MICROALGAE

MICROALGAE
Cultivation, Recovery of Compounds and Applications

Edited by

Charis M. Galanakis
Galanakis Laboratories, Chania, Greece

ELSEVIER

ACADEMIC PRESS
An imprint of Elsevier

Academic Press is an imprint of Elsevier
125 London Wall, London EC2Y 5AS, United Kingdom
525 B Street, Suite 1650, San Diego, CA 92101, United States
50 Hampshire Street, 5th Floor, Cambridge, MA 02139, United States
The Boulevard, Langford Lane, Kidlington, Oxford OX5 1GB, United Kingdom

Notices
Knowledge and best practice in this field are constantly changing. As new research and experience broaden our
understanding, changes in research methods, professional practices, or medical treatment may become
necessary.

Practitioners and researchers must always rely on their own experience and knowledge in evaluating and using any
information, methods, compounds, or experiments described herein. In using such information or methods they
should be mindful of their own safety and the safety of others, including parties for whom they have a professional
responsibility.

To the fullest extent of the law, neither the Publisher nor the authors, contributors, or editors, assume any liability
for any injury and/or damage to persons or property as a matter of products liability, negligence or otherwise, or
from any use or operation of any methods, products, instructions, or ideas contained in the material herein.

Library of Congress Cataloging-in-Publication Data
A catalog record for this book is available from the Library of Congress

British Library Cataloguing-in-Publication Data
A catalogue record for this book is available from the British Library

ISBN 978-0-12-821218-9

For information on all Academic Press publications
visit our website at https://www.elsevier.com/books-and-journals

Publisher: Charlotte Cockle
Acquisitions Editor: Patricia Osborn
Editorial Project Manager: Lena Sparks
Production Project Manager: Sreejith Viswanathan
Cover Designer: Vicky Pearson

Typeset by SPi Global, India

Contents

Contributors   ix

1. Cultivation techniques
LUISA FERNANDA RIOS PINTO, GABRIELA FILIPINI FERREIRA, AND MARIJA TASIC

1 Introduction   1
2 Laboratory cultivation techniques   3
3 Pilot cultivation techniques   5
4 Industrial cultivation techniques   13
5 Dark fermentation—Fermenters   13
6 Lowering cultivation costs   22
7 Conclusions   24
Statement   26
References   26

2. Photobioreactor design for microalgae culture
MARIANY COSTA DEPRÁ, IHANA AGUIAR SEVERO, ROSANGELA RODRIGUES DIAS, LEILA QUEIROZ ZEPKA, AND EDUARDO JACOB-LOPES

1 Introduction   35
2 System hydrodynamics   36
3 Parameters of environmental conditions in photobioreactors   43
4 Measuring the photobioreactors performance   48
5 Bottlenecks to achieve expansion of photobioreactors   50
6 Advances in the design of photobioreactors   54
7 Conclusions   56
Declaration of competing interest   57
References   57

3. Transport phenomena models affecting microalgae growth
JOSÉ REBOLLEDO-OYARCE, CÉSAR SÁEZ-NAVARRETE, AND LEONARDO RODRIGUEZ-CORDOVA

1 Introduction   63
2 Most important factors for the growth of a microalgae   64
3 Irradiation models   68
4 Growth models in microalgae   73
5 Momentum transfer models   75
6 Effect of shear stress on the growth of microalgae   77
7 Gas exchange and temperature effect   79
8 Energy consumption of a cultivation system   82
9 Conclusion   83
References   84

4. Edible bio-oil production from microalgae and application of nano-technology
NAZIA HOSSAIN, TAMAL CHOWDHURY, HEMAL CHOWDHURY, ASHFAQ AHMED, SABZOI NIZAMUDDIN, GREGORY GRIFFIN, TEUKU MEURAH INDRA MAHLIA, AND YOUNG-KWON PARK

1 Introduction   91
2 Suitable microalgae candidates for edible bio-oil and nanotechnology application for higher growth of microalgal species   92
3 Microalgae pretreatment   94
4 Methods of lipid extraction for edible bio-oil production   100
5 Conversion processes of bio-oil from microalgae   102
6 Bio-oil recovery, distillation, and purification   104
7 Integrated approaches   107
8 Environmental and socioeconomic impacts   108
9 Conclusions   111
References   111

5. Catalytic conversion of microalgae oil to green hydrocarbon

MIN-YEE CHOO, YI PIN PHUNG, AND JOON CHING JUAN

1 Introduction 117
2 Catalytic deoxygenation of microalgae oil, DO 131
3 Conclusion and future prospect 138
Acknowledgment 139
References 139

6. Biofuel production

NATHASKIA SILVA PEREIRA NUNES, MÔNICA ANSILAGO, NATHANYA NAYLA DE OLIVEIRA, RODRIGO SIMÕES RIBEIRO LEITE, MARCELO FOSSA DA PAZ, AND GUSTAVO GRACIANO FONSECA

1 General introduction 145
2 Main biofuels produced from microalgae 146
3 Other biofuels 161
4 Influence of cultivation conditions 162
5 Commercial application of these technologies 165
6 Perspectives 166
References 166

7. Emerging technologies for the clean recovery of antioxidants from microalgae

FRANCESCO DONSÌ, GIOVANNA FERRARI, AND GIANPIERO PATARO

1 Introduction 173
2 Extraction technologies for antioxidant compounds 175
3 Nonconventional extraction of bioactive compounds 177
4 Conclusions and future perspectives 197
References 198

8. Food applications

MARCO GARCIA-VAQUERO

1 Introduction 207
2 Composition of microalgae 208
3 Extraction of microalgal high-value compounds for food applications 211
4 The current market of microalgae and microalgal products 223
5 Legislation concerning microalgae as food 230
6 Future market and challenges of the use of microalgae as food 231
Acknowledgments 232
References 232

9. Microalgae as feed ingredients for livestock production and aquaculture

LUISA M.P. VALENTE, ANA R.J. CABRITA, MARGARIDA R.G. MAIA, INÊS M. VALENTE, SOFIA ENGROLA, ANTÓNIO J.M. FONSECA, DAVID MIGUEL RIBEIRO, MADALENA LORDELO, CÁTIA FALCÃO MARTINS, LUÍSA FALCÃO E CUNHA, ANDRÉ MARTINHO DE ALMEIDA, AND JOÃO PEDRO BENGALA FREIRE

1 Introduction 239
2 Microalgae in ruminants 240
3 Microalgae in swine 265
4 Microalgae in poultry 269
5 Microalgae in rabbit 278
6 Microalgae in diets for relevant species for aquaculture 283
7 Conclusion and perspectives 301
Acknowledgments 302
References 302

10. Cosmetics applications

ANDRESSA COSTA DE OLIVEIRA, ANA LUCÍA MOROCHO-JÁCOME, CIBELE RIBEIRO DE CASTRO LIMA, GABRIELA ARGOLLO MARQUES, MAÍRA DE OLIVEIRA BISPO, AMANDA BEATRIZ DE BARROS, JOÃO GUILHERME COSTA, TÂNIA SANTOS DE ALMEIDA, CATARINA ROSADO, JOÃO CARLOS MONTEIRO DE CARVALHO, MARIA VALÉRIA ROBLES VELASCO, AND ANDRÉ ROLIM BABY

1 Introduction 313
2 The necessity of products environmentally sustainable in cosmetics 314
3 Skin structure 315
4 Property of algae in skincare products 317
5 Natural dyes 319
6 Moisturizer agents 319
7 Antiaging agents 321
8 Anticellulite agents 323
9 Sunscreen/UV filter compounds 324
10 Skin-whitening agents 326
11 Haircare products: The benefits of algae 327
12 Formulation adjuvants 330
13 Conclusions and perspectives 331
References 332

11. Microalgal applications toward agricultural sustainability: Recent trends and future prospects

KSHIPRA GAUTAM, MEGHNA RAJVANSHI, NEERA CHUGH, RAKHI BAJPAI DIXIT, G. RAJA KRISHNA KUMAR, CHITRANSHU KUMAR, UMA SHANKAR SAGARAM, AND SANTANU DASGUPTA

1 Introduction   339
2 Biofertilizers   340
3 Plant biostimulants   351
4 Biopesticides   357
5 Symbiotic interaction of microalgae with higher plants   361
6 Microalgae in bioremediation and reclamation of degraded land   363
7 Conclusion and future prospects   368
References   368
Further reading   377

12. Microalgae biofilms for the treatment of wastewater

HASSIMI ABU HASAN, SITI NUR HATIKA ABU BAKAR, AND MOHD SOBRI TAKRIFF

1 Introduction   381

2 Species of microalgae applicable for wastewater treatment   382
3 Microalgae biofilms   386
4 Microalgae biofilm photobioreactors   391
5 Kinetics of microalgae biofilm photobioreactors   398
6 Conclusion   402
Acknowledgments   402
References   402

13. Techno-economic assessment of microalgae for biofuel, chemical, and bioplastic

YESSIE WIDYA SARI, KIKI KARTIKASARI, WIDYARANI, IRIANI SETYANINGSIH, AND DIANIKA LESTARI

1 Introduction   409
2 Microalgae for biofuels   410
3 Microalgae for chemicals   416
4 Microalgae for bioplastic   421
5 Microalgae: Integrated system   423
6 Conclusion   428
References   428

Index   433

Contributors

Ashfaq Ahmed Department of Chemical Engineering, COMSATS University Islamabad Lahore Campus, Lahore, Pakistan; School of Environmental Engineering, University of Seoul, Seoul, Republic of Korea

Mônica Ansilago Laboratory of Bioengineering, Faculty of Biological and Environmental Sciences, Federal University of Grande Dourados (FCBA/UFGD), Dourados, MS, Brazil

André Rolim Baby Department of Pharmacy, Faculty of Pharmaceutical Sciences, University of São Paulo, São Paulo, Brazil

Siti Nur Hatika Abu Bakar Department of Chemical & Process Engineering, Faculty of Engineering and Built Environment, Universiti Kebangsaan, Malaysia, Bangi, Selangor, Malaysia

Ana R.J. Cabrita ICBAS, Institute of Biomedical Sciences Abel Salazar; REQUIMTE, LAQV, ICBAS, Abel Salazar Biomedical Sciences Institute, University of Porto, Porto, Portugal

Min-Yee Choo Institute of Biological Sciences, Faculty of Science; Nanotechnology & Catalyst Research Centre (NANOCAT), University of Malaya, Kuala Lumpur, Malaysia

Hemal Chowdhury Department of Mechanical Engineering, Chittagong University of Engineering and Technology, Chattogram, Bangladesh

Tamal Chowdhury Department of Electrical and Electronic Engineering, Chittagong University of Engineering and Technology, Chattogram, Bangladesh

Neera Chugh Reliance Technology Group, Reliance Industries Limited, Reliance Corporate Park, Ghansoli, Maharashtra, India

João Guilherme Costa CBIOS—Universidade Lusófona's Research Center for Biosciences and Health Technologies, Lisbon, Portugal

Marcelo Fossa da Paz Laboratory of Enzymology and Fermentation Processes, Faculty of Biological and Environmental Sciences, Federal University of Grande Dourados (FCBA/UFGD), Dourados, MS, Brazil

Santanu Dasgupta Reliance Technology Group, Reliance Industries Limited, Reliance Corporate Park, Ghansoli, Maharashtra, India

André Martinho de Almeida LEAF Linking Landscape, Environment, Agriculture and Food, School of Agriculture, University of Lisbon, Lisbon, Portugal

Tânia Santos de Almeida CBIOS—Universidade Lusófona's Research Center for Biosciences and Health Technologies, Lisbon, Portugal

Amanda Beatriz de Barros Department of Pharmacy, Faculty of Pharmaceutical Sciences, University of São Paulo, São Paulo, Brazil

João Carlos Monteiro de Carvalho Department of Biochemical and Pharmaceutical Technology, Faculty of Pharmaceutical Sciences, University of São Paulo, São Paulo, Brazil

Cibele Ribeiro de Castro Lima Physics Institute, University of São Paulo, São Paulo, Brazil

Andressa Costa de Oliveira Department of Pharmacy, Faculty of Pharmaceutical Sciences, University of São Paulo, São Paulo, Brazil

Nathanya Nayla de Oliveira Laboratory of Bioengineering, Faculty of Biological and Environmental Sciences, Federal University of Grande Dourados (FCBA/UFGD), Dourados, MS, Brazil

Maíra de Oliveira Bispo Department of Pharmacy, Faculty of Pharmaceutical Sciences, University of São Paulo, São Paulo, Brazil

Mariany Costa Deprá Bioprocesses Intensification Group, Federal University of Santa Maria, UFSM, Santa Maria, RS, Brazil

Rosangela Rodrigues Dias Bioprocesses Intensification Group, Federal University of Santa Maria, UFSM, Santa Maria, RS, Brazil

Rakhi Bajpai Dixit Reliance Technology Group, Reliance Industries Limited, Reliance Corporate Park, Ghansoli, Maharashtra, India

Francesco Donsì Department of Industrial Engineering, University of Salerno, Fisciano, Italy

Sofia Engrola Centre of Marine Sciences (CCMAR), University of Algarve, Campus de Gambelas, Faro, Portugal

Luísa Falcão e Cunha LEAF Linking Landscape, Environment, Agriculture and Food, School of Agriculture, University of Lisbon, Lisbon, Portugal

Giovanna Ferrari Department of Industrial Engineering, University of Salerno; ProdAl Scarl, Fisciano, Italy

Gabriela Filipini Ferreira Chemical Engineering, Student at the Faculty of Chemical Engineering at the University of Campinas (UNICAMP), Campinas, Brazil

António J.M. Fonseca ICBAS, Institute of Biomedical Sciences Abel Salazar; REQUIMTE, LAQV, ICBAS, Abel Salazar Biomedical Sciences Institute, University of Porto, Porto, Portugal

Gustavo Graciano Fonseca Laboratory of Bioengineering, Faculty of Biological and Environmental Sciences, Federal University of Grande Dourados (FCBA/UFGD), Dourados, MS, Brazil

João Pedro Bengala Freire LEAF Linking Landscape, Environment, Agriculture and Food, School of Agriculture, University of Lisbon, Lisbon, Portugal

Marco Garcia-Vaquero School of Agriculture and Food Science, University College Dublin, Dublin 4, Ireland

Kshipra Gautam Reliance Technology Group, Reliance Industries Limited, Reliance Corporate Park, Ghansoli, Maharashtra, India

Gregory Griffin School of Engineering, RMIT University, Melbourne, VIC, Australia

Hassimi Abu Hasan Department of Chemical & Process Engineering, Faculty of Engineering and Built Environment, Universiti Kebangsaan, Malaysia, Bangi, Selangor, Malaysia

Nazia Hossain School of Engineering, RMIT University, Melbourne, VIC, Australia

Eduardo Jacob-Lopes Bioprocesses Intensification Group, Federal University of Santa Maria, UFSM, Santa Maria, RS, Brazil

Joon Ching Juan Nanotechnology & Catalyst Research Centre (NANOCAT), University of Malaya, Kuala Lumpur; School of Science, Monash University, Subang Jaya, Selangor, Malaysia

Kiki Kartikasari Carbon and Environmental Research Indonesia, Bogor, Indonesia

Chitranshu Kumar Reliance Technology Group, Reliance Industries Limited, Reliance Corporate Park, Ghansoli, Maharashtra, India

G. Raja Krishna Kumar Reliance Technology Group, Reliance Industries Limited, Reliance Corporate Park, Ghansoli, Maharashtra, India

Rodrigo Simões Ribeiro Leite Laboratory of Enzymology and Fermentation Processes, Faculty of Biological and Environmental Sciences, Federal University of Grande Dourados (FCBA/UFGD), Dourados, MS, Brazil

Dianika Lestari Department of Food Engineering, Faculty of Industrial Technology, Institut Teknologi Bandung, Bandung, Indonesia

Madalena Lordelo LEAF Linking Landscape, Environment, Agriculture and Food, School of Agriculture, University of Lisbon, Lisbon, Portugal

Teuku Meurah Indra Mahlia School of Information, Systems and Modelling, Faculty of Engineering and Information Technology, University of Technology Sydney, Sydney, NSW, Australia

Margarida R.G. Maia ICBAS, Institute of Biomedical Sciences Abel Salazar; REQUIMTE, LAQV, ICBAS, Abel Salazar Biomedical Sciences Institute, University of Porto, Porto, Portugal

Gabriela Argollo Marques Department of Pharmacy, Faculty of Pharmaceutical Sciences, University of São Paulo, São Paulo, Brazil

Cátia Falcão Martins LEAF Linking Landscape, Environment, Agriculture and Food, School of Agriculture, University of Lisbon, Lisbon, Portugal

Ana Lucía Morocho-Jácome Department of Pharmacy, Faculty of Pharmaceutical Sciences, University of São Paulo, São Paulo, Brazil

Sabzoi Nizamuddin Civil and Infrastructure Engineering, School of Engineering, RMIT University, Melbourne, VIC, Australia

Nathaskia Silva Pereira Nunes Laboratory of Bioengineering, Faculty of Biological and Environmental Sciences, Federal University of Grande Dourados (FCBA/UFGD), Dourados, MS, Brazil

Young-Kwon Park School of Environmental Engineering, University of Seoul, Seoul, Republic of Korea

Gianpiero Pataro Department of Industrial Engineering, University of Salerno, Fisciano, Italy

Yi Pin Phung Nanotechnology & Catalyst Research Centre (NANOCAT), University of Malaya, Kuala Lumpur, Malaysia

Meghna Rajvanshi Reliance Technology Group, Reliance Industries Limited, Reliance Corporate Park, Ghansoli, Maharashtra, India

José Rebolledo-Oyarce Departamento de Ingeniería Química y Bioprocesos, Pontificia Universidad Católica de Chile, Santiago, Chile

David Miguel Ribeiro LEAF Linking Landscape, Environment, Agriculture and Food, School of Agriculture, University of Lisbon, Lisbon, Portugal

Luisa Fernanda Rios Pinto Chemical Engineering, Research Fellow Position at the Faculty of Chemical Engineering at the University of Campinas (UNICAMP), Campinas, Brazil

Leonardo Rodriguez-Cordova Departamento de Ingeniería Química y Bioprocesos, Pontificia Universidad Católica de Chile, Santiago, Chile

Catarina Rosado CBIOS—Universidade Lusófona's Research Center for Biosciences and Health Technologies, Lisbon, Portugal

César Sáez-Navarrete Departamento de Ingeniería Química y Bioprocesos; Centro de Investigación en Nanotecnología y Materiales Avanzados CIEN-UC, Facultad de Física; UC Energy Research Center (CE-UC), Pontificia Universidad Católica de Chile, Santiago, Chile

Uma Shankar Sagaram Reliance Technology Group, Reliance Industries Limited, Reliance Corporate Park, Ghansoli, Maharashtra, India

Yessie Widya Sari Department of Physics, Faculty of Mathematics and Natural Sciences; Center for Transdisciplinary and Sustainability Sciences, Institute of Research and Community Services, IPB University, Bogor, Indonesia

Iriani Setyaningsih Department of Aquatic Product Technology, Faculty of Fisheries and Marine Sciences, IPB University, Bogor, Indonesia

Ihana Aguiar Severo Bioprocesses Intensification Group, Federal University of Santa Maria, UFSM, Santa Maria, RS, Brazil

Mohd Sobri Takriff Department of Chemical & Process Engineering, Faculty of Engineering and Built Environment, Universiti Kebangsaan Malaysia, Bangi, Selangor, Malaysia

Marija Tasic Chemical Engineering, Associate Professor at the Faculty of Technology at the University of Nis, Leskovac, Serbia

Inês M. Valente REQUIMTE, LAQV, ICBAS, Abel Salazar Biomedical Sciences Institute; REQUIMTE, LAQV, Department of Chemistry and Biochemistry, Faculty of Sciences of the University of Porto, Porto, Portugal

Luisa M.P. Valente CIIMAR, Interdisciplinary Centre of Marine and Environmental Research, University of Porto, Terminal de Cruzeiros do Porto de Leixões, Matosinhos; ICBAS, Institute of Biomedical Sciences Abel Salazar, University of Porto, Porto, Portugal

Maria Valéria Robles Velasco Department of Pharmacy, Faculty of Pharmaceutical Sciences, University of São Paulo, São Paulo, Brazil

Widyarani Research Unit for Clean Technology, Indonesian Institute of Sciences, Bandung, Indonesia

Leila Queiroz Zepka Bioprocesses Intensification Group, Federal University of Santa Maria, UFSM, Santa Maria, RS, Brazil

1

Cultivation techniques

Luisa Fernanda Rios Pinto[a], Gabriela Filipini Ferreira[b], and Marija Tasic[c]

[a]Chemical Engineering, Research Fellow Position at the Faculty of Chemical Engineering at the University of Campinas (UNICAMP), Campinas, Brazil
[b]Chemical Engineering, Student at the Faculty of Chemical Engineering at the University of Campinas (UNICAMP), Campinas, Brazil
[c]Chemical Engineering, Associate Professor at the Faculty of Technology at the University of Nis, Leskovac, Serbia

1 Introduction

Microalgae have considerable potential as a feedstock for different bioproducts (Borowitzka & Moheimani, 2013; Klok, Lamers, Martens, Draaisma, & Wijffels, 2014; Moheimani, McHenry, de Boer, & Bahri, 2015; Raheem, Prinsen, Vuppaladadiyam, Zhao, & Luque, 2018), from biofuels to edible oils and proteins, and can be cultivated in different environmental conditions and systems. As the human population increases, it is required the production of more food, energy, and water, thus producing more carbon dioxide (CO_2). For this reason, microalgae cultivation brings a new opportunity and solution for all these problems toward a sustainable bioeconomy (Bussa, Eisen, Zollfrank, & Röder, 2019; Ferreira et al., 2018; Rizwan, Mujtaba, Memon, Lee, & Rashid, 2018).

The growth characteristics and metabolites accumulation strongly depend on the type of cultivation: phototrophic (using light and CO_2), heterotrophic (without light and organic carbon), and mixotrophic (using light, CO_2, and organic carbon). Phototrophic cultivation is the most common type used in large-scale microalgae production (Suparmaniam et al., 2019).

One of the essential parameters in algae cultivation is the type of system used. Several companies in some countries have developed the commercial production of microalgae biomass and/or bioproducts, also coupled with CO_2 fixation (Zhou et al., 2017).

Cultivation systems can be divided into two broad groups: open systems and closed systems (Okoro, Azimov, Munoz, Hernandez, & Phan, 2019). The most commonly used open systems are ponds (cylindrical, rectangular, and elliptical-bottomed) or largely used raceway ponds. It is called open systems because the microalgae culture is in contact with atmospheric air. These systems are not sophisticated; the cultures are usually grown under natural conditions and have lower control on variables (e.g. temperature, light, and pH). Several challenges are confronting with the use of open ponds: water evaporation, climate variations, contamination, among others. These systems are generally used to obtain low added-value products. On the other hand, in closed systems, generally called photobioreactors (PBRs), the cultivation is developed to achieve higher yields. These systems can be achieved by the high luminosity surface and the control of growth and contamination variables within the system. In these systems, the atmospheric air has no direct contact with the microalgae culture. Several PBRs have been developed so far, among them, vertical cylindrical, vertical or horizontal flat panel, tubular panels, and vertical and horizontal serpentines. The material may be of glass, plastic, or polycarbonate.

Microalgae cultivation, in particular growing systems, were developed thinking in terms of volume production, so at first, it was quite rustic and simple. With the development of different products from microalgae biomass (high-value products), these systems have been advancing technologically, using engineering concepts in the construction of improved systems, always aiming to achieve high productivity and cost optimization (Narala et al., 2016). Both of actually systems for microalgae cultivation, in particular growing systems, have advantages and disadvantages. Notably, open systems have minimal capital and operating costs and lower energy but require large areas and come with contamination problems. By contrast, the closed systems are more expensive; they do not require large areas, and the contamination problems decrease.

Nowadays, microalgae cultivation is well stabilized for food, feed, and pigments. However, in some areas, it is still not feasible. Microalgae biorefinery is a possibility for becoming microalgae cultivation a viable economic process, producing high-value products for all the cell fractions (Chew et al., 2017). Currently, many companies work in microalgae large-scale production. Following is a few facilities that produce microalgae in large scale: (a) Earthrise Nutritionals is used to produce cyanobacterial biomass for food (Chisti, 2007), (b) Phycom is an industry that produces *Chlorella vulgaris* and *Chlorella sorokiniana* for feed and food, and (c) Algawise that is a part from Corbion, that produce algae oils, among others. More industrial facilities are listed in Table 3.

This chapter deals with conventional and emerging cultivation techniques for microalgae cultivation.

1.1 The history of microalgae cultivation system

The large-scale systems of microalgae cultivation were established during the first decades of the 20th century (Hamed, 2016). Before this, algae culturing was restricted to laboratory-scale operations. Notwithstanding, simple systems for food chain enhancements commanding to algal production and following growth of comestible organisms exist for centuries, and natural colonies of *Spirulina* have been initially collected for food in Africa and Mexico. The outdoor microalgae production began in around 1948. Fig. 1 depicted when a few cultivation systems were created to enhance production over time. The first large cultivation was performed by growing microalgae in an open pond for the food chain. These systems were initially used in Germany and Japan during 1940–50. The construction of German ponds was in a plastic-lined, with approximately 20 cm depth and using air to improve mixing and obtain a homogeneous culture. The Japan design was a single flat-bottomed of 22 m^2 surface area and bubbling CO_2 enriched air thought aeration pipes and 15–20 cm deep (Terry & Raymond, 1985). In the early 1950s, the Research Corporation of New York contacted the Stanford Research Institute (SRI) in Palo Alto, California, to conduct the study for the continuous culture of *Chlorella*, but the results were not conclusive enough, one of the reasons was the

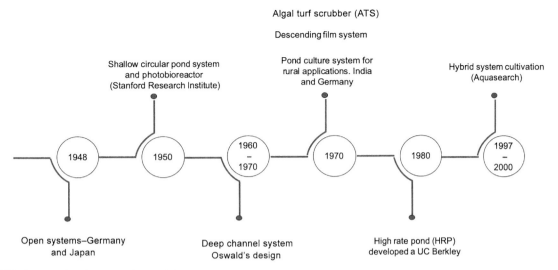

FIG. 1 Timeline for microalgae cultivation system.

contamination culture. The SRI was one of the first workers to recognize the need for a closed and sterilized system to maintain an uncontaminated algal culture (Lee, 1986).

Following the United States concentrated on microalgal-bacterial systems for wastewater treatment, even though during 1950–60, the application of microalgae for the conversion of CO_2 to oxygen in spacecraft and submarines was also explored. The more significant international effort in algal biomass culture with strongly mixed systems started in Eastern Europe in the 1960s, headed by researchers at the Czechoslovak Academy of Sciences and based at least in part on earlier Soviet work (Terry & Raymond, 1985).

Between the 1960s and 1970s, a deep channel system patterned after with Oswald's design, ponds of up to $100\,m^2$, was developed in California for the treatment (to remove nitrogenous) of runoff waters from agricultural tiling is in the San Joaquin Valley. Another model developed in the Czech Republic in the 1970s was a cascade of descending film.

The Commission of the European Communities began the algal biomass research for bioenergy in 1978. The project entitled "Mariculture on Land (MCL)," includes researchers in Germany, France, Italy, and Brazil. It proposed the utilization of arid coastal lands and seawater for the culture of micro and macroalgae for methane production. Later in 1979, the Solar Energy Research Institute (SERI), began a research program on the biomass production of aquatic plants for energy, which included the use of microalgae as a mechanism for the fixation of solar energy (by photosynthesis) as liquid biofuels (Terry & Raymond, 1985).

As already mentioned, the cultivation system will be selected among existing or new designs depending on the final product. This chapter provides a short review of the present systems and their applications. The cultivation system is a principal criterion after the selection of a microalgae species for optimal cultivation in order to be a cost-effective process.

2 Laboratory cultivation techniques

Generally, laboratory cultivations are used to study and optimize parameters, variables, and growth conditions. Lab cultivation is a mandatory step when the objective is a scale-up. Many of the research published in the literature are results of laboratory studies, through the propose of studying growth conditions for a specific microalga (Liu, Chen, Tao, & Wang, 2020; Zhou et al., 2019). Microalgae cultivation offers benefits in terms of climate change, but still has many challenges that need to overcome. The first factor that needs to be considered is the strain selection; the best strain is that one grows highly and has the metabolic accumulation necessary for the objective and can survive in severe conditions. The next challenge is energy consumption, as many steps in the microalgae cultivation process need power to work, including upstream and downstream. Closed PBRs need energy for mixing the culture, water pumping, gas bubbling (CO_2), harvesting/dewatering the culture, among others, that depend on the final product. Another challenge that needs to overcome is the availability of the nutrients and water, large amounts of water and nutrients are used for cultivation, but nowadays microalgae cultivation are overcoming these obstacles by using wastewater from industries, which have abundant nutrients and water often undervalued. The final challenge is the distribution of CO_2; the systems for the CO_2 distribution are problematic, as they demand high energy and require long pipelines to transport the gas.

The factors that affect the microalgae cultivation for biomass production are light intensity, photoperiod, temperature, nutrients, mixing, aeration, pH, and CO_2 absorption. Microalgae photosynthesize, which means that they assimilate inorganic carbon for conversion into organic matter. Light is a common source of energy, which drives this reaction. The light intensity is one of the most important variables, once the microalgae cell metabolism uses light as a source of energy for synthesizing the cell protoplasm. For lab cultivation, artificial illumination is necessary by incandescent lamps or without light (heterotrophic cultivation). Fluorescent tubes in the blue or red-light spectrum are ideal as these are the most active portions of the light spectrum for photosynthesis (Koc, Anderson, & Kommareddy, 2013). The depth and the density of the microalgal culture influence at the light intensity: at high depths and cell concentrations, the light intensity needs to be increased to access through the culture. The optimum light intensity depends on the microalgae strain; in the literature, it is possible to find that for *Desmodesmus* sp., a $98 \mu\,mol\,m^{-2}\,s^{-1}$ (Ji et al., 2013) source is the optimum, on the other hand, for *Dunaliella viridis* the optimum was $700 \mu\,mol\,m^{-2}\,s^{-1}$ (Gordillo, Goutx, Figueroa, & Niell, 1998). Some strains suffer photoinhibition when is increased the light intensity. It occurs with disruption of the chloroplast lamellae growth and inactivation of the enzymes involved in the carbon dioxide fixation. However, some strains can survive without light in heterotrophic cultivation.

The light intensity and duration affect photosynthesis of microalgae, and its biochemical composition, very low or very high luminescence, is not efficient for growth (Khan, Shin, & Kim, 2018). The range of lightness used in a lab-scale is $20–1000 \mu\,mol\,m^{-2}\,s^{-1}$ (ResearchGate, n.d.); being the most common, $150 \mu\,mol\,m^{-2}\,s^{-1}$. The lamps are placed, usually, at 25–30 cm of distance from the flasks.

Light cycles (dark:light) are important for the performance of the microalgae (Jacob-Lopes, Scoparo, Lacerda, & Franco, 2009). The most used photoperiod is 12:12 h (dark:light). Experimental studies report that the minimum photoperiod is (8:16) and the maximum (24:0). Many studies reported that 24 h of light exposed increased biomass productivity (Rai, Gautom, & Sharma, 2015). Biomass productivity reached for lab-scale is around $0.02–0.7\,g\,L^{-1}\,d^{-1}$ (Ferreira, Ríos Pinto, Maciel Filho, & Fregolente, 2019; Ishika, Moheimani, Laird, & Bahri, 2019; Rashid, Ryu, Jeong, Lee, & Chang, 2019).

The temperature influences the growth velocity, cell dimension, metabolites composition and accumulation, and nutrients requirements. The best temperature for cultivation also highly depends on the strain. The cultivation in a lab-scale is prepared in a cultivation room (indoor) with a stable temperature in a range of 16–27°C.

Some studies report that the optimum temperature is 18–24 °C but strongly depends on the microalgae strain.

In terms of pH, in a lab-scale is common uses a range of 6–9 (Qiu, Gao, Lopez, & Ogden, 2017; Song et al., 2020), but it is reported that the optimum is 8.2–8.7 for most microalgae strain. In order to control the pH value, it is strongly used the addition of carbon dioxide to reach acceptable levels (Valdés, Hernández, Catalá, & Marcilla, 2012; Vonshak & Coombs, 1985).

Another factor that influences in microalgae productivity is mixing. This factor is necessary to prevent cell sedimentation and the homogenization of all the cells. To good mixing granter, all the cells are exposed to the light and nutrients. In practice, among different ways to mix microalgae culturing, the most used are the mechanical and air bubbles. High velocities and turbulence can damage the cells; the optimum levels depend on the microalgae strain (Eriksen, 2008).

Finally, nutrient concentration affects the microalgae growth. Some nutrients are essential for the cultivation, principally macronutrients (phosphorus and nitrogen) (Zhuang et al., 2018). Many culture media are used for laboratory cultivation (BG 11 (Rippka, Deruelles, Waterbury, Herdman, & Stanier, 1979), f/2 Guillard (Guillard, 1975; Guillard & Ryther, 1962), CHU 13 (Chu, 1942), among others) that strongly depends on the microalgae strain, Chew et al. (2018) list the nutrients that are needed for different types of microalgae.

Lab cultivation generally is operating in a batch culture that consists in the inoculation of cells into a flask, following for a growth period of several days (7–40) and finally harvested. The scale-up in the laboratory is usually carried out by successive transfers of cultivation systems from small to larger (Pérez et al., 2017). The inoculum (2%–10%) is transferring to a bigger flask adding more cultural media (Cruz et al., 2018).

A few of lab systems are shown in Fig. 2, (a) Petri dishes that are used to isolate a microalga species, once microalgae usually grow in a colony that is easy to identify in a solid media, but the sample to be inoculated need to be in a deficient concentration of microalga cells. (b) Tubes are used for inoculum conservation, usually is the purest microalgae colonies that have in a lab, generally, from this sample, the experiments start. (c) Flasks are primarily used for lab cultivations; usually, the experiments start with a 250-mL flask. Various sizes and capacities are used for this purpose (100, 250, 500, 1000, and 2000–10,000 mL). Commonly is used 80% of the maximum volume, with 10% of inoculum, the remaining is cultivation medium. (d) Plastic bottles: there are several plastic bottles from water bottles to a gallon of water.

In laboratories, microalgae can also be cultivated in different systems; for example, PBRs, bags, and other technological systems. The cultivation technique is strongly influenced by the microalgae strain (Chew et al., 2018).

3 Pilot cultivation techniques

The cultivation parameters (Bux & Chisti, 2016; Gim et al., 2014; Najafabadi, Malekzadeh, Jalilian, Vossoughi, & Pazuki, 2015; Ruangsomboon, 2012) that are possible to manipulate at a pilot-scale are the same as laboratory-scale systems: nutrients concentration, nutrients source (wastewater and synthetic medium), carbon source (organic and inorganic), salinity, temperature, pH, light source (artificial and natural), light intensity, light spectra (Vadiveloo, Moheimani, Cosgrove, Parlevliet, & Bahri, 2017), photoperiod, and agitation. Among these parameters, the temperature ensures that the selected microorganism grows appropriately. The pH is also essential, as it interferes with cell growth. The agitation, on the other hand, has the role of ensuring a good transfer of mass and heat, keeping the medium homogeneous throughout the cultivation process as well as improving the access of algae cells to gases in the case air or CO_2 are

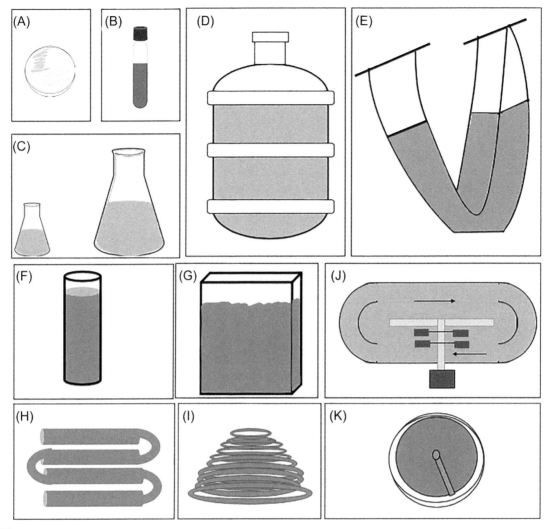

FIG. 2 Systems cultivation (A–D) Lab-scale systems, (E–I) closed systems, and (J and K) open systems.

supplemented. Considering that each parameter will be adjusted according to the need and characteristics of the process, stress conditions can be applied by manipulating these parameters to produce a specific metabolite. Nevertheless, better control is possible by using closed systems compared when scaling up PBRs.

Regardless of choosing a specific cultivation system to produce microalgae at a larger scale, it is desired to maintain the biomass productivity and its composition when scaling up the

production from laboratory to pilot to industrial scales. To achieve this goal, a few aspects must be taken into consideration, such as hydrodynamic and environmental conditions (Cuello, Hoshino, Kuwahara, & Brown, 2016). Additionally, it is essential to arrange a large pond or multiple tubes in a way that all culture volume is roughly under the same conditions and close to a complete mixture (homogeneous), with minimum dead zones, light incidence variation, among others. Thus larger-scale production of biomass must consider economic as well as technological limitations. In summary, the selection of a cultivation system will depend on the microalgal species selected, the cost of land, environment conditions, desired bioproduct (Lee, 2013) and process economic viability.

The microalgae cultivation technique is divided into two categories: open and closed systems (Klinthong, Yang, Huang, & Tan, 2015). Open systems (mostly open ponds) are commonly chosen to be used in larger scales due to lower construction and maintenance costs, thus resulting in a lower estimated cost of production (Zakariah, Abd. Rahman, Hamzah, & Md Jahi, 2015), direct exposure to outdoor sunlight and other technical advantages (Acién et al., 2017). However, these are more susceptible to contamination (fungi, bacteria, and other microalgae) and dependent on climate conditions. Therefore, robust cultures with high tolerance to adverse ambient conditions, such as pH and temperature variations, and microalgae prone to resist harmful invaders should be selected for this type of cultivation. To that purpose, it is essential to invest in upstream processes, such as isolation and selection of microalgae strains, as well as optimizing downstream processes (harvesting, dewatering, drying, cell disruption, and lipid extraction).

On the other hand, numerous PBRs with different shapes and sizes have been developed to achieve high productivity, which could be competitive to open systems if production costs are reduced, and revenue is also increased by targeting high-value bioproducts. Moreover, they allow axenic cultivation of monocultures, less water evaporation, lower CO_2 loss, and better control of conditions (El-Baz & Baky, 2018).

3.1 Photobioreactors

Closed algal cultures exhibit several benefits disregarding the economic barrier. They reduce contamination risks and limit CO_2 losses (Moheimani et al., 2015) in addition to the previously mentioned increase in biomass productivity. Although many economic studies show a higher price of construction and operation compared to open ponds, there are still process steps that could be explored to reduce costs. For example, expanses with nutrients and carbon sources could be significantly reduced if wastewater (Ayre, Moheimani, & Borowitzka, 2017; Chinnasamy, Bhatnagar, Hunt, & Das, 2010; Guldhe et al., 2017) and flue gases (Duarte, Fanka, & Costa, 2016; Hanifzadeh, Sarrafzadeh, & Tavakoli, 2012; Ji et al., 2017) were used during the cultivation step. Closed PBRs are usually divided by their geometrical configuration, such as tubular (cylindrical) and flat-plate (rectangular prism) types.

3.1.1 Tubular photobioreactors

These systems consist in glass or plastic tubes usually assembled vertically with upward flow obtained by using pumps and/or gas bubbling (air-lift), area-to-volume ratios of $80–100 \, m^{-1}$, diameters up to 10 cm to improve photosynthetic efficiency (Acién et al., 2017) and liquid velocities of $0.5 \, m \, s^{-1}$ (El-Baz & Baky, 2018). They are usually applied to obtain concentrated cultures and produce high-quality biomass for high-value applications such as pharmaceutical, food, and cosmetic industries. However, excess O_2 accumulation, above air saturation, is a disadvantage of this configuration as it can inhibit photosynthesis.

Another common tubular PBR is the "air-lift" or "CO_2-lift" type, which can achieve high-density cultures (Chiu, Tsai, Kao, Ong, & Lin, 2009). They consist of larger diameter tubes with two internal zones and gas bubbling that allows recirculation. The central zone, called a riser, is where gas is dispersed. Then the liquid is moved to the outer zone, called downcomer. This configuration differs from other tubular PBRs mainly due to the mixing mechanism that promotes better homogeneity and facilitates mass transfer. Also, they can have high photosynthetic efficiency due to the use of a LED light jacket covering most of the bioreactor's surface.

3.1.2 Flat-plate photobioreactors

The rectangular-shape PBRs were conceptualized to obtain optimal light absorbance as the rectangular panels, made of glass or a polymer, are transparent, have a high area-to-volume ratio ($400\,m^{-1}$) and a light path around 7 cm (Płaczek, Patyna, & Witczak, 2017). Consequently, cultures can achieve high volumetric biomass productivities (Acién et al., 2017). They show less O_2 accumulation than tubular PBRs but present a temperature variation challenge, overheating. Thus, as in some cases of tubular, flat-plate PBRs may require cooling. Similar mixing mechanisms are applied (pumps and gas bubbling).

3.1.3 Different designs of photobioreactors

Other PBR systems are (1) the oldest widely used serpentine and (2) manifold or the unusual (3) helical configurations (Acién et al., 2017; El-Baz & Baky, 2018). The first refers to several tubes usually displayed horizontally, connected by U-bends. They have been adapted and upgraded over the last decades to the recent model with improved mass transfer and mixing. The second is similar to serpentine as it consists of a series of straight, long, and parallel tubes, which are connected by two manifolds for distribution and collecting culture suspension. Finally, helical PBRs are smaller diameter circularly bent tubes coiled around an upright supporting structure with a separate or connected degasser. They may be advantageous in a space-limiting situation.

3.1.4 Comparison of photobioreactors

Photobioreactor selection must consider the best design to achieve high biomass productivity with a specific composition (target bioproduct), as well as the economic viability of the proposed system. Therefore, enclosed PBRs are often chosen for high-value metabolites production (Suparmaniam et al., 2019). Table 1 shows a collection of different microalgae cultivation using PBRs at a laboratory or pilot scale. Studies developed in recent years were chosen to compare biomass productivity using different systems. It can be seen that flat-plate configurations usually yield higher biomass production probably due to superior photosynthetic efficiency. On the other hand, tubular PBRs proved to be a suitable system for high-lipid production as well as wastewater treatment.

Another important parameter to be considered when scaling up PBRs is the cultivation regime, similar to chemical reactors, batch, semi-batch, and continuous. Batch reactors are the most used in industry when involving microorganisms due to easier control of the process variables and final product quality. However, a continuous process is desired to increase production and obtain constant growth throughout the year. Once defined the cultivation regime, a kinetic study of cell growth if often conducted to optimize production.

3.2 Open ponds

Open ponds are often divided into the raceway and circular configurations (Suparmaniam et al., 2019). They consist of cheap systems where the cultivation is usually conducted in outdoor conditions. Consequently, it depends on weather conditions being favorable such as abundant light incidence throughout the year; it has poor control of parameters such as temperature and pH; it is highly exposed to contamination (grazers, bacteria, fungi, viruses, and other microalgae) (Bux & Chisti, 2016). Although they usually achieve lower biomass productivity than closed PBRs due to these factors, and others, economic studies show that they still are the most viable option. For example, cost-efficient carbon supplies could improve biomass production and lipid accumulation by bubbling CO_2 from inexpensive flue gases.

TABLE 1 Literature review of recent microalgae cultivation studies using different photobioreactors.

PBR type	PBR system and scale	Microalgae and culture media	Main results	Reference
Tubular	Air-lift PBR (7 L) with 14-h light/10-h dark cycle and air bubbling Laboratory scale	*Chlorella vulgaris* Wastewater	Biomass productivity: $17.9\,\text{mg}\,\text{L}^{-1}\,\text{d}^{-1}$ Carbon content: 49.96%	Chaudhary, Dikshit, and Tong (2018)
		Scenedesmus obliquus Wastewater	Biomass productivity: $16.7\,\text{mg}\,\text{L}^{-1}\,\text{d}^{-1}$ Carbon content: 50.1%	
	Air-lift PBR (7 L) with a 14-h light/10-h dark cycle and CO_2 bubbling Laboratory scale	*Chlorella vulgaris* Wastewater	Biomass productivity: $94.1\,\text{mg}\,\text{L}^{-1}\,\text{d}^{-1}$ Carbon content: 50.4%	
		Scenedesmus obliquus Wastewater	Biomass productivity: $86.5\,\text{mg}\,\text{L}^{-1}\,\text{d}^{-1}$ Carbon content: 50.67%	
	Air-lift PBR (10 L) in fed-batch mode (addition of $0.11\,\text{g}\,\text{L}^{-1}\,\text{d}^{-1}$ acetate) at the outdoor tropical condition Laboratory scale	*Scenedesmus abundans* BBM	[a]Biomass productivity: $162.5\,\text{mg}\,\text{L}^{-1}\,\text{d}^{-1}$ Lipid content: 31.45%	Gupta, Pawar, Pandey, Kanade, and Lokhande (2019)
	Horizontal tubular PBR (800 L) in a greenhouse, with 400 L sunlight exposed part and 400 L dark part Pilot scale	*Nannochloropsis oceânica* f/2	Biomass productivity: $0.12\,\text{g}\,\text{L}^{-1}\,\text{d}^{-1}$ N uptake: $5.27\,\text{mg}$ (N) L^{-1}	Silkina, Zacharof, Ginnever, Gerardo, and Lovitt (2019)
		Nannochloropsis oceânica Wastewater	Biomass productivity: $0.11\,\text{g}\,\text{L}^{-1}\,\text{d}^{-1}$ N uptake: $7.37\,\text{mg}$ (N) L^{-1}	
		Scenedesmus quadricauda f/2	Biomass productivity: $0.11\,\text{g}\,\text{L}^{-1}\,\text{d}^{-1}$ N uptake: $6.82\,\text{mg}$ (N) L^{-1}	
		Scenedesmus quadricauda Wastewater	Biomass productivity: $0.09\,\text{g}\,\text{L}^{-1}\,\text{d}^{-1}$ N uptake: $7.17\,\text{mg}\,\text{L}^{-1}$	
	Multi-stage PBR (2000 L) enriched with gas from thermal power plants Pilot-scale	*Nephroselmis* sp. BBM	Biomass productivity: $0.163\,\text{g}\,\text{L}^{-1}\,\text{d}^{-1}$ Lipid content: 60.9%	Park et al. (2019)
	Tubular PBR (150 L). Pilot scale	*Chlorella vulgaris* BBM	Biomass productivity: $76\,\text{mg}\,\text{L}^{-1}\,\text{d}^{-1}$	Villaseñor Camacho, Fernández Marchante, and Rodríguez (2018)

Continued

TABLE 1   Literature review of recent microalgae cultivation studies using different photobioreactors—cont'd

PBR type	PBR system and scale	Microalgae and culture media	Main results	Reference
Flat-plate	Flat-plate PBR (25 L) with LED light Laboratory scale	*Chlorella vulgaris* (Andersen, Jacobson, & Sexton, 1991)	Biomass production: 1200–1350 mg L^{-1} Lipid productivity: up to 40 mg L^{-1} d^{-1}	Papapolymerou et al. (2018)
	Flat-plate PBRs (140 L) Pilot scale	*Nannochloropsis* sp. Seawater and f/2	Biomass productivity: 0.12–0.14 g L^{-1} d^{-1}	Nwoba, Parlevliet, Laird, Alameh, and Moheimani (2020)
	Flat-plate PBRs (10 in series, 30 L each) inclined 60 degrees Pilot scale	*Scenedesmus obtusiusculus* BG-11	Biomass productivity: 0.90 g L^{-1} d^{-1}	Koller, Wolf, Brück, and Weuster-Botz (2018)
		Scenedesmus ovalternus BG-11	Biomass productivity: 0.65 g L^{-1} d^{-1}	

[a] *Measured from days 4–6 of cultivation.*

3.2.1 Raceway ponds

One of the oldest developed open systems is raceway ponds, widely used for large-scale cultivation of microalgae (El-Baz & Baky, 2018). They have large areas (100–5000 m^2) and are divided into 2 or 4 channels for recirculation, the liquid velocity of 0.2 m s^{-1} to avoid settling of cells, short height to ensure light reaches to bottom (0.2–0.4 m) and low surface-area-to-volume (S/V) ratios (5–10 m^{-1}) (Acién et al., 2017). Culture circulation is usually obtained by using a paddle wheel or a propeller and a centered baffle. Deflector baffles may also be used to improve mixing by promoting a turbulent flow.

Besides fluid dynamics, mass transfer limitations are an issue of raceway ponds, as CO$_2$ biofixation and O$_2$ removal are also determining for system performance.

3.2.2 Circular pond

A rounded pond with a rotating arm usually has 20–30 cm depth and 40–50 m diameter. It is often employed in wastewater or other effluent treatments, especially in Asia (El-Baz & Baky, 2018), and is associated with a cheaper cost. However, due to high exposure to the surroundings, it can easily contaminate with other microorganisms.

3.2.3 Different designs of open systems

Inclined (cascade) systems allow the algae culture to flow down an angled surface of a few 100 m^2, be collected in a larger-volume recipient, and pumped back to the top (Borowitzka & Moheimani, 2013). During the day, the culture is continuously recirculated to increase light exposure and promote photosynthesis. They can achieve high production once the shallowest pond stores concentrating biomass cultures.

Thin-layer systems are employed to obtain high microalgal biomass concentrations by using a low-depth culture (<50 mm) and maximizing light-efficiency (Acién et al., 2017). They consist of inclined platforms, sloping cascades, or near-horizontal raceways with a high S/V ratio ($25–50\,\mathrm{m}^{-1}$), that enables the optimal sun incidence to achieve high productivity, in contrast with other open ponds or raceways. Thin-layer cascades can have more than $100\,\mathrm{m}^{-1}$ S/V, with biomass productivity and density higher than $30\,\mathrm{g\,m}^{-2}\,\mathrm{d}^{-1}$ and $10\,\mathrm{g\,L}^{-1}$, respectively (Grivalský et al., 2019).

Algal turf scrubbers consist of a similar system with a substrate that supports attached growth on a sloped surface, where algae absorb nutrients from the flowing wastewater (Hoffman, Pate, Drennen, & Quinn, 2017). They are simple in design and allow a more natural biomass harvest compared to other systems. Furthermore, they exhibit stable and promising biomass productivity, being one drawback of the high content of ash (Hess et al., 2019).

3.2.4 Comparison of open systems

Open ponds are the cheapest system to cultivate microalgae on a large scale; thus, they are mainly found in industrial production. The most common are raceway and circular ponds, but attention has been recently given to thin layer and microalgal turf scrubber to increase biomass productivity. Moreover, new designs have been developed, such as a flat-plate continuous open PBR (Luo et al., 2019), which was able to achieve $0.47\,\mathrm{g\,L}^{-1}\,\mathrm{d}^{-1}$ biomass productivity while treating a piggery biogas slurry. Table 2 compares different open systems used for microalgae cultivation at a laboratory or pilot scale. Studies developed in recent years were chosen to compare biomass productivity using different systems. Biomass productivities are overall lower compared to closed systems (Table 1), but promising results were obtained using different designs of unconventional open ponds with wastewater or another effluent as nutrient and carbon sources. Moreover, the use of flue gases instead of pure CO_2 can be a sustainable and economical option to improve biomass productivity, with improved mixing and mass transfer (Cheng, Yang, Ye, Zhou, & Cen, 2016).

3.3 Hybrid system

A hybrid system for microalgae cultivation involves the integration of different phases into a two-stage system (Płaczek et al., 2017). Usually, the initial growth takes place in a closed system to obtain high concentration cultures using a closed PBR. Then following larger-scale cultivations are performed using open ponds. This strategy is employed to benefit from the advantages of both systems (Rawat, Ranjith Kumar, Mutanda, & Bux, 2013), closed and open, by achieving a strong inoculum with minimum contamination and applying nutrient stress conditions toward the desired metabolites, respectively. These steps are not only applied to improve biomass production based on the advantages and avoiding the drawbacks of each system individually but also aim to understand and then manipulate the stages of cell growth. The first step usually implies in growing microalgae to reach a late exponential growth phase, and the second, where lipid accumulation occurs more expressively, for example, is conducted during the stationary phase, even though the culture will naturally have an adaptation and exponential phase when scaling up.

Liu, Chen, Wang, and Liu (2019) cultivated *Scenedesmus dimorphus* using an open pond–PBR hybrid system. They achieved a 46.3%–74.3% increase in biomass productivity using the hybrid system compared with the open pond and a 12.5% increase compared with the PBR. Similarly, Narala et al. (2016) indicated that a hybrid system could lead to significantly higher lipid production by combining exponential growth in an air-lift PBR followed by growth under nutrient depletion in open raceway ponds. Consequently, these systems cannot only be a solution to open the pond's lower biomass and lipid productions compared to closed systems, but also improve PBR's performance in some cases.

TABLE 2   Literature review of recent microalgae cultivation studies using different open pond systems.

Open system	Open pond system and scale	Microalgae and culture media	Main results	Reference
Raceway pond	Raceway pond (0.2 m^3) with paddlewheel Pilot-scale	*Nannochloropsis* sp. Seawater and f/2	Biomass productivity: 0.04 g L^{-1} d^{-1}	Nwoba et al. (2020)
	Raceway pond (15 cm depth, 1500 L). Cultivation in batch and semicontinuous modes Pilot-scale	*Chlorella* sp. and *Scenedesmus* sp. consortium Anaerobically digested piggery effluent	Biomass productivity: 60–61 mg L^{-1} d^{-1} Lipid productivity: 5.4–21 mg L^{-1} d^{-1}	Raeisossadati, Vadiveloo, Bahri, Parlevliet, and Moheimani (2019)
	Raceway pond (3 PBRs set with 0.11 m culture height and 7.2 m^2 total surface area each) with paddlewheel Pilot-scale	*Nannochloropsis gaditana* Seawater and centrate (from wastewater)	Biomass productivity: 0.27 g L^{-1} d^{-1}	Romero-Villegas, Fiamengo, Acién-Fernández, and Molina-Grima (2018)
	Raceway reactor (1000 L) with paddlewheel. Results at different pH levels Pilot-scale	*Graesiella* sp. BG11	Biomass productivity: 0.043–0.052 g L^{-1} d^{-1}	Wang et al. (2018)
	Raceway pond (200 m^2 culture area, 20 cm depth, 40,000 L) with two paddlewheels and CO_2 injection Pilot-scale	*Graesiella* sp. BG11 salts diluted in lake's water	Biomass productivity: 6.2–8.7 g m^{-2} d^{-1} Lipid productivity: 2.0–2.9 g m^{-2} d^{-1}	Wen et al. (2016)
	Raceway pond (1191 m^2, 26 cm depth, 310 m^3) with two paddlewheels and flue gas injection Pilot-scale	*Nannochloropsis oculate* Seawater	Biomass productivity: around 2–27 g m^{-2} d^{-1}	Cheng et al. (2015)
Circular pond	Circular bioreactor (4 m^3) in a greenhouse Pilot scale	*Chlorella vulgaris* (Andersen et al., 1991)	[a]Biomass production: 1200–500 mg L^{-1} [a]Lipid productivity: 22 mg L^{-1} d^{-1}	Papapolymerou et al. (2018)
	Circular culture pond (10 L) at the indoor artificial condition. Results at different pH levels Laboratory scale	*Graesiella* sp. BG11	Biomass productivity: 0.116–0.140 g L^{-1} d^{-1}	Wang et al. (2018)
Thin layer and turf scrubber	Thin-layer reactor (0.5 cm depth, 350 L). Cultivation in batch and semicontinuous modes Pilot-scale	*Chlorella* sp. and *Scenedesmus* sp. consortium Anaerobically digested piggery effluent	Biomass productivity: 31–45 mg L^{-1} d^{-1} Lipid productivity: 8.1–29 mg L^{-1} d^{-1}	Raeisossadati et al. (2019)
	Algal turf scrubber (12 lanes, 0.975 m^2 each) with 1% slope Pilot-scale	*Klebsormidium* sp. and *Stigeoclonium* spp. Wastewater	Biomass productivity: 13.5–63.3 mg L^{-1} d^{-1}	Liu, Danneels, Vanormelingen, and Vyverman (2016)

[a] *Obtained during summer.*

4 Industrial cultivation techniques

Increasing the scale of microalgae production from laboratory to industrial scales poses even more difficulties than pilot systems. Among the discussed cultivation techniques, open ponds lower costs and become more attractive. However, control of the environmental conditions and contamination is a greater challenge.

Although many research papers found in the literature are focused on the production of biodiesel from microalgal biomass, the existing sites for microalgae cultivation at an industrial scale aim the development of aquaculture and production of carotenoids, essential fatty acids and/or animal feed. Table 3 gathers details of specific commercial products from different species, being *Haematococcus pluvialis*, *Spirulina*, *Chlorella*, or marine microalgae cultivated in open ponds the most common. The data were collected from the company's websites, and the year is referred to as the date they were founded, not exclusively for algae products.

In general, any facility with enough sunlight incidence, cheap nutrient sources, and water availability are adequate for microalgae cultivation. However, the selection between open pods and PBRs should be based on the microalgae and target products as well as parameters other than solar radiation, temperature, and day length, such as wind and rainfall onsite (Schade & Meier, 2019). Also, CO_2 bubbling on growth media that enables carbon credit and improves both biomass production and lipid accumulation is another important factor. Typical locations are coastal areas, and the majority of existing sites for microalgae cultivation are found in the United States.

Both new companies and existing ones aiming to expand their product catalog are investing in the production of microalgae. The recent alarming news of climate change effects encourages the market to provide clean energy sources with minimum impact on the environment. In that sense, microalgae are promising biomass

regarding land use (Georgianna & Mayfield, 2012), CO_2 biofixation, biomass productivity, water consumption, and many other aspects compared to terrestrial plants (Nakamura & Li-Beisson, 2016).

5 Dark fermentation—Fermenters

Dark fermentation is the process cultivation of some microalgae strains in the dark, heterotrophic environment.

5.1 Heterotrophic microalgae strains

The dinoflagellates, the green algae, and the thraustochytrids strains can grow under heterotrophic conditions. The easiest way to determine if microalgae are heterotrophy is to use the test with Biolog, Inc. to sell microtiter plates. In this test, there are 96-well plates with different organic carbon compounds that can be used to determine if and on which carbon source microalgae grow.

The genera suitable to heterotrophically grown are *Amphora*, *Ankistrodesmus*, *Arthrospira*, *Chlamydomonas*, *Chlorella*, *Chlorococcum*, *Crypthecodinium*, *Cyclotella*, *Dunaliella*, *Euglena*, *Nannochloropsis*, *Nitzschia*, *Ochromonas*, and *Tetraselmis* (Behrens, 2005). However, the heterotrophic *Dunaliella* strains or mutants were not commercially developed (Ben-Amotz, 2007).

Most microalgae culture collection strains cannot grow under heterotrophic conditions since their maintenance is in the phototrophic regime. Therefore, the isolation of strains from organic material sources is a better alternative.

5.2 Heterotrophic cultivation

The custom heterotrophic cultivation conditions are media with glucose or acetate, nitrogen and phosphorus compounds, no light, agitation (200–480 rpm), pH (6.1–6.5), with or without oxygen (Chew et al., 2018). Besides glucose

TABLE 3 Existing sites for microalgae production at the industrial scale.

Company (year)	Production facility	Main target products
Algaetech (2004)	Cultivation of *Haematococcus pluvialis*, *Spirulina* using raceway ponds in Kuala Lumpur, Malaysia	Astaxanthin, *Spirulina* powder, and tablet
Algatech (1998)	Cultivation of *Haematococcus pluvialis*, *Phaeodactylum tricornutum*, *Porphyridium cruentum*, *Nannochloropsis*, *Euglena gracilis* using photobioreactors (glass tubes) in Eilot, Israel	Astaxanthin and fucoxanthin
Algenol (2006)	Cultivation of microalgae using closed systems in Florida, United States	Algae extracts, *Spirulina* powder, *Spirulina* protein isolates, natural colorant (blue), biofuels (ethanol), crude oils
BlueBioTech (2001)	Cultivation of microalgae (e.g., *Nannochloropsis*, *Spirulina*, and *Chlorella*) in Kaltenkirchen, Germany	Phycocyanin, astaxanthin, microalgae powder (EPA, DHA, and ARA)
Cellana (2004)	Cultivation of marine microalgae using photobioreactors coupled with open ponds in Hawaii, United States	Omega-3 EPA and DHA oils, animal feed, biofuel feedstocks
Cyanotech (1983)	Cultivation *Haematococcus pluvialis* and *Spirulina* in Kailua Kona, Hawaii, United States	Astaxanthin and *Spirulina* powder and tablet
Corbion (1919), purchases: TerraVia (2016) and Solazyme (2003)	Cultivation of microalgae mainly using fermenters in different locations: AlgaPrime (DHA) in Brazil; AlgaePür (oleic), ThriveAlgae (culinary), AlgaVia (whole algae), AlgaWise (oils), and Encapso (lubricant) in the United States	DHA, high stability high oleic algal oil, culinary algae oil, whole algae ingredient, algae oils, drilling lubricant
DIC Corporation (1908)	Cultivation of *Spirulina* in Tokyo, Japan	Food supplement and blue colorant
E.I.D.—Parry (1788)	Cultivation of *Spirulina* and *Chlorella* in Onnaiyur, South India	Nutraceuticals
Kent BioEnergy Corporation (1995)	Cultivation of microalgae in California, USA	Livestock feed additives, fertilizer, and biomass-derived energy
Pond Technologies (2007)	Cultivation of microalgae (*Chlorella vulgaris*, *Arthrospira platensis*, and *Haematococcus pluvialis*) using closed systems in Ontario, Canada	Astaxanthin, phycocyanin, aquaculture
Sapphire Energy (2007)	Cultivation of microalgae in open ponds in California, United States	Crude oil
Taiwan Chlorella (1964)	Cultivation of *Chlorella* in Taiwan, China	*Chlorella* tablets, extract powder
Veramaris (2018), partnerships: DSM (1902) and Evonik (2007)	Cultivation of *Schizochytrium* sp. in giant lakes in Nebraska, United States	Omega-3 (EPA and DHA)

Glossary: eicosapentaenoic acid (EPA), docosahexaenoic acid (DHA), arachidonic acid (ARA), research and development (R&D).

and acetate, fructose, sucrose, galactose, glycerol, and mannose can be a carbon source for microalgae fermentation, too (Liang, Sarkany, & Cui, 2009). As a source of nitrogen, ammonia and nitrate are most common. The agitation of heterotrophic cultures is necessary for uniform nutrient and oxygen distribution. Oxygen is involved in the reproduction and growth of heterotrophic algae. The limiting factor for most microalgal heterotrophic cultures growth is oxygen, which makes increasing operating cost. It is due to the low solubility of oxygen in aqueous media. After a strong algae colony is formed, heterotrophic algae may ferment under anaerobic conditions.

Recently, scientists investigate cheap organic carbon sources. The agricultural waste, cattle slurry, food waste, whey permeate are some of the studied cultivation media. The solid and liquid wastes from the treatment plants and industry are very common nowadays (Table 4). These substrates are rich in carbon, nitrogen, and phosphorus, so they supplement the wastewaters for better growth of microalgae (Gladue & Maxey, 1994; Guldhe et al., 2017). However, these waste substrates are rich in various contaminants and need purifying pretreatment. Fungi and bacteria can contaminate media, too. Thus, sterilization is essential. Marine microalgae must be capable of growth in low-salinity medium. Namely, a combination of high chloride levels and the high temperatures during cultivation and sterilization respectively will result in corrosion of stainless-steel vessels (Harel & Place, 2007).

High-energy density in the organic carbon source provides high cell densities ($>100\,\mathrm{g\,L^{-1}}$) in heterotrophic cultures of microalgae (Barclay, Kirk, & Dong, 2013). However, the maximum specific growth of heterotrophic cultures is generally lower (i.e., from 0.008 to $0.098\,\mathrm{h^{-1}}$ for *Spirulina platensis* and *Chlorella vulgaris*, respectively) than that in phototrophic cultivating regime (Lee, 2007). These lower growth values are the consequence of low affinity for organic components. Excess organic compounds concentrations inhibit cell growth instead of increasing maximum specific growth (Chew et al., 2018). The only exception is *Chlorella vulgaris*, whose photosynthetic and heterotrophic maximum specific growth rates are comparable.

Algae can successfully synthesize several products. The lipids (lipids in general and docosahexaenoic—DHA), pigments, and carotenoids are the most common heterotrophic microalgae aerobic products, whereas hydrogen, alcohols, acetate, lactate, and succinate are linked to anaerobic heterotrophic microalgae production.

Owing to the absence of light, the pigmentation on the algae is eliminated. Thus, lipid productivities increase and light-induced metabolites decrease (Chew et al., 2018). Even higher lipid productivities rates are associated with low levels of nitrogen or silicate (in diatoms) (Perez-Garcia, Escalante, de Bashan, & Bashan, 2011). A type of sugar determines the accumulation of specific types of lipids. Low temperatures favor the synthesis of polyunsaturated fatty acids. The carbon-limited but nitrogen-sufficient heterotrophic cultures of *Galdieria sulphurariam* and *Spirulina platensis* (Sloth, Wiebe, & Eriksen, 2006) accumulate a high concentration of pigments (C-phycocyanin). On the other hand, *Chlorella pyrenoidosa*, *Chlorella protothecoides*, *Chlorella zofingiensis*, *Haematococcus pluvialis*, and *Dunaliella* sp. under nitrogen-limited and at very high C/N ratios synthesize carotenoid—astaxanthin, while with glucose and urea in medium synthesize carotenoid—lutein.

Under anaerobic conditions in the dark, *Chlorella*, *Scenedesmus*, and *Chlamydomonas* produce formate, acetate, ethanol, glycerol with small amounts of H_2, and CO_2, from carbohydrates (mainly starch) from cells biomass Ethanol and hydrogen, are most common anaerobic products of microalgae heterotrophic cultivation. Microalgae such as *Chlamydomonas reinhardtii* synthesize ethanol ($66.7\,\mathrm{mg\,g^{-1}}$ biomass) due to the

TABLE 4   Productivity figures for heterotrophic microalgae cultures.

Species	Application	Medium	Cultivation techniques	Biomass	Product	Reference
Galdieria sulphuraria	C-phycocyanin	S+G[a]	Continuous	$83.3\,\mathrm{g\,L^{-1}}$	$50.0\,\mathrm{mg\,L^{-1}\,d^{-1}}$	Graverholt and Eriksen (2007)
Galdieria sulphuraria	C-phycocyanin	S+G	Fed-batch	$109\,\mathrm{g\,L^{-1}}$	17.50	Graverholt and Eriksen (2007)
Chlorella protothecoides	Lipids	S+G	Fed-batch	3.2	57.8%	Xiong, Li, Xiang, and Wu (2008)
Chlorella	DHA	S+G	Fed-batch	116.2	1.02	Wu and Shi (2007)
Crypthecodinium cohnii	DHA	S+G	Fed-batch	109	–	Swaaf, Sijtsma, and Pronk (2003)
Chlorella protothecoides	Lipids	Corn powder hydrolysate (CPH)	Fed-batch	$2\,\mathrm{g\,L^{-1}\,d^{-1}}$	$932\,\mathrm{mg\,L^{-1}\,d^{-1}}$	Xu, Miao, and Wu (2006)
Schizochytrium mangrovei; Chlorella pyrenoidosa	Lipids	Food waste hydrolysate obtained via fungal hydrolysis	Batch	$20\,\mathrm{g\,L^{-1}}$	$300\,\mathrm{mg\,g^{-1}}$	Pleissner, Lam, Sun, and Lin (2013)
Chlorella vulgaris	Lipids	Soya whey	Batch	2.5	$9.8\,\mathrm{g\,L^{-1}}$	Mitra, van Leeuwen, and Lamsal (2012)
Chlorella vulgaris	Lipids	Ethanol thin stillage	Batch	5.8	43%	Mitra et al. (2012)
Chlorococcum sp. RAP13	Lipids	Diary waste supplemented with 4% and 6% of glycerol	Batch	1.4–1.9	39–42%	Ummalyma and Sukumaran (2014)
Scenedesmus sp.	Lipids	S+G supplemented with nitrate	Batch	$3.46\,\mathrm{g\,L^{-1}}$	52.6%	Ren, Liu, Ma, Zhao, and Ren (2013)
Mixed culture	Lipids	Domestic wastewater	Batch	$1.69\,\mathrm{g\,L^{-1}}$	28.2%	Prathima Devi, Venkata Subhash, and Venkata (2012)
Mixed	Lipids	Concentrated municipal wastewater	Batch	$269\,\mathrm{mg\,L^{-1}\,d^{-1}}$	$77\,\mathrm{mg\,L^{-1}\,d^{-1}}$	Zhou et al. (2011)

TABLE 4 Productivity figures for heterotrophic microalgae cultures—cont'd

Species	Application	Medium	Cultivation techniques	Biomass	Product	Reference
Auxenochlorella protothecoides UMN 280	Lipids	Municipal wastewater	Batch	$1.12\,\mathrm{g\,L^{-1}}$	28.9%	Zhou et al. (2012)
Chlorella protothecoides	Lutein	S+G	Batch	$17.2\,\mathrm{g\,L^{-1}}$	$4.43\,\mathrm{mg\,g^{-1}}$	Shi and Chen (1999)
Chlorella protothecoides	Lutein	S+G supplemented with urea	Fed-batch	$46.9\,\mathrm{g\,L^{-1}}$	$22.67\,\mathrm{mg\,L^{-1}\,d^{-1}}$	Shi and Chen (2002)
Chlorella zofingiensis	Astaxanthin	S+G	Fed-batch	$53\,\mathrm{g\,L^{-1}}$	$32\,\mathrm{mg\,L^{-1}}$	Hu, Nagarajan, Zhang, Chang, and Lee (2018)

[a] *Synthetic medium supplemented with glucose.*

high-starch content via the dark fermentation process (Hirano, Ueda, Hirayama, & Ogushi, 1997). While in sulfur-deprived medium, the same strain can produce sustained quantities of H_2 (Kosourov, Seibert, & Ghirardi, 2003). In, acid-rich effluents, such as food wastes and wastewater, *Chlorella* sp. showed the dominant H_2 producing comparing to consortia of *Scenedesmus* sp. and Diatoms under starvation (Chandra & Venkata, 2011).

5.3 Fermenters

Fermenters or stirred-tank bioreactors are process vessels used to cultivate heterotrophic microalgae. Strictly speaking, eukaryotic microalgae grow in a bioreactor under lower mixing intensity, and prokaryotic microalgae grow in a fermenter. However, manufacturers and scientists still refer to both as fermenters. Laboratory-scale fermenters are mainly batch operation mode. This mode reduced growth and productivities, and so feed-batch or continuous mode is more implied. Where product formation is strictly associated with rapid biomass growth, the pulsed fed-batch strategy is applicable (Venkata Mohan, Rohit, Chiranjeevi, Chandra, & Navaneeth, 2015). The continuous fermenter modes are chemostat (Fig. 3) and perfusion culture systems (Fig. 4).

Chemostat is a static fermenter, which could maintain high-density algal cells. While fermenter is stocked with cells, tubes continually add nutrients and oxygen and remove the spend medium and CO_2. Therefore, intracellular molecules and ions concentrations, as well as cell populations, are maintained constant over an indefinite period. Chemostat can provide 100 times higher productivity comparing to the batch mode. For example, the highest obtained cell concentration of *C. reinhardtii* was $1.5\,\mathrm{g\,L^{-1}}$ using heterotrophic chemostat culture (Chen & Johns, 1996a).

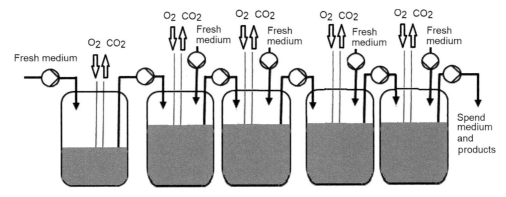

FIG. 3   Chemostat.

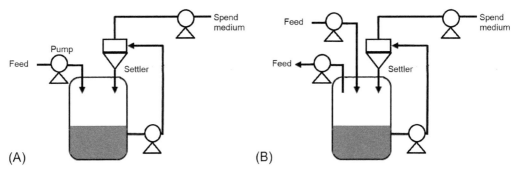

FIG. 4   (A) Perfusion and (B) perfusion-bleeding culture systems.

Perfusion culture systems (Chen & Johns, 1995) are applied when inhibitory metabolites need to be removed. Perfusion retains cells inside the fermenter while fresh media and waste and spend media are continuously provided and removed, respectively. Hollow fiber filtration or gravity settler can be used to achieve perfusion. However, clogging of the membrane is frequent and reduces cultivation efficiency. Therefore, the strategy of bleeding the cells, where the spend medium was removed while algae cells were continuously harvested, allowed higher biomass and product productivity than the single perfusion culture. For example, the highest productivity ($175\,mg\,L^{-1}\,d^{-1}$) of eicosapentaenoic acid (EPA) in microalgal cultures was reported for diatom *Nitzschia laevis* cultivated in perfusion–cell bleeding culture system (Wen & Chen, 2001).

Fermenters maintain a controlled environment to maximize productivity in strictly aseptic conditions. Owing to repeated sterilization, noncorrosive and nontoxic material must be used in fermenters design. Laboratory (research or bench-top, 1–50L) fermenters can be of borosilicate glass while pilot-plant vessels (50–1000L) are of stainless steel. Regardless of fermenter type, (1) the culture vessel, (2) input–output systems ports for nutrient media, gas, and product, and (3) control systems are common to all. Mixing is crucial for fermentation since it causes oxygen dispersion and homogenization of media. Also, during heterotrophic cultivation, high biomass levels can be expected, so proper mixing is critical (Barclay et al., 2013). Airlift, bladed turbine, and impellers are generally used mixing systems for laboratory, pilot plant- and industrial scale, respectively. For the design and construction of fermenters, sterilization must be included too.

Finally, fermenters are geometrically designed as packed-bed fermenters, bubble-column type, air-lift fermenter, and tank fermenters (Figs. 5–8). The fed-batch, chemostat, and membrane cell recycle systems were specially studied to overcome growth inhibition of low organic substrates concentrations (Chen, 1996; Hochfeld, 2006; Liu, Sun, & Chen, 2014). Very few processes for heterotrophic microalgal cultivation use membrane cell recycle systems to overcome the inhibitory effect products (Chen & Johns, 1995, 1996b).

Packed-bed fermenters are systems where microalgae are immobilized in beds (usually

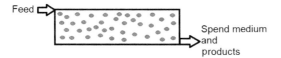

FIG. 5   Packed-bed fermenter.

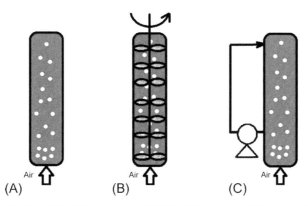

FIG. 6   (A) Bubble column, (B) stirred bubble column, and (C) bubble column with external loop.

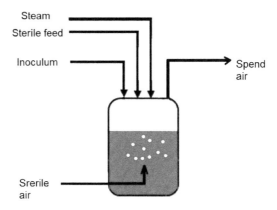

FIG. 7   Air-lift fermenter.

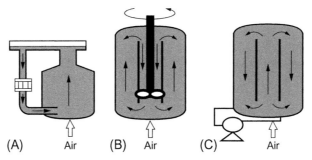

(A)   Air   (B)   Air   (C)   Air

FIG. 8   (A) Tank fermenter pressure cycle with external loop, (B) stirred tank with internal loop, and (C) tank fermenters pressure cycle with internal loop.

gel beds) to increase the productivity of biomass and products. Frequently immobilized genera of microalgae were *Chlorella* and *Scenedesmus* (Lebeau, Robert, & Subba Rao, 2006), mainly for wastewater treatment. A co-immobilized cell system can also improve the productivity of microalgae. For example, heterotrophic cultivation of *Chlorella* sp. co-immobilized with bacteria *Azospirillum brasilense* provides higher biomass productivity and higher total starch content per culture comparing to packed-bed fermenters with *Chlorella* sp. immobilized alone (Choix, de Bashan, & Bashan, 2012). The bacterial presence significantly prolonged the production of starch in this immobilized heterotrophic system.

Bubble-column type is aerobically fermenters because bubbles come from oxygen. The fermenter is a long tube with a sparging device at the bottom, which creates rising bubbles to mix algae content. During cultivation, cells grow and increase weight, settle out, and therefore, can be easily separated from the fermenter. However, bubbles cannot provide enough mixing across the whole length of the tube. Only lower tube parts maintain high densities algal cells since bubbles cannot pass faraway. Fermenting is fast and cells do not have enough time to produce desirable products. To overcome this problem, agitators or external pumps can be used in bubble column (Fig. 6B and C). Bioremediation with

heterotrophic *Phormidium* sp. was performed in a bubble column in the secondary treatment of cattle-slaughterhouse wastewater (Kalk & Langlykke, 1986).

Air-lift fermenter (Fig. 7) has a similar mechanism to a bubble column. Air from the bottom push the media and creates circulation differences in cell density over the unit. As well as the bubble column, the air-lift fermenter is not suitable for algae with both a high and low specific gravity. Algae with a high and low specific density tend to settle down and foam, respectively.

The tank fermenter pressure-cycle fermenter with an external loop is an air-lift unit (Fig. 8A). It is equipped with an external loop that provides heat removal for thermosensitive algal strains. Air is pumped from the bottom, and mixing can be enhanced by installing a mechanical mixer (Fig. 8B) or an external pump (Fig. 8C). Among these three modes of tank fermenter designs, stirred tank with the internal loop are most frequently utilized, since most of the commercially available fermenters are this construction type. The high-lipid content (55.2%) was reached from a microalga *C. protothecoids* ($15\,g\,L^{-1}$) through this type of cultivation technology (Xu et al., 2006). The highest phycocyanin production rate of $861\,mg\,L^{-1}\,d^{-1}$ was obtained in the continuous stirred tank with the internal loop flow culture of *Galdieria sulphuraria* 074G-G1 ($110\,g\,L^{-1}$) with glucose as substrate (Graverholt & Eriksen, 2007).

For microalgae technology, fermenters have several significant advantages over PBRs. The advantages and disadvantages of heterotrophic microalgae cultivation in fermenters, concerning possible commercialization, are summarized in Table 5.

5.4 Heterotrophic cultivation costs

A high cost derives from the fermentation media. Synthetic media can contain cheap components, such as glucose, sucrose, ammonium, or nitrate. However, the inclusion of yeast extract, trace metals, and vitamins increase the cost. Therefore, the desired characteristic of microalgae is the capability to grow on low-cost media, such as wastewaters.

Commercial microalgal fermentation already exists. First commercial heterotrophic fermentation production started in the United States, in the late 1980s with DHA and omega-3 fatty acids by *Crypthecodinium cohnii* algae. Nowadays, commercial success exists for various heterotrophic microalgae products (shown in Table 6).

Recent attempts to extend the commercial dark fermentation of biodiesel, renewable chemicals, and food ingredients were not with success.

Heterotrophic biomass production in fermenters is estimated to cost less than US$7/kg of dry biomass weight (Gladue & Maxey, 1994; Tredici, 2007). The cost of cultivation $45\,g\,L^{-1}$ $(20\,g\,L^{-1}\,d^{-1})$ *Nitzschia alba* in a conventional stirred-tank fermenter was estimated at US$12/kg of dry biomass (Gladue & Maxey, 1994). Doubling rates of this strain were 6.5h, which was like the one of *C. vulgaris*, and therefore, the price of dry weight was the same. Similarly, the cost of 20h doubling time algae would range from $160 to 600/kg dry (Kalk & Langlykke, 1986). However, it was reported even higher prices (US$43/kg) of dry *C. vulgaris* biomass produced for human nutrition (Brennan & Owende, 2010). In 1991, Cell Systems spray-dried *Tetraselmis suecica* and *Cyclotella cryptica* prices were approximately US$170/kg of dry weight (Gladue & Maxey, 1994). Production cost for DHA (*C. cohnii* or *Schizochytrium* sp.) is from US$25/kg (Harel & Place, 2007) to US$52/kg (Brennan & Owende, 2010), whereas astaxanthin, from *Haematococcus pluvialis*, is US$45/kg (Gladue & Maxey, 1994). The prices for natural β-carotene and phycobiliprotein range from

TABLE 5 Advantages and limitations of heterotrophic microalgae cultivation in fermenters.

Advantages	Limitations
Rapid growth	The maximum specific growth rate of heterotrophic algae is lower than of photosynthetic
High cell density	The microalgae cells also need to be unicellular (nonfilamentous)
Use of existing and standardized fermentation technology (both at the pilot plant and large-scale levels)	Not appropriate for culture collection microalgae
The ability to produce a high level of the target product	Less useful for producing light-induced metabolites
Lower fermentation cost	Axenic growth
Low harvesting cost	High costs of organic compounds substrate
Effectively recover the high-level product from the cell or the fermentation media	The necessary supply of nitrogen and phosphorus compounds

TABLE 6   Commercial heterotrophic microalgae processes.

Strain	Process	Product	Cell density	Annual capacity	Application	Country
Chlorella sp.	Glucose-limited, fed-batch fermentation	Dry biomass	$80\,\mathrm{g\,L^{-1}}$	1000 t	1. Powder or tablets nutritional supplement (36 €/kg)	Japan, Taiwan, Germany
Crypthecodinium cohnii	Fed-batch bubble-column fermenters	DHA	–	240 t of DHA	1. DHA	United States
Schizochytrium sp.	Conventional stainless-steel fermenters	1. Dry biomass 2. DHA	$200\,\mathrm{g\,L^{-1}}$	10 t of DHA	1. An aquaculture feed 2. DHA	United States

US$215/kg to US$2150/kg and US$215/kg to US$1790/kg, respectively.

Although heterotrophic culture conditions are nearly 18 times cheaper in terms of electricity consumption (Harel & Place, 2007) comparing to phototrophic processes, microalgal processes are still expensive since costs of production 450–$680\,\mathrm{g\,L^{-1}\,d^{-1}}$ baker's yeast is about US$1/kg of dry weight (Atkinson, 1991).

6 Lowering cultivation costs

Decreasing microalgae cultivation cost needs a reduction in (a) the cultivation vessel cost (simple design, smaller units, and low power consumption), (b) materials (use wastewater, saline aquifer or seawater, and flue gases), and (c) less labor. At the same time, microalgal biomass productivity should be higher than that of irrigated and fertilized agricultural plants.

Reducing cost in open pond cultivation design is orientated toward: (a) using concrete pond or dug in the ground, both with lower depth which will reduce amount of water but increase the biomass density; (b) using a clay soil or liner, as a cover, to prevent evaporation losses, rainfalls dilution, and temperature fluctuations; (c) using power plant waste heat for climate (mainly temperature) control; (d) reducing of the mean fluid velocity by substitution of paddle wheel with pumps or mixing only by wind or convection; and (e) using those microalgae strains that can grow under extreme conditions to avoid the contamination.

Generally, PBRs have higher operating costs than open ponds. Therefore, large-scale biomass production in raceway ponds uses small-scale PBRs for inoculum preparation (Grima, Fernández, & Medina, 2013). However, many companies are still investing in PBRs research because their area biomass production is seven times higher than open ponds. Except for contamination and evaporation, as in open systems, there are many problems to be solved before the commercialization of PBRs. Illumination as a limiting factor for phototrophic microalgae growth is one of the major problems. Outdoor PBRs installation can reduce the high cost of artificial light and temperature control. Nevertheless, natural illumination has one disadvantage—the sun does not shine 24 h a day. During nights or on heavily shaded occasions, PBRs can be supplement with fluorescent lamps. Another way is to use microalgae strains, which grow in light/night regime. Thus, temperature control by immersion in a water pool is already reported (Grima et al., 2013). Pump mixed cells and air bubbling are very efficient ways of mixing. Mixing and thermoregulation by wave energy in floating PBRs can be efficient, too (Wiley et al., 2013). However, weathering and storms over the open waters are significant concerns. Another auspicious PBRs designs are vertical or inclined flat PBRs, which works as a solar tracking collector's system (Qiang, Faiman, & Richmond, 1998). Air bubbling can be adopted in this design, but it might not be necessary since moving (tracking) turbulence plays a vital role in mixing.

Similar problems to those of PBRs exist for fermenters type of design. Here, illumination is excluded, but a high cost should be credited to the high-power consumption for mixing of the high viscosity fermentation broth of filamentous microalgae. Using unicellular microalgae strains can significantly reduce cultivation costs in fermenters. These unicellular strains should also be able to produce high cell densities in the fermenters at deficient dissolved oxygen concentrations. Fermenters allow full control, but due to space requirements, it should be able to operate outdoor.

6.1 Cultivation in wastewater

Reduction in cost associated with culture medium should be oriented to the contribution of the power plants waste gases (carbon dioxide) and wastewaters (organic carbon) (Gouveia et al., 2016).

The wastewater environment is extensive. An anaerobic fermentation waste, diluted industrial wastewater from olive-oil extraction, effluents from the electroplating, leather and textile industries, hog wastewater, the leachate of a suburban landfill, a marine sewage outfall, municipal "grey—water" wastewater, palm oil mill effluent, pig manure, the secondary effluent of a wastewater treatment stabilization pond, sewage outfall, and struvite (magnesium ammonium phosphate hexahydrate) as precipitates from wastewater, wastewater discharge from a steel plant, the waters of a domestic drainage canal are among common ones (Barclay & Apt,

2013; Brennan & Owende, 2010; Chew et al., 2018; Gressel, 2013).

Monfet and Unc (2017) give a pervasive literature review of various wastewaters used for algal cultivation, as well as their nutrient composition. Besides a few studies which use raw or untreated wastewater, as cultivation media, many wastewaters must be pretreated. Filtration, dilution, metal ions removal, electro flocculation (Hu et al., 2018), sterilization, UV treatment, settling, centrifugation, biologically treated, mixing with seawater, trace elements, and polymers, effluents of internal circulation reactor are usually implemented methods for wastewater pretreatment. Many studies have investigated different wastewaters pretreatment technologies and their impact on microalgae cultivation. Thermal pretreatment or sterilization is the most widely used method for providing axenic conditions in microalgae cultivation media. Sterilization has been employed extensively as an inoculum pretreatment, too, at both laboratory and industrial scale. However, selecting pretreatment should be guided with the lowering of energy costs. Settling and mixing with internal circulation reactor effluents (instead of dilution with freshwater) are the most promising energy-efficient techniques.

Different wastewaters have similar organic but different inorganic contents. That means that their most influenced nutrient ratios, such as nitrogen to phosphorus ratio (N:P), are different. Based on the N:P ratio, the most suitable wastewaters are from the dairy industry, domestic wastewaters, and anaerobically digested municipal sewage from the swine industry. However, the choice of wastewater does not only depend on the N:P ratio and product yield but also local availability.

Unique species that can be cultivated in all types of wastewaters do not exist. The most effective are the ones isolated from a wastewater environment in which will be later cultivated. A nutrient removal efficiency and biomass production depend on the number of species.

It was shown that consortiums do not give better results (Stockenreiter, Haupt, Seppälä, Tamminen, & Spilling, 2016). However, in wastewater treatment (Gonçalves, Pires, & Simões, 2017) as well as for increased biomass productivity, crop protection, and benefits of mutualism, improving the stability (Kazamia, Riseley, Howe, & Smith, 2014) microalgal consortia is shown to be very advantageous. The eight most applied microalgae genera are *Chlamydomonas, Chlorella, Euglena, Navicula, Nitzschia, Oscillatoria, Scenedesmus,* and *Stigeoclonium* (Chew et al., 2018).

6.2 Cultivation for high-value products

Owing to a great diversity in the chemical composition, microalgae are extremely attractive to produce a wide range of valuable chemicals. There is a long history of commercial use of microalgae for various high-value products, such as health food (Tamiya, 1957), β-carotene (Ben-Amotz, Shaish, & Avron, 1991), astaxanthin (Lorenz & Cysewski, 2000), and DHA (Kyle, 2010). Microalgae can produce vitamins such as vitamins C, E, and B12, too. However, their production costs are higher than the vitamin's production using bacteria, fungi, or higher plants. Although dating back to 2013, Borowitzka (2013) gave a very significant overview of high-value products from microalgae, summarized in Table 7.

A biorefinery is a refinery in which mechanical, biochemical, chemical, and thermochemical processes transform biomass (feedstock) to various bioproducts. Based on the feedstock used, there are sugar-, starch-, lignocellulosic- and oil-based crops, grasses, and marine biomass biorefineries. There are two types of bioproducts: energy (biofuels, heat, and power) and material (i.e., sugars and proteins from sugarcane biorefinery).

Among other types, sugarcane biorefinery is well established since it combines the conversion of both sugar and lignocellulose to a final bioproduct.

The fusing of biofuel production with wastewater (vinasse) microalgae cultivation and CO_2 mitigation from fermenters provides a reduction in the cost of biofuel as well as environmental benefits of sugarcane biorefinery. Vinasse can be a medium for *Arthrospira maxima* OF15, *Chlamydomonas reinhardtii, Micractinium* sp. ME05, *Chlamydomonas biconvexa* for potential biological new peptides, hydrogen, biodiesel, carbohydrates production, respectively (Buitrón, Carrillo-Reyes, Morales, Faraloni, & Torzillo, 2017; Engin, Cekmecelioglu, Yücel, & Oktem, 2018; Montalvo et al., 2019; Santana et al., 2017).

Recently, digestion (anaerobically digested) effluents gathered most concerns from the environmental point of view, and extensive research had been directed toward it. One option is anaerobically digested wastewater treatment by microalgae with the benefit of multi-product production, especially biofuels.

7 Conclusions

This review summarizes all microalgae cultivation techniques and technologies and discusses the factors that influence biomass production and metabolites accumulation.

Considering all those systems and techniques, it shows to be evident the potential of microalgae cultivation. However, there are still challenges to be overcome in terms of cost and scale-up. Future large-scale cultivation advances and new technologies could come to be the microalgae cultivation for bioproducts economically feasible.

The studies published in the literature shown that the PBRs are more expensive when compared with open systems, specifically raceway ponds that are the most used. The PBRs reach up to three times more biomass productivity and are easily scalable. Currently, hybrid configurations bring many advantages once that

TABLE 7 Summary of high-value products from microalgae.

Group of products	Specific high-value product	The best algal sources
Carotenoids	β-Carotene	*Dunaliella salina*
	Astaxanthin	*Haematococcus pluvialis, Chlorella zofigiensis*
	Lutein	*Muriellopsis* sp., *Scenedesmus almeriensis, Chlorella potothecoides, Chlorella zofigiensis*
	Canthaxanthin	*Chlorella zofigiensis, Scenedesmus komareckii, Dunaliella salina*
	Zeaxanthin	*Dunaliella salina*
	Echinenone	*Botryococcus braunii*
	Fucoxanthin	*Phaeodactylum tricornutum*
	Phytoene	*Dunaliella salina*
	Phytofluene	*Dunaliella salina*
Pigments	Phycocyanin	*Spirulina, Porphyridium,* Rhodella, *Galdieria, Cryptophyta, Glaucophyta*
	Phycoerythrin	
	Allophycocyanin	
Fatty acids	γ-Linolenic acid	*Spirulina platensis*
	Arachidonic acid	*Mortiriella alpine, Parietochloris incisa*
	EPA	*Schizochytrium, Nannochloropsis* *Ulkneia* sp. *Nitzschia alba* *Nitzschia laevis*
	DHA	*Crypthecodinium cohnii, Schizochytrium, Nannochloropsis*
Sterols	Brassicasterol	*Olisthodiscus luteus*
	Poriferasterol	
	Fucosterol	
	Squalene	*Botryococcus, Aurantiochytrium*
Polyhydroxyalkonates	poly-3-Hydroxybutyrate (P3HB)	*Spirulina, Synechocystis*
	poly-3-Hydroxybutyrate-*co*-3-hydroxyvalerate (P3HB-co-3HV)	*Nostoc muscorum*
Biologically active compounds	Antioxidant	*Tetraselmis suecica, Botryococcus braunii, Neochloris oleoabundans, Isochrysis* sp., *Chlorella vulgaris, Phaeodactylum tricornutum, Porphyridium purpureum*
	Antibiotic	*Gambierdiscus toxicus, Chlorococcum* strain HS-101, *Dunaliella primolecta*

Continued

TABLE 7   Summary of high-value products from microalgae—cont'd

Group of products	Specific high-value product	The best algal sources
	Antiviral	*Spirulina platensis*
	Anticancer	*Poteriochromonas malhamensis, Scytonema ocellatum*
	Antihypotensive	*Microcystis* spp., *Dunaliella salina*
	Dolastatin 10	*Symploca* sp.,
Proteins		*Tetraselmis* sp.
Food gels		*Spirulina, Haematococcus*
Amino acids	Shionorine	*Porphyra umbilicalis*
	Porphyra-334	
Food supplements	Biscuits enriched with ω-3 fatty acids	*Isochrysis galbana*

includes the advantages of both systems—different configurations are already proposed and used (Bose et al., 2019; García-Galán et al., 2018; Zhou et al., 2014). Generally, the cultivation starts in the PBR for accelerating biomass productivity and then scale-upping to open systems (Aziz et al., 2020).

From an application point of view, further research should also focus on heterotrophic cultivation systems using wastewaters. Also, the continuous cultivation techniques, such as airlift fermenters, tank fermenter pressure cycle with external loop, or tank fermenters pressure cycle with internal loop for heterotrophic microalgae cultures, are necessary to develop.

Statement

We declare that all figures and tables presented in this chapter were made by the authors and have not been published elsewhere.

References

Acién, F. G., Molina, E., Reis, A., Torzillo, G., Zittelli, G. C., Sepúlveda, C., et al. (2017). Photobioreactors for the production of microalgae. In *Microalgae-based biofuels and bioproducts* (pp. 1–44). Elsevier. https://doi.org/10.1016/B978-0-08-101023-5.00001-7.

Andersen, R. A., Jacobson, D. M., & Sexton, J. P. (1991). *Provasoli-Guillard Center for culture of marine phytoplankton catalog of strains 1991*: (p. 98). West Boothbay Harbor, ME, USA.

Atkinson, B. (1991). *Mavituna F.* Biochemical engineering and biotechnology handbook: Stockton Press.

Ayre, J. M., Moheimani, N. R., & Borowitzka, M. A. (2017). Growth of microalgae on undiluted anaerobic digestate of piggery effluent with high ammonium concentrations. *Algal Research, 24,* 218–226. https://doi.org/10.1016/j.algal.2017.03.023.

Aziz, M. M. A., Kassim, K. A., Shokravi, Z., Jakarni, F. M., Liu, H. Y., Zaini, N., et al. (2020). Two-stage cultivation strategy for simultaneous increases in growth rate and lipid content of microalgae: A review. *Renewable and Sustainable Energy Reviews, 119,* 109621. https://doi.org/10.1016/j.rser.2019.109621.

Barclay, W., & Apt, K. (2013). Strategies for bioprospecting microalgae for potential commercial applications. In *Handbook of microalgal culture* (pp. 69–79). John Wiley & Sons, Ltd. https://doi.org/10.1002/9781118567166.ch4.

Barclay, W., Kirk, A., & Dong, X. D. (2013). Commercial production of microalgae via fermentation. In *Handbook of microalgal culture* (pp. 134–145). (1st ed.). John Wiley & Sons, Ltd. https://doi.org/10.1002/9781118567166.

Behrens, P. W. (2005). Photobioreactors and fermentors: The light and dark sides of growing algae. In R. A. Andersen (Ed.), *Algal culturing techniques* (pp. 273–280). Elsevier.

Ben-Amotz, A. (2007). Industrial production of microalgal cell-mass and secondary products—Major industrial species: Dunaliella. In *Handbook of microalgal culture* (pp. 273–280). John Wiley & Sons, Ltd. https://doi.org/10.1002/9780470995280.ch13.

Ben-Amotz, A., Shaish, A., & Avron, M. (1991). The biotechnology of cultivating Dunaliella for production of β-carotene rich algae. *Bioresource Technology*, *38*, 233–235. https://doi.org/10.1016/0960-8524(91)90160-L.

Borowitzka, M. A. (2013). High-value products from microalgae—Their development and commercialisation. *Journal of Applied Phycology*, *25*, 743–756. (2013). https://doi.org/10.1007/s10811-013-9983-9.

Borowitzka, M. A., & Moheimani, N. R. (Eds.). (2013). *Algae for biofuels and energy*. Dordrecht: Springer Netherlands. https://doi.org/10.1007/978-94-007-5479-9.

Bose, A., Lin, R., Rajendran, K., O'Shea, R., Xia, A., & Murphy, J. D. (2019). How to optimise photosynthetic biogas upgrading: A perspective on system design and microalgae selection. *Biotechnology Advances*, *37*, 107444. https://doi.org/10.1016/j.biotechadv.2019.107444.

Brennan, L., & Owende, P. (2010). Biofuels from microalgae—A review of technologies for production, processing, and extractions of biofuels and co-products. *Renewable and Sustainable Energy Reviews*, *14*, 557–577. https://doi.org/10.1016/j.rser.2009.10.009.

Buitrón, G., Carrillo-Reyes, J., Morales, M., Faraloni, C., & Torzillo, G. (2017). Biohydrogen production from microalgae. (chapter 9)In C. Gonzalez-Fernandez, & R. Muñoz (Eds.), *Microalgae-based biofuels and bioproducts* (pp. 209–234). Woodhead Publishing. https://doi.org/10.1016/B978-0-08-101023-5.00009-1.

Bussa, M., Eisen, A., Zollfrank, C., & Röder, H. (2019). Life cycle assessment of microalgae products: State of the art and their potential for the production of polylactid acid. *Journal of Cleaner Production*, *213*, 1299–1312. https://doi.org/10.1016/j.jclepro.2018.12.048.

Bux, F., & Chisti, Y. (Eds.). (2016). *Algae biotechnology*. Cham: Springer International Publishing. https://doi.org/10.1007/978-3-319-12334-9.

Chandra, R., & Venkata, M. S. (2011). Microalgal community and their growth conditions influence biohydrogen production during integration of dark-fermentation and photo-fermentation processes. *International Journal of Hydrogen Energy*, *36*, 12211–12219. https://doi.org/10.1016/j.ijhydene.2011.07.007.

Chaudhary, R., Dikshit, A. K., & Tong, Y. W. (2018). Carbon-dioxide biofixation and phycoremediation of municipal wastewater using Chlorella vulgaris and Scenedesmus obliquus. *Environmental Science and Pollution Research*, *25*, 20399–20406. https://doi.org/10.1007/s11356-017-9575-3.

Chen, F. (1996). High cell density culture of microalgae in heterotrophic growth. *Trends in Biotechnology*, *14*, 421–426. https://doi.org/10.1016/0167-7799(96)10060-3.

Chen, F., & Johns, M. R. (1995). A strategy for high cell density culture of heterotrophic microalgae with inhibitory substrates. *Journal of Applied Phycology*, *7*, 43–46. https://doi.org/10.1007/BF00003548.

Chen, F., & Johns, M. R. (1996a). Heterotrophic growth of Chlamydomonas reinhardtii on acetate in chemostat culture. *Process Biochemistry*, *31*, 601–604. https://doi.org/10.1016/S0032-9592(96)00006-4.

Chen, F., & Johns, M. R. (1996b). High cell density culture of Chlamydomonas reinhardtii on acetate using fed-batch and hollow-fibre cell-recycle systems. *Bioresource Technology*, *55*, 103–110. https://doi.org/10.1016/0960-8524(95)00141-7.

Cheng, J., Yang, Z., Huang, Y., Huang, L., Hu, L., Xu, D., et al. (2015). Improving growth rate of microalgae in a 1191m2 raceway pond to fix CO2 from flue gas in a coal-fired power plant. *Bioresource Technology*, *190*, 235–241. https://doi.org/10.1016/j.biortech.2015.04.085.

Cheng, J., Yang, Z., Ye, Q., Zhou, J., & Cen, K. (2016). Improving CO2 fixation with microalgae by bubble breakage in raceway ponds with up–down chute baffles. *Bioresource Technology*, *201*, 174–181. https://doi.org/10.1016/j.biortech.2015.11.044.

Chew, K. W., Chia, S. R., Show, P. L., Yap, Y. J., Ling, T. C., & Chang, J.-S. (2018). Effects of water culture medium, cultivation systems and growth modes for microalgae cultivation: A review. *Journal of the Taiwan Institute of Chemical Engineers*, *91*, 332–344. https://doi.org/10.1016/j.jtice.2018.05.039.

Chew, K. W., Yap, J. Y., Show, P. L., Suan, N. H., Juan, J. C., Ling, T. C., et al. (2017). Microalgae biorefinery: High value products perspectives. *Bioresource Technology*, *229*, 53–62. https://doi.org/10.1016/j.biortech.2017.01.006.

Chinnasamy, S., Bhatnagar, A., Hunt, R. W., & Das, K. C. (2010). Microalgae cultivation in a wastewater dominated by carpet mill effluents for biofuel applications. *Bioresource Technology*, *101*, 3097–3105. https://doi.org/10.1016/j.biortech.2009.12.026.

Chisti, Y. (2007). Biodiesel from microalgae. *Biotechnology Advances*, *25*, 294–306. https://doi.org/10.1016/j.biotechadv.2007.02.001.

Chiu, S.-Y., Tsai, M.-T., Kao, C.-Y., Ong, S.-C., & Lin, C.-S. (2009). The air-lift photobioreactors with flow patterning for high-density cultures of microalgae and carbon dioxide removal. *Engineering in Life Sciences*, *9*, 254–260. https://doi.org/10.1002/elsc.200800113.

Choix, F. J., de Bashan, L. E., & Bashan, Y. (2012). Enhanced accumulation of starch and total carbohydrates in alginate-immobilized Chlorella spp. induced by Azospirillum brasilense: II. Heterotrophic conditions. *Enzyme and Microbial Technology*, *51*, 300–309. https://doi.org/10.1016/j.enzmictec.2012.07.012.

Chu, S. P. (1942). The influence of the mineral composition of the medium on the growth of planktonic algae: Part I. Methods and culture media. *Journal of Ecology*, *30*, 284–325. https://doi.org/10.2307/2256574.

Cruz, Y. R., Aranda, D. A. G., Seidl, P. R., Diaz, G. C., Carliz, R. G., Fortes, M. M., et al. (2018). Cultivation systems of microalgae for the production of biofuels. In *Biofuels state of development*. IntechOpen. https://doi.org/10.5772/intechopen.74957.

Cuello, J. L., Hoshino, T., Kuwahara, S., & Brown, C. L. (2016). Scale-up—Bioreactor design and culture optimization. *Biotechnology for Biofuel Production and Optimization*, 497–511. https://doi.org/10.1016/B978-0-444-63475-7.00019-4.

Duarte, J. H., Fanka, L. S., & Costa, J. A. V. (2016). Utilization of simulated flue gas containing CO2, SO2, NO and ash for Chlorella fusca cultivation. *Bioresource Technology*, *214*, 159–165. https://doi.org/10.1016/j.biortech.2016.04.078.

El-Baz, F. K., & Baky, H. H. A. E. (2018). Pilot scale of microalgal production using photobioreactor. In C. JCG, & L. GLL (Eds.), *Photosynthesis—From its evolution to future improvements in photosynthetic efficiency using nanomaterials*. InTech. https://doi.org/10.5772/intechopen.78780.

Engin, I. K., Cekmecelioglu, D., Yücel, A. M., & Oktem, H. A. (2018). Evaluation of heterotrophic and mixotrophic cultivation of novel Micractinium sp. ME05 on vinasse and its scale up for biodiesel production. *Bioresource Technology*, *251*, 128–134. https://doi.org/10.1016/j.biortech.2017.12.023.

Eriksen, N. T. (2008). The technology of microalgal culturing. *Biotechnology Letters*, *30*, 1525–1536. https://doi.org/10.1007/s10529-008-9740-3.

Ferreira, A., Marques, P., Ribeiro, B., Assemany, P., de Mendonça, H. V., Barata, A., et al. (2018). Combining biotechnology with circular bioeconomy: From poultry, swine, cattle, brewery, dairy and urban wastewaters to biohydrogen. *Environmental Research*, *164*, 32–38. https://doi.org/10.1016/j.envres.2018.02.007.

Ferreira, G. F., Ríos Pinto, L. F., Maciel Filho, R., & Fregolente, L. V. (2019). A review on lipid production from microalgae: Association between cultivation using waste streams and fatty acid profiles. *Renewable and Sustainable Energy Reviews*, *109*, 448–466. https://doi.org/10.1016/j.rser.2019.04.052.

García-Galán, M. J., Gutiérrez, R., Uggetti, E., Matamoros, V., García, J., & Ferrer, I. (2018). Use of full-scale hybrid horizontal tubular photobioreactors to process agricultural runoff. *Biosystems Engineering*, *166*, 138–149. https://doi.org/10.1016/j.biosystemseng.2017.11.016.

Georgianna, D. R., & Mayfield, S. P. (2012). Exploiting diversity and synthetic biology for the production of algal biofuels. *Nature*, *488*, 329–335. https://doi.org/10.1038/nature11479.

Gim, G. H., Kim, J. K., Kim, H. S., Kathiravan, M. N., Yang, H., Jeong, S.-H., et al. (2014). Comparison of biomass production and total lipid content of freshwater green microalgae cultivated under various culture conditions. *Bioprocess and Biosystems Engineering*, *37*, 99–106. https://doi.org/10.1007/s00449-013-0920-8.

Gladue, R. M., & Maxey, J. E. (1994). Microalgal feeds for aquaculture. *Journal of Applied Phycology*, *6*, 131–141. https://doi.org/10.1007/BF02186067.

Gonçalves, A. L., Pires, J. C. M., & Simões, M. (2017). A review on the use of microalgal consortia for wastewater treatment. *Algal Research*, *24*, 403–415. https://doi.org/10.1016/j.algal.2016.11.008.

Gordillo, F. J. L., Goutx, M., Figueroa, F. L., & Niell, F. X. (1998). Effects of light intensity, CO2 and nitrogen supply on lipid class composition of Dunaliella viridis. *Journal of Applied Phycology*, *10*, 135–144. https://doi.org/10.1023/A:1008067022973.

Gouveia, L., Graça, S., Sousa, C., Ambrosano, L., Ribeiro, B., Botrel, E. P., et al. (2016). Microalgae biomass production using wastewater: Treatment and costs: Scale-up considerations. *Algal Research*, *16*, 167–176. https://doi.org/10.1016/j.algal.2016.03.010.

Graverholt, O. S., & Eriksen, N. T. (2007). Heterotrophic high-cell-density fed-batch and continuous-flow cultures of Galdieria sulphuraria and production of phycocyanin. *Applied Microbiology and Biotechnology*, *77*, 69–75. https://doi.org/10.1007/s00253-007-1150-2.

Gressel, J. (2013). Transgenic marine microalgae: A value-enhanced fishmeal and fish oil replacement. In *Handbook of microalgal culture* (pp. 653–670). John Wiley & Sons, Ltd. https://doi.org/10.1002/9781118567166.ch35.

Grima, E. M., Fernández, F. G. A., & Medina, A. R. (2013). Downstream processing of cell mass and products. In *Handbook of microalgal culture* (pp. 267–309). John Wiley & Sons, Ltd. https://doi.org/10.1002/9781118567166.ch14.

Grivalský, T., Ranglová, K., da Câmara Manoel, J. A., Lakatos, G. E., Lhotský, R., & Masojídek, J. (2019). Development of thin-layer cascades for microalgae cultivation: Milestones (review). *Folia Microbiologia (Praha)*, *64*, 603–614. https://doi.org/10.1007/s12223-019-00739-7.

Guillard, R. R. L. (1975). Culture of phytoplankton for feeding marine invertebrates. In: W. L. Smith, & M. H. Chanley (Eds.), *Culture of marine invertebrates animals*, New York: Plenum Press, pp. 29–60. https://doi.org/10.1007/978-1-4615-8714-9_3.

Guillard, R. R. L., & Ryther, J. H. (1962). Studies of marine planktonic diatoms: I. Cyclotella Nana Hustedt, and Detonula Confervacea (cleve) gran. *Canadian Journal of Microbiology*, *8*, 229–239. https://doi.org/10.1139/m62-029.

Guldhe, A., Kumari, S., Ramanna, L., Ramsundar, P., Singh, P., Rawat, I., et al. (2017). Prospects, recent advancements and challenges of different wastewater streams for microalgal cultivation. *Journal of Environmental Management*, *203*, 299–315. https://doi.org/10.1016/j.jenvman.2017.08.012.

Gupta, S., Pawar, S. B., Pandey, R. A., Kanade, G. S., & Lokhande, S. K. (2019). Outdoor microalgae cultivation in airlift photobioreactor at high irradiance and temperature conditions: Effect of batch and fed-batch strategies, photoinhibition, and temperature stress. *Bioprocess and Biosystems Engineering*, *42*, 331–344. https://doi.org/10.1007/s00449-018-2037-6.

Hamed, I. (2016). The evolution and versatility of microalgal biotechnology: A review. *Comprehensive Reviews in Food Science and Food Safety*, *15*, 1104–1123. https://doi.org/10.1111/1541-4337.12227.

Hanifzadeh, M. M., Sarrafzadeh, M. H., & Tavakoli, O. (2012). Carbon dioxide biofixation and biomass production from flue gas of power plant using microalgae. In: *2012 Second Iranian conference on renewable energy and distributed generation*, pp. 61–64.

Harel, M., & Place, A. R. (2007). Heterotrophic production of marine algae for aquaculture. In *Handbook of microalgal culture* (pp. 513–524). John Wiley & Sons, Ltd. https://doi.org/10.1002/9780470995280.ch31.

Hess, D., Wendt, L. M., Wahlen, B. D., Aston, J. E., Hu, H., & Quinn, J. C. (2019). Techno-economic analysis of ash removal in biomass harvested from algal turf scrubbers. *Biomass and Bioenergy*, *123*, 149–158. https://doi.org/10.1016/j.biombioe.2019.02.010.

Hirano, A., Ueda, R., Hirayama, S., & Ogushi, Y. (1997). CO2 fixation and ethanol production with microalgal photosynthesis and intracellular anaerobic fermentation. *Energy*, *22*, 137–142. https://doi.org/10.1016/S0360-5442(96)00123-5.

Hochfeld, W. L. (2006). *Producing biomolecular substances with fermenters, bioreactors, and biomolecular synthesizers*. CRC Presshttps://doi.org/10.1201/9781420021318.

Hoffman, J., Pate, R. C., Drennen, T., & Quinn, J. C. (2017). Techno-economic assessment of open microalgae production systems. *Algal Research*, *23*, 51–57. https://doi.org/10.1016/j.algal.2017.01.005.

Hu, J., Nagarajan, D., Zhang, Q., Chang, J.-S., & Lee, D.-J. (2018). Heterotrophic cultivation of microalgae for pigment production: A review. *Biotechnology Advances*, *36*, 54–67. https://doi.org/10.1016/j.biotechadv.2017.09.009.

Ishika, T., Moheimani, N. R., Laird, D. W., & Bahri, P. A. (2019). Stepwise culture approach optimizes the biomass productivity of microalgae cultivated using an incremental salinity increase strategy. *Biomass and Bioenergy*, *127*, 105274. https://doi.org/10.1016/j.biombioe.2019.105274.

Jacob-Lopes, E., Scoparo, C. H. G., Lacerda, L. M. C. F., & Franco, T. T. (2009). Effect of light cycles (night/day) on CO2 fixation and biomass production by microalgae in photobioreactors. *Chemical Engineering and Processing Process Intensification*, *48*, 306–310. https://doi.org/10.1016/j.cep.2008.04.007.

Ji, F., Hao, R., Liu, Y., Li, G., Zhou, Y., & Dong, R. (2013). Isolation of a novel microalgae strain Desmodesmus sp. and optimization of environmental factors for its biomass production. *Bioresource Technology*, *148*, 249–254. https://doi.org/10.1016/j.biortech.2013.08.110.

Ji, M.-K., Yun, H.-S., Hwang, J.-H., Salama, E.-S., Jeon, B.-H., & Choi, J. (2017). Effect of flue gas CO2 on the growth, carbohydrate and fatty acid composition of a green microalga *Scenedesmus obliquus* for biofuel production. *Environmental Technology*, *38*, 2085–2092. https://doi.org/10.1080/09593330.2016.1246145.

Kalk, J. P., & Langlykke, A. F. (1986). Cost estimation for biotechnology projects. In A. L. Demain, & N. A. Solomon (Eds.), *Manual of industrial microbiology and biotechnology*. Washington, DC: Amer. Soc. Microbiol.

Kazamia, E., Riseley, A. S., Howe, C. J., & Smith, A. G. (2014). An engineered community approach for industrial cultivation of microalgae. *Industrial Biotechnology (New Rochelle, NY)*, *10*, 184–190. https://doi.org/10.1089/ind.2013.0041.

Khan, M. I., Shin, J. H., & Kim, J. D. (2018). The promising future of microalgae: Current status, challenges, and optimization of a sustainable and renewable industry for biofuels, feed, and other products. *Microbial Cell Factories*, *17*, 36. (2018). https://doi.org/10.1186/s12934-018-0879-x.

Klinthong, W., Yang, Y.-H., Huang, C.-H., & Tan, C.-S. (2015). A review: Microalgae and their applications in CO2 capture and renewable energy. *Aerosol and Air Quality Research*, *15*, 712–742. https://doi.org/10.4209/aaqr.2014.11.0299.

Klok, A. J., Lamers, P. P., Martens, D. E., Draaisma, R. B., & Wijffels, R. H. (2014). Edible oils from microalgae: Insights in TAG accumulation. *Trends in Biotechnology*, *32*, 521–528. https://doi.org/10.1016/j.tibtech.2014.07.004.

Koc, C., Anderson, G. A., & Kommareddy, A. (2013). Use of red and blue light-emitting diodes (LED) and fluorescent lamps to grow microalgae in a photobioreactor. *The Israeli Journal of Aquaculture – Bamidgeh*, *65*, 797.

Koller, A. P., Wolf, L., Brück, T., & Weuster-Botz, D. (2018). Studies on the scale-up of biomass production with Scenedesmus spp. in flat-plate gas-lift photobioreactors. *Bioprocess and Biosystems Engineering*, *41*, 213–220. https://doi.org/10.1007/s00449-017-1859-y.

Kosourov, S., Seibert, M., & Ghirardi, M. L. (2003). Effects of extracellular pH on the metabolic pathways in sulfur-deprived, H2-producing Chlamydomonas reinhardtii cultures. *Plant & Cell Physiology*, *44*, 146–155. https://doi.org/10.1093/pcp/pcg020.

Kyle, D. J. (2010). Future development of single cell oils. (chapter 20)In Z. Cohen, & C. Ratledge (Eds.), *Single cell oils* (pp. 439–451). (2nd ed.). AOCS Press. https://doi.org/10.1016/B978-1-893997-73-8.50024-4.

Lebeau, T., Robert, J.-M., & Subba Rao, D.-V. (2006). Biotechnology of immobilized micro algae: a culture technique for the future? In D. V. S. Rao (Ed.), *Algal cultures, analogues of blooms and applications* (pp. 801–837). Enfield, NH, USA: Science Publishers.

Lee, Y.-K. (1986). Enclosed bioreactors for the mass cultivation of photosynthetic microorganisms: The future trend. *Trends in Biotechnology*, 4, 186–189. https://doi.org/10.1016/0167-7799(86)90243-X.

Lee, Y.-K. (2007). Algal nutrition—Heterotrophic carbon nutrition. In *Handbook of microalgal culture* (pp. 116–124). John Wiley & Sons, Ltd. https://doi.org/10.1002/9780470995280.ch7.

Lee, J. W. (Ed.). (2013). *Advanced biofuels and bioproducts*. New York, NY: Springer New York. https://doi.org/10.1007/978-1-4614-3348-4.

Liang, Y., Sarkany, N., & Cui, Y. (2009). Biomass and lipid productivities of Chlorella vulgaris under autotrophic, heterotrophic and mixotrophic growth conditions. *Biotechnology Letters*, 31, 1043–1049. https://doi.org/10.1007/s10529-009-9975-7.

Liu, X., Chen, G., Tao, Y., & Wang, J. (2020). Application of effluent from WWTP in cultivation of four microalgae for nutrients removal and lipid production under the supply of CO2. *Renewable Energy*, 149, 708–715. https://doi.org/10.1016/j.renene.2019.12.092.

Liu, W., Chen, Y., Wang, J., & Liu, T. (2019). Biomass productivity of *Scenedesmus dimorphus* (Chlorophyceae) was improved by using an open pond–photobioreactor hybrid system. *European Journal of Phycology*, 54, 127–134. https://doi.org/10.1080/09670262.2018.1519601.

Liu, J., Danneels, B., Vanormelingen, P., & Vyverman, W. (2016). Nutrient removal from horticultural wastewater by benthic filamentous algae Klebsormidium sp., Stigeoclonium spp. and their communities: From laboratory flask to outdoor algal turf scrubber (ATS). *Water Research*, 92, 61–68. https://doi.org/10.1016/j.watres.2016.01.049.

Liu, J., Sun, Z., & Chen, F. (2014). Heterotrophic production of algal oils. (chapter 6)In A. Pandey, D.-J. Lee, Y. Chisti, & C. R. Soccol (Eds.), *Biofuels algae* (pp. 111–142). Amsterdam: Elsevier. https://doi.org/10.1016/B978-0-444-59558-4.00006-1.

Lorenz, R. T., & Cysewski, G. R. (2000). Commercial potential for Haematococcus microalgae as a natural source of astaxanthin. *Trends in Biotechnology*, 18, 160–167. https://doi.org/10.1016/S0167-7799(00)01433-5.

Luo, L., Lin, X., Zeng, F., Luo, S., Chen, Z., & Tian, G. (2019). Performance of a novel photobioreactor for nutrient removal from piggery biogas slurry: Operation parameters, microbial diversity and nutrient recovery potential. *Bioresource Technology*, 272, 421–432. https://doi.org/10.1016/j.biortech.2018.10.057.

Mitra, D., van Leeuwen, J. (. H.), & Lamsal, B. (2012). Heterotrophic/mixotrophic cultivation of oleaginous Chlorella vulgaris on industrial co-products. *Algal Research*, 1, 40–48. https://doi.org/10.1016/j.algal.2012.03.002.

Moheimani, N. R., McHenry, M. P., de Boer, K., & Bahri, P. A. (Eds.). (2015). *Biomass and biofuels from microalgae*. In Vol. 2. Cham: Springer International Publishing. https://doi.org/10.1007/978-3-319-16640-7.

Monfet, E., & Unc, A. (2017). Defining wastewaters used for cultivation of algae. *Algal Research*, 24, 520–526. https://doi.org/10.1016/j.algal.2016.12.008.

Montalvo, G. E. B., Thomaz-Soccol, V., Vandenberghe, L. P. S., Carvalho, J. C., Faulds, C. B., Bertrand, E., et al. (2019). Arthrospira maxima OF15 biomass cultivation at laboratory and pilot scale from sugarcane vinasse for potential biological new peptides production. *Bioresource Technology*, 273, 103–113. https://doi.org/10.1016/j.biortech.2018.10.081.

Najafabadi, H. A., Malekzadeh, M., Jalilian, F., Vossoughi, M., & Pazuki, G. (2015). Effect of various carbon sources on biomass and lipid production of Chlorella vulgaris during nutrient sufficient and nitrogen starvation conditions. *Bioresource Technology*, 180, 311–317. https://doi.org/10.1016/j.biortech.2014.12.076.

Nakamura, Y., & Li-Beisson, Y. (Eds.). (2016). *Lipids in plant and algae development*. In Vol. 86. Cham: Springer International Publishing. https://doi.org/10.1007/978-3-319-25979-6.

Narala, R. R., Garg, S., Sharma, K. K., Thomas-Hall, S. R., Deme, M., Li, Y., et al. (2016). Comparison of microalgae cultivation in photobioreactor, open raceway pond, and a two-stage hybrid system. *Frontiers in Energy Research*, 4, 29. (2016). https://doi.org/10.3389/fenrg.2016.00029.

Nwoba, E. G., Parlevliet, D. A., Laird, D. W., Alameh, K., & Moheimani, N. R. (2020). Pilot-scale self-cooling microalgal closed photobioreactor for biomass production and electricity generation. *Algal Research*, 45, 101731. https://doi.org/10.1016/j.algal.2019.101731.

Okoro, V., Azimov, U., Munoz, J., Hernandez, H. H., & Phan, A. N. (2019). Microalgae cultivation and harvesting: Growth performance and use of flocculants—A review. *Renewable and Sustainable Energy Reviews*, 115, 109364. https://doi.org/10.1016/j.rser.2019.109364.

Papapolymerou, G., Karayannis, V., Besios, A., Riga, A., Gougoulias, N., & SpIliotis, X. (2018). Scaling-up sustainable Chlorella vulgaris microalgal biomass cultivation from laboratory to pilot-plant photobioreactor, towards biofuel. *Global NEST Journal*, 21, 37–42. https://doi.org/10.30955/gnj.002777.

Park, S., Ahn, Y., Pandi, K., Ji, M.-K., Yun, H.-S., & Choi, J.-Y. (2019). Microalgae cultivation in pilot scale for biomass production using exhaust gas from thermal power plants. *Energies*, 12, 3497. https://doi.org/10.3390/en12183497.

Pérez, L., Salgueiro, J. L., González, J., Parralejo, A. I., Maceiras, R., & Cancela, Á. (2017). Scaled up from indoor to outdoor cultures of Chaetoceros gracilis and Skeletonema costatum microalgae for biomass and oil production. *Biochemical Engineering Journal*, *127*, 180–187. https://doi.org/10.1016/j.bej.2017.08.016.

Perez-Garcia, O., Escalante, F. M. E., de Bashan, L. E., & Bashan, Y. (2011). Heterotrophic cultures of microalgae: Metabolism and potential products. *Water Research*, *45*, 11–36. https://doi.org/10.1016/j.watres.2010.08.037.

Płaczek, M., Patyna, A., & Witczak, S. (2017). Technical evaluation of photobioreactors for microalgae cultivation. *E3S Web of Conferences*, *19*, 02032. https://doi.org/10.1051/e3sconf/20171902032.

Pleissner, D., Lam, W. C., Sun, Z., & Lin, C. S. K. (2013). Food waste as nutrient source in heterotrophic microalgae cultivation. *Bioresource Technology*, *137*, 139–146. https://doi.org/10.1016/j.biortech.2013.03.088.

Prathima Devi, M., Venkata Subhash, G., & Venkata, M. S. (2012). Heterotrophic cultivation of mixed microalgae for lipid accumulation and wastewater treatment during sequential growth and starvation phases: Effect of nutrient supplementation. *Renewable Energy*, *43*, 276–283. https://doi.org/10.1016/j.renene.2011.11.021.

Qiang, H., Faiman, D., & Richmond, A. (1998). Optimal tilt angles of enclosed reactors for growing photoautotrophic microorganisms outdoors. *Journal of Fermentation and Bioengineering*, *85*, 230–236. https://doi.org/10.1016/S0922-338X(97)86773-6.

Qiu, R., Gao, S., Lopez, P. A., & Ogden, K. L. (2017). Effects of pH on cell growth, lipid production and CO2 addition of microalgae Chlorella sorokiniana. *Algal Research*, *28*, 192–199. https://doi.org/10.1016/j.algal.2017.11.004.

Raeisossadati, M., Vadiveloo, A., Bahri, P. A., Parlevliet, D., & Moheimani, N. R. (2019). Treating anaerobically digested piggery effluent (ADPE) using microalgae in thin layer reactor and raceway pond. *Journal of Applied Phycology*, *31*, 2311–2319. https://doi.org/10.1007/s10811-019-01760-6.

Raheem, A., Prinsen, P., Vuppaladadiyam, A. K., Zhao, M., & Luque, R. (2018). A review on sustainable microalgae based biofuel and bioenergy production: Recent developments. *Journal of Cleaner Production*, *181*, 42–59. https://doi.org/10.1016/j.jclepro.2018.01.125.

Rai, M. P., Gautom, T., & Sharma, N. (2015). Effect of salinity, pH, light intensity on growth and lipid production of microalgae for bioenergy application. *OnLine Journal of Biological Sciences*, *15*, 260–267. https://doi.org/10.3844/ojbsci.2015.260.267.

Rashid, N., Ryu, A. J., Jeong, K. J., Lee, B., & Chang, Y.-K. (2019). Co-cultivation of two freshwater microalgae species to improve biomass productivity and biodiesel production. *Energy Conversion and Management*, *196*, 640–648. https://doi.org/10.1016/j.enconman.2019.05.106.

Rawat, I., Ranjith Kumar, R., Mutanda, T., & Bux, F. (2013). Biodiesel from microalgae: A critical evaluation from laboratory to large scale production. *Applied Energy*, *103*, 444–467. https://doi.org/10.1016/j.apenergy.2012.10.004.

ResearchGate (n.d.) A review of effect of light on microalgae growth. (Accessed 20 December 2019) https://www.researchgate.net/publication/263094270_A_Review_of_Effect_of_Light_on_Microalgae_Growth.

Rippka, R., Deruelles, J., Waterbury, J. B., Herdman, M., & Stanier, R. Y. (1979). Generic assignments, strain histories and properties of pure cultures of cyanobacteria. *Microbiology*, *111*, 1–61. https://doi.org/10.1099/00221287-111-1-1.

Rizwan, M., Mujtaba, G., Memon, S. A., Lee, K., & Rashid, N. (2018). Exploring the potential of microalgae for new biotechnology applications and beyond: A review. *Renewable and Sustainable Energy Reviews*, *92*, 394–404. https://doi.org/10.1016/j.rser.2018.04.034.

Romero-Villegas, G. I., Fiamengo, M., Acién-Fernández, F. G., & Molina-Grima, E. (2018). Utilization of centrate for the outdoor production of marine microalgae at the pilot-scale in raceway photobioreactors. *Journal of Environmental Management*, *228*, 506–516. https://doi.org/10.1016/j.jenvman.2018.08.020.

Ruangsomboon, S. (2012). Effect of light, nutrient, cultivation time and salinity on lipid production of newly isolated strain of the green microalga, Botryococcus braunii KMITL 2. *Bioresource Technology*, *109*, 261–265. https://doi.org/10.1016/j.biortech.2011.07.025.

Santana, H., Cereijo, C. R., Teles, V. C., Nascimento, R. C., Fernandes, M. S., Brunale, P., et al. (2017). Microalgae cultivation in sugarcane vinasse: Selection, growth and biochemical characterization. *Bioresource Technology*, *228*, 133–140. https://doi.org/10.1016/j.biortech.2016.12.075.

Schade, S., & Meier, T. (2019). A comparative analysis of the environmental impacts of cultivating microalgae in different production systems and climatic zones: A systematic review and meta-analysis. *Algal Research*, *40*, 101485. https://doi.org/10.1016/j.algal.2019.101485.

Shi, X. M., & Chen, F. (1999). Production and rapid extraction of lutein and the other lipid-soluble pigments from Chlorella protothecoides grown under heterotrophic and mixotrophic conditions. *Food Nahrung*, *43*, 109–113. https://doi.org/10.1002/(SICI)1521-3803(19990301)43:2<109::AID-FOOD109>3.0.CO;2-K.

Shi, X.-M., & Chen, F. (2002). High-yield production of lutein by the green microalga Chlorella protothecoidesin heterotrophic fed-batch culture. *Biotechnology Progress*, *18*, 723–727. https://doi.org/10.1021/bp0101987.

Silkina, A., Zacharof, M.-P., Ginnever, N. E., Gerardo, M., & Lovitt, R. W. (2019). Testing the waste based biorefinery concept: Pilot scale cultivation of microalgal species on spent anaerobic digestate fluids. *Waste and Biomass Valorization*. https://doi.org/10.1007/s12649-019-00766-y.

Sloth, J. K., Wiebe, M. G., & Eriksen, N. T. (2006). Accumulation of phycocyanin in heterotrophic and mixotrophic cultures of the acidophilic red alga Galdieria sulphuraria. *Enzyme and Microbial Technology*, 38, 168–175. https://doi.org/10.1016/j.enzmictec.2005.05.010.

Song, C., Han, X., Qiu, Y., Liu, Z., Li, S., & Kitamura, Y. (2020). Microalgae carbon fixation integrated with organic matters recycling from soybean wastewater: Effect of pH on the performance of hybrid system. *Chemosphere*, 126094. https://doi.org/10.1016/j.chemosphere.2020.126094.

Stockenreiter, M., Haupt, F., Seppälä, J., Tamminen, T., & Spilling, K. (2016). Nutrient uptake and lipid yield in diverse microalgal communities grown in wastewater. *Algal Research*, 15, 77–82. https://doi.org/10.1016/j.algal.2016.02.013.

Suparmaniam, U., Lam, M. K., Uemura, Y., Lim, J. W., Lee, K. T., & Shuit, S. H. (2019). Insights into the microalgae cultivation technology and harvesting process for biofuel production: A review. *Renewable and Sustainable Energy Reviews*, 115, 109361. https://doi.org/10.1016/j.rser.2019.109361.

Swaaf, M. E. D., Sijtsma, L., & Pronk, J. T. (2003). High-cell-density fed-batch cultivation of the docosahexaenoic acid producing marine alga Crypthecodinium cohnii. *Biotechnology and Bioengineering*, 81, 666–672. https://doi.org/10.1002/bit.10513.

Tamiya, H. (1957). Mass culture of algae. *Annual Review of Plant Physiology*, 8, 309–334. https://doi.org/10.1146/annurev.pp.08.060157.001521.

Terry, K. L., & Raymond, L. P. (1985). System design for the autotrophic production of microalgae. *Enzyme and Microbial Technology*, 7, 474–487. https://doi.org/10.1016/0141-0229(85)90148-6.

Tredici, M. R. (2007). Mass production of microalgae: Photobioreactors. In *Handbook of microalgal culture* (pp. 178–214). John Wiley & Sons, Ltd. https://doi.org/10.1002/9780470995280.ch9.

Ummalyma, S. B., & Sukumaran, R. K. (2014). Cultivation of microalgae in dairy effluent for oil production and removal of organic pollution load. *Bioresource Technology*, 165, 295–301. https://doi.org/10.1016/j.biortech.2014.03.028.

Vadiveloo, A., Moheimani, N. R., Cosgrove, J. J., Parlevliet, D., & Bahri, P. A. (2017). Effects of different light spectra on the growth, productivity and photosynthesis of two acclimated strains of Nannochloropsis sp. *Journal of Applied Phycology*, 29, 1765–1774. https://doi.org/10.1007/s10811-017-1083-9.

Valdés, F. J., Hernández, M. R., Catalá, L., & Marcilla, A. (2012). Estimation of CO2 stripping/CO2 microalgae consumption ratios in a bubble column photobioreactor using the analysis of the pH profiles. Application to Nannochloropsis oculata microalgae culture. *Bioresource Technology*, 119, 1–6. https://doi.org/10.1016/j.biortech.2012.05.120.

Venkata Mohan, S., Rohit, M. V., Chiranjeevi, P., Chandra, R., & Navaneeth, B. (2015). Heterotrophic microalgae cultivation to synergize biodiesel production with waste remediation: Progress and perspectives. *Bioresource Technology*, 184, 169–178. https://doi.org/10.1016/j.biortech.2014.10.056.

Villaseñor Camacho, J., Fernández Marchante, C. M., & Rodríguez, R. L. (2018). Analysis of a photobioreactor scaling up for tertiary wastewater treatment: Denitrification, phosphorus removal, and microalgae production. *Environmental Science and Pollution Research*, 25, 29279–29286. https://doi.org/10.1007/s11356-018-2890-5.

Vonshak, A., & Coombs, J. (1985). Micro-algae: Laboratory growth techniques and outdoor biomass production. (chapter 15)In D. O. Hall, S. P. Long, & J. M. O. Scurlock (Eds.), *Techniques in Bioproductivity and Photosynthesis* (pp. 188–200). (2nd ed.). Pergamon. https://doi.org/10.1016/B978-0-08-031999-5.50025-X.

Wang, Z., Wen, X., Xu, Y., Ding, Y., Geng, Y., & Li, Y. (2018). Maximizing CO2 biofixation and lipid productivity of oleaginous microalga Graesiella sp. WBG-1 via CO2-regulated pH in indoor and outdoor open reactors. *Science of the Total Environment*, 619–620, 827–833. https://doi.org/10.1016/j.scitotenv.2017.10.127.

Wen, Z.-Y., & Chen, F. (2001). A perfusion–cell bleeding culture strategy for enhancing the productivity of eicosapentaenoic acid by Nitzschia laevis. *Applied Microbiology and Biotechnology*, 57, 316–322. https://doi.org/10.1007/s002530100786.

Wen, X., Du, K., Wang, Z., Peng, X., Luo, L., Tao, H., et al. (2016). Effective cultivation of microalgae for biofuel production: A pilot-scale evaluation of a novel oleaginous microalga Graesiella sp. WBG-1. *Biotechnology for Biofuels*, 9, 123. https://doi.org/10.1186/s13068-016-0541-y.

Wiley, P., Harris, L., Reinsch, S., Tozzi, S., Embaye, T., Clark, K., et al. (2013). Microalgae cultivation using offshore membrane enclosures for growing algae (OMEGA). *Journal of Sustainable Bioenergy Systems*, 3, 18–32. https://doi.org/10.4236/jsbs.2013.31003.

Wu, Z., & Shi, X. (2007). Optimization for high-density cultivation of heterotrophic Chlorella based on a hybrid neural network model. *Letters in Applied Microbiology*, 44, 13–18. https://doi.org/10.1111/j.1472-765X.2006.02038.x.

Xiong, W., Li, X., Xiang, J., & Wu, Q. (2008). High-density fermentation of microalga Chlorella protothecoides in bioreactor for microbio-diesel production. *Applied Microbiology and Biotechnology*, 78, 29–36. https://doi.org/10.1007/s00253-007-1285-1.

Xu, H., Miao, X., & Wu, Q. (2006). High quality biodiesel production from a microalga Chlorella protothecoides by heterotrophic growth in fermenters. *Journal of Biotechnology*, 126, 499–507. https://doi.org/10.1016/j.jbiotec.2006.05.002.

Zakariah, N. A., Abd. Rahman, N., Hamzah, F., & Md Jahi, T. (2015). Cultivation system of green microalgae, Botryococcus braunii: A review. *Jurnal Teknologi, 5*, 76. https://doi.org/10.11113/jt.v76.5517.

Zhou, W., Chen, P., Min, M., Ma, X., Wang, J., Griffith, R., et al. (2014). Environment-enhancing algal biofuel production using wastewaters. *Renewable and Sustainable Energy Reviews, 36*, 256–269. https://doi.org/10.1016/j.rser.2014.04.073.

Zhou, W., Li, Y., Min, M., Hu, B., Chen, P., & Ruan, R. (2011). Local bioprospecting for high-lipid producing microalgal strains to be grown on concentrated municipal wastewater for biofuel production. *Bioresource Technology, 102*, 6909–6919. https://doi.org/10.1016/j.biortech.2011.04.038.

Zhou, W., Min, M., Li, Y., Hu, B., Ma, X., Cheng, Y., et al. (2012). A hetero-photoautotrophic two-stage cultivation process to improve wastewater nutrient removal and enhance algal lipid accumulation. *Bioresource Technology, 110*, 448–455. https://doi.org/10.1016/j.biortech.2012.01.063.

Zhou, W., Wang, J., Chen, P., Ji, C., Kang, Q., Lu, B., et al. (2017). Bio-mitigation of carbon dioxide using microalgal systems: Advances and perspectives. *Renewable and Sustainable Energy Reviews, 76*, 1163–1175. https://doi.org/10.1016/j.rser.2017.03.065.

Zhou, J., Wu, Y., Pan, J., Zhang, Y., Liu, Z., Lu, H., et al. (2019). Pretreatment of pig manure liquid digestate for microalgae cultivation via innovative flocculation-biological contact oxidation approach. *Science of the Total Environment, 694*, 133720. https://doi.org/10.1016/j.scitotenv.2019.133720.

Zhuang, L.-L., Yu, D., Zhang, J., Liu, F., Wu, Y.-H., Zhang, T.-Y., et al. (2018). The characteristics and influencing factors of the attached microalgae cultivation: A review. *Renewable and Sustainable Energy Reviews, 94*, 1110–1119. https://doi.org/10.1016/j.rser.2018.06.006.

2

Photobioreactor design for microalgae culture

Mariany Costa Deprá, Ihana Aguiar Severo, Rosangela Rodrigues Dias, Leila Queiroz Zepka, and Eduardo Jacob-Lopes

Bioprocesses Intensification Group, Federal University of Santa Maria, UFSM, Santa Maria, RS, Brazil

1 Introduction

The global context of natural resource depletion, pollution issues, and dependence on fossil energy has raised new concerns for society (Préat, Taelman, Meester, Allais, & Dewulf, 2020). Consequently, the emerging need to use renewable resources has induced public authorities to direct their research programs toward environmentally viable alternatives (Míguez, Porteiro, Pérez-Orozco, Patiño, & Gómez, 2020). Associated with these aspects, large-scale demand for biofuels and other biobased products is growing worldwide among industrial processes (Gayen, Bhowmick, & Maity, 2019). Thus, the use, development, and expansion of biotechnology of phototrophic microorganisms have gained prominence in recent decades due to their high potential for technological exploitation (Paladino & Neviani, 2020).

Microalgae are grown for a variety of purposes. Its main applications include food and feed supply, fine chemicals (Maroneze, Jacob-Lopes, Zepka, Roca, & Pérez-Gálvez, 2019; Zepka, Jacob-Lopes, & Roca, 2019) and bulk, wastewater treatment and deodorization (Santos, Fernandes, Wagner, Jacob-Lopes, & Zepka, 2016; Vieira et al., 2019), and bioenergy production (Severo, Barin, Wagner, Zepka, & Jacob-Lopes, 2018; Severo, Deprá, et al., 2018; Severo, Deprá, Zepka, & Jacob-Lopes, 2016). However, microalgae-based processes mainly depend on their productivity.

Photobioreactors are the core of microalgal bioprocess engineering. The design and operation of photobioreactors had a broad interest in biotechnological industrial tracking. Technically, these configurations, in current operation, are extrapolations from conventional chemical reactors (Jacob-Lopes, Zepka, Ramírez-Mérida, Maroneze, & Neves, 2016). These systems, as well as their variations, were developed and adapted to meet the peculiarities of microalgae bioprocesses. As a consequence, these reaction vessels are constructed through specific geometric designs under robust operating conditions to favor microalgae growth and achieve high biomass productivity (Moroni, Lorino, Cicci, & Bravi, 2019).

However, one of the main reasons for this slow industrial progress is the high cost of production associated mainly with materials. Besides, the construction and installation of photobioreactors must consider land availability and energy demand (Deprá, Mérida, de Menezes, Zepka, & Jacob-Lopes, 2019). Concomitantly, as these are constant biochemical reactions, it is essential to understand the key phenomena that limit the performance of microalgae cells, such as substrate delivery, pH, gas exchange (CO_2 addition/O_2 removal), temperature and lighting requirements (photo limitation, seasonal periods, and photoperiods) (Acién et al., 2018).

Therefore, this chapter aims to cover an overview of the operational parameters and critical details to be considered for the optimal process, as well as specific characteristics and challenges that will be critical to the success of photobioreactors applied to microalgae culture.

2 System hydrodynamics

Flow hydrodynamics is a crucial factor in influencing photobioreactor performance. This is because the hydrodynamic differences in such equipment can provide excessive damage to the physical properties of cellular biochemicals during microalgal cultivation (Heredia et al., 2018). More specifically, fluid hydrodynamics have the ability to interfere with nutrient delivery as well as direct cell movements in the reactor, allowing for continued light exposure (Gao, Kong, & Vigil, 2018).

Given this scenario, to improve the operational efficiency of photobioreactors, reduce optimization costs, build and scale-up, numerical evaluation instructions are performed (Guler, Deniz, Demirel, & Imamoglu, 2019). In short, modeling and simulation procedures are implemented to better understand the rheological and hydrodynamic parameters of fluids within photobioreactors (Guler, Deniz, Demirel, Oncel, & Imamoglu, 2019). In addition,

these techniques aim to explore the mixing properties, interphase mass transfer rates, and biological functions of microalgae (Guler, Deniz, Demirel, & Imamoglu, 2019).

Thus, an optimal photobioreactor design requires highly robust control systems. Therefore, to achieve this goal, the burden of experimentally investigating the critical parameters involved in photobioreactor performance must be considered. They will enhance and ease the transition from laboratory-scale to pilot and industrial plants. In this sense, the process engineering parameters and their respective technological advances are discussed below.

2.1 Superficial liquid velocity and superficial gas velocity

In recent decades, the biphasic liquid–gas flow has attracted considerable attention due to its full application in the engineering fields. As a result, the superficial velocity of the liquid and superficial velocity of the gas aroused interest in their real relationship with the productivity of microalgal photobioreactors (Zhang, Wang, Sun, & Yuan, 2019).

From an initial perspective, it should be noted that surface velocity differs from actual velocity since surface velocity reflects relative flow rates. A better understanding results in the thought that photobioreactors do not have a single-phase system as it is composed of the solid, liquid, and gas fraction (Li, Li, & Li, 2013). Thus, the need to evaluate the surface velocity of the gas, when in interaction with the surface velocity of the liquid is necessary, since they directly influence the system flow regime.

On the other hand, the superficial gas velocity is one of the main parameters used to determine the hydrodynamics of the flow in gas–solid fluidized beds. However, the interrelation of surface velocity provides variation depending on the arrangement of photobioreactors, interfering with data consistency.

Thus, when gas and liquid flow simultaneously in a pipe, several types of discharge can occur, differing from each other by the spatial distribution of the interface. These configurations are referred to as flow patterns and are considering several models or flow regimes in pipes. The classification of horizontal or vertical flow patterns can be understood in Table 1.

Additionally, to illustrate the gas–liquid relationship in reaction systems, Fig. 1 shows the correlation between liquid and gas surface velocities. If for a constant flow of liquid in a given pipe, a small flow of gas is added, the

TABLE 1 Classification of horizontal or vertical flow patterns.

Classification	Definition
Bubble flow	Gas bubbles scatter within the liquid, which is the continuous phase, and the bubbles are concentrated at the top
Plug flow	Elongated gas pockets are formed, which tend to move in the upper part of the tube, with or without small bubbles in the region just below the pockets
Stratified flow	The phases flow separated by a relatively smooth horizontal interface, gas at the top, and liquid at the bottom of the tube. This flow pattern is associated with slow velocities for both fluids and is not common to occur
Wavy flow	The phases flow apart, but with an irregular interface, showing ripples. These waves are associated with high gas velocities
Slug flow	Disruption of the liquid–gas interface occurs at some points, and liquid droplets are suspended in the gas phase; Liquid ridges can reach the top surface of everything. This flow tends to occur as gas velocity increases
Annular low	Liquid film forms on the pipe wall surrounding the gaseous phase flowing as a core, with the thickness of the liquid film at the bottom of the pipe being higher than at the top. The interface is unstable and has bubbles in the liquid film, as well as a considerable amount of liquid is kept suspended in the gas phase as filaments
Spray low	In this type of regime, the liquid dispersed in the gas phase in the form of droplets is dragged by the gas

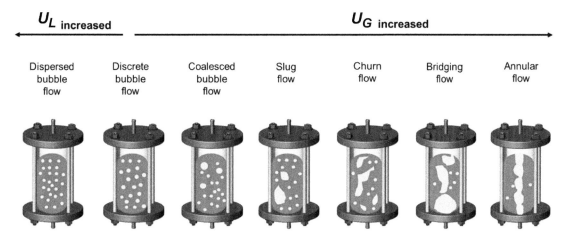

FIG. 1 Relation of superficial liquid velocity and superficial gas velocity associated with flow regimes. *Adapted from Zhang, J. P., Grace, J. R., Epstein, N., & Lim, K. S. (1997). Flow regime identification in gas-liquid flow and three-phase fluidized beds.* Chemical Engineering Science, 52(21 − 22), 3979–3992.

gas forms small bubbles that do not coalesce as they flow through the liquid (bubbled flow). As the proportion of gas is increased, the number and size of bubbles increase, and they tend to come together, forming alternating pockets of liquid and gas (buffered flow). In other proportions, gas may flow into the center of the pipe while liquid flows along the wall like a ring (annular flow). Increasing the gas flow will reach the point where the gas will become the continuous phase, and the liquid will be dispersed as a mist (dispersed flow). This occurs in both vertical and horizontal flow. However, in horizontal flow, a stratified flow can also occur, where the lighter phase occupies the upper part of the pipe and the denser the bottom (Sada, Yasunishi, Katoh, & Nishioka, 1978).

Thus, by knowing the physical properties of the phases, the surface velocity of the liquid (νL), the surface velocity of the gas (νG), and the cross-sectional area of the pipe, it is possible to predict in which region of the Baker map the liquid–biphasic flow, where the gas is found and hence determine the flow pattern through Eqs. (1) and (2):

$$U_G = \frac{Q}{A} \qquad (1)$$

$$U_L = \frac{Q}{A} \qquad (2)$$

where U_G is the superficial gas velocity (m/s), U_L is the superficial liquid velocity (m/s), Q is the phase volume flow rate (m³/s), and the cross-sectional area of the tube or porous medium into which fluid is flowing through (m²).

Also, at the level of example, Table 2 presents the studies regarding the main photobioreactor models and their respective gas superficial velocities.

2.2 Gas holdup

Liquid bubbles are commonly addressed in industrial biotechnological processes. This is because, in many multistage reaction systems (solid, liquid, gaseous), the reaction rate is often controlled by the interfacial area of liquid gas, associated with the diameter of the bubble and its gas holdup (Tian et al., 2020). In this regard, the projection and expansion of photobioreactors must be capable of generating a large number of fine bubbles to enhance mass transfer and accelerate the reaction in gas–liquid systems. Thus, the main parameter concerning these aspects can be determined by characterizing the gas retention.

The gas holdup consists of evaluating the dynamic properties of the bubble, such as bubble passage frequency, chord length, bubble velocity, and interfacial area. In another aspect, it may also be specified as the fraction of the reactor volume absorbed by the gas. This can be estimated as the volume of liquid displaced by gas (expansion of liquid volume) due to aeration. Associated with this, these principles directly affect the growth of microalgae, since carbon dioxide distribution usually occurs through contact bubbles (Ojha & Al-Dahhan, 2018).

TABLE 2   Main photobioreactor models and their respective gas superficial velocities.

Design	Superficial gas velocity (m/s)	Reference
Tubular photobioreactor	0.02	Ugwu, Ogbonna, and Tanaka (2002)
Bubble column photobioreactor	0.0082	Merchuk, Gluz, and Mukmenev (2000)
Flat-plate photobioreactor	0.009	Zhang, Kurano, and Miyachi (2002)
Airlift photobioreactor	0.055	Contreras, García, Molina, and Merchuk (1998)

On the other hand, gas holdup analysis becomes crucial because it is indirectly related to the circulation rate, the gas residence time, as well as the general mass transfer rate ($kL\alpha$) (which will be discussed in detail in Section 2.5). Thus, the literature reports many parameters that influence this value, and therefore characterizing them concomitantly is indispensable in light of the photobioreactor operating conditions (Leonard, Ferrasse, Lefevre, Viand, & Boutin, 2019).

Generally, these parameters are naturally assigned to bubble column reactor configurations. That is because, although these designers appear to be easily constructed, their characterization is substantially complex due to the instability involved in the flows and their dependence on the superficial gas velocity (Kannan, Naren, Buwa, & Dutta, 2019).

Empirically, the gas holdup is calculated using Eq. (3):

$$\varepsilon = \frac{U_G}{U_L} \tag{3}$$

where ε is the gas holdup (dimensionless), U_G is the superficial gas velocity (m/s), and U_L is the superficial liquid velocity (m/s) (Mirón, Camacho, Gomez, Grima, & Chisti, 2000).

Also, it is important to note that this parameter mainly depends on the diameter of the bubbles and their increasing velocity. In short, gas retention is high when the number of bubbles is significant, and its diameter is small (lower ascent rate).

2.3 Flow regime

The flow regime is commonly established as an evaluation parameter of photobioreactor design since it determines the behavior of fluids concerning several variables within the reaction vessels. Moreover, the main feature taken into consideration is directly related to the flow of a fluid and the particle's trajectory, as well as the relation of the dependence of the state of flow organization.

Reynolds's number can be calculated to ascertain the movement of the culture medium according to the different flow regimes (Peiran & Shizhu, 1990) (Eq. 4).

$$Re = \frac{\rho \times V \times D}{\mu} \tag{4}$$

where ρ is the specific mass of the fluid (kg/m^3), V is the average fluid velocity (m/s), D is the longitude characteristic of the flow (m), and μ is the dynamic viscosity of the fluid (kg/m/s).

In short, transition regimes can be verified within the photobioreactors: laminar and turbulent. The first can be defined as the regime in which fluid particles tend to travel parallel paths. In contrast, the second is characterized as the regime in which the particles have irregular and curvilinear trajectories, changing in direction, and generally forming swirls (Shariff & Chakraborty, 2017).

In practice, most industrial installations have fluid turbulence. Similarly, in photobioreactors, it is necessary to approximate fluid flow to a regime closer to turbulent fluctuations (Marshall & Sala, 2011). However, it is essential to note that while turbulent regimes are well developed and validated, the complexity of establishing turbulence within photobioreactors without causing cellular damage remains a barrier to project consolidation.

Approaching turbulent flows can be advantageous if used properly. Since microalgal cells need to be brought from poorly lit areas to be regularly exposed to well-lit areas, for robust microalgae, turbulence can stimulate and reduce residence time for cell growth to an optimal point. In contrast, studies have shown that some dinoflagellate microalgae strains appear to be excessively sensitive to hydrodynamic forces found in the turbulent regime typical of photobioreactors (Camacho et al., 2011).

Also, when considering the flow path, the turbulent regime can cause vertical mixing, which is defined as the mixture flow relevant to light

TABLE 3 Main photobioreactor models and their respective flow regimes.

Design	Re	Reference
Tubular photobioreactor	25,000	Gómez-Pérez, Oviedo, Ruiz, and Van Boxtel (2017)
Bubble column photobioreactor	6368	Kováts, Thévenin, and Zähringer (2018)
Flat-plate photobioreactor	1761	Hernández-Melchor, Cañizares-Villanueva, Terán-Toledo, López-Pérez, and Cristiani-Urbina (2017)
Biofilm photobioreactor	0.52	Yang et al. (2011)
Hybrid photobioreactor	2500	Deprá et al. (2019)

exposure. However, as this regime is in rapid flux, it may not be sufficiently mixed with the need for light exposure. Also, high hydrodynamic forces may limit the performance of cyclic movements of the liquid between the bottom (dark zone) and the surface (light zone) of the runways (Prussi et al., 2014). Therefore, it is necessary to understand that flow with a high turbulent number may not be a flow with a correspondingly high frequency of light/dark cell cycles. For example, a flow for which the whirlpool is far from the light and dark zone separation line cannot contribute to the light–dark cycles.

Thus, at the level of exemplification, Table 3 presents the studies regarding the main photobioreactor models and their respective flow regimes.

2.4 Mixing

Photobioreactor designers are continually investigating ways to increase microalgae productivity in photobioreactors (Qin & Wu, 2019). In the last decades, several models of mixers have been proposed (Cheng, Huang, &

Chen, 2016; Huang et al., 2014; Perner-Nochta & Posten, 2007; Ugwu et al., 2002; Ugwu, Ogbonna, & Tanaka, 2005), resulting in researchers encouragement to refine their studies. However, to date, the real understanding by which the mixing system can increase microalgae productivity has not yet been deepened.

To this end, mathematical modeling is recommended to evaluate the effectiveness of implementing these new models. Thus, variables such as mixing time and circulation time should be considered.

In short, the determination of the mixing time (TM) is understood as the time interval required to acquire a distinct difference from the thoroughly mixed state after bioreactor entry (Mirón, García, Camacho, Grima, & Chisti, 2004). Its relevance in a photobioreactor is given since the mixing effect caused by the mixer accelerates gas exchange between cells and the intensity must be appropriate to the workload. However, while it is crucial to improve mixing, it is also necessary to take into account the shear effect that can be generated by over mixing, causing a reduction in productivity (Hinterholz et al., 2019). The empirical method can be given by the acid tracer method that combines both acid injection and pH measurement over time (Eq. 5):

$$CT = [H^+] = \left(\frac{[H^+]_{\text{instantaneous}} - [H^+]_{\text{inicial}}}{[H^+]_{\text{final}} - [H^+]_{\text{inicial}}} \right) \quad (5)$$

where H^+ is the difference of hydrogen potential, corresponding to the required time to attain a 5% deviation from complete homogeneity after the injection of a tracer pulse is injected into the reactor.

Also, another way to ensure the right mix is through circulation time. This, in turn, can be understood as the time interval that the entire volume of fluid passes through the system (Chisti, Halard, & Moo-Young, 1988).

Thus, the methods perform fundamental functions to cells since they keep the algal cells suspended in the nutrient medium distribution and increase the gas/liquid/mass transfer, besides preserving the cellular integrity, they must ensure their suspension in constant illumination (El-Baz & El Baky, 2018). Additionally, both methods are applied concurrently to flow rate parameters. That is because, in bioreactors where mixing time and circulation time are shorter, greater workforces at the gas inlet should be designed and readjusted, which would result in higher energy demands.

2.5 Mass transfer

The successful design of photobioreactors is also evaluated from mass and heat transfer. CO_2 is usually the main carbon source for growing photosynthetic microalgae, which may be continuously or intermittently transferred from the gas phase to the liquid phase of the culture medium. Despite being a problem commonly encountered in these cultivation systems, these CO_2/O_2 exchange phenomena must be efficient to maintain high productivity and offset the high production expenses (Ugwu, Aoyagi, & Uchiyama, 2008).

The mass transfer characteristics that are applicable in photobioreactors include the overall volumetric mass transfer coefficient ($k_L\alpha$), mixing, liquid velocity, gas bubble velocity, and gas holdup. Besides, the $k_L\alpha$ depends on agitation rate, temperature, pH control, type of sparger, and surfactants/antifoam agents. These issues already discussed in detail above.

Regardless, CO_2 mass transfer from the gas phase to the cell phase comprises different stages, although the gas–liquid stage is the determining process in photobioreactors. Such a phenomenon may be described by Eq. (6) (Raeesossadati, Ahmadzadeh, McHenry, & Moheimani, 2014; Tobajas & Garcia-Calvo, 2000):

$$N_{CO_2} = K_L\alpha(CL* - CL) \qquad (6)$$

where N_{CO_2} is the carbon dioxide production rate (mg/L/s), k_L is the liquid-phase mass transfer coefficient (m/s), α is the specific available area for mass transfer (m^{-1}), $CL*$ is saturation concentration of dissolved oxygen (mg/L), and CL is the dissolved oxygen concentration (mg/L).

For a better understanding, Fig. 2 illustrates a CO_2 mass transfer scheme in a photobioreactor.

The reactivity of carbon dioxide in aqueous solutions establishes several balances in its contact with water. The first equilibrium refers to the dissolution of gas in water to form carbonic acid. Carbonic acid undergoes almost instantaneous dissociation into bicarbonate and carbonate ions, with the total inorganic carbon concentration given by the total of CO_3^{2-}, HCO_3^-, and CO_2 species (Cabello, Morales, & Revah, 2017).

In terms of solubility, CO_2 is much more water-soluble than O_2. However, due to the low solubility of both gases in aqueous solutions (low mass transfer coefficient), there is a need to provide these elements throughout the process. Thus, an efficient carbon dioxide transfer system is expected in photobioreactors. Carbon dioxide transfer efficiency is required to increase volumetric mass transfer coefficients $K_L\alpha$ allowing better performance in the transfer of gas to the liquid phase. The volumetric coefficients of mass transfer depend mainly on the physical properties of the fluid, the fluid flow, and the system and geometry of the gas injector. It is worth mentioning that to improve mass transfer, the reduction in bubble size leads to faster dissolution, slower rise, and high surface/volume ratio, and this corroborates the well-known Fick's Law (Pires, Alvim-Ferraz, & Martins, 2017).

Therefore, the CO_2 mass transfer into photobioreactors becomes a limiting factor of the processes, as the dissolved carbon dioxide concentration decreases with increasing temperature, as well as decreasing with increased concentration of dissolved salts. These factors

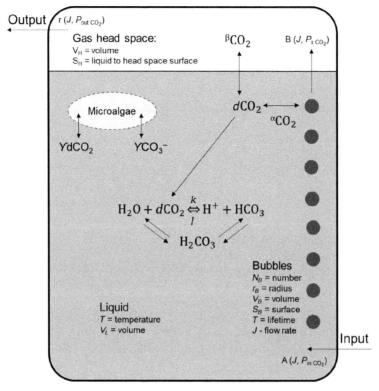

FIG. 2 Schematic view of CO_2 mass transfer in a photobioreactor. *Note*: CO_2-enriched air is supplied to the photobioreactor through a channel, forming the bubbles (N_B) of effective mean radius rB (m). The average bubble lifetime is T (s). The gaseous and liquid phases are assumed to be of the same temperature T (°C). The volume of the liquid phase is V_L (m³), and the surface between the liquid and the gaseous headspace is S_H (m²). The headspace volume is V_H (m³). The gas flow is constant at J (m³/s), entering with CO_2 partial pressure of P_{inCO_2} (Pa), and the resulting CO_2 mass transfer flux is A (mol s⁻¹). The CO_2 partial pressure of the gas flow from bubbles to the headspace is P_{XCO_2} (Pa), resulting in CO_2 flux of B (mol s⁻¹). The integral mass transfer flux of CO_2 from bubbles to the liquid medium is αCO_2 (mol s⁻¹), from the headspace to the liquid medium it is βCO_2 (mol s⁻¹), from the pool of dissolved CO_2 to microalgae it is $YdCO_2$ (mol s⁻¹), and from the bicarbonate pool to the microalgae $YHCO_3^-$ (mol s⁻¹). *Adapted from Nedbal, L., Červený, J., Keren, N., & Kaplan, A. (2010). Experimental validation of a nonequilibrium model of CO2 fluxes between gas, liquid medium, and algae in a flat-panel photobioreactor.* Journal of Industrial Microbiology & Biotechnology, 37(12), 1319–1326.

become relevant for microalgae CO_2 transfer and removal processes, considering the need for higher saturation concentration values (Jacob-Lopes, Scoparo, Queiroz, & Franco, 2010; Raeesossadati et al., 2014).

On the other hand, oxygen produced by water photolysis reactions can be accumulated in the medium at levels that may exceed 400% of the air saturation value. This phenomenon can damage cells (photo-oxidative damage),

leading to poor productivity. In tubular photobioreactors, for example, the accumulation of dissolved O_2 is typical along the tubes (Song, Li, He, Yao, & Huang, 2019). Then, degassing equipment (based on an airlift system) is normally integrated into the reactor to remove excess O_2. It is essential to couple degassers units to control this parameter because the high O_2 concentration favors the activity of the enzymes oxygenase, causing the O_2 absorption instead of CO_2 (photorespiration) and, consequently, reduces cell growth (Garcia-Ochoa & Gomez, 2009).

In this sense, the O_2 mass transfer from the medium to the atmosphere depends mainly on the diffusion coefficient and other secondary factors, such as turbulence. The addition of an O_2/CO_2 enriched airflow at appropriate concentrations can improve the driving force of mass transfer.

3 Parameters of environmental conditions in photobioreactors

3.1 Light

Microalgae cultivation presents a unique demand in biochemical engineering, which is the adequate luminosity supply to the cells. Photosynthesis depends exclusively on this process parameter, which varies in terms of quantity, intensity, quality, and type of light energy. These features substantially modify the photobioreactor design for microalgal cultures.

Photolimitation and photoinhibition phenomena are frequent in inadequately illuminated cultivations. In the first case, the problem occurs at a low light intensity. In the second case, light saturation occurs when cells are exposed to intensities that exceed the critical level. In both situations, inadequate lighting causes significant losses in kinetic performance (Grima, Sevilla, Pérez, & Camacho, 1996). Typical examples are outdoor photobioreactors that experience climate fluctuations. In winter, cultures are usually photo limited, while in summer they are photoinhibition. It should be mentioned that in addition to climate, the positioning of photobioreactors, whether horizontal, vertical or inclined, is a factor that affects the light availability. East–west vertical systems receive about 5% more solar energy compared to north–south horizontal systems (Fernández, Camacho, Pérez, Sevilla, & Grima, 1998). This orientation, of course, depends on latitude and longitude. Moreover, diffuse radiation or light dilution principle, as it is also known, will determine the incidence of direct, reflected, and scattered solar radiation. The light intensity in vertical systems arranged in parallel will be much lower than that which reaches horizontal systems in the same surface area. Thus, photosynthetic efficiency is higher due to diffuse light with subsaturant effect (Wondraczek et al., 2019).

Considering these aspects, the relationship between microalgae photosynthetic efficiency and luminosity is remarkable and, at the same time, somewhat complicated. Fig. 3A illustrates the effect of light intensity and cell growth rate, where incidences depend on cell type and phase. The interruption or decrease of growth dynamics occurs in the darkness period (i.e., when the light supply is below the critical intensity). In contrast, in a highly illuminated period, this becomes a problem. To achieve better kinetic performance, the photobioreactor must have a homogeneous external illumination. In practical terms, however, this is possible if the depth of cultivation is shallow, the moderate cell concentration, and the controlled light intensity (Fig. 3B). Both zones of limitation, saturation, light inhibition, and zone without cell growth can be seen concomitantly within the same photobioreactor (Ogbonna & Tanaka, 2000).

By subjecting the cultivation to mixing and agitation processes, for example, the cells are homogenized reducing photoinhibition in the surface zone and lower biomass productivity in the dark zone as well as biomass productivity in the shading zone, according to a study by Sivakaminathan, Hankamer, Wolf, and Yarnold (2018). Once the photobioreactor is improperly designed, this duality of the light–dark gradient will not be a well-defined operating parameter. The gradient of light intensity inside the photobioreactor as a function of surface distance can be represented by the empirical relationship of the Lambert–Beer's Law, mathematically described by Eq. (7) (Pires et al., 2017):

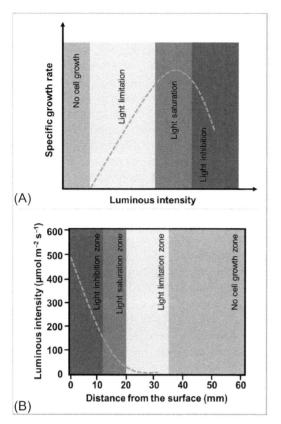

FIG. 3   Effect of light intensity on specific growth rate in photoautotrophic cultivation (A) and light distribution inside a photobioreactor (B). *Adapted from Ogbonna, J. C., & Tanaka, H. (2000). Light requirement and photosynthetic cell cultivation—Development of processes for efficient light utilization in photobioreactors.* Journal of Applied Phycology, 12, *207–218.*

$$I = I_0 \times e^{-k \times C \times d} \qquad (7)$$

where I is the light intensity at the desired surface ($\mu E/m^2/s$), I_0 is the luminosity intensity at the illuminated wall surface of the photobioreactor ($\mu E/m^2/s$), k is the attenuation constant due to the microalgae culture at concentration C, and d is the distance from the irradiated surface (m^{-1}). This Law means that the light intensity decreases exponentially through the pathway it travels.

Alternatively, the issue of the light gradient can be optimized through photoperiods. Alternations between light/dark cycles have been demonstrated, and biomass productivity proved (Jacob-Lopes, Scoparo, Lacerda, & Franco, 2009; Levasseur, Taidi, Lacombe, Perré, & Pozzobon, 2018). Maroneze et al. (2016) showed that photosynthetic efficiency increased during modulation of these cycles. Importantly, under dark conditions, respiration is the predominant metabolism for cell maintenance, causing losses close to 40% of daytime biomass productivity. In the photobioreactor design, light/dark cycles should be parameterized as a function of cycle time and light fraction. Studies based on computational fluid dynamics simulation in different photobioreactor configurations shows that long light/dark cycles can negatively affect microalgae growth, whose effect can be discarded when subjected to short cycles (fractions of seconds) (Barceló-Villalobos, Fernández-del Olmo, Guzmán, Fernández-Sevilla, & Fernández, 2019; Perner-Nochta & Posten, 2007; Pruvost, Cornet, & Legrand, 2008).

Another crucial factor in photobioreactor design is the light spectral quality. Although sunlight covers a wide range of the spectrum, only visible light with wavelengths between 400 and 700nm represents the photosynthetically active region (PAR) (responsible for 43%–50% of total solar radiation) (Wang, Lan, & Horsman, 2012). It can be converted from light energy to chemical energy in microalgae biomass. PAR is a function of microalgae natural pigments, which have characteristic colors and absorbances at specific wavelengths. Chlorophylls (mainly *a*) and the accessory pigments, such as carotenoids and phycobiliproteins, are involved in the light capture process and have their absorption bands in PAR. The ultraviolet (UV) and infrared (IR) regions generally do not participate in the light energy conversion process (Lehmuskero, Chauton, & Boström, 2018). Fig. 4 shows the relative photosynthetic efficiency as a function of wavelength.

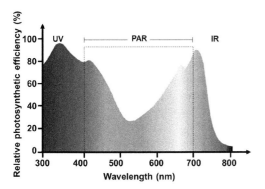

FIG. 4 Relative photosynthetic efficiency as a function of the wavelength of light.

Spectral bandwidth luminosity is the key factor in cell growth in photobioreactors, which must operate under optimal conditions. However, some cyanobacteria, such as *Acaryochloris marina*, may grow in IR radiation regions near

750 nm. But this phenomenon must be carefully controlled so that no productivity losses occur due to system overheating (see in detail in Section 3.2). On the other hand, UV radiation (<400 nm) causes cellular damage due to the ionization of materials (Nwoba, Parlevliet, Laird, Alameh, & Moheimani, 2019).

Additionally, the light energy represented in the PAR range is expressed in terms of photon flux density (PFD) or hemispherical incident light flux (q). The characterization of incident PFD in microalgae cultivation systems is commonly studied (Pfaffinger et al., 2019). As discussed earlier, the light intensity is not evenly distributed because of cell dissipation during cultivation. In this context, unlike classical bioprocesses, in microalgae cultivation, highly specific approaches to photobioreactor design and optimization are required. Fig. 5 demonstrates the relationship between light attenuation and microalgae growth in a photobioreactor.

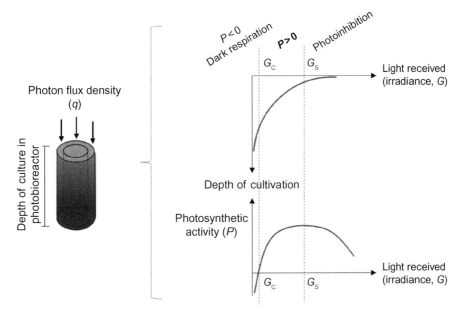

FIG. 5 Relationship between light attenuation and cell growth in microalgae photobioreactors. *Adapted from Pruvost, J. (2019). Cultivation of algae in photobioreactors for biodiesel production (Chapter 26). In: Biofuels: Alternative feedstocks and conversion processes for the production of liquid and gaseous biofuels (2nd ed.).* Biomass, biofuels, biochemicals, pp. 629–659. Academic Press.

It demonstrates photosynthetic activity in response to light supply, represented by the green curve. Such a curve shows a photosynthetic saturation with fluence rate G to a saturation fluence rate G_s (photoinhibition). In addition, to achieve satisfactory growth, a limiting value of the saturation fluence rate G_c, called the photosynthesis compensation point, is required (Pruvost, 2019).

Besides, PFD is expressed in quantitative terms of lumens in the PAR range and must be converted to photons (Eq. 8):

$$\sum_{\lambda=400}^{\lambda=700} I_{ph,nm} \Big/ \sum_{\lambda=400}^{\lambda=700} I_{ph,nm} = \Upsilon_{ph/lm} \qquad (8)$$

where $\Upsilon_{ph/lm}$ is the conversion factor which is calculated by the addition of the lumens (m^2/nm)$^{-1}$ ($l_{lm,nm}$) and the μmol-photons (m^2/s/nm)$^{-1}$ ($I_{ph,nm}$), in the range 400–700 nm and then further calculated by dividing the cumulative values (Blanken, Cuaresma, Wijffels, & Janssen, 2013). The different light intensities have been well documented in the literature, with values ranging from 141 to 1656 μmol/m^2/s. These variations depend on the type of cultivation system, the strain, location (indoor/outdoor), and the desired objective (Nwoba et al., 2019).

Supplying sunlight as natural illumination can affect the photosynthetic efficiency of outdoor cultivations, as we saw before. Artificial lighting can be used to circumvent this critical barrier to microalgal cultures, as it provides better operational control. Photobioreactor light sources may be provided by incandescent, fluorescent, high-intensity discharge, or light-emitting diode (LED) lamps (Schulze, Barreira, Pereira, Perales, & Varela, 2014; Schulze, Guerra, Pereira, Schüler, & Varela, 2017). According to research by the same authors, the use of LEDs has been considered as a method of intelligent light control. They address recent findings on providing light energy with better biomass yields. However, the critical point is

that in qualitative terms, including power consumption, PAR spectrum, and costs, the use of artificial illuminators can be very costly for the process (Janssen, 2016). Estimated electricity expenditure (~65%) associated with an investment in LEDs will add about USD 16/kg in costs for dry biomass production (Blanken et al., 2013).

Finally, the transparency of the materials used for photobioreactor construction, and the surface/volume ratio (S/V) should also be considered. As basic requirements in a photobioreactor design, the choice of material type should ensure the surface is sufficiently clear for light penetration (i.e., without the formation of biofilm layers) (which will be discussed in detail in Section 5.2). In contrast, in terms of S/V, there are currently different photobioreactor configurations designed to expand light distribution (which will be discussed in detail in Section 5.3). Moreover, internal illumination, such as the use of fiber optics, allows improving average irradiance and depth of incidence within the cultivation (Singh & Sharma, 2012; Wang et al., 2012).

3.2 Temperature

Temperature is another important factor associated with microalgal cultivation, especially concerning to cell morphology and physiology, influencing the oscillations of metabolic rates. According to temperature tolerance, microalgae can be classified in psychrophiles, mesophiles, and thermophiles microorganisms (Pires et al., 2017).

In general, the optimum cultivation temperature occurs in the mesophilic region (25–35 °C). However, some strains may tolerate values outside this range. For example, it has been shown that the genus *Chlorella* can grow over a wide range of 5–30 °C (Singh & Singh, 2014). Higher temperature values, in the range of 55–60 °C, may inhibit metabolic activity and insolubilize

CO_2 in the culture medium, impairing its assimilation (Patel, Matsakas, Rova, & Christakopoulos, 2019). On the other hand, values below 16 °C may retard growth kinetics. Nevertheless, thermophilic or thermotolerant strains for the removal of hot exhaust gases have been studied, as well as psychrophilic strains for the cold environments (Table 4).

Most cultivation systems, especially those outdoors, assume temperature variation as a result of environmental variation (seasonality and light/dark cycles). Additional heating and cooling mechanisms, such as the use of thermal blankets, coils, heat exchangers, sprays, and shading systems, may be installed photobioreactors depending on their design, size, and location (Wang et al., 2012). In temperate regions, the temperature can easily reach 40 °C, that is, substantial heat amount must be removed (about 18,000 GJ/ha/yr) through cooling systems. To avoid overheating, direct submersion in water reservoirs has been more widely accepted. But this route requires considerable freshwater volumes (from 2400 to 8000 m^3/ha/yr) to cool the surface of the photobioreactor, especially in large-scale cultivations. The most obvious solution would be to use seawater for this purpose, but it also has the capital and operational expenditures (Nwoba et al., 2019). In contrast, the use of greenhouses can be a heating method for systems located in low-temperature regions; however, extremely energy-intensive (Pruvost, 2019). In this sense, the choice of the best temperature regulation mechanism has been investigated as a function of the desired process design, according to the study by Huang, Jiang, Wang, and Yang (2017).

3.3 pH

The control of pH value has a very significant impact on microalgae cultivation. This variable needs to be kept close to its optimal value and strongly depends on the photosynthetic rate in the photobioreactor (Pawlowski, Guzmán, Berenguel, & Acién, 2019).

The optimal pH of microalgal cultures usually ranges from neutral (pH ~ 7) to alkaline (pH ~ 9). However, the growth of some species is favored at pH 4 (*Chlorococcum littorale*) and

TABLE 4 Microalgae cultivation in different photobioreactors at wide temperatures ranges.

Microalgae	Type of photobioreactor	Temperature (°C)	Reference
Chlorella sp. MT-7	Bubble column	40	Ong, Kao, Chiu, Tsai, and Lin (2010)
Thermosynechococcus sp. CL-1	Bubble column	50	Hsueh, Li, Chen, and Chu (2009)
Thermosynechococcus elongatus BP-1	Flat-plate	55	Bergmann and Trösch (2016)
Chlorella sorokiniana	Helical airlift	55–60	Roy, Kumar, Ghosh, and Das (2014)
Thermosynechococcus elongatus PKUAC-SCTE542	Tubular	60	Liang et al. (2019)
Chlamydomonas, Chlorella, Tetraselmis, Pseudopleurochloris, Nannochloropsis and *Phaeodactylum*	Bubble column	8–15	Schulze, Hulatt, Morales-Sánchez, Wijffels, and Kiron (2019)
Monoraphidium sp.	Thin-layer	10	Řezanka, Nedbalová, Lukavský, Střížek, and Sigler (2017)
Chlorella sorokiniana	Flat-plate	10	Oncel et al. (2015)

pH 8–10 (*Arthrospira platensis*). In high cell density cultures, for example, the pH value may increase to around 9, which can easily be corrected by CO_2-enriched aeration (pure CO_2 or from flue gas). This operational control is a dominant factor that will determine the pH of the medium (Wang et al., 2012).

Regardless, there is a complex relationship between the availability and assimilation of dissolved nutrients in the culture medium. In terms of carbon species, such as CO_2, HCO_3^-, and CO_3^{2-}, the equilibrium between gas phase mass transfer to the liquid phase and the CO_2 assimilation by the cells depends on the pH. In terms of nitrogen, NH_3^- and NH_4^+ are also strongly affected by the pH of the culture medium. This relationship extends to phosphate precipitation and trace elements (Pires et al., 2017). Raeesossadati et al. (2014) and Vasumathi, Premalatha, and Subramanian (2012) discuss in more detail the phenomena of CO_2 fixation and mass transfer concerning pH.

Thus, CO_2 supply at the photobioreactor inlet changes the acidity of the culture medium, which influences the pH and the CO_2 absorption by microalgae. If CO_2 is supplied in large quantities, the pH is rapidly reduced, causing damage to cells. In contrast, insufficient CO_2 supply limits growth (Pawlowski et al., 2019). In this regard, when designing a photobioreactor, an apparatus for efficient pH control should be coupled to it to facilitate maintenance of the culture medium, as this is a standard daily practice for larger-scale reactors.

4 Measuring the photobioreactors performance

Photobioreactors require precise tracking of various cultivation parameters to reflect process dynamics correctly. To this end, the estimation and optimization of microalgal yields require, as well as the development of suitable photobioreactor designs, specific growth modeling to provide useful information on the performance of cultivation systems. Thus, the use of mathematical models becomes crucial elements of simulations (Table 5).

As seen in Table 5, the main photobioreactor performance evaluation models consist of kinetic models as a function of the growth and substrate. These models generally prioritize cell productivity, as it takes into account the maximum biomass density and the time required to reach it (Nwoba et al., 2019). Besides, the main substrate to be used by microalgae is carbon dioxide. Thus, carbon dioxide fixation rates, as well as its elimination capacity and removal efficiency, become essential to determine how efficient the cell ratio and substrate conversion within the system are shown. Additionally, the application of carbon balance proves to be more robust modeling than most other sensors, as it addresses the main carbon conversion routes by photosynthetic microorganisms. Through this is possible to determine the solid, liquid, and volatile fraction of carbon, totaling the global conversion of carbon into photobioreactors.

Similarly, as these are biological reactions, parameters as a function of photo-oxidation require special attention, mainly concerning oxygen production. This is because its excess production, in photobioreactors, is known to be a limiting factor for photosynthesis.

In contrast, kinetic models as a function of light intensity are presented as pertinent but unusual models. This occurs because the corresponding methodology is somewhat primitive since the variables consider photon flux absorbed parameters and energy content of the biomass.

Therefore, gather parameters that have specific characteristics such as fluid dynamics, biochemical reactions, and light and mass transfers does not ensure the overall success of photobioreactors. However, it allows a better understanding of the biological system and its interactions with the environment. Also, it enables the

TABLE 5 Parameters of assessment photobioreactors performance.

Parameter	Mathematical models	Variables nomenclature	Reference
Kinetic models as a function of growth			
Cell productivity	$P_x = \frac{X_1 - X_0}{t_1 - t_0}$	where Px is the biomass productivity (mg/L/h), X_i is biomass concentration at time t_1 (mg/L), X_0 is the biomass concentration at time t_0 (mg/L), t is the residence time (h)	Weigand (1981)
Kinetic models as a function of substrate			
Carbon fixation rate	$Rc = C_C \times \left(\frac{M_{CO_2}}{M_C}\right)$	where Rc is the carbon fixation rate (mg/L/h), C_c is the percent carbon in the biomass (%), M_{CO_2} is the molecular weight of CO_2 (mg), and M_C molecular weight of carbon (mg)	Yun, Lee, Park, Lee, and Yang (1997)
CO_2 elimination capacity	$EC = \frac{(C_i - C_o)}{V_R} \times Q$	where EC is the CO_2 elimination capacity (mg/L/min), C_o and C_i correspond to the outlet and inlet CO_2 concentration (mg/L/min), Q is the gas flow (L/min), and V_R is the reactor volume (L)	Jacob-Lopes, Lacerda, and Franco (2008)
CO_2 removal efficiency	$RE = \frac{(C_i - C_o)}{C_i} \times 100$	where RE is the CO_2 removal efficiency (%), C_o and C_i correspond to the outlet and inlet CO_2 concentration (mg/L/min)	Jacob-Lopes, Lacerda, and Franco (2008)
Carbon mass balance	$C_{in} - C_{out} = V_L M_C \left(\frac{d[C_{biomass}]}{dt} + \frac{d[DIC]}{dt}\right)$ $+ V_g M_C \left(\frac{d[C_{headspace}]}{dt}\right)$	where C_{in} is the mass flow rate of carbon (mg/d), C_{out} is the mass flow rate of carbon out of the photobioreactor (mg/d), V_L is the volume of the media (L), V_g is the volume of the headspace (L), M_C is the atomic weight of carbon (mg), (d$[C_{biomass}]$)/dt is the rate of change in concentration of carbon in the biomass (mg/L/d), (d[DIC])/dt is the rate of change in dissolved inorganic carbon concentration (mg/L/d), and (d$[C_{headspace}]$)/dt is the rate of change in carbon concentration in the headspace (mg/L/d)	Deprá et al. (2019)
Kinect models as a function of photo-oxidation			
Production capacity oxygen	$PC = \frac{(C_i - C_o)}{V_R} \times Q$	where PC is the production capacity O_2 (mg/L/min), C_o and C_i correspond to the outlet and inlet O_2 concentration (mg/L/min), Q is the gas flow (L/min), and V_R is the reactor volume (L)	Jacob-Lopes, Lacerda, and Franco (2008)
Kinect models as a function of light intensity			
Quantum yield	$\psi_E = \frac{P_b}{F_{vol}}$	where Ψ_E is the quantum yield (%), P_b represents the volumetric productivity of biomass (g/L/d) and, F_{vol} is the photon flux absorbed in the unit of volume ($\mu E/m^2/s$). This coefficient can be converted into energy units, commonly observed qk, taking into account the average light energy used (kJ/E)	(Grima, Camacho, Pérez, Fernández, & Sevilla, 1997)
Bioenergetic yield	$\psi = \psi_{KJ} \times Q_b$	where Ψ is the bioenergetic yield (%), Ψ_{KJ} is the product of the system (kJ/E), and Q_b is the biomass combustion heat (kJ/g)	Lee and Tay (1991)

development of systems with greater operational control, as well as assisting in the design process and expansion of photobioreactors to ensure high microalgal productivity rates.

5 Bottlenecks to achieve expansion of photobioreactors

5.1 Power consumption

The microalgae have become prominent in biotechnological processes, because of their potential application as a source of renewable energy, it is imperative to state that these processes produce more energy than they consume. However, it is important to note that higher power consumption is a common problem in the design of the photobioreactor (Pirouzi, Nosrati, Shojaosadati, & Shakhesi, 2014). As a result, continuous efforts are deliberate in building models with low-energy to ensure the economic viability of microalgal bioprocesses (Qin & Wu, 2019).

Indeed, while there are many energy demands involved in microalgal photobioreactors, substantial improvements have been made through new reactor geometries and optimization of mixing and aeration strategies. However, under certain circumstances, the high energy supply to provide cultures mixing and circulation is generally the dominant factor (Jones, Louw, & Harrison, 2017).

In this way, to assertively determine the total energy requirements of photobioreactors, theoretical energy consumptions are estimated for a known range of flow rates in order to characterize equipment performance (Hulatt & Thomas, 2011).

Often, energy consumption is given through Eq. (9):

$$W = \Delta P \times V \qquad (9)$$

where W is energy consumption (Nm/s), ΔP is pressure drop measurement (Nm), and V is volumetric flow (m^3/s).

Another energy parameter that has stood out as an energy measure is the net energy ratio (NER). By definition, the NER can be understood as the ratio of the total energy produced (energy content resulting from the process) to the energy required for photobioreactor operation (Eq. 10).

$$NER = \frac{E_{out}}{E_{inp}} \qquad (10)$$

where NER is net energy ratio (dimensionless), E_{out} is output energy (MJ), and E_{inp} is input energy (MJ).

Furthermore, for us to consider the net energy ratio of a viable process, the NER must be theoretically greater than 1. This is because this value means that there is a net usable energy gain. On the other hand, a net energy ratio of less than one means that there is a general energy loss. However, most of the industrial processes in operation do not meet these requirements, boosting their optimization to achieve these parameters. In this sense, at an exemplary level, Table 6

TABLE 6 Energy requirements and net energy ratio of main photobioreactor configurations.

Design	Energy requirements (W/m³)	Net energy ratio	Reference
Tubular photobioreactor	2500	0.08	Jorquera, Kiperstok, Sales, Embiruçu, and Ghirardi (2010)
Bubble column photobioreactor	78	1.1	Hulatt and Thomas (2011)
Flat-plate photobioreactor	53	1.86	Jorquera et al. (2010)
Biofilm photobioreactor	–	6.01	Ozkan, Kinney, Katz, and Berberoglu (2012)
Hybrid photobioreactor	52	2.49	Deprá et al. (2019)

shows energy demands and energy ratios between major photobioreactor configurations.

5.2 Material quality and investment cost

Currently, a considerable effort has been put into the construction of photobioreactors, aiming at the application of high-quality materials and concomitantly aiming at reducing their investment costs in their developments (Zhu, Chi, Bi, Zhao, & Cai, 2019).

The raw material used in the ideal construction of such equipment, especially for the photo-phase, plays a crucial role in decision making, as well as in the costs and maintenance of them. Thus, the materials used must have the following characteristics: (i) high transparency, i.e., not allow the reduction of light transmittance after outdoor exposure, (ii) flexible and durable properties, (iii) nontoxic, (iv) resistance to chemicals and metabolites produced by microorganisms, as well as resistance to external weather, and (v) low cost (Dasgupta et al., 2010).

Thus, among the most commonly used materials in construction, today include glass, polymethyl methacrylate (PMMA), polycarbonate (PC), low-density polyethylene (LDPE), and

other polymers (Table 7). All of these materials meet the requirements of transparency and mechanical robustness of photobioreactor manufacturing. However, each material has its pros and cons for building specific types of photobioreactors.

As can be seen from Table 7, the main materials used have similar characteristics. However, polycarbonate stands out against the other potentialities, since it presents itself as a high strength material and does not require substantial efforts for cleaning, offering intermediate cost values. In contrast, glass is preferred for applications requiring high quality. However, its handling and processing make its use difficult (Posten, 2012).

In addition, the combination of materials such as steel frame and glass are often found in photobioreactor manufacture. These materials have advantages as the design is simple and can be functioned better as a shading device. However, as disadvantages, the thickness between glass and steel is narrow, making it difficult to make models with many connections. Moreover, the high propensity of cell mortality when there is a cellular allocation in the corners can jeopardize the photobioreactor cellular

TABLE 7 Comparison of main materials for microalgae photobioreactor design.

Material	[a]Material price (USD/kg)	Light transmittance (%)	Material density (kg/m^3)	Material lifespan (year)	Energy content (kJ/kg)	Lifespan-weighted energy content (kJ/m^2 year)
Glass	0.43–1.07	80–90	2400–2800	20	25,000	15,500
Polymethyl methacrylate (PMMA)	3.0–6.20	92	1180	20	131,000	72,800
Polycarbonate (PC)	2.67–4.28	89	1250	20	113,000	66,500
Low-density polyethylene (LDPE)	0.53–5.94	88	920	3	78,000	13,500

[a] *These costs data were corrected for inflation when needed, and thus, provide a reasonable estimate of the current costs.*

Adapted from Uyar, B. (2016). Bioreactor design for photofermentative hydrogen production. Bioprocess and Biosystems Engineering, *39(9), 1331–1340.*

productivity, besides the difficult maintenance of the equipment.

Another possibility of material association can be found between the acrylic molding and steel pipes. These materials are easy to adjust, have a limited probability of leakage, and increase the rate of reduction in cell mortality. On the other hand, as disadvantages are costly for molding acrylic, they require supporting structures in addition to high scratch propensity, which would result in the possibility of cracking in photobioreactors (Ardiani, Koerniawan, Martokusumo, Suyono, & Poerbo, 2019).

Finally, it should be emphasized that the choice and cost of materials for photobioreactor construction, particularly for large-scale systems, and the costs of plant managers, need to be carefully considered for the appropriate price. Besides, certain principles, such as the level of cleanliness, will determine the life cycle of a photobioreactor. Therefore, measures such as smoothing the inner surface, minimizing the quantity and surface area of the inner parts and folds, and having an ample internal space have been introduced to increase the cleanability of these types of equipment. Thus, the best photobioreactor configuration should have a simple structure, practical operation, easy temperature control, low capital cost, low operating cost (including low power consumption and convenient cleaning and maintenance), long service life, and so on.

5.3 Scale-up

Inherently, expansion is considered a complex and detailed process (Guler, Deniz, Demirel, Oncel, et al., 2019). The design of phototrophic bioprocesses can be extremely challenging, especially when reactor types on the considered scales differ (Pfaffinger et al., 2019). Based on this principle, for industrial microalgae production systems, it is essential to select the most suitable photobioreactor type and conditions to operate the workloads required in large-scale processes (Guler, Deniz, Demirel, & Imamoglu, 2020). For this, different parameters must be considered, which are directly associated with the selection of the microorganism and its ideal growth conditions, resistance to variations in environmental conditions, in addition to the product value and total system cost (Riveros et al., 2018).

Briefly, photobioreactor scalability involves the transition of production volumes: (I) laboratory vials, (II) bench photobioreactors, (III) pilot-scale, (IV) industrial scale, to increase the microalgal production volume. Given this scenario, the design of a broad range of photobioreactors requires the identification, not only of the basic principles of microalgal cell biochemical reactions but fundamentally of the primordial engineering concepts involved in the construction and operation of photobioreactors. Among the criteria required is the requirement of a homogeneous light source of the reaction volume, alternating light–dark cycles and intermittent light effects associated with uniform light intensity at different depths of the reaction vessels (Jacob-Lopes, Scoparo, et al., 2009). Besides, the mixing system employed must ensure the removal of oxygen formed to prevent cellular inhibition. Also, equipment monitoring controls must ensure temperature adjustments, in addition to regulation and injection availability from external carbon sources. Finally, the structure of the equipment must ensure that the variation in layer thickness is easily accessible for sterilization procedures (Socher, Löser, Schott, Bley, & Steingroewer, 2016).

Thus, to assess the scalability of photobioreactors, it is imperative to evaluate the optimal surface/volume ratio. This is because this parameter includes the main concern, which determines the amount of light that enters the system per unit of reactor volume. Thus, this criterion improves the efficiency of light utilization of the system, besides increasing the biomass yield, theoretically providing the reduction of production costs (Banerjee & Ramaswamy,

2019). Therefore, the higher the surface/volume ratio, the higher the cell concentration, and consequently, the higher the photobioreactor volumetric productivity.

Numerical tools are known to allow analysis and optimization of the processes under evaluation. Not unlike, since scaling up requires prior experiments, which require significant time and resources to perform, it is pertinent to estimate the surface/volume ratio using this feature as a derivation of guidelines for structural design engineering (Perez-Castro, Sanchez-Moreno, & Castilla, 2017).

Typically, extensively proposed photobioreactors for industrial-scale use are based on specific configurations, which include bubble, tubular and flat plate columns. Therefore, to determine the surface-volume ratio of each model, an individual equation model is required.

The steps for sizing bubble photobioreactors can be followed by Eqs. (11)–(14):

$$C_f = 0.0791 \times Re^{-0.25} \tag{11}$$

$$U_L = 2 \left[\frac{g \times d_b}{1.8} \right]^{0.5} \tag{12}$$

$$V = \frac{Q_L \times \left([O_2]_{in} - [O_2]_{out}\right)}{K_L \times a_L \left([O_2]_{sat} - [O_2]\right)(1 - \varepsilon)} \tag{13}$$

$$D = \sqrt{\frac{4 \times Q_L}{\pi \times U_b}} \tag{14}$$

where C_f is fanning friction factor (dimensionless), U_L is the liquid velocity (m/s), g is the gravitational acceleration (m/s^2), d_b is bubble diameter (m), V is the volume of bubble column (m^3), Q_L liquid flow rate entering bubble column (m/s), $[O_2]_{in}$ and $[O_2]_{out}$ is oxygen concentration in the inlet and outlet (mg/L), $[O_2]_{sat}$ and $[O_2]$ is the driving force for the transport of oxygen from liquid to gas phase (mg/L), $K_L a_L$ is the volumetric mass transfer efficiency (dimensionless), ε is the gas holdup

(dimensionless), D is the minimum diameter of the bubble column (m), and U_b bubble rise velocity (m/s).

Additionally, it is essential to note that, in practical terms, some specifications are assigned in the geometrically cylindrical reaction vessel configurations. Height-to-diameter ratios (H/D) are generally up to 1:10, allowing for high surface areas for light energy uptake by cells. Also, usually, the surface velocity of the gas used ranges from 1 to 2.8 cm/s, providing an increase in the frequency of bubble passage by up to 62%. Another consideration is directly associated with bubble chord length distributions, which were observed to be longer when arranged at higher gas velocities. Besides, the length of the bubble chord ranges from 0.164 to 0.221 cm, while the rate of rising of the bubble is recorded from 64 to 107 cm/s (Ojha & Al-Dahhan, 2018). On the other hand, 500 μm microbubbles have particularities of greater application interest, as they may enhance mass transfer and photobioreaction (Zhao, Wang, Han, & Yu, 2019).

However, for the extension of tubular photobioreactors, the following Eqs. (15)–(18) may be used to drive the scaling up:

$$V = \frac{1}{t_f + t_d} \tag{15}$$

$$U_{RL} = f \alpha U_{RS} \tag{16}$$

$$\alpha = 1 - \varphi L1 - \varphi S \tag{17}$$

$$\varphi = \frac{V_f}{V_f + V_d} \tag{18}$$

where V is cycle frequency (s^{-1}), t_f, t_d is the light and dark period (s), U_{RL} is the velocity at large-scale (m/s), U_{RS} is the small-scale (m), f is the ratio of the tube diameters at both the larger and smaller scales (m), α is retrieved from the φ values at the two scales, V_f illuminated volume of period t_f, and V_d is the dark volume of period t_d.

As tubular photobioreactors consist of a series of transparent tubes with a diameter of

0.1–0.3 m arranged horizontally or vertically, the surface/volume ratio is generally high. Thus, in large workloads, this configuration is limited by factors such as concentration gradients along the tubes, biofilm formation on the tube walls, and requiring high soil area.

Conversely, the sizing of flat plate photobioreactors emphasizes the proportion of productivity rates. Thus, the steps for expansion can be followed by the following Eqs. (19) and (20):

$$P_r = P \times \frac{V}{S_t} \qquad (19)$$

$$n = \frac{S_t}{S} \qquad (20)$$

where P_r is the real biomass productivity (g/L/d), V is the volume (m^3), S_t is total surface area (m^2) of flat panels, and S is the surface area of each flat plate (m^2).

Among some factors that interfere with the scaling of this model are some dimensions generally considered. Several arrangements have been built to maximize your workloads. However, although they have shown satisfactory kinetic performance, some impasses regarding the structure of the equipment must be respected (Jacob-Lopes, Revah, Hernández, Shirai, & Franco, 2009). Thus, the separation of the plates must necessarily vary between 2 and 4 cm. In addition, the width and height usually have values in the order of 2.5 and 1.5 m, respectively. Thus, the V/S ratio is from 40 to 70 L/m^2, making it difficult to expand and operate the equipment (Slegers, Wijffels, Van Straten, & Van Boxtel, 2011).

However, although these configurations have industrial application potentials, to date, as seen above, the ideal scaling model has not been consolidated. In this sense, several theoretical studies direct the new reactor designs to achieve the proper surface-volume ratio, resulting in H/D values of 1.0–1.5. Given these aspects, the hybrid models developed in recent years aim to compensate for the inconvenience caused by the limitation of the S/V ratio and scaling up of conventional photobioreactors (Severo, Deprá, Zepka, & Jacob-Lopes, 2019). At the level of exemplification, Deprá et al. (2019) developed a hybrid photobioreactor model with potential application characteristics that require high workloads. This is because in the proposed design, considering only the area of the illuminated platform presented a surface of 0.22 m^2/m^3. Besides, it is crucial to take into account that the surface required by the bubble column (dark volume) has not been accounted for, as this part of the equipment does not require exposure to light, allowing its placement underground. On the other hand, when compared to the already consolidated models, the tubular and flat plate photobioreactors present values in the order of 14.28 and 14.42 m^2/m^3, respectively.

Also, the bottlenecks encountered in sizing photobioreactors must be taken into account in the light of economic issues. Studies by several authors show that the increase in the area required represents, at a commercial level, a significant capital increase and, consequently, the operating costs of the process, since the cost of land acquisition and improvement is estimated at around 12.0 USD/m^2 (Irwin, Irwin, Martin, & Aracena, 2018; Li, Zhu, Niu, Shen, & Wang, 2011). Yet, this low area requirement enhances the scalability of the photobioreactor, especially under conditions of low land availability.

6 Advances in the design of photobioreactors

Notably, one of the leading scientific tools underway in the field of microalgal research is associated with advances in configurations and new photobioreactor designs for industrial applications.

As seen earlier, key technical challenges prevent the selection of an ideal industrial-scale reactor model. Thus, numerous photobioreactor models were designed in error attempts to create a standard and robust configuration of such equipment (Kunjapur & Eldridge, 2010).

However, enhancing new photobioreactor designs for custom applications means learning and imitating nature. The responsibility and assurance of reproducing equipment that meets the environmental requirements and special conditions for specific microalgal strains for particular end products are far from being achieved. Alternatively, a potential strategy is to establish models of individually tailored equipment to meet the specific demands of industrial segments (Morweiser, Kruse, Hankamer, & Posten, 2010).

As a result, the various designed models do not always appear to have advanced features to previous models. Some of them have lower robustness due to their decrease practicality resulting from different hydrodynamic and scalability factors, as well as structural hygiene and maintenance problems.

In short, only a small portion of photobioreactor configurations can be employed for the mass cultivation of microalgae. Among them, Table 8 addresses and lists the main parameters and their advantages and disadvantages associated with these models.

The different photobioreactor configurations allowed a significant improvement in biomass productivity. Evidently, biofilm photobioreactors present a potential opportunity, since beyond the high microalgal productivity rates

TABLE 8 Main advances of photobioreactors models and their respective advantages and drawbacks.

Design photobioreactor	Cell productivity (g/L/d)	Advantages	Drawbacks
Column photobioreactors	0.06–0.21	Good biomass growth, high efficiency of photosynthesis, high potential of scalability, limited photoinhibition, and photo-oxidization Cheap, compact and easy to maintain The potential of exposition to alternating dark and light cycles Small prerequisites of space demanded cultivation Suitable for outdoor cultivation Low energy use Suitability for algae immobilization	A small area of light exposition that is additionally reduced with the increase of column diameter Low ratio of reactor surface to its volume Possibility of biofilm formation on reactor walls
Tubular photobioreactors	0.05–1.9	High mass transfer Good mixing Potential for scalability Easy to sterilize Least land use Reduced photoinhibition	Small illumination area Sophisticated construction materials Support costs Modest scalability
Flat-plate photobioreactors	0.16–4.3	High biomass productivity Large illumination surface area Suitable for outdoor cultures Uniform distribution of light Low power consumption	Difficult to scale-up Difficult temperature control Fouling Photoinhibition Shear damage from aeration

Continued

TABLE 8 Main advances of photobioreactors models and their respective advantages and drawbacks—cont'd

Design photobioreactor	Cell productivity (g/L/d)	Advantages	Drawbacks
Biofilm photobioreactor	0.015–7.07	Low cost of microalgae harvesting Reduced light limitation Low footprint Low water consumption Efficient CO_2 mass transfer	Formation of gradients Scaling-up
Hybrid photobioreactor	0.05–2.8	Low use of the land area Low operating costs High stability	Limited to a few microalgal strains

Adapted from Gupta, P. L., Lee, S. M., & Choi, H. J. (2015). A mini review: Photobioreactors for large scale algal cultivation. World Journal of Microbiology and Biotechnology, 31(9), 1409–1417; Płaczek, M., Patyna, A., & Witczak, S. (2017). Technical evaluation of photobioreactors for microalgae cultivation. In E3S web of conferences, Vol. 19, p. 02032. EDP Sciences; Maroneze, M. M., & Queiroz, M. I. (2018). Microalgal production systems with highlights of bioenergy production. In Energy from microalgae, pp. 5–34. Springer, Cham.

(0.015–7.07 g/L) compared to the typical concentrations of suspended algae found in photobioreactors (0.5–5.0 g/L), have a solid content ranging from 10% to 20% (Gross, Jarboe, & Wen, 2015). However, they do not have adequate supports for industrial demands, since their structural configuration does not allow large-scale increases.

On the other hand, the hybrid models, although presenting more modest values of microalgal productivities, their configuration aims to compensate for the drawbacks caused by the limitation of the S/V ratio and scale-up of conventional photobioreactors (Jacob-Lopes et al., 2016; Ramirez-Merida, Zepka, & Jacob-Lopes, 2015). Also, this layout would generate higher volumetric production, reducing capital, and operational expenditure.

Finally, regardless of the model, to date, microalgal photobioreactors do not perform optimally for large industrial photobioreactors because they produce very low biomass yields in any geometry compared to other microbial processes (yeast, bacteria) (50–100 g/L) (Bosma, Vermue, Tramper, & Wijffels, 2010). Thus, beyond having to fit the basic rules of hydrodynamics and energy performance, ideally,

photobioreactors should at least achieve the productivity of around 15–30 g/L so that microalgae-based processes can be consolidated.

7 Conclusions

Until now, mass culture in large-scale photobioreactors has undoubtedly been the hottest topic in algal biotechnology. Thus, in order to circumvent these bottlenecks, research has rapidly evolved to meet industrial production. As a result, a wide range of photobioreactor models has been developed over the last decades. These developments were based exclusively on aspects related to hydrodynamic parameters. However, it is well known that techniques rooted in biology, such as the assessment of environmental parameters, can have a substantial impact on microalgal yields and raise the overall efficiency of photobioreactors to a new level in the near future.

Therefore, the critical technical challenges that hinder the optimal selection of a commercial-scale reactor require solutions based on the integration of engineering knowledge associated with biological aspects to consolidate the use of photobioreactors.

Besides, although significant technological improvements have been made, the consolidation of microalgae production for bulk commodities is not yet feasible. Thus, in order to circumvent the economic bottlenecks, current forecasts show that microalgae production should assume characteristic biorefinery models. Therefore, the absolute valorization of all components present in microalgal biomass, especially those of fine chemicals, will facilitate the transition from current small-scale production to large-scale microalgae production in future.

Declaration of competing interest

The authors declare that all figures and tables are self-authorship and not published elsewhere.

References

Acién, F. G., Molina, E., Reis, A., Torzillo, G., Zittelli, G. C., Sepúlveda, C., et al. (2018). Photobioreactors for the production of microalgae. In *Microalgae-based biofuels and bioproducts—From feedstock cultivation to end-products* (pp. 1–44). Woodhead Publishing Series in Energy.

Ardiani, N. A., Koerniawan, M. D., Martokusumo, W., Suyono, E. A., & Poerbo, H. W. (2019). Feasibility of algae photobioreactor as Façade in the office building in Indonesia. In: *IOP conference series: Earth and environmental science, Vol. 322 (1)*, IOP Publishing, 012020.

Banerjee, S., & Ramaswamy, S. (2019). Comparison of productivity and economic analysis of microalgae cultivation in open raceways and flat panel photobioreactor. *Bioresource Technology Reports*, 100328.

Barceló-Villalobos, M., Fernández-del Olmo, P., Guzmán, J. L., Fernández-Sevilla, J. M., & Fernández, F. A. (2019). Evaluation of photosynthetic light integration by microalgae in a pilot-scale raceway reactor. *Bioresource Technology*, *280*, 404–411.

Bergmann, P., & Trösch, W. (2016). Repeated fed-batch cultivation of *Thermosynechococcus elongatus* BP-1 in flat-panel airlift photobioreactors with static mixers for improved light utilization: Influence of nitrate, carbon supply and photobioreactor design. *Algal Research*, *17*, 79–86.

Blanken, W., Cuaresma, M., Wijffels, R. H., & Janssen, M. (2013). Cultivation of microalgae on artificial light comes at a cost. *Algal Research*, *2*(4), 333–340.

Bosma, R., Vermue, M. H., Tramper, J., & Wijffels, R. H. (2010). Towards increased microalgal productivity in photobioreactors. *International Sugar Journal*, *112*(1334), 74–85.

Cabello, J., Morales, M., & Revah, S. (2017). Carbon dioxide consumption of the microalga *Scenedesmus obtusiusculus* under transient inlet CO_2 concentration variations. *Science of the Total Environment*, *584*, 1310–1316.

Camacho, F. G., Rodríguez, J. G., Mirón, A. S., Belarbi, E. H., Chisti, Y., & Grima, E. M. (2011). Photobioreactor scale-up for a shear-sensitive dinoflagellate microalga. *Process Biochemistry*, *46*(4), 936–944.

Cheng, W., Huang, J., & Chen, J. (2016). Computational fluid dynamics simulation of mixing characteristics and light regime in tubular photobioreactors with novel static mixers. *Journal of Chemical Technology & Biotechnology*, *91*(2), 327–335.

Chisti, Y., Halard, B., & Moo-Young, M. (1988). Liquid circulation in airlift reactors. *Chemical Engineering Science*, *43*, 451–457.

Contreras, A., García, F., Molina, E., & Merchuk, J. C. (1998). Interaction between CO_2-mass transfer, light availability, and hydrodynamic stress in the growth of *Phaeodactylum tricornutum* in a concentric tube airlift photobioreactor. *Biotechnology and Bioengineering*, *60*(3), 317–325.

Dasgupta, C. N., Gilbert, J. J., Lindblad, P., Heidorn, T., Borgvang, S. A., Skjanes, K., et al. (2010). Recent trends on the development of photobiological processes and photobioreactors for the improvement of hydrogen production. *International Journal of Hydrogen Energy*, *35*(19), 10218–10238.

Deprá, M. C., Mérida, L. G., de Menezes, C. R., Zepka, L. Q., & Jacob-Lopes, E. (2019). A new hybrid photobioreactor design for microalgae culture. *Chemical Engineering Research and Design*, *144*, 1–10.

El-Baz, F. K., & El Baky, H. H. A. (2018). Pilot scale of microalgal production using Photobioreactor. In *Photosynthesis: From its evolution to future improvements in photosynthetic efficiency using nanomaterials* (pp. 53–67).

Fernández, F. A., Camacho, F. G., Pérez, J. S., Sevilla, J. F., & Grima, E. M. (1998). Modeling of biomass productivity in tubular photobioreactors for microalgal cultures: Effects of dilution rate, tube diameter, and solar irradiance. *Biotechnology and Bioengineering*, *58*(6), 605–616.

Gao, X., Kong, B., & Vigil, R. D. (2018). Multiphysics simulation of algal growth in an airlift photobioreactor: Effects of fluid mixing and shear stress. *Bioresource Technology*, *251*, 75–83.

Garcia-Ochoa, F., & Gomez, E. (2009). Bioreactor scale-up and oxygen transfer rate in microbial processes: An overview. *Biotechnology Advances*, *27*(2), 153–176.

Gayen, K., Bhowmick, T. K., & Maity, S. K. (Eds.). (2019). *Sustainable downstream processing of microalgae for industrial application*. CRC Press.

Gómez-Pérez, C. A., Oviedo, J. E., Ruiz, L. M., & Van Boxtel, A. J. B. (2017). Twisted tubular photobioreactor fluid dynamics evaluation for energy consumption minimization. *Algal Research, 27*, 65–72.

Grima, E. M., Camacho, F. G., Pérez, J. S., Fernández, F. A., & Sevilla, J. F. (1997). Evaluation of photosynthetic efficiency in microalgal cultures using averaged irradiance. *Enzyme and Microbial Technology, 21*(5), 375–381.

Grima, E. M., Sevilla, J. F., Pérez, J. S., & Camacho, F. G. (1996). A study on simultaneous photolimitation and photoinhibition in dense microalgal cultures taking into account incident and averaged irradiances. *Journal of Biotechnology, 45*(1), 59–69.

Gross, M., Jarboe, D., & Wen, Z. (2015). Biofilm-based algal cultivation systems. *Applied Microbiology and Biotechnology, 99*(14), 5781–5789.

Guler, B. A., Deniz, I., Demirel, Z., & Imamoglu, E. (2019). Computational fluid dynamics simulation in scaling-up of airlift photobioreactor for astaxanthin production. *Journal of Bioscience and Bioengineering, 129*(1), 86–92.

Guler, B. A., Deniz, I., Demirel, Z., Oncel, S. S., & Imamoglu, E. (2019). Comparison of different photobioreactor configurations and empirical computational fluid dynamics simulation for fucoxanthin production. *Algal Research, 37*, 195–204.

Guler, B. A., Deniz, I., Demirel, Z., & Imamoglu, E. (2020). Computational fluid dynamics simulation in scaling-up of airlift photobioreactor for astaxanthin production. *Journal of Bioscience and Bioengineering, 129*(1), 86–92.

Heredia, J. C. R., García, A. N., Marin, A. R., Lopez, Y. C., Loria, J. D. C. Z., & Rivero, J. C. S. (2018). Effect of hydrodynamic conditions of photobioreactors on lipids productivity in microalgae. *Microalgal Biotechnology*, 39–57.

Hernández-Melchor, D. J., Cañizares-Villanueva, R. O., Terán-Toledo, J. R., López-Pérez, P. A., & Cristiani-Urbina, E. (2017). Hydrodynamic and mass transfer characterization of flat-panel airlift photobioreactors for the cultivation of a photosynthetic microbial consortium. *Biochemical Engineering Journal, 128*, 141–148.

Hinterholz, C. L., Trigueros, D. E. G., Módenes, A. N., Borba, C. E., Scheufele, F. B., Schuelter, A. R., et al. (2019). Computational fluid dynamics applied for the improvement of a flat-plate photobioreactor towards high-density microalgae cultures. *Biochemical Engineering Journal, 151*, 107257.

Hsueh, H. T., Li, W. J., Chen, H. H., & Chu, H. (2009). Carbon bio-fixation by photosynthesis of *Thermosynechococcus* sp. CL-1 and *Nannochloropsis oculta. Journal of Photochemistry and Photobiology B: Biology, 95*(1), 33–39.7.

Huang, Q., Jiang, F., Wang, L., & Yang, C. (2017). Design of photobioreactors for mass cultivation of photosynthetic organisms. *Engineering, 3*(3), 318–329.

Huang, J., Li, Y., Wan, M., Yan, Y., Feng, F., Qu, X., et al. (2014). Novel flat-plate photobioreactors for microalgae cultivation with special mixers to promote mixing along the light gradient. *Bioresource Technology, 159*, 8–16.

Hulatt, C. J., & Thomas, D. N. (2011). Productivity, carbon dioxide uptake and net energy return of microalgal bubble column photobioreactors. *Bioresource Technology, 102*(10), 5775–5787.

Irwin, N. B., Irwin, E. G., Martin, J. F., & Aracena, P. (2018). Constructed wetlands for water quality improvements: Benefit transfer analysis from Ohio. *Journal of Environmental Management, 206*, 1063–1071.

Jacob-Lopes, E., Lacerda, L. M. C. F., & Franco, T. T. (2008). Biomass production and carbon dioxide fixation by *Aphanothece microscopica Nägeli* in a bubble column photobioreactor. *Biochemical Engineering Journal, 40*(1), 27–34.

Jacob-Lopes, E., Revah, S., Hernández, S., Shirai, K., & Franco, T. T. (2009). Development of operational strategies to remove carbon dioxide in photobioreactors. *Chemical Engineering Journal, 153*(1–3), 120–126.

Jacob-Lopes, E., Scoparo, C. H. G., Lacerda, L. M. C. F., & Franco, T. T. (2009). Effect of light cycles (night/day) on CO_2 fixation and biomass production by microalgae in photobioreactors. *Chemical Engineering and Processing: Process Intensification, 48*(1), 306–310.

Jacob-Lopes, E., Scoparo, C. H. G., Queiroz, M. I., & Franco, T. T. (2010). Biotransformations of carbon dioxide in photobioreactors. *Energy Conversion and Management, 51*(5), 894–900.

Jacob-Lopes E., L.Q. Zepka, L. Ramírez-Mérida, M.M. Maroneze, Neves, C. (2016) Bioprocess for the conversion of carbon dioxide from industrial emissions, bioproducts use thereof and hybrid photobioreactor, WO 2016/041028 A1, March 24.

Janssen, M. (2016). Microalgal photosynthesis and growth in mass culture. *Advances in chemical engineering* (pp. 185–256). Vol. 48 (pp. 185–256). Academic Press.

Jones, S. M., Louw, T. M., & Harrison, S. T. (2017). Energy consumption due to mixing and mass transfer in a wave photobioreactor. *Algal Research, 24*, 317–324.

Jorquera, O., Kiperstok, A., Sales, E. A., Embiruçu, M., & Ghirardi, M. L. (2010). Comparative energy life-cycle analyses of microalgal biomass production in open ponds and photobioreactors. *Bioresource Technology, 101*(4), 1406–1413.

Kannan, V., Naren, P. R., Buwa, V. V., & Dutta, A. (2019). Effect of drag correlation and bubble-induced turbulence closure on the gas hold-up in a bubble column reactor. *Journal of Chemical Technology & Biotechnology, 94*(9), 2944–2954.

Kováts, P., Thévenin, D., & Zähringer, K. (2018). Characterizing fluid dynamics in a bubble column aimed for the determination of reactive mass transfer. *Heat and Mass Transfer, 54*(2), 453–461.

Kunjapur, A. M., & Eldridge, R. B. (2010). Photobioreactor design for commercial biofuel production from microalgae. *Industrial & Engineering Chemistry Research, 49*(8), 3516–3526.

Lee, Y. K., & Tay, H. S. (1991). High CO_2 partial pressure depresses productivity and bioenergetic growth yield of *Chlorella pyrenoidosa* culture. *Journal of Applied Phycology*, *3*(2), 95–101.

Lehmuskero, A., Chauton, M. S., & Boström, T. (2018). Light and photosynthetic microalgae: A review of cellular-and molecular-scale optical processes. *Progress in Oceanography*, *168*, 43–56.

Leonard, C., Ferrasse, J. H., Lefevre, S., Viand, A., & Boutin, O. (2019). Gas hold up in bubble column at high pressure and high temperature. *Chemical Engineering Science*, *200*, 186–202.

Levasseur, W., Taidi, B., Lacombe, R., Perré, P., & Pozzobon, V. (2018). Impact of seconds to minutes photoperiods on *Chlorella vulgaris* growth rate and chlorophyll a and b content. *Algal Research*, *36*, 10–16.

Li, H. L., Li, Y. D., & Li, D. X. (2013). Influence of superficial gas velocity on the dynamic characteristics parameters of CFB. *Advanced materials research* (pp. 953–957). Vol. 663 (pp. 953–957). Trans Tech Publications.

Li, J., Zhu, D., Niu, J., Shen, S., & Wang, G. (2011). An economic assessment of astaxanthin production by large scale cultivation of *Haematococcus pluvialis*. *Biotechnology Advances*, *29*(6), 568–574.

Liang, Y., Tang, J., Luo, Y., Kaczmarek, M. B., Li, X., & Daroch, M. (2019). *Thermosynechococcus* as a thermophilic photosynthetic microbial cell factory for CO_2 utilisation. *Bioresource Technology*, *278*, 255–265.

Maroneze, M. M., Jacob-Lopes, E., Zepka, L. Q., Roca, M., & Pérez-Gálvez, A. (2019). Esterified carotenoids as new food components in cyanobacteria. *Food Chemistry*, *287*, 295–302.

Maroneze, M. M., Siqueira, S. F., Vendruscolo, R. G., Wagner, R., de Menezes, C. R., Zepka, L. Q., et al. (2016). The role of photoperiods on photobioreactors—A potential strategy to reduce costs. *Bioresource Technology*, *219*, 493–499.

Marshall, J. S., & Sala, K. (2011). A stochastic Lagrangian approach for simulating the effect of turbulent mixing on algae growth rate in a photobioreactor. *Chemical Engineering Science*, *66*(3), 384–392.

Merchuk, J. C., Gluz, M., & Mukmenev, I. (2000). Comparison of photobioreactors for cultivation of the red microalga Porphyridium sp. *Journal of Chemical Technology & Biotechnology: International Research in Process, Environmental & Clean Technology*, *75*(12), 1119–1126.

Míguez, J. L., Porteiro, J., Pérez-Orozco, R., Patiño, D., & Gómez, M. Á. (2020). Biological systems for CCS: Patent review as a criterion for technological development. *Applied Energy*, *257*, 114032.

Mirón, A. S., Camacho, F. G., Gomez, A. C., Grima, E. M., & Chisti, Y. (2000). Bubble-column and airlift photobioreactors for algal culture. *AIChE Journal*, *46*(9), 1872–1887.

Mirón, A. S., García, M. C. C., Camacho, F. G., Grima, E. M., & Chisti, Y. (2004). Mixing in bubble column and airlift reactors. *Chemical Engineering Research and Design*, *82*(10), 1367–1374.

Moroni, M., Lorino, S., Cicci, A., & Bravi, M. (2019). Design and bench-scale hydrodynamic testing of thin-layer wavy photobioreactors. *Water*, *11*(7), 1521.

Morweiser, M., Kruse, O., Hankamer, B., & Posten, C. (2010). Developments and perspectives of photobioreactors for biofuel production. *Applied Microbiology and Biotechnology*, *87*(4), 1291–1301.

Nwoba, E. G., Parlevliet, D. A., Laird, D. W., Alameh, K., & Moheimani, N. R. (2019). Light management technologies for increasing algal photobioreactor efficiency. *Algal Research*, *39*, 101433.

Ogbonna, J. C., & Tanaka, H. (2000). Light requirement and photosynthetic cell cultivation—Development of processes for efficient light utilization in photobioreactors. *Journal of Applied Phycology*, *12*, 207–218.

Ojha, A., & Al-Dahhan, M. (2018). Local gas holdup and bubble dynamics investigation during microalgae culturing in a split airlift photobioreactor. *Chemical Engineering Science*, *175*, 185–198.

Oncel, S. S., Kose, A., Faraloni, C., Imamoglu, E., Elibol, M., Torzillo, G., et al. (2015). Biohydrogen production from model microalgae *Chlamydomonas reinhardtii*: A simulation of environmental conditions for outdoor experiments. *International Journal of Hydrogen Energy*, *40*(24), 7502–7510.

Ong, S. C., Kao, C. Y., Chiu, S. Y., Tsai, M. T., & Lin, C. S. (2010). Characterization of the thermal-tolerant mutants of *Chlorella* sp. with high growth rate and application in outdoor photobioreactor cultivation. *Bioresource Technology*, *101*(8), 2880–2883.

Ozkan, A., Kinney, K., Katz, L., & Berberoglu, H. (2012). Reduction of water and energy requirement of algae cultivation using an algae biofilm photobioreactor. *Bioresource Technology*, *114*, 542–548.

Paladino, O., & Neviani, M. (2020). Scale-up of photobioreactors for microalgae cultivation by π-theorem. *Biochemical Engineering Journal*, *153*, 107398.

Patel, A., Matsakas, L., Rova, U., & Christakopoulos, P. (2019). A perspective on biotechnological applications of thermophilic microalgae and cyanobacteria. *Bioresource Technology*, *278*, 424–434.

Pawlowski, A., Guzmán, J. L., Berenguel, M., & Acién, F. G. (2019). Control system for pH in raceway photobioreactors based on wiener models. *IFAC-PapersOnLine*, *52*(1), 928–933.

Peiran, Y., & Shizhu, W. (1990). A generalized Reynolds equation for non-Newtonian thermal elastohydrodynamic lubrication. *Journal of Tribology*, *112*(4), 631–636.

Perez-Castro, A., Sanchez-Moreno, J., & Castilla, M. (2017). PhotoBioLib: A Modelica library for modeling and simulation of large-scale photobioreactors. *Computers & Chemical Engineering, 98*, 12–20.

Perner-Nochta, I., & Posten, C. (2007). Simulations of light intensity variation in photobioreactors. *Journal of Biotechnology, 131*(3), 276–285.

Pfaffinger, C. E., Severin, T. S., Apel, A. C., Göbel, J., Sauter, J., & Weuster-Botz, D. (2019). Light-dependent growth kinetics enable scale-up of well-mixed phototrophic bioprocesses in different types of photobioreactors. *Journal of Biotechnology, 297*, 41–48.

Pires, J. C., Alvim-Ferraz, M. C., & Martins, F. G. (2017). Photobioreactor design for microalgae production through computational fluid dynamics: A review. *Renewable and Sustainable Energy Reviews, 79*, 248–254.

Pirouzi, A., Nosrati, M., Shojaosadati, S. A., & Shakhesi, S. (2014). Improvement of mixing time, mass transfer, and power consumption in an external loop airlift photobioreactor for microalgae cultures. *Biochemical Engineering Journal, 87*, 25–32.

Posten, C. (2012). Design and performance parameters of photobioreactors. *TATuP-Zeitschrift für Technikfolgenabschätzung in Theorie und Praxis, 21*(1), 38–45.

Préat, N., Taelman, S. E., Meester, S. D., Allais, F., & Dewulf, J. (2020). Identification of microalgae biorefinery scenarios and development of mass and energy balance flowsheets. *Algal Research, 45*, 101737.

Prussi, M., Buffi, M., Casini, D., Chiaramonti, D., Martelli, F., Carnevale, M., et al. (2014). Experimental and numerical investigations of mixing in raceway ponds for algae cultivation. *Biomass and Bioenergy, 67*, 390–400.

Pruvost, J. (2019). Cultivation of algae in photobioreactors for biodiesel production. (Chapter 26)*Biofuels: Alternative feedstocks and conversion processes for the production of liquid and gaseous biofuels* (pp. 629–659). (2nd ed.). *Biomass, biofuels, biochemicals*Academic Press.

Pruvost, J., Cornet, J. F., & Legrand, J. (2008). Hydrodynamics influence on light conversion in photobioreactors: An energetically consistent analysis. *Chemical Engineering Science, 63*(14), 3679–3694.

Qin, C., & Wu, J. (2019). Influence of successive and independent arrangement of Kenics mixer units on light/dark cycle and energy consumption in a tubular microalgae photobioreactor. *Algal Research, 37*, 17–29.

Raeesossadati, M. J., Ahmadzadeh, H., McHenry, M. P., & Moheimani, N. R. (2014). CO_2 bioremediation by microalgae in photobioreactors: Impacts of biomass and CO_2 concentrations, light, and temperature. *Algal Research, 6*, 78–85.

Ramirez-Merida, L. G., Zepka, L. Q., & Jacob-Lopes, E. (2015). Current status, future developments and recent patents on photobioreactor technology. *Recent Patents on Engineering, 9*(2), 80–90.

Řezanka, T., Nedbalová, L., Lukavský, J., Střížek, A., & Sigler, K. (2017). Pilot cultivation of the green alga Monoraphidium sp. producing a high content of polyunsaturated fatty acids in a low-temperature environment. *Algal Research, 22*, 160–165.

Riveros, K., Sepulveda, C., Bazaes, J., Marticorena, P., Riquelme, C., & Acién, G. (2018). Overall development of a bioprocess for the outdoor production of Nannochloropsis gaditana for aquaculture. *Aquaculture Research, 49*(1), 165–176.

Roy, S., Kumar, K., Ghosh, S., & Das, D. (2014). Thermophilic biohydrogen production using pre-treated algal biomass as substrate. *Biomass and Bioenergy, 61*, 157–166.

Sada, E., Yasunishi, A., Katoh, S., & Nishioka, M. (1978). Bubble formation in flowing liquid. *The Canadian Journal of Chemical Engineering, 56*(6), 669–672.

Santos, A. B., Fernandes, A. S., Wagner, R., Jacob-Lopes, E., & Zepka, L. Q. (2016). Biogeneration of volatile organic compounds produced by *Phormidium autumnale* in heterotrophic bioreactor. *Journal of Applied Phycology, 28*(3), 1561–1570.

Schulze, P. S., Barreira, L. A., Pereira, H. G., Perales, J. A., & Varela, J. C. (2014). Light emitting diodes (LEDs) applied to microalgal production. *Trends in Biotechnology, 32*(8), 422–430.

Schulze, P. S., Guerra, R., Pereira, H., Schüler, L. M., & Varela, J. C. (2017). Flashing LEDs for microalgal production. *Trends in Biotechnology, 35*(11), 1088–1101.

Schulze, P. S., Hulatt, C. J., Morales-Sánchez, D., Wijffels, R. H., & Kiron, V. (2019). Fatty acids and proteins from marine cold adapted microalgae for biotechnology. *Algal Research, 42*, 101604.

Severo, I. A., Barin, J. S., Wagner, R., Zepka, L. Q., & Jacob-Lopes, E. (2018). Biofuels from microalgae: Photobioreactor exhaust gases in oxycombustion systems. In *Energy from microalgae* (pp. 271–290). Cham: Springer.

Severo, I. A., Deprá, M. C., Barin, J. S., Wagner, R., de Menezes, C. R., Zepka, L. Q., et al. (2018). Bio-combustion of petroleum coke: The process integration with photobioreactors. *Chemical Engineering Science, 177*, 422–430.

Severo, I. A., Deprá, M. C., Zepka, L. Q., & Jacob-Lopes, E. (2016). Photobioreactors and oxycombustion: A mini-review on the process integration. *Journal of Chemical Engineering & Process Technology, 7*, 310.

Severo, I. A., Deprá, M. C., Zepka, L. Q., & Jacob-Lopes, E. (2019). Carbon dioxide capture and use by microalgae in photobioreactors. (Chapter 8)In *Bioenergy with carbon capture and storage* (pp. 151–171). Academic Press.

Shariff, S., & Chakraborty, S. (2017). Two-scale model for quantifying the effects of laminar and turbulent mixing on algal growth in loop photobioreactors. *Applied Energy*, *185*, 973–984.

Singh, R. N., & Sharma, S. (2012). Development of suitable photobioreactor for algae production—A review. *Renewable and Sustainable Energy Reviews*, *16*(4), 2347–2353.

Singh, S. P., & Singh, P. (2014). Effect of CO_2 concentration on algal growth: A review. *Renewable and Sustainable Energy Reviews*, *38*, 172–179.

Sivakaminathan, S., Hankamer, B., Wolf, J., & Yarnold, J. (2018). High-throughput optimisation of light-driven microalgae biotechnologies. *Scientific Reports*, *8*(1), 11687.

Slegers, P. M., Wijffels, R. H., Van Straten, G., & Van Boxtel, A. J. B. (2011). Design scenarios for flat panel photobioreactors. *Applied Energy*, *88*(10), 3342–3353.

Socher, M. L., Löser, C., Schott, C., Bley, T., & Steingroewer, J. (2016). The challenge of scaling up photobioreactors: Modeling and approaches in small scale. *Engineering in Life Sciences*, *16*(7), 598–609.

Song, B. Y., Li, M. J., He, Y., Yao, S., & Huang, D. (2019). Electrochemical method for dissolved oxygen consumption on-line in tubular photobioreactor. *Energy*, *177*, 158–166.

Tian, H., Pi, S., Feng, Y., Zhou, Z., Zhang, F., & Zhang, Z. (2020). One-dimensional drift-flux model of gas holdup in fine-bubble jet reactor. *Chemical Engineering Journal*, *368*, 121222.

Tobajas, M., & Garcia-Calvo, E. (2000). Comparison of experimental methods for determination of the volumetric mass transfer coefficient in fermentation processes. *Heat and Mass Transfer*, *36*(3), 201–207.

Ugwu, C. U., Aoyagi, H., & Uchiyama, H. (2008). Photobioreactors for mass cultivation of algae. *Bioresource Technology*, *99*(10), 4021–4028.

Ugwu, C., Ogbonna, J., & Tanaka, H. (2002). Improvement of mass transfer characteristics and productivities of inclined tubular photobioreactors by installation of internal static mixers. *Applied Microbiology and Biotechnology*, *58*(5), 600–607.

Ugwu, C. U., Ogbonna, J. C., & Tanaka, H. (2005). Characterization of light utilization and biomass yields of *Chlorella sorokiniana* in inclined outdoor tubular photobioreactors equipped with static mixers. *Process Biochemistry*, *40*(11), 3406–3411.

Vasumathi, K. K., Premalatha, M., & Subramanian, P. (2012). Parameters influencing the design of photobioreactor for the growth of microalgae. *Renewable and Sustainable Energy Reviews*, *16*(7), 5443–5450.

Vieira, K. R., Pinheiro, P. N., Santos, A. B., Cichoski, A. J., de Menezes, C. R., Wagner, R., et al. (2019). The role of microalgae-based systems in the dynamics of odors compounds in the meat processing industry. *Desalination and Water Treatment*, *150*, 282–292.

Wang, B., Lan, C. Q., & Horsman, M. (2012). Closed photobioreactors for production of microalgal biomasses. *Biotechnology Advances*, *30*(4), 904–912.

Weigand, W. A. (1981). Maximum cell productivity by repeated fed-batch culture for constant yield case. *Biotechnology and Bioengineering*, *23*(2), 249–266.

Wondraczek, L., Gründler, A., Reupert, A., Wondraczek, K., Schmidt, M. A., Pohnert, G., et al. (2019). Biomimetic light dilution using side-emitting optical fiber for enhancing the productivity of microalgae reactors. *Scientific Reports*, *9*(1), 9600.

Yang, Y., Liao, Q., Zhu, X., Wang, H., Wu, R., & Lee, D. J. (2011). Lattice Boltzmann simulation of substrate flow past a cylinder with PSB biofilm for bio-hydrogen production. *International Journal of Hydrogen Energy*, *36*(21), 14031–14040.

Yun, Y. S., Lee, S. B., Park, J. M., Lee, C. I., & Yang, J. W. (1997). Carbon dioxide fixation by algal cultivation using wastewater nutrients. *Journal of Chemical Technology & Biotechnology: International Research in Process, Environmental and Clean Technology*, *69*(4), 451–455.

Zepka, L. Q., Jacob-Lopes, E., & Roca, M. (2019). Catabolism and bioactive properties of chlorophylls. *Current Opinion in Food Science*, *26*, 94–100.

Zhang, K., Kurano, N., & Miyachi, S. (2002). Optimized aeration by carbon dioxide gas for microalgal production and mass transfer characterization in a vertical flat-plate photobioreactor. *Bioprocess and Biosystems Engineering*, *25*(2), 97–101.

Zhang, S., Wang, Z., Sun, B., & Yuan, K. (2019). Pattern transition of a gas–liquid flow with zero liquid superficial velocity in a vertical tube. *International Journal of Multiphase Flow*, *118*, 270–282.

Zhao, Y., Wang, H. P., Han, B., & Yu, X. (2019). Coupling of abiotic stresses and phytohormones for the production of lipids and high-value by-products by microalgae: A review. *Bioresource Technology*, *274*, 549–556.

Zhu, C., Chi, Z., Bi, C., Zhao, Y., & Cai, H. (2019). Hydrodynamic performance of floating photobioreactors driven by wave energy. *Biotechnology for Biofuels*, *12*(1), 54.

Transport phenomena models affecting microalgae growth

José Rebolledo-Oyarce[a], César Sáez-Navarrete[a,b,c], and Leonardo Rodriguez-Cordova[a]

[a]Departamento de Ingeniería Química y Bioprocesos, Pontificia Universidad Católica de Chile, Santiago, Chile [b]Centro de Investigación en Nanotecnología y Materiales Avanzados CIEN-UC, Facultad de Física, Pontificia Universidad Católica de Chile, Santiago, Chile [c]UC Energy Research Center (CE-UC), Pontificia Universidad Católica de Chile, Santiago, Chile

1 Introduction

Photosynthetic microorganisms use energy delivered by the sun and carbon dioxide from the environment to produce high-value compounds such as proteins, lipids, pigments, biofuels, food supplements, drugs, among other things. Likewise, there are species organisms that can be used to treat wastewater or gaseous waste using a filter adapted to absorb harmful compounds for the environment (Borowitzka, 2013; Kumar, Santhanam, Park, & Kim, 2016; Leong, Lim, Lam, Uemura, & Ho, 2018; Markou, Wang, Ye, & Unc, 2018; Vo et al., 2019).

Due to these industrial applications, the study of the cultivation of these organisms and their subsequent use to manufacture high-value compounds has become widespread in recent years and has allowed many compound purification and extraction techniques to be improved and perfected.

However, due to the great difficulty in producing these organisms on large scales is the application these advances have been hindered because the production of these compounds is the starting point to produce different applications (Fig. 1).

In the 1950s, the industry that was engaged in the production of large-scale microalgae in large culture pools began to face great problems when trying to optimize their cultivation processes, with this the first closed photobioreactors were developed that were used in the food industry (Vo et al., 2019). These systems provided greater control of environmental factors such as temperature, the number of gases exchanged, the surface area irradiated with light, among other factors.

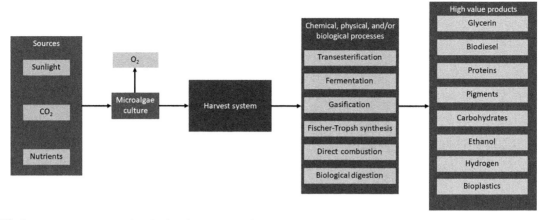

FIG. 1   Macro stages to produce high-value compounds using microalgae.

With the passage of time, when it was necessary to understand the behavior of these reactors in order to optimize these processes, theories of matter and energy exchange began to be applied (Darvehei, Bahri, & Moheimani, 2018; Vo et al., 2019; Wang, Lan, & Horsman, 2012).

These theories, in the beginning, were very vague approximations of reality, but that over time were evolving to highly complex and precise models that involve even the dynamics of the fluid applying numerical methods (Bitog et al., 2011), and all this evolution can be explained due to the increase of the capacity of computers and a better understanding of the physicochemical processes present in these systems.

Of these developed theories, the one that has always had a high interest is the law that governs the distribution of light inside the photobioreactors (Molina Grima et al., 1994), due to the impact that light has on the growth of microalgae and consequently on the production of compounds high commercial value (Naderi, Znad, & Tade, 2017). Likewise, it has been determined that the most limiting factor when carrying out both closed and open cultures is the amount of energy delivered by the lighting system (Brindley, Jiménez-Ruíz, Acién, & Fernández-Sevilla,

2016; Naderi et al., 2017; Wágner, Valverde-Pérez, & Plósz, 2018); therefore, developing an adequate understanding of this phenomenon allows for better elaboration, mixing systems, and substrate feed.

Due to the aforementioned, it is that, in this chapter, the various irradiance models, their impact on microalgae growth, and fluid transfer models applied in the understanding of photobioreactors will be analyzed (Li et al., 2010; Naderi, Tade, & Znad, 2015; Perner-Nochta & Posten, 2007; Pilon, Berberoğlu, & Kandilian, 2011; Rosello Sastre, Csögör, Perner-Nochta, Fleck-Schneider, & Posten, 2007; Zhang et al., 2016).

2 Most important factors for the growth of a microalgae

To achieve the optimization of microalgae cultures, it is important to consider the following factors: temperature, cultivation method, nutrients, pH of the environment, intensity of light among others (Chia et al., 2018). These factors directly impact the growth rate of the microalgae biomass and the composition of this same biomass. Therefore, finding a balance or an optimal area of these factors for each microalga is essential to reduce the operating costs of closed farming systems.

2.1 Type of reactor

Despite the years, the type of reactor remains one of the biggest challenges of these organisms, that is, the problem is in the design and optimization of the reactors (photobioreactor), especially in energy consumption, stating that the commercial success of photobioreactors will be achieved when their energy consumption is understood (Xu, Lv, Huo, & Li, 2018). Therefore, the two types of photobioreactors are presented below: open and closed.

2.1.1 Open photobioreactors

Open photobioreactors are the first biological reactors that were used to produce large-scale microalgae biomass since the first microalgae culture (*Chlorella vulgaris*) presented by Beyerinck (1890).

The development of these reactors began in 1948 in the United States (Stanford), Germany (Essen), and Japan (Tokyo) (Borowitzka, 1999; Burlew, 1953). But it was not until the 1960s that the Nihon Chlorella company, in Japan, managed to develop the first large-scale open reactor giving rise to a whole new branch of study on microalgae that involves its process of growth and harvesting and processing of the biomass (Borowitzka, 1999; Mata, Martins, & Caetano, 2010; Spolaore, Joannis-Cassan, Duran, & Isambert, 2006).

After the impulses of this Japanese company, many companies started a process of research and implementation of these culture systems reaching, in 1980, a total of 46 companies throughout Asia that could produce more than 1000 kg of microalgae year. And for the year 1996, this production exceeded 2000 tons (Borowitzka, 1999; Lee, 1997). Among the most emblematic companies of that era is the company Sosa Texcoco S.A., which was located in Lake Texcoco in Mexico and that in 2009 ceased

operations due to economic problems and the company Dainippon Ink and Chemicals Corporation (DIC Corporation) that, since 1977, produces *Spirulina* for the manufacture of food and that until Today (year 2020) is still in operation.

In these more than 60 years of development of these reactors, three types of reactors have been mainly implemented: (1) raceway pond, (2) shallow circular ponds with a rotating arm to mix the culture, and (3) shallow sloping ponds (Borowitzka, 1999; Doucha & Lívanský, 2009; Mata et al., 2010).

The raceway pond mainly consists of large pools with depths ranging from 15 to 40 cm, which at the beginning were rectangular without agitation, but over time, they modified systems with infinite loops in which the culture moves through of the pond by means of stirring blades that also help to homogenize the system. Circular ponds are also shallow containers and that, like raceway ponds, the culture is kept homogeneous by a rotating arm. Also, the shallow sloping ponds correspond to a system developed in the Czech Republic, where it takes advantage of the force of gravity to keep the system agitated. In addition, these latter ponds have a lower depth than the other two, with a range of depth ranging from 6 to 8 mm (Doucha & Lívanský, 2006, 2009).

Among the three types of reactors mentioned above, the most used by the industry are raceway ponds and shallow inclined ponds, due to their easy and economical construction and low operating cost (Christenson & Sims, 2011; Doucha & Lívanský, 2009).

These systems, being open to the environment, do not have a culture temperature control; therefore, when selecting the microalgae used for cultivation, the ability of the microorganism to grow in the climate of the area must be considered. However, these systems have the advantage of not subjecting microalgae to great hydrodynamic stress, so that the fall in growth due to this factor is avoided. Also, these open systems have expanded due to easy scalability (Christenson & Sims, 2011).

2.1.2 Closed photobioreactors

These systems, unlike the previous ones, are not open to the environment; therefore, great control of the culture is obtained allowing the growth of any microalgae independent of the climate of the area (Borowitzka, 1999). This advantage gives them a commercial interest in the ability to produce both low-value compounds (fuel and food) and high-value compounds (β-carotene and polysaccharides) independent of climatic conditions.

Like open systems, various designs of these reactors have been explored and these reactors can be divided into two large groups: (1) flat plate reactors and (2) tubular reactors (Dasgupta et al., 2010).

The flat plate reactors are characterized by being thin-shaped orthopedic systems, in which the lighting of these systems is carried out by one of the wide faces, achieving the maximum use of area vs volume allowing for this configuration. As for the tubular reactors, these can be divided into cylindrical tubular and helical tubular; these reactors are characterized by greater efficiency in the use of light and be the ones that have presented the best results in the biomass production process (Adeniyi, Azimov, & Burluka, 2018).

With respect to tubular reactors, as mentioned above, they have presented better yields in biomass production, due to a better distribution of light in the system accompanied by an adequate homogenization of the reactor allowing microalgae to receive light properly and the resources needed to grow such as carbon dioxide (CO_2).

Therefore, in general, it can be summarized that when designing microalgae culture systems, we have that between open and closed systems; open photobioreactors have advantages when scaling them for an industry, but they have serious problems in the process of harnessing the light, generated that the systems are limited by the light and the amount of CO_2 available (Christenson & Sims, 2011).

In contrast, closed photobioreactors solve these problems of light availability and CO_2 availability, but present serious problems at the time of scaling causing operating costs to be more than twice as expensive as open systems, which is why one of the most prominent branches of studies of closed reactors is the optimization of these systems, to reduce operating costs, especially in tubular closed systems, since they have the highest yields in biomass production (Borowitzka, 1999; Mata et al., 2010).

2.2 Temperature

Temperature is one of the factors that have been determined to affect both the growth rate of microalgae and its biochemical composition, provided there is no CO_2 limitation or limitation of light availability (Pulz, 2001), having an influence on these factors from 7% to 9% (Converti, Casazza, Ortiz, Perego, & Del Borghi, 2009).

With this, a large temperature range has been investigated in order to determine the impact of temperature on various microalgae, being able to determine that for a large number of microalgae its optimal operating range is between 20°C and 25°C (Chia et al., 2018). However, there are microalgae that can survive in low-temperature environmental conditions, such as *Chlorella vulgaris* that can grow at a temperature of 4°C (Bartosh & Banks, 2007), an important characteristic if what we are looking for is the reduction of operating costs.

2.3 pH

The pH of the culture medium does not have a significant impact on the biomass composition of the microalgae, but it does have an important effect on the survival of the microalgae (Chia et al., 2018).

The effect of pH occurs in the biological processes performed by the microalgae, such as the absorption of CO_2, ions, and nutrients, intracellular and cell wall functions associated with photosynthetic enzymes (Juneja, Ceballos, & Murthy, 2013). For example, the acidic pH alters the absorption capacity of nutrients, while very alkaline pH reduces the affinity for CO_2 and delays the completion of the cell cycle (Juneja et al., 2013). Due to these effects, it is important to keep microalgae in the optimum pH and temperature ranges in the reactors (Table 1 shows optimal ranges of some microalgae).

2.4 Available nutrients

The availability of nutrients in the culture medium is essential so that the microalgae can perform the biological cycles since they use these nutrients as raw material to convert them into carbohydrates and lipids; the latter are those that are desired to be produced to obtain fuels.

In the case of the microalgae mentioned in Table 1, which correspond to photosynthetic organisms, the compound used as a carbon source is carbon dioxide (CO_2). This compound is added to the culture medium either from the atmosphere or through a controllable flow injection system; this second alternative is generally implemented because the atmospheric air does not have enough CO_2 to maintain the medium (Christenson & Sims, 2011). Additionally, with the CO_2 injection system, it is possible to perform the removal of gaseous oxygen (O_2), a compound that is a product of photosynthesis, which generates photooxidation when there is the presence of light (Carvalho & Meireles, 2006).

TABLE 1   Tested and optimal ranges of temperature and pH of various strains of microalgae.

Strain	Temperature range tested (°C)	Optimum temperature range (°C)	pH range tested	Optimum pH range	References
Dunaliella salina	15–30	20–25	6–9	7–7.5	Çelekli and Dönmez (2006), Wu, Duangmanee, Zhao, and Ma (2016), and Ying, Gilmour, and Zimmerman (2014)
Dunaliella tertiolecta	12–28	20–25	7.3–9.3	7.4–7.8	Goldman, Riley, and Dennett (1982), Rukminasari (2013), and Sosik and Mitchell (1994)
Chlorella vulgaris	20–30	25–30	3–10	7.5–8	Gong, Feng, Kang, Luo, and Yang (2014), Kessler (1985), Rachlin and Grosso (1991), Serra-Maia, Bernard, Gonçalves, Bensalem, and Lopes (2016), and Sharma, Singh, and Sharma (2012)
Isochrysis galbana	10–25	18–22	6–9	7.5–8	Molina Grima, Sánchez Pérez, García Sánchez, García Camacho, and López Alonso (1992)
Nannochloropsis salina	10–35	25–30	5–10	7.5–8	Bartley, Boeing, Dungan, Holguin, and Schaub (2014) and Van Wagenen et al., (2012)
Spirulina platensis	16–50	35–40	7–11	9–9.5	Belkin and Boussiba (1991), Göksan, Zekeriyaoğlu, and Ak (2007), and Ismaiel, El-Ayouty, and Piercey-Normore (2016)

Together with these gases, there are micronutrients that must be added to the culture medium in order to maintain and/or enhance the growth of microalgae, such as iron (Fe), magnesium (Mg), calcium (Ca), sodium (Na), potassium (K), copper (Cu), manganese (Mn), among other nutrients (Chia et al., 2018; Özgür et al., 2010; Uyar, Schumacher, Gebicki, & Modigell, 2009).

2.5 Light intensity

Finally, the intensity of light is one of the most studied factors due to the direct impact this factor has on photosynthesis (Chia et al., 2018). In fact, the intensity of light and the efficient use of light are fundamental parameters when designing the reactor in which the microalgae is going to be cultivated since it determines both the economic and the production parameters.

This lighting in closed photobioreactors can be delivered both by sunlight and by an artificial lighting system; however, in recent years, the latter systems have gained popularity due to the flexibility they present as it is possible to supply colors specific, and the decrease in the consumption of these systems, becoming a feasible alternative (Pulz, 2001).

This illumination, independent of the type chosen, must maintain a balance between the saturation and inhibition conditions, since, below the light saturation condition, the rate of photosynthesis and biomass growth increase with increasing the intensity of the light, but at some point, the phenomenon of light saturation begins to occur in which, the photon absorption is equal to the exchange of electrons by photosynthesis. Beyond saturation, there is an inhibition of light that causes irreversible damage to the photosynthetic apparatus and produces a decrease in biomass.

3 Irradiation models

To model the transfer of irradiance into a system with microalgae, it is necessary to understand the interactions of the microalgae with the incident light. In general terms, interactions are reduced to absorption, reflection, refraction, and transmission (Pilon & Kandilian, 2016). These last three phenomena, generally, are grouped into one called scattering (Fig. 2).

These interactions and their transfer have been studied in many situations; however, over the years, the models that have prevailed over time are:

- Beer-Lambert law
- Two-flux approximation (TFA)
- Radiative transfer equation (RTE)

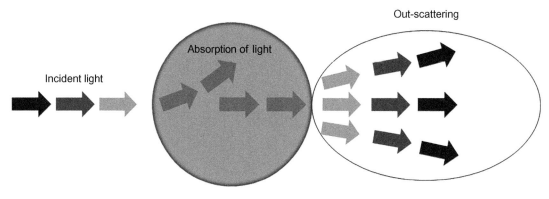

FIG. 2   Scheme of the basic interactions inside a microalga.

These three methodologies are mentioned from the lowest level of complexity to the highest. This increase in complexity is given by the assumptions and phenomena considered.

3.1 Beer-Lambert law

If we refer to the modeling of the irradiance inside a photobioreactor, the Beer-Lambert law has been preferred by many researchers to explain how light interacts with its surroundings due to its easy resolution and its high level of adjustment with experimental data (Lee, Jalalizadeh, & Zhang, 2015; Naderi et al., 2017; Zhang et al., 2016).

This law considers that a monochromatic light source is transferred into a single transverse direction to that focus. And this transferred energy is only affected by the phenomenon of absorption of the microorganism, that is, the phenomena of light scattering are neglected (Cornet, Dussap, & Dubertret, 1992; Mosorov, 2017; Wágner et al., 2018; Zhang, Dechatiwongse, & Hellgardt, 2015), as shown in Eq. (1)

$$\frac{dI_\lambda}{dr} = -\kappa_\lambda \cdot X \cdot I_\lambda \tag{1a}$$

$$I_\lambda(r) = I_0 \cdot \exp\left(-\kappa_\lambda \cdot X \cdot r\right) \tag{1b}$$

where I_λ corresponds to the irradiance of a wavelength λ measured in terms of $W\,m^{-2}$. r corresponds to the path traveled by light in a linear manner (m). κ_λ represents the absorption made by a specific microorganism of a certain wavelength λ, since it is assumed that the intensity will be converted into internal energy of matter along the line of sight of the beam without being redirected to other directions due to scattering, that is, there is no scattering (Howell, Siegel, & Mengüç, 2010), and this parameter has units of $m^2\,g^{-1}$. Also, the term X corresponds to the concentration of the microorganism used in the culture ($g\,m^{-3}$).

However, despite its simplicity, it has allowed us to understand the phenomenon of

radiation in a fairly wide range of biomass concentrations with a maximum cell density equal to $2\,g\,L^{-1}$ (Molina Grima, Acién Fernández, García Camacho, & Chisti, 1999; Pruvost, Legrand, Legentilhomme, & Muller-Feuga, 2002).

It is important to note that the assumption that the light travels in a linear fashion is possible to fulfill it in reality using either open photobioreactors or closed planar photobioreactors; in both cases, there is a flat face perpendicular to the light source forcing that the light transfers mostly linearly. But this equation is not very applicable in tubular-closed photobioreactors, where the surface facing the culture is not a flat surface.

In addition, Eq. (1) assumes that the term radiation does not vary with respect to time, that there is a steady state. However, as biomass grows over time, irradiance also varies over time; therefore, Eq. (1) depends on time indirectly.

3.2 Two-flux approximation

Going up in the complexity level, the two-flux approximation model proposed by Schuster in 1905 and first applied to a photobioreactor by Cornet et al. (1992) is presented.

In this model, like Beer-Lambert's law, a monochromatic light source is emitted that emits radially, but in addition, said light has the following characteristics: (1) the light field that develops inside the reactor is isotropic and (2) absorption and scattering phenomena are present, where the latter is performed by the suspended particle (microalgae) (Cornet et al., 1992; Cornet, Dussap, & Gros, 1994).

As can be assumed, the scattering can be in all directions, which is why an approximation is made of the number of flows that are dispersed by the particle ranging from two to eight flows, achieving a numerical solution for each of them (Cornet et al., 1994). However, after comparative analysis, it was possible to determine that the approximation of two flows is sufficiently

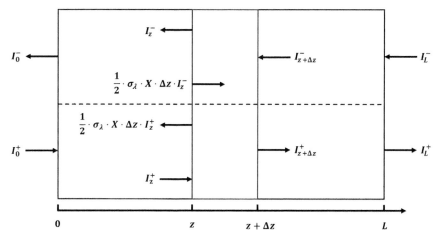

FIG. 3   Global radiation balance performed by two-flux approximation.

accurate to obtain the radiation profile (Cornet et al., 1994).

This approximation of two flows considers that the scattering occurs in two directions opposite to each other and parallel to the direction of emission. Fig. 3 shows the differential balance, allowing us to obtain Eqs. (2) and (3).

$$I^+|_{z,\lambda} - I^+|_{z+\Delta z,\lambda} + \frac{1}{2} \cdot \sigma_\lambda \cdot X \cdot \Delta z \cdot I^-|_{z,\lambda}$$

$$= \kappa_\lambda \cdot X \cdot \Delta z \cdot I^+|_{z,\lambda} + \frac{1}{2} \cdot \sigma_\lambda \cdot X \cdot \Delta z \cdot I^+|_{z,\lambda} \quad (2)$$

$$I^-|_{z+\Delta z,\lambda} - I^-|_{z,\lambda} + \frac{1}{2} \cdot \sigma_\lambda \cdot X \cdot \Delta z \cdot I^+|_{z,\lambda}$$

$$= \kappa_\lambda \cdot X \cdot \Delta z \cdot I^-|_{z,\lambda} + \frac{1}{2} \cdot \sigma_\lambda \cdot X \cdot \Delta z \cdot I^-|_{z,\lambda} \quad (3)$$

where the terms $I^-|_{z,\lambda}$ and $I^+|_{z,\lambda}$ correspond to the irradiance of the wavelength λ in the positive and negative direction to the z-axis, respectively (W m^{-2}). The term σ_λ corresponds to the scattering term generated by the suspended particle with respect to the wavelength λ(m^{2}g^{-2}).

With respect to the scattering, it is possible to determine that the coefficient $1/2$ responds to the consideration that the scattering occurs equitably both forward and backward.

From Eqs. (2) and (3) dividing by Δz and having said term zero, it is possible to obtain the following differential equations:

$$\frac{dI_z^+}{dz} = -\kappa_\lambda \cdot X \cdot I_z^+ + \frac{1}{2} \cdot \sigma_\lambda \cdot X \cdot \left(I_z^- - I_z^+\right) \quad (4)$$

$$\frac{dI_z^-}{dz} = \kappa_\lambda \cdot X \cdot I_z^- + \frac{1}{2} \cdot \sigma_\lambda \cdot X \cdot \left(I_z^- - I_z^+\right) \quad (5)$$

The system of differential equations presented in Eqs. (4) and (5) can be solved by providing adequate edge conditions. For example, if the reactor is illuminated on only one side ($z = 0$) and on the other side ($z = L$) it is obscured with a material that absorbs light, it is possible to say that the edge conditions are:

$$z = 0 \quad I_z^+ = I_0 \quad (6)$$

$$z = L \quad I_z^- = 0 \quad (7)$$

With Eqs. (4)–(7) it is possible to obtain the total radiation flux (I_λ) by making the vector sum of the positive and negative irradiance:

$$I_\lambda = I_\lambda^+ - I_\lambda^- \quad (8)$$

Finally, it is important to note that from Eqs. (4)–(8), it is possible to recover the Beer-Lambert law by making the scattering term equal to zero, and that, like the Beer-Lambert law, it has the limitation that it is difficult to include the geometry of the reactor in the equations and therefore, are valid for containers, in general, flat (Cornet et al., 1994; Cornet, Dussap, Gros, Binois, & Lasseur, 1995). However, there are some adaptations to apply them in other geometries, for example, a toroid (Pottier et al., 2005).

3.3 Radiative transfer equation (RTE)

At the highest level of difficulty is the Radiative Transfer Equation that corresponds to the generalized equation of heat transfer by radiation, which could consider the geometry of the problem and properly incorporate scattering phenomena.

RTE has been very useful both in the world of physics, chemistry, and engineering, but it was only a few years ago when it began to be implemented in photobioreactors due to its high computational costs (Pareek, Chong, Tadé, & Adesina, 2008).

Although this equation in very few texts is explained in detail, it is known that it is a simplification of the Boltzmann transfer equation (developed by Ludwig Boltzmann in 1872) and is made possible because the photons do not interact with each other, but if interact linearly with the medium through which it spreads (Menguc & Viskanta, 1985; Pareek et al., 2008; Selçuk & Ayrancı, 2003).

$$\frac{1}{c} \cdot \frac{\partial I_\lambda}{\partial t} + \mu \cdot \frac{\partial I_\lambda}{\partial x} + \eta \cdot \frac{\partial I_\lambda}{\partial y} + \xi \cdot \frac{\partial I_\lambda}{\partial z}$$
$$= -\kappa_\lambda \cdot X \cdot I_\lambda - \sigma_\lambda \cdot X \cdot I_\lambda$$
$$+ \frac{\sigma_\lambda \cdot X}{4\pi} \int_{4\pi} \Phi(\Omega', \Omega) \cdot I_\lambda(s, \Omega') \, d\Omega' \qquad (9)$$

where the term $\Phi(\Omega', \Omega)$ corresponds to phase function that determines how the radiation is scattered by the suspended and/or medium particle. The term c corresponds to the constant of the speed of light in a vacuum (m s^{-1}). Also, the terms μ, η, and ξ are the cosine directions of the solid angle for the x, y, and z coordinates, respectively, where:

$$\mu = \cos\theta \quad \eta = \sin\theta \cdot \sin\phi \quad \xi = \sin\theta \cdot \cos\phi \qquad (10)$$

From Eq. (9), we have that the terms on the left side represent the temporal and spatial distribution, respectively. Also, on the right side are, in order, the terms of absorption, scattering out of the particle and into the particle of the wavelength λ made by the microalgae.

Among the most important advantages of RTE is that it is possible to incorporate the geometry of the problem. For example, in cylindrical enclosures, the following formulation is obtained (Ben Salah, Askri, Slimi, & Ben Nasrallah, 2004; Menguc & Viskanta, 1986):

$$\frac{1}{c} \cdot \frac{\partial I_\lambda}{\partial t} + \frac{1}{r} \cdot \frac{\partial}{\partial r}(r \cdot \mu \cdot I_\lambda)|_{(\phi, z, \theta, \psi)}$$
$$- \frac{1}{r} \cdot \frac{\partial}{\partial \psi}(\eta \cdot I_\lambda)|_{(r, \phi, z, \theta)} + \frac{\partial}{\partial z}(\xi \cdot I_\lambda)|_{(r, \phi, \theta, \psi)}$$
$$= -\kappa_\lambda \cdot X \cdot I_\lambda - \sigma_\lambda \cdot X \cdot I_\lambda$$
$$+ \frac{\sigma_\lambda \cdot X}{4\pi} \int_{4\pi} \Phi(\Omega', \Omega) \cdot I_\lambda(s, \Omega') \, d\Omega' \qquad (11)$$

As in the case of the rectangular enclosure configuration, the terms μ, η, and ξ are the cosine directions of the solid angle:

$$\mu = \sin\theta \cdot \cos\psi \quad \eta = \sin\theta \cdot \sin\psi \quad \xi = \cos\theta \qquad (12)$$

Although from Eq. (11) it is possible to obtain a solution in a large majority of cases, it is possible to assume that the radiative transport is symmetrical in the ψ direction; therefore, the term of variation in that direction is negligible. Additionally, since the speed of light is a much larger parameter than the rest of the terms in the equation (approximately seven orders of greater magnitude), it is commonly assumed that the temporal variation is negligible (Cornet et al., 1994, 1995), obtaining the following equation:

$$\frac{1}{r} \cdot \frac{\partial}{\partial r}(r \cdot \mu \cdot I_\lambda)|_{(\phi, z, \theta, \psi)} + \frac{\partial}{\partial z}(\xi \cdot I_\lambda)|_{(r, \phi, \theta, \psi)}$$
$$= -\kappa_\lambda \cdot X \cdot I_\lambda - \sigma_\lambda \cdot X \cdot I_\lambda$$
$$+ \frac{\sigma_\lambda \cdot X}{4\pi} \int_{4\pi} \Phi(\Omega', \Omega) \cdot I_\lambda(s, \Omega') \, d\Omega' \qquad (13)$$

It is important to note that from Eq. (13) it is possible to recover Beer-Lambert's law and the two-flux approximation.

In the case of Beer-Lambert's law, it is possible to obtain it assuming that the scattering made by the culture is negligible compared to the absorption term and that the light source is uniform and infinite in the z-direction, therefore, there is not variation in that direction.

Also, in the case of the two-flux approximation, it can be derived from Eq. (13) considering that the scattering occurs in an isotropic medium, that is, $\Phi(\Omega',\Omega)=1$ and that this occurs in only two directions opposite each other and parallel to the r-axis.

As you have seen in the equations that present the variation in irradiance with respect to the special position inside a photobioreactor, there are advantages and disadvantages of using one equation over the other. For example, Beer-Lambert's law is interesting since a simple equation to solve has proven useful for predicting the behavior of light inside a photobioreactor. However, RTE has determined in a concrete way the effect of different colors on the growth of microalgae.

Additionally, the ability of RTE to be able to apply to different geometries represents an advantage toward the future to allow the cultivation systems to be scaled in a more reliable way to that obtained in the laboratory. That is why the challenge in this area is to be able to develop more efficient and less expensive computational resolution methods to allow the application of these systems in the scaling process. It is also important to be able to develop a methodology to empirically determine the phase function of the culture medium with the microalgae and allow its application in future models.

3.3.1 Phase function: Meaning and numerical approximation

The phase function term, $\Phi(\Omega',\Omega)$, is a function that expresses how the scattering term in the system occurs. This term can be determined empirically as shown in the publication of Kandilian, Jesus, Legrand, Pilon, and Pruvost (2017). But alternatively, they have been developed in different ways in order to solve the radiation equation numerically.

The first approach is presented by Fiveland (1984), which proposes that the phase function can be decomposed into a constant term and a variable term that depends exclusively on the type of scattering that the medium presents, distinguishing cases where there is 100% scattering forward ($a=1$), 100% backward scattering ($a=-1$), and the base case of isotropic scattering ($a=0$). This approach is as follows:

$$\Phi(\Omega',\Omega) = 1 + a \cdot \cos\Theta \qquad (14a)$$

$$-1 \le a \le 1 \qquad (14b)$$

The second approach proposes that the phase function can be represented as a series of Legendre polynomials (P_n) (Fiveland, 1988), which is expressed as follows:

$$\Phi(\Omega',\Omega) = \sum_{n=0}^{N} (2n+1)a_n P_n(\cos\Theta) \qquad (15)$$

where a_n is the coefficient of a Legendre series. Also, in both equations, we have the term Θ that represents the angle between the incoming and outgoing intensity of the system.

The third expression corresponds to that carried out by Henyey and Greenstein (1941) in an empirical manner that has the following form:

$$\int_{4\pi} \Phi(\Omega',\Omega)\,d\Omega' = 2\pi \int_{-1}^{1} \Phi(t)\,dt = 1 \qquad (16a)$$

$$\Phi(\Omega',\Omega) = \Phi_{HGg}(t)$$
$$= \frac{1-g^2}{4\pi \cdot (1+g^2 - 2 \cdot g \cdot t)^{3/2}} \quad t := \hat{\omega} \cdot \omega \in [-1, 1]$$

$$(16b)$$

$$-1 < g < 1 \qquad (16c)$$

where the term g as in Eq. (14) determines the type of scattering performed by the system, being isotropic scattering when the term g is equal to zero, forward scattering with $g > 0$ and backward scattering with $g < 0$ (Han, Long, Cong, Intes, & Wang, 2017; Henyey & Greenstein, 1941).

These three approaches, in the case of biological tissues and photosynthesizing organisms (microalgae), should be applied considering a strongly forward scattering due to the size of the cells compared to the wavelength (Han et al., 2017; Kandilian, Lee, & Pilon, 2013; Pilon & Kandilian, 2016), always resulting in a dispersion of Mie (Lock & Gouesbet, 2009; Mishchenko, 2009).

4 Growth models in microalgae

Together with the light intensity distribution model, it is necessary to have a model that allows us to relate how irradiance affects the growth of microalgae and incorporates it into Eq. (17) (Zhang et al., 2015a, 2015b; Zhang et al., 2016).

$$\frac{dX}{dt} = \mu \cdot X - \mu_d \cdot X^2 \tag{17}$$

where X is the concentration of the biomass of the microalgae (g m^{-3}). μ is the specific growth rate of the microalgae in question (day^{-1}). μ_d corresponds to the specific death rate at the present environmental conditions (m^{3}g^{-1}day^{-1}).

The Eq. (17) presented above can be derived from the following development of the equation that represents the concentration of the microalgae biomass with respect to time:

$$X(t) = \frac{X_{max}}{1 + \exp[-\mu(t - t_0)]}$$

if we derive the previous equation, it is possible to obtain the variation of the biomass with respect to time:

$$\frac{dX(t)}{dt} = \frac{X_{max} \, \mu \exp[-\mu(t - t_0)]}{(1 + \exp[-\mu(t - t_0)])^2}$$

$$\frac{dX(t)}{dt} = \frac{X_{max} \, \mu \, (\exp[-\mu(t - t_0)] + 1) - X_{max} \, \mu}{(1 + \exp[-\mu(t - t_0)])^2}$$

$$\frac{dX(t)}{dt} = \frac{X_{max} \, \mu}{1 + \exp[-\mu(t - t_0)]} - \frac{X_{max}^2}{(1 + \exp[-\mu(t - t_0)])^2} \frac{\mu}{X_{max}}$$

$$\frac{dX(t)}{dt} = \mu X(t) - \frac{\mu}{X_{max}} X^2(t) = \mu X(t) - \mu_d X^2(t)$$

For this, it is necessary to know that for microalgae there are three categories of models that allow estimating the specific growth rate of the microalgae (μ) (Darvehei et al., 2018; Lee et al., 2015). These categories are as follows:

- Growth models based on a single substrate
- Growth models based on irradiance
- Multifactor growth models

In addition to the models mentioned in the previous table, Lee et al. (2015) and Anwar, Lou, Chen, Li, and Hu (2019) perform a complete review of the vast majority of existing models; however, in Table 2 are the most used models in the literature and without more direct to apply in a scaling-up process.

As can be seen in Table 2, there are different ways of modeling the growth rate of microalgae. However, the equation to be used will depend exclusively on which or what are the limiting parameters in the cultivation system that are planned to be studied in a mathematical model. The reason behind this is that the limiting variable will have high sensitivity with respect to the growth of the microalgae, that is, that the positive or negative effect on the magnitude of one of the variables also affects the other variable.

Likewise, it is possible to appreciate that as we need to improve the accuracy of the model we lose in degrees of freedom since more variables are necessary to estimate the growth of the microalgae.

Finally, it is possible to raise the challenge of developing a growth equation that allows

TABLE 2 Growth models for microalgae.

Type of model	Model	Formula	References
Growth models based on a single substrate	Andrews	$\mu = \mu_{max} \dfrac{S}{K_S + S + \dfrac{S^2}{K_{S,i}}}$	Kunikane and Kaneko (1984), Kurano and Miyachi (2005); Moya, Sánchez-Guardamino, Vilavella, and Barberà (1997)
	Martínez et al.	$\mu = \dfrac{\mu_{m1}S + \mu_{m2}K_S + \dfrac{\mu_{m3}S^2}{K_{S,i}}}{K_S + S + \dfrac{S^2}{K_{S,i}}}$	Martínez, Jiménez, and Yousfi (1999) and Sancho, Castillo, and Yousfi (1997)
Growth models based on irradiance	Aiba	$\mu = \mu_{max} \dfrac{I}{K_I + I + \dfrac{I^2}{K_{I,L}}}$	Lee et al. (2015), Lee, Erickson, and Yang, (1987), Rebolledo-Oyarce, Mejía-López, García, Rodríguez-Córdova, and Sáez-Navarrete (2019), and Zhang, Dechatiwongse, del Rio-Chanona, et al. (2015a, 2015b)
	Steele	$\mu = \mu_{max} \cdot \dfrac{I}{I_{opt}} e^{\left(1 - \frac{I}{I_{opt}}\right)}$	Kurano and Miyachi (2005), Ranganathan, Amal, Savithri, and Haridas (2017), Rebolledo-Oyarce et al. (2019)
	Grima et al.	$\mu = \mu_{max} \dfrac{I_{av}^{b+\frac{c}{I}}}{\left[I_K + \left(\dfrac{I}{K_{I,L}}\right)^a\right]^{b+\frac{c}{I}} + I_{av}^{b+\frac{c}{I}}}$	Fernández, Camacho, Pérez, Sevilla, and Grima (1998), Lee et al. (2015), and Molina Grima, Fernández Sevilla, Sánchez Pérez, and García Camacho (1996)
Multifactor growth models	Franz et al.	$\mu = \mu_{max} \left(\dfrac{I}{K_I + I}\right)\left(\dfrac{S_{CO_2}}{K_{S,CO_2} + S_{CO_2}}\right)$ $\left(\dfrac{S_{nu}}{K_{S,nu} + + S_{nu}}\right) f(T)$	Franz, Lehr, Posten, and Schaub (2012)
	He et al.	$\mu = \mu_{max} \left(\dfrac{I_{av}}{K_I + I_{av}}\right)$ $\left(\dfrac{S_{CO_2}}{K_{CO_2} + S_{CO_2} + \dfrac{S_{CO_2}^2}{K_{I,CO_2}}}\right)$	He, Subramanian, and Tang (2012)

modeling of the growth process of microalgae with oscillation in biomass concentration over time. This situation is quite interesting to attack because, in industrial processes, temperature, pH, and feed flow controllers have a lag, whether large or small, with respect to the situation that triggers the action of the controller; this causes an oscillation in the culture medium that is not easy to simulate and model, so if a simple equation to characterize this situation is developed, it would be possible to better understand the large cultivation systems.

4.1 Important equations of biomass growth

4.1.1 Aiba model

This model is proposed by Aiba in 1982 as a modification of the first equation that related the growth of microalgae with irradiance, the equation of Tamiya et al. (Darvehei et al., 2018; Lee et al., 2015). Tamiya's equation corresponds to a simile of the classical Monod equation of cell growth dependent on a substrate, but applied to irradiance and which is stated as follows:

$$\mu = \mu_{max} \cdot \frac{I}{K_I + I}$$

where μ and μ_{max} correspond to the specific growth rate (day^{-1}) and the maximum specific growth rate (day^{-1}), respectively. The term I corresponds to the average irradiance inside the bioreactor (W·m^{-2}). Finally, the term K_I corresponds to a parameter of photo saturation (W·m^{-2}).

From the previous equation and the modification to the Monod equation proposed by Andrews (1968), Aiba proposed the following equation:

$$\mu = \mu_{max} \cdot \frac{I}{K_I + I + \dfrac{I^2}{K_{i,L}}}$$

In this equation, an additional adjustment term appears that considers the microalgae photoinhibition, $K_{i,L}$ (W·m^{-2}), an important factor since, when the microalgae are illuminated with an irradiance rate above the necessary damage is generated in the cells causing a decrease in their growth rate.

This model has been useful over the years to explain the behavior of fresh and saltwater microalgae such as *Isochrysis galbana* (Molina Grima et al., 1996), *Spirulina platensis* (Lee et al., 1987), *Cyanothece* sp. (Zhang, Dechatiwongse, & Hellgardt, 2015), among others (Zhang, Dechatiwongse, del Rio-Chanona, et al., 2015a, 2015b).

4.1.2 Steele model

For another length, the Steele model, proposed in 1962, was the first model to incorporate photoinhibition of photosynthetic microorganisms. This effect is presented in the following equation:

$$\mu = \mu_{max} \cdot \frac{I}{I_{opt}} e^{\left(1 - \frac{I}{I_{opt}}\right)}$$

where the term I_{opt} represents the optimal growth irradiance that allows μ to be equal to μ_{max} and is measured in W·m^{-2}.

The previous equation, like the Aiba model, has been used to explain the behavior of different microalgae, with great success in explaining the effect of photoinhibition (Bello, Ranganathan, & Brennan, 2017; Ranganathan et al., 2017; Wágner et al., 2018). However, the same author (Steele) reports high model deviation in shallow culture; therefore, its use in such culture is not recommended (Steele, 1962).

5 Momentum transfer models

The momentum transfer models allow us to understand how the movement inside a photobioreactor is. This understanding allows us to reduce the shear stress, improve the delivery of nutrients, and allow the adequate transfer of gases into the system (Gao, Kong, & Vigil, 2018; Wang & Lan, 2018).

Due to this, the simulations of computational dynamics of fluids (CFD) have played a fundamental role in the construction of new reactors and the scaling of them in industrial processes (Bitog et al., 2011). The CFD technique consists in solving the Navier-Stokes equations using different mathematical methods such as finite volumes or finite elements (Zienkiewicz & Taylor, 2000).

By solving these equations, it is possible to study the influence of certain hydrodynamic factors on the behavior of the photobioreactor and its consequent impact on the growth of microalgae. These factors can be the geometry of the reactor, the speed of the circulating gas, the average size of the bubbles used to aerate the system, the pressure, the speed of the mechanical mixing system in the case of existing, among other things (Bitog et al., 2011; Gao et al., 2018; Gao, Kong, & Dennis Vigil, 2017a, 2017b; Wu & Merchuk, 2004).

5.1 Three phase model

In a photobioreactor, there are three phases living together: gas (CO_2 and oxygen exchanged between the microalgae and the environment), liquid (provided by the culture medium), and solid (the microalgae in suspension) (Gao et al., 2018).

For the general case of an airlift reactor, the equations of conservation of mass and momentum, from the Eulerian point of view, are:

$$\frac{\partial}{\partial t}(\alpha_k \rho_k) + \nabla \cdot \left(\alpha_k \rho_k \vec{u}_k \right) = 0 \qquad (18a)$$

$$\frac{\partial}{\partial t}\left(\alpha_k \rho_k \vec{u}_k \right) + \nabla \cdot \left(\alpha_k \rho_k \vec{u}_k \vec{u}_k \right)$$
$$= -\alpha_k \nabla p + \nabla \cdot \left(\overline{\overline{\tau_k}} + \overline{\overline{\tau^{Re}}} \right) + \alpha_k \rho_k \vec{g} + \vec{F}_{lk} \quad (18b)$$

where α_k and $\vec{u}_k$ are the fractions of phase volume and liquid velocities ($k = l$), solids ($k = s$) and gases ($k = g$), correspondingly. $\overline{\overline{\tau_k}}$ and $\overline{\overline{\tau^{Re}}}$ are the phase stress and Reynolds stress tensors, respectively. These terms depend on the turbulence model used to solve the problem (Gao et al., 2017b). ρ_k corresponds to the density of phase k. p is the system pressure. Finally, the term $\vec{F}_{lk}$ is the exchange of momentum between the liquid phase and the k phase; this term is broken down into: drag, lift, virtual mass, wall lubrication, and turbulent dispersion forces (Gao et al., 2017b).

Together with the general balance equation, the constitutive equations of the system can be simplified in the following equations:

- Drag model for gas-liquid (Tomiyama, Kataoka, Zun, & Sakaguchi, 1998)

$$\overrightarrow{F_{Dg}} = -\overrightarrow{F_{Dl}} = \frac{3}{4}\alpha_g \alpha_l \frac{\rho_l}{d_b} C_D \left| \vec{u}_g - \vec{u}_l \right| \left(\vec{u}_g - \vec{u}_l \right) \quad (19)$$

$$C_D = \max \left(\left(\min \left(\frac{24}{Re_D}\left(1 + 0.15Re_b^{0.687}\right), \frac{72}{Re_b} \right), \frac{8}{3}\frac{E_O}{E_O + 4} \right) \right) \quad (20)$$

- Lift force for gas-liquid (Gao et al., 2018)

$$\overrightarrow{F_{Lg}} = -\overrightarrow{F_{Ll}} = -C_L \alpha_g \rho_l \left(\vec{u}_l - \vec{u}_g \right) \times \left(\nabla \times \vec{u}_l \right) \quad (21)$$

- Virtual mass force for gas-liquid (Antal, Lahey, & Flaherty, 1991)

$$\overrightarrow{F_{VMg}} = -\overrightarrow{F_{VMl}}$$
$$= -C_{VM}\alpha_g \rho_l \left(\frac{d_l \vec{u}_l}{dt} - \frac{d_g \vec{u}_g}{dt} \right) \times \left(\nabla \times \vec{u}_l \right)$$
$$(22)$$

$$\frac{d_l \vec{u}_l}{dt} = \frac{\partial_l \vec{u}_l}{\partial t} + \vec{u}_l \cdot \nabla \vec{u}_l \qquad (23)$$

$$\frac{d_g \vec{u}_g}{dt} = \frac{\partial_g \vec{u}_g}{\partial t} + \vec{u}_g \cdot \nabla \vec{u}_g \qquad (24)$$

- Lubrication wall force for gas-liquid (Antal et al., 1991):

$$\overrightarrow{F_{Wg}} = -\overrightarrow{F_{Wl}}$$
$$= -\frac{\alpha_g \rho_l \left| \vec{u}_g - \vec{u}_l \right|^2}{d_b} \max \left[0, \left(C_1 + C_2\frac{d_b}{y} \right) \right] \vec{n}_r$$
$$(25)$$

- Turbulent dispersion force for gas-liquid (Behzadi, Issa, & Rusche, 2004):

$$\overrightarrow{F_{Tg}} = -\overrightarrow{F_{Tl}} = -\rho_l \overrightarrow{u_g'} \overrightarrow{u_l'} \nabla \alpha_g = C_T \alpha_l k_k \nabla \alpha_g \quad (26)$$

- Drag model for liquid-solid (Gao et al., 2018):

$$\overrightarrow{F_{Ds}} = -\overrightarrow{F_{Dl}} = \frac{3}{4}\alpha_s \alpha_l \frac{\rho_l}{d_s} C_D \left| \vec{u}_s - \vec{u}_l \right| \left(\vec{u}_s - \vec{u}_l \right) \quad (27)$$

$$C_D = \frac{24}{Re_s}\left(1 + 0.15Re_s^{0.687}\right) \qquad (28)$$

In these equations, d_s and d_g correspond to the diameters of the microalgae and the bubble, respectively.

5.2 Models applied in photobioreactors

Applying the equations and using mathematical models, it is possible to understand the behavior of different types of photobioreactors, whether open or closed, as summarized in Table 3.

As can be seen in Table 3, the model is $k - \epsilon$ is used in the most usual reactors (flat plane, bubble column, cylinder reactor, raceway). Even the $k - \epsilon$ models have been modified to consider bubble-induced turbulence (BIT) at higher gas flow rates (Li, Zhao, Cheng, Du, & Liu, 2013). However, for rotary flow reactors such as in a Torus or Taylor-Couette reactor, it is advisable to use the turbulence models of $k - \omega$ (Gao et al., 2017a; Gao et al., 2017b; Pottier et al., 2005; Pruvost et al., 2008).

6 Effect of shear stress on the growth of microalgae

Another factor that is related to the hydrodynamics of photobioreactors is the effect that shear stress has on the growth of microalgae.

The shear stress caused by the mixing and/or pumping system of nutrients in the culture system exceeding a certain tolerance threshold, which depends on each microalgae, can cause severe damage to the microalgae cell affecting the growth process and consequently affecting the production of high-value compounds (Merchuk, 1991; Wang & Lan, 2018).

This shear stress is a function of the shear rate and fluid viscosity, as shown in the following equation:

$$\tau = \dot{\gamma} \mu \qquad (29)$$

where τ is the shear stress and $\dot{\gamma}$ is the shear rate. Also, the term μ corresponds to the apparent viscosity.

This shear stress depends on many factors, either of the microalgae or environmental factors. Among the most important environmental factors is the viscosity of the system.

As mentioned in other publications, microalgae cultures depend on their concentration to determine their hydrodynamic behavior (Wang & Lan, 2018). Low-density culture systems have a Newtonian behavior, while high-density culture systems have a non-Newtonian behavior (Bernaerts et al., 2017).

This change in viscosity in the culture system directly impacts the mixing of the system, since the viscosity of the system increases, the nonidealities within the photobioreactor increase. Within the nonidealities, there are dead zones; these zones are volumes of zero movements of the fluid in question; this can cause the decay of the microalgae and consequently the death of these due to the lack of lighting to perform photosynthesis.

Additional to the dead zones are the microeddies, these eddies cause an increase in shear stress causing damage to the cells of the microalgae if this eddy is smaller than the diameter of the microalgae (Sobczuk, Camacho, Grima, & Chisti, 2006).

Another external factor that affects the shear stress is the temperature since there are microalgae that, due to changes in the temperatures, adapt to be able to survive in a said environment allowing it, in addition, to be able to increase its resistance capacity toward the shear stress (Mitsuhashi, Hosaka, Tomonaga, Muramatsu, & Tanishita, 1995).

With respect to the mathematical formulation that relates the growth of microalgae with shear stress, the classical formulation assumes that the effect of shear stress on the growth of microalgae is negligible when this parameter is below a certain critical value and that above this value the effect begins to manifest itself exponentially incrementally, as can be seen in the following equation (Wu & Merchuk, 2002, 2004):

$$\frac{dX}{dt} = \mu X - \mu_d X^2 - Me\,X \qquad (30a)$$

$$Me = \begin{cases} \overline{Me}\,e^{k_m(\tau - \tau_c)} & (\tau > \tau_c) \\ \overline{Me} & (\tau \leq \tau_c) \end{cases} \qquad (30b)$$

TABLE 3   Different models applied to solve the Navier-Stokes equations.

Type of system	Type of PBR	Phase model	Turbulence model	CFD code	Focus of the study	References
Open system	Raceway	L-G	$k-\omega$	FLUENT	Flow patterns of different configurations of a raceway	Xu, Li, and Waller (2014)
	Raceway	L-G-S	N.A.	OpenFoam	Vertical mixing, cell trajectory tracking	Prussi et al. (2014)
	Raceway	L-S	$k-\epsilon$	CFX	Improved flow characteristic using flow modifiers including baffles and deflector	Zhang, Dechatiwongse, and Hellgardt (2015)
	Raceway	L	$k-\epsilon$, LES	CFX	Effects of different types of paddle wheels	Zeng et al. (2016)
Closed system	Flat Plate	L-G	$k-\epsilon$	COMSOL	Improve the mixing system to allow high-density cultures	Hinterholz et al. (2019)
	Flat Plate	L-S	$k-\epsilon$	FLUENT	Study on the destabilization mixing in the flat plate PBR	Su, Kang, Shi, Cong, and Cai (2010)
	Flat Plate	L-G	$k-\epsilon$	CFX	Structural optimization and cultivation performance	Wang, Tao, and Mao (2014)
	Draft-tube airlift	L-G	$k-\epsilon$	FLUENT	Optimize the inner structure	Xu, Liu, Wang, and Liu (2012)
	Airlift	L-G-S	$k-\epsilon$ with BIT	CFX	Multiphase CFD model development	Luo and Al-Dahhan (2012)
	Bubble column	L-G	$k-\epsilon$	FLUENT	Effect of bubbling on the flow pattern	Akhtar, Pareek, and Tadé (2007)
	Bubble column	L-G	$k-\epsilon$	FLUENT	Study the effect of microalgae growth on fluid dynamics equations and irradiance transfer	Nauha and Alopaeus (2013)
	Torus	L-S	$k-\omega$	FLUENT	Track the trajectory of a cell influenced by the presence of light	Pruvost, Cornet, and Legrand (2008)
	Torus	L	$k-\omega$	FLUENT	Flow pattern inside a Torus	Pruvost, Pottier, and Legrand (2006)
	Taylor–Couette	L-G-S	$k-\omega$	FLUENT	Cell trajectory tracking, the effect of mixing on biomass production	Gao et al. (2017a)
	Twisted Tubular	L	$k-\epsilon$	COMSOL	Study the trajectory inside the twisted tubular photobioreactor to minimize energy consumption and maximize biomass production	Gómez-Pérez, Espinosa Oviedo, Montenegro Ruiz, and van Boxtel (2017)

where Me corresponds to maintenance term. $\overline{Me}$ is the maintenance term without the effect of shear stress. k_m corresponds to the extinction coefficient for shear stress, and τ_c is critical shear stress.

In this equation, it is necessary to cut the stress calculated by assuming that the energy inside the system can be defined by the isothermal expansion of the gas and that energy is transferred through the gas-liquid interface. This assumption facilitates us to determine the overall term of shear stress using the following formula (Merchuk & Ben-Zvi(Yona), 1992):

$$\gamma' = \left(\frac{p_1 J_G \ln\left(\frac{p_1}{p_2}\right)}{a L_R^2 \kappa} \right)^{1/n} \quad (31)$$

where p_1 and p_2 are the pressures at the top and bottom of the system to be modeled. J_G is the superficial gas velocity, and L_R is the length of the reactor. The term a is the interfacial area between the liquid and the gas. κ is the fluid consistency index. Using this formulation, it is possible to obtain the global term shear stress ($\dot{\gamma}$) by applying the following equation:

$$\dot{\gamma} = \gamma' \left(\frac{V_r \phi}{Q_G} \right) \quad (32)$$

where γ' is the global shear stress per unit of time and V_r is the culture volume. Q_G is the gas flow rate, and the term ϕ is the gas holdup (Wu & Merchuk, 2002).

With respect to the term a, it is possible to calculate it using the formulation presented by Deckwer (Wu & Merchuk, 2002, 2004):

$$a = 4.65 \times 10^{-2} \left(\frac{J_G}{\kappa} \right)^{0.51} \quad (33)$$

Finally, shear stress can be calculated using Eq. (29). With the equations presented above, it is possible to clearly appreciate the effect of shear stress on the growth of microalgae. This effect should ideally be minimized to the maximum to avoid cell damage and/or avoid slowing down the culture time.

Due to the above, it is that one of the challenges in this area of work is to be able to develop models in a nonstationary state to determine the variation in shear stress in the system and to deliver an indicator to the culture system operator to increase or decrease system agitation. Additionally, it should be borne in mind that this term always contrasts with the purpose of the mixing system that tries to reduce the times in which the cells are not exposed to the lighting system, so it becomes interesting to be able to determine to what extent it compensates to maintain the system mixing over the possible cell damage caused to the cell due to shear stress.

7 Gas exchange and temperature effect

Finally, inside a photobioreactor, there is an exchange between ambient gases and gases generated by microalgae (CO_2, N_2, O_2, etc.). This gas exchange allows the cell to feed and grow, preventing death due to the lack of nutrients such as carbon and/or nitrogen and photooxidation due to excess oxygen in the culture medium (Subramanian, Dineshkumar, & Sen, 2016).

To reduce cell death inside a photobioreactor, it is important to have a good mixing system that allows the rapid escape of O_2 from the system and the adequate injection of CO_2 into the culture medium (Eriksen, 2008; Ugwu, Aoyagi, & Uchiyama, 2008). Therefore, fluid dynamics models become important to understand and implement adequate mixing systems that allow this gas exchange as well as preventing the death of microalgae due to hydrodynamic stress caused by the mixing system (Wang & Lan, 2018).

In addition to the photobioreactor mixing system, gas exchange depends on the temperature of the system. This temperature can be rising and/or falling due to the biological processes of the microalgae and the corresponding reactor agitation system. Due to these variations, a heat exchange system is coupled to the photobioreactors to maintain the stable temperature of the system. This temperature control is not only due to the fact that microalgae develops better at a certain temperature of the culture medium but also because the gases and organic compounds present in a said medium are affected in their transfer from one phase to another due to the temperature.

The temperature of the system impacts different abiotic and biotic factors, for example, in the solubility. Solubility is an important factor in the growth of microalgae because it allows nitrogen and phosphorus compounds to be incorporated into the culture medium and gases such as CO_2 or O_2 are transferred to the culture medium for later release or incorporation (Scheufele et al., 2019).

With respect to solubility, the simplest way to model the relationship between the gaseous compound dissolved in a liquid is Henry's law (Scheufele et al., 2019), as shown in the following equation:

$$H_i = \frac{c_i}{P_i} \qquad (34)$$

where H_i corresponds to Henry's constant of compound i. c_i is the concentration of the species i in the aqueous phase, and P_i is the partial pressure of the species i in the gas under equilibrium conditions.

However, because Henry's constant depends on the temperature, it is possible to use Van't Hoff's law to determine the value of that constant, using the following equation (Scheufele et al., 2019):

$$CO_{2(g)} \overset{HCO2}{\longleftrightarrow} {}_{H_{CO_2}} CO_{2(aq)}$$

$$\frac{d \ln H_i}{d(1/T)} = -\frac{\Delta H_{Sol}}{R} \qquad (35)$$

where T is the system temperature, R is the universal gas constant, and H_{Sol} is the enthalpy of dissociation.

If we assume that ΔH_{Sol} does not vary with temperature (in a limited range of working temperature), it is possible to obtain the final equation that governs Henry's constant with respect to temperature:

$$H_i(T) = H^0 \exp\left[-\frac{\Delta H_{Sol}}{R}\left(\frac{1}{T} - \frac{1}{T_0}\right)\right] \qquad (36)$$

Many of Henry's constants under standard conditions (20–25°C and 1 atm) for different compounds dissolved in pure water and seawater are available in Sander (2015).

This change in Henry's constant of a compound with respect to temperature can be seen in a change in solubility of the worked substance. Also, this change in solubility directly impacts in the mass transfer of gases of interest to microalgae, such as CO_2 and O_2, since the general equation that models the variation of concentration of a compound G in the culture medium (Marsullo et al., 2015) is:

$$\frac{dG}{d} = F_G + \mu X Y_{AM} - k_{M,L}\alpha(G - G^*) \qquad (37)$$

where F_G is the flux term of compound G introduced or expelled from the culture system; this term is positive if the compound is introduced into the system and negative if it is expelled from it. Y_{AM} corresponds to the mass of compound G, either generated or consumed by the microalgae per unit mass of microalgae produced. The last term of the equation corresponds to the equilibrium that exists between the two phases present in the culture medium (liquid and gas), where $k_{M,L}\alpha$ corresponds to the mass transfer term given for the element in question. Finally, G^* is the saturation concentration of the compound in the culture medium at a given temperature.

In this equation, it can appreciate the effect that temperature has on the transfer of gases in a bioculture, since if the temperature changes, so does Henry's constant and therefore the global coefficient of matter transfer. This overall mass transfer coefficient is related to Henry's constant as follows (Liu et al., 2015):

$$\frac{1}{K_{M,L}} = \frac{1}{k_{M,L}\alpha} = \frac{1}{k_L} + \frac{1}{k_G H} \qquad (38)$$

where k_L is the transfer coefficient from the liquid phase to the gas phase. k_G is the transfer coefficient from the gas phase to the liquid phase of the compound in question. Also, the term H corresponds to Henry's constant that depends on the temperature.

In addition to the effect of mass transfer, temperature also affects the growth rate of the microalgae. As mentioned above, temperature is a factor that can limit the growth of microalgae (Chia et al., 2018) to such an extent that it can cause cell death if the temperature is below the minimum temperature at which the microalgae can survive or if the temperature is above the maximum temperature allowed by the microorganism.

In practical terms, we have seen that the maximum growth rate of the microalgae studied depends directly proportional to the temperature of the system, as can be seen in the following equation:

$$\mu_{max}(T) = \mu_{max}\Theta(T) \qquad (39)$$

where the term μ_{max} corresponds to the maximum growth rate of the microalgae at a reference temperature. Also, the term $\Theta(T)$ directly depends on the minimum temperature (T_{min}) at which the microalgae can grow naturally, the optimal temperature (T_{opt}), at which the microalgae grow in a greater proportion considering the constant environmental conditions and the maximum temperature (T_{max}) at which the microalgae can survive (Pires, Alvim-Ferraz, & Martins, 2017).

With these parameters, it is possible to construct the following mathematical relationship to determine the variation of the parameter $\Theta(T)$:

This equation in graphic terms can be represented in the following schematic curve (Fig. 4), where the critical points in microalgae growth are possible:

In general, in microalgae, it has an optimal and maximum temperature very close to each other, so the variation curve of $\Theta(T)$; in general, it has a long tail on the left side of the optimum temperature and a very drastic drop in growth from the right side of the optimum temperature.

In contrast to the approach presented above, there is another mathematical way of relating temperature to microalgae growth, which is the one presented by Zhang, Dechatiwongse, and Hellgardt (2015). This approach consists of using the Arrhenius equation to determine this multiplier at the microalgae growth rate:

$$\mu_{max}(T) = \mu_{max}\phi(T) \qquad (41)$$

This approach is possible to realize it has already worked well to estimate the effect of temperature on bacteria (Alagappan & Cowan, 2004; Vargas, Mariano, Corrêa, & Ordonez, 2014) and microalgae (Zhang et al., 2016), as in *Haematococcus pluvialis*. This new term $\phi(T)$ is formulated as follows:

$$\phi(T) = A e^{-\frac{E_a}{RT}} - B e^{-\frac{E_b}{RT}} \qquad (42)$$

where A and B correspond to pre-exponential coefficients, R corresponds to the universal gas constant, T is the system temperature, and E_a and E_b correspond to the activation and deactivation energy, respectively.

Finally, due to the aforementioned, it is that the photobioreactors are incorporated with a temperature control system, usually a steam or hot water jacket that allows the system to be maintained at a stable and oscillating

$$\Theta(T) = \begin{cases} \dfrac{(T - T_{max})(T - T_{min})^2}{(T_{opt} - T_{min})\left[(T_{opt} - T_{min})(T - T_{opt}) - (T_{opt} - T_{max})(T_{opt} + T_{min} - 2T)\right]} & \text{if } T \in [T_{min}, T_{max}] \\ 0 & \text{otherwise} \end{cases}$$

$$(40)$$

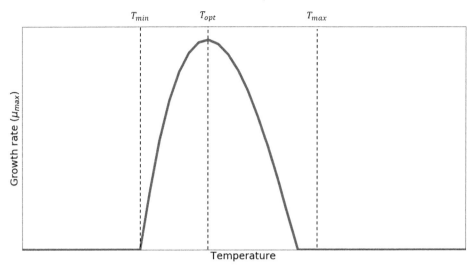

FIG. 4   Representation of the variation in the growth rate of the microalgae with respect to the system temperature.

temperature at the optimum temperature of the system. However, due to the incorporation of this heat exchange system, the energy costs of photobioreactors increase due to the expense required to keep the temperature regulated.

Therefore, one of the main challenges for the future is to be able to incorporate it into the transfer models that affect the growth of microalgae, a section on the energy expenditure that exists to keep the system running optimally, and how that expense is changing depends on the dimensions of the reactor.

Additionally, it is always important to remember that, when incorporating a heat exchange system, it is important to keep in mind the maintenance costs and know the ease to clean the system of each of the temperature regulation equipment or devices to avoid future inconveniences at the time of keeping the system operating.

8 Energy consumption of a cultivation system

Another important aspect when modeling photobioreactors and the impact of microalgae growth is to consider the energy consumption of the process of bubbling, mixing, harvesting, and filling the culture system.

To maintain the culture systems in a steady state of growth, the culture system must be kept in constant pumping of CO_2 and/or other nutrients necessary for the microalgae to grow and produce the compounds of interest. To determine the energy expenditure of the bubbling system, the classical equation of electrical consumption of a pump must be used:

$$E_{bubbling} = P_{pump, bubbling} \ t_{bubbling} \qquad (43)$$

where $t_{bubbling}$ corresponds to the bubbling or feeding time of the culture system. $P_{pump, bubbling}$ corresponds to the power of the pump used to inject the substance of interest into the culture system. This pump energy can be calculated as follows:

$$P_{pump, bubbling} = \frac{\rho g Q \Delta h}{\eta_{pump}} \qquad (44)$$

where ρ corresponds to the density of the fluid displaced by the pump, either an aqueous mixture or a gas. g corresponds to the acceleration of gravity. Q is the volumetric flow rate of the displaced fluid. Δh corresponds to the height difference between the point before the pump and the exit point of the pump. η_{pump} is the efficiency of the pump used to displace the fluid.

Similarly, in the case of the energy used to mix, we have that the equation to determine the electrical consumption is like that presented for the bubbling system, but the power term is calculated as follows (Marsullo et al., 2015):

$$P_{mix} = \frac{\rho g Q \Delta h}{\eta_{mixing\ equipment}} \tag{45}$$

where $\eta_{mixing\ equipment}$ is the efficiency of the equipment used to mix the culture either a hydraulic paddle or an air pumping system.

Finally, in the case of the energy of the harvesting and filling system, the energy consumption equation is (Marsullo et al., 2015):

$$E_{harv/refil} = P_{pump}\ t_{harv/refil} \tag{46}$$

where $t_{harv/refil}$ corresponds to the time used to harvest and cool the culture medium. However, we can see that the power of the pump can vary over time as the microalgae grow since the density of the system varies as the microalgae grow and release compounds to the culture medium.

Due to the aforementioned, one of the important challenges in this area is to determine the optimum harvest and fill point of the system. For this, an interesting perspective is to be able to combine an energy consumption model with a microalgae growth model since in this way it is possible to appreciate if this factor has a limitation when scaling the cultivation system in the future because if the consumption energy increases excessively as the system grows, the process of scaling said microalgae is useless due to the economic unfeasibility of maintaining the system.

9 Conclusion

By reviewing the wide range of existing models to describe the behavior of microalgae, it can be determined that as our understanding of various mathematical models and the computational capacity of computers has improved; all these models are possible.

Likewise, it is clearly observed that the greatest challenges in future research correspond to multiphysical modeling, that is, to incorporate the phenomena of irradiance, growth, and movement and, on the other hand, are to create support tools for the scaling process of these reactors. For example, a tool that allows determining the effect of increasing the speed of the circulating gas inside the system or that allows predicting in a simple way the effect of increasing the amount of light radiated to the system was developed.

Additionally, one of the most interesting challenges is to be able to use the equations of mass, irradiance, and momentum transfer to be able to determine dimensionless numbers to consider when scaling up culture systems.

The latter has become critical in recent years because it does not matter how sophisticated models become to understand the behavior of microalgae if it is not possible to find a relationship between these equations and their respective large-scale systems will never be. It is not possible to obtain an economically viable microalgae culture system without connecting the results obtained in the laboratory with those obtained in large industrial reactors.

Among the most interesting dimensionless parameters to determine is the relationship between the amount of irradiance entered into the system and the growth of the microalgae. This relationship would be critical at the time of scaling up a culture system because, through a simple relationship, it would be possible to determine whether the scaling up of the proposed system is viable or that the situation is irrational.

Finally, despite the wide variety of equations that exist to model the behavior of microalgae, there is still no clarity in which situations certain equations are better than others due to either the behavior of the microalgae or the behavior of the photobioreactor. Therefore, one of the important improvements in the area of modeling of microalgae is to develop an adequate working methodology to determine the use of one model over the other.

References

Adeniyi, O. M., Azimov, U., & Burluka, A. (2018). Algae biofuel: Current status and future applications. *Renewable and Sustainable Energy Reviews, 90*(August 2017), 316–335. https://doi.org/10.1016/j.rser.2018.03.067.

Akhtar, A., Pareek, V., & Tadé, M. (2007). CFD simulations for continuous flow of bubbles through gas-liquid columns: Application of VOF method. *Chemical Product and Process Modeling. 2*(1) https://doi.org/10.2202/1934-2659.1011.

Alagappan, G., & Cowan, R. M. (2004). Effect of temperature and dissolved oxygen on the growth kinetics of *Pseudomonas putida* F1 growing on benzene and toluene. *Chemosphere, 54*(8), 1255–1265. https://doi.org/10.1016/j.chemosphere.2003.09.013.

Andrews, J. F. (1968). A mathematical model for the continuous culture of microorganisms utilizing inhibitory substrates. *Biotechnology and Bioengineering, 10*(6), 707–723. https://doi.org/10.1002/bit.260100602.

Antal, S. P., Lahey, R. T., & Flaherty, J. E. (1991). Analysis of phase distribution in fully developed laminar bubbly two-phase flow. *International Journal of Multiphase Flow, 17*(5), 635–652. https://doi.org/10.1016/0301-9322(91)90029-3.

Anwar, M., Lou, S., Chen, L., Li, H., & Hu, Z. (2019). Recent advancement and strategy on bio-hydrogen production from photosynthetic microalgae. *Bioresource Technology, 292*(August), 121972. https://doi.org/10.1016/j.biortech.2019.121972.

Bartley, M. L., Boeing, W. J., Dungan, B. N., Holguin, F. O., & Schaub, T. (2014). pH effects on growth and lipid accumulation of the biofuel microalgae *Nannochloropsis salina* and invading organisms. *Journal of Applied Phycology, 26*(3), 1431–1437. https://doi.org/10.1007/s10811-013-0177-2.

Bartosh, Y., & Banks, C. J. (2007). Algal growth response and survival in a range of light and temperature conditions: Implications for non-steady-state conditions in waste stabilisation ponds. *Water Science and Technology, 55*(11), 211–218. https://doi.org/10.2166/wst.2007.365.

Behzadi, A., Issa, R. I., & Rusche, H. (2004). Modelling of dispersed bubble and droplet flow at high phase fractions. *Chemical Engineering Science, 59*(4), 759–770. https://doi.org/10.1016/j.ces.2003.11.018.

Belkin, S., & Boussiba, S. (1991). Resistance of spirulina platensis to ammonia at high pH values. *Plant and Cell Physiology, 32*(7), 953–958. https://doi.org/10.1093/oxfordjournals.pcp.a078182.

Bello, M., Ranganathan, P., & Brennan, F. (2017). Dynamic modelling of microalgae cultivation process in high rate algal wastewater pond. *Algal Research, 24*, 457–466. https://doi.org/10.1016/j.algal.2016.10.016.

Ben Salah, M., Askri, F., Slimi, K., & Ben Nasrallah, S. (2004). Numerical resolution of the radiative transfer equation in a cylindrical enclosure with the finite-volume method. *International Journal of Heat and Mass Transfer, 47* (10 − 11), 2501–2509. https://doi.org/10.1016/j.ijheatmasstransfer.2003.11.023.

Bernaerts, T. M. M., Panozzo, A., Doumen, V., Foubert, I., Gheysen, L., Goiris, K., et al. (2017). Microalgal biomass as a (multi)functional ingredient in food products: Rheological properties of microalgal suspensions as affected by mechanical and thermal processing. *Algal Research, 25*(December 2016), 452–463. https://doi.org/10.1016/j.algal.2017.05.014.

Beyerinck, M. W. (1890). Culturversuche mit Zoochlorellen, Lichenengonidien und anderen niederen Algen. *Botanische Zeitung, 47*, 725–785. https://doi.org/10.1017/CBO9781107415324.004.

Bitog, J. P., Lee, I. B., Lee, C. G., Kim, K. S., Hwang, H. S., Hong, S. W., et al. (2011). Application of computational fluid dynamics for modeling and designing photobioreactors for microalgae production: A review. *Computers and Electronics in Agriculture, 76*(2), 131–147. https://doi.org/10.1016/j.compag.2011.01.015.

Borowitzka, M. A. (1999). Commercial production of microalgae: Ponds, tanks, tubes and fermenters. *Journal of Biotechnology, 70*(1–3), 313–321. https://doi.org/10.1016/S0168-1656(99)00083-8.

Borowitzka, M. A. (2013). High-value products from microalgae-their development and commercialisation. *Journal of Applied Phycology, 25*(3), 743–756. https://doi.org/10.1007/s10811-013-9983-9.

Brindley, C., Jiménez-Ruíz, N., Acién, F. G., & Fernández-Sevilla, J. M. (2016). Light regime optimization in photobioreactors using a dynamic photosynthesis model. *Algal Research, 16*, 399–408. https://doi.org/10.1016/j.algal.2016.03.033.

Burlew, J. S. (1953). In J. S. Burlew (Ed.), *Algae culture from laboratory to pilot plant.* Carnegie Institution of Washington.

Carvalho, A. P., & Meireles, L. A. (2006). Microalgae reactors: A review of enclosed systems and performances. *Biotechnology Progress, 3*(1), 1490–1506.

Çelekli, A., & Dönmez, G. (2006). Effect of pH, light intensity, salt and nitrogen concentrations on growth and β-carotene accumulation by a new isolate of Dunaliella sp. *World Journal of Microbiology and Biotechnology, 22*(2), 183–189. https://doi.org/10.1007/s11274-005-9017-0.

Chia, S. R., Ong, H. C., Chew, K. W., Show, P. L., Phang, S. M., Ling, T. C., et al. (2018). Sustainable approaches for algae utilisation in bioenergy production. *Renewable Energy*, *129*, 838–852. https://doi.org/10.1016/j.renene.2017.04.001.

Christenson, L., & Sims, R. (2011). Production and harvesting of microalgae for wastewater treatment, biofuels, and bioproducts. *Biotechnology Advances*, *29*(6), 686–702. https://doi.org/10.1016/j.biotechadv.2011.05.015.

Converti, A., Casazza, A. A., Ortiz, E. Y., Perego, P., & Del Borghi, M. (2009). Effect of temperature and nitrogen concentration on the growth and lipid content of *Nannochloropsis oculata* and *Chlorella vulgaris* for biodiesel production. *Chemical Engineering and Processing: Process Intensification*, *48*(6), 1146–1151. https://doi.org/10.1016/j.cep.2009.03.006.

Cornet, J. F., Dussap, C. G., & Dubertret, G. (1992). A structured model for simulation of cultures of the cyanobacterium *Spirulina platensis* in photobioreactors: I. Coupling between light transfer and growth kinetics. *Biotechnology and Bioengineering*, *40*(7), 817–825. https://doi.org/10.1002/bit.260400709.

Cornet, J. F., Dussap, C. G., & Gros, J. B. (1994). Conversion of radiant light energy in photobioreactors. *AICHE Journal*, *40*(6), 1055–1066. https://doi.org/10.1002/aic.690400616.

Cornet, J. F., Dussap, C. G., Gros, J. B., Binois, C., & Lasseur, C. (1995). A simplified monodimensional approach for modeling coupling between radiant light transfer and growth kinetics in photobioreactors. *Chemical Engineering Science*, *50*(9), 1489–1500. https://doi.org/10.1016/0009-2509(95)00022-W.

Darvehei, P., Bahri, P. A., & Moheimani, N. R. (2018). Model development for the growth of microalgae: A review. *Renewable and Sustainable Energy Reviews*, *97*(September), 233–258. https://doi.org/10.1016/j.rser.2018.08.027.

Dasgupta, C. N., Jose Gilbert, J., Lindblad, P., Heidorn, T., Borgvang, S. A., Skjanes, K., et al. (2010). Recent trends on the development of photobiological processes and photobioreactors for the improvement of hydrogen production. *International Journal of Hydrogen Energy*, *35*(19), 10218–10238. https://doi.org/10.1016/j.ijhydene.2010.06.029.

Doucha, J., & Lívanský, K. (2006). Productivity, CO2/O2 exchange and hydraulics in outdoor open high density microalgal (Chlorella sp.) photobioreactors operated in a Middle and Southern European climate. *Journal of Applied Phycology*, *18*(6), 811–826. https://doi.org/10.1007/s10811-006-9100-4.

Doucha, J., & Lívanský, K. (2009). Outdoor open thin-layer microalgal photobioreactor: Potential productivity. *Journal of Applied Phycology*, *21*(1), 111–117. https://doi.org/10.1007/s10811-008-9336-2.

Eriksen, N. T. (2008). The technology of microalgal culturing. *Biotechnology Letters*, *30*(9), 1525–1536. https://doi.org/10.1007/s10529-008-9740-3.

Fernández, F. G. A., Camacho, F. G., Pérez, J. A. S., Sevilla, J. M. F., & Grima, E. M. (1998). Modeling of biomass productivity in tubular photobioreactors for microalgal cultures: Effects of dilution rate, tube diameter, and solar irradiance. *Biotechnology and Bioengineering*, *58*(6), 605–616. https://doi.org/10.1002/(SICI)1097-0290(19980620)58:6<605::AID-BIT6>3.0.CO;2-M.

Fiveland, W. A. (1984). Discrete-ordinates solutions of the radiative transport equation for rectangular enclosures. *ASME Journal of Heat Transfer*, *106*(4), 699–706. https://doi.org/10.1115/1.3246741.

Fiveland, W. A. (1988). Three-dimensional radiative heat-transfer solutions by the discrete-ordinates method. *Journal of Thermophysics and Heat Transfer*, *2*(4), 309–316. https://doi.org/10.2514/3.105.

Franz, A., Lehr, F., Posten, C., & Schaub, G. (2012). Modeling microalgae cultivation productivities in different geographic locations—Estimation method for idealized photobioreactors. *Biotechnology Journal*, *7*(4), 546–557. https://doi.org/10.1002/biot.201000379.

Gao, X., Kong, B., & Dennis Vigil, R. (2017a). Comprehensive computational model for combining fluid hydrodynamics, light transport and biomass growth in a Taylor vortex algal photobioreactor: Eulerian approach. *Algal Research*, *24*, 1–8. https://doi.org/10.1016/j.algal.2017.03.009.

Gao, X., Kong, B., & Dennis Vigil, R. (2017b). Comprehensive computational model for combining fluid hydrodynamics, light transport and biomass growth in a Taylor vortex algal photobioreactor: Eulerian approach. *Bioresource Technology*, *224*, 523–530. https://doi.org/10.1016/j.algal.2017.03.009.

Gao, X., Kong, B., & Vigil, R. D. (2018). Multiphysics simulation of algal growth in an airlift photobioreactor: Effects of fluid mixing and shear stress. *Bioresource Technology*, *251*, 75–83. https://doi.org/10.1016/j.biortech.2017.12.014.

Göksan, T., Zekeriyaoğlu, A., & Ak, I. (2007). The growth of Spirulina platensis in different culture systems under greenhouse condition. *Turkish Journal of Biology*, *31*(1), 47–52.

Goldman, J. C., Riley, C. B., & Dennett, M. R. (1982). The effect of ph in intensive microalgal cultures. ii. Species competition. *Journal of Experimental Marine Biology and Ecology*, *5*(4799), 15–24.

Gómez-Pérez, C. A., Espinosa Oviedo, J. J., Montenegro Ruiz, L. C., & van Boxtel, A. J. B. (2017). Twisted tubular photobioreactor fluid dynamics evaluation for energy consumption minimization. *Algal Research*, *27*(February), 65–72. https://doi.org/10.1016/j.algal.2017.08.019.

Gong, Q., Feng, Y., Kang, L., Luo, M., & Yang, J. (2014). Effects of light and pH on cell density of *Chlorella vulgaris*. *Energy Procedia, 61*, 2012–2015. https://doi.org/10.1016/j.egypro.2014.12.064.

Han, W., Long, F., Cong, W., Intes, X., & Wang, G. (2017). Radiative transfer with delta-Eddington-type phase functions. *Applied Mathematics and Computation, 300*, 70–78. https://doi.org/10.1016/j.amc.2016.12.001.

He, L., Subramanian, V. R., & Tang, Y. J. (2012). Experimental analysis and model-based optimization of microalgae growth in photo-bioreactors using flue gas. *Biomass and Bioenergy, 41*, 131–138. https://doi.org/10.1016/j.biombioe.2012.02.025.

Henyey, L. G., & Greenstein, J. L. (1941). Diffuse radiation in the galaxy. *The Astrophysical Journal, 93*, 70–83.

Hinterholz, C. L., Trigueros, D. E. G., Módenes, A. N., Borba, C. E., Scheufele, F. B., Schuelter, A. R., et al. (2019). Computational fluid dynamics applied for the improvement of a flat-plate photobioreactor towards high-density microalgae cultures. *Biochemical Engineering Journal, 151* (June), 107257. https://doi.org/10.1016/j.bej.2019.107257.

Howell, J. R., Siegel, R., & Mengüç, M. P. (2010). Radiative energy loss and gain along a line-of-sight. In R. H. Bedford (Ed.), *Thermal radiation heat transfer* (pp. 31–36). CRC Press.

Ismaiel, M. M. S., El-Ayouty, Y. M., & Piercey-Normore, M. (2016). Role of pH on antioxidants production by Spirulina (Arthrospira) platensis. *Brazilian Journal of Microbiology, 47*(2), 298–304. https://doi.org/10.1016/j.bjm.2016.01.003.

Juneja, A., Ceballos, R. M., & Murthy, G. S. (2013). Effects of environmental factors and nutrient availability on the biochemical composition of algae for biofuels production: A review. *Energies, 6*(9), 4607–4638. https://doi.org/10.3390/en6094607.

Kandilian, R., Jesus, B., Legrand, J., Pilon, L., & Pruvost, J. (2017). Light transfer in agar immobilized microalgae cell cultures. *Journal of Quantitative Spectroscopy and Radiative Transfer, 198*, 81–92. https://doi.org/10.1016/j.jqsrt.2017.04.027.

Kandilian, R., Lee, E., & Pilon, L. (2013). Radiation and optical properties of *Nannochloropsis oculata* grown under different irradiances and spectra. *Bioresource Technology, 137*, 63–73. https://doi.org/10.1016/j.biortech.2013.03.058.

Kessler, E. (1985). Upper limits of temperature for growth in Chlorella (Chlorophyceae). *Plant Systematics and Evolution, 151*(1–2), 67–71. https://doi.org/10.1007/BF02418020.

Kumar, S. D., Santhanam, P., Park, M. S., & Kim, M. K. (2016). Development and application of a novel immobilized marine microalgae biofilter system for the treatment of shrimp culture effluent. *Journal of Water Process Engineering, 13*, 137–142. https://doi.org/10.1016/j.jwpe.2016.08.014.

Kunikane, S., & Kaneko, M. (1984). Growth and nutrient uptake of green alga, *Scenedesmus dimorphus*, under a wide range of nitrogen/phosphorus ratio—II. Kinetic model. *Water Research, 18*(10), 1313–1326. https://doi.org/10.1016/0043-1354(84)90037-X.

Kurano, N., & Miyachi, S. (2005). Selection of microalgal growth model for describing specific growth rate-light response using extended information criterion. *Journal of Bioscience and Bioengineering, 100*(4), 403–408. https://doi.org/10.1263/jbb.100.403.

Lee, Y. K. (1997). Commercial production of microalgae in the Asia-Pacific rim. *Journal of Applied Phycology, 9*(5), 403–411. https://doi.org/10.1023/A:1007900423275.

Lee, H. Y., Erickson, L. E., & Yang, S. S. (1987). Kinetics and bioenergetics of light-limited photoautotrophic growth of *Spirulina platensis*. *Biotechnology and Bioengineering, 29*(7), 832–843. https://doi.org/10.1002/bit.260290705.

Lee, E., Jalalizadeh, M., & Zhang, Q. (2015). Growth kinetic models for microalgae cultivation: A review. *Algal Research, 12*, 497–512. https://doi.org/10.1016/j.algal.2015.10.004.

Leong, W. H., Lim, J. W., Lam, M. K., Uemura, Y., & Ho, Y. C. (2018). Third generation biofuels: A nutritional perspective in enhancing microbial lipid production. *Renewable and Sustainable Energy Reviews, 91*(July 2017), 950–961. https://doi.org/10.1016/j.rser.2018.04.066.

Li, D., Xiong, K., Li, W., Yang, Z., Liu, C., Feng, X., et al. (2010). Comparative study in liquid-phase heterogeneous photocatalysis: Model for photoreactor scale-up. *Industrial and Engineering Chemistry Research, 49*(18), 8397–8405. https://doi.org/10.1021/ie100277g.

Li, Q., Zhao, X., Cheng, K., Du, W., & Liu, D. (2013). Simulation and experimentation on the gas holdup characteristics of a novel oscillating airlift loop reactor. *Journal of Chemical Technology and Biotechnology, 88*(4), 704–710. https://doi.org/10.1002/jctb.3888.

Liu, X., Guo, Z., Roache, N. F., Mocka, C. A., Allen, M. R., & Mason, M. A. (2015). Henry's law constant and overall mass transfer coefficient for formaldehyde emission from small water pools under simulated indoor environmental conditions. *Environmental Science and Technology, 49*(3), 1603–1610. https://doi.org/10.1021/es504540c.

Lock, J. A., & Gouesbet, G. (2009). Generalized Lorenz-Mie theory and applications. *Journal of Quantitative Spectroscopy and Radiative Transfer, 110*(11), 800–807. https://doi.org/10.1016/j.jqsrt.2008.11.013.

Luo, H. P., & Al-Dahhan, M. H. (2012). Airlift column photobioreactors for Porphyridium sp. culturing: Part II. Verification of dynamic growth rate model for reactor performance evaluation. *Biotechnology and Bioengineering, 109*(4), 942–949. https://doi.org/10.1002/bit.24362.

Markou, G., Wang, L., Ye, J., & Unc, A. (2018). Using agroindustrial wastes for the cultivation of microalgae and duckweeds: Contamination risks and biomass safety concerns. *Biotechnology Advances*, *36*(4), 1238–1254. https://doi.org/10.1016/j.biotechadv.2018.04.003.

Marsullo, M., Mian, A., Ensinas, A. V., Manente, G., Lazzaretto, A., & Marechal, F. (2015). Dynamic modeling of the microalgae cultivation phase for energy production in open raceway ponds and flat panel photobioreactors. *Frontiers in Energy Research*, *3*(September), 1–18. https://doi.org/10.3389/fenrg.2015.00041.

Martínez, M. E., Jiménez, J. M., & Yousfi, F. E. (1999). Influence of phosphorus concentration and temperature on growth and phosphorus uptake by the microalga *Scenedesmus obliquus*. *Bioresource Technology*, *67*(3), 233–240. https://doi.org/10.1016/S0960-8524(98)00120-5.

Mata, T. M., Martins, A. A., & Caetano, N. S. (2010). Microalgae for biodiesel production and other applications: A review. *Renewable and Sustainable Energy Reviews*, *14* (1), 217–232. https://doi.org/10.1016/j.rser.2009.07.020.

Menguc, M. P., & Viskanta, R. (1985). Radiative transfer in three-dimensional rectangular enclosures containing inhomogeneous, anisotropically scattering media. *Journal of Quantitative Spectroscopy and Radiative Transfer*, *33*(6), 533–549.

Menguc, M. P., & Viskanta, R. (1986). Radiative transfer in axisymmetric, finite cylindrical enclosures. *ASME Journal of Heat Transfer*, *108*(May 1986), 271–277. https://doi.org/10.1115/1.3246915.

Merchuk, J. C. (1991). Shear effects on suspended cells. *Advances in Biochemical Engineering/Biotechnology*, *44*(July 1988), 65–95. https://doi.org/10.1007/bfb0000748.

Merchuk, J. C., & Ben-Zvi(Yona), S. (1992). A novel approach to the correlation of mass transfer rates in bubble columns with non-Newtonian liquids. *Chemical Engineering Science*, *47*(13), 3517–3523. https://doi.org/10.1016/0009-2509(92)85065-J.

Mishchenko, M. I. (2009). Gustav Mie and the fundamental concept of electromagnetic scattering by particles: A perspective. *Journal of Quantitative Spectroscopy and Radiative Transfer*, *110*(14–16), 1210–1222. https://doi.org/10.1016/j.jqsrt.2009.02.002.

Mitsuhashi, S., Hosaka, K., Tomonaga, E., Muramatsu, H., & Tanishita, K. (1995). Effects of shear flow on photosynthesis in a dilute suspension of microalgae. *Applied Microbiology and Biotechnology*, *42*(5), 744–749. https://doi.org/10.1007/BF00171956.

Molina Grima, E., Acién Fernández, F. G., García Camacho, F., & Chisti, Y. (1999). Photobioreactors: Light regime, mass transfer, and scaleup. *Journal of Biotechnology*, *70* (1–3), 231–247. https://doi.org/10.1016/S0168-1656(99)00078-4.

Molina Grima, E., Fernández Sevilla, J. M., Sánchez Pérez, J. A., & García Camacho, F. (1996). A study on simultaneous photolimitation and photoinhibition in dense microalgal cultures taking into account incident and averaged irradiances. *Journal of Biotechnology*, *45*(1), 59–69. https://doi.org/10.1016/0168-1656(95)00144-1.

Molina Grima, E., Garcia Carnacho, F., Sanchez Perez, J. A., Fernandez Sevilla, J. M., Acien Fernandez, F. G., & Contreras Gomez, A. (1994). A mathematical model of microalgal growth in light-limited chemostat culture. *Journal of Chemical Technology & Biotechnology*, *61*, 167–173. https://doi.org/10.1002/jctb.280610212.

Molina Grima, E., Sánchez Pérez, J. A., García Sánchez, J. L., García Camacho, F., & López Alonso, D. (1992). EPA from Isochrysis galbana. Growth conditions and productivity. *Process Biochemistry*, *27*(5), 299–305. https://doi.org/10.1016/0032-9592(92)85015-T.

Mosorov, V. (2017). The Lambert-Beer law in time domain form and its application. *Applied Radiation and Isotopes*, *128*, 1–5. https://doi.org/10.1016/j.apradiso.2017.06.039.

Moya, M. J., Sánchez-Guardamino, M. L., Vilavella, A., & Barberà, E. (1997). Growth of Haematococcus lacustris: A contribution to kinetic modelling. *Journal of Chemical Technology and Biotechnology*, *68*(3), 303–309. https://doi.org/10.1002/(SICI)1097-4660(199703)68:3<303::AID-JCTB639>3.0.CO;2-1.

Naderi, G., Tade, M. O., & Znad, H. (2015). Modified photobioreactor for biofixation of carbon dioxide by *Chlorella vulgaris* at different light intensities. *Chemical Engineering and Technology*, *38*(8), 1371–1379. https://doi.org/10.1002/ceat.201400790.

Naderi, G., Znad, H., & Tade, M. O. (2017). Investigating and modelling of light intensity distribution inside algal photobioreactor. *Chemical Engineering and Processing: Process Intensification*, *122*(December 2016), 530–537. https://doi.org/10.1016/j.cep.2017.04.014.

Nauha, E. K., & Alopaeus, V. (2013). Modeling method for combining fluid dynamics and algal growth in a bubble column photobioreactor. *Chemical Engineering Journal*, *229*, 559–568. https://doi.org/10.1016/j.cej.2013.06.065.

Özgür, E., Mars, A. E., Peksel, B., Louwerse, A., Yücel, M., Gündüz, U., et al. (2010). Biohydrogen production from beet molasses by sequential dark and photofermentation. *International Journal of Hydrogen Energy*, *35*(2), 511–517. https://doi.org/10.1016/j.ijhydene.2009.10.094.

Pareek, V., Chong, S., Tadé, M., & Adesina, A. A. (2008). Light intensity distribution in heterogenous photocatalytic reactors. *Asia-Pacific Journal of Chemical Engineering*, *3*, 171–201. https://doi.org/10.1002/apj.

Perner-Nochta, I., & Posten, C. (2007). Simulations of light intensity variation in photobioreactors. *Journal of Biotechnology*, *131*(3), 276–285. https://doi.org/10.1016/j.jbiotec.2007.05.024.

Pilon, L., Berberoğlu, H., & Kandilian, R. (2011). Radiation transfer in photobiological carbon dioxide fixation and fuel production by microalgae. *Journal of Quantitative*

Spectroscopy and Radiative Transfer, *112*(17), 2639–2660. https://doi.org/10.1016/j.jqsrt.2011.07.004.

Pilon, L., & Kandilian, R. (2016). Interaction between light and photosynthetic microorganisms. (chapter 2) C. E. Legrand (Ed.), *Photobioreaction engineering* (pp. 107–149). Vol. 48(pp. 107–149). Academic Press. https://doi.org/10.1016/bs.ache.2015.12.002.

Pires, J. C. M., Alvim-Ferraz, M. C. M., & Martins, F. G. (2017). Photobioreactor design for microalgae production through computational fluid dynamics: A review. *Renewable and Sustainable Energy Reviews*, *79*(May), 248–254. Elsevier Ltd. (2017). https://doi.org/10.1016/j.rser.2017.05.064

Pottier, L., Pruvost, J., Deremetz, J., Cornet, J. F., Legrand, J., & Dussap, C. G. (2005). A fully predictive model for one-dimensional light attenuation by *Chlamydomonas reinhardtii* in a torus photobioreactor. *Biotechnology and Bioengineering*, *91*(5), 569–582. https://doi.org/10.1002/bit.20475.

Prussi, M., Buffi, M., Casini, D., Chiaramonti, D., Martelli, F., Carnevale, M., et al. (2014). Experimental and numerical investigations of mixing in raceway ponds for algae cultivation. *Biomass and Bioenergy*, *67*, 390–400. https://doi.org/10.1016/j.biombioe.2014.05.024.

Pruvost, J., Cornet, J.-F., & Legrand, J. (2008). Hydrodynamics influence on light conversion in photobioreactors: An energetically consistent analysis. *Chemical Engineering Science*, *63*(14), 3679–3694. https://doi.org/10.1016/j.ces.2008.04.026.

Pruvost, J., Legrand, J., Legentilhomme, P., & Muller-Feuga, A. (2002). Simulation of microalgae growth in limiting light conditions: Flow effect. *AICHE Journal*, *48*(5), 1109–1120. https://doi.org/10.1002/aic.690480520.

Pruvost, J., Pottier, L., & Legrand, J. (2006). Numerical investigation of hydrodynamic and mixing conditions in a torus photobioreactor. *Chemical Engineering Science*, *61*(14), 4476–4489. https://doi.org/10.1016/j.ces.2006.02.027.

Pulz, O. (2001). Photobioreactors: Production systems for phototrophic microorganisms. *Applied Microbiology and Biotechnology*, *57*(3), 287–293. https://doi.org/10.1007/s002530100702.

Rachlin, J. W., & Grosso, A. (1991). The effects of pH on the growth of *Chlorella vulgaris* and its interactions with cadmium toxicity. *Archives of Environmental Contamination and Toxicology*, *20*, 505–508. https://doi.org/10.1007/BF01065839.

Ranganathan, P., Amal, J. C., Savithri, S., & Haridas, A. (2017). Experimental and modelling of *Arthrospira platensis* cultivation in open raceway ponds. *Bioresource Technology*, *242*, 197–205. https://doi.org/10.1016/j.biortech.2017.03.150.

Rebolledo-Oyarce, J., Mejía-López, J., García, G., Rodríguez-Córdova, L., & Sáez-Navarrete, C. (2019). Novel photobioreactor design for the culture of *Dunaliella tertiolecta*—Impact of color in the growth of microalgae. *Bioresource Technology*, *289*(June), 121645. https://doi.org/10.1016/j.biortech.2019.121645.

Rosello Sastre, R., Csögör, Z., Perner-Nochta, I., Fleck-Schneider, P., & Posten, C. (2007). Scale-down of microalgae cultivations in tubular photo-bioreactors—A conceptual approach. *Journal of Biotechnology*, *132*(2), 127–133. https://doi.org/10.1016/j.jbiotec.2007.04.022.

Rukminasari, N. (2013). Effect of temperature and nutrient limitation on the growth and lipid content of three selected microalgae (*Dunaliella tertiolecta*, *Nannochloropsis* sp. and *Scenedesmus* sp.) for biodiesel production. *International Journal of Marine Science*, *3*(17), 135–144. https://doi.org/10.5376/ijms.2013.03.0017.

Sancho, M. E. M., Castillo, J. M. J., & Yousfi, F. E. (1997). Influence of phosphorus concentration on the growth kinetics and stoichiometry of the microalga Scenedesmus obliquus. *Process Biochemistry*, *32*(8), 657–664. https://doi.org/10.1016/S0032-9592(97)00017-4.

Sander, R. (2015). Compilation of Henry's law constants (version 4.0) for water as solvent. *Atmospheric Chemistry and Physics*, *15*(8), 4399–4981. https://doi.org/10.5194/acp-15-4399-2015.

Scheufele, F. B., Hinterholz, C. L., Zaharieva, M. M., Najdenski, H. M., Módenes, A. N., Trigueros, D. E. G., et al. (2019). Complex mathematical analysis of photobioreactor system. *Engineering in Life Sciences*, *19*(12), 844–859. https://doi.org/10.1002/elsc.201800044.

Selçuk, N., & Ayrancı, I. (2003). The method of lines solution of the discrete ordinates method for radiative heat transfer in enclosures containing scattering media. *Numerical Heat Transfer, Part B: Fundamentals*, *43*(2), 179–201. https://doi.org/10.1080/713836169.

Serra-Maia, R., Bernard, O., Gonçalves, A., Bensalem, S., & Lopes, F. (2016). Influence of temperature on *Chlorella vulgaris* growth and mortality rates in a photobioreactor. *Algal Research*, *18*, 352–359. https://doi.org/10.1016/j.algal.2016.06.016.

Sharma, R., Singh, G. P., & Sharma, V. K. (2012). Effects of culture conditions on growth and biochemical profile of *Chlorella vulgaris*. *Journal of Plant Pathology & Microbiology*, *3*(5), 100131. https://doi.org/10.4172/2157-7471.1000131.

Sobczuk, T. M., Camacho, F. G., Grima, E. M., & Chisti, Y. (2006). Effects of agitation on the microalgae *Phaeodactylum tricornutum* and *Porphyridium cruentum*. *Bioprocess and Biosystems Engineering*, *28*(4), 243–250. https://doi.org/10.1007/s00449-005-0030-3.

Sosik, H. M., & Mitchell, B. G. (1994). Effects of temperature on growth, light absorption, and quantum yield in *Dunaliella tertiolecta* (Chlorophyceae). *Journal of Phycology*, *30*(5), 833–840. https://doi.org/10.1111/j.0022-3646.1994.00833.x.

Spolaore, P., Joannis-Cassan, C., Duran, E., & Isambert, A. (2006). Commercial applications of microalgae. *Journal of Bioscience and Bioengineering, 101*(2), 87–96. https://doi.org/10.1263/jbb.101.87.

Steele, J. H. (1962). Environmental control of photosynthesis in the sea. *Limnology and Oceanography, 7*(2), 137–150. https://doi.org/10.1016/B978-075064637-6/50022-8.

Su, Z., Kang, R., Shi, S., Cong, W., & Cai, Z. (2010). Study on the destabilization mixing in the flat plate photobioreactor by means of CFD. *Biomass and Bioenergy, 34*(12), 1879–1884. https://doi.org/10.1016/j.biombioe.2010.07.025.

Subramanian, G., Dineshkumar, R., & Sen, R. (2016). Modelling of oxygen-evolving-complex ionization dynamics for energy-efficient production of microalgal biomass, pigment and lipid with carbon capture: An engineering vision for a biorefinery. *RSC Advances, 6*(57), 51941–51956. https://doi.org/10.1039/c6ra08900c.

Tomiyama, A., Kataoka, I., Zun, I., & Sakaguchi, T. (1998). Drag coefficients of single bubbles under normal and micro gravity conditions. *JSME International Journal Series B, 41*(2), 472–479. https://doi.org/10.1299/jsmeb.41.472.

Ugwu, C. U., Aoyagi, H., & Uchiyama, H. (2008). Photobioreactors for mass cultivation of algae. *Bioresource Technology, 99*(10), 4021–4028. https://doi.org/10.1016/j.biortech.2007.01.046.

Uyar, B., Schumacher, M., Gebicki, J., & Modigell, M. (2009). Photoproduction of hydrogen by rhodobacter capsulatus from thermophilic fermentation effluent. *Bioprocess and Biosystems Engineering, 32*(5), 603–606. https://doi.org/10.1007/s00449-008-0282-9.

Van Wagenen, J., Miller, T. W., Hobbs, S., Hook, P., Crowe, B., & Huesemann, M. (2012). Effects of light and temperature on fatty acid production in *Nannochloropsis salina*. *Energies, 5*(3), 731–740. https://doi.org/10.3390/en5030731.

Vargas, J. V. C., Mariano, A. B., Corrêa, D. O., & Ordonez, J. C. (2014). The microalgae derived hydrogen process in compact photobioreactors. *International Journal of Hydrogen Energy, 39*(18), 9588–9598. https://doi.org/10.1016/j.ijhydene.2014.04.093.

Vo, H. N. P., Ngo, H. H., Guo, W., Nguyen, T. M. H., Liu, Y., Liu, Y., et al. (2019). A critical review on designs and applications of microalgae-based photobioreactors for pollutants treatment. *Science of the Total Environment, 651*, 1549–1568. https://doi.org/10.1016/j.scitotenv.2018.09.282.

Wágner, D. S., Valverde-Pérez, B., & Plósz, B. G. (2018). Light attenuation in photobioreactors and algal pigmentation under different growth conditions—Model identification and complexity assessment. *Algal Research, 35*(July), 488–499. https://doi.org/10.1016/j.algal.2018.08.019.

Wang, C., & Lan, C. Q. (2018). Effects of shear stress on microalgae—A review. *Biotechnology Advances, 36*(4), 986–1002. https://doi.org/10.1016/j.biotechadv.2018.03.001.

Wang, B., Lan, C. Q., & Horsman, M. (2012). Closed photobioreactors for production of microalgal biomasses. *Biotechnology Advances, 30*(4), 904–912. https://doi.org/10.1016/j.biotechadv.2012.01.019.

Wang, L., Tao, Y., & Mao, X. (2014). A novel flat plate algal bioreactor with horizontal baffles: Structural optimization and cultivation performance. *Bioresource Technology, 164*, 20–27. https://doi.org/10.1016/j.biortech.2014.04.100.

Wu, Z., Duangmanee, P., Zhao, P., & Ma, C. (2016). The effects of light, temperature, and nutrition on growth and pigment accumulation of three *Dunaliella salina* strains isolated from saline soil. *Jundishapur Journal of Microbiology, 9*(1), 1–9. https://doi.org/10.5812/jjm.26732.

Wu, X., & Merchuk, J. C. (2002). Simulation of algae growth in a bench-scale bubble column reactor. *Biotechnology and Bioengineering, 80*(2), 156–168. https://doi.org/10.1002/bit.10350.

Wu, X., & Merchuk, J. C. (2004). Simulation of algae growth in a bench scale internal loop airlift reactor. *Chemical Engineering Science, 59*(14), 2899–2912. https://doi.org/10.1016/j.ces.2004.02.019.

Xu, B., Li, P., & Waller, P. (2014). Study of the flow mixing in a novel ARID raceway for algae production. *Renewable Energy, 62*, 249–257. https://doi.org/10.1016/j.renene.2013.06.049.

Xu, L., Liu, R., Wang, F., & Liu, C.-Z. (2012). Development of a draft-tube airlift bioreactor for Botryococcus braunii with an optimized inner structure using computational fluid dynamics. *Bioresource Technology, 119*, 300–305. https://doi.org/10.1016/j.biortech.2012.05.123.

Xu, K., Lv, B., Huo, Y. X., & Li, C. (2018). Toward the lowest energy consumption and emission in biofuel production: Combination of ideal reactors and robust hosts. *Current Opinion in Biotechnology, 50*, 19–24. https://doi.org/10.1016/j.copbio.2017.08.011.

Ying, K., Gilmour, D. J., & Zimmerman, W. (2014). Effects of CO2 and pH on growth of the microalga *Dunaliella salina*. *Journal of Microbial & Biochemical Technology, 6*(3), 167–173. https://doi.org/10.4172/1948-5948.1000138.

Zeng, F., Huang, J., Meng, C., Zhu, F., Chen, J., & Li, Y. (2016). Investigation on novel raceway pond with inclined paddle wheels through simulation and microalgae culture experiments. *Bioprocess and Biosystems Engineering, 39*(1), 169–180. https://doi.org/10.1007/s00449-015-1501-9.

Zhang, D., Dechatiwongse, P., del Rio-Chanona, E. A., Maitland, G. C., Hellgardt, K., & Vassiliadis, V. S. (2015a). Modelling of light and temperature influences on cyanobacterial growth and biohydrogen production. *Algal Research, 9*, 263–274. https://doi.org/10.1016/j.algal.2015.03.015.

Zhang, D., Dechatiwongse, P., del Rio-Chanona, E. A., Maitland, G. C., Hellgardt, K., & Vassiliadis, V. S. (2015b). Dynamic modelling of high biomass density cultivation and biohydrogen production in different scales of flat plate photobioreactors. *Biotechnology and Bioengineering*, *112*(12), 2429–2438. https://doi.org/10.1002/bit.25661.

Zhang, D., Dechatiwongse, P., & Hellgardt, K. (2015). Modelling light transmission, cyanobacterial growth kinetics and fluid dynamics in a laboratory scale multiphase photo-bioreactor for biological hydrogen production.

Algal Research, *8*, 99–107. https://doi.org/10.1016/j.algal.2015.01.006.

Zhang, D., Wan, M., del Rio-Chanona, E. A., Huang, J., Wang, W., Li, Y., et al. (2016). Dynamic modelling of Haematococcus pluvialis photoinduction for astaxanthin production in both attached and suspended photobioreactors. *Algal Research*, *13*, 69–78. https://doi.org/10.1016/j.algal.2015.11.019.

Zienkiewicz, O. C., & Taylor, R. L. (2000). *The finite element method* (5th ed.). (Vols. 1–3). Butterworth Heinemann.

Edible bio-oil production from microalgae and application of nano-technology

Nazia Hossain[a], Tamal Chowdhury[b], Hemal Chowdhury[c], Ashfaq Ahmed[d,g], Sabzoi Nizamuddin[e], Gregory Griffin[a], Teuku Meurah Indra Mahlia[f], and Young-Kwon Park[g]

[a]School of Engineering, RMIT University, Melbourne, VIC, Australia,[b]Department of Electrical and Electronic Engineering, Chittagong University of Engineering and Technology, Chattogram, Bangladesh,[c]Department of Mechanical Engineering, Chittagong University of Engineering and Technology, Chattogram, Bangladesh,[d]Department of Chemical Engineering, COMSATS University Islamabad Lahore Campus, Lahore, Pakistan,[e]Civil and Infrastructure Engineering, School of Engineering, RMIT University, Melbourne, VIC, Australia,[f]School of Information, Systems and Modelling, Faculty of Engineering and Information Technology, University of Technology Sydney, Sydney, NSW, Australia,[g]School of Environmental Engineering, University of Seoul, Seoul, Republic of Korea

1 Introduction

Increasing food demand all over the world based on agricultural products has become a global concern and edible oil is one of these high demanding food substances. Edible oil is used to process most of the fried foods and snacks from small to large scale as well as increases the shelf life of processed foods. From household to food industries, fulfilling the extensive demand of edible oil from vegetable sources has turned a major challenge all over the world. Therefore, researchers are looking for third-generation oil-producing resources nowadays and microalgae are one of them (Xue et al., 2018). Microalgae are unicellular/multicellular microorganisms that can grow on fresh, saline, or wastewater and the biomass productivity is much higher than other terrestrial oleaginous crops. Microalgae contain very less water, land, and carbon footprint, simultaneously accumulate heavy metal wastewater, adsorb, and sequestrate CO_2 from the environment, grow at a much faster rate than other crops (Hossain, Zaini, & Indra Mahlia, 2019; Hossain, Zaini, Mahlia, & Azad, 2019).

Microalgae. https://doi.org/10.1016/B978-0-12-821218-9.00004-9

Microalgae have been preferred for edible bio-oil production due to the superiority of oil content over the terrestrial oleaginous crops (such as sunflower, olive, oil palm, corn, and others), enriched lipid composition, abundant omega-3 polyunsaturated fatty acid (PUFA), antioxidant components, therapeutic components (e.g., sterols, pigments), and vitamins (e.g., astaxanthin, a type of beta carotene) necessary for overall human health, especially for muscle recovery, cardiovascular, antiaging, and nutrition benefits. Due to the nutrient compounds, microalgae oil also has been applied to produce high-quality cosmetics and nutrition supplements (Huang, Zhang, Xue, Wang, & Cong, 2016; Xue et al., 2018).

A commercial microalgae-producing company in Brunei Darussalam, Mitsubishi Corporation Biotech Sdn Bhd. has been producing astaxanthin (beta carotene containing 550 times higher antioxidative than vitamin E) from red microalgae, *Haematococcus pluvialis* since 2016. Hence, microalgae oil has been considered as functional oil with great commercial potential (Kaneda, 2016). Besides, oil containment, microalgae are also rich in protein, carbohydrate, and other nutrients under specific conditions. Therefore, nonedible bio-oil, biofuels such as biodiesel, bioethanol, biobutanol, solid fuel (microalgae char), and feed for fuel cells can be produced simultaneously with integrated approach during bio-oil generation from microalgae (Hossain & Mahlia, 2019; Hossain, Mahlia, & Saidur, 2019; Hossain & Morni, 2019; Hossain, Zaini, & Mahlia, 2019; Hossain, Zaini, Mahlia, & Azad, 2019).

Microalgal bio-oil has been produced in a laboratory scale, yet not expanded through industrial scale due to the economic feasibility. Previous studies have presented that microalgal bio-oil contained higher market price compared to vegetable oils due to the significant investment for cultivation and harvesting in a photobioreactor, expensive extraction, and purification processes as well as lack of integrated applications. Since microalgae bio-oil production still is not flourished immensely, and research is ongoing;

industries have not come forward to establish commercial bio-oil (Huang et al., 2016; Klok, Lamers, Martens, Draaisma, & Wijffels, 2014). However, commercial bio-diesel from crude algal bio-oil and bioethanol from microalgae for fuel purposes has schemed for the commercial application so far (Hossain, Bhuiyan, Pramanik, Nizamuddin, & Griffin, 2020; Hossain, Mahlia, Zaini, & Saidur, 2019; Hossain, Zaini, & Indra Mahlia, 2019). If comprehensive techno-economic studies can be conducted on the most optimized conditions in the future, microalgal bio-oil would have a strong possibility to be implemented on a large scale.

The main objective of this chapter is to emphasize on the mechanism of microalgae biomass conversion to edible bio-oil and oil recovery from crude bio-oil. This study comprehensively elaborated various pretreatment methods prior to bio-oil extraction, the effect of different pretreatment methods on different microalgae species, the most prominent bio-oil extraction methods, bio-oil yield from different microalgae species via several extraction methods and techniques for bio-oil recovery and purification. Along with the detailed bio-oil production process, this chapter also presented an integrated approach of production of nonedible bio-oil, biodiesel, bioethanol (from alcohol recovery), and some other high value-added products simultaneously from this process. Moreover, the possible impacts on the economy, society, ecology, and environment have been delineated extensively in this study.

2 Suitable microalgae candidates for edible bio-oil and nanotechnology application for higher growth of microalgal species

Overuse and gradual consumption of edible oil in daily life may result in a quick depletion of oil. Since microalgae have a similar concentration of fatty acid like vegetable oils, it is the most encouraging alternative to vegetable oil. Microalgae possess a high amount of polyunsaturated fatty acids (PFA), the main source of edible bio-oil and PFA is beneficial to human health (Hossain, Mahlia, & Saidur, 2019; Nguyen, Moon, Bui, Oh, & Lee, 2019). Table 1 presents some suitable microalgae candidates for edible bio-oil production, water type for cultivation, the origin of species, and fatty acid contents.

However, acceptable candidates require to be single out to obtain edible oil. Several factors such as the strength of light, temperature, the composition of medium, and growth phase play important role in the selection of microalgae (Farooq, Lee, Huh, & Lee, 2016; Pádrová et al., 2014; Sarma et al., 2014). It is reported that during the normal growth phase, the medium that contains microalgae does not have enough nutrient. So, microalgae do not present significant growth (De Windt, Aelterman, & Verstraete, 2005;

Radzun et al., 2015). To solve this hurdle, researchers have applied metallic and hybrid nanomaterials in the medium (Kang et al., 2014). They observed that metallic nanomaterials act as hindrance for the growth of fungi and another microbial organism which are competitors of microalgae. Besides, hybrid nanomaterials tend to speed up the capture and absorption of CO_2 which results in higher supply of CO_2 in cultivation media. It is necessary to have an adequate supply of CO_2 in cultivation media since it helps in the photosynthesis process of microalgae (Choi, Kim, & Lee, 2012; Kang et al., 2014; Kim et al., 2016). Kadar et al. evaluated the aftermath of three kinds of manufactured nanoscale zerovalent iron (nZVI) nanoparticles (NP) on three marine microalgae (*Pavlova lutheri*, *Isochrysis galbana*, and *Tetraselmis suecica*). The expansion in the percentage of lipid in *T. suecica* and

TABLE 1 Suitable microalgae candidates for edible bio-oil production, growth condition, the origin of species, and fatty acid contents.

Microalgae	Growth condition (water type)	Origin of species	Fatty acid content	References
Bacillariophyceae sp.	Marine	Indonesia	C16:0 (38%–48%), C16:1 (0.5%–19%), C18:2 (1%–10%), C18:3 (0.02%–26%)	Widianingsih, Hartati, Endrawati, and Mamuaja (2013)
Cyanophyceae sp.	Marine	Indonesia	C16:0 (47%–66%), C18:0 (9%–14%), C16:1 (4%–12%), C18:1 (5%–25%)	Widianingsih et al. (2013)
Prasinophyceae sp.	Marine	Indonesia	C16:0 (48%–54%), C18:0 (9%–13%)	Widianingsih et al. (2013)
Nannochloropsis sp.	Marine	Indonesia	C16:0 (32%), C18:0 (14.4%), C16:1 (22.4%), C20:0 (13.4%)	Widianingsih et al. (2013)
Spirulina platensis	Marine	Indonesia	C16:0 (66.05%)	Widianingsih et al. (2013)
Phaeodactylum tricornutum	Marine	China	C16:0, C16:1, C20:5n3	Shen et al. (2016)
Nanochloropsis oceanica	Marine	China	C16:0, C20:5n3	Shen et al. (2016)
Chlorella pyrenoidosa	Marine	China	C18:1, C18:3	Shen et al. (2016)

Note: Palmitic acid-C16:0, Palmiotolic acid-C16:1, Linoleic acid-C18:2, Linolenic acid-C18:3, Stearic acid-C18:0, Eicosanoic acid-C20:0, Eicosapentaenoic acid-C20:5n3.

P. lutheri improved with uncoated nZVI and inorganically covered nZVI powder (Choi et al., 2012). Padrova et al. has also observed a similar outcome and added that the growth in microalgae was due to the iron supplement in the medium (Pádrová et al., 2014). Effect of using carbon nanotubes (CNTs), nano Fe_2O_3, and nano MgO in microalgae cultivation medium was observed by He et al. (2017). This research identified that the growth of microalgae was only enhanced by nano Fe_2O_3. $MgSO_4$ NPs were used in glycerol-based media to observe the development of *Chlorella vulgaris*. The utilization of $MgSO_4$ NPs resulted in the deduction of availability of light to microalgae and thus bringing about an increment in chlorophyll content (Sarma et al., 2014). Metzler et al. have also observed a similar phenomenon (Metzler, Erdem, Tseng, & Huang, 2012). The assessment of TiO_2 NPs on the development of microalgae was also evaluated (Aruoja, Dubourguier, Kasemets, & Kahru, 2009; Li et al., 2015; Tang, Li, Qiao, Wang, & Li, 2013). At optimal condition (concentration of TiO_2, 0.1 g/L and 2 days UV exposure), *C. vulgaris* (UTEX 265) experienced growth in lipid and fame content. A little consolidation of TiO_2, the maximum extreme development rate, and last cell thickness of microalgae were not influenced (Li et al., 2015). Jeon et al. added SiO_2 NPs in the cultivation media, bringing about a prompt 34% improvement in FAME efficacy (Jeon, Park, Ahn, & Kim, 2017). Deng et al. have presented that the development of the diatom was upgraded when treated by low CeO_2 NPs (≤ 5 mg/L). Additionally, the chlorophyll α content and dissolvable sugar substance of *P. tricornutum* were improved after treatment by nanoTiO_2 and nano-CeO_2 (Deng et al., 2017). CdSe NPs have been applied in two species of microalgae, for example, *P. cruentum* and *Spirulina platensis*. The first species showed enhanced biomass percentage while the other species did not present the same while CdSe NPs were used (Rudic et al., 2011). The use of NPs also brings about a drastic improvement in light-absorption productivity with phycoerythrin which delineates the invigorating effect of CdSe NPs (Rudic et al., 2012). 3-aminopropyl-functionalized magnesium phyllosilicate (aminoclays (ACs) which is a combination of an organic and inorganic compound have shown extraordinary interaction with molecules and cells (Datta, Achari, & Eswaramoorthy, 2013; Lee et al., 2013; Lee et al., 2013). ACs have rich-essential amine bunches that initiate adsorption onto cells of microalgae and by generating ROS create oxidative stress conditions. In addition, under high CO_2 fixations, the amine group of MgAC's can fasten CO_2 assimilation in water (Holmström, Patil, Butler, & Mann, 2007; Mann, 2009). Farooq eat al. have conducted research which states that MgACs have the capability to capture CO_2 that brings about rich HCO_3^- content in cultivated medium of microalgae (Farooq et al., 2016). Table 2 signifies the application of different NPs that have been employed in the metabolic accumulation and growth of microalgae.

3 Microalgae pretreatment

3.1 Cell disruption methods of microalgae

Thermochemical, biological and chemical methods are generally used to convert biomass to produce energy. Compared to nonconventional sources, raw biomass does not have high calorific value and high bulk density. Besides moisture found in raw biomass is quite high. These factors have a negative effect on production capacity and due to this, signification reduction in yield occurred (Tumuluru, 2016). These issues need to be solved. The concept of previous studies provided that biorefineries can operate at their maximum capacities and hence pretreatment is necessary.

Cellular pretreatment is an indispensable part that is mandatory to disintegrate the cell membranes of microalgae to ease the extraction process of lipid. Lipid found in microalgae cells is usually encircled by a monolayer of

TABLE 2 Effect of NP$_S$ on microalgae metabolic accumulation and growth.

Microalgae Species	Density (g/L)	Nanoparticles			Performances			
		Types of NPs	Dosage (mg/L)	Size (nm)	Efficiency (%)	Time (days)	Improved target	References
Desmodesmus subspicatus	1×10^5 cells/mL	nZVI	5.1	50	58.33	9	Lipid	Nguyen et al. (2019) and Pádrová et al. (2014)
Dunaliella salina	1×10^5 cells/mL	nZVI	5.1	50	33.33	9	Lipid	Nguyen et al. (2019) and Pádrová et al. (2014)
Scenedesmus obliquus	5×10^6 cells/mL	MgO	40	<50	18.5	6	Lipid	Nguyen et al. (2019) and He et al. (2017)
	1×10^7 cells/mL	CNTs	5	<2	8.9	8	Lipid	Nguyen et al. (2019) and He et al. (2017)
	1×10^7 cells/mL	Fe$_2$O$_3$	≤ 20	<30	10	7	Growth rate	Nguyen et al. (2019) and He et al. (2017)
	1×10^7 cells/mL	Fe$_2$O$_3$	5	<30	39.6	7	Lipid	Nguyen et al. (2019) and He et al. (2017)
Phaeodactylum tricornutum	2×10^6 cells/mL	CeO$_2$	≤ 5	10–30	~10	4	Growth rate	Nguyen et al. (2019) and Rudic et al. (2012)
Porphyridium cruentum	0.01	CdSe	Quantum dot form; 6 mg/L	3.5	47.5	10	Biomass	Nguyen et al. (2019), Rudic et al. (2012), and Guo, Yiyun, and Chen (2015)
Microcystis aeruginosa	1×10^6 cells/mL	TiO$_2$	0.8	–	~50	11	Growth rat	Guo et al. (2015) and Nguyen et al. (2019)
Chromochloris zofingiensis	0.2	CNTs	80	Single-wall	119	6	Biomass	Nguyen et al. (2019) and Wang and Yang (2013)

Note: Reactive oxygen species (ROS), polyunsaturated fatty acids (PFA), nanoscale zero-valent iron (nZVI), nanoparticles (NPs), carbon nanotubes (CNTs).

phospholipids which is covered by protein (Walther & Farese Jr., 2009). Different components such as cellulose and hemicellulose, mineral, and glycoprotein build up the robust structure of microalgae's wall (Scholz et al., 2014). And so, the solvent cannot penetrate the microalgae's thick membrane. To ease the lipid extraction, disruption of cell structure is necessary and hence pretreatment is required which can ensure the high yield within a shorter amount of time. So, suitable pretreatment method needs to be executed before the extraction process to lessen the solvent and energy consumption. The common methods that are used for pretreatment purposes are summarized below.

3.1.1 Bead beating

Bead beating (BB) is a process in which the membrane of microalgae is disrupted by the action of fast-moving spinning beads. This process usually takes several minutes and can be applicable to all species of microalgae without any arrangement (Chen, Wang, Qiu, & Ge, 2018). The two most accepted bed beating process are agitated beads and shaking vessels (Lee, Lewis, & Ashman, 2012). Shaking vessels usually contain some containers on a pulsating platform and disruption is generally done by quivering the vessels on the pulsating platform. Being employed in research facilities, this method is only applicable to numerous samples that require different treatment conditions. Similarly, in agitated bead type, a vessel is generally filled up by beads where agitator rotates, and that leads to better cell disruption efficacies. But during the time of disruption process, heat is generated by the agitator and so to protect the heat-sensitive biomolecules from heat, a cooling jacket must be provided (Kim et al., 2013). In a nutshell, this method has two advantages; the simple arrangement of the equipment and the quickness of the treatment process whereas the only disadvantage of the process is the requirement of additional equipment for cooling purposes (Kim et al., 2013).

3.1.2 High-pressure homogenization

High-pressure homogenization (HPH) (French process) uses a hydraulic shear force to disrupt the cell wall. This force is obtained when the biomass is held at high pressure and gushed through a small tube (Menegazzo & Fonseca, 2019). When compared with another process, this treatment does not require a huge amount of energy. Besides the cost associated with operation and risk involved in thermal degradation is quite low in this process (Lee, Yoo, Jun, Ahn, & Oh, 2010). However, the efficacy of HPH tends to vary according to species and the rigidity of the cell walls may cause efficiency to drop (Ursu et al., 2014). Several studies have applied this process with a view to treating microalgae. To extract intracellular ingredients of *Nannochloropsis* sp., HPH has been applied and it shows high efficiency although high consumption of energy is also reported (Halim, Harun, Danquah, & Webley, 2012). Another study notified that when HPH has been applied in *Scenedesmus acutus*, the cell was completely damaged, but retrieval of fatty acids was observed to be less than 80% (Dong, Knoshaug, Pienkos, & Laurens, 2016).

3.1.3 Pressing

To demolish the thick membrane of microalgae to release the oil content, the mechanical force can be used. The use of mechanical force to release the lipid content of microalgae is called Pressing (Mubarak, Shaija, & Suchithra, 2015). It has been applied extensively in the factory-based extraction process of seed oils such as soybean and sunflower. In this process, pollutants from external sources hardly have any effect on microalgae and preservice of the chemical purity of the substance is embraced (Halim, Danquah, & Webley, 2012). Screw press, extruder, and biomass spraying are the main parts of the mechanical pressing process (Rawat, Ranjith Kumar, Mutanda, & Bux, 2013). It is very difficult to achieve lipid from microalgae biomass if there is moisture in it and it is an absolute necessity to boost the efficiency of this technique since the efficacy of the technique is reported to be 75% (Mubarak et al., 2015). Pressing is generally applied both for large scale and small scale to extract lipids to produce biodiesel.

3.1.4 Microwave method

Being an electromagnetic wave, a microwave has the frequency range from 300 MHz to 300 GHz. Its frequency is generally lesser than the infrared ray and higher than the radio wave. This technology has been applied in the extraction of target particles in different fields, especially in microalgae (Balasubramanian, Allen, Kanitkar, & Boldor, 2011). First microalgae species are subjected to microwave radiation. When they are subjected to radiation, a fast oscillation is generated by cell molecules in the rotating electric field which results in friction in the molecules. Due to inter and intramolecular movements, a substantial amount of heat is generated and this vaporizes water, which significantly ruptures the cell membrane (Amarni & Kadi, 2010). This method has the advantage of less reaction time and low cost in the whole operating system. All species of microalgae can be efficiently extracted by this method. This method faced the disadvantage of employing an additional cooling system which is required to save the target compounds from heat (Ranjith Kumar, Hanumantha Rao, & Arumugam, 2015).

3.1.5 Chemical method

The cracking of the cell takes place when chemicals are exploited to raise the permeability of the cell to a specific level (Vaara, 1992). It was proclaimed that the degradation of chemicals linkage of microalgae cells occurred through acids, alkalis, and surfactants (Sathish & Sims, 2012). The main benefit of this treatment is that it does not consume much energy. However, large quantities of chemicals are required in this method which questions the financial sustainability of this treatment. Besides, chemicals can attack the microalgae by causing corrosion to

the reactor and thus ruining the whole process (Kim et al., 2013).

The chemical method is an promising approach to speed up the extraction of lipids from the cells of microalgae since it tears apart the bond between the biomass matrix and the lipids (Dong et al., 2016). The cell disruption was performed on a commixture of *Chlorella* sp. and *Scenedesmus* sp., with 1 M H_2SO_4, 1 M, and 5 M NaOH at 90°C and the duration of the experiment was 30 min. After dissolving chlorophyll, have 0.5 M H_2SO_4 1 M solution was used to convert the free fatty acids (Xue et al., 2018).

3.1.6 Enzymatic disruption

To degrade the cell walls of microalgae, enzymatic disruption is the most promising approach. A single or mixture of enzymes has been applied for the disruption of the cell membrane of microalgae (Suslick & Flannigan, 2008). To extract lipid from microalgae, Wu et al. have applied four enzymes named lysozyme, cellulase, protease, and pectinase and yielded lipid of 73.11% (Wu et al., 2017). This process is different from the chemical method because the condition required for this treatment course is mild. To enhance the performance of disruption course and for the better economical process, enzymatic disruption is combined with another process. The main obstacle that hinders the development of this treatment in large scale application is the long duration time and high cost associated with the treatment. So, more researches are needed to solve these issues.

3.1.7 Ultrasonication

To get higher productivity from disruption of cells in a short duration, ultrasonication has been applied widely (Ferreira, Dias, Silva, & Costa, 2016). In this method, sonic waves are applied to fluid cultures which consequently generate microbubbles. Owing to the generation of microbubbles, cavitation is induced and results in an acting force on the cell of microalgae and thus rupturing the cell (Kita et al., 2010). This process is more intense in lower frequency than higher frequency and several factors such as temperature, type of wall, viscosity affect this process. To stop the temperature rising owing to heat dissipation, cooling the medium is a prerequisite for the effectiveness of this matter and due to this low temperature is favorable. Since the cooling process and the high power of ultrasound is required, the consumption of energy is high in this process (Prabakaran & Ravindran, 2011).

Various studies reported the suitability of ultrasonication in the pretreatment process of microalgae. Three species such as *Chlorella* sp., *Nostoc* sp., and *Tolypothrix* sp. were subjected to diverse pretreatment courses in a study and this study suggested ultrasonication as a suitable pretreatment method for yielding of lipid [25]. Agitation and ultrasonication were both applied in pretreatment of *Chlorella pyrenoidosa* and no significant difference was observed in the extraction of lipid (D'oca et al., 2011). However, two studies experimented ultrasonication on *Botryococcus* sp., *Chlorella vulgaris*, and *Schizochytrium* sp. S34, and notified that this method is not very efficient in cracking the microalgae's wall (Petkov & Garcia, 2007).

3.2 Selection of cell disruption methods

The true objective of cellular treatment is to disintegrate the cell walls of microalgae to ease the extraction process of lipid. Different pretreatment methods are delineated in Table 3. High energy decay takes place due to the collaboration of several issues such as pressure and temperature involved in the extraction process. This issue is directly conjunct with the structure of the cell wall. So, during the selection of pretreatment method, specific requirements such as total process energy requirement, the financial and environmental impact of solvent, difficulties in scale-up operation must be considered (D'Alessandro & Antoniosi Filho, 2016; Halim, Harun, et al., 2012; Kim et al., 2013; Mubarak et al., 2015). Table 3 compares the requirements for different pretreatment methods.

Table 4 presents the yield percentage of lipids from various species of microalgae. These species had undergone a drying process after their cell disruption. After the drying process, they were subjected to solvent extraction. All species may need different treatment methods. Single treatment methods should not be considered good for all microalgae species excluding *Spirulina* sp. Nevertheless, it was foreseen that

TABLE 3   Comparison of different methods for biodiesel production (D'Alessandro & Antoniosi Filho, 2016; Halim, Harun, et al., 2012; Kim et al., 2013; Mubarak et al., 2015).

Method of cell disruption	Scale-up	Energy consumption	Operational cost	Increased capacity	Solvent utilization
HPH	High	Low-high	Medium-High	Low	Low
Bead beating	High	High	Low-medium	Low	Medium-high
Ultrasonication	Low-medium	High	Medium-high	Low	Medium-high
Microwave method	Medium-high	Medium-high	Medium-high	Low	Medium-high
Enzymatic disruption	High	Low-medium	High	Medium-high	Low
Chemical method	High	Low	Medium-high	Medium-high	Medium-high
Pressing	High	Low	Low-medium	Low	Low

TABLE 4 Effect of cell disruption process on the extraction of lipid (Lee et al., 2010; Pohndorf et al., 2016; Prabakaran & Ravindran, 2011; Zheng et al., 2011).

Microalgae species for lipid extraction	Cell disruption techniques	Drying process requirement	Lipid yield (%)
Chlorella vulgaris	Auto-flocculation	Yes (solar)	3
Chlorella vulgaris	Wet biomass pressing with silica powder	Yes (solar)	4.7
Chlorella vulgaris	Dry biomass pressing with silica powder	Yes (solar)	6
Chlorella vulgaris	Pressing with liquid nitrogen	Yes (solar)	29
Chlorella vulgaris	Ultrasound (600 W, 20 min)	Yes (solar)	14
Chlorella vulgaris	Ball mill (1500 rpm, 20 min)	Yes (solar)	9.6
Chlorella vulgaris	Snailase (37°C, 2 h)	–	6.8
Chlorella vulgaris	Lysozyme (55°C, 10 h)	–	24
Chlorella vulgaris	Cellulase (55°C, 10 h)	–	22
Chlorella vulgaris	Microwave (2450 MHz, 100°C, 5 min)	Yes (solar)	17
Spirulina sp.	Auto-flocculation	Yes	5.86
Spirulina sp.	Ball mill (600 rpm, 2 h)	Yes	5.82
Spirulina sp.	Microwave (2450 MHz, 2 min)	Yes	5.7
Spirulina sp.	Autoclaving (0.2 MPa, 30 min)	Yes	5.85
Chlorella sp.	Auto-flocculation	Yes	15
Chlorella sp.	Autoclaving (15 lbs, 121°C, 5 min)	Yes	24
Chlorella sp.	Ball mill (3500 rpm, 5 min)	Yes	30
Chlorella sp.	Microwave (2450 MHz, 100°C, 5 min)	Yes	36
Chlorella sp.	Ultrasound (50 kHz, 15 min)	Yes	38
Nostoc sp.	Auto-flocculation	Yes	14.8
Nostoc sp.	Autoclaving (15 lbs, 121°C, 5 min)	Yes	20
Nostoc sp.	Ball mill (3500 rpm, 5 min)	Yes	26
Nostoc sp.	Microwave (2450 MHz, 100°C, 5 min)	Yes	32
Nostoc sp.	Ultrasound (50 kHz, 15 min)	Yes	35
Tolypothrix sp.	Auto-flocculation	Yes	6
Tolypothrix sp.	Autoclaving (15 lbs, 121°C, 5 min)	Yes	18
Tolypothrix sp.	Ball mill (3500 rpm, 5 min)	Yes	28
Tolypothrix sp.	Microwave (2450 MHz, 100°C, 5 min)	Yes	32

Continued

TABLE 4   Effect of cell disruption process on the extraction of lipid (Lee et al., 2010; Pohndorf et al., 2016; Prabakaran & Ravindran, 2011; Zheng et al., 2011)—cont'd

Microalgae species for lipid extraction	Cell disruption techniques	Drying process requirement	Lipid yield (%)
Tolypothrix sp.	Ultrasound (50 kHz, 15 min)	Yes	28
Botryococcus sp.	Auto-flocculation	Yes	7
Botryococcus sp.	Autoclaving (1.5 MPa, 125°C, 5 min)	Yes	11
Botryococcus sp.	Ball mill (2800 rpm, 5 min)	Yes	28
Botryococcus sp.	Microwave (2450 MHz, 100°C, 5 min)	Yes	28.5
Botryococcus sp.	Ultrasound (10 kHz, 5 min)	Yes	8
Chlorella vulgaris	Auto-flocculation	Yes	5
Chlorella vulgaris	Autoclaving (1.5 MPa, 125°C, 5 min)	Yes	10
Chlorella vulgaris	Ball mill (2800 rpm, 5 min)	Yes	8
Chlorella vulgaris	Microwave (2450 MHz, 100°C, 5 min)	Yes	10
Chlorella vulgaris	Ultrasound (10 kHz, 5 min)	Yes	5.5
Scenedesmus sp.	Auto-flocculation	Yes	2
Scenedesmus sp.	Autoclaving (1.5 MPa, 125°C, 5 min)	Yes	5
Scenedesmus sp.	Ball mill (2800 rpm, 5 min)	Yes	9
Scenedesmus sp.	Microwave (2450 MHz, 100°C, 5 min)	Yes	10
Scenedesmus sp.	Ultrasound (10 kHz, 5 min)	Yes	7

microwave and ultrasonication delineated the highest efficacy of lipid yield (Macías-Sánchez et al., 2015).

4 Methods of lipid extraction for edible bio-oil production

A huge scale mechanical press has been globally utilized to extract oils from crops including soybeans, wheat, and peanuts. However, this method is not acceptable for microalgae due to the characteristic differences in morphology (Xue et al., 2018). Besides, the water present in the microalgae has a significant impact on oil production. Compared to this method, supercritical fluid extraction, organic solvent extraction, and solvent-free extraction provide a more satisfactory result in extracting oil from wet microalgae.

4.1 Supercritical fluid extraction

Being an ascending green innovative technology, supercritical fluid extraction (SFE) has drawn huge attention in the recent world for its high selectivity and the utilization of elements that have the characteristics of fluids and gases when subjected to high pressure and temperature (Griffiths, Van Hille, & Harrison, 2010; Zhang, Li, Zhang, & Tan, 2015). When these parameters of a fluid attained a cynical value, it went to the supercritical (SC) region and thus enabling the fluid to behave as solvent extraction. And because of that, when temperature and pressure come down to room temperature and baroscopic pressure, no debris is left behind. It has been depicted that SFE has the potentiality to replace the traditional approaches of lipid yielding by organic solvents (Mendes et al., 1995). An arrangement for squeezing and conveying the liquid CO_2 to the extraction vessel is incorporated in this process. A heating valve whose function is to depressurize the SC-CO_2 is also found in this process. As soon as the furnace is warmed up, the compressed CO_2 gets into the SC stage and yields lipid from microalgae. The real advantage of this process is that this process has low toxicity and large salvation power. Mass transfer occurs due to viscous property of fluid and this process produces lipids that are solvent-free (Mendes et al., 1995; Taher, Sulaiman, Al-Marzouqi, Haik, & Farid, 2014). The main obstacle to this process is the high cost related to operation and infrastructure (Mubarak et al., 2015).

SC-CO_2 has been applied for biofuel production from *Chlorococcum* sp. and a comparison has been made with Soxhlet and hexane extraction on their lipid yield (Halim, Danquah, & Webley, 2012). This study reported that SC-CO_2 has a higher lipid yield than both these extraction methods. Lipid has also been extracted from *Chlorella vulgaris* utilizing SC-CO_2 (Mendes et al., 1995). This research delineated that when the pressure increases, it has a positive effect on lipid yield. However, this method is not suitable for all microalgae species.

4.2 Solvent extraction method

Solvents are exploited in this course to extract oil from lipid. Diverse issues such as species of microalgae, cost of solvents, toxicity contribute to the selection of solvents (Rawat, Ranjith Kumar, Mutanda, & Bux, 2011). The two methods that are applied for oil extraction are Soxhlet extraction and Bligh and Dyer's method. The former process uses hexane as a solvent for

extracting purposes while the latter utilizes a mixture of methanol and chloroform as solvents (Pragya, Pandey, & Sahoo, 2013). Ether and benzene are also used as a solvent but hexane has gained popularity in the present world due to its cost effectiveness.

4.2.1 Soxhlet extraction

Soxhlet extraction (SE) utilizes either hexane or combination of hexane with oil expeller/ press method to extract oil from algae. First, the expeller extracts oils and after that cyclohexane is used to yield oil from the remaining pulp. Oil is dissolved into hexane and the pulp is filtered out. After that, distillation is used to isolate oil and solvent. If both hexane solvent and cold press are used, it can yield up to more than 95% oil from algae (Pragya et al., 2013). The main disadvantage of this process is the possible hazards involved in dealing with chemicals such as the use of Benzene can lead to an explosion (Pragya et al., 2013).

4.2.2 Bligh and Dyer's method

A mixture of methanol, chloroform, and water is utilized in Bligh and Dyer's technique as solvents and the proportion to their mix up is 2:1:1.8. First, the solvents and culture are blended to their given ratios and are assimilated to create a monophasic system. After that, they are re-assimilated with the addition of the same percentage of chloroform. Therefore, the proportion of methanol, chloroform, and water is 2:1:1.8 and that of solvent to tissue is [(3 + 1):1] [38]. For the wet route, the ratio should be considered is 2:1:1.8 and for the dry route, [(3 + 1):1] needs to be investigated since the content of water is less with respect to biomass. At last, fractional distillation is applied to isolate or free lipids from chloroform (Iverson, Lang, & Cooper, 2001). It has been reported that Bligh & Dyer's method exhibits high efficacy and the percentage of lipids can be extracted is above 95% (Lam & Lee, 2012; Young, Nippgen, Titterbrandt, & Cooney, 2010). So, for high lipid yield, this method is the most suitable method when compared with Soxhlet

TABLE 5 Performance of various solvent extraction methods.

Microalgae species	Solvent extraction method	Bio-oil recovery (%)	References
Nannochloropsis sp.	SC-CO$_2$	25	Andrich, Nesti, Venturi, Zinnai, and Fiorentini (2005)
Spirulina platensis	SC-CO$_2$	77.9	Andrich, Zinnai, Nesti, and Venturi (2006)
Chlorococcum sp.	SC-CO$_2$	81.7	Halim, Danquah, and Webley (2012)
Chlorella vulgaris	Bligh and Dyer's method	10.6	Kim et al. (2012)
Chlorococcum sp.	Soxhlet	45	Halim, Danquah, and Webley (2012)

extraction. Table 5 presents the performance of various solvent extraction methods from different microalgae species.

4.3 Solvent-free extraction

Some studies have considered not utilizing solvent to extract lipids from microalgae for improving the economic and environmental aspects of oil production (Xue et al., 2018). A research was conducted on *Aurantiochytrium* sp. reported that without using solvent, the yield percentage of lipid was 77.37% (Xue et al., 2018). Aqueous enzymatic extraction (AEE) is a proven technique that has been utilized strongly to obtain lipids from different oil seeds. This procedure is largely reliable on oilseed squashing, the insolubility of oil in water, and the thickness distinction between water and oil. Enzymes perform a significant role in disordering the lipoprotein and lipopolysaccharide to speed up the lipid release. Organic solvent consumption can be wiped out effectively by this strategy and this study does not contribute to generating noxious waste. Moreover, in a medium where liquidity prevails, phospholipids get isolated from the oil and thus diminishing the need for the distillation process (Xue et al., 2018). An investigation was performed on lipid extraction by Chen et al. to observe the outcome of AEE with a combined thermal lysis process. They used wet microalgae as raw materials and reported the

extraction efficiency of 88.3% (Xue et al., 2018). Another study recovered lipid using AEE combining with sonication-enzyme treatment on *Chlorella vulgaris* (Chen, Li, Ren, & Liu, 2016). So, for yielding of oil from wet microalgae, AEE could be recognized as a popular candidate.

5 Conversion processes of bio-oil from microalgae

Microalgal bio-oil can be applied as both edible and nonedible while by-products: biochar and gaseous products can be used as direct biofuel. Bio-char can be stored underground as a carbon sink for hundreds of years for the next generation. Microalgae bio-oil and biochar are carbon-rich inherent products of thermo-chemical (e.g., slow pyrolysis, fast pyrolysis, hydrogenation, torrefaction, catalytic hydro-pyrolysis, hydrodeoxygenation, and others) and hydrothermal (e.g., liquefaction, carbonization, wet impregnation, and others) processes. Fig. 1 presents the overall bio-oil production process flow from microalgae (Chang et al., 2015; Hossain & Mahlia, 2019; Nam et al., 2017; Torri et al., 2011; Xie et al., 2015).

5.1 Hydrothermal liquefaction

Hydrothermal liquefaction (HTL) is a type of hydrothermal processing where any biomass such as microalgae can be converted into crude

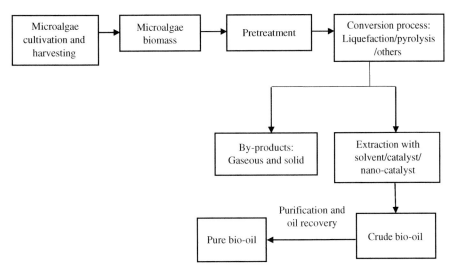

FIG. 1 Bio-oil production process flow from microalgae (Chang, Duan, & Xu, 2015; Hossain & Mahlia, 2019; Nam, Kim, Capareda, & Adhikari, 2017; Torri et al., 2011; Xie et al., 2015).

bio-oil utilizing water at super or subcritical pressures and temperatures (Panga et al., 2018). The yield of bio-oil from biomass can be determined by Eq. (1) (Hossain & Mahlia, 2019; Nizamuddin et al., 2018).

$$\text{Bio}-\text{oil yield (wt\%)} = \frac{m_{\text{bio}-\text{oil}}}{m_{\text{biomass}}} \times 100\% \qquad (1)$$

Water has been used as the most suitable and cheaper solvent to carry out HTL for bio-oil production. Besides water, some other solvents such as ethanol, isopropanol, and others have experimented in laboratory-scale albeit water remained the best solvent among all due to the characteristics like nontoxicity, readily available, cheap, nonhazardous, ecologically safe and environmentally benign. Therefore, water was the key factor to run HTL for bio-oil extraction as water performs as both catalyst and reactant (Demirbas, 2011; Nizamuddin et al., 2015; Nizamuddin et al., 2017; Savage, 2009). To produce crude bio-oil via HTL, conditions: temperature range 280–380°C and pressure, 7–30 MPa in the liquid state has been used in previous experimental studies. The main products obtained from HTL are bio-crude (bio-oil mixed

with aqueous phase) and the by-products are solid particles (biochar), water-soluble products, and gases. Bio-oil extracted from microalgae via HTL contained a high amount of oxygen and nitrogen. The characteristics of algal bio-oil have been proven more suitable for the purposes of aromatics, polymers lubricants and others, but not very favorable for the petroleum blend. Through HTL, both dry or wet microalgae can be converted into bio-crude. Microalgae biomass contains a large amount of moisture content and pretreatment especially drying requires high heat and energy to evaporate moisture content (Toor, Rosendahl, & Rudolf, 2011). In addition, the HTL process contains negligible SO_x emission for bio-oil production which manifested this process much environment-friendly than other processing methods (Elliott, Biller, Ross, Schmidt, & Jones, 2015). An experimental study on microalgal bio-oil via HTL presented that three different microalgae species: *Chlorella vulgaris, Botryococcus braunii*, and *Scenedesmus quadricauda* were processed with different water concentrations (1:6, 1:7, 1:8, 1:9 and 1:10) via HTL at 200–320°C and 60 bar pressure. *Chlorella vulgaris,*

Botryococcus braunii, and *Scenedesmus quadricauda* yielded a maximum 12.4%, 16.3%, and 18% (wt), respectively. The gas chromatography analysis presented the presence of furan, phenol, acid, and ester derivatives in crude bio-oil. The Fourier Transform Infrared Spectroscopy (FTIR) analysis of bio-oil showed high containment of aliphatic, phenolic, alcoholic, carboxylic, and hydroxylic groups in the solid residues (Panga et al., 2018).

5.2 Slow and fast pyrolysis

Pyrolysis is a thermal process to heat the biomass with high temperature, 450–550°C or above at an anoxic environment. The main product through pyrolysis is pyrolyzed oil or bio-oil and the by-products are biochar and gaseous products which can be applied for fuel purposes directly (Chaiwong, Kiatsiriroat, Vorayos, & Thararax, 2013; Nizamuddin et al., 2018). An experimental study on algal bio-oil presented that different algal species such as *Polytrichum commune, Sphagnum palustre, Dicranum scoparium, Thuidium tamarascinum, Drepanocladua revolvens, Chlorella prototheocoides,* and *Cladophora fracta* yielded bio-oil 34.3%–53.3% at 775°C with the heating rate, 10 K/s via fast pyrolysis (Demirbaş, 2006). Another experimental study presented that *Spirulina* sp. yielded 32.42% via slow pyrolysis in a fixed-bed reactor with the optimum temperature range 500–550°C. This bio-oil contains aromatic, phenolic, amide, heterocyclic, indole, alkane, and nitrile groups (Chaiwong et al., 2013).

5.3 Hydrothermal decarboxylation, hydrogenation, and others

Various techniques such as catalytic cracking, hydrodeoxygenation (HDO), decarboxylation have been implemented to extract crude bio-oil from microalgae for oxygen reduction without hydrogen requirement. Hydrodeoxygenation process removed oxygen from bio-oil as water and decarboxylation and decarbonylation eliminated oxygen through CO_2 and CO formation. These different processes can be applied simultaneously in a single hydro-processing reaction. The catalysts and catalytic cracking process include alkalines (NaOH, MgO, CaO), acids (Al_2O_3, $AlCl_3$), zeolites (MCM-41, SAPO11, SAPO5, HBEA, USY, HZSM-5) (Elliott, Hart, Neuenschwander, Rotness, & Zacher, 2009; Yang et al., 2016). An experimental study demonstrated that a novel catalyst, Pt/C presented decarbonylation activities and strong stability of continuous operation while performing deoxygenation of algal triglycerides. Moreover, this catalyst was also very cheap and presented 90% selectivity (Fu, Lu, & Savage, 2011). HDO has been carried out with model compounds e.g., guaiacol, sorbitol, vanillin, phenol, acetic acid, cresol, eugenol, and methyl heptonate (Yang et al., 2016). Based on the previous experimental results, *Scenedesmus* sp. yielded approximated 50% bio-oil with the presence of Ni-Al catalysts at 300°C under H_2 pressure at 4 MPa; *Nannochloropsis salina* presented nearly 98.7% complete conversion of algal oil with sulfided NiMo/γ-Al_2O_3 at 3.45 Mpa H_2 pressure and 360°C via hydrodeoxygenation; *Nannochloropsis salina* also showed nearly 62.7%–76.5% algal oil (hydrocarbons) with Pt/Al_2O_3, Rh/Al_2O_3, and presulfided NiMo/Al_2O_3 at 2–3.44 Mpa H_2 pressure and 310–360°C via hydrodeoxygenation (Yang et al., 2016).

6 Bio-oil recovery, distillation, and purification

Various requirements need to be maintained for the final product to be called bio-oil. Impurities found in the oil can reduce the product value and impure bio-oil cannot be utilized as food substances. Impure bio-oil can only be implemented for other fuel purposes. These impurities usually originate from feedstocks such as surplus alcohol remaining after the reaction, glycerol, water, and catalyst residues (Paiva

Pinheiro Pires et al., 2019). These contaminating particles need to be eliminated to maintain the standard of oil. Several methods such as supercritical fluid equipment application, liquid-liquid extraction, distillation, membrane extraction, precipitation and others, and washing are available to eliminate these impurities (Paiva Pinheiro Pires et al., 2019; Saleh, Dubé, & Tremblay, 2010). To eliminate unstable scums from a liquid mixture, distillation is generally applied (Bateni, Saraeian, & Able, 2017). Different types of distillation techniques such as conventional (vacuum, organic and steam distillation), extractive, evaporative and molecular distillation are available. To remove water and remaining impurities form crude bio-oil, evaporation and conventional distillation methods are the most familiar methods that have been widely used (Kouzu & Hidaka, 2013). In molecular distillation which is generally done under high vacuum, the particles free path is higher than the evaporation and condenser exterior area and so a huge amount of the scattered atoms reach at the consolidating surface without being redirected on crash with overseas gas atoms, bringing about a larger separation yield (Bateni, Bateni, Able, & Noori, 2018; Bateni & Karimi, 2016). Molecular distillation has been applied to purify bio-oil and yielded 98% segregation at the 120°C temperature (evaporator temperature) (Wang et al., 2010).

6.1 Supercritical fluid separation

In the processing of biomass, supercritical fluid separation (SFS) is regarded as a safe and favorable process. Being a mass exchange process, it is operated at the working conditions (temperature and pressure) over the cynical point of the solvent (Manjare & Dhingra, 2019). This process is a faster separation process. This process is based on density for liquid (bio-oil) that can penetrate the solid matrix not permitted to access through different density liquids because of different viscosity and surface tension. This process can be implemented as an alternative to organic solvent for commercial and laboratory-based bio-oil separation. This application can be applied for microalgal bio-oil purification for higher purity (Manjare & Dhingra, 2019).

6.2 Liquid-liquid extraction

Liquid-liquid extraction (LLE) which is also known as solvent extraction generally includes all the techniques of wet washing. For extracting components from the liquid feed, a suitable solvent is used. Water is the most common solvent used in this process and its temperature, pressure plays an effective role in the purification of bio-oil. It has experimented that when the temperature is increased; it results in an improvement in glycerol diffusivity and brings about a better volumetric mass transfer coefficient and thus resulting in higher biodiesel purification. However, increment in temperature also raises the solubility of water which affects the standard properties of bio-oil since the water might be present in bio-oil. From a modeling perspective, it was also observed that biodiesel obtained after multistage wet washing can maintain the standard of oil since it reduces glycerol content. But the formulation of emulsions and loss in yielding of the final product is the main setback of this multistage wet washing (Abbott, Cullis, Gibson, Harris, & Raven, 2007). So, researchers have been searching for the appropriate solvents and techniques which can reduce the water content in bio-oil. Acidified water has been used in this purpose since the quantity of water required to nullify the remaining homogeneous alkali catalyst in bio-oil is less (Baptista, 2018). Sulfuric acid, phosphoric acid, and hydrochloric acid are the most favorable acids that have been used for this purpose. But using acids has also some disadvantages, for example, the acidity content in final biodiesel may be acute. Organic solvents are applied to remove this setback (Baptista, 2018). To purify the bio-oil in this process, crude bio-oil is mixed in the organic solvents such as petroleum ether and *n*-hexane and then exposed to the water washing process. But the use of water washing in this process brings about the higher costs in the purification process. Also, this process is also energy-intensive (Baptista, 2018). So, the focus was also given to investigate more waterless purification method and researchers have proposed to use ionic liquids (ILs) in the purification of bio-oil. Although these liquids are expensive, they are also nonflammable, reusable, and nonvolatile and can dissolve organic and inorganic compounds. A series of ammonium salts have been investigated in bio-oil purification and this study reported that [ClEt-Me$_3$N]Cl, [EtNH$_4$]Cl, and choline chloride showed better performance in impurity removal (Abbott et al., 2007). Ethylene glycol, methyl triphenyl phosphonium bromide, and 2,2,2-trifluoroacetamide had been used with choline chloride to enhance the performance of impurity removal (Shahbaz, Mjalli, Hashim, & Alnashef, 2010; Shahbaz, Mjalli, Hashim, & Alnashef, 2011).

6.3 Membrane extraction

Ceramic membranes have the chemical and physical properties to isolate impurities from bio-oil. They can also be effectively used in separating impurities (Ellingboe & Runnels, 1966; López Granados, Alba-Rubio, Vila, Martín Alonso, & Mariscal, 2010; Wang et al., 2009). Two mono-channel ceramic layers having a pore diameter of 0.05 and 0.1 μm have been used to purify bio-oil (Ferrero, Almeida, Alvim-Ferraz, & Dias, 2014). Bio-oil was made to propagate (2.11 mL/min) through the membranes of the tubes. The bio-oil was cross filtered by the tubes along with the flux through the ceramic layer of 26.4 L/m^2/h. The propagation was made to stop when the volumetric concentration factor reached 3. The outcome reported that the two membranes have low calcium removal efficacy. The membrane with the highest porosity showed a better result (30%) while the other membrane result was low (22%). The outcome of this study was validated by an experimental study which states that ceramic membranes have low calcium removal efficacy (Wang et al., 2009).

6.4 Precipitation

Precipitating agents are utilized to abolish calcium ions from bio-oil. When these agents react with bio-oil containing calcium ions, a compound (insoluble) is formed when the reaction is stirred. After that, filtration can be applied to remove the precipitate. The better precipitation process is largely dependent on many constituents such as pH, temperature, and concentration of calcium ions (Veljković, Banković-Ilić, & Stamenković, 2015). The following equation represents the reactions between biodiesel containing calcium ions and precipitating agents (oxalic and citric acid).

$$Ca^{2+} + H_2C_2O_4 + H_2O = CaC_2O_4 \cdot H_2O \downarrow + 2H^+$$

$$3CaO + 2H_3C_6H_5O_7 = Ca_3(C_6H_5O_7)_2 \downarrow + 6H^+$$

Oxalic and citric acid have been tested to purify biodiesel which was obtained from Jatropha oil and for this purpose; various agents were used to calcium molar ratios by keeping the similar precipitating conditions (45°C, 20 min and stirring) (Zhu et al., 2006). At the molar ratio of 1:1, the reported calcification efficacy for oxalic and citric acid was 92.2% and 96.2%, respectively and the acquired refined bio-oil was 90.7% and 95.5%, respectively. When the number of agents was halved, the calcification efficacy came down to half but the change in calcification efficacy was not observed when the quantity of water changed. Oxalic acid has the disadvantage of getting quickly dispersed in biodiesel and resulting in the formation of an emulsion. So, phase separation gets difficult.

7 Integrated approaches

Although the production of nonedible bio-oil from microalgae as a renewable source for fuel purpose has been initiated primarily a few decades ago, yet edible bio-oil strived to catch attention among edible oil researchers. Edible bio-oil can contribute greatly beside other vegetable oils to mitigate the high oil demand for food consumption by individuals and food processing industries but attaining economic feasibility prior to the application is a mandatory attempt. Since research for cost-effective bio-oil extraction technologies is still ongoing, an integrated approach with simultaneous applications of co-products and by-products may achieve the desired goal. Hence, it could be considered as a major industrial challenge to overcome to introduce a new source of edible oil to the oil industries. Fig. 2 presents a possible integrated approach of microalgae bio-oil.

Microalgal bio-oil is extracted from the cellular unsaturated fatty acid, the remaining cell debris contains protein, carbohydrates, and other biochemical compounds which can be raw materials to produce biodiesel, bioethanol, biobutanol, biogas, jet-fuel, and other nutritious and pharmaceutical products (Bwapwa, Anandraj, & Trois, 2018; Goh, 2019; Hossain, Haji Zaini, & Mahlia, 2017; Hossain & Mahlia, 2019; Hossain, Mahlia, & Saidur, 2019). To enhance the economic feasibility, biomass residues can be separated after edible bio-oil extraction, then residues can be converted into different value-added products and recovered pure products through biorefineries. Protein components of residual microalgae biomass can be converted into biodiesel via the esterification or transesterification process. Besides pure

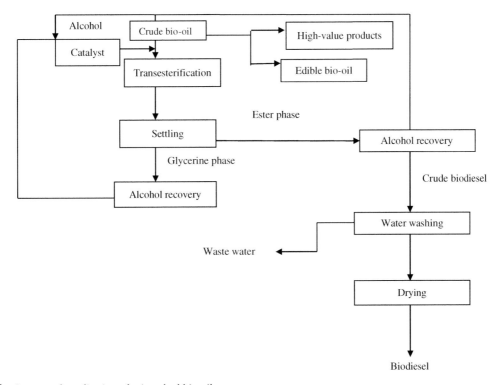

FIG. 2   Integrated application of microalgal bio-oil.

biodiesel obtainment, crude glycerol can be produced simultaneously. Crude glycerol can be used as feedstock for value-added products as medical, pharmaceutical, and personal care applications. Purified glycerol can be used as solvent, sweetener, humectant, food preservatives for foods and beverages; processing tools of low-fat foods as well as a thickening agent for liqueurs (Yang, Hanna, & Sun, 2012). High-value product after microalgal bio-oil extraction, phospholipid removed from bio-oil can be employed as additives (at maximum concentration, 0.57%) with cocoa butter (almost 14% of the total weight of commercial chocolate) in chocolate industries mainly for dark chocolate processing to prevent bloom formation as well as remove crystallization (Parsons & Keeney, 1969). Phospholipids also can be used commercially as taste enhancers for breadmaking and wheat dough (Xue et al., 2018). Extraction of chlorophyll II from the residue can be employed as an organic food coloring agent in food industries as well as an additive in cosmetics and pharmaceuticals due to its' antimutagenic and antioxidant properties (Hosikian, Lim, Halim, & Danquah, 2010). Phycobiliproteins can be implemented as fluorescent dyes either research or clinical immunology laboratories. Vitamins and minerals (e.g., astaxanthin, betacarotenoid, chlorophylls, and other vitamins) are applied as nutrition supplements, cosmetics, and pharmaceuticals for antiaging, muscle recovery, cardiovascular improvement, and improving eye-sight (Kaneda, 2016; Xue et al., 2018). After processing all the valuable products, the microbial slurry is still enriched with phosphorus and nitrogen component. Therefore, the slurry can be utilized for harvesting further batch of microalgae as well as organic fertilizers (Hossain, Mahlia, Zaini, & Saidur, 2019; Xue et al., 2018).

8 Environmental and socioeconomic impacts

The consistent rise in the population of the world is pressing high at the need to develop unconventional thoughts to produce edibles from the nonedible natural resources. The utilization of microalgae to produce edible oil is one of the interesting possibilities being explored for the said purpose. Microalgae have proven their potential suitability to produce many chemicals, energy applications, and feedstock (Giwa et al., 2018; Kebelmann, Hornung, Karsten, & Griffiths, 2013; Raheem, Wan Azlina, Taufiq Yap, Danquah, & Harun, 2015). Presently edible oils are being produced from the vegetable oils extracted from the food crops such as sunflower, canola, and soybean. These crops are needed to be grown on the large land acres requiring the well-established agricultural formalities and inputs. The rise in demand for edible oils coupled with the cost of production of the feedstock crops is the main cause of the rise in the cost and affecting the communities in the developing world. Traditional methods require large fertile land areas and a lot of effort to grow the crops needed to produce edible oils. The introduction of the concept to utilize underutilized feedstocks such as microalgae can bring a revolution and ease to produce edible oil to meet the daily needs of living communities. Still, a lot of efforts are needed to mature this possibility to reality especially in terms of the cost-effectiveness of the process both at the microalgae cultivation and edible oil production sides (Milledge & Heaven, 2013). The maturity of the concept will impart numerous socioeconomic and environmental impacts on the communities (Draaisma et al., 2013). The benefits coupled with the microalgae include their higher growth rates as compared to the growth of the crops. Microalgae are considered under the category of vegetable harvests as they grow by consuming carbon dioxide from the environment thus offering tremendous benefits to the environment helping in the reduction of greenhouse gases. Microalgae are called micro plants as well based on their growth resulting from the consumption of carbon dioxide and light. Carbon dioxide is believed to be the main responsible greenhouse gas contributing to global warming. With the growth of microalgae for edible oil production, we can grab two benefits including the reduction in the carbon dioxide in the environment and the availability of feedstock other than the conventional seeds for the edible oil production (Giwa et al., 2018). They can be grown even in the nonarable and underutilized lands offering more availability of the fertile land for the growth of direct food crops such as wheat and rice (Menegazzo & Fonseca, 2019). They are capable to produce the edible proteins, carbohydrates, and lipids in different forms (Klok et al., 2014). Although microalgae are an attractive source, their production costs are still believed to be high and the main challenge for their utilization. The cultivation of microalgae for edible oil production can pave the way to educate and sensitize the communities for about their usefulness for the biofuels production as well. Biofuels are commonly considered as offering numerous needs, including supportability, a decrease of ozone-depleting substance discharges, local advancements, social structure and horticulture, the security of supply. Simultaneous production of edible oils from the microalgae and biofuels from the resulting residues by the applications of different energy conversion processes such as thermochemical conversion and biochemical conversion processes could be another advantage in line with the production of edible oil from the microalgae (Andrade, Batista, Lira, Barrozo, & VIEIRA, 2018; Phusunti, Phetwarotai, & Tekasakul, 2018). The utilization of the microalgae for value-added products will offer attractive socioeconomic benefits to the societies by providing new investment and economic opportunities. The challenges associated with the increase in the growth rate, cultivation, dewatering, and pretreatment of the biomass from microalgae should be considered as opportunities. The thought of edible oil production from microalgae can attract investment in multiple sectors including the agricultural, engineering, and research and development sectors. Utilizing microalgae to produce edible oil and biofuels can lessen the expense of traditional

crude materials essential to produce edible oil and biofuels, when contrasted with customary sources, and by gathering and utilizing, its expense can be additionally decreased. The concept of producing edible oil from microalgae can impart direct positive impacts on the developing countries but indirectly effects the world. It mainly results in the decreased prices of the edible commodities. On the other hand it will reduce the requirements of agricultural land to grow crops which are traditionally being used to produce the edible oil which also results in an increase in commodity prices and affect the poor community of the world drastically (Van Der Horst & Vermeylen, 2011). However, water scarcity is another issue that must be kept in mind and addressed in the present century.

Besides bio-oil production, the integrated approach with biofuels (such as biodiesel, bioethanol, aviation fuel, and others) is also considered as offering numerous needs, including supportability, a decrease of ozone-depleting substance discharges, local advancements, social structure and horticulture, the security of supply. Biofuels are noncontaminating, locally accessible, available, reasonable, and are a solid fuel got from inexhaustible sources. The emanation of harmful hydrocarbons, for example, benzene has declined in Brazil, notwithstanding the discharge of sulfur and CO (Ahmed, Abu Bakar, Azad, Sukri, & Mahlia, 2018; Ahmed, Abu Bakar, Azad, Sukri, & Phusunti, 2018; Ahmed et al., 2018; Cremonez et al., 2015; Hossain, Mahlia, & Saidur, 2019). Bio-oil—regardless of whether unadulterated or mixed outcomes in lower outflows of most contaminations' comparative with diesel, including altogether lower emanation of particulates, sulfur, hydrocarbons, CO, poisons. Outflows change with a motor plan, state of vehicles, and nature of the fuel. In bio-oil–diesel mixes, potential decreases of most contaminations increment straight as the portion of biodiesel increments, except for NO_x outflow. Bioethanol utilized as an oxygenate can lessen the outflow of a few contaminations especially in more established vehicles (Hossain & Jalil, 2015; Hossain, Zaini, Jalil, & Mahlia, 2018; Mahlia, Ismail, Hossain, Silitonga, & Shamsuddin, 2019). Anyway, the utilization of oxygenates, for example, ethanol (and biodiesel), to modify the fuel to oxygen proportion won't positively affect emanation if a vehicle's air-to-fuel proportion is set low or if an excessive amount of ethanol is added to gas in a vehicle with a fixed air-to-fuel proportion. It is the situation, oxygenate can expand NO_x discharges and cause "lean fizzle" expanding hydrocarbon emanations. Indeed, it was contending by experimental studies that ethanol has no emanation related favorable circumstances over reformulated fuel other than the decrease of CO_2 (Hossain et al., 2017; Hossain & Jalil, 2015; Hossain, Mahlia, & Saidur, 2019; Kusumo et al., 2017; Kusumo, Silitonga, Ong, Masjuki, & Mahlia, 2017).

Utilizing microalgae in an integrated approach to produce edible oil and biofuels can lessen the expense of traditional crude materials essential to produce edible oil and biofuels when contrasted with customary sources, and by gathering and utilizing, its expense can be additionally decreased. This examination includes the plan and improvement of a recreation model to break down the expenses and emanations related to the production of edible oil from microalgae. Discharges related to nonrenewable energy source burning incorporate carbon monoxide, carbon dioxide, nitrogen oxides, sulfur dioxide, particulate issues, and unpredictable natural mixes. These toxins cause various destructive effects, for example, respiratory issues and the fermentation of crisp water sources. For every one of the reasons expressed over, the utilization of oil-based energies for transportation displays a test to supporting portability, nature, and human well-being. Choices to oil should be investigated to decide a progressively maintainable arrangement. Choices have grown, for example, hydrogen and biofuels, anyway their wide-scale appropriation have been obstructed. One of the significant

misfortunes to these options is the cost (Hossain et al., 2019; Hossain & Mahlia, 2019; Hossain, Mahlia, & Saidur, 2019; Silitonga et al., 2018). For instance, hydrogen and power modules for vehicles and the infrastructural necessity to supply it are excessively costly with current innovation, hence not financially reasonable. In any case, there are arrangements that can contend with the customary transportation vitality and waste techniques as far as financial matters and furthermore give outside advantages to the earth and human wellbeing. There are transportation fuel choices that don't require new frameworks. Biodiesel is a fuel that can be mixed into phospholipids diesel or utilized without anyone else to decrease the rate at which oil-based fuel is devoured. This decrease in the utilization rate can consider more opportunity to contemplate elective fuel sources and innovation while diminishing outflows related to the transportation business. A present strategy for diminishing expenses is to use microalgae as a modest feedstock to produce edible oil (Giwa et al., 2018; Solomon, 2010).

Three significant financial issues related to extended biofuels improvement can be recognized: little scale financing, monetary improvement and work age, and well-being and gender orientation implications. It can be normal that best in class biofuels advancements, for example, cellulosic ethanol, particularly little scale, will experience issues acquiring financing in certain locales and nations because of high capital necessities and saw high hazard given the emanant idea of the market. However, the forthright cost boundary can be overwhelmed by using low-or no-cost build-ups for feedstock, ease obligation financing, and joining into a biorefinery platform. Moreover, such activities would appear to be customized for subsidizing from the "Global Environment Facility and the Energy Efficiency" (Solomon, 2010).

The concept of producing edible oil from microalgae can impart direct positive impacts on the developing countries but indirectly effects the world. It mainly results in the decreased prices of edible commodities. On the other hand, it will reduce the requirements of agricultural land to grow crops which are traditionally being used to produce the edible oil which also results in an increase in commodity prices and affect the poor community of the world drastically (Van Der Horst & Vermeylen, 2011). Therefore, microalgae could be a great and sustainable source for the production of edible oils we cannot ignore the fact that biomass is still the largest water consumer leading to water crisis (Cremonez et al., 2015). So, a good mix of well-understood policies formed on the eve of the best technologies will be helpful in the utilization of microalgae for edible oil production.

9 Conclusions

Microalgae are one of the most promising third-generation sources for edible bio-oil production. Though processing stages and extraction technologies are still under research, bio-oil yield in laboratory scale is still adequate to project for commercialization in the coming decades. Applications of different catalysts, nano-catalysts, and solvents utilization with the bio-oil extraction process have enhanced the overall yield. Along with the improvement of extraction techniques, the determination of suitable microalgal strain and processing techniques is expected in the coming future for scalable applications because of the emphasis on alternative and more sustainable oil sources globally. Therefore, this study proposed an integrated approach clarifying various processing routes of high-value product generation simultaneously with bio-oil. If this approach can be implemented, economic feasibility might be achieved. To achieve this goal, a comprehensive techno-economic analysis of the integrated process is recommended. A detailed life-cycle assessment and energy balance analysis can suggest the energy and heat requirement for the overall system as well as determine the exact water, land, and carbon footprint. This chapter concludes that if microalgal bio-oil production with value-added products can achieve economic feasibility, a new window will be opened in the oil, food, beverages, and pharmaceutical industries worldwide.

References

Abbott, A. P., Cullis, P. M., Gibson, M. J., Harris, R. C., & Raven, E. (2007). Extraction of glycerol from biodiesel into a eutectic based ionic liquid. *Green Chemistry, 9*, 868–872.

Ahmed, A., Abu Bakar, M. S., Azad, A. K., Sukri, R. S., & Mahlia, T. M. I. (2018). Potential thermochemical conversion of bioenergy from Acacia species in Brunei Darussalam: A review. *Renewable and Sustainable Energy Reviews, 82*, 3060–3076.

Ahmed, A., Abu Bakar, M. S., Azad, A. K., Sukri, R. S., & Phusunti, N. (2018). Intermediate pyrolysis of Acacia cincinnata and Acacia holosericea species for bio-oil and biochar production. *Energy Conversion and Management, 176*, 393–408.

Ahmed, A., Hidayat, S., Abu Bakar, M. S., Azad, A. K., Sukri, R. S., & Phusunti, N. (2018). Thermochemical characterisation of Acacia auriculiformis tree parts via proximate, ultimate, TGA, DTG, calorific value and FTIR spectroscopy analyses to evaluate their potential as a biofuel resource. *Biofuels*, 1–12.

Amarni, F., & Kadi, H. (2010). Kinetics study of microwave-assisted solvent extraction of oil from olive cake using hexane: Comparison with the conventional extraction. *Innovative Food Science & Emerging Technologies, 11*, 322–327.

Andrade, L. A., Batista, F. R. X., Lira, T. S., Barrozo, M. A. S., & VIEIRA, L. G. M. (2018). Characterization and product formation during the catalytic and non-catalytic pyrolysis of the green microalgae Chlamydomonas reinhardtii. *Renewable Energy, 119*, 731–740.

Andrich, G., Nesti, U., Venturi, F., Zinnai, A., & Fiorentini, R. (2005). Supercritical fluid extraction of bioactive lipids from the microalga Nannochloropsis sp. *European Journal of Lipid Science and Technology, 107*, 381–386.

Andrich, G., Zinnai, A., Nesti, U., & Venturi, F. (2006). Supercritical fluid extraction of oil from microalga Spirulina (Arthrospira) platensis. *Acta Alimentaria, 35*, 195–203.

Aruoja, V., Dubourguier, H. C., Kasemets, K., & Kahru, A. (2009). Toxicity of nanoparticles of CuO, ZnO and TiO_2 to microalgae Pseudokirchneriella subcapitata. *Science of the Total Environment, 407*, 1461–1468.

Balasubramanian, S., Allen, J. D., Kanitkar, A., & Boldor, D. (2011). Oil extraction from Scenedesmus obliquus using a continuous microwave system—design, optimization, and quality characterization. *Bioresource Technology, 102,* 3396–3403.

Baptista, A. (2018). Biodiesel—from production to combustion. In *Biodiesel production systems: Operation, process control and troubleshooting.* Switzerland: Springer International Publishing (chapter 3).

Bateni, H., Bateni, F., Able, C., & Noori, M. S. (2018). Biorefinery of safflower seeds in a sequential process for effective use of the substrate for biofuel production. *Waste and Biomass Valorization, 9,* 2145–2155.

Bateni, H., & Karimi, K. (2016). Biodiesel production from castor plant integrating ethanol production via a biorefinery approach. *Chemical Engineering Research and Design, 107,* 4–12.

Bateni, H., Saraeian, A., & Able, C. (2017). A comprehensive review on biodiesel purification and upgrading. *Biofuel Research Journal, 4,* 668–690.

Bwapwa, J. K., Anandraj, A., & Trois, C. (2018). Microalgae processing for jet fuel production. *Biofuels, Bioproducts and Biorefining, 12,* 522–535.

Chaiwong, K., Kiatsiriroat, T., Vorayos, N., & Thararax, C. (2013). Study of bio-oil and bio-char production from algae by slow pyrolysis. *Biomass and Bioenergy, 56,* 600–606.

Chang, Z., Duan, P., & Xu, Y. (2015). Catalytic hydropyrolysis of microalgae: Influence of operating variables on the formation and composition of bio-oil. *Bioresource Technology, 184,* 349–354.

Chen, L., Li, R., Ren, X., & Liu, T. (2016). Improved aqueous extraction of microalgal lipid by combined enzymatic and thermal lysis from wet biomass of Nannochloropsis oceanica. *Bioresource Technology, 214,* 138–143.

Chen, Z., Wang, L., Qiu, S., & Ge, S. (2018). Determination of microalgal lipid content and fatty acid for biofuel production. *BioMed Research International, 2018,* 17.

Choi, W., Kim, G., & Lee, K. (2012). Influence of the CO_2 absorbent monoethanolamine on growth and carbon fixation by the green alga Scenedesmus sp. *Bioresource Technology, 120,* 295–299.

Cremonez, P. A., Feroldi, M., De Oliveira, C. D. J., Teleken, J. G., Alves, H. J., & Sampaio, S. C. (2015). Environmental, economic and social impact of aviation biofuel production in Brazil. *New Biotechnology, 32,* 263–271.

D'Alessandro, E. B., & Antoniosi Filho, N. R. (2016). Concepts and studies on lipid and pigments of microalgae: A review. *Renewable and Sustainable Energy Reviews, 58,* 832–841.

D'oca, M. G. M., Viêgas, C. V., Lemões, J. S., Miyasaki, E. K., Morón-Villarreyes, J. A., Primel, E. G., et al. (2011). Production of FAMEs from several microalgal lipidic extracts and direct transesterification of the Chlorella pyrenoidosa. *Biomass and Bioenergy, 35,* 1533–1538.

Datta, K. K. R., Achari, A., & Eswaramoorthy, M. (2013). Aminoclay: A functional layered material with multifaceted applications. *Journal of Materials Chemistry A, 1,* 6707–6718.

De Windt, W., Aelterman, P., & Verstraete, W. (2005). Bioreductive deposition of palladium (0) nanoparticles on Shewanella oneidensis with catalytic activity towards reductive dechlorination of polychlorinated biphenyls. *Environmental Microbiology, 7,* 314–325.

Demirbaş, A. (2006). Oily products from mosses and algae via pyrolysis. *Energy Sources, Part A: Recovery, Utilization, and Environmental Effects, 28,* 933–940.

Demirbas, A. (2011). Competitive liquid biofuels from biomass. *Applied Energy, 88,* 17–28.

Deng, X.-Y., Cheng, J., Hu, X.-L., Wang, L., Li, D., & Gao, K. (2017). Biological effects of TiO_2 and CeO_2 nanoparticles on the growth, photosynthetic activity, and cellular components of a marine diatom Phaeodactylum tricornutum. *Science of the Total Environment, 575,* 87–96.

Dong, T., Knoshaug, E. P., Pienkos, P. T., & Laurens, L. M. L. (2016). Lipid recovery from wet oleaginous microbial biomass for biofuel production: A critical review. *Applied Energy, 177,* 879–895.

Draaisma, R. B., Wijffels, R. H., Slegers, P. M., Brentner, L. B., Roy, A., & Barbosa, M. J. (2013). Food commodities from microalgae. *Current Opinion in Biotechnology, 24,* 169–177.

Ellingboe, J. L., & Runnels, J. H. (1966). Solubilities of sodium carbonate and sodium bicarbonate in acetone-water and methanol-water mixtures. *Journal of Chemical & Engineering Data, 11,* 323–324.

Elliott, D. C., Biller, P., Ross, A. B., Schmidt, A. J., & Jones, S. B. (2015). Hydrothermal liquefaction of biomass: Developments from batch to continuous process. *Bioresource Technology, 178,* 147–156.

Elliott, D. C., Hart, T. R., Neuenschwander, G. G., Rotness, L. J., & Zacher, A. H. (2009). Catalytic hydroprocessing of biomass fast pyrolysis bio-oil to produce hydrocarbon products. *Environmental Progress & Sustainable Energy, 28,* 441–449.

Farooq, W., Lee, H. U., Huh, Y. S., & Lee, Y.-C. (2016). Chlorella vulgaris cultivation with an additive of magnesium-aminoclay. *Algal Research, 17,* 211–216.

Ferreira, A. F., Dias, A. P. S., Silva, C. M., & Costa, M. (2016). Effect of low frequency ultrasound on microalgae solvent extraction: Analysis of products, energy consumption and emissions. *Algal Research, 14,* 9–16.

Ferrero, G. O., Almeida, M. F., Alvim-Ferraz, M. C. M., & Dias, J. M. (2014). Water-free process for eco-friendly purification of biodiesel obtained using a heterogeneous Ca-based catalyst. *Fuel Processing Technology, 121,* 114–118.

Fu, J., Lu, X., & Savage, P. E. (2011). Hydrothermal decarboxylation and hydrogenation of fatty acids over Pt/C. *ChemSusChem, 4,* 481–486.

Giwa, A., Adeyemi, I., Dindi, A., Lopez, C. G.-B., Lopresto, C. G., Curcio, S., et al. (2018). Techno-economic assessment of the sustainability of an integrated biorefinery from microalgae and Jatropha: A review and case study. *Renewable and Sustainable Energy Reviews, 88*, 239–257.

Goh, B. (2019). Sustainability of direct biodiesel synthesis from microalgae biomass: A critical review. *Renewable and Sustainable Energy Reviews, 107*, .

Griffiths, M. J., Van Hille, R. P., & Harrison, S. T. (2010). Selection of direct transesterification as the preferred method for assay of fatty acid content of microalgae. *Lipids, 45*, 1053–1060.

Guo, R., Yiyun, L., & Chen, J. (2015). Toxic effect of nano-TiO_2 and nano-carbon on microcystis aeruginosa. In S. Chen, & S. Zhou (Eds.), *Proceedings of the international conference on advances in energy, environment and chemical engineering*. Vol. 23. Atlantis Press.

Halim, R., Danquah, M. K., & Webley, P. A. (2012). Extraction of oil from microalgae for biodiesel production: A review. *Biotechnology Advances, 30*, 709–732.

Halim, R., Harun, R., Danquah, M. K., & Webley, P. A. (2012). Microalgal cell disruption for biofuel development. *Applied Energy, 91*, 116–121.

He, M., Yan, Y., Pei, F., Wu, M., Gebreluel, T., Zou, S., et al. (2017). Improvement on lipid production by Scenedesmus obliquus triggered by low dose exposure to nanoparticles. *Scientific Reports, 7*, 15526.

Holmström, S., Patil, A., Butler, M., & Mann, S. (2007). Influence of polymer co-intercalation on guest release from aminopropyl-functionalized magnesium phyllosilicate mesolamellar nanocomposites. *Journal of Materials Chemistry, 17*.

Hosikian, A., Lim, S., Halim, R., & Danquah, M. K. (2010). Chlorophyll extraction from microalgae: A review on the process engineering aspects. *International Journal of Chemical Engineering, 2010*, 11.

Hossain, N., Bhuiyan, M. A., Pramanik, B. K., Nizamuddin, S., & Griffin, G. (2020). Waste materials for wastewater treatment and waste adsorbents for biofuel and cement supplement applications: A critical review. *Journal of Cleaner Production, 255*, 120261.

Hossain, N., Haji Zaini, J., & Mahlia, T. M. I. (2017). A review of bioethanol production from plant-based waste biomass by yeast fermentation. *International Journal of Technology, 8*, 5–18.

Hossain, N., & Jalil, R. (2015). Sugar and bioethanol production from oil palm trunk (OPT). *Asia Pacific Journal of Energy and Environment, 2*, 89–92.

Hossain, N., & Mahlia, T. M. I. (2019). Progress in physicochemical parameters of microalgae cultivation for biofuel production. *Critical Reviews in Biotechnology, 39*, 835–859.

Hossain, N., Mahlia, T. M. I., & Saidur, R. (2019). Latest development in microalgae-biofuel production with nano-additives. *Biotechnology for Biofuels, 12*, 125.

Hossain, N., Mahlia, T. M. I., Zaini, J., & Saidur, R. (2019). Techno-economics and sensitivity analysis of microalgae as commercial feedstock for bioethanol production. *Environmental Progress & Sustainable Energy, 38*, 13157.

Hossain, N., & Morni, N. A. H. (2019). Co-pelletization of microalgae-sewage sludge blend with sub-bituminous coal as solid fuel feedstock. *BioEnergy Research 13*, 618–629.

Hossain, N., Razali, A. N., Mahlia, T. M. I., Chowdhury, T., Chowdhury, H., Ong, H. C., et al. (2019). Experimental investigation, techno-economic analysis and environmental impact of bioethanol production from Banana stem. *Energies, 12*, 3947.

Hossain, N., Zaini, J., & Indra Mahlia, T. M. (2019). Life cycle assessment, energy balance and sensitivity analysis of bioethanol production from microalgae in a tropical country. *Renewable and Sustainable Energy Reviews, 115*, 109371.

Hossain, N., Zaini, J., Jalil, R., & Mahlia, T. M. I. (2018). The efficacy of the period of Saccharification on oil palm (Elaeis guineensis) trunk sap hydrolysis. *International Journal of Technology, 9*, 652–662.

Hossain, N., Zaini, J., & Mahlia, T. M. I. (2019). Experimental investigation of energy properties for Stigonematales sp. microalgae as potential biofuel feedstock. *International Journal of Sustainable Engineering, 12*, 123–130.

Hossain, N., Zaini, J., Mahlia, T. M. I., & Azad, A. K. (2019). Elemental, morphological and thermal analysis of mixed microalgae species from drain water. *Renewable Energy, 131*, 617–624.

Huang, Y., Zhang, D., Xue, S., Wang, M., & Cong, W. (2016). The potential of microalgae lipids for edible oil production. *Applied Biochemistry and Biotechnology, 180*, 438–451.

Iverson, S. J., Lang, S. L., & Cooper, M. H. (2001). Comparison of the Bligh and Dyer and Folch methods for total lipid determination in a broad range of marine tissue. *Lipids, 36*, 1283–1287.

Jeon, H.-S., Park, S. E., Ahn, B., & Kim, Y.-K. (2017). Enhancement of biodiesel production in Chlorella vulgaris cultivation using silica nanoparticles. *Biotechnology and Bioprocess Engineering, 22*, 136–141.

Kaneda, Y. (2016). *Why MC biotech chose Brunei* (2016 ed.). Brunei: The Worldfolio.

Kang, N. K., Lee, B., Choi, G.-G., Moon, M., Park, M. S., Lim, J., et al. (2014). Enhancing lipid productivity of Chlorella vulgaris using oxidative stress by TiO_2 nanoparticles. *Korean Journal of Chemical Engineering, 31*, 861–867.

Kebelmann, K., Hornung, A., Karsten, U., & Griffiths, G. (2013). Intermediate pyrolysis and product identification by TGA and Py-GC/MS of green microalgae and their extracted protein and lipid components. *Biomass and Bioenergy, 49*, 38–48.

Kim, Y.-H., Choi, Y.-K., Park, J., Lee, S., Yang, Y.-H., Kim, H. J., et al. (2012). Ionic liquid-mediated extraction of lipids from algal biomass. *Bioresource Technology, 109*, 312–315.

Kim, B., Praveenkumar, R., Lee, J., Nam, B., Kim, D. M., Lee, K., et al. (2016). Magnesium aminoclay enhances lipid production of mixotrophic chlorella sp. KR-1 while reducing bacterial populations. *Bioresource Technology*, *219*, 608–613.

Kim, J., Yoo, G., Lee, H., Lim, J., Kim, K., Kim, C. W., et al. (2013). Methods of downstream processing for the production of biodiesel from microalgae. *Biotechnology Advances*, *31*, 862–876.

Kita, K., Okada, S., Sekino, H., Imou, K., Yokoyama, S., & Amano, T. (2010). Thermal pre-treatment of wet microalgae harvest for efficient hydrocarbon recovery. *Applied Energy*, *87*, 2420–2423.

Klok, A. J., Lamers, P. P., Martens, D. E., Draaisma, R. B., & Wijffels, R. H. (2014). Edible oils from microalgae: Insights in TAG accumulation. *Trends in Biotechnology*, *32*, 521–528.

Kouzu, M., & Hidaka, J.-S. (2013). Purification to remove leached CaO catalyst from biodiesel with the help of cation-exchange resin. *Fuel*, *105*, 318–324.

Kusumo, F., Silitonga, A. S., Masjuki, H. H., Ong, H. C., Siswantoro, J., & Mahlia, T. M. I. (2017). Optimization of transesterification process for Ceiba pentandra oil: A comparative study between kernel-based extreme learning machine and artificial neural networks. *Energy*, *134*, 24–34.

Kusumo, F., Silitonga, A. S., Ong, H. C., Masjuki, H. H., & Mahlia, T. M. I. (2017). A comparative study of ultrasound and infrared transesterification of Sterculia foetida oil for biodiesel production. *Energy Sources, Part A: Recovery, Utilization, and Environmental Effects*, *39*, 1339–1346.

Lam, M. K., & Lee, K. T. (2012). Immobilization as a feasible method to simplify the separation of microalgae from water for biodiesel production. *Chemical Engineering Journal*, *191*, 263–268.

Lee, Y.-C., Huh, Y. S., Farooq, W., Chung, J., Han, J.-I., Shin, H.-J., et al. (2013). Lipid extractions from docosahexaenoic acid (DHA)-rich and oleaginous chlorella sp. biomasses by organic-nanoclays. *Bioresource Technology*, *137*, 74–81.

Lee, Y.-C., Jin, E., Jung, S. W., Kim, Y.-M., Chang, K. S., Yang, J.-W., et al. (2013). Utilizing the algicidal activity of aminoclay as a practical treatment for toxic red tides. *Scientific Reports*, *3*, 1292.

Lee, A. K., Lewis, D. M., & Ashman, P. J. (2012). Disruption of microalgal cells for the extraction of lipids for biofuels: Processes and specific energy requirements. *Biomass and Bioenergy*, *46*, 89–101.

Lee, J.-Y., Yoo, C., Jun, S.-Y., Ahn, C.-Y., & Oh, H.-M. (2010). Comparison of several methods for effective lipid extraction from microalgae. *Bioresource Technology*, *101*, S75–S77.

Li, F., Liang, Z., Zheng, X., Zhao, W., Wu, M., & Wang, Z. (2015). Toxicity of nano-TiO$_2$ on algae and the site of reactive oxygen species production. *Aquatic Toxicology*, *158*, 1–13.

López Granados, M., Alba-Rubio, A. C., Vila, F., Martín Alonso, D., & Mariscal, R. (2010). Surface chemical

promotion of ca oxide catalysts in biodiesel production reaction by the addition of monoglycerides, diglycerides and glycerol. *Journal of Catalysis*, *276*, 229–236.

Macías-Sánchez, M. D., Robles-Medina, A., Hita-Peña, E., Jiménez-Callejón, M. J., Estéban-Cerdán, L., González-Moreno, P. A., et al. (2015). Biodiesel production from wet microalgal biomass by direct transesterification. *Fuel*, *150*, 14–20.

Mahlia, T. M. I., Ismail, N., Hossain, N., Silitonga, A. S., & Shamsuddin, A. H. (2019). Palm oil and its wastes as bioenergy sources: A comprehensive review. *Environmental Science and Pollution Research International*.

Manjare, S. D., & Dhingra, K. (2019). Supercritical fluids in separation and purification: A review. *Materials Science for Energy Technologies*, *2*, 463–484.

Mann, S. (2009). Self-assembly and transformation of hybrid nano-objects and nanostructures under equilibrium and non-equilibrium conditions. *Nature Materials*, *8*, 781–792.

Mendes, R. L., Coelho, J. P., Fernandes, H. L., Marrucho, I. J., Cabral, J. M. S., Novais, J. M., et al. (1995). Applications of supercritical CO$_2$ extraction to microalgae and plants. *Journal of Chemical Technology & Biotechnology*, *62*, 53–59.

Menegazzo, M. L., & Fonseca, G. G. (2019). Biomass recovery and lipid extraction processes for microalgae biofuels production: A review. *Renewable and Sustainable Energy Reviews*, *107*, 87–107.

Metzler, D., Erdem, A., Tseng, Y., & Huang, C. P. (2012). Responses of algal cells to engineered nanoparticles measured as algal cell population, chlorophyll a, and lipid peroxidation: Effect of particle size and type. *Journal of Nanotechnology*, *2012*.

Milledge, J. J., & Heaven, S. (2013). A review of the harvesting of micro-algae for biofuel production. *Reviews in Environmental Science and Bio/Technology*, *12*, 165–178.

Mubarak, M., Shaija, A., & Suchithra, T. V. (2015). A review on the extraction of lipid from microalgae for biodiesel production. *Algal Research*, *7*, 117–123.

Nam, H., Kim, C., Capareda, S. C., & Adhikari, S. (2017). Catalytic upgrading of fractionated microalgae bio-oil (Nannochloropsis oculata) using a noble metal (Pd/C) catalyst. *Algal Research*, *24*, 188–198.

Nguyen, M. K., Moon, J.-Y., Bui, V. K. H., Oh, Y.-K., & Lee, Y.-C. (2019). Recent advanced applications of nanomaterials in microalgae biorefinery. *Algal Research*, *41*, 101522.

Nizamuddin, S., Baloch, H. A., Griffin, G. J., Mubarak, N. M., Bhutto, A. W., Abro, R., et al. (2017). An overview of effect of process parameters on hydrothermal carbonization of biomass. *Renewable and Sustainable Energy Reviews*, *73*, 1289–1299.

Nizamuddin, S., Jaya Kumar, N. S., Sahu, J. N., Ganesan, P., Mubarak, N. M., & Mazari, S. A. (2015). Synthesis and characterization of hydrochars produced by hydrothermal carbonization of oil palm shell. *The Canadian Journal of Chemical Engineering*, *93*, 1916–1921.

Nizamuddin, S., Siddiqui, M. T. H., Baloch, H. A., Mubarak, N. M., Griffin, G., Madapusi, S., et al. (2018). Upgradation of chemical, fuel, thermal, and structural properties of rice husk through microwave-assisted hydrothermal carbonization. *Environmental Science and Pollution Research, 25*, 17529–17539.

Nizamuddin, S., Siddiqui, M. T. H., Mubarak, N. M., Baloch, H. A., Mazari, S. A., Tunio, M. M., et al. (2018). Advanced nanomaterials synthesis from pyrolysis and hydrothermal carbonization: A review. *Current Organic Chemistry, 22*, 446–461.

Pádrová, K., Lukavský, J., Nedbalová, L., Čejková, A., Cajthaml, T., Sigler, K., et al. (2014). Trace concentrations of iron nanoparticles cause overproduction of biomass and lipids during cultivation of cyanobacteria and microalgae. *Journal of Applied Phycology, 27*.

Paiva Pinheiro Pires, A., Arauzo, J., Fonts, I., Dómine, M., Arroyo, A., García-Pérez, M., et al. (2019). Challenges and opportunities for bio-oil refining: A review. *Energy & Fuels, 33*, 4683–4720.

Panga, K., Saranga, V., Verma, K., Pooja, K., Dheeravath, D., Srilatha, K., et al. (2018). Bio oil production from microalgae via hydrothermal liquefaction technology under subcritical water conditions. *Journal of Microbiological Methods, 153*.

Parsons, J. G., & Keeney, P. G. (1969). Phospholipid concentration in cocoa butter and its relationship to viscosity in dark chocolate. *Journal of the American Oil Chemists' Society, 46*, 425–427.

Petkov, G., & Garcia, G. (2007). Which are fatty acids of the green alga chlorella? *Biochemical Systematics and Ecology, 35*, 281–285.

Phusunti, N., Phetwarotai, W., & Tekasakul, S. (2018). Effects of torrefaction on physical properties, chemical composition and reactivity of microalgae. *Korean Journal of Chemical Engineering, 35*, 503–510.

Pohndorf, R. S., Camara, Á. S., Larrosa, A. P. Q., Pinheiro, C. P., Strieder, M. M., & Pinto, L. A. A. (2016). Production of lipids from microalgae Spirulina sp.: Influence of drying, cell disruption and extraction methods. *Biomass and Bioenergy, 93*, 25–32.

Prabakaran, P., & Ravindran, A. D. (2011). A comparative study on effective cell disruption methods for lipid extraction from microalgae. *Letters in Applied Microbiology, 53*, 150–154.

Pragya, N., Pandey, K. K., & Sahoo, P. K. (2013). A review on harvesting, oil extraction and biofuels production technologies from microalgae. *Renewable and Sustainable Energy Reviews, 24*, 159–171.

Radzun, K. A., Wolf, J., Jakob, G., Zhang, E., Stephens, E., Ross, I., et al. (2015). Automated nutrient screening system enables high-throughput optimisation of microalgae production conditions. *Biotechnology for Biofuels, 8*, 65.

Raheem, A., Wan Azlina, W. A. K. G., Taufiq Yap, Y. H., Danquah, M. K., & Harun, R. (2015). Optimization of the microalgae Chlorella vulgaris for syngas production using central composite design. *RSC Advances, 5*, 71805–71815.

Ranjith Kumar, R., Hanumantha Rao, P., & Arumugam, M. (2015). Lipid extraction methods from microalgae: A comprehensive review. *Frontiers in Energy Research, 2*, .

Rawat, I., Ranjith Kumar, R., Mutanda, T., & Bux, F. (2011). Dual role of microalgae: Phycoremediation of domestic wastewater and biomass production for sustainable biofuels production. *Applied Energy, 88*, 3411–3424.

Rawat, I., Ranjith Kumar, R., Mutanda, T., & Bux, F. (2013). Biodiesel from microalgae: A critical evaluation from laboratory to large scale production. *Applied Energy, 103*, 444–467.

Rudic, V., Cepoi, L., Gutsul, T., Ludmila, R., Chiriac, T., Miscu, V., et al. (2012). Red algae Porphyridium cruentum growth stimulated by CdSe quantum dots covered with Thioglycerol. *Journal of Nanoelectronics and Optoelectronics, 7*, 681–687.

Rudic, V., Cepoi, L., Rudi, L., Chiriac, T., Nicorici, A., Todosiciuc, A., et al. (2011). Synthesis of CdSe nanoparticles and their effect on the antioxidant activity of Spirulina platensis and Porphyridium cruentum cells. In: *ICNBME-2011: International conference on nanotechnologies and biomedical engineering; German-Moldovan workshop on novel nanomaterials for electronic, photonic and biomedical applications proceedings*, Moldova: Republic of. Technical University of Moldova460.

Saleh, J., Dubé, M. A., & Tremblay, A. Y. (2010). Effect of soap, methanol, and water on glycerol particle size in biodiesel purification. *Energy & Fuels, 24*, 6179–6186.

Sarma, S. J., Das, R. K., Brar, S. K., Le Bihan, Y., Buelna, G., Verma, M., et al. (2014). Application of magnesium sulfate and its nanoparticles for enhanced lipid production by mixotrophic cultivation of algae using biodiesel waste. *Energy, 78*, 16–22.

Sathish, A., & Sims, R. C. (2012). Biodiesel from mixed culture algae via a wet lipid extraction procedure. *Bioresource Technology, 118*, 643–647.

Savage, P. E. (2009). A perspective on catalysis in sub- and supercritical water. *The Journal of Supercritical Fluids, 47*, 407–414.

Scholz, M. J., Weiss, T. L., Jinkerson, R. E., Jing, J., Roth, R., Goodenough, U., et al. (2014). Ultrastructure and composition of the Nannochloropsis gaditana cell wall. *Eukaryotic Cell, 13*, 1450–1464.

Shahbaz, K., Mjalli, F., Hashim, M., & Alnashef, I. (2010). Using deep eutectic solvents for the removal of glycerol from palm oil-based biodiesel. *Journal of Applied Sciences, 24*, 3349–3354.

Shahbaz, K., Mjalli, F. S., Hashim, M. A., & Alnashef, I. M. (2011). Eutectic solvents for the removal of residual palm oil-based biodiesel catalyst. *Separation and Purification Technology, 81*, 216–222.

Shen, P.-L., Wang, H.-T., Pan, Y.-F., Meng, Y.-Y., Wu, P.-C., & Xue, S. (2016). Identification of characteristic fatty acids to quantify triacylglycerols in microalgae. *Frontiers in Plant Science, 7.*

Silitonga, A. S., Masjuki, H. H., Ong, H. C., Sebayang, A. H., Dharma, S., Kusumo, F., et al. (2018). Evaluation of the engine performance and exhaust emissions of biodiesel-bioethanol-diesel blends using kernel-based extreme learning machine. *Energy, 159,* 1075–1087.

Solomon, B. D. (2010). Biofuels and sustainability. *Annals of the New York Academy of Sciences, 1185,* 119–134.

Suslick, K. S., & Flannigan, D. J. (2008). Inside a collapsing bubble: Sonoluminescence and the conditions during cavitation. *Annual Review of Physical Chemistry, 59,* 659–683.

Taher, H., Sulaiman, A.-Z., Al-Marzouqi, A., Haik, Y., & Farid, M. (2014). Mass transfer modeling of Scenedesmus sp. lipids extracted by supercritical CO_2. *Biomass and Bioenergy, 70.*

Tang, Y., Li, S., Qiao, J., Wang, H., & Li, L. (2013). Synergistic effects of nano-sized titanium dioxide and zinc on the photosynthetic capacity and survival of Anabaena sp. *International Journal of Molecular Sciences, 14,* 14395–14407.

Toor, S. S., Rosendahl, L., & Rudolf, A. (2011). Hydrothermal liquefaction of biomass: A review of subcritical water technologies. *Energy, 36,* 2328–2342.

Torri, C., Samorì, C., Adamiano, A., Fabbri, D., Faraloni, C., & Torzillo, G. (2011). Preliminary investigation on the production of fuels and bio-char from Chlamydomonas reinhardtii biomass residue after bio-hydrogen production. *Bioresource Technology, 102,* 8707–8713.

Tumuluru, J. S. (2016). Effect of deep drying and Torrefaction temperature on proximate, ultimate composition, and heating value of 2-mm Lodgepole pine (Pinus contorta) grind. *Bioengineering (Basel), 3.*

Ursu, A.-V., Marcati, A., Sayd, T., Sante-Lhoutellier, V., Djelveh, G., & Michaud, P. (2014). Extraction, fractionation and functional properties of proteins from the microalgae Chlorella vulgaris. *Bioresource Technology, 157,* 134–139.

Vaara, M. (1992). Agents that increase the permeability of the outer membrane. *Microbiological Reviews, 56,* 395–411.

Van Der Horst, D., & Vermeylen, S. (2011). Spatial scale and social impacts of biofuel production. *Biomass and Bioenergy, 35,* 2435–2443.

Veljković, V. B., Banković-Ilić, I. B., & Stamenković, O. S. (2015). Purification of crude biodiesel obtained by heterogeneously-catalyzed transesterification. *Renewable and Sustainable Energy Reviews, 49,* 500–516.

Walther, T. C., & Farese, R. V., Jr. (2009). The life of lipid droplets. *Biochimica et Biophysica Acta, 1791,* 459–466.

Wang, Y., Nie, J., Zhao, M., MA, S., Kuang, L., Han, X., et al. (2010). Production of biodiesel from waste cooking oil via a two-step catalyzed process and molecular distillation. *Energy & Fuels, 24,* 2104–2108.

Wang, Y., Wang, X., Liu, Y., Ou, S., Tan, Y., & Tang, S. (2009). Refining of biodiesel by ceramic membrane separation. *Fuel Processing Technology, 90,* 422–427.

Wang, Y., & Yang, K. (2013). Toxicity of single-walled carbon nanotubes on green microalga Chromochloris zofingiensis. *Chinese Journal of Oceanology and Limnology, 31,* 306–311.

Widianingsih, W., Hartati, R., Endrawati, H., & Mamuaja, J. (2013). Fatty acid composition of marine microalgae in Indonesia. *Research Journal of Soil Biology, 10,* 75–82.

Wu, C., Xiao, Y., Lin, W., Li, J., Zhang, S., Zhu, J., et al. (2017). Aqueous enzymatic process for cell wall degradation and lipid extraction from Nannochloropsis sp. *Bioresource Technology, 223,* 312–316.

Xie, Q., Addy, M., Liu, S., Zhang, B., Cheng, Y., Wan, Y., et al. (2015). Fast microwave-assisted catalytic co-pyrolysis of microalgae and scum for bio-oil production. *Fuel, 160,* 577–582.

Xue, Z., Wan, F., Yu, W., Liu, J., Zhang, Z., & Kou, X. (2018). Edible oil production from microalgae: A review. *European Journal of Lipid Science and Technology, 120,* 1700428.

Yang, F., Hanna, M. A., & Sun, R. (2012). Value-added uses for crude glycerol—A byproduct of biodiesel production. *Biotechnology for Biofuels, 5,* 13.

Yang, C., Li, R., Cui, C., Liu, S., Qiu, Q., Ding, Y., et al. (2016). Catalytic hydroprocessing of microalgae-derived biofuels: A review. *Green Chemistry, 18,* 3684–3699.

Young, G., Nippgen, F., Titterbrandt, S., & Cooney, M. J. (2010). Lipid extraction from biomass using co-solvent mixtures of ionic liquids and polar covalent molecules. *Separation and Purification Technology, 72,* 118–121.

Zhang, Y., Li, Y., Zhang, X., & Tan, T. (2015). Biodiesel production by direct transesterification of microalgal biomass with co-solvent. *Bioresource Technology, 196,* 712–715.

Zheng, H., Yin, J., Gao, Z., Huang, H., Ji, X., & Dou, C. (2011). Disruption of Chlorella vulgaris cells for the release of biodiesel-producing lipids: A comparison of grinding, ultrasonication, bead milling, enzymatic lysis, and microwaves. *Applied Biochemistry and Biotechnology, 164,* 1215–1224.

Zhu, H., Wu, Z., Chen, Y., Zhang, P., Duan, S., Liu, X., et al. (2006). Preparation of biodiesel catalyzed by solid super base of calcium oxide and its refining process. *Chinese Journal of Catalysis, 27,* 391–396.

Catalytic conversion of microalgae oil to green hydrocarbon

Min-Yee Choo[a,b], Yi Pin Phung[b], and Joon Ching Juan[b,c]

[a]Institute of Biological Sciences, Faculty of Science, University of Malaya, Kuala Lumpur, Malaysia
[b]Nanotechnology & Catalyst Research Centre (NANOCAT), University of Malaya, Kuala Lumpur,
Malaysia [c]School of Science, Monash University, Subang Jaya, Selangor, Malaysia

1 Introduction

1.1 Background

Over the past decades, an energy crisis has arose because of the excessive utilization of global oil reserves by the constantly growing world population and production of various materials such as fine chemicals, detergents, pharmaceuticals, plastics, synthetic fibers, fuel, pesticides, lubricants, solvents, fertilizers, waxes, coke, asphalt, etc. (Naik, Goud, Rout, & Dalai, 2010). The United States Environmental Protection Agency also known as USEPA has mentioned that the transportation sector alone has consumed almost 40% of the primary energy. Besides, it also contributed to nearly 71% of the greenhouse gas (GHG) emissions in 2010 and the value increased to 72% in 2017 (Agency, 2019; Mohan, Pittman, & Steele, 2006; Pham, Lee, & Kim, 2016). Currently, fossil fuels were estimated to contribute about 90% of the world's energy demand (Yen et al., 2012). Several worrying issues related to the high utilization of nonrenewable fuels such as environmental issues, fluctuation of petroleum fuel prices, and most importantly the deterioration of the health standards (Ho, Ye, Hasunuma, Chang, & Kondo, 2014).

Combustion of fossil fuels produces a high amount of carbon dioxide gasses which indirectly is the leading contributor to the thinning of the ozone layer that propagated global warming (Ahove & Bankole, 2018). With the exhaustion of fossil fuel reserves and the harmful effects of these GHG emissions to the environment has led to the quest to seek for an environmentally benign and sustainable resource, commonly known as the renewable resources. Recently, renewable hydrocarbon fuel demand has been growing rapidly and it is getting difficult to meet the demands of the power generation and transportation sector.

Furthermore, as predicted by the Global Energy Forum, fossil fuel was estimated to be exhausted in less than 100 years due to the rate of natural products is 5 magnitude lower than the consumption of said fuel (Banković-Ilić, Miladinović, Stamenković, & Veljković, 2017). Originally, researchers diverted out to renewable resources of energy such as solar, wind, and hydro energy but due to the high cost and the fact that solar, wind, and hydro energies are highly unpredictable due to direct influence by weather, these ideas were not suitable replacements for fossil fuel (Kaygusuz & Sekerci, 2016). Therefore, enormous efforts have been devoted in developing sustainable alternatives that can prevail over the growing need for a novel source of hydrocarbon, which is usable in various industrial applications to overcome the impending environmental destruction. One of the alternatives that show promise is the use of biomass for the production of biofuels. Biomass contains a high amount of carbohydrates that can be most certainly converted into carbon-based fuel. The fuel of choice and method of production are still uncertain. There were heated debates over whether the conversion of biomass should be thermochemical (as in using heat and some metal catalyst) or biological (which utilizes enzymes and microorganisms). Different processes will produce different types of fuel, for example, if the biomass were to undergo a thermochemical

process, it will produce synthetic fuel or also known as biofuel (Schmidt & Dauenhauer, 2007). There are several advantages of using biofuels compared to fossil fuels. These advantages include extremely low carbon dioxide emissions, biodegradability, nontoxicity, and sustainability (Sajjadi, Raman, Parthasarathy, & Shamshirband, 2016). In Table 1, the evolution of biofuel starting from the first generation to the fourth generation can be seen clearly. The first-generation biofuel was adapted using edible oil crops such as palm, soybean, corn, coconut, sunflower, rapeseed, etc., meanwhile, the fourth-generation biofuel was more focused with the development of engineered microalgae (Rahimi, Mohammadi, Basiri, Parsamoghadam, & Masahi, 2016; Zhang, Fang, & Wang, 2015).

Being from abundant oil crops, several benefits come from the first-generation biofuels. One such benefit is the better carbon dioxide emission control due to the plants' own ability to convert carbon dioxide that was produced into nutrients for their own growth via photosynthesis. It was expected that the carbon dioxide used for photosynthesis could offset the carbon dioxide amount produced from the combustion of biofuel (Naik et al., 2010). On the other hand,

TABLE 1 Biofuel generations and their pros and cons.

Biofuel generation	Feedstock	Advantages	Disadvantages
First	Rapeseed, corn, etc.	• Reduced GHG emissions • Simple and low cost	• Compatibility and storage issues • Competition with food production
Second	Waste cooking oil, lignocellulosic feedstock	• Utilize food and agricultural waste • No competition with food crop	• High processing cost • High advanced technology needed
Third	Microalgae	• Higher growth rate • High versatility	• Low lipid content • Contamination problem in open pond system
Fourth	Engineered microalgae	• High biomass and lipid productivity • High CO_2 sequestration	• High initial investment • Ongoing research

Reproduced from Choo, M.-Y., Oi, L.E., Show, P.L., Chang, J.-S., Ling, T.C., Ng, E.-P., et al. (2017). Recent progress in catalytic conversion of microalgae oil to green hydrocarbon: A review. Journal of the Taiwan Institute of Chemical Engineers, 79, 116–124.

several issues arise from the use of these first-generation crops as feedstock for biofuel production. The main issue that sparked numerous debates in recent years is the predominant impact of first-generation biofuel on the ecosystem. Scientists fear that the constant demand for more biofuel might lead to the devastation of the environment and cause an imbalance as immense areas of arable land are required for the planting and cultivating of such terrestrial oil crops (Parmar, Singh, Pandey, Gnansounou, & Madamwar, 2011).

Another drawback related to the first-generation biofuel is the engine compatibility and storage issues. These drawbacks were one of the major limitations to the application of the first-generation biofuel. In addition, the cost price of the feedstock (about 80% of the overall cost) has contributed significantly to the increase in the biodiesel price and vice-versa because the increase in first-generation biofuel production will also lead to the increase in the cost price of the feedstock (Rahimi et al., 2016). Due to this, a different approach was made which is to use various nonfood feedstock to produce biofuel. Numerous nonfood based feedstock such as waste cooking oils (WCO) and agricultural wastes was being tested and were successfully converted into second-generation biofuels which are bioethanol or biodiesel (Sahar et al., 2018). As usual, there were some debates with the use of such waste as feedstock. The main factor that leads to the reconsideration of these feedstocks is the requirement to segregate the wastes as well as separating the unusable wastes and the raw materials from the feedstock. The segregation of these materials takes too much time and requires a lot of effort and manpower, therefore, causing the secondary-generation biofuel to be limited and made it difficult to meet the global demand (Li et al., 2016; Li, Cheng, Huang, Zhou, & Cen, 2015; Liu & Lien, 2016; Ullah, Bustam, & Man, 2015).

After much consideration, an alternative was determined which is the use of microalgae as a feedstock for the production of bio-fuel. Microalgae was chosen as one of the most reassuring substitutes (known as the third-generation feedstock) due to their tremendously rapid growth rate, and photosynthetic ability. Microalgae do not possess a vascular system for the transportation of nutrient, unlike higher plants which utilizes this method of nutrient transfer. Every microalgae cell is photoautotrophic which means that each cell is capable of synthesizing its own nutrient using inorganic sources in the presence of light. The most crucial winning point is that it does not require much arable land. With only simple environmental conditions, the microalgae can survive and reproduce, these microalgae also use up the carbon dioxide for photosynthesis which indirectly offsets the carbon dioxide emission which is produced by the fuel (Chen et al., 2013). In comparison to many other carbon capture and storage (CCS) methods, the carbon dioxide fixation using microalgae shows many benefits such as low cost of operation, good environmental adaptability, a rapid growth rate of microalgae due to good supply of carbon dioxide, and high photosynthesis rate due to the increase in microalgae amount which was caused by the rapid growth rate. It was also found that the photosynthesis rate of microalgae is approximately 50 times higher than terrestrial plants (Suali & Sarbatly, 2012). Furthermore, chemical compounds found in microalgae can be assimilated into cosmetics as a source of valuable molecules known as polyunsaturated fatty acid oils which can also be incorporated into nutritional supplements and baby formulas (Spolaore, Joannis-Cassan, Duran, & Isambert, 2006). However, even with all the benefits that come from microalgae, but the high growth rate is often accompanied by low lipid content and vice versa. Hence, engineered microalgae with both high growth rate and lipid content have become the fourth-generation biofuel feedstock (Tan, Show, Chang, Ling, & Lan, 2015).

The aim of the fourth-generation biofuel was to induce to increase the lipid contents of microalgae in order to grasp the full potential of microalgae not only in the biofuel industry but as well as other industries that rely on high unsaturated fatty acid oils. Therefore, these unique features of microalgae led it to be the future source of sustainable nonedible oil which will be transmuted into green hydrocarbons.

1.1.1 Advantages and disadvantages

Microalgae was introduced as a promising source for the production of green hydrocarbons. The potential stems from their high lipid content (fourth generation biofuel—engineered microalgae) and immense growth rate. Certain species of microalgae contain up to 60% of lipid content by their dry cell weight and an average of 35 wt% (Feng, Deng, Fan, & Hu, 2012; Ho, Chang, Lai, & Chen, 2014; Karpagam, Preeti, Ashokkumar, & Varalakshmi, 2015). Lipid content and lipid productivity of the microalgae are influenced by the species and phenotypic factors, for example, culture conditions (e.g., adding adequate nitrate content into their cultivating environment) (Peng, Yao, Zhao, & Lercher, 2012; Zhao, Brück, & Lercher, 2013). The yield of terrestrial oil crops such as corn, coconut, and palm are on average below 1000 gallons per acre. On the other hand, the oil yield from microalgae (~5000 gallons per acre) is almost 5 times more than those of terrestrial crops (Galadima & Muraza, 2014). Judging from the yield and space factors, microalgae shows better prospect as it can generate more yield within the same amount of space given as compared to terrestrial oil crops. In addition, microalgae biomass production can be accelerated with the use of industry flue gas and wastewater to supply carbon dioxide and nitrogen, respectively (Cheah et al., 2016; Kuo et al., 2015; Marjakangas et al., 2015; Sutherland, Howard-Williams, Turnbull, Broady, & Craggs, 2015). These approaches not only aid in the cultivation of microalgae but also help to reduce the waste generated being released into the environment, which indirectly leads to the lowering of environmental pollution. By channeling the carbon dioxide from flue gas into the cultivation chamber of the microalgae, it can help reduce the risk of carbon dioxide being released into the atmosphere, which causes a reduction of GHG emission (Clarens, Resurreccion, White, & Colosi, 2010). By average, the consumption of CO_2 to microalgae growth is in a weight ratio of 1.83:1 (Jiang et al., 2013). The rate of photosynthesis of these microalgae can be more than 6.9×10^4 cells/mL/h. The average estimation was done based on the *Chlorella vulgaris* sp. *cell* number of 5.7×10^7 cells/mL in a controlled media for 34 days (Scragg, Illman, Carden, & Shales, 2002). Nascimento et al. (2015) demonstrated that 2.52 tonnes of carbon dioxide per ton of microalgae can be fixed by the *Botryococcus terribilis* sp. Microalgae biomass is an ideal oil feedstock for green hydrocarbon production due to the ability to produce high lipid yield, to capture a high amount of carbon dioxide, to achieve high biomass, to grow on wastewater, and the ease of cultivation on nonarable space.

Different strains of microalgae affect the chemical composition as well as the cultivation conditions required. Commonly, the composition of microalgae cells comprises 20%–40% of lipids, 0%–20% of carbohydrates, 0%–5% of nucleic acids, and 30%–50% of proteins (Zhao et al., 2013). Focusing on the lipid content, it can be further classified into two major groups, such as structural lipids (also known as polar lipids) and storage lipids (known as neutral lipids). The core function of the structural lipids is to control the permeability of the membrane. The storage lipids are made up of triglycerides (a glycerol molecule that has three fatty acid chains coupled onto it). The most common fatty acids that can be found in microalgae cells are vaccenic acids, palmitic acids, linoleic acids, palmitoleic acids, and stearic acids (Table 2).

TABLE 2 Fatty acid composition of microalgae oil and microalgae species.

Fatty acid	Microalgae oil[a]	Microalgae oil[b]	Chlamydomonas planctogloea	Chlorella vulgaris	Parachlorella kessleri	Marine Chlorella sp.	Nannochloropsis sp.	Nitzschia cf. ovalis	Scenedesmus obliquus	Tetraseknus sp.
Saturated fatty acids (SFA)										
$C_{12:0}$[c]	0.32	—	—	—	—	0.68	0.46	—	5.32	—
$C_{14:0}$	4.59	0.04	3.5	6.5	6.6	3.49	2.89	2.67	3.45	1.31
$C_{16:0}$	22.34	4.41	22.4	38.2	14.7	72.04	70.18	13.25	25.12	12.56
$C_{17:0}$	—	—	0.7	—	0.3	0.68	0.74	—	4.85	—
$C_{18:0}$	0.67	4.41	1.1	0.9	3.8	13.39	13.91	0.45	16.58	0.45
$C_{20:0}$	—	0.43	0.1	1.2	0.3	0.82	0.69	—	1.85	—
$C_{22:0}$	—	0.44	—	—	—	—	—	—	—	—
$C_{24:0}$	—	0.36	—	—	—	—	—	—	—	—
Monounsaturated fatty acids (MUFA)										
$C_{16:1}$	21.88	—	11.5	1.2	0.3	3.24	2.62	17.12	—	31.47
$C_{18:1}$	—	32.2	22.0.3	0.7	44	2.56	4.69	0.59	34.44	0.41
$C_{22:1}$	—	0.97	—	—	—	—	—	—	—	—
Polyunsaturated fatty acids (PUFA)										
$C_{16:2}$	1.13	—	2.4	—	1.0	—	—	2.45	—	7.45
$C_{16:3}$	—	—	2.5	—	0.6	—	—	7.20	—	2.84
$C_{18:2}$	14.79	56.2	5.4	5.4	13.3	2.50	2.69	0.43	5.68	2.0
$C_{18:3}$	3.94	—	19.7	44.1	11.4	0.56	1.02	0.37	2.44	1.10
$C_{20:4}$	—	0.07	—	—	—	—	—	4.40	—	0.12
$C_{20:5}$	30.32	—	—	—	—	—	—	26.67	—	16.65

Continued

TABLE 2 Fatty acid composition of microalgae oil and microalgae species—cont'd

Fatty acid	Microalgae oil[a]	Microalgae oil[b]	Chlamydomonas planctogloea	Chlorella vulgaris	Parachlorella kessleri	Marine Chlorella sp.	Nannochloropsis sp.	Nitzschia cf. ovalis	Scenedesmus obliquus	Tetraseknus sp.
$C_{22:4}$	–	0.19	–	–	–	–	–	–	–	–
$C_{22:6}$	–	0.13	–	–	–	–	–	4.20	–	1.33
Ref	Kandel, Anderegg, Nelson, Chaudhary, and Slowing (2014)	Peng, Yao, et al. (2012), Peng, Yuan, Zhao, and Lercher (2012), and Song, Zhao, and Lercher (2013)	Soares, da Costa, Vieira, and Antoniosi Filho (2019)	Abedini Najafabadi, Malekzadeh, Jalilian, Vossoughi, and Pazuki (2015)	Soares et al. (2019)	Cheirsilp and Torpee (2012)	Cheirsilp and Torpee (2012)	Patil, Källqvist, Olsen, Vogt, and Gislerød (2007)	Abd El Baky, El-Baroty, Bouaid, Martinez, and Aracil (2012)	Patil et al. (2007)

[a] Supplied by Solix Biofuels, Inc.
[b] Supplied by Verfahrenstechnik Schwedt GmbH.
[c] The nomenclature shows the number of carbon atoms and the number of unsaturations in the chains: for example, the present sample contained 12C atoms and no double bonds.

These fatty acids have a carbon chain length of about 12–18 which made them fairly suitable for the production of green hydrocarbon. As shown in Table 3, *Tribonema minus* sp., *Desmodesmus* sp., *Chlorella protothecoides* sp., and *Scenedesmus* sp. have high lipid content of about 47.4%–64.1% and fairly high lipid productivity which is about 224.1–384.7 mg/L/day. However, the lipid content and lipid productivity may vary as the microalgae species mentioned are cultured under different cultivation conditions.

The extraction of these microalgae oil can be done chemically or mechanically. This chemical extraction process often involves the use of solvents such as hexanes or alcohols to extract the organic oil from microalgae. According to literature, the most efficient extraction method is called the supercritical fluid extraction technique (Couto et al., 2010). There are several other techniques such as ultrasonic extraction, expeller presses, Soxhlet extractors, and electromechanical methods. The extraction purpose is to obtain oil from the algae cells in order to convert them into biofuel or other agricultural products via thermochemical or biochemical means (Demirbaş, 2008). The conventional method of extraction involves the dewatering process before extracting as moisture level in microalgae

TABLE 3 The lipid content of microalgae.

Microalgae species	Lipid content (% dry weight)	References
Botryococcus braunii	25–75	Nautiyal, Subramanian, and Dastidar (2014)
Chlorella protothecoides	55.8	Li et al. (2014)
Crypthecodinium cohnii	20	Swaaf et al. (2001)
Cylindrotheca sp.	16–37	Cruz et al. (2018)
Desmodesmus sp.	64.1	Ho, Chang, et al. (2014)
Dunaliella primolecta	23	Cruz et al. (2018)
Isochrysis sp.	25–33	Liu and Lin (2001)
Monallanthus salina	>20	Weldy and Huesemann (2007)
Monoraphidium sp.	47.4	Zhao et al. (2016)
Nannochloris sp.	>20	Andruleviciute, Makareviciene, Skorupskaite, and Gumbyte (2014)
Nannochloropsis sp.	37–60	Ma, Chen, Yang, Liu, and Chen (2016)
Neochloris oleoabundans	35–54	Tornabene, Holzer, Lien, and Burris (1983)
Nitzschia sp.	45–47	Renaud, Zhou, Parry, Thinh, and Woo (1995)
Phaeodactylum tricornutum	20–30	Haro, Sáez, and Gómez (2017)
Scenedesmus sp.	47.4	Ren et al. (2014)
Schizochytrium sp.	50–77	Nautiyal et al. (2014)
Tetraselmis sueica	33.72	Huang, Wei, Huang, and Yan (2014)
Tribonema minus	61.8	Hui, Wenjun, Wentao, Lili, and Tianzhong (2016)

biomass may up to 78.4%. It was reported that the microalgae biomass can be directly converted into fuel oil via thermochemical liquefaction at 300–360°C and 10 MPa. However, the yield is much lower compared to those extracted from the dewatering method (Ikenaga, Ueda, Matsui, Ohtsuki, & Suzuki, 2001). The dewatering process is normally conducted using an expeller or by using a press. Heating or steaming under high pressure before transferring the biomass into a mechanical press to extract the oil. The pressing method is best suited for microalgae oil extraction because the micron-sized microalgae make it difficult to apply the expeller method. Another upside of this method is the fact that it doesn't require any chemicals to be added and yet it can yield 70%–75% of oil (Shuping et al., 2010). As for supercritical-fluid extraction, the use of carbon dioxide as a solvent to extract the oil allows the oil to be free from contamination by toxic solvents and thermal degradation. The oil extracted is of high quality however, this method takes time thus leading it to be less efficient for commercial or large-scale production.

The conversion of microalgae into biofuel is broken down into two processes, namely biochemical conversion process and thermochemical conversion process. The biochemical conversion process refers to the transesterification and fermentation methods, which produce biodiesel and ethanol as main products, respectively. Thermochemical processes can be further categorized as liquefaction, hydrogenation, gasification, and pyrolysis. All these methods produce bio-oil fuel as their main product, whereas hydrogenation is used to improve the biofuel or feedstock property and gasification is used to produce syngas. The main issue that arises from the conversion process is that the downstream management of the formation of tar, especially during gasification and pyrolysis methods. The tar produced can lead to possible polymerization into complex structures that are unfavorable (Ni, Leung, Leung, & Sumathy,

2006). Microalgae is well known for the production of biodiesel. This product stems from the transesterification of glycerol as a sideproduct. Based on reports, 73.4% of lipid content accumulates in microalgae, majority of the lipid component exists as triglycerides. Therefore, biodiesel can be produced from these lipid content via transesterification. The process takes place in a reactor where blended methanol and catalyst were added with the microalgae triglycerides.

Another famous conversion method is the pyrolysis process. Pyrolysis is an anaerobic heating procedure that occurs at high temperatures between 200°C and 750°C and does not involve oxidation. There are two main categories of pyrolysis which are fast and slow pyrolysis. Fast pyrolysis results in the production of bio-oil and biochar. Slow pyrolysis on the other hand yields pyrolysis gas (comprising of methane and carbon dioxide) and biochar. The rate of pyrolysis of microalgae biomass can achieve up to 87% of the conversion rate at 300°C to about 600°C. Another benefit of this process is that the pyrolysis of microalgae produces bio-oil that is more stable compared to bio-oil produced from the pyrolysis of other crops. Catalytic conversion of microalgae oil into biofuel is also possible with the help of certain metal oxide catalyst using the deoxygenation (DO) process. This conversion shows promise as the yield of the conversion is high and the use of catalyst further promotes better conversion rate (Chisti, 2008).

Previously, biofuel production was mostly made from the extraction of lipids from the targeted microalgae which is then accompanied by the transesterification process to produce fatty acid methyl ester (FAME) (Chen et al., 2015; Yen, Yang, Chen, Jesisca, & Chang, 2015). The extraction method that was applied only works on microalgae strains with high lipid content. Even though the transesterification process was a properly developed method of conversion, the converted FAME contained a huge amount of oxygenates that will deteriorate the value of the fuel produced. Several drawbacks were suffered from FAME such as the low heating value, high oil acidity, and high viscosity, which prevents it to be used directly as a drop-in fuel (Bautista, Vicente, & Garre, 2012). As an alternative, bio-oils can be manufactured from microalgae with low lipid content through two thermochemical conversion methods known as pyrolysis and hydrothermal liquefaction (Bai, Duan, Xu, Zhang, & Savage, 2014). These two methods are widely used due to its cost-effectiveness. On the contrary, the disadvantage of this route of conversion is the low selectivity because unwanted products mostly oxygenate will be produced. Hence, an upgrade is necessary for this process to remove all the unwanted oxygenates from the final product. After the removal of these, oxygenates from biofuels hydrocarbon fuels will be obtained which resembled petroleum-derived fuels that are in the range of diesel, jet fuels, and gasoline which are commonly known as green-diesel, green jet fuel, and green gasoline, respectively.

There are many benefits of using microalgae as an alternative renewable resource for the production of biofuel. Unlike fossil fuels, microalgae are renewable, they grow rapidly under proper conditions. Certain strains of microalgae can produce a high amount of oil and grow easily can produce a sufficient amount of biodiesel. According to an article published by GTM research in 2011, it is expected that these biodiesels produced by the microalgae can replace about 17% of oil imported for the transportation sector (Drevense, 2011). The cultivation of microalgae can also reduce the space necessary for the cultivation compared to terrestrial plants such as corn and palm oil. Another plus point is that microalgae is not a major food source thus this does not lead to a reduction of the cultivated food source. Although there are many benefits that microalgae can offer, there are still some complications that come with such cultivations. The first complication that arise with such cultivation is the cost of cultivation. In order to cultivate specific species or high lipid content

microalgae, the cost required to prepare the suitable cultivation as well as proper infrastructure is extremely high (Acién, Fernández, Magán, & Molina, 2012). Second, the biofuel produced is yet to be compatible with a lot of the current transportation sector machinery. The most successful fuel produced thus far is the cellulose ethanol (a kind of biofuel), which requires expensive enzymes to break down the cellulose into cellulose ethanol. In a study conducted by Dan Edmunds and Philip Reed, the fuel efficiency of cellulose ethanol is lower than that of gasoline (13.5 and 18.3 mpg, respectively) (Markings, 2017).

Other than just the incompatibility, there were several other aspects that have to be reviewed as well. One such aspect is the fact that most pipelines which are built are incompatible with the biofuel produced due to the biofuels affinity to water and corrosion. To modify the entire pipeline would incur another cost, which is extremely expensive. This would also lead to a higher cost of transports through the use of railways and tankers. In terms of storage and shelf life, the normal gasoline blends that don't possess ethanol can be stored for many years without the fear of contamination. Biofuel, on the other hand, are hygroscopic, which makes them absorb 50 times more water than conventional fuels more readily. This resulted in reduced shelf life for ethanol fuels (Hassan & Kalam, 2013).

1.2 Catalyst and catalysis

In order to understand the DO process of the microalgae, one must first understand the catalysis process that is taking place. Catalysis is known as the process of adding a substance known as a catalyst to increase the rate of a chemical reaction (Vignarooban et al., 2015). The catalyst is not consumed in the process and can be used repeatedly. Due to the regenerative property of the catalyst, only a small amount of it is needed to completely alter the reaction rate. Generally, a reaction has its very own reaction time for the complete chemical reaction to take place. In order to overcome the barrier of time, the presence of catalyst provides an alternative reaction pathway with lower activation energy compared to the noncatalyzed mechanism to produce the same end product. During the reaction, the catalyst will react to form an intermediate which will then regenerate the original catalyst in a cyclic process. As for the original reactions (the reaction pathway before the use of catalyst), it will maintain the same and still produce end products. If the substance added is used up during the reaction, then it is not considered as a catalyst. The substance is known as a reaction inhibitor as it is actually a limiting factor (Laidler & Meiser, 2006).

Catalysts can be classified as three sorts of catalysts, such as the homogeneous catalyst, heterogeneous catalyst, and biocatalyst. Homogeneous catalysts are catalysts that are in the same chemical phase as the reactant. For example, a liquid catalyst being used in a liquid phase reaction. A heterogeneous catalyst is a catalyst that exists as a different phase as compared to the reactant's chemical phase. For example, a solid catalyst being used in a liquid phase reaction (Hävecker et al., 2012). During a reaction, catalysis takes place to reduce the requirement of activation energy to achieve the respective transition state. The total free energy of the reactant to products doesn't change. Some catalysts can propagate several chemical transformations. In the field of chemistry, the activation energy is defined as the energy that has to be supplied for a certain reaction in order to convert into the end product. The activation energy (E_a) of a reaction is denoted in kilocalories per mole (kcal/mol) or joules (J) or kilojoules per mole (kJ/mol).

According to the Arrhenius equation, the formula for the calculation of activation energy is,

$$k = Ae^{(-E_a/RT)}$$

where,

k = reaction rate coefficient
A = preexponential factor of the reaction

E_a = activation energy
R = universal gas constant
T = absolute temperature (unit: kelvins).

With this equation, an estimation of the activation energy can be calculated to determine the energy required for the reaction that's taking place.

$$\ln k = -E_a/RT \times \ln A$$

Based on the equation it can be seen that temperature plays a major role in affecting the reaction forward towards the production of the end product. The higher the temperature, the higher the rate of reaction. Based on the theory, reactions occur when particles collide. If a substance is heated up, the particles move faster. As the temperature increases, the particles are provided with a sufficient amount of energy to agitate the particles to move about. This allows the molecules to collide more frequently thus enabling the reaction to occur at a higher rate.

1.2.1 Types of catalysts

Generally, catalysts react with one or more reactants and turn into intermediates which will subsequently produce the final reaction product (Bautista et al., 2012). During the process, the catalyst will regenerate itself and continue to produce even more intermediates and final products. By labeling the catalyst as "c," the reaction mechanism is as shown below,

$$B + c \rightarrow Ac$$
$$A + Bc \rightarrow ABc$$
$$ABc \rightarrow cD$$
$$cD \rightarrow c + D$$

Even though the catalyst was consumed at some points of the reaction, the catalyst will subsequently be reproduced at the end of the reaction and can be reused to continue the catalytic reaction.

There are three kinds of catalysts, namely homogeneous catalysts, heterogeneous catalysts, and last but not least the biocatalyst. The types of catalyst to be chosen for a reaction depends greatly on different factors such as ease of catalyst removal, types of desired intermediates and end products, rate of reaction, removal of specific trace elements, and so on. During the selection of catalysts, one must consider multiple factors as mentioned above. For example, if the reaction that was to take place produces liquid products, it would be recommended to use solid catalysts to ease in the separation of a catalyst via filtration or centrifugation from the reaction medium. The heterogeneous catalysts will be preferred in this situation.

Heterogeneous catalysts work in a different phase as compared to the reactants. Most heterogeneous catalysts comprised of solid catalysts that work on liquid or gaseous reaction mixtures. Adsorption plays a vital role in this type of catalytic reaction. The effectiveness depends greatly on the total surface area of the solid. The smaller the particle size of the catalyst, the more surface area it possesses. Active sites on the catalyst in which the reaction takes place are exposed to the reactants (Kishida, Hanaoka, Hayashi, Tashiro, & Wakabayashi, 1998). This allows the reactants to adsorb onto the surface and undergo chemisorption which will result in the dissociation into adsorbed atomic species and form new bonds with atoms within close proximity. An example of such a reaction is the use of vanadium (V) oxide for the production of sulfuric acid. A supported heterogeneous catalyst is catalysts that have been dispersed on a secondary material that further enhances the effectiveness of the catalyst and minimizes the cost. These support not only reduce and sometimes prevent agglomeration, but they also aid in increasing the surface area of the catalysts. Some supports are merely binding agents to hold the catalysts in place, increasing their surface area and exposing their active sites for the desired particles. Porous catalysts have increased surface area as well. Some of the well-known porous catalytic supports are alumina and various kinds of activated carbons (Wu et al., 2019).

Homogeneous catalysts are catalysts that function in the same phase as the reactants. The mechanism of this catalytic conversion is similar to those of heterogeneous catalytic reactions. A homogeneous catalyst is dissolved into a solvent with the substrates. An example of such reaction is the formation of alkyl aldehydes in which carbon monoxide was added to an alkene product. This process is also known as hydroformylation. Another example is the use of organic molecules that exhibit catalytic property such as organometallic compounds to catalyze reactions (Cornils & Herrmann, 2003).

Biocatalyst is mostly enzymes or protein-based catalysts used in metabolism and catabolism. Some nonprotein-based biomolecules can also exhibit catalytic properties such as synthetic deoxyribozymes. Biocatalyst can be classified as the intermediate between homogeneous and heterogeneous catalysts. In biocatalyst, the parameters dictating the activity of the catalysts are pH, the concentration of substrate, the concentration of enzyme, and temperature. In this catalytic reaction, enzymes are used to prepare commodity chemicals such as acrylamide and high-fructose corn syrup.

Catalysts play a significant role in the major productions of chemical products. Due to the reduction of time and cost, catalysts were employed in many industries such as energy processing, bulk chemical production, food industry, and more. The use of catalysts significantly improves the production rate of the items but it also has its impact on the environment. One noticeable catalytic reaction that affects the environment is the role of free radical chlorine, which breaks down the ozone layer. The free radicals were formed from the reaction of ultraviolet radiation on chlorofluorocarbons (also known as CFCs). Catalysts can be used for the conversion of microalgae oil into biofuel. The conversion rate can be enhanced thus producing more biofuel with less time consumption. The use of the catalyst in biofuel production has been heavily studied and heterogeneous catalysts were selected for the conversion. Most of these metallic catalysts aids in increasing the conversion rate and produce biofuel of a specific carbon number. In order to understand the mechanism behind the conversion of microalgae into biofuel via catalytic conversion, the catalytic reaction plays a major role in providing insight to understand the logic behind the mechanism.

1.3 Catalytic deoxygenation

1.3.1 Introduction

In order to convert microalgae oil into green hydrocarbon, catalytic DO can be utilized. The reaction pathway that was proposed for the conversion of microalgae oil into hydrocarbons and the thermodynamic data are all shown in Table 4. Several catalytic reactions can take place to convert the microalgae into biofuel. Some such ways are the use of the DO method, namely hydrodeoxygenation (HDO), decarboxylation/decarbonylation (DCO). The above-mentioned

routes are the majority of the conversion routes that microalgae oil undergoes to convert into hydrocarbons in liquid form. The principal reactions which take place in the vapor phase are water gas shift (WGS) and methanation. DO is the chemical reaction that involves the removal of oxygen atoms from the starting molecule. This is also used to refer to the process to remove oxygen from gases and solvents. The DO process for microalgae conversion is typically related to the breaking of the hydrocarbon chain. The hydrocarbon chain is being broken and the oxygen is being removed in the form of CO_2/CO via DCO reactions. The green diesel that is produced presented similar properties to conventional petroleum-derived fuels. DO is also an economical process due to the fact that some methods of DO do not require the use of hydrogen gas (H_2) for example the DCO process. Many well-established refineries use a DO process known as HDO. HDO is the process which involved an exothermic reaction that removes oxygen in the form of water (H_2O) in the

TABLE 4   Thermodynamic data of individual steps in deoxygenation of fatty acid in microalgae oil.

Liquid phase	Reaction	ΔG°_{573K} (kJ/mol)
In the presence of hydrogen		
Hydrodeoxygenation	$R-COOH + 3H_2 \rightarrow R-CH_3 + 2H_2O$	−86.1
Decarbonylation	$R-COOH + H_2 \rightarrow R-H + CO + H_2O$	−67.6
Can occur with or without the presence of hydrogen		
Decarbonylation	$R-COOH \rightarrow R'-H + CO + H_2O$	−17.0
Decarboxylation	$R-COOH \rightarrow R-H + CO_2$	−83.5
Gas phase	**Reaction**	ΔG°_{573K} **(kJ/mol)**
Water gas shift	$CO + H_2O \rightarrow CO_2 + H_2$	−17.6
Methanation	$CO + 3H_2 \rightarrow CH_4 + H_2O$	−78.8
Methanation	$CO_2 + 4H_2 \rightarrow CH_4 + 2H_2O$	−61.2

R, saturated carbon chain; *R'*, unsaturated carbon chain.
Modified from Choo, M.-Y., Oi, L.E., Show, P.L., Chang, J.-S., Ling, T.C., Ng, E.-P., et al. (2017). Recent progress in catalytic conversion of microalgae oil to green hydrocarbon: A review. Journal of the Taiwan Institute of Chemical Engineers, 79, 116–124.

presence of hydrogen gas (H_2). HDO is almost the same as the existing hydrotreating technology that was utilized by the petroleum refinery sector. This process allows the utilization of currently preexisting petroleum refinery infrastructures. This is actually an additional advantage of this HDO technology because less cost is incurred due to the compatibility of the technique with the existing infrastructures (Ayodele, Farouk, Mohammed, Uemura, & Daud, 2015; Kumar, Yenumala, Maity, & Shee, 2014). It is a feasible technique in the removal of oxygen from the carbon chain but due to the requirement of hydrogen gas, it is actually fairly costly and dangerous. HDO process takes place in the form of,

$$C_{17}H_{35}COOH + 3H_2 \rightarrow C_{18}H_{38} + 2H_2O$$

in which the reaction removes oxygen in the form of water.

DCO remove the oxygen in the form of carbon dioxide and carbon monoxide, respectively. An example of such a process is,

$$C_{17}H_{35}COOH \rightarrow C_{17}H_{36} + CO_2 \qquad (1)$$

$$C_{17}H_{35}COOH + H_2 \rightarrow C_{17}H_{36} + CO + H_2O \qquad (2)$$

Please refer to Table 4 for the simplified chemical equation.

Where process reaction (1) represents the decarboxylation process and process reaction (2) represents the decarbonylation process. Decarboxylation is an exothermic reaction whereas decarbonylation is described to be an endothermic reaction (Jęczmionek & Porzycka-Semczuk, 2014). In some cases, due to the water gas shift activity of the catalyst, it is kind of difficult to establish whether are the observed carbon monoxide and carbon dioxide were produced by decarboxylation. There is some decarbonylation routes that can produce similar end products which were proposed in the literature (Donnis, Egeberg, Blom, & Knudsen, 2009).

There are some differences between the HDO process and the DCO process. One such example is the difference in the carbon number of the final outcome. In the HDO process, oxygen is removed as water and the end product has the same number of carbons as the feedstock. DCO, on the other hand, produces subsequent hydrocarbons (paraffin and olefin) with one less carbon number than the feedstock that was used. The HDO process normally involves the removal of oxygen atom by reacting with the triglycerides and free fatty acids (FFAs) under the constant supply of hydrogen gas thus producing *n*-paraffin and water. DCO, on the other hand, removes oxygen in the form of carbon dioxide or carbon monoxide with little to no hydrogen gas supply during the process. The only drawback from DCO is the fact that the end product will lose one carbon atom hence bringing down the energy value of the produced hydrocarbon by a little. The carbon dioxide and carbon monoxide produced during DCO have a tendency of being adsorbed onto the active sites of the catalyst which leads to the catalyst being deactivated or poisoned (Mäki-Arvela, Snåre, Eränen, Myllyoja, & Murzin, 2008).

Theoretically, the hydrocarbon yield of a complete reaction of HDO and DCO pathway is approximately 85% and 80%, respectively (Zhou & Lawal, 2015). The two reaction pathways might even simultaneously during the reaction and the favored reaction route can then be determined by measuring the ratio of HDO/DCO. For example, if the ratio of HDO/DCO is 3, this means that the HDO route is three times more preferred as compared to the DCO route. HDO and DCO will both produce biofuels that possess hydrocarbon of high purity which does share a lot of similarities in terms of properties with the existing petroleum fuels. Higher selectivity of target hydrocarbons, minimal or no hydrogen is required, and lower temperature requirement which is around 250–300°C are all advantages of the DCO process. Hydrogen gas requirement and consumption for each route can be ranked as HDO > decarbonylation > decarboxylation routes (Srifa, Faungnawakij,

Itthibenchapong, & Assabumrungrat, 2015). This technology is able to be deployed on a smaller magnitude due to the lesser hydrogen gas requirement and consumption. Additionally, microalgae do not need to be transported to a centralized facility for processing. Therefore, the catalytic DO process can achieve a much more cost-effective and energy-efficient outlook for the conversion of microalgae oil into green hydrocarbons.

1.3.2 Reaction pathway

A general idea of the reaction pathway for the conversion of microalgae oil into hydrocarbon can be seen clearly in Fig. 1. The reaction pathway starts with the metal-catalyzed induced hydrogenation to eliminate the double bonds found in the microalgae oil to produce saturated triglycerides. This process then proceeds with the hydrogenolysis through β-elimination. After triglyceride eliminates one of its fatty acids, the

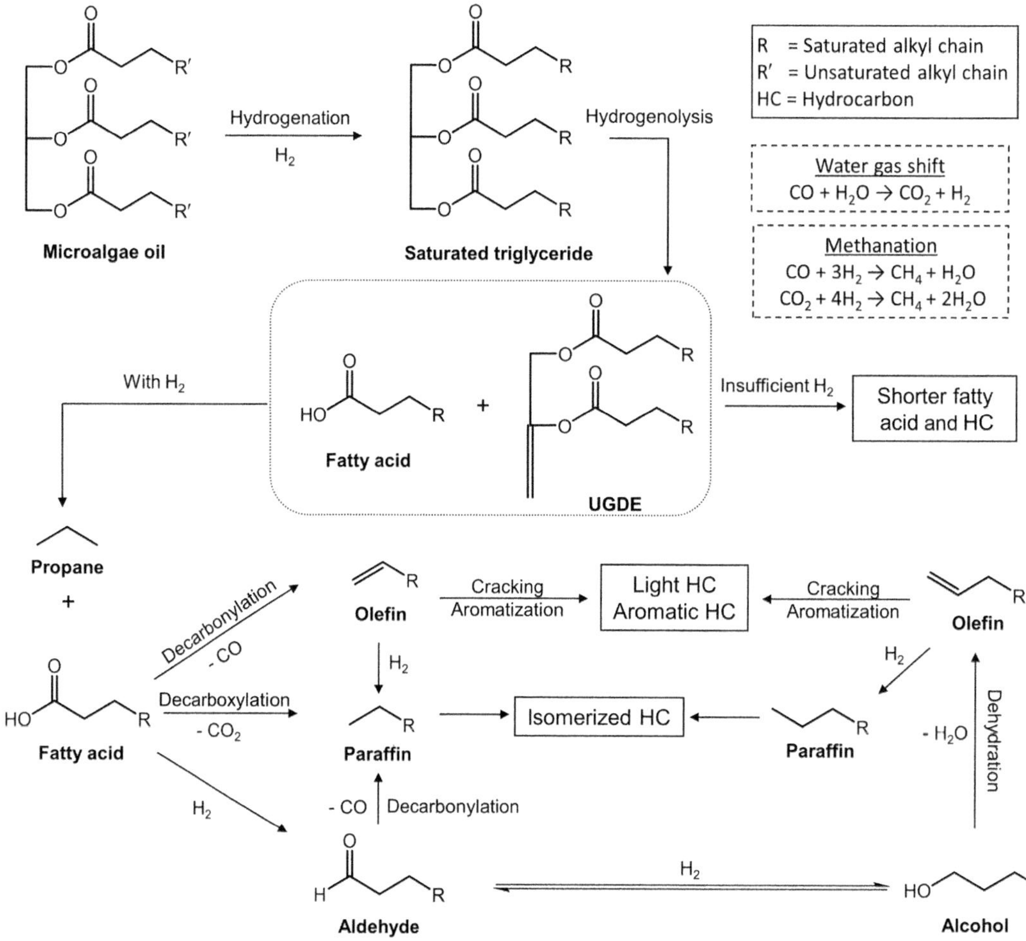

FIG. 1   Reaction pathway in catalytic deoxygenation of microalgae oil. *From Choo, M.-Y., Oi, L.E., Show, P.L., Chang, J.-S., Ling, T.C., Ng, E.-P., et al. (2017). Recent progress in catalytic conversion of microalgae oil to green hydrocarbon: A review.* Journal of the Taiwan Institute of Chemical Engineers, 79, 116–124.

unsaturated glycol di-fatty esters (UGDE) were produced. Subsequently, the β-elimination of unsaturated glycol di-fatty acid ester will not happen because of the lack of hydrogen in the system. At high temperature, the cleaving or cracking of the formed unsaturated glycol di-fatty acid ester to form shorter fatty acids and alkanes will occur. During the presence of hydrogen gas or by using hydrotreating catalysts, β-elimination can occur, thus producing fatty acids and propane. Peng et al. clearly illustrated that the fatty acids act as the intermediate in the DO of microalgae (Peng, Yao, et al., 2012). It was determined by the author that the yield of stearic acid reaches almost 80 wt% after close to 60 min. The yield of stearic acid will then slowly decrease as the hydrocarbon yield will increase with time. Without the presence of constant hydrogen gas supply, the fatty acid produced will either undergo decarboxylation to form carbon dioxide and paraffin or the fatty acids will undergo decarbonylation to produce carbon monoxide and olefins. Whereas in the presence of constant hydrogen gas supply, the hydrogenation of these carboxylic groups, for example, fatty acid, tends to occur. This will then lead to the production of aldehydes. Aldehydes will then go through further decarbonylation to produce carbon dioxide and paraffin. It is also possible to take on the hydrogenation route to produce alcohol. The alcohol can partake in the metal-catalyzed hydrogenation and acid-catalyzed dehydration, which produces paraffin as the end products. The decarbonylation of fatty acids and the dehydration of alcohol produced olefins which will then engage in the cracking and cyclization process with the accompaniment of acid support to produce lighter hydrocarbons and cycloalkanes. This reaction happened due to the C=C double bond is not as stable thus are more susceptible to the cracking process compared to singly bonded C—C. On another note, the adsorption of C=C double bond on the active sites of the catalysts might contribute to the formation of coke, which

will lead to the production of undesired by-products such as aromatic compounds (Gosselink et al., 2013). Nevertheless, this disadvantage is not as extreme in the hydrogen gas-filled condition. Mostly due to the hydrogenation that took place on the unsaturated compounds that converted them into the saturated compounds which are not as susceptible to cracking or cleaving. Paraffins, on the other hand, might engage in the isomerization process to produce isomerized hydrocarbons. This is actually beneficial in the process of aviation fuel production. During the water gas shift (WGS) reaction, carbon monoxide formed via the decarbonylation process might react with water to form carbon dioxide gas and hydrogen gas. Moreover, the carbon dioxide and carbon monoxide from the DCO process can react with the hydrogen gas to form methane via the methanation process.

2 Catalytic deoxygenation of microalgae oil, DO

DO of microalgae oil is a process of oxygen removal from the hydrocarbon chain as can be seen in Table 4. This process involves several routes, each has its unique chemical reaction which will eventually produce the same final product which is a hydrocarbon chain with no oxygen atom with an exception of DCO route which will produce a hydrocarbon chain with one carbon less compared to the precursor feedstock. Their pathway all derived from the DO process and are different methods of DO via different reaction conditions with the aid of catalysts. The DO process that was discussed consists of HDO, and DCO. The pathway shown in Fig. 1 summarized the reaction pathway of all the processes. Microalgae oil which is the main biomass stock feed is made up of unsaturated fatty acids which differ in percentage by weight for different species of microalgae. The lipid content in % dry weight is shown in Table 2.

Based solely on lipid content, microalgae show high lipid content which is beneficial for the conversion of the lipid contents into saturated hydrocarbons. Due to the fact that microalgae contain several types of fatty acids in its lipid composition as shown in Table 2, the catalytic DO process aims to produce only saturate hydrocarbons. This means that the unsaturated fatty acids have to be converted into saturated fatty acids. The main reason why the conversion is important is that double bonds are not favorable as the final product of DO process because it is prone to unwanted oxidative damage (Evans, 2012). By using catalysts such as precious-metal-based catalysts, nonprecious metal-based catalysts, catalysts with mesoporous structures, and etc., allows the microalgae oil to undergo a higher rate of conversion into biofuel. The catalyst that was used in the DO process can be varied for the enhancement of the conversion process.

Precious metal-based catalysts, for instance, can be used in the conversion of microalgae oil into biofuel. Precious metals are naturally occurring rare metal elements with high economic value. Examples of precious metals are gold, silver, platinum, palladium, and etc. The precious metals tend to be more chemically inert and not as reactive as compared to most elements and they have high luster and are mostly ductile (Na et al., 2012).

In chemistry, some precious metals are also known as noble metals such as silver, platinum, and gold. These noble metals are most well known for their resistance to corrosion and oxidation in air with high moisture content. The noble metals are highly sought after for their applications in ornamentation, metallurgy as well as their use in frontier technology. Noble metals are also known to be very expensive due to their rarity in the earth's crust. The use of noble metals in the production of biofuel has been studied and was determined that noble metals were able to produce high yields oh hydrocarbon chains.

The majority of the fatty acids found in microalgae oil are stearic acid, palmitic acid, linoleic acid, vaccenic acid, and palmitoleic acid. The supported noble metal catalysts are extremely active in the DO of these fatty acids (Snåre, Kubičková, Mäki-Arvela, Eränen, & Murzin, 2006). In one of the studies conducted by Deutschmann et al., it was actually determined that by using just 5 wt% of Pd/C catalyst, it was able to produce 100% C17 hydrocarbons from the C18 stearic acid (Deutschmann, Knözinger, Kochloefl, & Turek, 2009). Researchers found that 5 wt% of Pt/C was the most suitable catalyst for the DO of oleic acid (Table 5, under DCO only category entry no. 2) (Evans, 2012). The large surface area of activated carbon made it achieve better dispersion compared to silica support with the same amount of loading. Another study that uses multiwall carbon nanotubes (MWCNT) with about 5 wt% platinum supported, discovered that the catalyst support does not contribute to the conversion of stearic acid. Both the catalysts used which are 5 wt% Pt/MWCNT and 5 wt% Pt/C showed a similar conversion rate (about 52.4%). However, the selectivity towards *n*-heptadecane by Pt/MWCNT (around 97.0%) was far better than those of Pt/C (around 57.0%) (Yang, Nie, Fu, Hou, & Lu, 2013). The results concluded that Pt/MWCNT was actually capable of decarboxylation of different fatty acids found in microalgae oil.

Nonprecious metals or more widely known as base-metals are any nonferrous (contains no iron elements) metals which are neither precious metals nor noble metals. The most common nonprecious metals are nickel, aluminum, zinc, tin, lead, and copper. They're more common than precious and noble metals and are more readily extracted. Pure base metals tend to oxidize fairly easily except for copper. Nonprecious metals are used in a wide variety of applications. Copper is commonly used in electrical wiring due to its high conductivity (Elkonin & Sokolowski, 1991) and ductile strength. Lead on the other hand is a reliable source for batteries. Nickel is

TABLE 5 Shows the different routes of catalytic conversion for crude microalgae oil, microalgae-based bio oils, and free fatty acids (FFAs).

No.	Process and catalysts	Feedstock	T (°C)	P (bar)	t (h)	Conversion (%)	Selectivity (%)			Reference
							HDO	DCO	Isomerisation	
HDO and DCO										
1	10 wt% Ni/HBEA (Si/Al=180)	Crude microalgae oil	250	40	8	68	54	5.1	3.8	Peng, Yao, et al. (2012)
	10 wt% Ni/HBEA (Si/Al=180)	Crude microalgae oil	260	15	8	55	23	17	11	Peng, Yao, et al. (2012)
	10 wt% Ni/HBEA (Si/Al=180)	Crude microalgae oil	260	40	8	75	54	8.9	6.3	Peng, Yao, et al. (2012)
	10 wt% Ni/HBEA (Si/Al=180)	Crude microalgae oil	260	60	8	78	61	7.1	3.9	Peng, Yao, et al. (2012)
	10 wt% Ni/HBEA (Si/Al=180)	Crude microalgae oil	270	40	8	78	48	13	11	Peng, Yao, et al. (2012)
2	Fe-MSN	Crude microalgae oil	290	30	6	67	Ratio (6 HDO:1 DCO)		–	Kandel et al. (2014)
3	S-Ni-Mo/γ-Al$_2$O$_3$	Nannochloropsis salina	360	34.47	–	98.7	Ratio (6.4 HDO:1 DCO)		–	Zhou and Lawal (2015)
4	10 wt% Ni/HZSM-5 90	Stearic acid	260	–	6	100	41	9.2	0.4	Li et al. (2015)
	10 wt% Ni/HZSM-5240	Stearic acid	260	–	6	65	67	8.8	–	Li, Zhang, et al. (2015)
	10 wt% Ni/HZSM-5400	Stearic acid	260	–	6	60	80	6.1	–	Li, Zhang, et al. (2015)
DCO only										
1	5 wt% Pt/MWCNT	Stearic acid	330	–	0.5	52.4	–	97	–	Lestari et al. (2009)
2	5 wt% Pt/C	Stearic acid	330	–	0.5	52.4	–	57	–	Lestari et al. (2009)
3	5 wt% Ru/C	Stearic acid	330	–	0.5	42.3	–	42.9	–	Lestari et al. (2009)
4	10 wt% Ni/ZrO$_2$	Microalgae oil	260	6	8	–	–	26	–	Peng, Yuan, et al. (2012)

Continued

TABLE 5 Shows the different routes of catalytic conversion for crude microalgae oil, microalgae-based bio oils, and free fatty acids (FFAs)—cont'd

No.	Process and catalysts	Feedstock	T (°C)	*P* (bar)	*t* (h)	Conversion (%)	Selectivity (%) HDO	DCO	Isomerisation	Reference
	10 wt% Ni/ZrO_2	Microalgae oil	260	40	8	–	–	66	–	Peng, Yuan, et al. (2012)
	10 wt% Ni/ZrO_2	Microalgae oil	260	70	8	–	–	59	–	Peng, Yuan, et al. (2012)
	10 wt% Ni/ZrO_2	Microalgae oil	270	40	8	76	–	68	–	Peng, Yuan, et al. (2012)
5	Ni-Al LDH	*Scenedesmus* sp. algal lipid	260	40	1	65	–	Majority	–	Santillan-Jimenez, Morgan, Loe, and Crocker (2015)
	Ni-Al LDH	*Scenedesmus* sp. algal lipid	260	40	4	55	–	Majority	–	Santillan-Jimenez et al. (2015)
6	5% Pd/SiO_2	Stearic acid	300	–	6	89	–	100	–	Ping, Wallace, Pierson, Fuller, and Jones (2010)
7	Pd/SBA-15	Stearic acid	300	–	6	100	–	100	–	Ping et al. (2010)
8	5% Pd/C	Stearic acid	300	–	6	100	–	100	–	Maier, Roth, Thies, and Schleyer (1982)

often used to harden and strengthen metal alloys (Na et al., 2012).

These nonprecious metals are preferable to be adapted as catalysts as compared to noble metals because of their cheaper cost and abundance. Nevertheless, the catalytic performance of both precious and nonprecious metals was still barely comprehended. Thus, Snåre et al. (2006) have researched a series of metals and their DO properties on fatty acids. The authors came to a conclusion stating that the DO activity of the metals is arranged in a specific order which is, $Pd > Pt > Ni > Rh > Rh > Ir > Ru > Os$. Even though Pd and Pt are more intrinsically active, the result proved that Ni is a more promising alternative as a substitute for noble metals. Based on Peng, Yuan, et al. (2012) findings, the microalgae oil can be converted into alkanes at the conditions of 270°C and 40 bar hydrogen gas with 10 wt % Ni/ZrO_2 (can be seen in Table 5, under DCO only, entry no. 4). A total of 76% yield of liquid alkanes and the selectivity of C17 hydrocarbon (which is about 66%) were obtained after a 4 h reaction. This finding actually explains that DCO process is being favored by group 10 metal catalysts. The Ni/ZrO_2 catalyst has higher stability and it was able to maintain its catalytic activity even after 72 h. Based on the examples aforementioned above, this goes to show that catalytic DO of microalgae is capable of producing high yield products but the system behind the conversion has to be planned properly as the pathway of the conversion is fairly specific.

2.1 Hydrodeoxygenation process

HDO process is one of the methods of catalytic conversion of microalgae oil into biofuel. The fundamental requirement of this method of conversion is the need to constantly supply hydrogen gas at a fixed pressure in order to direct the reaction pathway to remove oxygen in the form of water. There are several factors that affect the reaction pathway. In order for the pathway to be selected catalyst loading plays

a major role. For example, in a study conducted it was determined that the selectivity towards the HDO process will decrease by almost half when Si/Al ratio decreases from 200 to 45 (Peng, Yao, et al., 2012). This is mostly due to the density of the acid sites on the catalyst. Reportedly, the HDO/cracking ratio of HBEA with the Si/Al catalyst of ratio 45 and 180, achieved the ratio of 2.49 and 8.57, respectively. This proves that the cracking process is more prone to occur in HBEA with a low Si/Al ratio. It was found that when nickel loading increases, the acid site density of HBEA will reduce but the selectivity of HDO products will increase.

In order to solve the diffusional issue, an idea of using mesoporous material arises. The idea was to convert microalgae oil using mesoporous zeolites also known as hierarchical zeolite (pore size between 2 and 50 nm) with the characteristic micropores of zeolite still well intact. According to the International Union of Pure and Applied Chemistry (IUPAC) nomenclature, materials containing pores are separated into different categories depending on the pore size. Materials having pores larger than 50 nm are known as microporous materials, those within the range of 2–50 nm are known as mesoporous materials, last but not least, those with pore sizes smaller than 2 nm are known as microporous materials. Examples of mesoporous materials are alumina, zirconium, titanium, and silica (McCusker, 2005).

Zeolite has been determined to be an effective catalyst in the conversion of microalgae oil due to the presence of Brönsted acid sites even though the presence of micropores has to lead it to face some blunders. This was most probably due to the bigger reactants were unable to diffuse freely into and out of the micropore of zeolite. As for smaller molecules, the diffusion is also limited which greatly deteriorates the catalytic performance. Hence the idea of using hierarchical zeolite that offers mesopore to enhance the diffusional of bulky molecules while preserving the zeolitic characteristics. Hierarchical zeolites work as solid acid support which has the tendency to produce HDO products due to the fast dehydration step of alcohol (Kumar et al., 2014). Hierarchical zeolites have another advantage which is that it possesses stronger acid sites (due to the lower Si/Al ratio) compared to other ordered mesoporous materials such as MCM-41, HMS and more. An example is when Nickel-Molybdenum was supported onto HZSM-5, it was able to convert 98% of the microalgae oil and 78.5% of the yield was of jet fuel that possessed an average i/n ratio of 2.5 (Verma, Kumar, Rana, & Sinha, 2011).

An extensive study on the different preparation methodologies of Ni/HBEA and its effect on the conversion of microalgae oil using the HDO reaction pathway was done by Song et al. (2013). In another study, they discovered that the conversion of microalgae oil into biofuels using Ni/HBEA catalyst and by using the impregnation method could only achieve 16% conversion rate whereas by using another 3 different methods which are Nanoparticles (NP), Ion-exchanged/Precipitation (IP), and Deposition/Precipitation (DP) were all able to convert 61% of the stearic acid into C18 hydrocarbons. The 3 methods mentioned are all methods of nickel dispersion. The ion-exchange/Precipitation (IP) method allows the zeolite to exchange its ions with nickel. This allows good metal dispersion but the method is limited by the zeolite's exchange capacity. The Deposition/Precipitation method makes use of the basification method on the nickel salt suspension. The basification agent that was used is urea. Lastly, the Nanoparticle method uses a two-step synthesis that prepares the monodispersed nickel nanoparticles separately which are then grafted onto a support by using mechanical mixing (Song et al., 2013). Based on the results obtained from the methods mentioned, this goes to show that smaller and highly dispersed nickel nanoparticles are able to catalyze several different steps in the HDO reaction process which includes the breaking of the double bonds, hydrogenolysis of triglyceride,

hydrogenation of fatty acids, and decarbonylation of aldehydes. On the contrary, the C18/C17 ratio generated by this catalyst remained in the range of 1.57–1.83 which is actually much lower than that of the impregnated catalyst (which is at the ratio of 2.96).

The HDO process is actually a fairly promising method of conversion for microalgae oil and the catalysts required for this reaction process are also readily available and can be manufactured fairly easily. This pathway also guarantees a product with a similar carbon number as the feedstock which makes it easier to control. The only problem arises from the use of the HDO process is the hazards that come with the system set up. Hydrogen gas can be dangerous when exposed to heat as it can lead to an explosion.

2.2 Decarboxylation and decarbonylation process

In these processes, each of the processes produces different end products. DCO process will be able to produce a hydrocarbon of the one less carbon number as its feedstock. Both processes remove oxygen in the form of carbon monoxides or carbon dioxide. This route is actually temperature-dependent. In order to obtain a high yield from the DCO process, it is often required to hike the process temperatures to extreme degrees (Ding et al., 2015). Undesirable side reactions such as aromatization, cracking, and dehydrogenation normally occurs at high temperatures (Santillan-Jimenez et al., 2015). As the temperature is increased from 250°C to 270°C, the DCO product yield, as well as isomerized products, will increase from 5.1% and 3.8% to 13% and 11%, respectively (Table 5, under HDO & DCO, entry no. 1). With higher activation energy requirement and longer contact time of intermediates of decarbonylation, the temperature has to be hiked up in order for the DCO route to proceed (Peng, Yao, et al., 2012).

DCO can proceed with or without the flow of hydrogen gas. In Peng's study, the effect of hydrogen gas on the conversion rate of microalgae oil via the DCO route was studied (Peng, Yuan, et al., 2012). The results obtained showed that at a higher pressure of hydrogen gas, the more final products are formed from the DCO process. The DCO process was able to increase its yield significantly from 26 wt% to 66 wt% as the hydrogen pressure increases from 6 to 40 bar. As the hydrogen pressure increases up to 90 bar, the n-heptadecane yield drops from 66 wt% to 59 wt%. This indicates that the decarbonylation reaction is actually negated at high hydrogen gas pressure (Table 5, under DCO only, entry no. 4). It was determined that the highest selectivity for DCO products can be achieved by peaking the pressure of hydrogen gas at 40 bar. Previous literature has shown similar observations in which DCO reaction route is still more favorable at low hydrogen gas pressure whereas high hydrogen gas pressure actually favors HDO process routes (Ding et al., 2015). Another type of catalyst which is Ni-Al LDH shows the good conversion rate of microalgae lipids where the majority of the DO process favors the DCO route. The problem faced by this metal catalyst is when the contact time increases from 1 to 4 h, the catalyst's stability drops which leads it to reduce in its conversion rate (from 65% conversion rate at 1 h to 55% conversion rate at 4 h). This can be seen in Table 5 under DCO only, entry no. 5 (Choo et al., 2017).

Generally, DCO route is more prone in group 10 element catalysts such as platinum, nickel, palladium, and etc. These catalysts can aid the DCO process to yield high amounts of C_{n-1} hydrocarbons (Gosselink et al., 2013; Na et al., 2012; Peng, Yuan, et al., 2012). However, there is an exception of Nickel in which the Ni/HBEA catalyst favors the HDO route more due to the presence of Brönsted acid sites as discussed in the segment above (Song et al., 2013; Zhao et al., 2013).

In some cases, the conversion route of microalgae oil will undergo both HDO and DCO routes simultaneously. This can be seen in the study conducted by Zhou et al., where he used sulfided Ni-Mo/γ-Al$_2$O$_3$ to almost completely convert Nannochloropsis salina into biofuel (a high conversion rate of 98.2%). This process is named hydrotreating in which sulfided catalysts are preferred as it provides good conversion rate with a route selectivity of ratio 6 HDO:1 DCO. A drawback of using these kinds of catalysts for the process is the deactivation of the catalyst caused by the reverse Mars van Krevelen mechanism and contamination of water.

2.3 Deactivation of catalyst

Despite the fact that the catalytic reaction greatly aids in the conversion of microalgae biomass into biofuel, the catalytic reaction also faces some significant drawbacks from the constant requirements of catalyst replacements. One of the problems that arose from the usage of catalysts in the DO process is the deactivation of the catalyst that is being used. This brings about the reduction of effectiveness (the loss of catalytic activity) of the future catalytic reactions. Examples of major causes of HDO deactivation process are carbon deposition, solid-state transformation, sintering, coke formation, and poisoning (Chen et al., 2001). Coke formation, on the other hand, is mostly caused by the polycondensation as well as polymerization process reactions. It was determined that furans and phenols are the predominant precursors for the formation of coke due to their strong interaction with the surface of the catalyst (Furimsky & Massoth, 1999).

In the study conducted by González-Borja and Resasco, it was reported that the deactivation of the catalyst was more severe in compounds that contain two oxygen atoms compared to compounds that only has a singular oxygen atom. The report shows that the catalyst used to convert guaiacol via the HDO process shows more coke formation and reduction of catalytic quality as compared to catalysts used to convert anisole (González Borja & Resasco, 2011; Popov et al., 2010). Another factor that affects the formation of coke is the catalyst property. An example of such property is the acidity of the catalyst. This can be seen via the study that was reported by Bui et al. According to their findings, it was determined that during the process of HDO of guaiacol in the presence of CoMoS catalyst, the use of alumina support resulted in the formation of heavy by-products which caused the catalyst to be deactivated whereas catalysts with lower acidity produced lighter by-products (Bui, Laurenti, Delichère, & Geantet, 2011). Other than the catalyst, the condition in which the reaction takes place (such as hydrogen gas pressure and temperature) plays a major role in controlling the amount of coke formation. For example, the temperature was reported by Li et al. to be a prominent factor in determining the severity of coke formation. As the temperature increases, coke formation increases despite the amelioration of the HDO process (Li, Cheng, et al., 2015).

According to Zanuttini et al., carbon deposition favors the Brønsted acid site compared to the lewis acid sites. This was reported in their study using Pt/Al_2O_3 and Pt/SiO_2 catalysts for the HDO reaction of m-cresol. The results showed a higher carbon deposition rate in the Silica catalyst as compared to the Alumina catalyst. This goes to show that Brønsted and Lewis acid sites play a major role in the carbon deposition selectivity. As for their contribution to coke formation is debatable (Zanuttini, Dalla Costa, Querini, & Peralta, 2014).

3 Conclusion and future prospect

In conclusion, microalgae are a promising feedstock for green hydrocarbon production due to its high lipid content, its lack of competition from global food production, pairing up with a high sequestration ability towards carbon dioxide. In order to effectively utilize microalgae, the catalytic DO process has proven to be one of the more effective alternatives to remove the unwanted oxygenated compounds from the microalgae feedstocks. With the removal of oxygenates via HDO, DCO process, accompanied by several catalytic processes (mostly isomerization and cracking), green hydrocarbon can be produced to serve multiple purposes (green jet fuel, green diesel, and green gasoline). The catalytic DO of microalgae oil employed the supported metal catalysts method. These supports vary from reducible oxide (Zirconium Oxide), Solid acid support (Hierarchical zeolite, HBEA, HZSM-5), refractory oxide (silicon oxide, aluminum oxide) to neutral support (activated carbon, CNT). Different metal sites have their own advantages and disadvantages, for example, precious metal such as platinum and palladium suffered from cost-effectiveness, sulfided metals such as cobalt and molybdenum suffers from catalytic deactivation, and low-cost base metals such as nickel and iron suffer from lower activity compared to noble metals.

The use of the catalyst in the production of green hydrocarbon has proven to be efficient and the catalyst selection has to be specific for different processes. In the production of green diesel (C14–C18 hydrocarbon) which is obtainable through the DO process by using the catalyst such as Ni-Al LDH and Ni/ZrO_2 are capable of producing diesel with the hydrocarbon range of C15–C18 hydrocarbons. The cracking of the products into shorter chains of hydrocarbon which are actually beneficial for green gasoline was made possible with the help of the brönsted acid site on Fe-MSN and Ni/HBEA. Moreover, the use of hierarchical zeolite such as the mesoporous ZSM-5 which combined the capability of both microporous zeolite and mesoporous support to perform isomerization, DO, and cracking simultaneously really helped in the process of producing green jet fuel (C9–C15 hydrocarbon with high i/n ratio). Instead of the long winding process of DO, cracking, and isomerization, respectively, this catalyst is able to combine all the steps mentioned into one single step reaction which indirectly helps to reduce the time taken for the green jet fuel production.

It is crucial to understand the fundamental chemistry and mechanism behind the catalytic DO process in order to extract its benefits and utilize it in the field of production. To be able to grasp the physiochemical properties of catalyst (acidity, catalyst dispersion, metal particle size, stability, etc.), the impact of the reaction conditions (such as temperature, pressure, catalyst loading, etc.) and the kinetics, as well as reaction mechanism, adds value to the production of green hydrocarbon. With the understanding of these aspects, a design for the production can be tailor-made to meet the requirements as well as to produce an efficient, highly integrated and cost-effective process for the conversion of biofuel from microalgae. In order to propagate the use of microalgae as an alternative renewable resource to substitute fossil fuel, multiple factors need to be factored in to weigh out the benefits as well as the problems that might arise from the substitution. With the understanding of the different pros and cons of using microalgae as feedstock, this allows researchers to grasp the current condition and overcome any possible complications that might arise. Therefore, continuous effort should be deployed to expand upstream and downstream technology in order to accelerate the commercialization of green hydrocarbon from microalgae oil in hopes that 1-day microalgae can be the solution to the depletion problem of fossil fuels.

Acknowledgment

The study was supported by the Fundamental Research Grant Scheme (FP029-2017A), SATU Joint Research Scheme (ST009-2017), Research University Grants (RU018D-2016), and IPPP Postgraduate Research Grant (PG013-2015B).

References

Abd El Baky, H. H., El-Baroty, G. S., Bouaid, A., Martinez, M., & Aracil, J. (2012). Enhancement of lipid accumulation in Scenedesmus obliquus by optimizing CO_2 and Fe^{3+} levels for biodiesel production. *Bioresource Technology*, *119*, 429–432.

Abedini Najafabadi, H., Malekzadeh, M., Jalilian, F., Vossoughi, M., & Pazuki, G. (2015). Effect of various carbon sources on biomass and lipid production of Chlorella vulgaris during nutrient sufficient and nitrogen starvation conditions. *Bioresource Technology*, *180*, 311–317.

Acién, F. G., Fernández, J. M., Magán, J. J., & Molina, E. (2012). Production cost of a real microalgae production plant and strategies to reduce it. *Biotechnology Advances*, *30*(6), 1344–1353.

Agency, E. E. (2019). Greenhouse gas emissions from transport in Europe: EEA. [updated 17 December 2019]. Available from:(2019). https://www.eea.europa.eu/data-and-maps/indicators/transport-emissions-of-greenhouse-gases/transport-emissions-of-greenhouse-gases-12.

Ahove, M. A., & Bankole, S. I. (2018). Petroleum industry activities and climate change: Global to national perspective. In P. E. Ndimele (Ed.), *The political ecology of oil and gas activities in the nigerian aquatic ecosystem* (pp. 277–292). Academic Press (chapter 18).

Andruleviciute, V., Makareviciene, V., Skorupskaite, V., & Gumbyte, M. (2014). Biomass and oil content of Chlorella sp., Haematococcus sp., Nannochloris sp. and Scenedesmus sp. under mixotrophic growth conditions in the presence of technical glycerol. *Journal of Applied Phycology*, *26*(1), 83–90.

Ayodele, O. B., Farouk, H. U., Mohammed, J., Uemura, Y., & Daud, W. M. A. W. (2015). Hydrodeoxygenation of oleic acid into n- and iso-paraffin biofuel using zeolite supported fluoro-oxalate modified molybdenum catalyst: Kinetics study. *Journal of the Taiwan Institute of Chemical Engineers*, *50*, 142–152.

Bai, X., Duan, P., Xu, Y., Zhang, A., & Savage, P. E. (2014). Hydrothermal catalytic processing of pretreated algal oil: A catalyst screening study. *Fuel*, *120*, 141–149.

Banković-Ilić, I. B., Miladinović, M. R., Stamenković, O. S., & Veljković, V. B. (2017). Application of nano CaO–based catalysts in biodiesel synthesis. *Renewable and Sustainable Energy Reviews*, *72*, 746–760.

Bautista, L. F., Vicente, G., & Garre, V. (2012). Biodiesel from microbial oil. (chapter 8)In R. Luque, & J. A. Melero (Eds.), *Advances in biodiesel production* (pp. 179–203). Woodhead Publishing.

Bui, V., Laurenti, D., Delichère, P., & Geantet, C. (2011). Hydrodeoxygenation of guaiacol: Part II: Support effect for CoMoS catalysts on HDO activity and selectivity. *Applied Catalysis B: Environmental*, *101*, 246–255.

Cheah, W. Y., Ling, T. C., Juan, J. C., Lee, D.-J., Chang, J.-S., & Show, P. L. (2016). Biorefineries of carbon dioxide: From carbon capture and storage (CCS) to bioenergies production. *Bioresource Technology*, *215*, 346–356.

Cheirsilp, B., & Torpee, S. (2012). Enhanced growth and lipid production of microalgae under mixotrophic culture condition: Effect of light intensity, glucose concentration and fed-batch cultivation. *Bioresource Technology*, *110*, 510–516.

Chen, C.-L., Huang, C.-C., Ho, K.-C., Hsiao, P.-X., Wu, M.-S., & Chang, J.-S. (2015). Biodiesel production from wet microalgae feedstock using sequential wet extraction/transesterification and direct transesterification processes. *Bioresource Technology, 194*, 179–186.

Chen, D., Lødeng, R., Omdahl, K., Anundskås, A., Olsvik, O., & Holmen, A. (2001). A model for reforming on ni catalyst with carbon formation and deactivation. J. J. Spivey, G. W. Roberts, & B. H. Davis (Eds.), *Studies in surface science and catalysis 139* (pp. 93–100). Elsevier.

Chen, C.-Y., Zhao, X.-Q., Yen, H.-W., Ho, S.-H., Cheng, C.-L., Lee, D.-J., et al. (2013). Microalgae-based carbohydrates for biofuel production. *Biochemical Engineering Journal, 78*, 1–10.

Chisti, Y. (2008). Biodiesel from microalgae beats bioethanol. *Trends in Biotechnology, 26*(3), 126–131.

Choo, M.-Y., Oi, L. E., Show, P. L., Chang, J.-S., Ling, T. C., Ng, E.-P., et al. (2017). Recent progress in catalytic conversion of microalgae oil to green hydrocarbon: A review. *Journal of the Taiwan Institute of Chemical Engineers, 79*, 116–124.

Clarens, A. F., Resurreccion, E. P., White, M. A., & Colosi, L. M. (2010). Environmental life cycle comparison of algae to other bioenergy feedstocks. *Environmental Science & Technology, 44*(5), 1813–1819.

Cornils, B., & Herrmann, W. A. (2003). Concepts in homogeneous catalysis: The industrial view. *Journal of Catalysis, 216*(1), 23–31.

Couto, R. M., Simões, P. C., Reis, A., Da Silva, T. L., Martins, V. H., & Sánchez-Vicente, Y. (2010). Supercritical fluid extraction of lipids from the heterotrophic microalga *Crypthecodinium cohnii. Engineering in Life Sciences, 10*(2), 158–164.

Cruz, Y., Aranda, D. A. G., Seidl, P., Diaz, G., Garliz, R., Fortes, M., et al. (2018). Cultivation systems of microalgae for the production of biofuels. In K. Biernat (Ed.), *Biofuels – State of development* (pp. 199–207). IntechOpen.

Demirbaş, A. (2008). Production of biodiesel from algae oils. *Energy Sources, Part A: Recovery, Utilization, and Environmental Effects, 31*(2), 163–168.

Deutschmann, O., Knözinger, H., Kochloefl, K., & Turek, T. (2009). Heterogeneous catalysis and solid catalysts. *In* Ullmann's encyclopedia of industrial chemistry. Wiley VCH

Ding, R., Wu, Y., Chen, Y., Liang, J., Liu, J., & Yang, M. (2015). Effective hydrodeoxygenation of palmitic acid to diesel-like hydrocarbons over MoO_2/CNTs catalyst. *Chemical Engineering Science, 135*, 517–525.

Donnis, B., Egeberg, R., Blom, P., & Knudsen, K. (2009). Hydroprocessing of bio-oils and oxygenates to hydrocarbons. Understanding the reaction routes. *Topics in Catalysis, 52*, 229–240.

Drevense, S. (2011). *Algae can replace 17 percent of U.S fuel imports, study says.* Green Tech Media. updated 15 April

2011. Available from:(2011). https://www.greentechmedia.com/articles/read/algae-can-replace-17-percent-of-u-s-fuel-study-says.

Elkonin, B. V., & Sokolowski, J. S. (1991). Simple technique for increasing the conductivity of copper for current lead conductors. *Cryogenics, 31*(12), 1053–1054.

Evans, R. (2012). 2 - Selection and testing of metalworking fluids. In V. P. Astakhov, & S. Joksch (Eds.), *Metalworking fluids (MWFs) for cutting and grinding* (pp. 22–78). Woodhead Publishing.

Feng, P., Deng, Z., Fan, L., & Hu, Z. (2012). Lipid accumulation and growth characteristics of Chlorella zofingiensis under different nitrate and phosphate concentrations. *Journal of Bioscience and Bioengineering, 114*(4), 405–410.

Furimsky, E., & Massoth, F. E. (1999). Deactivation of hydroprocessing catalysts. *Catalysis Today, 52*(4), 381–495.

Galadima, A., & Muraza, O. (2014). Biodiesel production from algae by using heterogeneous catalysts: A critical review. *Energy, 78*, 72–83.

González Borja, M., & Resasco, D. (2011). Anisole and guaiacol hydrodeoxygenation over monolithic Pt–Sn catalysts. *Energy & Fuels, 25*, .

Gosselink, R., Hollak, S., Chang, S.-W., Haveren, J., De Jong, K., Bitter, J. H., et al. (2013). Reaction pathways for the deoxygenation of vegetable oils and related model compounds. *ChemSusChem, 6*.

Haro, P., Sáez, K., & Gómez, P. I. (2017). Physiological plasticity of a Chilean strain of the diatom Phaeodactylum tricornutum: The effect of culture conditions on the quantity and quality of lipid production. *Journal of Applied Phycology, 29*(6), 2771–2782.

Hassan, M. H., & Kalam, M. A. (2013). An overview of biofuel as a renewable energy source: Development and challenges. *Procedia Engineering, 56*, 39–53.

Hävecker, M., Wrabetz, S., Kröhnert, J., Csepei, L.-I., Naumann d'Alnoncourt, R., Kolen'ko, Y. V., et al. (2012). Surface chemistry of phase-pure M1 MoVTeNb oxide during operation in selective oxidation of propane to acrylic acid. *Journal of Catalysis, 285*(1), 48–60.

Ho, S.-H., Chang, J.-S., Lai, Y.-Y., & Chen, C.-N. N. (2014). Achieving high lipid productivity of a thermotolerant microalga Desmodesmus sp. F2 by optimizing environmental factors and nutrient conditions. *Bioresource Technology, 156*, 108–116.

Ho, S.-H., Ye, X., Hasunuma, T., Chang, J.-S., & Kondo, A. (2014). Perspectives on engineering strategies for improving biofuel production from microalgae—A critical review. *Biotechnology Advances, 32*, .

Huang, X., Wei, L., Huang, Z., & Yan, J. (2014). Effect of high ferric ion concentrations on total lipids and lipid characteristics of Tetraselmis subcordiformis, Nannochloropsis oculata and Pavlova viridis. *Journal of Applied Phycology, 26*(1), 105–114.

Hui, W., Wenjun, Z., Wentao, C., Lili, G., & Tianzhong, L. (2016). Strategy study on enhancing lipid productivity of filamentous oleaginous microalgae Tribonema. *Bioresource Technology*, *218*, 161–166.

Ikenaga, N.-o., Ueda, C., Matsui, T., Ohtsuki, M., & Suzuki, T. (2001). Co-liquefaction of micro algae with coal using coal liquefaction catalysts. *Energy & Fuels*, *15*(2), 350–355.

Jęczmionek, Ł., & Porzycka-Semczuk, K. (2014). Hydrodeoxygenation, decarboxylation and decarbonylation reactions while co-processing vegetable oils over a NiMo hydrotreatment catalyst. Part I: Thermal effects—Theoretical considerations. *Fuel*, *131*, 1–5.

Jiang, Y., Zhang, W., Wang, J., Chen, Y., Shen, S., & Liu, T. (2013). Utilization of simulated flue gas for cultivation of Scenedesmus dimorphus. *Bioresource Technology*, *128*, 359–364.

Kandel, K., Anderegg, J. W., Nelson, N. C., Chaudhary, U., & Slowing, I. I. (2014). Supported iron nanoparticles for the hydrodeoxygenation of microalgal oil to green diesel. *Journal of Catalysis*, *314*, 142–148.

Karpagam, R., Preeti, R., Ashokkumar, B., & Varalakshmi, P. (2015). Enhancement of lipid production and fatty acid profiling in Chlamydomonas reinhardtii, CC1010 for biodiesel production. *Ecotoxicology and Environmental Safety*, *121*, 253–257.

Kaygusuz, K., & Sekerci, T. (2016). Biomass for efficiency and sustainability energy utilization in Turkey. *Journal of Engineering Research and Applied Science*, *5*(1), 332–341.

Kishida, M., Hanaoka, T., Hayashi, H., Tashiro, S., & Wakabayashi, K. (1998). Novel preparation method for supported metal catalysts using microemulsion—Control of catalyst surface area. B. Delmon, P. A. Jacobs, R. Maggi, J. A. Martens, P. Grange, & G. Poncelet (Eds.), *Studies in surface science and catalysis* (pp. 265–268). *118*(pp. 265–268). Elsevier.

Kumar, P., Yenumala, S. R., Maity, S. K., & Shee, D. (2014). Kinetics of hydrodeoxygenation of stearic acid using supported nickel catalysts: Effects of supports. *Applied Catalysis A: General*, *471*, 28–38.

Kuo, C.-M., Chen, T.-Y., Lin, T.-H., Kao, C.-Y., Lai, J.-T., Chang, J.-S., et al. (2015). Cultivation of Chlorella sp. GD using piggery wastewater for biomass and lipid production. *Bioresource Technology*, *194*, 326–333.

Laidler, K. J., & Meiser, J. H. (2006). *Physical chemistry* (2nd ed.). Menlo Park, CA: Benjamin/Cummings Pub. Co., ©1982.

Lestari, S., Mäki-Arvela, P., Simakova, I., Beltramini, J., Lu, G. Q. M., & Murzin, D. Y. (2009). Catalytic deoxygenation of stearic acid and palmitic acid in semibatch mode. *Catalysis Letters*, *130*(1), 48–51.

Li, T., Cheng, J., Huang, R., Zhou, J., & Cen, K. (2015). Conversion of waste cooking oil to jet biofuel with nickel-based mesoporous zeolite Y catalyst. *Bioresource Technology*, *197*, 289–294.

Li, W., Ding, Y., Zhang, W., Shu, Y., Zhang, L., Yang, F., et al. (2016). Lignocellulosic biomass for ethanol production and preparation of activated carbon applied for supercapacitor. *Journal of the Taiwan Institute of Chemical Engineers*, *64*, 166–172.

Li, Y., Han, F., Xu, H., Mu, J., Chen, D., Feng, B., et al. (2014). Potential lipid accumulation and growth characteristic of the green alga Chlorella with combination cultivation mode of nitrogen (N) and phosphorus (P). *Bioresource Technology*, *174*, 24–32.

Li, Y., Zhang, C., Liu, Y., Hou, X., Zhang, R., & Tang, X. (2015). Coke deposition on Ni/HZSM-5 in bio-oil hydrodeoxygenation processing. *Energy & Fuels*, *29*, 150310062419000.

Liu, Y.-K., & Lien, P.-M. (2016). Bioethanol production from potato starch by a novel vertical mass-flow type bioreactor with a co-cultured-cell strategy. *Journal of the Taiwan Institute of Chemical Engineers*, *62*, 162–168.

Liu, C.-P., & Lin, L.-P. (2001). Ultrastructural study and lipid formation of Isochrysis sp. CCMP1324. *Botanical Bulletin of Academia Sinica*, *42*, .

Ma, X.-N., Chen, T.-P., Yang, B., Liu, J., & Chen, F. (2016). Lipid production from Nannochloropsis. *Marine Drugs*, *14*(4), 61.

Maier, W. F., Roth, W., Thies, I., & Schleyer, P. V. R. (1982). Hydrogenolysis, IV. Gas phase decarboxylation of carboxylic acids. *Chemische Berichte*, *115*(2), 808–812.

Mäki-Arvela, P., Snåre, M., Eränen, K., Myllyoja, J., & Murzin, D. (2008). Continuous decarboxylation of lauric acid over Pd/C catalyst. *Fuel*, *87*, 3543–3549.

Marjakangas, J. M., Chen, C.-Y., Lakaniemi, A.-M., Puhakka, J. A., Whang, L.-M., & Chang, J.-S. (2015). Selecting an indigenous microalgal strain for lipid production in anaerobically treated piggery wastewater. *Bioresource Technology*, *191*, 369–376.

Markings, S. (2017). *The disadvantages of cellulose biofuel*. Sciencing News. updated 25 April 2017. Available from:(2017). https://sciencing.com/can-tanks-stored-inside-buildings-7853526.html.

McCusker, L. (2005). IUPAC nomenclature for ordered microporous and mesoporous materials and its application to non-zeolite microporous mineral phases. *Reviews in Mineralogy & Geochemistry*, *57*, 1–16.

Mohan, D., Pittman, C., & Steele, P. (2006). Pyrolysis of wood/biomass for bio-oil: A critical review. *Energy*, *20*, 848.

Na, J.-G., Yi, B. E., Han, J. K., Oh, Y.-K., Park, J.-H., Jung, T. S., et al. (2012). Deoxygenation of microalgal oil into hydrocarbon with precious metal catalysts: Optimization of reaction conditions and supports. *Energy*, *47*(1), 25–30.

Naik, S. N., Goud, V. V., Rout, P. K., & Dalai, A. K. (2010). Production of first and second generation biofuels: A comprehensive review. *Renewable and Sustainable Energy Reviews*, *14*(2), 578–597.

Nascimento, I. A., Cabanelas, I. T. D., Santos, J. N., Nascimento, M. A., Sousa, L., & Sansone, G. (2015). Biodiesel yields and fuel quality as criteria for algal-feedstock selection: Effects of CO$_2$-supplementation and nutrient levels in cultures. *Algal Research, 8*, 53–60.

Nautiyal, P., Subramanian, K. A., & Dastidar, M. G. (2014). Recent advancements in the production of biodiesel from algae: A review. In *Reference module in earth systems and environmental sciences*. Elsevier.

Ni, M., Leung, D., Leung, M. K. H., & Sumathy, K. (2006). An overview of hydrogen production from biomass. *Fuel Processing Technology, 87*, 461–472.

Parmar, A., Singh, N. K., Pandey, A., Gnansounou, E., & Madamwar, D. (2011). Cyanobacteria and microalgae: A positive prospect for biofuels. *Bioresource Technology, 102*(22), 10163–10172.

Patil, V., Källqvist, T., Olsen, E., Vogt, G., & Gislerød, H. (2007). Fatty acid composition of 12 microalgae for possible use in aquaculture feed. *Aquaculture International,*.

Peng, B., Yao, Y., Zhao, C., & Lercher, J. (2012). Towards quantitative conversion of microalgae oil to diesel-range alkanes with bifunctional catalysts. *Angewandte Chemie International Edition in English, 51*, 2072–2075.

Peng, B., Yuan, X., Zhao, C., & Lercher, J. (2012). Stabilizing catalytic pathways via redundancy: Selective reduction of microalgae oil to alkanes. *Journal of the American Chemical Society, 134*, 9400–9405.

Pham, T.-H., Lee, B.-K., & Kim, J. (2016). Novel improvement of CO$_2$ adsorption capacity and selectivity by ethylenediamine-modified nano zeolite. *Journal of the Taiwan Institute of Chemical Engineers, 66*.

Ping, E. W., Wallace, R., Pierson, J., Fuller, T. F., & Jones, C. W. (2010). Highly dispersed palladium nanoparticles on ultra-porous silica mesocellular foam for the catalytic decarboxylation of stearic acid. *Microporous and Mesoporous Materials, 132*(1), 174–180.

Popov, A., Kondratieva, E., Goupil, J.-M., Mariey, L., Bazin, P., Gilson, J.-P., et al. (2010). Bio-oils hydrodeoxygenation: Adsorption of phenolic molecules on oxidic catalyst supports. *Journal of Physical Chemistry C, 114*, 15661–15670.

Rahimi, M., Mohammadi, F., Basiri, M., Parsamoghadam, M. A., & Masahi, M. M. (2016). Transesterification of soybean oil in four-way micromixers for biodiesel production using a cosolvent. *Journal of the Taiwan Institute of Chemical Engineers, 64*, 203–210.

Ren, H.-Y., Liu, B.-F., Kong, F., Zhao, L., Xie, G.-J., & Ren, N.-Q. (2014). Enhanced lipid accumulation of green microalga Scenedesmus sp. by metal ions and EDTA addition. *Bioresource Technology, 169*, 763–767.

Renaud, S. M., Zhou, H. C., Parry, D. L., Thinh, L.-V., & Woo, K. C. (1995). Effect of temperature on the growth, total lipid content and fatty acid composition of recently isolated tropical microalgae Isochrysis sp., Nitzschia closterium, Nitzschia paleacea, and commercial species

Isochrysis sp. (clone T.ISO). *Journal of Applied Phycology, 7*(6), 595–602.

Sahar, Sadaf, S., Iqbal, J., Ullah, I., Bhatti, H. N., Nouren, S., et al. (2018). Biodiesel production from waste cooking oil: An efficient technique to convert waste into biodiesel. *Sustainable Cities and Society, 41*, 220–226.

Sajjadi, B., Raman, A. A. A., Parthasarathy, R., & Shamshirband, S. (2016). Sensitivity analysis of catalyzed-transesterification as a renewable and sustainable energy production system by adaptive neuro-fuzzy methodology. *Journal of the Taiwan Institute of Chemical Engineers, 64*, 47–58.

Santillan-Jimenez, E., Morgan, T., Loe, R., & Crocker, M. (2015). Continuous catalytic deoxygenation of model and algal lipids to fuel-like hydrocarbons over Ni–Al layered double hydroxide. *Catalysis Today, 258*, 284–293.

Schmidt, L. D., & Dauenhauer, P. J. (2007). Hybrid routes to biofuels. *Nature, 447*, 914–915. Available from:(2007). https://www.nature.com/articles/447914a.

Scragg, A. H., Illman, A. M., Carden, A., & Shales, S. W. (2002). Growth of microalgae with increased calorific values in a tubular bioreactor. *Biomass and Bioenergy, 23*(1), 67–73.

Shuping, Z., Yulong, W., Mingde, Y., Kaleem, I., Chun, L., & Tong, J. (2010). Production and characterization of bio-oil from hydrothermal liquefaction of microalgae Dunaliella tertiolecta cake. *Energy, 35*(12), 5406–5411.

Snåre, M., Kubičková, I., Mäki-Arvela, P., Eränen, K., & Murzin, D. (2006). Heterogeneous catalytic deoxygenation of stearic acid for production of biodiesel. *Industrial & Engineering Chemistry Research, 45*.

Soares, A. T., da Costa, D. C., Vieira, A. A. H., & Antoniosi Filho, N. R. (2019). Analysis of major carotenoids and fatty acid composition of freshwater microalgae. *Heliyon, 5*(4), e01529.

Song, W., Zhao, C., & Lercher, J. (2013). Importance of size and distribution of ni nanoparticles for the hydrodeoxygenation of microalgae oil. *Chemistry (Weinheim an der Bergstrasse, Germany), 19*.

Spolaore, P., Joannis-Cassan, C., Duran, E., & Isambert, A. (2006). Commercial applications of microalgae. *Journal of Bioscience and Bioengineering, 101*(2), 87–96.

Srifa, A., Faungnawakij, K., Itthibenchapong, V., & Assabumrungrat, S. (2015). Roles of monometallic catalysts in hydrodeoxygenation of palm oil to green diesel. *Chemical Engineering Journal, 278*, 249–258.

Suali, E., & Sarbatly, R. (2012). Conversion of microalgae to biofuel. *Renewable and Sustainable Energy Reviews, 16*(6), 4316–4342.

Sutherland, D. L., Howard-Williams, C., Turnbull, M. H., Broady, P. A., & Craggs, R. J. (2015). Enhancing microalgal photosynthesis and productivity in wastewater treatment high rate algal ponds for biofuel production. *Bioresource Technology, 184*, 222–229.

Swaaf, M. E., Grobben, G. J., Eggink, G., Rijk, T., Meer, P., & Sijtsma, L. (2001). Characterisation of extracellular poly-saccharides produced by Crypthecodinium cohnii. *Applied Microbiology and Biotechnology, 57*, 395–400.

Tan, C. H., Show, P. L., Chang, J.-S., Ling, T. C., & Lan, J. C.-W. (2015). Novel approaches of producing bioenergies from microalgae: A recent review. *Biotechnology Advances, 33*(6, Part 2), 1219–1227.

Tornabene, T. G., Holzer, G., Lien, S., & Burris, N. (1983). Lipid composition of the nitrogen starved green alga Neochloris oleoabundans. *Enzyme and Microbial Technology, 5*(6), 435–440.

Ullah, Z., Bustam, M. A., & Man, Z. (2015). Biodiesel production from waste cooking oil by acidic ionic liquid as a catalyst. *Renewable Energy, 77*, 521–526.

Verma, D., Kumar, R., Rana, B., & Sinha, A. (2011). Aviation fuel production from lipids by a single-step route using hierarchical mesoporous zeolites. *Energy & Environmental Science, 4*, 1667–1671.

Vignarooban, K., Lin, J., Arvay, A., Kolli, S., Kruusenberg, I., Tammeveski, K., et al. (2015). Nano-electrocatalyst materials for low temperature fuel cells: A review. *Chinese Journal of Catalysis, 36*(4), 458–472.

Weldy, C. S., & Huesemann, M. (2007). Lipid production by Dunaliella salina in batch culture: Effects of nitrogen limitation and light intensity. *Journal of Undergraduate Research, 7*, .

Wu, J., Wang, X., Wang, Q., Lou, Z., Li, S., Zhu, Y., et al. (2019). Nanomaterials with enzyme-like characteristics (nanozymes): Next-generation artificial enzymes (II). *Chemical Society Reviews, 48*(4), 1004–1076.

Yang, C., Nie, R., Fu, J., Hou, Z., & Lu, X. (2013). Production of aviation fuel via catalytic hydrothermal decarboxylation of fatty acids in microalgae oil. *Bioresource Technology, 146*, 569–573.

Yen, H.-W., Hu, I. C., Chen, C.-Y., Ho, S.-H., Lee, D.-J., & Chang, J.-S. (2012). Microalgae-based biorefinery—From biofuels to natural products. *Bioresource Technology, 135*.

Yen, H.-W., Yang, S.-C., Chen, C.-H., Jesisca, & Chang, J.-S. (2015). Supercritical fluid extraction of valuable compounds from microalgal biomass. *Bioresource Technology, 184*, 291–296.

Zanuttini, M. S., Dalla Costa, B. O., Querini, C. A., & Peralta, M. A. (2014). Hydrodeoxygenation of m-cresol with Pt supported over mild acid materials. *Applied Catalysis A: General, 482*, 352–361.

Zhang, F., Fang, Z., & Wang, Y.-T. (2015). Biodiesel production directly from oils with high acid value by magnetic Na_2SiO_3@Fe_3O_4/C catalyst and ultrasound. *Fuel, 150*, 370–377.

Zhao, C., Brück, T., & Lercher, J. (2013). Catalytic deoxygenation of microalgae oil to green hydrocarbons. *Green Chemistry, 15*.

Zhao, Y., Li, D., Ding, K., Che, R., Xu, J.-W., Zhao, P., et al. (2016). Production of biomass and lipids by the oleaginous microalgae Monoraphidium sp. QLY-1 through heterotrophic cultivation and photo-chemical modulator induction. *Bioresource Technology, 211*, 669–676.

Zhou, L., & Lawal, A. (2015). Evaluation of presulfided NiMo/γ-Al_2O_3 for hydrodeoxygenation of microalgae oil to produce green diesel. *Energy & Fuels, 29*(1), 262–272.

Biofuel production

*Nathaskia Silva Pereira Nunes[a], Mônica Ansilago[a],
Nathanya Nayla de Oliveira[a], Rodrigo Simões Ribeiro Leite[b],
Marcelo Fossa da Paz[b], and Gustavo Graciano Fonseca[a]*

[a]Laboratory of Bioengineering, Faculty of Biological and Environmental Sciences, Federal University
of Grande Dourados (FCBA/UFGD), Dourados, MS, Brazil
[b]Laboratory of Enzymology and Fermentation Processes, Faculty of Biological and Environmental
Sciences, Federal University of Grande Dourados (FCBA/UFGD), Dourados, MS, Brazil

1 General introduction

The global need for energy is growing apace, and dependence on fossil fuels is under threat since supplies are unsustainable and finite. Thus, the production of food and fuel depends on obtaining sustainable resources (Talebi et al., 2013). Microalgae offer a promising solution to meet energy and fuel needs. These microscopic unicellular and photoautotrophic organisms found primarily in aquatic environments are at the bottom of the food chain and are considered one of the oldest living organisms on the planet (Kumar & Sharma, 2014).

The main components of microalgae biomass are carbohydrates, proteins, lipids, and pigments. Each microalgal species can produce different levels of these components, and its metabolism may change according to changes in the chemical composition of the environment and other conditions such as temperature and light (Nascimento, Nascimento, & Fonseca, 2020). Microalgae can be used as a source for the synthesis of a variety of bioproducts, such as fuels, chemicals, materials, cosmetics, animal feed, and food supplements.

Algal biomass can potentially be used in various energy processes. It has considerable advantages over traditional raw materials, such as high productivity (usually 10–100 times higher than traditional crops), highly efficient carbon capture, high content of lipids, or starch (that can be used to produce biodiesel or ethanol, respectively). It is also cultivable in freshwater, seawater, brackish water, or even wastewater and can be grown on nonagricultural land (Cazzaniga et al., 2014; Xu, Wim Brilman, Withag, Brem, & Kersten, 2011).

Microalgae can be harvested continuously throughout the year since there is no off-season. Another aspect that favors the integrated and sequential production of various products and reduces logistics costs in biorefinery facilities is that microalgae can be grown and processed at the same site. The techniques currently used in the large-scale production of microalgae include raceway ponds, which are large open ponds, and closed tubular photobioreactors. There are numerous possibilities for the production of biofuels from algal biomass, offering opportunities for the development of a sustainable microalgae-based industry whose productivity is independent of soil fertility and less dependent on water purity.

Biofuels produced from first-generation feedstock (soybeans, sunflowers, corn) and second-generation feedstock (forest product wastes and other materials) have a limited ability to meet commercial targets for biofuel production, as well as the low potential for carbon mitigation (Mata, Martins, & Caetano, 2010). In contrast, microalgae, a third-generation raw material (Fig. 1), use both inorganic (CO_2) and organic carbon as carbon sources for the formation of fatty acids, and hence, lipids, whose amount in each cell differs among species. Increasing interest has, therefore, focused on the use of microalgae for the production of biofuel precursors, such as triacylglycerol and starch lipids, which can be transformed into biodiesel and bioethanol, respectively, and on the production of biogas from algal biomass (Fig. 2). Moreover, some species can produce biohydrogen, an attractive and clean fuel that helps reduce carbon emissions.

2 Main biofuels produced from microalgae

2.1 Biodiesel

Among the various applications of algal biomass, its use in biodiesel production has gained increasing importance, since microalgae require much less land area, less time to mature (24–48h), and yield about 30 times or higher volume in oil than terrestrial oilseed crops (Kumar & Sharma, 2014).

The literature reports lipid contents of microalgae ranging from 5% to 75% in terms of dry biomass (Ohse et al., 2015; Sharma, Schuhmann, & Schenk, 2012). The composition

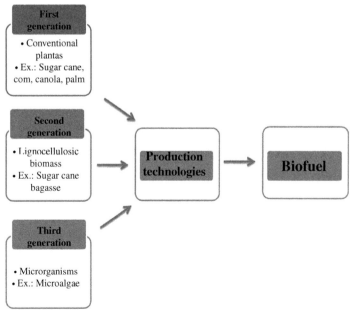

FIG. 1 Biofuel production through different sources of resources.

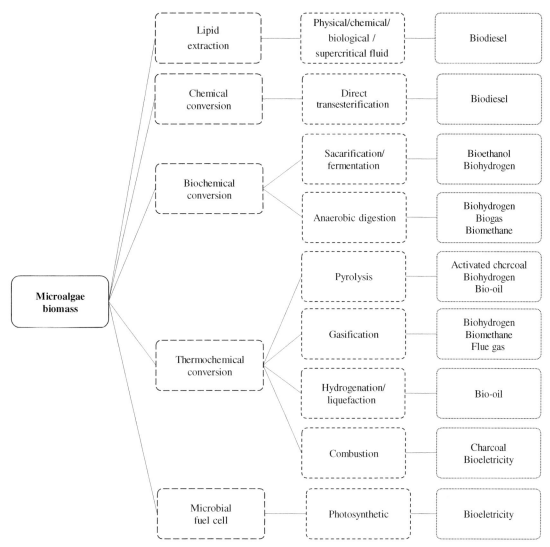

FIG. 2 Scheme of biofuels produced from microalgae.

of microalgae includes fatty acids with 14–22 carbon atoms, which is similar to the vegetable oils used in biodiesel production. Moreover, their physicochemical characteristics are similar to those of traditional biomasses and diesel, however, their environmental impact is significantly lower than that of petrodiesel (Adesanya, Cadena, Scott, & Smith, 2014).

Current studies around the world focus mainly on accumulating the lipid content in microalgae under various conditions of growth for increased oil production. Lipid accumulation in microalgae occurs when a nutrient from the environment is depleted or becomes the limiting factor for growth (Brennan & Owende, 2010). Thus, biotechnological interventions can result in significant advances in the commercial-scale production of biofuel from microalgae.

To enhance lipid production, researchers have resorted to genetic improvement, enabling direct control of these microorganisms through mutagenesis or metabolic engineering by introducing transgenes that adjust the cell's metabolic pathways to trigger metabolite production (Bt Md Nasir, Aminul Islam, Anuar, & Yaakob, 2019). Genetic modifications in microalgae provide new means for the precise control of target mechanisms, increasing the accumulation of cell lipids under normal growth conditions (Lim & Schenk, 2017; Xue et al., 2015).

Cultivation conditions also directly influence the concentration of lipids produced by each algal species. Therefore, subjecting cells to stressors such as nutrient depletion and variations in light intensity, temperature, salinity, and pH are conventionally employed to increase the accumulation of lipids within the biological limits of cells (Bartley, Boeing, Dungan, Holguin, & Schaub, 2014; Suyono et al., 2015).

The biomass of *Chlorella* sp. obtained from cultures with 150 ppm $MgSO_4$, 12.5% salinity, and low light intensity presented high lipid content (32.5%). On the other hand, when *Chlorella* sp. was cultivated with a lower concentration of $MgSO_4$ and higher salinity and light intensity, its lipid content was lower (12.5%) (Shekh et al., 2015). Belotti, Bravi, de Caprariis, de Filippis, and Scarsella (2013) analyzed stressful conditions aimed at increasing the lipid productivity of *Chlorella vulgaris* and found that the lack of nitrogen leads to only a slight improvement. However, the combined effect of lack of nitrogen and phosphorus was to increase the overall productivity of nonpolar lipids.

Lipids are the biofuel feedstock most easily extracted from algae, but their storage is hampered by the presence of polyunsaturated fatty acids (PUFAs), which cause oxidation reactions and high moisture content (Brennan & Owende, 2010). The high content of polar lipids, a class of molecules that negatively affect the process of biodiesel production from microalgae crude oil, is generally not analyzed (Belotti et al., 2013). The transesterification of triacylglycerol (TAG) is significantly influenced by several operating parameters, including the oil-biomass-alcohol ratio, catalyst loading, reaction time and temperature, and reagent purity grade (Menegazzo & Fonseca, 2019).

2.1.1 Production methods

Biodiesel is traditionally produced using oils that are converted through the transesterification reaction, using short-chain alcohols such as methanol and ethanol, which are responsible for providing the radicals methyl and ethyl, respectively, and a base or acid catalyst. As the reaction progresses, glycerol and a set of fatty acid esters are generated as products (Wahlen, Willis, & Seefeldt, 2011). The direct transesterification of microalgae has been adopted as an alternative to the classic biodiesel production process, using solutions with alcohol, catalyst, and a solvent, together with biomass.

Traditional transesterification takes place in two steps. In the first step, triacylglycerides (TAG) are extracted from microalgal lipids by the solvent extraction method. Solvents such as chloroform, methanol, hexane, or isopropanol usually are used for extraction, either individually or in mixed proportions. In the second step, the TAGs are transformed into fatty acid methyl ester (FAME) in the presence of a monohydroxy alcohol (such as methanol) and a catalyst (alkaline or acid) with glycerol as a byproduct (Park, Nguyen, & Jin, 2019). Albeit well-established, the conventional transesterification process has some disadvantages since it generates large amounts of contaminants with solvent in wastewater, causing severe recycling problems, and handling and storing the large volume of these solvents can cause environmental and health risks.

Direct transesterification is an alternative for one-step biodiesel production from algal biomass (Fig. 3). The algal biomass is mixed with

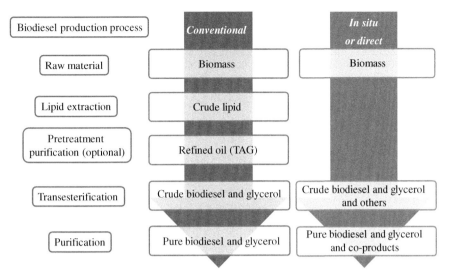

FIG. 3   Flowchart of biodiesel production methods.

methanol in the presence of a catalyst for a certain period and at a given temperature. The methanol used in this process is quickly recovered and can be reused several times. This process, therefore, minimizes the use of large amounts of solvents, shortens processing times, and reduces environmental and health hazards. However, it requires proper optimization of physicochemical parameters, such as temperature, catalyst concentration, and reaction time, to obtain higher biodiesel yields (Ghosh, Banerjee, & Das, 2017).

Efforts have, therefore, focused on eliminating the process of extraction and purification of lipids from biomass (Ghosh et al., 2017; Talebi et al., 2013; Torres, Acien, García-Cuadra, & Navia, 2017; Velasquez-Orta, Lee, & Harvey, 2012). Based on these characteristics, the direct transesterification of microalgal biomass may reach similar values in the recovery of fatty acids as those that would be achieved by methods that involve extraction followed by transesterification (Table 1). However, these results are strongly dependent on the species in question (Menegazzo & Fonseca, 2019). *Chlorella vulgaris* yielded higher FAME levels by the conventional transesterification process (Nascimento et al., 2012; Talebi et al., 2013; Velasquez-Orta et al., 2012), while *Nannochloropsis gaditana* yielded higher FAME levels via direct transesterification (Jazzar et al., 2015; Torres et al., 2017).

Triacylglycerol molecules react with short-chain alcohol, methanol, or ethanol, although methanol is most commonly used. It is sold with a high degree of purity and is nonhygroscopic, however, as mentioned earlier, it is highly toxic and of fossil origin. Therefore, the use of ethanol as a solvent would contribute to rendering the process 100% renewable, although the presence of water in hydrous alcohol makes the synthesis of biodiesel difficult. Hydrolysis reduces the catalyst's activity, increases its consumption, increases saponification, and decreases the biodiesel yield (Rahman, Aziz, Al-khulaidi, Sakib, & Islam, 2017).

Makareviciene, Gumbyte, and Sendzikiene (2019) evaluated the use of ethanol in the direct transesterification of biodiesel from the microalga *Ankistrodesmus* sp. These researchers found that the ideal processing conditions were a temperature of 42°C, a molar ratio of ethanol to the

TABLE 1 Comparison of FAME yield generated in conventional transesterification and direct transesterification.

Method	Condition	Microalgae species	SFA	MUFA	PUFA	References
Conventional transesterification	NaOH in methanol, followed by methylation with BF3 catalyst (12% in methanol)	*Ankistrodesmus falcatus*	41.39	28.41	30.20	Nascimento et al. (2012)
		Ankistrodesmus fusiformis	37.33	22.43	40.24	
		Kirchneriella lunaris	32.06	23.11	44.83	
		Coelastrum microporum	45.87	38.03	16.10	
		Desmodesmus brasiliensis	34.54	44.08	21.38	
		Scenedesmus obliquus	70.83	21.71	7.46	
		Pseudokirchneriella subcapitata	35.39	47.36	17.25	
		Chlorella vulgaris	52.15	37.51	10.33	
	Hexane/chloroform (4: 1, v / v)	*Achnanthes* sp.	40.2	45.9	14.3	Doan, Sivaloganathan, and Obbard (2011)
		Heterosigma sp.	45.4	31	23.7	
		Nannochloropsis sp	47.5	41.8	10.7	
Direct transesterification	Methanol and 2% sulfuric acid in 4 mg microalgae (2 h, 80°C, 750 rpm). 300 μL of 0,9% NaCl solution and 300 μL of hexane; centrifuged (3000 g, 20°C, 3 min)	*Ankistrodesmus* sp.	23.43	23.27	37.16	Talebi et al. (2013)
		Dunaliella salina (UTEX)	22.77	22.89	34.47	
		Scenedesmus sp.	18.59	26.86	30.00	
		Chlorella emersonii	24.55	17.01	38.37	
		Chlorella protothecoides	22.79	19.23	36.19	
		Chlorella salina	29.34	18.52	40.63	
		Chlorella vulgaris	25.0	24.80	45.90	
	Methanol and HCl catalyst of 1 g algal biomass	*Chlorella* sp.	28.5	40.6	28.2	Ghosh et al. (2017)
	Methanol ratio 600:1, catalyst sulfuric acid 0.35:1	*Chlorella vulgaris*	18.6	42.9	20.9	Velasquez-Orta et al. (2012)
	Methanol ratio 600:1, catalyst alkaline sodium hydroxide 0.35:1	*Chlorella vulgaris*	15.3	41.86	31	
	Methanol, catalyst acid sulfuric, stirring for 2 h at 95–100°C, hexane, stirring for 30 min; biodiesel was distilled twice	*Nannochloropsis gaditana*	29.8	20.49	14.79	Torres et al. (2017)

SFA: saturated fatty acids (%), MUFA: monounsaturated fatty acids (%), PUFA: polyunsaturated fatty acids (%).

oil of 8:1, of the biocatalyst lipozyme TL IM, 9.6% (of oil mass), and processing time of 12 h. These conditions resulted in 97.69% of oil transesterification, with a final product containing 6.8% of ethyl esters from microalgae oil mixed with diesel oil, which met the requirements of the European diesel fuel specifications. Hence, ethanol can be used instead of methanol to produce biodiesel and make the process eco-efficient, requiring only the proper catalysts to promote the best performance of this solvent in the direct transesterification process.

2.1.2 Relevant characteristics

Biodiesel FAME profile

Lipid biomolecules are composed of carbon, hydrogen, and oxygen. These lipids include phospholipids, glycolipids, mono-, di-, and triacylglycerols, among others (Greenwell, Laurens, Shields, Lovitt, & Flynn, 2010), but only triglycerides are easily converted to biodiesel by transesterification. Microalgae are also composed of triglycerides and polyunsaturated fatty acids that are not traditionally used to produce biodiesel because they are prone to undergo undesirable oxidation reactions (Adesanya et al., 2014). A large amount of polyunsaturated fatty acids limits the speed of the reaction, which, in turn, increases the activation energy (Ghosh et al., 2017).

Most marine microalgae do not meet the parameters required by international standards in terms of their fatty acid composition because of their considerably high contents of unsaturated acids, especially tri- and polyunsaturated acids. Thus, given the potential of biodiesel production from marine microalgae, it is estimated that the physicochemical properties of biodiesel derived from freshwater species will be of better quality. According to the parameters established by the EN 14214 standard, the maximum limit for linolenic acid or, in the case of microalgae, tri-unsaturated fatty acids, is 12%, while the maximum fatty acid content with more than three double bonds (PUFA) is 1%.

Unsaturation reduces the viscosity of biodiesel and improves filter clogging point properties. However, the presence of unsaturated fatty acids, especially PUFAs, reduces the oxidative stability of biofuel. The predominance of oleic and palmitoleic acids, among unsaturated acids and low di-, tri-, and polyunsaturated contents, favor some of the physical properties of biodiesel, such as its clogging point and viscosity. Although the saturation and fatty acid profile of microalgae have only a minor influence on the production of biodiesel by transesterification, they can affect the properties of biofuel.

Biodiesel properties

Many literature reviews describe the lipid and fatty acid content obtained by cultivating different microalgal species and strains, but few link the influence of production methods with the quality of biodiesel produced (Goh et al., 2019). Islam, Ayoko, Brown, Stuart, and Heimann (2013), who investigated the influence of each fatty acid on biodiesel properties, reached the following conclusions: the higher the polyunsaturated fatty acid content, the higher the iodine value (IV) and the lower the cetane number (CN); viscosity is strongly associated with saturated fatty acids; a higher heating value increases the saturated fatty acids; and IV is a parameter that indicates the degree of saturation of fuel, which affects fuel viscosity and the cold filter plugging point.

The properties of biodiesel produced from *Chlorella* sp. by conventional transesterification (CT) were evaluated by Gumbytė, Makareviciene, Skorupskaite, Sendzikiene, and Kondratavicius (2018) and by direct transesterification (DT) by Ghosh et al. (2017) (Table 2). Fuel density and viscosity affect engine performance and its emission characteristics. Density values of 833 kg/m^3 and 886 kg/m^3 and viscosity values of 3.82 and 4.6 were obtained by CT and DT, respectively. The number of acids describes the quality of biodiesel and measures its corrosion potential. This property was evaluated only in DT, showed a

TABLE 2 Biodiesel properties.

Reaction conditions biodiesel	Microalgae species	ID	CN	SV	IV	LCSF	CFPP	CP	AV	D	V	FP	References
Conventional transesterification	Ankistrodesmus falcatus	88.81	50.52	201.97	101.33	1.69	−10.43	–	–	–	–	–	Nascimento et al. (2012)
	Ankistrodesmus fusiformis	102.91	48.00	199.85	113.81	1.78	−10.14	–	–	–	–	–	
	Kirchneriella lunaris	112.77	42.47	202.21	136.97	1.94	−9.62	–	–	–	–	–	
	Coelastrum microporum	70.23	52.95	205.63	88.42	1.98	−9.52	–	–	–	–	–	
	Desmodesmus brasiliensis	86.84	53.28	205.46	87.05	1.23	−0.55	–	–	–	–	–	
	Scenedesmus obliquus	36.63	63.63	216.04	35.28	1.23	−11.87	–	–	–	–	–	
	Pseudokirchneriella subcapitata	81.86	53.94	207.68	82.83	1.95	−9.60	–	–	–	–	–	
	Chlorella vulgaris	58.17	61.83	199.37	52.63	1.57	−10.81	–	–	–	–	–	
	Chlorella protothecoides	–	51.32	–	–	–	–	246.0	0.224	0.874	4.88*	183.9	Batista, Lucchesi, Carareto, Costa, and Meirelles (2018)
	Chlorella sp.	–	52	–	–	–	–	–	–	833**	3.82**	58	Gumbytė et al. (2018)
	Acutodesmus obliquus	–	–	165.66	156.40	–	–	–	74.41	–	33.33**	–	Escorsim et al. (2018)
Direct transesterification	Ankistrodesmus sp.	97.59	52.45	170.60	114.88	5.22	−0.08	3.55	–	–	–	–	Talebi et al. (2013)
	Dunaliella salina (UTEX)	91.83	55.40	170.56	108.58	4.85	−1.24	3.60	–	–	–	–	
	Scenedesmus sp.	86.86	59.57	152.99	99.57	3.05	−6.91	3.22	–	–	–	–	
	Chlorella emersonii	93.75	54.24	162.28	114.18	6.37	3.55	2.77	–	–	–	–	
	Chlorella protothecoides	91.60	54.57	163.37	111.75	4.93	−0.99	3.51	–	–	–	–	
	Chlorella salina	99.78	49.93	180.97	117.92	6.07	2.58	6.32	–	–	–	–	
	Chlorella vulgaris	116.59	44.0	194	135.26	6.71	4.60	2.66	–	–	–	–	
	Chlorella sp. MJ 11/11	–	56.1	244.8	80.6	–	–	−2.2	0.7	886*	4.6**	113	Ghosh et al. (2017)
	Ankistrodesmus sp.	–	52	–	–	–	–	–	–	844**	4.45**	59	Makareviciene et al. (2019)
	Nannochloropsis gaditana	–	–	–	86.5	–	–	–	2.7	852**	3.76**	–	Torres et al. (2017)

ID: degree of unsaturation; CN: cetane number; SV: saponification value (mg g⁻¹); IV: iodine value (g I$_2$ 100 g⁻¹); LCSF: long chain saturated factor; CFPP: cold filter plugging point (°C); CP: cloud point (°C); AV: acidity value (mg KOH g⁻¹); D: density (g cm⁻³) at *0°C, **15°C; V: viscosity (mm² s⁻¹) at *60°C, **40°C; FP: flash point (°C).

value of 0.7. The iodine value, which indicates the chemical stability of biodiesel, showed a value of 80.6, also evaluated only in DT. The saponification value, which indicates the number of alkalis required for the hydrolysis of fatty acids in glycerides, showed a value of 244.8 in DT. Another critical characteristic of fuel is its CN, which is the ability of a fuel to ignite rapidly. Fuel produced from *Chlorella* sp. yielded CN values of 52 and 56.1 in CT and DT, respectively.

Gumbytė et al. (2018) obtained a transesterification yield of 98% with an ethanol-to-oil molar ratio of 4.54:1, and the low content of PUFAs was a positive property for biofuel production, as polyunsaturated acids tend to become oxidized. During oxidation, secondary oxidation products generate rubber, which clogs the injectors and prevents the fuel from reaching the engine. On the other hand, using a molar ratio of methanol-to-microalgae biomass of 5:1, Ghosh et al. (2017) achieved a maximum lipid conversion of 95%.

The FAME profile analyzed by Nascimento et al., 2012 suggested that the best approach to generate high-quality biodiesel from microalgae is by blending oils from different cell cultures. That is because *Kirchneriella lunaris*, *Ankistrodesmus fusiformis*, and *Ankistrodesmus falcatus* showed the highest levels of polyunsaturated FAME, resulting in the production of biodiesel with the lowest CN, the highest iodine value (IV), and the lowest oxidation stability. Conversely, higher levels of saturated FAME in *Scenedesmus obliquus* oil indicate it is a source of biodiesel with high oxidation stability, high CN, and low IV. Biodiesel produced from microalgae may have high IV values since it is composed mainly of unsaturated fatty acid methyl esters (Torres et al., 2017).

Biodiesel composed of long-chain saturated fatty acid esters will have a high CN, which favors the quality of biofuel. The CN property, which measures combustion quality, is related to ignition timing. A high CN indicates good engine operation, minimizing the number of pollutant emissions. This property is influenced by the degree of saturation and length of the carbon chain. The minimum CN of biodiesel is 51 in Europe, 47 in the US, and 45 in Brazil. Microalgae biodiesel meets all the CN requirements, and this is an additional indication of the feasibility of this biofuel as an alternative to biodiesel from edible vegetable oils (Batista et al., 2018).

Thus, there is no definitive barrier against the future of microalgae biofuel production, although significant biological and engineering innovations are needed to meet the demand for resources. It is essential to identify the most efficient method for each species, in terms of the biodiesel production process, direct or conventional transesterification, and the solvent and catalyst to be used. These parameters determine the quality of the profile of extracted fatty acids, and, hence, the properties of the biodiesel (Menegazzo & Fonseca, 2019).

2.2 Bioethanol

The world population increases each year significantly, thus intensifying the demand for energy. This, in turn, leads to environmental concerns regarding the sources, renewability, and sustainability of the raw materials employed to produce biofuels (Harun & Danquah, 2011).

Much effort has focused on research into plant and microbial biomass for biofuel generation. In regard to microbial biomass, microalgae are promising candidates because they meet these prerequisites and because that their synthesized and accumulated bioproducts are, e.g., lipids and carbohydrates (Harun, Danquah, & Forde, 2010; Kim, Choi, Kim, Wi, & Bae, 2014), which can be used for the production of biodiesel and bioethanol, respectively.

Bioethanol produced with microalgae is a so-called third-generation biofuel (Lakatos et al., 2019), since it originates from the fermentation of carbohydrate contained in the biomass of a

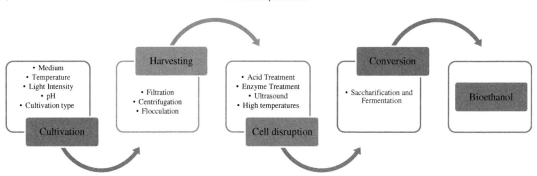

FIG. 4   Flowchart of bioethanol production process variables.

microorganism, in this case, microalgae. Microalgae bioethanol is an excellent substitute for conventional gasoline refined from petroleum (Doan, Moheimani, Mastrangelo, & Lewis, 2012).

The advantages of using microalgal biomass over other types of biomass are that high volumes of biomass can be grown in smaller areas, microalgae can be cultivated in unproductive areas, their cell disruption, and carbohydrate release are natural, and unlike first-generation biofuels, they do not represent a food conflict (Dutta, Daverey, & Lin, 2014; Phwan et al., 2018).

Till date, the disadvantage of bioethanol derived from biomass is that the variables of the production process (cultivation, harvesting, cell disruption, conversion) are not all optimized so far (Fig. 4) (Lee, Seong, Lee, & Lee, 2015). Moreover, although enzymes are incredibly efficient, they represent a high industrial cost. However, this may be circumvented using higher investments in genetically modified microalgae (Lakatos et al., 2019).

2.2.1 Production methods

Cell disruption

Cell disruption methods may be physical, chemical, or biological. Physical methods are shaking, ultrasound, autoclaving, microwave, and others. These methods cause mechanical breakdown through contact, pressure, or denaturation (Lakatos et al., 2019). One of the advantages of these methods is that they are efficient

and generate no waste, but their disadvantage is that they require energy.

On the other hand, chemical methods are characterized by the use of acid or alkaline chemical reagents for cell disruption. The advantage of this process is that it simplifies two processes since cell disruption and polysaccharide hydrolysis can be achieved by using chemical reagents (mainly sulfuric acid) (Chng, Lee, & Chan, 2017). The disadvantages are that waste is generated and that the process represents operational risks because sulfuric acid is highly corrosive.

The biological method, in turn, is characterized by the use of enzymes for both cell disruption and polysaccharide hydrolysis. The advantage of enzymes is that they are not harmful to the environment. However, they are expensive, and the process depends on external factors, such as pH, temperature, among others (Harun & Danquah, 2011).

Saccharification processes

Saccharification processes are indispensable due to the diversity of polysaccharides found in microalgal biomass. The sugars accumulated in microalgae biomass include xylose, mannose, arabinose, mannose, galactose, and glucose (Kim et al., 2014; Kim, Oh, & Bae, 2017).

The most common saccharification processes for the release and hydrolysis of polysaccharides contained in microalgae are chemical and enzymatic hydrolysis. Chemical hydrolysis is usually performed with sulfuric acid, but some studies have reported using hydrochloric acid and even acetic acid (Chng et al., 2017; Phwan et al., 2019).

Thus, the microalgal biomass is mixed with an acid solution (the concentration of acid may vary from one species to another). The temperature is then raised to trigger hydrolysis (Khan, Lee, Shin, & Kim, 2017; Zhou, Zhang, Wu, Gong, & Wang, 2011). After the hydrolysis, the pH must be adjusted for subsequent or simultaneous fermentation (Ho et al., 2013).

Enzymatic hydrolysis, in turn, involves the use of enzymes from different microorganisms. Such enzymes include amyloglucosidase (Chng et al., 2017), β-glucosidase, cellulase (Onay, 2019), and lysozyme (Khan et al., 2017). This process produces exceptionally high yields, but the more purified enzymes are used, the higher the production costs.

Fermentation

Bioethanol is produced by fermenting polysaccharides found in microalgal biomass (Dutta et al., 2014). Concerning the fermentation process, the choice may be to perform separate hydrolysis and fermentation (SHF) or to perform simultaneous saccharification and fermentation (SSF). The SSF process may be preceded by physical treatment, or it may involve the use of enzymes to perform cell disruption.

Most studies have found that the bioethanol yield in the SSF process is higher than in SHF because there is a synergistic effect between carbohydrate digestion (with the enzyme or acid) and bioethanol production (Chng et al., 2017; Phwan et al., 2019; Kim, Im, & Lee, 2015). The SHF process involves two operations, i.e., hydrolysis followed by fermentation.

In both processes, fermentation occurs using two main microorganisms, *Zymomonas mobilis* (Ho, Huang, et al., 2013, Ho, Kondo, Hasunuma, & Chang, 2013) and *Saccharomyces cerevisiae* (Kim et al., 2014; Phwan et al., 2019). To confirm

the efficiency of carbohydrate conversion into ethanol, we use the following equation described by Mussatto et al. (2010):

$$\text{Theoretical ethanol yield;\%}$$
$$= \frac{\text{Ethanol}\,(g)}{\text{Total sugar}\,(g) \times 0.511 \times 100}$$

where 0.511 is the maximum theoretical conversion of sugar into ethanol.

As can be seen in Table 3, each step of the production process is significant, from the choice of strain to the method of hydrolysis and the type of fermentation, as they will directly influence the final ethanol productivity achieved. Therefore, it is essential to evaluate both hydrolysis methods and verify the efficiency of each one, as well as the fermentation methods, for each microalga and hence for the hydrolysis of its polysaccharides.

Genetic recombination could potentially be used to minimize the steps of the production process and modify the microalga so that its metabolism increasingly resembles that of yeast.

2.2.2 *Relevant characteristics*

The amount of sugar found in biomass varies from one species to another and can also be influenced by other factors such as temperature, light intensity, pH, carbon dioxide supplementation, macronutrients (nitrogen, phosphorus), among others (Buono, Langellotti, Martello, Rinna, & Fogliano, 2014; Ho, Chen, & Chang, 2012; Lakatos et al., 2019; Silva & Fonseca, 2020).

To exemplify, in the study by Kim et al. (2014), when stress was induced by limiting the availability of nitrogen and sulfur nutrients in the cultivation of *Chlorella vulgaris*, the total carbohydrate content showed an increase of 7% and 40% (w/w), respectively, over that found in normal conditions.

Ho et al. (2012) associated the influence of light intensity and nutrient limitation on the production process of biomass and the product of the microalga *Scenedesmus obliquus*. They found

TABLE 3 Influence of strain, hydrolysis method, and fermentation type on the final ethanol yield obtained.

Microalgae	Hydrolysis method	Type of fermentation	Yeast	Bioethanol production, g bioethanol/g biomass	Bioethanol theoretical yield, %	References
Scenedesmus dimorphus	4% *v*/v Sulfuric acid	SHF	*Saccharomyces cerevisiae*	0.178	80.3	Chng et al. (2017)
Scenedesmus dimorphus	Enzymatic (AGL)	SHF	*Saccharomyces cerevisiae*	0.183	84.3	
Scenedesmus obliquus	2.5% Sulfuric acid	SHF	*Zymomonas mobilis*	0.195	–	Ho, Kondo, et al. (2013)
Chlorella sp.	1% Sulfuric acid	SSF	*Saccharomyces cerevisiae*	0.137	–	Phwan et al. (2019)
Chlorella sp.	3% Sulfuric acid	SSF	*Saccharomyces cerevisiae*	0.169	–	
Chlorella sp.	5% Sulfuric acid	SSF	*Saccharomyces cerevisiae*	0.281	–	
Chlorella sp.	7% Sulfuric acid	SSF	*Saccharomyces cerevisiae*	0.154	–	
Chlorella sp.	9% Sulfuric acid	SSF	*Saccharomyces cerevisiae*	0.084	–	
Chlorella sp.	1% Acetic acid	SSF	*Saccharomyces cerevisiae*	0.110	–	
Chlorella sp.	3% Acetic acid	SSF	*Saccharomyces cerevisiae*	0.133	–	
Chlorella sp.	5% Acetic acid	SSF	*Saccharomyces cerevisiae*	0.230	–	
Chlorella sp.	7% Acetic acid	SSF	*Saccharomyces cerevisiae*	0.134	–	
Chlorella sp.	9% acetic acid	SSF	*Saccharomyces cerevisiae*	0.080	–	
Porphyridium cruentum (SPC)	Enzymatic (CEL and PEC)	SSF	*Saccharomyces cerevisiue*	–	65.4	Kim et al. (2017)
Porphyridium cruentum (FPC)	Enzymatic (CEL and PEC)	SSF	*Saccharomyces cerevisiae*	–	70.3	
Scenedesmus obliquus CNW-N	2.0% sulfuric acid	SHF	*Zymomonas mobilis*	0.213	99.8	Ho, Kondo, et al. (2013)
Chlorella vulgaris	Enzymatic (AMY and CEL)	SHF	*Zymomonas mobilis*	0.178	89.3	Ho, Huang, et al. (2013)

TABLE 3 Influence of strain, hydrolysis method, and fermentation type on the final ethanol yield obtained—cont'd

Microalgae	Hydrolysis method	Type of fermentation	Yeast	Bioethanol production, g bioethanol/g biomass	Bioethanol theoretical yield, %	References
Chlorella vulgaris	Enzymatic (AMY and CEL)	SSF	*Zymomonas mobilis*	0.214	92.3	
Chlorella vulgaris	Sulfuric acid	SHF	*Zymomonas mobilis*	0.233	87.59	

SPC: seawater *Porphyridium cruentum*; FPC: Freshwater *Porphyridium cruentum*; AGL: amyloglucosidases; AMY: amylases; CEL: cellulases; PEC: pectinases.

that as the light intensity increased, the amount of biomass produced was also more significant, and carbon fixation was greater. However, the maximum increase for this species under these conditions was $540\,\mu mol\,m^{-2}\,s^{-1}$, and the highest carbohydrate content was 22.4% (w/w), after 5 days of nitrogen limitation.

Therefore, in terms of the final production cost, all these variables must be analyzed, since they also directly influence the amount of biomass produced. Their optimization may thus result in cost reductions due to high biomass productivity plus high product productivity (Ho, Huang, et al., 2013; Ho, Kondo, et al., 2013).

2.3 Biohydrogen

In recent years, the ever-growing demand for energy and the marked increase in the consumption of nonrenewable fuel reserves have prompted researchers to focus particular attention on technologies for the production of fuels derived from renewable energy sources. These include biohydrogen, which is considered a potentially cleaner energy source and a promising alternative to other conventional fossil fuels (Khosravitabar, 2019; Show et al., 2019). Biohydrogen releases only water as an end product of its production process, owing to which it has been considered a sustainable alternative to replace gasoline and/or diesel, generating less negative environmental impacts (Szwaja & Grab-Rogalinski, 2009).

Hydrogen, which is not freely available in the environment but can be found in combination with other elements, is considered a common source of energy. Hydrogen can contain three times more energy than hydrocarbon-based fuels (Khosravitabar, 2019). Furthermore, according to Khosravitabar (2019), the hydrogen production process requires several external energy inputs, such as light, heat, or even electricity. The challenge posed by the use of hydrogen is the sustainability of the production and storage phases of this fuel (Ghimire et al., 2015).

Biohydrogen production from microalgae is a new clean technology that offers several benefits. These include high CO_2 capture when compared to other higher plants, simplified methods of cultivation of this microorganism, whose wastes can be used as a carbon source, accelerated production and high carbohydrate content, without the presence of lignocellulosic material, enabling it to be used as a raw material in the production of third-generation fuels (Callegari, Bolognesi, Cecconet, & Capodaglio, 2020; Saratale, Saratale, Banu, & Chang, 2019).

The substrates commonly used for the production of microalgae include unsalable or unwanted waste materials, some of which are environmentally harmful. Among the materials that can be used as a substrate for microalgae production is agricultural waste such as ligno-cellulosic bagasse, livestock and swine manure, palm oil, olive mill waste, municipal organic waste, and other industrial waste (Ghimire et al., 2015).

2.3.1 Production methods

Hydrogen can be produced in different ways, using both renewable and non-renewable energy sources. Hydrogen production techniques from non-renewable energy sources include gas steam reforming, gasification, pyrolysis, and other techniques involving water, such as electrolysis, and microorganisms, such as fermentation, and biophotolysis (Demirbas, 2009). However, technologies for hydrogen production from fossil fuels are still harmful to the environment, requiring the study and application of clean and environmentally sustainable technologies (Show et al., 2019). The techniques for hydrogen production using renewable sources include hydroelectric energy, wind energy, wave energy, solar energy, geothermal energy, biomass energy, and even non-renewable

energy such as coal, natural gas, and nuclear energy (Abe, Popoola, Ajenifuja, & Popoola, 2019).

Biohydrogen from microalgae can be produced employing photobiological systems in bioreactors, which are used for biomass production under controlled conditions such as light, temperature, pH, aeration, among others (Chandra, Iqbal, Vishal, Lee, & Nagra, 2019; Hallenbeck, 2005). These processes include photofermentation, anaerobic fermentation, microbial electrolysis, pyrolysis, and gasification (Aziz, 2015; Saratale et al., 2019), as illustrated in Fig. 5. Anaerobic fermentation occurs in the absence of light, producing hydrogen, while photofermentation involves direct and indirect electrolysis and fermentation in the presence of light energy (Saratale et al., 2019). The process of photolysis involves photosynthesis and accumulation of carbohydrates, which are later converted to hydrogen. In the process of microbial electrolysis, electrons originating from the metabolism of microorganisms flow from the anode to the cathode, generating

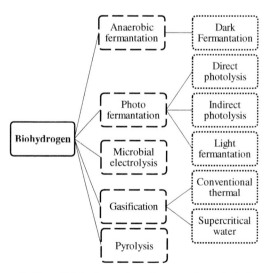

FIG. 5   Methodologies for the production of biohydrogen by microorganisms, including biochemical processes and thermochemical production processes.

energy, thereby converting chemical energy into electrical energy (Monasterio, Mascia, & Lorenzo, 2017).

Microorganisms produce biohydrogen through autotrophy and heterotrophy, and the chemical reactions involved in these processes illustrated in Fig. 6. In autotrophic conversion, solar energy is converted into biohydrogen directly through photosynthesis, which occurs in a way similar to that in higher plants. In heterotrophic conversion, organic substrates converted to biohydrogen with concomitant formation of other less complicated organic compounds (Ghimire et al., 2015).

The advantages and disadvantages of each method vary according to the sustainability and practicality of the technology involved and the energy efficiency of the process. Despite their promising potential for production, microalgae biofuel technologies are still under research and require large-scale studies. The notable advantage of microbial electrolysis stems from the method's high energy efficiency compared to that of others, as well as the widespread availability of intracellular material such as lipids. On the other hand, the disadvantage of this method is the difficulty of expanding it on a larger scale (Monasterio et al., 2017).

The disadvantage of biochemical conversion is that it is a slow process, and its energy conversion is lower than that of thermochemical conversion. Conventional and supercritical methods are generally used in thermochemical conversion through gasification. The disadvantage of the conventional method is that it requires maximum drying of microalgae biomass, thereby generating high energy consumption (Aziz, Oda, & Kashiwagi, 2014), which is not necessary for the supercritical method because gasification is performed in an aqueous medium. However, the supercritical gasification technique also involves high energy consumption at the moment when the microalgae biomass reaches a supercritical state (Aziz, 2015).

Biophotolysis, which involves biological activity based on luminous energy to dissociate a substrate into hydrogen and oxygen, has the advantage of using natural resources that are more readily available in the environment, such as light energy and water. However, the

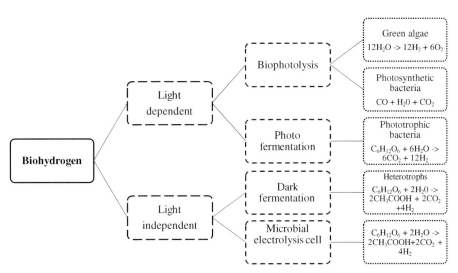

FIG. 6 Schematic of the chemical reactions involved in the production of biohydrogen from the presence and absence of light energy.

technique is limited by low biofuel production (Singh & Das, 2018). In dark fermentation, the microbial metabolism does not require specific energy to transport glucose into the cell, nor does it use adenosine triphosphate (ATP) to produce glucose-6-phosphate (Shobana et al., 2017).

2.3.2 Relevant characteristics

Hydrogen produced from biological agents such as microalgae and bacteria is known as biohydrogen. These microorganisms can decompose organic matter and transform it into carbon dioxide and hydrogen (Callegari et al., 2020). Microalgal hydrogen production is a promising technology, given the numerous benefits and properties of these microorganisms, and requires small areas for the production of microalgae biomass, unlike other biofuels that require a large area of arable land (Kruse & Hankamer, 2010). The enzymes involved in the biohydrogen production process are hydrogenases and nitrogenases. Hydrogenases use protons, which they reduce to produce hydrogen, while nitrogenases are involved in fixing nitrogen, reducing it to ammonia (Singh & Das, 2018).

Table 4 lists studies in which microalgae were used in the production of biohydrogen through different processes and in various culture media, both synthetic, such as BG-11 medium, and natural, such as starch industry wastes.

TABLE 4 Biohydrogen productivity by different microalgae species.

Microalgae species	Feedstock	Reaction conditions	Batch culture mode	Yield	References
Chlorococcum sp.	BG-11 medium	Anaerobic acetone-butano fermentation	Mini-reactors	8.4 and 6.1 mmol L^{-1} of medium/day	Chernova and Kiseleva (2017)
Scenedesmus sp.	TAP media (with S)	Photo fermentation	Rectangular glass reactor	Maximum evolution of 17.72% v/v H_2 of total gases	Dasgupta et al. (2015)
Chlamydomonas reinhardtii	TAP media	Photo fermentation	Tubular PBR	3121.5 ± 178.9 mL	Giannelli and Torzillo (2012)
Gloeocystis sp.	BG-11 devoid of glucose	Light conditions / Dark conditions	Septum vials/ septum vials wrapped in aluminum foil	346 µmol H_2 (mg chl a $h^{-1})^{-1}$/108.5 µmol H_2 (mg chl a $h^{-1})^{-1}$	Junaid, Khanna, Lindblad, and Ahmed (2019)
Chlorella vulgaris	—	Pyrolysis (900°C and 4 Mpa)	Entrained flow reactor	88.01 vol%	Maliutina, Tahmasebi, and Yu (2018)
Scenedesmus sp. and anaerobic sludge	Starch wastewater	Dark fermentation	Reactors where the air was removed from the headspace by argon gas	1508.3 mL L^{-1}	Ren, Liu, Kong, Zhao, and Ren (2015)
Scenedesmus sp.	Corn starch	Dark fermentation	Anaerobic sequencing batch reactor (ASBR)	811.1 mL L^{-1} d^{-1}	Ren et al. (2019)
Chlorella sp.	Crude gycerol	Anaerobic condition	Serum bottle/ bioreactor	10.31 ± 0.05 mL L^{-1}; 11.65 ± 0.65 mL L^{-1}	Sengmee, Cheirsilp, Suksaroge, and Prasertsan (2017)

3 Other biofuels

Less studied but no less critical are other biofuels obtained from algal biomass, such as bio-oil, flue gas, biomethane, charcoal-based electricity, biogas, and bioelectricity.

3.1 Bio-oil

Bio-oil is produced by algal biomass pyrolysis. It has similar characteristics to diesel, so bio-oil could replace diesel and solve the fuel demand and supply conflict problem. However, the oxygen content of microalgae bio-oil is not the same as diesel, which restricts the industrial use of this material. Therefore, to improve its quality, microalgae bio-oil needs to be deoxygenated and made suitable for the fuel industry. Thus, Guo et al. (2019) used CaO hydroxide calcination (HCCaO) to deoxygenate the pyrolysis of bio-oil from *Nannochloropsis* sp. Their results indicated that HCCaO significantly deoxygenated the bio-oil, improving its performance through direct fixation of "quasi-CO_2 active intermediates."

3.2 Flue gas

The application of microalgae technology to capture CO_2 from flue gases has gained significant attention. Although it cannot be treated as a long-term CO_2 mitigation method (unless CO_2 is stored), it is still a carbon-neutral method because the amount of CO_2 released during the combustion of algal biofuels is no higher than CO_2 captured during algae cultivation (Vuppaladadiyam et al., 2018). Carrying out a review of the technical aspects and the feasibility of combined carbon fixation and cultivation of microalgae for carbon reuse, Vuppaladadiyam et al. (2018) concluded that carbon dioxide fixation of flue gases depends mainly on the microalgae strain and the combustion gases and their concentrations. Despite technological innovations, their feasibility for commercial application still requires further research.

3.3 Biomethane

Anaerobic digestion of microalgae for biomethane production has great potential due to its ease of cultivation and low lignin content. However, according to Bishop and Rahman (2017), many obstacles remain in the path of commercial expansion, such as prohibitively high water consumption, variable quality/quantity of gaseous products, and inhibition phenomena. To overcome these problems, more research is needed related to the diversity and function of microorganisms within the digester and the development of a robust approach to cultivation. A new process (SunCHem) was described elsewhere for biomethane production by hydrothermal gasification (Haiduc et al., 2009). In this process, the nutrients, water, and CO_2 produced are recycled, indicating that it is a promising process for efficient production of methane by catalytic gasification of microalgae biomass, despite that the process savings should be improved.

3.4 Bioelectricity

Bioelectricity can be generated by the use of photosynthetic microalgae microbial fuel cells. In this system, microalgae act using light (artificial or natural) to generate electrical energy. Thus, they perform the role of biocatalysts in oxidation or reduction reactions and maybe in both the anode and cathode compartments (Bazdar, Roshandel, Yaghmaei, & Mardanpour, 2018; Gouveia, Neves, Sebastião, Nobre, & Matos, 2014; Kumar, Basu, Gupta, Sharma, & Bishnoi, 2019).

3.5 Biochar

The biochar obtained from microalgae biomass is approximately 10–100 μm in size, with porosity around 1 μm. The biochar can be produced by the pyrolysis process or by the combustion process. In the pyrolysis process, to obtain activated charcoal, which determines the success of the technique, is the time and temperature employed (Maliutina et al., 2018). Pyrolysis occurs in the absence of oxygen at 350–700°C. One of the applications of activated carbon is in the use of this material as a biofertilizer, and also as a carbon dioxide sequester from the atmosphere (El-Naggar et al., 2019; Rizwan, Mujtaba, Memon, Lee, & Rashid, 2018). When pyrolysis occurs quickly, there is a production of biochar, but with the generation of by-products, such as bio-oil, while pyrolysis, that takes a longer time, presents a higher yield only of biochar (Yu et al., 2017). In the carbonization process, which is a chemical process of combustion, microalgae biomass undergoes a thermal degradation process that occurs at moderate temperatures, thus taking longer to complete (Rodrigues & Braghini Junior, 2019).

3.6 Biogas

Biogas is one of the products of decomposition of organic matter. Coal, agro-industrial waste, algal biomass, petroleum coke, among others, can be used as raw material for the production of biogas (Toro, Pérez, & Alzate, 2018). Among the processes for obtaining biogas are gasification and anaerobic digestion. Obtaining biogas through the gasification process involves several steps, such as feeding the gasification system with the microalgae biomass, with the temperature increase up to 1200°C. As the temperature of this process increases, the water evaporates, thus causing partial oxidation of the biomass, eventually producing biogas, which in this process can also be called synthesis gas (Patinvoh, Osadolor, Chandolias, Horváth, & Taherzadeh, 2016). Biogas obtained through anaerobic digestion can be produced with different substrates such as organic matter and biomass. For the process of obtaining biogas occur, chemical reactions take place at different stages, such as hydrolysis, acidification, acetate production, and methane production. From this chemical process, the biogas formed consists of methane and carbon dioxide (Toro et al., 2018). However, other traces of different gases can be found, such as hydrogen sulfide, ammonia, hydrogen, carbon monoxide, among others.

4 Influence of cultivation conditions

4.1 Algae metabolism

Each cultivation condition provides different sources of nutrients and energy sources, resulting in a variation in the content of carbohydrates, lipids, proteins, and biomass productivity. Microalgae have different forms of energy metabolism: (1) photoautotrophic, when light is utilized to produce chemical energy through the process of photosynthesis, e.g., sunlight, as an energy source, and CO_2 as an inorganic carbon source (2) heterotrophic: when energy and carbon are obtained from an external organic source, usually by the oxidation of sugars; (3) mixotrophic: when photosynthesis and oxidation of organic compounds occur concurrently, and (4) photoheterotrophic: when the energy source is light, and the carbon source is an organic compound. Choosing the best condition for production will depend on the purpose of cultivation, as well as the characteristics of the microalgae, to achieve a higher content of the intended compound.

Traditionally, the cultivation of microalgae has explored its photoautotrophic metabolism. However, studies have pointed out the advantages of biomass production by other metabolic pathways. The knowledge of the metabolism of each species is relevant to optimize the production of compounds on a large scale since these metabolic pathways influence the growth characteristics, as well as the products to be extracted and consequently their energy applications (Amaro, Guedes, & Malcata, 2011; Andruleviciute, Makareviciene, Skorupskaite, & Gumbyte, 2014; Liang, 2013).

Microalgae in heterotrophic cultivation do not perform photosynthesis, so they do not need light, only an external organic source to obtain energy and carbon. However, photoautotrophic microalgae are also capable of oxidizing organic compounds, originated from the internal process of photosynthesis to obtain energy, whereas, in heterotrophic ones, organic carbon captured from the external environment (Perez-Garcia, Escalante, de-Bashan, & Bashan, 2011).

With yields significantly higher than those of photoautotrophic cultivation, heterotrophic cultivation avoids problems associated with light limitation, using organic carbon both as an energy source and as a carbon source for biomass production (Huang, Chen, Wei, Zhang, & Chen, 2010; Liang, 2013). On the other hand, a carbon source must be added to the medium. The carbon sources that have been most used are glucose, glycerol, and acetate, applied in heterotrophic and mixotrophic cultures (Andruleviciute et al., 2014; Feng, Li, & Zhang, 2011; Zheng, Chi, Lucker, & Chen, 2012).

The entry of organic carbon in the cell is one of the most determining characteristics of heterotrophic metabolism (Azma, Mohamed, Mohamad, Rahim, & Ariff, 2011). The use of an organic carbon source permits to reduce the cost of cultivating microalgae, reusing the excess of nutrients found in either industrial or domestic wastewater, in addition to the environmental aspect. Thus, some researchers have combined the cultivation of microalgae with the treatment of wastewater, with excellent production of lipids and biomass, as well as the assimilation of some elements (bioremediation). Using microalgae in an alcohol distillery wastewater, Solovchenko, Pogosyan, Chivkunova, and Selyakh (2014) observed a decline in nitrate (95%), phosphate (77%), and sulfate (35%), in addition to an increase in the composition of chlorophyll and fatty acids from microalgae biomass. Higgins et al. (2016) studied the coculture of microalgae with *Auxenochlorella protothecoides* and *Escherichia coli* under mixotrophic

conditions, which led to a 2–6-fold increase in microalgae growth, a doubling of the lipid content, in addition to high nutrient removal rates. These studies are examples of the simultaneous wastewater treatment and production of bioproducts from microalgae.

The carbon source and its concentration utilized in the culture medium influence the biochemical composition of the algal biomass. Liang, Sarkany, and Cui (2009) reported different biomass compositions of *Chlorella vulgaris*, depending on the culture medium. Cells presented 44% carbohydrates, 21% lipids, 32% protein, and 3% ash in 1% glucose; 23% carbohydrates, 31% lipids, 42% proteins, and 4% ash in 1% acetate; 29% carbohydrates, 22% lipids, 45% proteins, and 4% ash in 1% glycerol; and 34% carbohydrates, 32% lipids, 30% proteins, and 3% ash in 2% glycerol-based media. Thus, the choice of carbon source for the heterotrophic culture medium must be according to the compound that one wants to be obtained.

Barros et al. (2019) observed values of maximum specific growth rate (μ_{max}) of 1.24 and 1.28 day^{-1} for *Chlorella vulgaris'* cultivations at heterotrophic and photoautotrophic conditions, respectively. The authors suggested combining the two cultivation modes for efficient production in two stages, based on heterotrophic growth to obtain a high concentration of cells, and then operate main bioreactors (flat panels) under photoautotrophy. The cells cultured heterotrophically led to a 100-fold decrease in the volume needed to inoculate a flat panel, compared to inoculants obtained using photoautotrophic conditions, without any delay caused by the metabolic change to the new trophic conditions.

Another study reported that the best growth parameters for *Chlorella sorokiniana* were obtained under mixotrophic conditions when compared to photoautotrophic and heterotrophic conditions. For glucose at 4 g L^{-1} as the carbon source, the μ_{max} was 3.40 d^{-1} and the dry weight biomass (Dw, 3.55 g L^{-1}) in mixotrophic culture. These values were 1.8 and 2.4 times higher than the heterotrophic culture, and 5.4 and 5.2 times higher than the photoautotrophic culture. In the mixotrophic culture, there was a substantial increase in the lipid content, reaching 45% of the dry biomass in comparison to the 13% obtained in the heterotrophic culture (Li et al., 2016).

To evaluate the effect of temperature and nitrogen concentration on the lipid productivity and fatty acid composition in three *Chlorella* strains (Ördög et al., 2016), and the morphology, growth, biochemical composition, and photosynthetic performance of *Chlorella vulgaris* under low and high nitrogen supplies (Li et al., 2016), these authors concluded that to maximize the production of lipids in microalgae it is essential to regulate the supply of nitrogen and to be aware that each species responds differently to the received supplies.

Comparing the productivity of the different microalgae cultivation systems is not trivial, as there is an excellent variation in the methodologies used, as well as in the cultivation conditions to obtain the data—the addition of nutrients in the medium results in a higher cost. Wastewater is an option to make cultivation more economical and sustainable. However, only a few species survive in these cultivation conditions. Heterotrophic and photoautotrophic cultures are the most used so that heterotrophic cultivation produces microalgae with higher lipid content, but microalgae are easily contaminated, especially in open culture systems, while photoautotrophic cultivation is ecological, but the lipid content produced is low (Chew et al., 2018). These perspectives for the cultivation of microalgae serve as a basis and incentive for further research to be developed, creating an impulse to improve this technology for a new era of biofuel generation.

4.2 Algal cultivation systems

Several systems are used for the production of microalgae, from open ponds to controlled automatized bioreactors. Each system has its advantages and disadvantages, with different cultivation volumes and characteristics. Production control can be an advantage given the optimization of production. However, the increase in energy and operating costs ends up being less attractive. Thus, the choice of each system must be based on the intrinsic needs of each microorganism used, the climatic conditions of the region, and the costs that will be spent on the system (Brennan & Owende, 2010).

The closed systems have as a characteristic non-direct contact with atmospheric air, which allows the control of several cultivation conditions, such as temperature, light, pH, amount of nutrients, and cell density, thus reducing the risk of contamination (Bitog et al., 2011). This system is generally used to obtain products with higher added value, such as pharmaceuticals and food supplements.

Closed photobioreactors have different configurations, and among them, we can mention flat panel, tubular, and spiral. The flat panel photobioreactor consists of two transparent glass plates arranged in a rectangular shape connected in a cascade that can be arranged vertically and in an inclined manner (BITOG et al., 2011). Presenting a higher contact surface, which allows for a higher brightness of the culture. However, it has as a disadvantage the possibility of biofilm formation on the internal surface of the glass plates (Chew et al., 2018). Romero-Villegas, Fiamengo, Acién Fernández, and Molina Grima (2018) observed that the *Geitlerinema* sp. reached $47.7 \, \text{g m}^{-2}$ of dry biomass, in addition to the removal of 82%, 85%, and 100% of carbon, nitrogen, and phosphorus, respectively when cultivated in 100 L flat panel photobioreactors under non-aseptic cultivation conditions.

Tubular photobioreactors, on the other hand, are a series of transparent tubes connected, which can be arranged in an inclined, horizontal, or vertical manner, being divided into columns of bubbles, and airlift (Demirel, Imamoglu, & Conk Dalay, 2015). In the airlift, configuration occurs the introduction of CO_2 by dispersers. The physical agitation provoked by the bubbles into the bioreactor carries out transport to the surface of the bioreactor. When it reaches the surface, the gas loosens and returns to the system flow (Singh & Sharma, 2012). As an example, it was reported elsewhere that *Cylindrotheca closterium* cultivated with an F/2 medium in an airlift bioreactor presented productivities of $0.356 \, \mathrm{g} \, \mathrm{L}^{-1}$ for biomass and $7.364 \, \mathrm{mg} \, \mathrm{L}^{-1}$ for lipids (Demirlel et al., 2015).

In bubble columns, the bubbles move randomly, often by mechanical means, while the airlift has a flow of cyclic bubbles. Usually, the bubble column bioreactor has twice the height of the cylinder diameter, light is provided externally, and the movement of the bubbles leads to greater cellular mobility, thus improving photosynthesis (Chew et al., 2018).

Under natural conditions, microalgae are grown in open systems, with little or no control of growing conditions. The cultivation ponds are open, thus being in contact with atmospheric air, but they present a higher risk of external contamination. Generally, the ponds used are shallow to be more efficient in the use of light (Brennan & Owende, 2010).

The open systems of microalgae production, generally of the raceways type, were developed, aiming at higher production of algal biomass with lower operational cost. However, when this cultivation is being carried out outdoors, it can alter the biochemical composition and the productivities of the algal cells, due to seasonal, temperature and light variations, and contamination (Chew et al., 2018). Open systems are generally applied to produce algal biomass with less added value and can be applied in the feeding of aquatic organisms and the production of biofuels (He, Yang, & Hu, 2016), among others.

Adesanya et al. (2014) combined tubular photobioreactors of the airlift type in the first stage and raceways in a second stage for the cultivation of *Chlorella vulgaris*. In this study, the authors assessed the environmental impact of biodiesel produced using microalgae and fossil fuels. Thus, a reduction in global warming potential of 76% and a reduction in fossil energy requirements of 75% were observed. It was reported elsewhere lipid productivities of $5.15 \, \mathrm{g} \, \mathrm{m}^{-2} \, \mathrm{d}^{-1}$ and $4.06 \, \mathrm{g} \, \mathrm{m}^{-2} \, \mathrm{d}^{-1}$ for *Chlorella* sp., and of $5.35 \, \mathrm{g} \, \mathrm{m}^{-2} \, \mathrm{d}^{-1}$ and $3.00 \, \mathrm{g} \, \mathrm{m}^{-2} \, \mathrm{d}^{-1}$ for and *Monoraphidium dybowskii* Y2 when cultivated in the semi-continuous feeding mode and the batch mode, respectively (He et al., 2016).

Despite being economical and simple, the open systems have some disadvantages for algae growth, such as the losses by evaporation that result in low yield and contamination in the culture medium. On the other hand, photobioreactors are predominantly designed to grow algae at ideal conditions that eliminate most of the problems encountered in open systems (Chew et al., 2018).

5 Commercial application of these technologies

One of the most striking features that attract commercial interest with the study and commercialization of microalgae is its ability to accumulate different compounds that, in turn, can be utilized as raw material for different coproducts of a production process (Zhan, Rong, & Wang, 2017). The concept of biorefinery refers to the maximization of the production process to convert microalgal biomass into different materials, i.e., it is a way to extract different products and raw materials for different applications, based in this case on the microalgae biomass composition (Chen, Li, & Wang, 2019; Menegazzo & Fonseca, 2019).

Thus, the production of biofuels can be associated with other processes, whose objective is the generation of compounds of medium to high added value (antioxidants, cosmetics, supplements, among others) (Chia et al., 2018). Besides, it is essential to emphasize that it must be integrated, either by reusing water or using effluents for algal cultivation (Deprá et al., 2018), which will reduce production costs in terms of nutrient supply, turning the process more sustainable.

Therefore, in the context of biofuel production, for the process to be economically viable, microalgae must be a versatile raw material. This strategy has already been adopted by pilot plants and some companies that already produce biofuel from microalgae, e.g., Sapphire Energy Inc., Cellana Inc. (Dickinson et al., 2017), which has been generating biofuel products, feed for aquaculture, and omega-three supplement from the microalgae cultivation. Cellana Inc., like Sapphire Energy Inc., has been targeting the joint production of biofuels, feed for aquaculture, and omega-three supplements (Cellana Inc., 2015). Algenol Biotech LLC, on the other hand, has its production process in closed bioreactors focused on the production of biofuels, such as ethanol, and biodiesel, beyond microalgae-based supplements, and personal care products (Algenol Biotech, 2011). Joule Unlimited Inc. and Synthetic Genomics Inc. have targeted on the production of biofuels, using genetic engineering to improve strains (Dickinson et al., 2017).

6 Perspectives

Notwithstanding the current problems in microalgae production, there is no definitive barrier to the future of microalgae biofuel production, given the significant engineering innovations under development to meet the demand for resources, as well as the various methods under analysis to increase the production of biofuels from algal biomass. Concerning the future contribution of biofuels, Kagan and Bradford (2009) are very optimistic, predicting that algae feedstock would reach 37% (40 billion gallons) of the world's total biofuel production by 2022. Other researchers believe that the amount of microalgal-derived biofuel may correspond to

6000 gals/acre/year and that this amount could be increased to up to 10,000 gal/acre/year if existing limiting factors are adequately addressed (Tabatabaei, Tohidfar, Jouzani, Safarnejad, & Pazouki, 2011; Waltz, 2009). It is known that microalgae offer a potentially sustainable source of biofuels, requiring only the methodological adaptation of production technologies to each species to potentiate biofuel generation.

References

Abe, J. O., Popoola, A. P. I., Ajenifuja, E., & Popoola, O. M. (2019). Hydrogen energy, economy and storage: Review and recommendation. *International Journal of Hydrogen Energy*, *44*(29), 15072–15086. https://doi.org/ 10.1016/j.ijhydene.2019.04.068.

Adesanya, V. O., Cadena, E., Scott, S. A., & Smith, A. G. (2014). Life cycle assessment on microalgal biodiesel production using a hybrid cultivation system. *Bioresource Technology*, *163*, 343–355. https://doi.org/10.1016/j. biortech.2014.04.051.

Algenol Biotech, L. L. C. (2011). *About algenol*. (2011). http:// www.algenol.com/. Accessed 28 January 2020.

Amaro, H. M., Guedes, A. C., & Malcata, F. X. (2011). Advances and perspectives in using microalgae to produce biodiesel. *Applied Energy*, *88*, 3402–3410.

Andruleviciute, V., Makareviciene, V., Skorupskaite, V., & Gumbyte, M. (2014). Biomass and oil content of *Chlorella* sp., *Haematococcus* sp., *Nannochlopsis* sp. and *Scenedesmus* sp. under mixotrophic growth conditions in the presence of technical glycerol. *Journal of Applied Phycology*, *26*, 83–90. https://doi.org/10.1007/s10811-013-0048-x.

Aziz, M. (2015). Integrated hydrogen production and power generation from microalgae. *International Journal of Hydrogen Energy*, *41*, 104–112. https://doi.org/10.1016/ j.ijhydene.2015.10.115.

Aziz, M., Oda, T., & Kashiwagi, T. (2014). Advanced energy harvesting from macroalgae-innovative integration of drying, gasification and combined cycle. *Energies*, *7*, 8217–8235. https://doi.org/10.3390/en7128217.

Azma, M., Mohamed, M. S., Mohamad, R., Rahim, R. A., & Ariff, A. B. (2011). Improvement of medium composition for heterotrophic cultivation of green microalgae, *Tetraselmis suecica*, using response surface methodology. *Biochemical Engineering Journal*, *53*, 187–195. https://doi. org/10.1016/j.bej.2010.10.010.

Barros, A., Pereira, H., Campos, J., Marques, A., Varela, J., & Silva, J. (2019). Heterotrophy as a tool to overcome the long and costly autotrophic scale-up process for large scale production of microalgae. *Scientific Reports*, 9, 13935. https://doi.org/10.1038/s41598-019-50206-z.

Bartley, M. L., Boeing, W. J., Dungan, B. N., Holguin, F. A., & Schaub, T. (2014). pH effects on growth and lipid accumulation of the biofuel microalgae *Nannochloropsis salina* and invading organisms. *Journal of Applied Phycology*, 26, 1431–1437. https://doi.org/10.1007/s10811-013-0177-2.

Batista, F. R. M., Lucchesi, K. W., Carareto, N. D. D., Costa, M. C. D., & Meirelles, A. J. A. (2018). Properties of microalgae oil from the species *Chlorella protothecoides* and its ethylic biodiesel. *Brazilian Journal of Chemical Engineering*, 35, 1383–1394. https://doi.org/10.1590/0104-6632.20180354s20170191.

Bazdar, E., Roshandel, R., Yaghmaei, S., & Mardanpour, M. M. (2018). The effect of different light intensities and light/dark regimes on the performance of photosynthetic microalgae microbial fuel cell. *Bioresource Technology*, 261, 350–360. https://doi.org/10.1016/j.biortech.2018.04.026.

Belotti, G., Bravi, M., de Caprariis, B., de Filippis, P., & Scarsella, M. (2013). Effect of nitrogen and phosphorus starvations on *Chlorella vulgaris* lipids productivity and quality under different trophic regimes for biodiesel production. *American Journal of Plant Sciences*, 4, 44–51. https://doi.org/10.4236/ajps.2013.412A2006.

Bishop, T., & Rahman, P. (2017). The production of biomethane from the anaerobic digestion of microalgae: Production technologies for biofuels. *Advances in Biofeedstocks and Biofuels*, 177–200. https://doi.org/10.1002/9781119117551.ch7.

Bitog, J. P., Lee, I.-B., Lee, C.-G., Kim, K.-S., Hwang, H.-S., Hong, S.-W., et al. (2011). Application of computational fluid dynamics for modeling and designing photobioreactors for microalgae production: A review. *Computers and Electronics in Agriculture*, 76, 131–147. https://doi.org/10.1016/j.compag.2011.01.015.

Brennan, L., & Owende, P. (2010). Biofuels from microalgae—a review of technologies for production, processing, and extractions of biofuels and co-products. *Renewable and Sustainable Energy Reviews*, 14, 557–577. https://doi.org/10.1016/j.rser.2009.10.009.

Bt Md Nasir, N.-A. N., Aminul Islam, A. K. M., Anuar, N., & Yaakob, Z. (2019). Genetic improvement and challenges for cultivation of microalgae for biodiesel: A review. *Mini-Reviews in Organic Chemistry*, 16, 277–289. https://doi.org/10.2174/1570193X15666180627115502.

Buono, S., Langellotti, A. L., Martello, A., Rinna, F., & Fogliano, V. (2014). Functional ingredients from microalgae. *Food & Function*, 5, 1669–1685.

Callegari, A., Bolognesi, S., Cecconet, D., & Capodaglio, A. G. (2020). Production technologies, current role, and future prospects of biofuels feedstocks: A state-of-the-art review. *Critical Reviews in Environmental Science and Technology*, 50, 384–436. https://doi.org/10.1080/10643389.2019.1629801.

Cazzaniga, S., Dall'osto, L., Szaub, J., Scibilia, L., Ballottari, M., Purton, S. E., et al. (2014). Domestication of the green alga *Chlorella sorokiniana*: Reduction of antenna size improves light-use efficiency in a photobioreactor. *Biotechnology for Biofuels*, 7, 157. https://doi.org/10.1186/s13068-014-0157-z.

Cellana Inc. (2015). *Alduo technology. (2015)*. http://cellana.com/technology/core-technology/. Accessed 28 January 2020.

Chandra, R., Iqbal, H. M. N., Vishal, G., Lee, H. S., & Nagra, S. (2019). Algal biorefinery: A sustainable approach to valorize algal-based biomass towards multiple product recovery. *Bioresource Technology*, 278, 346–359. https://doi.org/10.1016/j.biortech.2019.01.104.

Chen, H., Li, T., & Wang, Q. (2019). Ten years of algal biofuel and bioproducts: Gains and pains. *Planta*, 249, 195–219. https://doi.org/10.1007/s00425-018-3066-8.

Chernova, N. I., & Kiseleva, S. V. (2017). Microalgae biofuels: Induction of lipid synthesis for biodiesel production and biomass residues into hydrogen conversion. *International Journal of Hydrogen Energy*, 42, 2861–2867. https://doi.org/10.1016/j.ijhydene.2016.05.302.

Chew, K. W., Chia, S. R., Show, P. L., Yap, Y. J., Ling, T. C., & Chang, J.-S. (2018). Effects of water culture medium, cultivation systems and growth modes for microalgae cultivation: A review. *Journal of the Taiwan Institute of Chemical Engineers*, 91, 332–344. https://doi.org/10.1016/j.jtice.2018.05.039.

Chia, S. R., Chew, K. W., Show, P. L., Yap, Y. J., Ong, H. C., Ling, T. C., et al. (2018). Analysis of economic and environmental aspects of microalgae biorefinery for biofuels production: A review. *Biotechnology Journal*, 13, 1700618. https://doi.org/10.1002/biot.201700618.

Chng, L. M., Lee, K. T., & Chan, D. J. C. (2017). Synergistic effect of pretreatment and fermentation process on carbohydrate-rich *Scenedesmus dimorphus* for bioethanol production. *Energy Conversion and Management*, 141, 410–419.

Dasgupta, C. N., Suseela, M. R., Mandotra, S. K., Kumar, P., Pandey, M. K., Toppo, K., et al. (2015). Dual uses of microalgal biomass: An integrative approach for biohydrogen and biodiesel production. *Applied Energy*, 146, 202–208. https://doi.org/10.1016/j.apenergy.2015.01.070.

Demirbas, A. (2009). Biorefineries: Current activities and future developments. *Energy Conversion and Management*, 50(11), 2782–2801. https://doi.org/10.1016/j.enconman.2009.06.035.

Demirel, Z., Imamoglu, E., & Conk Dalay, M. (2015). Fatty acid profile and lipid content of *Cylindrotheca Closterium* cultivated in air-lift photobioreactor. *Journal of Chemical*

Technology & Biotechnology, 90, 2290–2296. https://doi.org/10.1002/jctb.4687.

Deprá, M. C., dos Santos, A. M., Severo, I. A., Santos, A. B., Zepka, L. Q., & Jacob-Lopes, E. (2018). Microalgal biorefineries for bioenergy production: Can we move from concept to industrial reality? *Bioenergy Research, 11,* 727–747. https://doi.org/10.1007/s12155-018-9934-z.

Dickinson, S., Mientus, M., Frey, D., Amini-Hajibashi, A., Ozturk, S., Shaikh, F., et al. (2017). A review of biodiesel production from microalgae. *Clean Technologies and Environmental Policy, 19,* 637–668. https://doi.org/10.1007/s10098-016-1309-6.

Doan, Q. C., Moheimani, N. R., Mastrangelo, A. J., & Lewis, D. M. (2012). Microalgal biomass for bioethanol fermentation: Implications for hypersaline systems with an industrial focus. *Biomass and Bioenergy, 46,* 79–88. https://doi.org/10.1016/j.biombioe.2012.08.022.

Doan, Y., Sivaloganathan, B., & Obbard, J. (2011). Screening of marine microalgae for biodiesel feedstock. *Biomass and Bioenergy, 35,* 2534–2544. https://doi.org/10.1016/j.biombioe.2011.02.021.

Dutta, K., Daverey, A., & Lin, J. G. (2014). Evolution retrospective for alternative fuels: First to fourth generation. *Renewable Energy, 69,* 114–122. https://doi.org/10.1016/j.renene.2014.02.044.

El-Naggar, A., Lee, S. S., Rinklebe, J., Farooq, M., Song, H., Sarmah, A. K., et al. (2019). Biochar application to low fertility soils: A review of current status, and future prospects. *Geoderma, 337,* 536–554. https://doi.org/10.1016/j.geoderma.2018.09.034.

Escorsim, A., Rocha, G., Vargas, J., Mariano, A., Pereira Ramos, L., Corazza, M., et al. (2018). Extraction of *Acutodesmus obliquus* lipids using a mixture of ethanol and hexane as solvent. *Biomass and Bioenergy, 108,* 470–478. https://doi.org/10.1016/j.biombioe.2017.10.035.

Feng, Y., Li, C., & Zhang, D. (2011). Lipid production of *Chlorella vulgaris* cultured in artificial wastewater medium. *Bioresource Technology, 102,* 101–105. https://doi.org/10.1016/j.biortech.2010.06.016.

Ghimire, A., Frunzo, L., Pirozzi, F., Trably, E., Escudie, R., Lens, P. N. L., et al. (2015). A review on dark fermentative biohydrogen production from organic biomass: Process parameters and use of by-products. *Applied Energy, 144,* 73–95. https://doi.org/10.1016/j.apenergy.2015.01.045.

Ghosh, S., Banerjee, S., & Das, D. (2017). Process intensification of biodiesel production from *Chlorella* sp. MJ 11/11 by single step transesterification. *Algal Research, 27,* 12–20. https://doi.org/10.1016/j.algal.2017.08.021.

Giannelli, L., & Torzillo, G. (2012). Hydrogen production with the microalga *Chlamydomonas reinhardtii* grown in a compact tubular photobioreactor immersed in a scattering light nanoparticle suspension. *International Journal of Hydrogen Energy, 37*(22), 16951–16961. https://doi.org/10.1016/j.ijhydene.2012.08.103.

Goh, B. H. H., Ong, H. C., Cheah, M. Y., Chen, W. H., Yu, K. L., & Mahlia, T. M. I. (2019). Sustainability of direct biodiesel synthesis from microalgae biomass: A critical review. *Renewable and Sustainable Energy Reviews, 107,* 59–74. https://doi.org/10.1016/j.rser.2019.02.012.

Gouveia, L., Neves, C., Sebastião, D., Nobre, B. P., & Matos, C. T. (2014). Effect of light on the production of bioelectricity and added-value microalgae biomass in a photosynthetic alga microbial fuel cell. *Bioresource Technology, 154,* 171–177. https://doi.org/10.1016/j.biortech.2013.12.049.

Greenwell, H. C., Laurens, L. M. L., Shields, R. J., Lovitt, R. W., & Flynn, K. J. (2010). Placing microalgae on the biofuels priority list: A review of the technological challenges. *Journal of the Royal Society, Interface the Royal Society, 7,* 703–726. https://doi.org/10.1098/rsif.2009.0322.

Gumbytė, M., Makareviciene, V., Skorupskaite, V., Sendzikiene, E., & Kondratavicius, M. (2018). Enzymatic microalgae oil transesterification with ethanol in mineral diesel fuel media. *Renewable and Sustainable Energy, 10,* 013105. https://doi.org/10.1063/1.5012939.

Guo, Q., Yang, L., Wang, K., Xu, X., Wu, M., & Zhang, X. (2019). Effect of hydration-calcination CaO on the deoxygenation of bio-oil from pyrolysis of *Nannochloropsis* sp. *International Journal of Green Energy, 16*(14), 1179–1188. https://doi.org/10.1080/15435075.2019.1671393.

Haiduc, A. G., Brandenberger, M., Suquet, S., Vogel, F., Bernier-Latmani, R., & Ludwig, C. (2009). SunCHem: An integrated process for the hydrothermal production of methane from microalgae and CO_2 mitigation. *Journal of Applied Phycology, 21,* 529–541. https://doi.org/10.1007/s10811-009-9403-3.

Hallenbeck, P. (2005). Fundamentals of the fermentative production of hydrogen. *Water Science and Technology, 52,* 21–29. https://doi.org/10.2166/wst.2005.0494.

Harun, R., & Danquah, M. K. (2011). Enzymatic hydrolysis of microalgal biomass for bioethanol production. *Chemical Engineering Journal, 168,* 1079–1084. https://doi.org/10.1016/j.cej.2011.01.088.

Harun, R., Danquah, M. K., & Forde, G. M. (2010). Microalgal biomass as a fermentation feedstock for bioethanol production. *Journal of Chemical Technology & Biotechnology, 85,* 199–203. https://doi.org/10.1002/jctb.2287.

He, Q., Yang, H., & Hu, C. (2016). Culture modes and financial evaluation of two oleaginous microalgae for biodiesel production in desert area with open raceway pond. *Bioresource Technology, 218,* 571–579. https://doi.org/10.1016/j.biortech.2016.06.137.

Higgins, B. T., Gennity, I., Samra, S., Kind, T., Fiehn, O., & Vandergheynst, J. S. (2016). Cofactor symbiosis for enhanced algal growth, biofuel production, and wastewater treatment. *Algal Research, 17,* 308–315. https://doi.org/10.1016/j.algal.2016.05.024.

Ho, S. H., Chen, C. Y., & Chang, J. S. (2012). Effect of light intensity and nitrogen starvation on CO_2 fixation and lipid/carbohydrate production of an indigenous microalga *Scenedesmus obliquus* CNW-N. *Bioresource Technology, 113*, 244–252. https://doi.org/10.1016/j.biortech.2011.11.133.

Ho, S. H., Huang, S. W., Chen, C. Y., Hasunuma, T., Kondo, A., & Chang, J. S. (2013). Bioethanol production using carbohydrate-rich microalgae biomass as feedstock. *Bioresource Technology, 135*, 191–198. https://doi.org/10.1016/j.biortech.2012.10.015.

Ho, S. H., Kondo, A., Hasunuma, T., & Chang, J. S. (2013). Engineering strategies for improving the CO_2 fixation and carbohydrate productivity of *Scenedesmus obliquus* CNW-N used for bioethanol fermentation. *Bioresource Technology, 143*, 163–171. https://doi.org/10.1016/j.biortech.2013.05.043.

Huang, G., Chen, F., Wei, D., Zhang, X., & Chen, G. (2010). Biodiesel production by microalgal biotechnology. *Applied Energy, 87*, 38–46. https://doi.org/10.1016/j.apenergy.2009.06.016.

Islam, M. A., Ayoko, G., Brown, R., Stuart, D., & Heimann, K. (2013). Influence of fatty acid structure on fuel properties of algae derived biodiesel. *Procedia Engineering, 56*, 591–596. https://doi.org/10.1016/j.proeng.2013.03.164.

Jazzar, S., Olivares, C. P., De los Ríos, A., Marzouki, M. N., Acien, G., Fernandez-Sevilla, J. M., et al. (2015). Direct supercritical methanolysis of wet and dry unwashed marine microalgae (*Nannochloropsis gaditana*) to biodiesel. *Applied Energy, 148*, 210–219. https://doi.org/10.1016/j.apenergy.2015.03.069.

Junaid, S., Khanna, N., Lindblad, P., & Ahmed, M. (2019). Multifaceted biofuel production by microalgal isolates from Pakistan. *Biofuels, Bioproducts & Biorefining, 13*, 1187–1201. https://doi.org/10.1002/bbb.2009.

Kagan, J., & Bradford, T. (2009). *Biofuels 2010: Spotting the next wave.* The Prometheus Institute.

Khan, M. I., Lee, M. G., Shin, J. H., & Kim, J. D. (2017). Pretreatment optimization of the biomass of *Microcystis aeruginosa* for efficient bioethanol production. *AMB Express, 7*, 19. https://doi.org/10.1186/s13568-016-0320-y.

Khosravitabar, F. (2019). Microalgal biohydrogen photoproduction: Scaling up challenges and the ways forward. *Journal of Applied Phycology, 1–13.* https://doi.org/10.1007/s10811-019-01911-9.

Kim, K. H., Choi, I. S., Kim, H. M., Wi, S. G., & Bae, H. J. (2014). Bioethanol production from the nutrient stress-induced microalga *Chlorella vulgaris* by enzymatic hydrolysis and immobilized yeast fermentation. *Bioresource Technology, 153*, 47–54. https://doi.org/10.1016/j.biortech.2013.11.059.

Kim, B., Im, H., & Lee, J. W. (2015). In situ transesterification of highly wet microalgae using hydrochloric acid. *Bioresource Technology, 185*, 421–425. https://doi.org/10.1016/j.biortech.2015.02.092.

Kim, H. M., Oh, C. H., & Bae, H. J. (2017). Comparison of red microalgae (*Porphyridium cruentum*) culture conditions for bioethanol production. *Bioresource Technology, 233*, 44–50. https://doi.org/10.1016/j.biortech.2017.02.040.

Kruse, O., & Hankamer, B. (2010). Microalgal hydrogen production. *Current Opinion in Biotechnology, 21*, 238–243. https://doi.org/10.1016/j.copbio.2010.03.012.

Kumar, S. S., Basu, S., Gupta, S., Sharma, J., & Bishnoi, N. R. (2019). Bioelectricity generation using sulphate-reducing bacteria as anodic and microalgae as cathodic biocatalysts. *Biofuels, 10*, 81–86. https://doi.org/10.1080/17597269.2018.1426161.

Kumar, M., & Sharma, M. P. (2014). Potential assessment of microalgal oils for biodiesel production: A review. *Journal of Materials and Environmental Science, 5*, 757–766.

Lakatos, G. E., Ranglová, K., Manoel, J. C., Grivalský, T., Kopecký, J., & Masojídek, J. (2019). Bioethanol production from microalgae polysaccharides. *Folia Microbiologica, 1–18.* https://doi.org/10.1007/s12223-019-00732-0.

Lee, O. K., Seong, D. H., Lee, C. G., & Lee, E. Y. (2015). Sustainable production of liquid biofuels from renewable microalgae biomass. *Journal of Industrial and Engineering Chemistry, 29*, 24–31. https://doi.org/10.1016/j.jiec.2015.04.016.

Li, T., Xu, J., Gao, B., Xiang, W., Li, A., & Zhang, C. (2016). Morphology, growth, biochemical composition and photosynthetic performance of *Chlorella vulgaris* (Trebouxiophyceae) under low and high nitrogen supplies. *Algal Research, 16*, 481–491. https://doi.org/10.1016/j.algal.2016.04.008.

Liang, Y. (2013). Production liquid transportation fuels from heterotrophic microalgae. *Applied Energy, 104*, 860–868. https://doi.org/10.1016/j.apenergy.2012.10.067.

Liang, Y., Sarkany, N., & Cui, Y. (2009). Biomass and lipid productivities of Chlorella vulgaris under autotrophic, heterotrophic and mixotrophic growth conditions. *Biotechnology Letters, 31*, 1043–1049. https://doi.org/10.1007/s10529-009-9975-7.

Lim, D. K. Y., & Schenk, P. M. (2017). Microalgae selection and improvement as oil crops: GM vs non-GM strain engineering. *AIMS Bioengineering, 4*, 151–161. https://doi.org/10.3934/bioeng.2017.1.151.

Makareviciene, V., Gumbyte, M., & Sendzikiene, E. (2019). Simultaneous extraction of microalgae *Ankistrodesmus* sp. oil and enzymatic transesterification with ethanol in the mineral diesel medium. *Food & Bioproducts Processing: Transactions of the Institution of Chemical Engineers Part C, 116*, 89–97. https://doi.org/10.1016/j.fbp.2019.05.002.

Maliutina, K., Tahmasebi, A., & Yu, J. (2018). Pressurized entrained-flow pyrolysis of microalgae: Enhanced production of hydrogen and nitrogen-containing compounds. *Bioresource Technology, 256*, 160–169. https://doi.org/10.1016/j.biortech.2018.02.016.

Mata, T. M., Martins, A. A., & Caetano, N. S. (2010). Microalgae for biodiesel production and other applications: A review. *Renewable and Sustainable Energy Reviews, 14*, 217–232. https://doi.org/10.1016/j.rser.2009.07.020.

Menegazzo, M. L., & Fonseca, G. G. (2019). Biomass recovery and lipid extraction processes for microalgae biofuels production: A review. *Renewable and Sustainable Energy Reviews, 107*, 87–107. https://doi.org/10.1016/j.rser.2019.01.064.

Monasterio, S., Mascia, M., & Lorenzo, M. D. (2017). Electrochemical removal of microalgae with an integrated electrolysis-microbial fuel cell closed-loop system. *Separation and Purification Technology, 183*, 373–381. https://doi.org/10.1016/j.seppur.2017.03.057.

Mussatto, S. I., Dragone, G., Guimarães, P. M. R., Silva, J. P. A., Carneiro, L. M., Roberto, I. C., Vicente, A., Domingues, L., & Teixeira, J. A. (2010). Technological trends, global market, and challenges of bio-ethanol production-R1. *Biotechnology Advances, 28*(6), 817–830. https://doi.org/10.1016/j.biotechadv.2010.07.001.

Nascimento, I., Marques, S., Teles, I., Pereira, S., Druzian, J., Oliveira de Souza, C., et al. (2012). Screening microalgae strains for biodiesel production: Lipid productivity and estimation of fuel quality based on fatty acids profiles as selective criteria. *Bioenergy Research, 6*, 1–13. https://doi.org/10.1007/s12155-012-9222-2.

Nascimento, V. M., Nascimento, K. M., & Fonseca, G. G. (2020). Biotechnological potential of *Pseudokirchneriella subcapitata, Scenedesmus spinosus* and *Scenedesmus acuminatus. Acta Alimentaria, 49*, 154–162. https://doi.org/10.1556/066.2020.49.2.4.

Ohse, S., Derner, R., Ozório, R., Corrêa, R., Badiale-Furlong, E., & Cunha, P. (2015). Lipid content and fatty acid profiles in ten species of microalgae. *Idesia (Arica), 33*, 93–101. https://doi.org/10.4067/S0718-34292015000100010.

Onay, M. (2019). Bioethanol production via different saccharification strategies from *H. tetrachotoma* ME03 grown at various concentrations of municipal wastewater in a flat-photobioreactor. *Fuel, 239*, 1315–1323. https://doi.org/10.1016/j.fuel.2018.11.126.

Ördög, V., Stirk, W. A., Bálint, P., Aremu, A. O., Okem, A., Lovász, C., et al. (2016). Effect of temperature and nitrogen concentration on lipid productivity and fatty acid composition in three *Chlorella* strains. *Algal Research, 16*, 141–149. https://doi.org/10.1016/j.algal.2016.03.001.

Park, S., Nguyen, T. H. T., & Jin, E. (2019). Improving lipid production by strain development in microalgae: Strategies, challenges and perspectives. *Bioresource Technology, 292*, 121953. https://doi.org/10.1016/j.biortech.2019.121953.

Patinvoh, R. J., Osadolor, O. A., Chandolias, K., Horváth, I. S., & Taherzadeh, M. J. (2016). Innovative pretreatment strategies for biogas production. *Bioresource Technology, 224*, 13–24. https://doi.org/10.1016/j.biortech.2016.11.083.

Perez-Garcia, O., Escalante, F. M. E., de-Bashan, L. E., & Bashan, Y. (2011). Heterotrophic cultures of microalgae: Metabolism and potential products. *Water Research, 45*, 11–36. https://doi.org/10.1016/j.watres.2010.08.037.

Phwan, C. K., Chew, K. W., Sebayang, A. H., Ong, H. C., Ling, T. C., Malek, M. A., et al. (2019). Effects of acids pre-treatment on the microbial fermentation process for bioethanol production from microalgae. *Biotechnology for Biofuels, 12*, 191. https://doi.org/10.1186/s13068-019-1533-5.

Phwan, C. K., Ong, H. C., Chen, W. H., Ling, T. C., Ng, E. P., & Show, P. L. (2018). Overview: Comparison of pretreatment technologies and fermentation processes of bioethanol from microalgae. *Energy Conversion and Management, 173*, 81–94. https://doi.org/10.1016/j.enconman.2018.07.054.

Rahman, M. A., Aziz, M. A., Al-khulaidi, R. A., Sakib, N., & Islam, M. (2017). Biodiesel production from microalgae *Spirulina maxima* by two step process: Optimization of process variable. *Journal of Radiation Research and Applied Science, 10*, 140–147. https://doi.org/10.1016/j.jrras.2017.02.004.

Ren, H. Y., Kong, F., Zhao, L., Ren, N.-Q., Ma, J., Nan, J., et al. (2019). Enhanced co-production of biohydrogen and algal lipids from agricultural biomass residues in long-term operation. *Bioresource Technology, 289*, 121774. https://doi.org/10.1016/j.biortech.2019.121774.

Ren, H. Y., Liu, B. F., Kong, F., Zhao, L., & Ren, N. (2015). Hydrogen and lipid production from starch wastewater by co-culture of anaerobic sludge and oleaginous microalgae with simultaneous COD, nitrogen and phosphorus removal. *Water Research, 85*, 404–412. https://doi.org/10.1016/j.watres.2015.08.057.

Rizwan, M., Mujtaba, G., Memon, S. A., Lee, K., & Rashid, N. (2018). Exploring the potential of microalgae for new biotechnology applications and beyond: A review. *Renewable and Sustainable Energy Reviews, 92*, 394–404. https://doi.org/10.1016/j.rser.2018.04.034.

Rodrigues, T., & Braghini Junior, A. (2019). Charcoal: a discussion on carbonization kilns. *Journal of Analytical and Applied Pyrolysis*, 104670. https://doi.org/10.1016/j.jaap.2019.104670.

Romero-Villegas, G. I., Fiamengo, M., Acién Fernández, F. G., & Molina Grima, E. (2018). Utilization of centrate for the outdoor production of marine microalgae at pilot-scale in flat-panel photobioreactors. *Journal of Biotechnology, 284*, 102–114. https://doi.org/10.1016/j.jbiotec.2018.08.006.

Saratale, G. D., Saratale, R. G., Banu, J. R., & Chang, J.-S. (2019). Biohydrogen production from renewable biomass resources. In *Biohydrogen* (pp. 247–277). Elsevier. https://doi.org/10.1016/B978-0-444-64203-5.00010-1.

Sengmee, D., Cheirsilp, B., Suksaroge, T. T., & Prasertsan, P. (2017). Biophotolysis-based hydrogen and lipid production by oleaginous microalgae using crude glycerol as exogenous carbon source. *International Journal of Hydrogen Energy, 42*, 1971–1976. https://doi.org/10.1016/j.ijhydene.2016.10.089.

Sharma, K. K., Schuhmann, H., & Schenk, P. M. (2012). High lipid induction in microalgae for biodiesel production. *Energies*, *5*(5), 1532–1553. https://doi.org/10.3390/en5051532.

Shekh, A. Y., Shrivastava, P., Gupta, A., Krishnamurthi, K., Devi, S. S., & Mudliar, S. N. (2015). Biomass and lipid enhancement in *Chlorella* sp. with emphasis on biodiesel quality assessment through detailed FAME signature. *Bioresource Technology*, *201*, 276–286. https://doi.org/10.1016/j.biortech.2015.11.058.

Shobana, S., Saratale, G. D., Pugazhendhi, A., Arvindnarayan, S., Periyasamy, S., Kumar, G., et al. (2017). Fermentative hydrogen production from mixed and pure microalgae biomass: Key challenges and possible opportunities. *International Journal of Hydrogen Energy*, *42*, 26440–26453. https://doi.org/10.1016/j.ijhydene.2017.07.050.

Show, K. Y., Yan, Y., Zong, C., Guo, N., Chang, J. S., & Lee, D. J. (2019). State of the art and challenges of biohydrogen from microalgae. *Bioresource Technology*, *289*, 121747. https://doi.org/10.1016/j.biortech.2019.121747.

Silva, A. A., & Fonseca, G. G. (2020). Influence of luminosity, carbon source, and concentration of salts in the physiology of *Chlorella sorokiniana*. *Environmental Technology*, *41*, 719–729. https://doi.org/10.1080/09593330.2018.1509889.

Singh, H., & Das, D. (2018). *Biofuels from microalgae: Biohydrogen*: (pp. 201–228). Cham: Springer. https://doi.org/10.1007/978-3-319-69093-3_10.

Singh, R. N., & Sharma, S. (2012). Development of suitable photobioreactor for algae production—a review. *Renewable and Sustainable Energy Reviews*, *16*, 2347–2353. https://doi.org/10.1016/j.rser.2012.01.026.

Solovchenko, A., Pogosyan, S. I., Chivkunova, O. B., & Selyakh, I. (2014). Phycoremediation of alcohol distillery wastewater with a novel *Chlorella sorokiniana* strain cultivated in a photobioreactor monitored on-line via chlorophyll fluorescence. *Algal Research*, *6*, 234–241. https://doi.org/10.1016/j.algal.2014.01.002.

Suyono, E. A., Haryadi, W., Zusron, M., Nuhamunada, M., Rahayu, S., & Nugroho, P. (2015). The effect of salinity on growth, dry weight and lipid content of the mixed microalgae culture isolated from Glagah as biodiesel substrate. *Journal of Life Sciences*, *9*, 229–233. https://doi.org/10.17265/1934-7391/2015.05.006.

Szwaja, S., & Grab-Rogalinski, K. (2009). Hydrogen combustion in a compression ignition diesel engine. *International Journal of Hydrogen Energy*, *34*, 4413–4421. https://doi.org/10.1016/j.ijhydene.2009.03.020.

Tabatabaei, M., Tohidfar, M., Jouzani, G. S., Safarnejad, M., & Pazouki, M. (2011). Biodiesel production from genetically engineered microalgae: Future of bioenergy in Iran. *Renewable and Sustainable Energy Reviews*, *15*, 1918–1927. https://doi.org/10.1016/j.rser.2010.12.004.

Talebi, A. F., Mohtashami, S. K., Tabatabaei, M., Tohidfar, M., Bagheri, A., Zeinalabedini, M., et al. (2013). Fatty acids profiling: A selective criterion for screening microalgae strains for biodiesel production. *Algal Research*, *2*, 258–267. https://doi.org/10.1016/j.algal.2013.04.003.

Toro, J. C. S., Pérez, Y. C., & Alzate, C. A. C. (2018). Evaluation of biogas and syngas as energy vectors for heat and power generation using lignocellulosic biomass as raw material. *Electronic Journal of Biotechnology*. *33*, https://doi.org/10.1016/j.ejbt.2018.03.005.

Torres, S., Acien, G., García-Cuadra, F., & Navia, R. (2017). Direct transesterification of microalgae biomass and biodiesel refining with vacuum distillation. *Algal Research*, *28*, 30. https://doi.org/10.1016/j.algal.2017.10.001.

Velasquez-Orta, S. B., Lee, J. G. M., & Harvey, A. (2012). Alkaline *in situ* transesterification of *Chlorella vulgaris*. *Fuel*, *94*, 544–550. https://doi.org/10.1016/j.fuel.2011.11.045.

Vuppaladadiyam, A. K., Yao, J. G., Florin, N., George, A., Wang, X., Labeeuw, L., et al. (2018). Impact of flue gas compounds on microalgae and mechanisms for carbon assimilation and utilization. *ChemSusChem*, *11*, 334–355. https://doi.org/10.1002/cssc.201701611.

Wahlen, B., Willis, R., & Seefeldt, L. (2011). Biodiesel production by simultaneous extraction and conversion of total lipids from microalgae, cyanobacteria, and wild mixed-cultures. *Bioresource Technology*, *102*, 2724–2730. https://doi.org/10.1016/j.biortech.2010.11.026.

Waltz, E. (2009). Biotech's green gold? *Nature Biotechnology*, *27*, 15–18. https://doi.org/10.1038/nbt0109-15.

Xu, L., Wim Brilman, D. W., Withag, J. A., Brem, G., & Kersten, S. (2011). Assessment of a dry and a wet route for the production of biofuels from microalgae: Energy balance analysis. *Bioresource Technology*, *102*, 5113–5122. https://doi.org/10.1016/j.biortech.2011.01.066.

Xue, J., Niu, Y. F., Huang, T., Yang, W. D., Liu, J. S., & Li, H. Y. (2015). Genetic improvement of the microalga *Phaeodactylum tricornutum* for boosting neutral lipid accumulation. *Metabolic Engineering*, *27*, 1–9. https://doi.org/10.1016/j.ymben.2014.10.002.

Yu, K. L., Show, P. L., Ong, H. C., Ling, T. C., Chi-Wei Lan, J., Chen, W.-H., et al. (2017). Microalgae from wastewater treatment to biochar—Feedstock preparation and conversion technologies. *Energy Conversion and Management*, *150*, 1–13. https://doi.org/10.1016/j.enconman.2017.07.060.

Zhan, J., Rong, J., & Wang, Q. (2017). Mixotrophic cultivation, a preferable microalgae cultivation mode for biomass/bioenergy production, and bioremediation, advances and prospect. *International Journal of Hydrogen Energy*, *42*, 8505–8517. https://doi.org/10.1016/j.ijhydene.2016.12.021.

Zheng, Y., Chi, Z., Lucker, B., & Chen, S. (2012). Two-stage heterotrophic and phototrophic culture strategy for algal biomass and lipid production. *Bioresource Technology*, *103*, 484–488. https://doi.org/10.1016/j.biortech.2011.09.122.

Zhou, N., Zhang, Y., Wu, X., Gong, X., & Wang, Q. (2011). Hydrolysis of *Chlorella* biomass for fermentable sugars in the presence of HCl and $MgCl_2$. *Bioresource Technology*, *102*, 10158–10161. https://doi.org/10.1016/j.biortech.2011.08.051.

Emerging technologies for the clean recovery of antioxidants from microalgae

Francesco Donsì[a], Giovanna Ferrari[a,b], and Gianpiero Pataro[a]

[a]Department of Industrial Engineering, University of Salerno, Fisciano, Italy
[b]ProdAl Scarl, Fisciano, Italy

1 Introduction

Microalgae encompass extremely diverse groups of microorganisms in freshwater and saltwater systems (Fu et al., 2017). They refer to eukaryotic or prokaryotic microorganisms, which have drawn considerable attention as a prosperous and sustainable source of a gamut of natural bioactive compounds with potential applications, different industrial segments such as food, feed, pharmaceutical, and cosmetic (Geada et al., 2018; Ruiz et al., 2016; Zhang et al., 2019). These bioactive compounds not only include proteins, carbohydrates, lipids but also fatty acids, carotenoids phycobiliproteins, chlorophylls, polyphenols, and vitamins, which are of great commercial interest mainly due to their therapeutic properties (Fu et al., 2017; Ruiz et al., 2016). For example, microalgae represent an alternative source of natural carotenoids, which are a class of lipid-soluble pigments of different color hues and intensities, ranging from brilliant yellow, orange, and red (Poojary et al., 2016). In microalgae, carotenoids are generally localized in the inner cell bodies, such as chloroplasts or vesicles, or are contained in the cytoplasmic matrix or bound to membranes and other macromolecules in the intracellular space. They can be categorized into two groups, namely primary and secondary carotenoids. Primary carotenoids, comprising, for example, α-carotene, β-carotene, fucoxanthin, lutein, neoxanthin, violaxanthin, and zeaxanthin, represent essential structural and functional elements of the photosynthetic apparatus. Secondary carotenoids include extraplastidic pigments, such as astaxanthin, canthaxanthin, and echinenone, which are accumulated in microalgal cells in large amounts upon exposure to specific environmental stimuli (Poojary et al., 2016).

Microalgae, such as *Dunaliella salina, Chlorella vulgaris, Haematococcus pluvialis,* and *Nannochloropsis oculata,* are an example of microalgae species-rich in carotenoid compounds (Fu et al., 2017; Geada et al., 2018). Although synthetic forms are cheaper, carotenoids of microalgal

Microalgae. https://doi.org/10.1016/B978-0-12-821218-9.00007-4

173

origin offer the advantage of supplying natural colorants, with superior nutritional and therapeutic value. More specifically, they exhibit unusual biological activities, such as antioxidant properties, cardiovascular protection, anticancer, antidiabetic, and anti-obesity characteristics, as recently reviewed (Fu et al., 2017).

Likewise to carotenoids, chlorophylls are lipid-soluble pigments of commercial interest, characterized by low polarity and green color (Henriques, Silva, & Rocha, 2007). Chlorophylls also exhibit significant health benefits and are used as nutraceutical ingredients agents because of their antioxidant, antiinflammatory, antimutagenic, and antimicrobial activity (da Silva & Sant'Anna, 2017).

A third necessary type of photosynthetic pigments is represented by phycobiliproteins, especially C-phycocyanin, assembled in the thylakoid membranes of the chloroplast. These water-soluble proteins can be recovered for commercial use from the cyanobacterium *Arthrospira platensis* (spirulina) and rhodophyte *Porphyridium* (Román, Alvárez-Pez, Fernández, & Grima, 2002; Viskari & Colyer, 2003). The primary use of phycobiliproteins is as natural dyes; however, a growing interest is recently emerging towards their health-beneficial properties, especially those associated with the decreased risk of degenerative, neural, and renal diseases (Fernández-Rojas, Hernández-Juárez, & Pedraza-Chaverri, 2014; Li et al., 2015; Memije-Lazaro, Blas-Valdivia, Franco-Colín, & Cano-Europa, 2018; Raja, Hemaiswarya, Ganesan, & Carvalho, 2016), as well as with their potential application as fluorescent biomarkers (Fernández-Rojas et al., 2014; Martínez, Luengo, Saldaña, Álvarez, & Raso, 2017).

In addition to pigments, microalgae are also a source of polyphenol compounds, which range from phenolic acids and other polyphenolic compounds with relatively simple chemical structures to the more complex structures of phlorotannins (Ibáñez, Herrero, Mendiola, & Castro-Puyana, 2012). In addition to their potent antioxidant activity, polyphenols from microalgae also play beneficial activities, including chemopreventive (Kang et al., 2003), UV-protective (Artan et al., 2008), and antiproliferative effects (Kang et al., 2003).

Microalgae also produce a large variety of essential polyunsaturated fatty acids (PUFAs), namely, ω-3 eicosapentaenoic acids (EPA) and docosahexaenoic acid (DHA), as well as ω-6 γ-linoleic acid, which may be involved in the formulation of infant products and as nutritional supplements (Fu et al., 2017). Furthermore, PUFAs are thought to play an essential role in cardio-circulatory and coronary heart diseases, as well as on the treatment of atherosclerosis, hypertension, cholesterol, and cancer (Geada et al., 2018).

Bioactive compounds produced by algae are located within the cells. Thus their recovery is hindered by the presence of the cell envelope, which significantly limits their rate of mass transfer during extraction. As illustrated in Fig. 1, microalgae are characterized by complex cell envelope structures and composition, significantly varying from species to species, ranging from tiny membranes to complex, multilayered structures. As an example, among the microalgae species most used for biotechnological applications, the cell envelope of *Dunaliella* consists of a single phospholipid bilayer cell membrane, while that of *Spirulina*, *Chlorella*, *Haematococcus*, *Scenedesmus*, or *Nannochloropsis* consist of a cell membrane surrounded by a complex cell wall of different composition and hence, of different rigidity (Martínez, Delso, Álvarez, & Raso, 2020). Moreover, internal organelles, such as chloroplasts (photosynthetic lamellae, discs, or thylakoids), confined within membranes, may have different structures and contain membrane-bound photosynthetic pigments (chlorophylls, carotenoids, and phycobiliproteins). Therefore, the recovery of antioxidant compounds from microalgae must be supported by efficient methods for their selective extraction, optimized for the biological diversity, as well as the different localization of bioactive compounds in the intracellular space.

Conventional extraction processes of antioxidant compounds are often conducted following

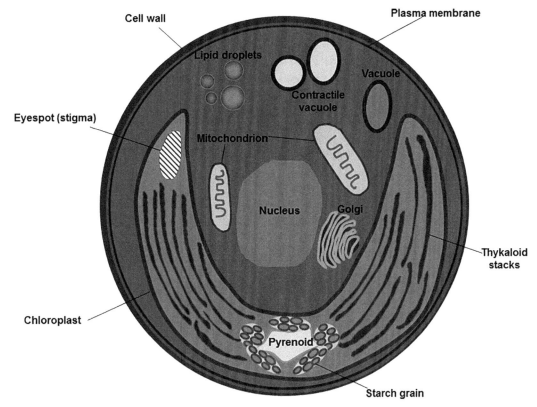

FIG. 1 Illustration detailing cell envelop and internal organelles present in a typical microalgae cell.

the dry or wet route using solvent extraction (e.g., maceration, Soxhlet extraction) with organic or aqueous solvents, depending on the polarity of the target compounds (Carullo et al., 2018). These methods offer a straightforward approach but typically suffer for several drawbacks such as high solvent consumption, long extraction times, relatively high temperature as well as low yield and selectivity (purity), which have promoted the pursuit of novel, sustainable, and ecofriendly extracting technologies (Günerken et al., 2015; Poojary et al., 2016; Zhang, Parniakov, et al., 2019). Therefore, this chapter critically analyzes the characteristics of the most promising emerging technologies for the recovery of bioactive compounds from microalgae, highlighting advantages and limitations, given their future implementation in

microalgae biorefinery. The discussed innovative techniques are based on electrotechnologies, including pulsed electric field (PEF), moderate electric field (MEF), and high-voltage electric discharges (HVED), pressurized liquid extraction (PLE), supercritical fluid extraction (SFE), microwave-assisted extraction (MAE), ultrasound-assisted extraction (UAE) and high-pressure homogenization (HPH).

2 Extraction technologies for antioxidant compounds

Between 20% and 30% of the total cost of production of algal biomass is associated with the preliminary harvesting and concentration processes (Molina Grima, Belarbi, Acién Fernández, Robles

Medina, & Chisti, 2003). This is quite a relevant issue because depending on the extraction process, different stable concentrations are required. Microalgal culture broths are significantly diluted (generally at <0.5 kg/m^3 dry biomass in commercial production), hence requiring the handling of large volumes for biomass recovery (Molina Grima et al., 2003). Owing to the small cell size (3–30 μm diameter), sedimentation, also when aided by flocculation, might not be sufficient to reach the required level of concentration of algal biomass (5%–15% dry solids). Therefore, filtration or centrifugation is often used. Filtration is a slow process, of limited use with large volumes. Centrifugation is suitable for obtaining less diluted biomass but is used mainly for high-value products (Molina Grima et al., 2003). Moreover, before extraction is carried out, cell disruption is necessary to further reduce the mass transfer resistances from the different cell structures, where the compounds of interest are located.

Mechanical disruption processes, for example, by HPH, bead milling, or ultrasonication, are discussed in the following sections. Chemical treatments can also be applied, such as the treatment with alkali, which is useful in achieving the lysis of the cell wall, despite it might cause degradation of sensitive products (e.g., proteins) (Molina Grima et al., 2003). In other cases, it is required that the microalgal biomass is wholly dried (e.g., by spray-drying, freeze-drying, or sun-drying). For example, conventional solvent extraction has been proven to be more useful for the recovery of intracellular metabolites when applied to dry biomass rather than wet biomass (Molina Grima et al., 2003), especially for what concerns the recovery of oils (Belarbi, Molina, & Chisti, 2000). Therefore, extraction technology should be selected not only based on its natural capital and operating costs but also regarding the impact on the whole process, from the perspective that a biorefinery approach is nowadays necessary for the economically viable full exploitation of all the products produced by microalgae (Vanthoor-Koopmans, Wijffels, Barbosa, & Eppink, 2013).

The main goal of a biorefinery approach is to integrate different technologies, which apply to a wide variety of end products for their recovery with suitable quality and sufficient quantities, without damaging one or more of the product fractions (Vanthoor-Koopmans et al., 2013).

2.1 Conventional solvent extraction methods

Conventional solvent extraction from microalgae is generally carried out on dry biomass, using aqueous or organic solvents, depending on their chemical structure, polarity, and distribution in the matrix (Ameer, Shahbaz, & Kwon, 2017; Monteiro et al., 2020). Especially for antioxidant compounds, the solvent choice is a critical factor, which determines the solubilization of the antioxidant compounds and hence the extraction yields result in antioxidant activity of the algal extracts (Monteiro et al., 2020). In particular, for polyphenols recovery from algae and microalgae, polar solvents are frequently used, such as aqueous mixtures, preferentially with methanol and ethanol (Goiris et al., 2012; Safafar, Van Wagenen, Møller, & Jacobsen, 2015; Zhang, Sun, Sun, Chen, & Chen, 2014), but also with acetone, or ethyl acetate (Dai & Mumper, 2010). Chloroform, diethyl ether, ethanol, hexane, or their combination (e.g., chloroform-methanol 2:1 (v/v) (Barba, Grimi, & Vorobiev, 2014)) are, instead, used to extract fatty acids such as EPA, DHA, and arachidonic acid (AA) from various microalgae (Barba et al., 2014; Molina Grima et al., 2003). Acetone, dichloromethane, dimethyl ether, diethyl ether, hexane, and octane have been used in the extraction of different carotenoids (Kanda, Kamo, Machmudah, Wahyudiono, & Goto, 2014; Mojaat, Foucault, Pruvost, & Legrand, 2008; Poojary et al., 2016; Sarkar, Das, Bhagawati, & Goswami, 2012). Increasing interest has gathered, in the last years, owing to the use of green solvents, especially ethanol and limonene (Castro-Puyana, Herrero, Urreta, et al., 2013; Castro-Puyana, Herrero, Mendiola, et al., 2013).

The most commonly used solvent extraction processes include maceration, percolation, counter-current extraction, thermal or pressurized liquid extraction, and Soxhlet (Barba et al., 2014; Poojary et al., 2016). The selection of the optimal process should also be based on the microalgae cellular structural properties. Microalgae have, in fact, different cell wall structures (Fig. 1), with most of them having rigid cell walls, which exert a significant resistance against mass transfer, slowing down the solvent extraction process (Barba et al., 2014; Wiltshire, Boersma, Möller, & Buhtz, 2000). Therefore, to enhance the mass transfer rate, in conventional solvent extraction processes large amounts of solvents are used, to ensure a significant driving force, and high temperatures to reduce viscosity, enhance diffusivity, and permeability, with an effect on the mass transfer coefficients (Barba et al., 2014), as well as to cause the lysis of the cell membranes. However, at high temperatures (>100°C), the risk of thermal denaturation or transformation of the molecules of interest is hugely significant (Barba et al., 2014; Wang & Weller, 2006).

However, the conventional solvent extraction technology for the recovery of specific antioxidant compounds lack selectivity and generally requires multiple separation steps, based on filtration followed by various chromatographic methods depending on the target metabolites, typically carotenoids, PUFAs, and polyphenols (Lim, Lee, Lee, Haam, & Kim, 2002), whereas proteins are usually purified using ion-exchange chromatography (Román et al., 2002). Therefore, the full industrial exploitation of conventional solvent extraction processes is limited by extraction efficiency, selectivity, and high solvent consumption.

3 Nonconventional extraction of bioactive compounds

Table 1 reports a survey of literature data about the use of emerging extraction technologies, such as electrotechnologies (PEF, MEF, HVED), supercritical fluid extraction (SFE), microwave-assisted extraction (MAE), ultrasound-assisted extraction (UAE) and use of high-pressure homogenization (HPH) as a pretreatment to extraction, for the recovery of antioxidant compounds.

3.1 Electrotechnologies

In the last decades' electrotechnologies, such as PEF, MEF, and HVED, have been used as cell disruption techniques for improving the extraction efficiency of different valuable compounds from a wide range of biomaterial, including agro-industrial food wastes and suspensions of microbial or algae cells (Donsì, Ferrari, & Pataro, 2010; Golberg et al., 2016; Kulshrestha & Sastry, 2003; Vorobiev & Lebovka, 2010). Although these technologies are all based on the direct exposure of wet biomaterial to an external electric field, they differ for the nature of the electric flow (i.e., alternating or direct, and pulsed or nonpulsed), the electric field strength (V/cm), and the extension of heat dissipation (Geada et al., 2018). These parameters influence not only the design and operational requirements of the corresponding equipment but also the effects induced in terms of cells damages, which may range from mild (PEF, MEF) to complete disruption (HVED) of biological cells, thus affecting the extraction efficiency of intracellular compounds of interest.

3.1.1 Pulsed electric field (PEF)-assisted extraction

PEF is one of the most advanced nonthermal processing methods, which has received considerable attention during the last decades due to its potential to create valuable and sustainable alternatives to conventional methods in biotechnological processing in which partial or total disintegration of biological (microbial, algae, plant, animal) cells is a crucial step (Raso et al., 2016).

TABLE 1 Summary of emerging extraction technologies to enhance antioxidants recovery from microalgae.

Technology	Operating conditions	Extraction conditions	Improvement in antioxidants recovery	Specific energy	Ref.
Arthrospira platensis					
UAE	50–165 W (167 W/cm², 8 min, 20 kHz	n-Heptane, diethyl ether and hexane at 10–50°C (10–60 g/L solvent)	Enhanced β-carotene recovery (up to 0.8–1.0 mg/g)	792 kJ/L of culture	Dey and Rathod (2013)
SFE	60 bar, 150°C, 0.83 h	CO_2 and 26.7% ethanol, air-dried and milled sample	Enhanced carotenoid recovery (up to 283 mg/g algae)	–	Esquivel-Hernández et al. (2016)
MAE	400 W, 50°C, 1 bar, 100 mL. 15 min	methanol/ethyl acetate/light petroleum (1:1:1, v/v)	Enhanced carotenoid recovery with respect to SFE (629 μg/g)	3600 kJ/L of culture	Esquivel-Hernández et al. (2016)
Chlamydomonas reinhardtii					
UAE	2200 W, 0.17–0.5 min, 20 kHz	Water (1.5 g$_{dw}$/L)	Enhanced carotenoids extraction	80 kJ/kg dw	Gerde, Montalbo-Lomboy, Yao, Grewell, and Wang (2012)
Chlorella pyrenoidosa					
UAE	0–19 W/cm², 0–1000 W, 15–45 kg/h, 0–6 h, 20–24 Hz	Subcritical CO_2 at 5–35 MPa and 15–33°C	Enhanced lutein recovery (up to 87–124 mg/100 g)	130–400 kJ/kg dw	Fan, Hou, Huang, Qiu, and Jiang (2015)
Chlorella sorokiniana					
HPH	100 MPa, 3 passes	30 mg of dried biomass in 10 mL of water	Enhanced carotenoid recovery (up to 1.97 mg/g)	–	Taucher et al. (2016)
Chlorella vulgaris					
UAE	56 W/cm², 60–240 min, 35 kHz	Ethanol (90%), 37°C (31 mg/L)	Enhanced lutein recovery (up to 3.16–3.36 mg/g wet weight, with enzymatic pretreatment)	–	Deenu, Naruenartwongsakul, and Kim (2013)

Method	Conditions	Solvent/sample	Result	Energy	Reference
SFE	80 bar, 500°C, 3 h	CO_2 and 7.5% ethanol	Enhanced Lutein ($\geq$1.8 mg/g) and β-carotene recovery (up to 0.2 $\geq$ mg/g)	—	Kitada et al. (2009)
SFE	40 bar, 300°C	CO_2 and 5% ethanol, crushed sample	Enhanced carotenoid recovery (up to 0.299% wt)	—	Gouveia et al. (2007)
SFE	40 bar, 400°C, 0.75 h	CO_2 and ethanol	Enhanced lutein recovery	-	Ruen-ngam, Shotipruk, Pavasant, Machmudah, and Goto (2012)
PEF	15 kV/cm	n/a, 3% dw	Enhanced carotenoid recovery (+525% compared to conventional ball milling)	100 kJ/kg dw	Töpfl (2006)
PEF	10–25 kV/cm	96% ethanol, 20°C, 1 h	Enhanced carotenoid recovery (up to 1.04 mg/g dw)	0.6–93 kJ/L of culture	Luengo, Condón-Abanto, Álvarez, and Raso (2014)
PEF	1–40 ms pulses, 3.5–5 kV/cm 3 μs pulses, 10–25 kV/cm	96% ethanol, 20°C, 1 h	Enhanced carotenoid recovery (up to 1.58 mg/L)	9–150 kJ/L of culture	Luengo, Martínez, Coustets, et al. (2015)
PEF	10–40°C, 10–25 kV/cm	96% ethanol, 20°C, 1 h	Enhanced lutein recovery (up to 0.753 mg/g dw)	1.5–93 kJ/L of culture	Luengo, Martínez, Bordetas, Álvarez, and Raso (2015)

Chlorococcum littorale

Method	Conditions	Solvent/sample	Result	Energy	Reference
SFE	60 bar, 300°C, 1–3 h	CO_2 and 10 mol% ethanol, freeze-dried sample	Enhanced carotenoid recovery (up to 0.094%–0.21% dw)	—	Ota et al. (2009)

Chromochloris zofingiensi

Method	Conditions	Solvent/sample	Result	Energy	Reference
HPH	100 MPa, 3 passes	30 mg of dried biomass in 10 mL of water	Enhanced lutein and astaxanthin recovery (up to 2.87 mg/g of total carotenoids)	—	Taucher et al. (2016)

Cylindrotheca closterium

Method	Conditions	Solvent/sample	Result	Energy	Reference
UAE	4.3–12.2 W, 3–15 min	Acetone	Enhanced fucoxanthin recovery (up to 3.5 - 4.5 mg/g)	25–350 kJ/kg dw	Pasquet et al. (2011)
MAE	25–100 W, 3–15 min, 30 mL	Acetone	Fucoxanthin recovery comparable to conventional heat treatment (4.24 mg/g)	—	Pasquet et al. (2011)

Continued

TABLE 1 Summary of emerging extraction technologies to enhance antioxidants recovery from microalgae—cont'd

Technology	Operating conditions	Extraction conditions	Improvement in antioxidants recovery	Specific energy	Ref.
Desmodesmus sp.					
HPH	280 MPa, 4 passes, 2.0–90 g/L	Water	Enhanced carotenoid recovery for conventional methods	10 MJ/kg microalgae	Xie et al. (2016)
Dunaliella salina					
UAE	20–60 W for 30 min at 20–1146 kHz	Water, 15–20°C	Enhanced carotenoid recovery and high cell inactivation efficiency	5.4 kJ/kg dw	Yamamoto, King, Wu, Mason, and Joyce (2015)
SFE	60 bar, 400°C, 3 h	CO_2 homogenized sample	Enhanced carotenoid recovery (up to 12.17 mg/g algae dw). Higher selectivity than UAE	—	Macías-Sánchez, Mantell, et al. (2009)
SFE	9.8 bar, 443°C, 1.6 h	CO_2 freeze-dried sample	Enhanced carotenoid recovery (up to 6.72% wt)	—	Jaime et al. (2007)
SFE	40–60 bar, 100–500°C, 3 h	CO_2 and 5% ethanol, freeze-dried	Enhanced carotenoid recovery (up to 13 mg/g)	—	Macías-Sánchez, Serrano, Rodríguez, and Martínez de la Ossa (2009)
SFE	50 bar, 300°C, 3 h	CO_2 and 5% ethanol, freeze-dried sample	Enhanced carotenoid recovery (up to 9.629 mg/g algae dw)	—	Macías-Sánchez et al. (2008)
Dunaliella tertiolecta					
UAE	30 mL treated at 4–12 W for 3–15 min	Water, 8.5°C	Enhanced β-carotene recovery (up to 5 mg/g)	25–350 kJ/kg dw	Pasquet et al. (2011)
MAE	50 W	Acetone	β-Carotene recovery comparable to conventional heat treatment		Pasquet et al. (2011)
Haematococcus pluvialis					
UAE	200 W, 10–20 min, 40 Hz	Ethanol and ethyl acetate at 30–50°C (50 g_{dw}/L)	Enhanced astaxanthin recovery (up to 27 mg/g)	120–240 kJ/kg dw	Tang-Bin, Jia, Li, Wang, and Wu (2013)
SFE	300 bar, 60°C	CO_2 and 9.4% ethanol, sample crushed in dry ice	Enhanced astaxanthin recovery (>97% of total)	—	Valderrama, Perrut, and Majewski (2003)

UAE	18.4W, 0–90min, 38.5 kHz	Methanol, ethanol, acetonitrile, acetone at 30–60°C (0.1 g/30mL)	Enhanced astaxanthin recovery (up to 73% of total)	2000 kJ/kg dw	Ruen-ngam, Shotipruk, and Pavasant (2011)
SFE	70bar, 500°C, 4h	CO_2, dried sample	Enhanced astaxanthin recovery (up to 23.04mg/g dw[e])	—	Thana et al. (2008)
SFE	70bar, 550°C, 4h	CO_2 and 5% ethanol, freeze-dried and ball-milled sample	Enhanced astaxanthin recovery (up to 61mg/g dw)	—	Machmudah, Shotipruk, Goto, Sasaki, and Hirose (2006)
SFE	70bar, 400°C, 5h	CO_2 and 10% olive oil, dried sample	Enhanced astaxanthin recovery (up to 51% of total)		Krichnavaruk, Shotipruk, Goto, and Pavasant (2008)
SFE	60bar and 300°C	CO_2 and 10% ethanol, freeze-dried and ball-milled sample	Enhanced carotenoid recovery (up to 92% of total), especially lutein, astaxanthin, β-carotene, and canthaxanthin		Nobre et al. (2006)
SFE	65bar, 435°C, 3.5h	CO_2 and ethanol (2.3mL/g sample), freeze-dried and powdered sample	Enhanced astaxanthin recovery (up to 87.4% of total)		Wang, Yang, Yan, and Yao (2012)
MAE	720W, 100mL, 5min	Acetone	Enhanced astaxanthin recovery (up to 74% of total)	432 kJ/L of culture	Ruen-ngam et al. (2011)
MAE	141W, 9.8mL, 83 s	Ethanol and ethyl acetate (2:1, v/v)	Optimized astaxanthin recovery (5.94 µg/mg dw)	1170 kJ/L of culture	Zhao, Chen, Zhao, and Hu (2009)
HPH	100MPa, 3 passes	30mg of dried biomass in 10mL of water	Enhanced carotenoid recovery (up to 4.21mg/g)	—	Taucher et al. (2016)
HPH	70MPa, 1 pass	water	Enhanced astaxanthin recovery	—	Olaizola (2000)
HPH	100MPa, 3 passes	water	Enhanced carotenoid recovery (up to 4.21mg/g dw)	—	Taucher et al. (2016)

Heterochlorella luteoviridis

MEF	0–180V, 60Hz, 10min, <35°C	25%–75% ethanol, 50min, 30°C	Enhanced carotenoid recovery (up to 1.21mg/g dw)	—	Jaeschke, Menegol, Rech, Mercali, and Marczak (2016)

Continued

TABLE 1 Summary of emerging extraction technologies to enhance antioxidants recovery from microalgae—cont'd

Technology	Operating conditions	Extraction conditions	Improvement in antioxidants recovery	Specific energy	Ref.
Monoraphidium sp.					
SFE	60 bar, 200°C, 1 h	CO_2 and ethanol, freeze-dried sample	Enhanced astaxanthin recovery (up to 2.02 mg/g dw)	—	Fujii (2012)
Nannochloropsis sp.					
SFE	40 bar, 300°C, >1 h	CO_2 and 20% ethanol, ball-milled sample	Enhanced carotenoid yield	—	Nobre et al. (2013)
PEF	20 kV/cm, 1–4 ms	1% (w/w) in distilled water, 3 h, 50°C, pH = 8.5–11	Enhanced carotenoid recovery (up to 0.2 mg/g dw)	13.3–53.1 kJ/kg	Parniakov et al. (2015a)
HPH	150 MPa, 1–10 passes	Water	Enhanced chlorophylls and carotenoids recovery for electrotechnologies	150–1500 kJ/kg	Grimi et al. (2014)
HPH	150 MPa, 1–10 passes	Water	Enhanced carotenoids recovery	150-y/kg	Olaizola (2000)
Nannochloropsis gaditana					
SFE	70 bar, 550°C, 4 h	CO_2, freeze dried sample	Enhanced total carotenoid recovery (up to 0.343 mg/g dw)	—	Macı´as-Sánchez et al. (2005)
SFE	40–60 bar, 100–500°C, 3 h	CO_2 and 5% ethanol, freeze-dried	Enhanced carotenoid recovery (up to 3 mg/g)	—	Macías-Sánchez, Serrano, et al. (2009)
SFE	60 bar, 500°C, 3 h	CO_2 and 5% ethanol, freeze-dried sample	Enhanced carotenoid recovery (up to 2.893 mg/g algae dw)	—	Macías-Sánchez et al. (2008)
Nannochloropsis oculata					
SFE	50 bar, 350°C	CO_2 and 16.7 wt% ethanol	Enhanced carotenoid recovery (up to 7.61 mg/g dw)	—	Liau et al. (2010)
SFE	50 bar, 350°C	CO_2 and ethanol, freeze-dried and powdered sample	Enhanced zeaxanthin recovery (13.17 mg/g)	—	Liau et al. (2011)

Porphyridium cruentum

HPH	27–270 MPa, 1–3 passes	Water	Enhanced B-Phycoerythrin	—	Jubeau et al. (2013)

Scenedesmus sp.

SFE	70 bar, 400°C, 1h	CO_2 and ethanol (30 mol %), freeze-dried sample	Enhanced lutein recovery (up to 2.210 mg/g)	—	Yen, Chiang, and Sun (2012)

Scenedesmus almeriensis

SFE	60 bar, 400°C, 5h	CO_2, freeze-dried and powdered sample	Enhanced lutein (up to 0.0466 mg/g dw) and β-carotene (up to 1.5 mg/g dw) recovery	—	Macías-Sánchez, Fernandez-Sevilla, Fernández, García, and Grima (2010)

Synechococcus sp.

SFE	500 bar, 60°C, 4h	CO_2 (freeze-dried sample)	Enhanced carotenoid (β-carotene, β-cryptoxanthin, zeaxanthin) recovery (up to 2.76 mg/g dw)	—	Montero et al. (2005)
SFE	50 bar, 300°C, 3h	CO_2, freeze-dried sample	Enhanced carotenoid recovery (up to 1.511 mg/g algae dw)	—	Macías-Sánchez et al. (2007)
SFE	40–60 bar, 100–500°C, 3h	CO_2 and 5% ethanol, freeze-dried	Enhanced carotenoid recovery (up to 1.2 mg/g)	—	Macías-Sánchez, Serrano, et al. (2009)
SFE	50 bar, 300°C, 3h	CO_2 and 5% ethanol, freeze-dried sample	Enhanced carotenoid recovery (up to 1.86 mg/g algae dw)	—	Macías-Sánchez et al. (2008)
SFE	40–60 bar, 200–400°C, 3h	CO_2 and ethanol	Enhanced β-carotene recovery (up to 0.70 mg/g algae dw) and Zeaxanthin recovery (up to 0.70 mg/g algae dw)	—	Cardoso, Serrano, Rodríguez, de la Ossa, and Lubián (2012)

PEF processing involves the application of intermittent short-duration pulses (from several nanoseconds to several milliseconds) of low, moderate, or high-intensity electric fields ($E = 0.1$–$80\,kV/cm$) to a biomaterial placed between two electrodes in either batch or continuous flow treatment chamber (Poojary et al., 2016). The pulses commonly used in PEF treatments are unipolar or bipolar, with either exponential or square-wave shape. Depending on the treatment intensity, size, and morphological characteristics of biological cells, the application of electric pulses may cause reversible or irreversible pore formation on the cell membranes (e.g., cytoplasmic and intracellular membrane structure), referred as electroporation or electropermeabilization (Kotnik, Kramar, Pucihar, Miklavčič, & Tarek, 2012). The main parameters that affect electroporation are electric field strength, treatment time, total specific energy input, and processing temperature (Raso et al., 2016). In general, increasing the intensity of these parameters enhances electroporation (Pataro et al., 2017). For the intensification of mass transfer processes, the setting of these parameters should be such that irreversible electroporation of cell membranes is achieved.

In general, PEF pretreatment enables the selective release of either water-soluble or nonpolar compounds utilizing an appropriate solvent. In particular, several authors have studied the ability of PEF to enhance the extraction yields of polyphenols and pigments with antioxidant properties such as carotenoids, chlorophylls, and phycobiliproteins, even though the first research work led to controversial results. A PEF pretreatment at $27\,kV/cm$ and $100\,kJ/kg$ increased the extraction yield of phenolic compounds in water by 5.7 times as compared to control extraction (Pataro et al., 2017). On the other hand, when cell suspensions of microalgae *Nannochloropsis sp.* were PEF ($20\,kV/cm$, 13.3–$53.1\,kJ/kg$) treated before extraction of carotenoid compounds in the same solvent, it was not possible to found any

detectible amount of carotenoids (Grimi et al., 2014). Likely, the use of a polar solvent, such as water, to extract hydrophobic substances such as carotenoids, along with the thick cell wall structure of *Nannochloropsis* sp (Scholz et al., 2014), which is likely not affected by PEF treatment, could be the reason for the observed scarce extraction efficiency. This explains the remarkable extraction yield of carotenoids (42%) and chlorophyll-a (54%) observed in the supernatant of PEF ($20\,kV/cm$, $75\,\mu s$) treated *C. vulgaris* cell suspensions using 96% ethanol as extraction solvent. Furthermore, it is worth noting that to promote extraction of carotenoids, which are bounded to chloroplasts, more intense treatment conditions are likely required for the permeabilization of smaller internal organelles like chloroplasts (Esser, Smith, Gowrishankar, Vasilkoski, & Weaver, 2010), besides the electroporation of the cytoplasmic membrane. It is known that the critical electric field required to trigger electroporation is inversely related to cell size (Kotnik et al., 2012; Luengo, Martínez, Bordetas, et al., 2015). In this line, the application of either nanosecond pulses with the high-intensity electric field around $100\,kV/cm$ or longer microsecond pulses of lower intensity has been reported to cause the electroporation of intracellular organelles (Esser et al., 2010; Schoenbach, Beebe, & Buescher, 2001).

Besides the influence of PEF treatment parameters and solvent polarity, the extractability of target intracellular compounds also depends on the elapsed time between the application of the electrical treatment and the extraction process. As an example, a PEF pretreatment of *Nannochloropsis oceanica* suspensions at $10\,kV/cm$ and $100\,kJ/kg$ followed by $1\,h$ incubation in the water at room temperature, increased the recovery yield of hydrophobic compounds such as total carotenes and chlorophyll-a by 1.6 and 1.4 times, respectively, as compared with untreated sample (Pataro, Carullo, & Ferrari, 2019). Similarly, the extraction yields for carotenoids, and chlorophylls a and b from *Chlorella vulgaris* increased by 1.2, 1.6, and 2.1 times, respectively, when the extraction was conducted 1 h after the application of PEF (20 kV/cm, 75 µs) treatment to the algae suspension, while undetectable amounts could be observed when the extracts were analyzed just after the electrical treatment (Luengo et al., 2014). These results suggested that, since PEF merely electroporates plasma membrane, the increase in the yield of pigments upon the application of PEF pretreatment may be attributed to subsequent plasmolysis of the chloroplast during the maceration time due to osmolytic disequilibrium in the cytoplasmatic space, as a consequence of the electroporation of the cytoplasmic membrane of the algae cells (Luengo et al., 2014; Pataro et al., 2019).

The possibility of selectively intensifying the extractability of either water-soluble intracellular compounds (e.g., polyphenols and proteins) or hydrophobic pigments (e.g., chlorophylls a and b and carotenoids) from microalgae by PEF-assisted multistep extraction procedure was recently reported. For example, using this approach, it has been shown that the combination of PEF and solvent extractions at various pH and the usage of a biphasic mixture of organic solvents, can remarkably improve the recovery yield of low water solubility carotenoids (Parniakov, Barba, et al., 2015a, 2015b). Correctly, it has been observed that the application of PEF to *Nannochloropsis* spp. suspension at pH 8.5, followed by extraction at pH 11, led to a noticeable increase in the concentrations of carotenoids in the aqueous extracts (Parniakov, Barba, et al., 2015a). In a further study, the same authors efficiently recovered proteins, phenolic compounds, carotenoids, and other pigments with high antioxidant capacity from *Nannochloropsis* spp. with the application of biphasic mixtures of organic solvents (i.e., dimethyl sulfoxide [DMSO] and ethanol [EtOH]) and water. It was demonstrated that a two-stage PEF(20 kV/cm)-assisted extraction process, consisting of extraction in the water at the first step

followed by extraction in the binary mixture at the second step, enabled high recovery yields of both water-soluble compounds (i.e., proteins and polyphenols) at the first step and hydrophobic components (i.e., pigments) at the second step using less concentrated mixtures of organic solvents with water (Parniakov, Barba, et al., 2015b).

The use of PEF in a hurdle approach with moderate heating of microalgae suspension has also been suggested as a sustainable strategy to enhance the extractability of valuable compounds and reduce energy consumption (Postma et al., 2016). To this regard, it has been found that a PEF treatment of *C. vulgaris* suspension at 25 kV/cm for 75 µs increased the concentration of lutein (753 µg/g dw of *C. vulgaris* culture) in the supernatant up to 4.5-fold higher, in comparison with control extraction, when carried out at 40°C, whereas a lower increase in the extraction yield up to 2.3- and 3.2-fold was detected when the processing temperature was set at 10°C and 25°C, respectively (Luengo, Martínez, Bordetas, et al., 2015). Similarly, it has been found that the increment of the processing temperature during PEF treatment up to 40°C remarkably increased the extraction yield of *C-phycocyanin* from *A. platensis* cells (Martínez et al., 2017). Interestingly, in these works, it has also been observed that mild heating of algae suspension contributes in making the cell membrane more susceptible for a breakdown under the PEF treatment and decreased the field strength and treatment time to achieve the desired yields, consequently reducing the total specific energy required for the treatment.

Finally, the recent discovery of the triggering effect of enzymatic activity of cells after electroporation and incubation opens up the possibility of new applications of PEF for the improvement of the extractability of compounds that are bounded or assembled in intracellular structures. Specifically, when Martínez, Gojkovic, et al. (2019) and Martínez, Delso, Álvarez, and Raso (2019) studied the extraction of water-soluble phycobiliproteins, such as phyco-erythrin from *Porphyridium cruentuma* and β-phycoerythrin from *Porphyridium cruentum*, they observed that the release of these compounds requires not only the diffusion of the pigments through the cell membrane but also the disassembling of the molecule from the cell organization. In this regard, it was postulated that PEF could promote the release of hydrolytic enzymes from internal organelles that would lyse the bonds between the pigment and the internal cell structure, thus enabling the water-pigment complex to diffuse through the membrane, carried by a concentration gradient.

As a final point, PEF processing is particularly interesting due to its low energy consumption, easy scale-up, and implementation in a continuous flow to processing capacities in the magnitudes of thousands of liters per hour. Furthermore, the possibility of combining it with other methods makes PEF technology suitable for the integration in a multistage or cascade biorefinery approach designed for the full exploitation of microalgal biomass (Martínez et al., 2020). However, further advancement is required before the full commercialization of PEF technology, which should include the development of more reliable and affordable pulse generation systems, as well as the optimization of the overall PEF system design.

3.1.2 Moderate electric field (MEF)-assisted extraction

MEF processing appears as a prospective technology to enhance the extraction of different compounds from various biological matrices.

MEF is a nonpulsed electrotechnology where the electric current flows periodically, reverse direction (AC) without interruption for a significant period (Rocha et al., 2018). The MEF-assisted extraction process involves the application (from seconds to minutes) of relatively low electric fields typically under1000 V generally applied in the form of a sinusoidal or bipolar square wave in the range of Hz up to tens of kHz, with or without heating, to biomaterials in direct contact with two electrodes of a batch or a continuous flow treatment chamber (Jaeschke, Menegol, Rech, Mercali, & Marczak, 2016b). MEF processing is particularly interesting because it is far less expensive than PEF while achieving many of the same effects. Interestingly, despite the relatively low field strength applied, it has been shown that MEF treatment can promote electroporation of the cell membranes and microbial inactivation also at room temperature, thus indicating that not only temperature but also electricity, are responsible for cell inactivation (Geada et al., 2018). However, besides field strength and temperature, also frequency is a key parameter in the MEF application. Although the additional effects induced by the electric field in MEF have been reported to be more effective at low frequencies (Nair et al., 2014; Shynkaryk, Ji, Alvarez, & Sastry, 2010), they have also been observed at high frequencies (i.e., kHz range) and proven effective in other applications such as cancer cell disruption (Kirson, 2004). Besides, the use of high frequencies (above 15–20) effectively eliminates or minimizes the occurrence of electrochemical reactions that otherwise could lead to undesired effects such as electrode corrosion and consequent product contamination (Pataro, Barca, Pereira, & Vicente, 2014).

Either alone or combined with thermal effects, the potential of MEF in the recovery of valuable intracellular compounds has been widely reported, mainly from plant material (Kulshrestha & Sastry, 2003; Vorobiev & Lebovka, 2010), while studies evaluating bioactive compounds extraction from microalgae using this technology are still limited. In a recent study, Jaeschke et al. (2016) showed that carotenoid extraction yield at a controlled temperature between 30°C and 35°C increased with electrical field strength and ethanol concentration (25%–75%, v/v). Correctly, a two-step process consisting of 10 min MEF (180 V, 60 Hz, 25%, v/v) process followed by a conventional extraction phase using ethanol as solvent (75%, v/v, 50 min) enabled remarkable recovery yields (up to 73%) of carotenoids from the *Heterochlorella luteoviridis* microalga biomass. The analyses of the extracts revealed that all-*trans*-lutein (856 µg/g), all-*trans*-zeaxanthin (244 µg/g), and all-*trans*-β-carotene (185 µg/g) were present in significant quantities, while all-*trans*-α-carotene, 9-13-15-*cis*-β-carotene, *cis*-violaxanthin, all-*trans*-violaxanthin, and 13-13′-*cis*-lutein were detected in minor quantities (Nezammahalleh, Ghanati, Adams, Nosrati, & Shojaosadati, 2016). Moreover, it is worth noting that the use of ethanol alone (75%, v/v) was found insufficient for the extraction of carotenoids, thus supporting the hypothesis that the improved extractability of these compounds could be related to the cell membranes permeabilization effect induced by MEF treatment.

In subsequent research, it was evaluated the effect of mild temperature (30–60°C) during MEF-assisted extraction process of carotenoids from *Heterochlorella luteoviridis* (Jaeschke, Merlo, Mercali, Rech, & Marczak, 2019). Results showed that MEF pretreatment under mild heating at 30°C, 40°C, and 50°C promoted the intensification of the extraction leading to higher carotenoid extraction yields in comparison with the control experiments. A total carotenoids extraction yield up to 86% was detected under mild heating conditions (40–50°C), indicating the existence of a synergistic effect between electric field and temperature. The thermal degradation of carotenoids was instead observed at a processing temperature of 60°C.

Although MEF is still less referred in current literature for extraction purposes than PEF, the relatively less severe processing conditions of MEF, as well as the associated heating features and the availability of reliable and affordable electric generators and control systems, may contribute to a facilitated implementation of this technology in the biorefinery process of microalgae, compared with the more challenging high pulsed power applications (Rocha et al., 2018).

3.1.3 High voltage electric discharges (HVED)-assisted extraction

HVED is a cell disintegration technique of wet biomaterial based on the phenomenon of electrical breakdown of water. During the HVED process, high energy is directly released into the aqueous medium placed between two (needle-plate) electrodes of a batch treatment chamber through a plasma channel formed by a short duration (2–5 µs) high-current/high-voltage electrical discharge (40–60 kV; 10 kA) (Vorobiev & Lebovka, 2010). Although the mechanisms of HVED are not yet well-understood, the combination of electrical breakdown with different secondary phenomena (high-amplitude pressure shock waves, bubbles cavitation, creation of liquid turbulence, heat dissipation, etc.) occurring during the treatment, are likely the cause of structural cell damages, including cell wall disruption, which makes this electrotechnology a more effective cell disintegration technique than PEF and MEF. However, the nonselective extraction and release of all cellular material may create operational problems and increased downstream purification costs. Moreover, the demand for technological solutions for reliable high pulse power generators along with process limitations (such as operation only in batch mode) is the main drawback of the method (Rocha et al., 2018).

Nevertheless, HVED is one of the most referred electrotechnologies used for the extraction of bioactive compounds from biomaterial of plant origin (Rocha et al., 2018), while there is still a lack of information about the effect of HVED on the extractability of intracellular compounds from microalgae. However, the first attempts have shown that this technology is useful to achieve the extraction of water-soluble intracellular compounds, even of high molecular weight (e.g., proteins). In contrast, HVED appears not so useful for the extraction of pigments (e.g., chlorophylls or carotenoids), which instead requires the use of organic solvents or the application of harsher, mechanical homogenization techniques, such as UAE or HPH (Grimi et al., 2014).

As per the literature survey, it can be concluded that electrotechnologies (PEF, MEF, and HVED) offer considerable potential for improving the extraction of intracellular compounds from microalgae. In particular, while PEF and MEF could easily be integrated into the first disintegration steps of a biorefinery process, HVED could be used more downstream for the recovery of the residual amount of intracellular compounds. However, further studies are required to deeply understand the mechanisms of electrical disintegration of the cell envelops, as well as to optimize the process parameters, depending on the compounds of interest and morphology of algae cells.

3.2 Pressurized liquid extraction (PLE)

PLE uses mainly water as the extraction solvent and therefore, is also known as pressurized water extraction, subcritical water extraction, superheated water extraction, and pressurized hot-water extraction. However, other (green) solvents can also be used. The critical concept of PLE is that extraction is carried using liquid solvents at elevated temperature and pressure, but under—subcritical conditions, generally in a temperature range comprised between 50°C and 200°C and in a pressure range from 35 to 200 bar (Osorio-Tobón, Meireles, Osorio-Tobón, & Meireles, 2013; Plaza & Turner, 2015). Increasing temperature above the boiling point at atmospheric pressure causes a significant reduction in solvent viscosity and surface tension, enhancing the solubility and mass transfer of intracellular compounds, contributing to reduce both extraction times and solvent consumption (Gao, Haglund, Pommer, & Jansson, 2015; Iqbal & Theegala, 2013; Li et al., 2006; Santos, Veggi, & Meireles, 2012). In particular, under typical PLE conditions, the polarity of solvents considered as green, such as water

and ethanol, becomes similar to the polarity of organic solvents, expanding the range of compounds that can be efficiently extracted (Plaza & Turner, 2015). Although water is extensively used as the polar solvent of choice for PLE, methanol, ethanol, acetone, dichloromethane, ethyl acetate, n-hexane, and propane have also been used, as along with ionic liquids, or aqueous solutions containing surfactants (Poojary et al., 2016).

Despite this, till date, PLE has been widely investigated for the recovery of many different intracellular compounds from plant sources (Mustafa & Turner, 2011; Ong, Cheong, & Goh, 2006). Its use for the recovery of antioxidant compounds from microalgae is still limited, and however, are focused on carotenoids, which are considered to be the main antioxidant compounds in several microalgae, such as *Phormidium sp.* (Rodríguez-Meizoso et al., 2008) and *Phaeodactylum tricornutum* (Gilbert-López, Barranco, Herrero, Cifuentes, & Ibáñez, 2017). Pioneering studies showed that the use of PLE enabled a more efficient extraction of carotenoids from microalgae *Dunaliella salina* and *Hematococcus pluvialis*, with shorter extraction times and lower solvent consumption than conventional solvent extraction (Denery, Dragull, Tang, & Li, 2004). Remarkably, due to the shorter extraction times associated with, the higher PLE temperatures (between 100°C and 160°C), no detrimental effect was observed during PLE on thermolabile compounds, such as carotenoids (Cha et al., 2010; Herrero, Jaime, Martin-Alvarez, Cifuentes, & Ibanez, 2006; Koo, Cha, Song, Chung, & Pan, 2012). However, above 100°C, a slight degradation of fucoxanthin during PLE of edible seaweed *Eisenia bicyclis* was observed (Shang, Kim, Lee, & Um, 2011). In contrast, for *Haematococcus pluvialis* it was observed that higher temperatures (200°C) are more suitable to recover carotenoids with high yields. In comparison, a higher antioxidant activity of the extracts retained at lower temperatures (50°C) (Jaime et al., 2010), probably also to the preservation of more thermolabile compounds, such as PUFAs, as observed for PLE from *Phaeodactylum tricornutum* (Gilbert-López et al., 2017). Other works report an increase of antioxidant activity at higher temperatures (200°C) of PLE from *Phormidium* sp., due to Maillard reaction compounds produced by thermal degradation of the sample (Rodríguez-Meizoso et al., 2008). Therefore, PLE was found suitable for the recovery of bioactive compounds from *C. vulgaris* (Cha et al., 2010) as well as from *Phaeodactylum tricornutum* (Kim et al., 2012), with higher yields in comparison with conventional maceration or UAE.

In the case of carotenoids, water-ethanol binary mixtures have been successfully used in PLE from *C. vulgaris*. For example, a 90% aqueous-ethanol solution led to better carotenoid yields than acetone, *n*-hexane, and water (Cha, Lee, et al., 2010), whereas pure ethanol gave better recovery of fucoxanthin and EPA than binary mixtures with water in the extracts from *Phaeodactylum tricornutum* (Gilbert-López et al., 2017). Moreover, it was found that by adjusting the extraction temperature, the selectivity towards different carotenoids can be regulated, with β-carotene being preferably extracted at a higher temperature (>100°C) and lutein at a lower temperature (<50°C, and longer extraction times) (Cha, Lee, et al., 2010). Similarly, from *Phormidium* sp., higher carotenoids yields were obtained with hexane or ethanol, rather than with water (Rodríguez-Meizoso et al., 2008).

In addition to the extract yield, also the selectivity can be regulated by appropriate solvent selection. In the case of PLE of carotenoids from *Haematococcus pluvialis*, the ethanol extracts were more abundant in lutein, neoxanthin, and β-carotene, whereas the *n*-hexane extracts were more abundant in astaxanthin (Jaime et al., 2010). In the case of PLE from *C. vulgaris*, acetone performed significantly better in the recovery of α-carotene, β-carotene, neoxanthin, and violaxanthin, in comparison with ethanol and water (Plaza et al., 2012). For *C. ellipsoidea*, zeaxanthin yield was higher with ethanol than with *n*-hexane or isopropanol (Koo et al., 2012). For *Neochloris oleoabundans*, pure ethanol provided higher carotenoid yields than its combination with limonene (Castro-Puyana, Herrero, Urreta, et al., 2013). In the case of PLE of carotenoids from *H. pluvialis*, better yields obtained with dichloromethane than with acetone, ethanol, ethyl acetate, and *n*-hexane (Taucher et al., 2016). It must be remarked that likewise conventional solvent extraction, also for PLE, any mechanical pretreatment process, such as HPH and ball-mill disruption, might significantly affect the extraction yields, as shown for astaxanthin from *Chromochloris zofingiensis* (Taucher et al., 2016) and fucoxanthin from *Phaeodactylum tricornutum* (Gilbert-López et al., 2017).

Given the different cell structures and associated mass transfer resistances, a common approach for the determination of optimal solvent composition, PLE temperature, and time, is the use of statistical experimental design. For instance, the central composite design has been applied to the optimization of zeaxanthin from *C. ellipsoidea* (Koo et al., 2012), or the multilevel factorial design, which was used for PLE of *Neochloris oleoabundans* (Castro-Puyana, Herrero, Urreta, et al., 2013) or *Phaeodactylum tricornutum* (Gilbert-López et al., 2017) to recover carotenoids.

3.3 Supercritical fluid extraction (SFE)

SFE is the most extensively studied emerging extraction technique for the selective recovery of valuable compounds from microalgae. SFE has been considered as a "green" alternative to conventional extraction methods, where organic solvents are replaced by supercritical fluids, i.e., fluids at a temperature and pressure above its critical limit. Moreover, since supercritical fluids possess low viscosity and high diffusivity, they provide better solvating and transport properties than liquids (Lang & Wai, 2001). CO_2 is mostly preferred solvent in SFE technique (referred to as supercritical CO_2 extraction [$scCO_2$]) since it has several advantages, including low toxicity, flammability and cost, and high purity when compared to other fluids.

During scCO$_2$ extraction process the biomaterial, typically in dried form, is placed in contact with CO$_2$ for a particular time (from minutes to hours) in a batch, semi-batch, or continuous system, where the solvating power (polarity) of supercritical fluid can be adjusted by either manipulating the temperature and pressure of the fluid or by adding a cosolvent, such as ethanol, allowing the extraction of a wide range of target compounds of different polarity (Kitada et al., 2009).

Nowadays, many laboratories and industries are replacing conventional extraction techniques with SFE using CO$_2$ due to several advantages, such as low energy consumption, absence of toxic solvents, recovery of heat-sensitive compounds, biocompounds without thermal degradation, high throughput, and reduction of the downstream costs (Molino et al., 2020). Furthermore, scCO$_2$ extraction process does not require complicated and costly separation and purification steps of the extract from the organic solvent, since CO$_2$ is a gas at room temperature and pressure (Molino et al., 2019). However, it is worth noting that, in some cases, extracts with relatively low purity are obtained, and the cost of supercritical fluids and the associated equipment makes it difficult to compete with conventional solvent extraction methods. Moreover, the scCO$_2$ extraction process is often conducted upon drying of algal biomass that requires a significant amount of energy (Cheng et al., 2011) and may cause losses of valuable food compounds (Golberg et al., 2016).

Several investigations have shown the potential of scCO$_2$ extraction for the extractability of a wide range of bioactive compounds, including carotenoids, chlorophylls, polyunsaturated fatty acid, and polyphenols from different microalgae species. In general, the scCO$_2$ extraction process of these intracellular compounds from microalgae is reported to involve several steps including (i) solubilization of CO$_2$ in the external medium, (ii) diffusion of CO$_2$ through the cell wall/membranes, (iii) penetration of CO$_2$ inside the algae cell, (iv) solubilization of target compounds, (v) diffusion of solubilized compounds through the cell wall/membrane, and vi) subsequent mass transfer in the external medium into the bulk of the supercritical solvent. Most of these steps will not occur consecutively, but rather take place simultaneously in a very complex and interrelated manner (Garcia-Gonzalez et al., 2007). However, based on the available literature data, it may be concluded that the relative importance of each of these step on the recovery yields and composition of the extracts is greatly affected by extraction operational conditions including temperature, pressure, contact time, cosolvent, flow rate, and pretreatment (Molino et al., 2020; Poojary et al., 2016).

In general, in scCO$_2$ extraction of bioactive compounds, the extraction efficiency increases with CO$_2$ pressure and temperature up to an optimal level (Aravena & del Valle, 2012; Kitada et al., 2009; Macías-Sánchez et al., 2010; Nobre et al., 2006; Yen et al., 2012), even though this trend is dependent on the combined effect of pressure and temperature (Macías-Sánchez et al., 2007; Macías-Sánchez et al., 2010; Macı´as-Sánchez et al., 2005; Macías-Sánchez, Mantell, et al., 2009; Roh, Uddin, & Chun, 2008). For example, Pataro et al. (2019) reported that the amount of extracted pigments (total carotenes and chlorophyll a) from *Nannochloropsis oceanica* cell suspension at a fixed temperature (35°C), increased when pressure increased from 8.0 to 14.0 MPa. However, further increments of the pressure did not cause any significant increase in the extraction yields of the two compounds. Similarly, Macías-Sánchez et al. (2010) and Macı´as-Sánchez et al. (2005) observed that the maximum extraction yield of extraction carotenoids and chlorophyll a from *Nannochloropsis gaditana* and *Scenedesmus almeriensis* obtained at an intermediate pressure within the investigated range (10–60 MPa). This behavior can be mainly attributed to the two effects that the increase in pressure has on CO_2 when the temperature is maintained constant. In particular, increasing the pressure causes an increase in the CO_2 density and, consequently, in the solvation power of the fluids, which increases the solubility of the compounds and extraction yield. However, when pressure increases, the diffusion coefficient decreases, thus reducing the penetration capacity of the solvent into the matrix and, hence, the extraction yield (Macı́as-Sánchez et al., 2005; Pataro et al., 2019).

On the other hand, an increase in temperature at constant pressure increases the vapor pressure resulting in improved solubility of intracellular compounds. In contrast, an increase in temperature results in a decrease in the fluid density, which, in turn, results in lower solubility of pigments. Moreover, a higher temperature may cause the thermal degradation of compounds. Therefore, it may be concluded that the recovery of carotenoids is highly dependent on the complex interaction of pressure and temperature, which significantly affects density, viscosity, and vapor pressure in the system. The predominance of one or other effects is responsible for the extraction efficiency (Macías-Sánchez et al., 2010).

Several researchers have used cosolvents such as ethanol (Bustamante, Roberts, Aravena, & Del Valle, 2011; Cardoso et al., 2012; Esquivel-Hernández et al., 2016; Fujii, 2012; Kitada et al., 2009; Liau et al., 2010; Machmudah et al., 2006; Macías-Sánchez et al., 2008; Macías-Sánchez, Serrano, et al., 2009; Nobre et al., 2006; Ota et al., 2009; Saravana et al., 2017; Valderrama et al., 2003; Wang et al., 2012; Yen et al., 2012), acetone (Kitada et al., 2009), and vegetable oil (Krichnavaruk et al., 2008; Liau et al., 2011; Saravana et al., 2017) as polarity modifiers for the efficient recovery of carotenoids of different polarity (e.g., β-carotene (Cardoso et al., 2012; Kitada et al., 2009), lutein (Kitada et al., 2009; Yen et al., 2012), astaxanthin (Bustamante et al., 2011; Fujii, 2012; Goto, Kanda, Wahyudiono, & Machmudah, 2015; Krichnavaruk et al., 2008;

Machmudah et al., 2006; Nobre et al., 2006; Valderrama et al., 2003; Wang et al., 2012), and zeaxanthin (Cardoso et al., 2012)) and fatty acids (Molino et al., 2018; Molino et al., 2019; Poojary et al., 2016). Moreover, the addition of cosolvents can cause swelling (Fahmy, Paulaitis, Johnson, & McNally, 1993) of algal cells, thus enhancing the mass transfer of intracellular compounds from the algal biomass (Bustamante et al., 2011; Nobre et al., 2006). The choice of the cosolvent should be determined by its affinity with the extractable target bioactive compounds (Molino et al., 2020), as well as considering its potential toxicity. In this regard, ethanol has been widely used as a cosolvent in the extraction of bioactive compounds from biomaterial due to less toxicity, since it is considered as a food-grade solvent (Kitada et al., 2009; Macías-Sánchez, Mantell, et al., 2009; Pataro et al., 2019). For example, when Yen et al. (Yen et al., 2012) tested different cosolvents with varied polarity during $scCO_2$ extraction of lutein from *Scenedesmus* sp., they reported that ethanol was a more effective cosolvent compared with methanol, propanol, butanol, and acetone. Similar results were achieved by Kitada et al. (2009), who found that ethanol was a more suitable cosolvent for the $scCO_2$ extraction of lutein and β-carotene from *Chlorella vulgaris* concerning acetone.

However, in another study Krichnavaruk et al. (2008) showed that the use of 10% olive oil as a cosolvent for the $scCO_2$ extraction at 70°C and 400 bar increased the recovery yield of astaxanthin from *H. pluvialis* up to 51%, which was equivalent to that obtained using ethanol as a cosolvent.

The initial pretreatment of algal biomass with physical or mechanical cell disintegration techniques, such as crushing, sonication, and ball-milling, has been shown to play a crucial role in the intensification of the extractability of bioactive compounds from microalgae during $scCO_2$ extraction (Aravena & del Valle, 2012; Crampon, Mouahid, Toudji, Lépine, & Badens, 2013; Gouveia et al., 2007; Mendes, Nobre,

Cardoso, Pereira, & Palavra, 2003; Nobre et al., 2006; Valderrama et al., 2003). In particular, this hurdle approach could be, especially crucial in the case of "hard structured" microalgal cells, where the cell wall/membrane system dramatically limits the mass transfer phenomena of solvent and analytes, thus requiring the application of intense processing conditions (high pressure, contact time and temperature for $scCO_2$) to recover substantial amounts of valuable intracellular compounds (Pataro et al., 2019).

As an example, Valderrama et al. (2003) found that the crushing pretreatment of microalga *H. pluvialis* before $scCO_2$ (60°C at 300 bar) extraction enhanced the accessibility of supercritical fluids to the carotenoids bound to the cell organelles, leading to a significant increase in the extraction yield of astaxanthin. More recently, the use of PEF treatment before $scCO_2$ extraction has been successfully used to intensify the extractability of pigments (total carotenes and chlorophyll a) from an aqueous suspension of fresh *Nannochloropsis oceanica* microalgae (Pataro et al., 2019). Interestingly, it was found that the combination of PEF (10 kV/cm; 100 kJ/kg) and $scCO_2$ extraction (14.0 MPa, 35°C, 7 min) added with 10% ethanol as cosolvent, resulted in a marked increase in the extraction yield of total carotenes (36%) and chlorophyll-a (52%) for the single $scCO_2$ extraction process, showing an apparent synergistic effect. This preliminary study highlighted that the application of PEF pretreatment followed by $scCO_2$ extraction process could be suggested for efficient processing of fresh biomass since the electroporation of the cell membranes could enhance the penetration capacity of $scCO_2$ inside the algae cells, thus intensifying the extractability of valuable compounds with lower processing conditions and avoiding the need for energy-intensive drying.

In general, an application of SFE using CO_2 to recover bioactive components from microalgae is somewhat useful, especially when a cosolvent like ethanol is added, and results in high-quality extracts. Future development of the SFE extraction technique of bioactive compounds from specific microalgae species could closely be related to the optimization of processing parameters (pressure, temperature, contact time, cosolvents) as well as the selection of appropriate cell disintegration technique for efficient and selective recovery of target analytes. Preliminary drying or lyophilization of raw biomass required to accelerate the kinetics of hydrophobic compounds are a rather expensive procedure that could limit the exploitation of this technology. However, recent findings on the combination of PEF and $scCO_2$ for the recovery of hydrophobic compounds from wet microalgae open up the possibility of a new processing approach that could allow high extraction efficiency and energy saving. Finally, high equipment costs and scale-up problems should be faced before the commercial application of the SFE technique.

3.4 Microwave-assisted extraction

MAE, consisting of the application of microwave irradiation during solvent extraction, is carried out for quicker and more efficient heating of the extraction medium than through indirect heating, contributing to a better and faster dissolution and mass transfer of intracellular compounds. Moreover, microwaves are reported to cause the evaporation of water inside the microalgae, with the development of considerable pressures, which contribute to rupturing the cell membranes and thereby, enhancing the mass transfer of solvent into the cells and of intracellular compounds into the solvent (Poojary et al., 2016). Heating is caused by the dipole rotation of molecules and ionic conduction in the medium induced by microwaves, which are defined as non-ionizing electromagnetic radiations with a characteristic frequency in the range from 300 MHz to 300 GHz. MAE can be carried out either in open or in closed vessels. Open vessels are more suitable for low-temperature and atmospheric-pressure extraction, while closed vessels are frequently used for high-temperature and high-pressure operations, likewise PLE.

Recent studies have demonstrated that MAE can be successfully used to recover antioxidant compounds, such as carotenoids, from microalgae, with the operating conditions, and, in particular, total energy delivered per kilogram of biomass and per solvent volume, as well as the polarity of the solvent, which depends on algal cell structure and targeted compounds. More specifically, when the microalgae exhibit complex exopolysaccharide envelopes, more intense microwave treatments are required (Poojary et al., 2016). For example, in the case of MAE of fucoxanthin from *Cylindrotheca closterium* in acetone, microwaves contributed to the disruption of the frustule structure, contributing to significantly reduce the extraction time, by enhancing mass transfer efficiency, in comparison with conventional extraction (3–5 min vs. 60 min) to reach similar fucoxanthin yields (Pasquet et al., 2011). However, in the case of a more straightforward cell wall structure, as for *Dunaliella tertiolecta*, no significant difference was observed between MAE and conventional extraction in terms of β-carotene recovery (Pasquet et al., 2011). Remarkably, when a closed vessel is used, the combination of high pressure and temperature contributes to enhancing the mass transfer not only by rupturing the cell wall but also through better solubilization and reduction in viscosity and interfacial tension. For example, closed-vessel MAE of *H. pluvialis* to recover astaxanthin resulted in shorter treatment time in comparison with conventional extraction or UAE (5 min vs. 60 min) (Ruen-ngam et al., 2011).

The optimal MAE energy treatment level also depends on the potential degradation of the extracted compounds. Most of the carotenoids are temperature-sensitive and might rapidly degrade or undergo isomerization at high temperatures (Poojary & Passamonti, 2015a; Poojary & Passamonti, 2015b). In particular, astaxanthin is reported to rapidly degrade above 75°C, with a significant reduction in yields at higher microwave power, as observed for *H. pluvialis* (Ruen-ngam et al., 2011; Zhao et al., 2009) or from *Arthrospira platensis* (Esquivel-Hernández et al., 2016). However, fucoxanthin is generally reported to exhibit no significant dependence on temperature, also enabling high-intensity MAE treatments (Pasquet et al., 2011; Xiao, Si, Yuan, Xu, & Li, 2012).

Different solvents have been used in MAE. For example, astaxanthin recovery from *H. pluvialis* carried out with acetone (Ruen-ngam et al., 2011) or ethanol and ethyl acetate (Zhao et al., 2009). Pure ethanol resulted in being optimal for recovery of fucoxanthin, as well as of PUFAs from *Phaeodactylum tricornutum*, although a higher recovery was attained with PLE than with MAE (Gilbert-López et al., 2017). Acetone alone or in a mixture with hexane was used for the recovery of β-carotene from Arthrospira maxima (Giorgis et al., 2017). Remarkably, only carotenoids were responsible for the observed antioxidant activity, while no polyphenols detected in the extracts (Giorgis et al., 2017).

3.5 Ultrasound-assisted extraction (UAE)

Ultrasound (US) is a particular type of sound wave beyond human hearing in the range between 20 kHz and 1 MHz with intensities higher than $1\,W/cm^2$, which can induce disruption of biological (e.g., microbial, algae, and plant) cells, depending on the frequency utilized (Meullemiestre, Breil, Abert-Vian, & Chemat, 2015). Most literature data dealt with the application of the so-called low-frequency US, defined between 18–200 kHz, in the extraction of valuable compounds from biomaterial, whereas only recently it has been reported the application of high-frequency US standing waves between 400 kHz–2 MHz to enhance separation of particulate from biomass (Juliano, Augustin, Xu, Mawson, & Knoerzer, 2017).

The application of the US to the biomaterial, including microalgae in a liquid medium, mainly utilizes the process of cavitation to induce cell disintegration. Specifically, cavitation is a complex phenomenon that involves nucleation, growth, and the transient impulsive collapse of tiny bubbles in the liquid driven by bulk pressure variation due to US waves (Vanthoor-Koopmans et al., 2013). Moreover, cavitation results in the physical effects of micro-turbulence and release of large amounts of heat and shockwaves, leading localized temperature increases up to 5000 K and pressure jets from strong bubble implosions due to unstable cavitation, which induce cell damages and ultimately lead to cell wall disruption (Barba et al., 2014; Meullemiestre et al., 2015). Therefore, low-frequency UAE has been extensively explored for the extraction of high-value compounds from biomaterial, including microalgal biomass, due to its ability to trigger unstable cavitation phenomena that allows to improve the penetration of the solvent in the intracellular space and to accelerate the subsequent diffusion of analytes (Dolatowski & Stasiak, 2012).

During UAE, the biomaterial is typically treated in batch or continuous flow mode using four types of equipment: (i) ultrasonic bath, (ii) ultrasonic probe, (iii) US plates, and (iv) tubular devices populated with small transducer ceramics (Günerken et al., 2015; Poojary et al., 2016). However, nowadays only large scale 18–30 kHz ultrasonication devices are in use due to energy consumption concerns, which typically range from 0.06 kWh/kg to 100 kWh/kg, depending on the equipment, type, and structure of algal cells and biomass concentration, among others (Günerken et al., 2015). In UAE, several parameters such as high power, frequency, intensity, duty cycles, the geometry of the ultrasonic reactor, solvent type, temperature, presence of dissolved gases, and external pressure greatly influence the extraction efficiency (recently reviewed by Chemat et al., 2017). Moreover, control of the processing temperature of the biomass, which unavoidably increases during the treatment, is crucial to extract thermally labile bioactive compounds such as carotenoids (Poojary et al., 2016).

In recent years, UAE has been successfully employed to extract lipids, phenolic compounds, and pigments from suspensions of either fresh or dry microalgae. Several works published in the available literature were devoted to the UAE of lipids from microalgae (Adam, Abert-Vian, Peltier, & Chemat, 2012; Araujo et al., 2013; Bermúdez Menéndez et al., 2014; Cravotto et al., 2008; Keris-Sen, Sen, Soydemir, & Gurol, 2014; Qv, Zhou, & Jiang, 2014). For example, Cravotto et al. (2008) compared the effects of US (19–300 kHz) and conventional extraction on the extraction yield of lipids from *Crypthecodinium cohnii* and found that the disruption by the US of the tough algal cell wall considerably improved the oil extraction yield from 4.8% in Soxhlet to 25.9% after UAE. Moreover, they did not observe significant differences in the fatty acids composition after UAE or conventional extraction, being DHA (C22:6) (39.3%) and palmitic acid (C16:0) (37.3%), the predominant fatty acids in the recovered oil. Similarly, other scientists demonstrated that UAE resulted in a significant increase in oil extraction from *Chlorella vulgaris* (52.5% w/w) (Araujo et al., 2013) and *Dunaliella tertiolecta* (45.9%) under the optimum conditions of high power (Qv et al., 2014).

Many efforts were also devoted to the UAE of different bioactive compounds such as polyphenols, carotenoids, and chlorophylls from microalgae (Deenu et al., 2013; Dey & Rathod, 2013; Fan et al., 2015; Gerde et al., 2012; Macías-Sánchez, Mantell, et al., 2009; Parniakov, Apicella, et al., 2015; Ruen-ngam et al., 2011; Tang-Bin et al., 2013; Yamamoto et al., 2015). For example, Macías-Sánchez, Mantell, et al. (2009) investigated the efficiency of UAE for the recovery of total carotenoids from *D. salina* suspensions, prepared by suspending the lyophilized microalgae in either methanol or

N,N-dimethylformamide (DMF). According to their results, UAE performed using DMF as solvent led to the highest recovery yield (up to 27.7 µg/mg dw) of carotenoids, which was also significantly higher than the yield obtained by SFE (up to 14.92 µg/mg dw). On the other hand, the UAE had a lower selectivity for carotenoids when compared to SFE (Macías-Sánchez, Mantell, et al., 2009). In another study, Parniakov, Apicella, et al. (2015) tested the UAE of phenolic and pigments (chlorophylls) from the microalgae *Nannochloropsis spp* using either special green solvents (water, ethanol (EtOH), dimethyl sulfoxide (DMSO)) or a binary mixture of solvents (water-DMSO and water-EtOH). The extraction efficiency of the UAE was found higher than that observed upon the conventional extraction procedure. US-processing time had a significant influence on the recovery of antioxidant compounds, thus being necessary to optimize this parameter to prevent any degradation of these valuable compounds. When using the single solvents, the authors found that the recovery efficiency of phenolic compounds and chlorophylls decreased in the order of DMSO > EtOH > H_2O. However, the maximum recovery of phenolic compounds and chlorophylls was found after US (400 W, 5 min) + binary mixtures of solvents (water-DMSO and water-EtOH) at 25%–30% and microalgae concentration of 10%. In several other studies, UAE was successfully used for the recovery of different carotenoids such as fucoxanthin from *Cylindrotheca closterium* (Pasquet et al., 2011) and *Phaeodactylum tricornutum* (Kim et al., 2012), β-carotene from *Dunaliella tertiolecta* (Pasquet et al., 2011), violaxanthin, neoxanthin, β-carotene and lutein from *C. vulgaris* (Plaza et al., 2012), which have been recently reviewed by (Poojary et al., 2016). However, it has been found that the yield and composition of the extracts are greatly affected by a large number of process parameters and need to be optimized for the efficient recovery of bioactive compounds. In this line, Dey and Rathod (2013) studied the effects of various UAE parameters such as intensity, extraction time, temperature, solvents, solid to solvent ratio, probe immersion length, duty cycles, and pretreatment. They defined the optimal set of values for the extraction of β-carotene from *Spirulina platensis*. In another study, Tang-Bin et al. (2013) defined the optimal value of different parameters such as US irradiation power (200 W), frequency (40 kHz), solvent composition (48.0% ethanol in ethyl acetate), liquid-to-solid ratio (20:1, mL/g), extraction time (16.0 min), and temperature (41.1°C) to achieve the maximum recovery yield of astaxanthin (27.58 mg/g) from *Haematococcus pluvialis*.

To further improve the extraction efficiency and reduce energy consumption, UAE has also been used as a pretreatment step or in a hurdle approach with other extraction methods. In this regard, Fan et al. (2015) successfully used the combination of UAE with subcritical CO_2 extraction for the extraction of lutein from the microalga, *Chlorella pyrenoidosa* (Fan et al., 2015). The combination of UAE and enzyme-assisted extraction for the extraction of lutein from *C. vulgaris* was described by Aree et al. (Deenu et al., 2013), who found that the enzymatic pretreatment using cell wall degrading enzymes facilitated the recovery of lutein in subsequent UAE.

The major drawback of ultrasonication of microalgal biomass is the relatively low cell disruption efficiency for some microalgal species and the local and overall heat production, which could lead to degradation of heat-sensitive compounds. Therefore, temperature control during treatment is crucial to improve product quality, even though the effectiveness of cell disruption might decrease significantly. However, the possibility of combining UAE with different solvent systems or other disruption methods to increase the efficiency and decrease the energy demand remains interesting given its potential use for the future microalgal biorefinery.

3.6 Cell disruption by high-pressure homogenization (HPH)

The disruption of the cell wall of microalgae before extraction significantly contributes to reducing the resistance to mass transfer, favoring the non-selective recovery of the intracellular compounds. Moreover, if the cell wall disruption is carried out by purely physical or mechanical pretreatments, avoiding the use of chemicals and high temperature, and is applied to the wet biomass, without the need for its preliminary drying, providing more substantial advantages not only for enhanced extraction yields but also for energy saving (Taucher et al., 2016).

The wet-milling techniques that have the capability of disrupting microalgal cells of micrometric size include HPH and colloid milling (CM). However, CM is suitable only for microalgal cells with low cell wall resistance and large size, as in the case of *Arthrospira platensis*. HPH is a wet milling process capable of disrupting submicrometric cell, particle, or droplet suspension, as it is based on the high-intensity fluid-mechanical stresses generated in a liquid when high-pressure energy (typically in the range between 50 and 400 MPa) is converted into kinetic energy in a specially designed homogenization valve chamber (Donsì, Ferrari, Lenza, & Maresca, 2009; Poojary et al., 2016). In the homogenization valve, therefore, intense fluid-mechanical stresses are developed, including high-pressure gradients, elongational and shear stresses, turbulence, cavitation, as well as impact with the valve surfaces (Clarke, Prescott, Khan, & Olabi, 2010; Donsì, Annunziata, & Ferrari, 2013; Donsì, Ferrari, Lenza, & Maresca, 2009; Lee, Lewis, & Ashman, 2012). Many different HPH units are available in the market, each with its proprietary disruption valve design. Overall, valve types can be classified as piston valves, orifice valves, and microfluidic channels, as discussed in detail elsewhere (Coccaro, Ferrari, & Donsì, 2018; Donsì et al., 2013; Donsì, Ferrari, & Maresca, 2009).

As a significant part of the energy, the pressure is dissipated as heat in the fluid. HPH processes are associated with a nonnegligible temperature increase in the range of 15–20°C/100 MPa.

In comparison with nonmechanical methods, which are less destructive and more biologically specific, HPH requires shorter processing times, low or no chemical costs, no risks associated with degradation or denaturation of compounds, as well as of toxicity of solvents and undesired reactions. However, HPH, because of the generation of high temperatures, of rapid pressure changes, and shear stresses developed, might cause a deterioration of the extracted intracellular compounds, and their contamination with several impurities (Günerken et al., 2015; Lee, Cho, Chang, & Oh, 2017; Stirk et al., 2020).

HPH is especially promising for the pretreatment of microalgae because it can be used equally efficiently on freshly-harvested samples, or concentrated biomass (Yap, Dumsday, Scales, & Martin, 2015), without the need for energy-intensive drying steps (Samarasinghe, Fernando, Lacey, & Faulkner, 2012). Previous data have demonstrated no significant changes in disruption efficiency when increasing total solids concentration up to 9% (w/w) for *Desmodesmus* sp. (Xie et al., 2016) and up to 25% (w/w) for *Nannochloropsis* sp. (Yap et al., 2015). The HPH required energy depends on the volume of treated fluid, rather than on the concentration of suspended solids until the solid concentration is such to affect the fluid macroscopic properties (e.g., viscosity) or to exert a protecting role on the suspended cells, which is a significant advantage in terms of process intensification. For these reasons, HPH is generally considered to be more efficient not only than other mechanical disruption systems, such as a ball or colloidal milling, freeze-drying or thawing (Taucher et al., 2016) and the US (Grimi et al., 2014; Lee, Kang, Lee, & Lee, 2013; Mulchandani, Kar, & Singhal, 2015; Taucher et al., 2016) but also than PEF and HVED (Grimi et al., 2014) and MAE

(Lee et al., 2013), especially at higher biomass concentration (Zhang, Grimi, Marchal, Lebovka, & Vorobiev, 2019).

HPH has been tested in the recovery of lipids (Choi & Lee, 2016; Halim, Rupasinghe, Tull, & Webley, 2013; Lee et al., 2013; Mulchandani et al., 2015; Olmstead, Kentish, Scales, & Martin, 2013), carotenoids (Taucher et al., 2016; Xie et al., 2016), pigments, proteins, and sugars (Mulchandani et al., 2015; Shene, Monsalve, Vergara, Lienqueo, & Rubilar, 2016). The main operating parameters are the pressure level and the number of passes, with the former being reported to be the determining variable (Choi & Lee, 2016; Grimi et al., 2014; Halim et al., 2013; Lee et al., 2013; Mulchandani et al., 2015; Olmstead et al., 2013; Spiden et al., 2013; Taucher et al., 2016; Wang, Li, Hu, Su, & Zhong, 2015; Yap et al., 2015). However, lower operating pressures (e.g., 100 MPa vs. 250 MPa) and a higher number of passes, despite not disrupting the cell wall, might be sufficient to damage the cell membranes and enhance the release of intracellular compounds, with the additional advantage of not degrading thermolabile compounds (Bernaerts, Gheysen, Foubert, Hendrickx, & Van Loey, 2019). Another important factor is represented by valve geometry, which is a proprietary design of the manufacturer; not many data are available on microalgae, but from analogy with the observed disruption efficiency on microorganisms, it can be assumed that piston valves generally ensure a more efficient disruption than orifice ones (Donsì et al., 2013).

Since HPH is an on-off fluid-mechanical disruption process (Coccaro et al., 2018), and therefore, it causes the complete disintegration of the cell wall structure, it can be profitably framed within the biorefinery concept, where multiple high value-added products are recovered, through cascading processes of different intensity (Kulkarni & Nikolov, 2018). More specifically, HPH is suitable as the final treatment step to enable the extraction of all the remaining intracellular compounds, after preliminary milder extraction processes, such as PEF (Carullo et al., 2018). The main disadvantage of HPH is represented by the lack of selectivity, which makes downstream separation and purification processes more difficult. (Donsì, Ferrari, & Maresca, 2009).

4 Conclusions and future perspectives

Microalgae represent a valuable source of natural antioxidant compounds, of commercial and industrial interest for a more sustainable and bio-based economy. However, the state-of-the-art industrial technologies for the recovery of the intracellular compounds from microalgal biomass are still based on vastly energy-consuming processes (e.g., drying and milling) and the extensive use of organic solvents. The development of greener processes requires a higher investment in emerging technologies, which can reduce the mass transfer rates of the intracellular content, increase the process selectivity towards the desired target compounds, and/or enhance the rate of heat transfer, for more energy- and resource-efficiency approaches.

Different innovative technologies have emerged, in recent years, for their capability of improving the extraction process, as a pretreatment or in integrated processes, such as the electrotechnologies based on PEF, MEF, and HVED, microwave (MAE), ultrasounds (UAE), and HPH, or PLE and SFE, which ensure higher efficiency and selectivity than conventional processes, as well as a significant reduction in treatment time or solvent consumption.

MAE, UAE, and HPH are suitable for enhancing the rate of extraction, either by making the heating process more efficient or disrupting the cell structure, reducing the mass transfer resistances. However, MAE is not desirable for the recovery of thermolabile compounds while UAE and HPH lack of selectivity, hence implying higher purification costs. PLE may provide a significant contribution towards the reduction of solvent consumption, but its use is limited by the high capital costs and the high temperatures of operation. Electrotechnologies, and in particular PEF, are, instead, suitable for low-temperature extraction of thermolabile antioxidant compounds, such as many carotenoids, but results to be scarcely efficient for microalgae with a complex cell wall structure. Another technology of great interest is based on supercritical fluids, such as CO_2 as a solvent (SFE), which enables the reduced use of organic solvents (often, the use of ethanol as cosolvents is sufficient for most polar compounds), and simplifies the purification processes of the target compounds. However, it still has many drawbacks, including the capital costs of the pressurized plant, the challenging setting of the extraction conditions, and the loss of CO_2 during operation. Moreover, it requires the preliminary drying of the biomass, which is a highly energy-intensive process, not required for the other technologies under consideration, using water-based solvents.

In the future, the scientific research should focus on the enhancement of energy and cost efficiency of the novel technologies to overcome their main limitations and improve yields and selectivities, especially from the perspective of industrial scale-up for commercial applications. These developments could contribute to streamlining the food, feed, pharmaceutical, nutraceutical, and cosmeceutical industries towards greener, more sustainable processes, and products.

References

Adam, F., Abert-Vian, M., Peltier, G., & Chemat, F. (2012). "Solvent-free" ultrasound-assisted extraction of lipids from fresh microalgae cells: A green, clean and scalable process. *Bioresource Technology, 114*, 457–465.

Ameer, K., Shahbaz, H. M., & Kwon, J. H. (2017). Green extraction methods for polyphenols from plant matrices and their byproducts: A review. *Comprehensive reviews in food science and food safety* (pp. 295–315). Vol. 16 (pp. 295–315). Blackwell Publishing Inc.

Araujo, G. S., Matos, L. J. B. L., Fernandes, J. O., Cartaxo, S. J. M., Gonçalves, L. R. B., Fernandes, F. A. N., et al. (2013).

Extraction of lipids from microalgae by ultrasound application: Prospection of the optimal extraction method. *Ultrasonics Sonochemistry, 20*(1), 95–98.

Aravena, R. I., & del Valle, J. M. (2012). Effect of microalgae preconditioning on supercritical CO_2 extraction of astaxanthin from *Haematococcus pluvialis*. In: *10th International Conference of Supercritical Fluids. USA.*

Artan, M., Li, Y., Karadeniz, F., Lee, S. H., Kim, M. M., & Kim, S. K. (2008). Anti-HIV-1 activity of phloroglucinol derivative, 6,6'-bieckol, from Ecklonia cava. *Bioorganic and Medicinal Chemistry, 16*(17), 7921–7926.

Barba, F. J., Grimi, N., & Vorobiev, E. (2014). New approaches for the use of non-conventional cell disruption technologies to extract potential food additives and nutraceuticals from microalgae. *Food Engineering Reviews, 7*(1), 45–62.

Belarbi, E. H., Molina, E., & Chisti, Y. (2000). A process for high yield and scaleable recovery of high purity eicosapentaenoic acid esters from microalgae and fish oil. *Enzyme and Microbial Technology, 26*(7), 516–529.

Bermúdez Menéndez, J. M., Arenillas, A., Menéndez Díaz, J. Á., Boffa, L., Mantegna, S., Binello, A., et al. (2014). Optimization of microalgae oil extraction under ultrasound and microwave irradiation. *Journal of Chemical Technology and Biotechnology, 89*(11), 1779–1784.

Bernaerts, T. M. M., Gheysen, L., Foubert, I., Hendrickx, M. E., & Van Loey, A. M. (2019). Evaluating microalgal cell disruption upon ultra high pressure homogenization. *Algal Research, 42.*

Bustamante, A., Roberts, P., Aravena, R., & Del Valle, J. M. (2011). Supercritical extraction of astaxanthin from *H. pluvialis* using ethanol-modified CO_2. Experiments and modeling. In *11th International Conference of Eng Food, Athens.*

Cardoso, L. C., Serrano, C. M., Rodríguez, M. R., de la Ossa, E. J. M., & Lubián, L. M. (2012). Extraction of carotenoids and fatty acids from microalgae using supercritical technology. *American Journal of Analytical Chemistry, 03*(12), 877–883.

Carullo, D., Abera, B. D., Casazza, A. A., Donsì, F., Perego, P., Ferrari, G., et al. (2018). Effect of pulsed electric fields and high pressure homogenization on the aqueous extraction of intracellular compounds from the microalgae *Chlorella vulgaris*. *Algal Research, 31.*

Castro-Puyana, M., Herrero, M., Mendiola, J. A., Suárez-Alvarez, S., Cifuentes, A., & Ibáñez, E. (2013). Extraction of new bioactives from neochloris oleoabundans using pressurized technologies and food grade solvents. In: *III Iberoamerican Conference on Supercritical Fluids Cartagena de Indias (Combodia).*

Castro-Puyana, M., Herrero, M., Urreta, I., Mendiola, J. A., Cifuentes, A., Ibáñez, E., et al. (2013). Optimization of clean extraction methods to isolate carotenoids from the microalga *Neochloris oleoabundans* and subsequent chemical characterization using liquid chromatography tandem mass spectrometry. *Analytical and Bioanalytical Chemistry*, 405(13), 4607–4616.

Cha, K. H., Kang, S. W., Kim, C. Y., Um, B. H., Na, Y. R., & Pan, C. H. (2010). Effect of pressurized liquids on extraction of antioxidants from *Chlorella vulgaris*. *Journal of Agricultural and Food Chemistry*, 58(8), 4756–4761.

Cha, K. H., Lee, H. J., Koo, S. Y., Song, D. G., Lee, D. U., & Pan, C. H. (2010). Optimization of pressurized liquid extraction of carotenoids and chlorophylls from *Chlorella vulgaris*. *Journal of Agricultural and Food Chemistry*, 58(2), 793–797.

Chemat, F., Rombaut, N., Sicaire, A.-G., Meullemiestre, A., Fabiano-Tixier, A.-S., & Abert-Vian, M. (2017). Ultrasound assisted extraction of food and natural products. Mechanisms, techniques, combinations, protocols and applications. A review. *Ultrasonics Sonochemistry*, 34, 540–560.

Cheng, C.-H., Du, T.-B., Pi, H.-C., Jang, S.-M., Lin, Y.-H., & Lee, H.-T. (2011). Comparative study of lipid extraction from microalgae by organic solvent and supercritical CO_2. *Bioresource Technology*, 102(21), 10151–10153.

Choi, W. Y., & Lee, H. Y. (2016). Effective production of bioenergy from marine *Chlorella* sp. by high-pressure homogenization. *Biotechnology and Biotechnological Equipment*, 30(1), 81–89.

Clarke, A., Prescott, T., Khan, A., & Olabi, A. G. (2010). Causes of breakage and disruption in a homogeniser. *Applied Energy*, 87(12), 3680–3690.

Coccaro, N., Ferrari, G., & Donsì, F. (2018). Understanding the break-up phenomena in an orifice-valve high pressure homogenizer using spherical bacterial cells (Lactococcus lactis) as a model disruption indicator. *Journal of Food Engineering*, 236, 60–71.

Crampon, C., Mouahid, A., Toudji, S. A. A., Lepîne, O., & Badens, E. (2013). Influence of pretreatment on supercritical CO_2 extraction from *Nannochloropsis oculata*. *Journal of Supercritical Fluids*, 79, 337–344. https://doi.org/10.1016/j.supflu.2012.12.022.

Cravotto, G., Boffa, L., Mantegna, S., Perego, P., Avogadro, M., & Cintas, P. (2008). Improved extraction of vegetable oils under high-intensity ultrasound and/or microwaves. *Ultrasonics Sonochemistry*, 15(5), 898–902.

da Silva, F. V., & Sant'Anna, C. (2017). Impact of culture conditions on the chlorophyll content of microalgae for biotechnological applications. *World Journal of Microbiology and Biotechnology*, 33(1).

Dai, J., & Mumper, R. J. (2010). Plant phenolics: Extraction, analysis and their antioxidant and anticancer properties. *Molecules* (pp. 7313–7352)Vol. 15, (pp. 7313–7352).

Deenu, A., Naruenartwongsakul, S., & Kim, S. M. (2013). Optimization and economic evaluation of ultrasound extraction of lutein from *Chlorella vulgaris*. *Biotechnology and Bioprocess Engineering*, 18(6), 1151–1162.

Denery, J. R., Dragull, K., Tang, C., & Li, Q. X. (2004). Pressurized fluid extraction of carotenoids from *Haematococcus pluvialis* and *Dunaliella salina* and kavalactones from *Piper methysticum*. *Analytica Chimica Acta*, 501(2), 175–181.

Dey, S., & Rathod, V. K. (2013). Ultrasound assisted extraction of β-carotene from *Spirulina platensis*. *Ultrasonics Sonochemistry*, 20(1), 271–276.

Dolatowski, Z. J., & Stasiak, D. M. (2012). Ultrasonically assisted diffusion processes. In F. Lebovka, N. Vorobiev, & E. Chemat (Eds.), *Enhancing extraction processes in the food industry* (pp. 123–144). Boca Raton: CRC Press.

Donsì, F., Annunziata, M., & Ferrari, G. (2013). Microbial inactivation by high pressure homogenization: Effect of the disruption valve geometry. *Journal of Food Engineering*, 115(3), 362–370.

Donsì, F., Ferrari, G., Lenza, E., & Maresca, P. (2009). Main factors regulating microbial inactivation by high-pressure homogenization: Operating parameters and scale of operation. *Chemical Engineering Science*, 64(3), 520–532.

Donsì, F., Ferrari, G., & Maresca, P. (2009). High-pressure homogenization for food sanitization. *Global Issues in Food Science and Technology*, 309–352.

Donsì, F., Ferrari, G., & Pataro, G. (2010). Applications of pulsed electric field treatments for the enhancement of mass transfer from vegetable tissue. *Food Engineering Reviews*, 2(2), 109–130.

Esquivel-Hernández, D., López, V., Rodríguez-Rodríguez, J., Alemán-Nava, G., Cuéllar-Bermúdez, S., Rostro-Alanis, M., et al. (2016). Supercritical carbon dioxide and microwave-assisted extraction of functional lipophilic compounds from *Arthrospira platensis*. *International Journal of Molecular Sciences*, 17(5), 658.

Esser, A. T., Smith, K. C., Gowrishankar, T. R., Vasilkoski, Z., & Weaver, J. C. (2010). Mechanisms for the intracellular manipulation of organelles by conventional electroporation. *Biophysical Journal*, 98(11), 2506–2514.

Fahmy, T. M., Paulaitis, M. E., Johnson, D. M., & McNally, M. E. P. (1993). Modifier effects in the supercritical fluid extraction of solutes from clay, soil, and plant materials. *Analytical Chemistry*, 65(10), 1462–1469.

Fan, X. D., Hou, Y., Huang, X. X., Qiu, T. Q., & Jiang, J. G. (2015). Ultrasound-enhanced subcritical CO_2 extraction of lutein from *Chlorella pyrenoidosa*. *Journal of Agricultural and Food Chemistry*, 63(18), 4597–4605.

Fernández-Rojas, B., Hernández-Juárez, J., & Pedraza-Chaverri, J. (2014). Nutraceutical properties of phycocyanin. *Journal of Functional Foods*, 11, 375–392.

Fu, W., Nelson, D. R., Yi, Z., Xu, M., Khraiwesh, B., Jijakli, K., et al. (2017). Bioactive compounds from microalgae: Current development and prospects. *Studies in Natural Products Chemistry*, 54(December), 199–225.

Fujii, K. (2012). Process integration of supercritical carbon dioxide extraction and acid treatment for astaxanthin extraction from a vegetative microalga. *Food and Bioproducts Processing, 90*(4), 762–766.

Gao, Q., Haglund, P., Pommer, L., & Jansson, S. (2015). Evaluation of solvent for pressurized liquid extraction of PCDD, PCDF, PCN, PCBz, PCPh and PAH in torrefied woody biomass. *Fuel, 154*, 52–58.

Garcia-Gonzalez, L., Geeraerd, A. H., Spilimbergo, S., Elst, K., Van Ginneken, L., Debevere, J., et al. (2007). High pressure carbon dioxide inactivation of microorganisms in foods: The past, the present and the future. *International Journal of Food Microbiology, 117*(1), 1–28.

Geada, P., Rodrigues, R., Loureiro, L., Pereira, R., Fernandes, B., Teixeira, J. A., et al. (2018). Electrotechnologies applied to microalgal biotechnology—Applications, techniques and future trends. *Renewable and Sustainable Energy Reviews, 94*(July), 656–668. https://doi.org/10.1016/j.rser.2018.06.059.

Gerde, J. A., Montalbo-Lomboy, M., Yao, L., Grewell, D., & Wang, T. (2012). Evaluation of microalgae cell disruption by ultrasonic treatment. *Bioresource Technology, 125*, 175–181.

Gilbert-López, B., Barranco, A., Herrero, M., Cifuentes, A., & Ibáñez, E. (2017). Development of new green processes for the recovery of bioactives from *Phaeodactylum tricornutum. Food Research International, 99*, 1056–1065.

Giorgis, M., Garella, D., Cena, C., Boffa, L., Cravotto, G., & Marini, E. (2017). An evaluation of the antioxidant properties of *Arthrospira maxima* extracts obtained using nonconventional techniques. *European Food Research and Technology, 243*(2), 227–237.

Goiris, K., Muylaert, K., Fraeye, I., Foubert, I., De Brabanter, J., & De Cooman, L. (2012). Antioxidant potential of microalgae in relation to their phenolic and carotenoid content. *Journal of Applied Phycology, 24*(6), 1477–1486.

Golberg, A., Sack, M., Teissie, J., Pataro, G., Pliquett, U., Saulis, G., et al. (2016). Energy-efficient biomass processing with pulsed electric fields for bioeconomy and sustainable development. *Biotechnology for Biofuels, 9*, 94.

Goto, M., Kanda, H., Wahyudiono, & Machmudah, S. (2015). Extraction of carotenoids and lipids from algae by supercritical CO_2 and subcritical dimethyl ether. *Journal of Supercritical Fluids, 96*, 245–251.

Gouveia, L., Nobre, B. P., Marcelo, F. M., Mrejen, S., Cardoso, M. T., Palavra, A. F., et al. (2007). Functional food oil coloured by pigments extracted from microalgae with supercritical CO_2. *Food Chemistry, 101*(2), 717–723.

Grimi, N., Dubois, A., Marchal, L., Jubeau, S., Lebovka, N. I., & Vorobiev, E. (2014). Selective extraction from microalgae *Nannochloropsis* sp. using different methods of cell disruption. *Bioresource Technology, 153*, 254–259. https://doi.org/10.1016/j.biortech.2013.12.011.

Günerken, E., D'Hondt, E., Eppink, M. H. M., Garcia-Gonzalez, L., Elst, K., & Wijffels, R. H. (2015). Cell disruption for microalgae biorefineries. *Biotechnology advances* (pp. 243–260). Vol. 33(pp. 243–260). Elsevier Inc.

Halim, R., Rupasinghe, T. W. T., Tull, D. L., & Webley, P. A. (2013). Mechanical cell disruption for lipid extraction from microalgal biomass. *Bioresource Technology, 140*, 53–63.

Henriques, M., Silva, A., & Rocha, J. (2007). Extraction and quantification of pigments from a marine microalga: A simple and reproducible method. In A. Méndez-Vilas (Ed.), *Communicating current research and educational topics and trends in applied microbiology* (pp. 586–593). Formatex.

Herrero, M., Jaime, L., Martin-Alvarez, P. J., Cifuentes, A., & Ibanez, E. (2006). Optimization of the extraction of antioxidants from *Dunaliella salina* microalga by pressurized liquids. *Journal of Agricultural and Food Chemistry, 54*(15), 5597–5603.

Ibáñez, E., Herrero, M., Mendiola, J. A., & Castro-Puyana, M. (2012). Extraction and characterization of bioactive compounds with health benefits from marine resources: Macro and micro algae, cyanobacteria, and invertebrates. In *Marine bioactive compounds: Sources, characterization and applications* (pp. 55–98). USA: Springer.

Iqbal, J., & Theegala, C. (2013). Optimizing a continuous flow lipid extraction system (CFLES) used for extracting microalgal lipids. *GCB Bioenergy, 5*(3), 327–337.

Jaeschke, D. P., Menegol, T., Rech, R., Mercali, G. D., & Marczak, L. D. F. (2016). Carotenoid and lipid extraction from *Heterochlorella luteoviridis* using moderate electric field and ethanol. *Process Biochemistry*, 1–8.

Jaeschke, D. P., Merlo, E. A., Mercali, G. D., Rech, R., & Marczak, L. D. F. (2019). The effect of temperature and moderate electric field pre-treatment on carotenoid extraction from *Heterochlorella luteoviridis. International Journal of Food Science and Technology, 54*(2), 396–402.

Jaime, L., Mendiola, J. A., Ibáñez, E., Martin-Álvarez, P. J., Cifuentes, A., Reglero, G., et al. (2007). β-Carotene isomer composition of sub- and supercritical carbon dioxide extracts. Antioxidant activity measurement. *Journal of Agricultural and Food Chemistry, 55*(26), 10585–10590.

Jaime, L., Rodríguez-Meizoso, I., Cifuentes, A., Santoyo, S., Suarez, S., Ibáñez, E., et al. (2010). Pressurized liquids as an alternative process to antioxidant carotenoids' extraction from *Haematococcus pluvialis* microalgae. *LWT-Food Science and Technology, 43*(1), 105–112.

Jubeau, S., Marchal, L., Pruvost, J., Jaouen, P., Legrand, J., & Fleurence, J. (2013). High pressure disruption: A two-step treatment for selective extraction of intracellular components from the microalga *Porphyridium cruentum. Journal of Applied Phycology, 25*(4), 983–989.

Juliano, P., Augustin, M. A., Xu, X. Q., Mawson, R., & Knoerzer, K. (2017). Advances in high frequency ultrasound separation of particulates from biomass. *Ultrasonics Sonochemistry, 35*(Pt B), 577–590.

Kanda, H., Kamo, Y., Machmudah, S., Wahyudiono, E. Y., & Goto, M. (2014). Extraction of fucoxanthin from raw macroalgae excluding drying and cell wall disruption by liquefied dimethyl ether. *Marine Drugs, 12*(5), 2383–2396.

Kang, K., Park, Y., Hye, J. H., Seong, H. K., Jeong, G. L., & Shin, H. C. (2003). Antioxidative properties of brown algae polyphenolics and their perspectives as chemopreventive agents against vascular risk factors. *Archives of Pharmacal Research, 26*(4), 286–293.

Keris-Sen, U. D., Sen, U., Soydemir, G., & Gurol, M. D. (2014). An investigation of ultrasound effect on microalgal cell integrity and lipid extraction efficiency. *Bioresource Technology, 152*, 407–413.

Kim, S. M., Jung, Y. J., Kwon, O. N., Cha, K. H., Um, B. H., Chung, D., et al. (2012). A potential commercial source of fucoxanthin extracted from the microalga *Phaeodactylum tricornutum*. *Applied Biochemistry and Biotechnology, 166*(7), 1843–1855.

Kitada, K., Machmudah, S., Sasaki, M., Goto, M., Nakashima, Y., Kumamoto, S., et al. (2009). Supercritical CO$_2$ extraction of pigment components with pharmaceutical importance from *Chlorella vulgaris*. *Journal of Chemical Technology and Biotechnology, 84*(5), 657–661.

Koo, S. Y., Cha, K. H., Song, D. G., Chung, D., & Pan, C. H. (2012). Optimization of pressurized liquid extraction of zeaxanthin from *Chlorella ellipsoidea*. *Journal of Applied Phycology, 24*(4), 725–730.

Kotnik, T., Kramar, P., Pucihar, G., Miklavčič, D., & Tarek, M. (2012). Cell membrane electroporation—Part 1: The phenomenon. *IEEE Electrical Insulation Magazine, 28*(5), 14–23.

Krichnavaruk, S., Shotipruk, A., Goto, M., & Pavasant, P. (2008). Supercritical carbon dioxide extraction of astaxanthin from *Haematococcus pluvialis* with vegetable oils as co-solvent. *Bioresource Technology, 99*(13), 5556–5560.

Kulkarni, S., & Nikolov, Z. (2018). Process for selective extraction of pigments and functional proteins from *Chlorella vulgaris*. *Algal Research, 35*, 185–193.

Kulshrestha, S., & Sastry, S. (2003). Frequency and voltage effects on enhanced diffusion during moderate electric field (MEF) treatment. *Innovative Food Science and Emerging Technologies, 4*(2), 189–194.

Lang, Q., & Wai, C. M. (2001). Supercritical fluid extraction in herbal and natural product studies—A practical review. *Talanta, 53*(4), 771–782.

Lee, S. Y., Cho, J. M., Chang, Y. K., & Oh, Y. K. (2017). Cell disruption and lipid extraction for microalgal biorefineries: A review. *Bioresource technology* (pp. 1317–1328). Vol. 244(pp. 1317–1328). Elsevier Ltd.

Lee, C. G., Kang, D. H., Lee, D. B., & Lee, H. Y. (2013). Pretreatment for simultaneous production of total lipids and fermentable sugars from marine alga, *Chlorella* sp. *Applied Biochemistry and Biotechnology, 171*(5), 1143–1158.

Lee, A. K., Lewis, D. M., & Ashman, P. J. (2012). Disruption of microalgal cells for the extraction of lipids for biofuels: Processes and specific energy requirements. *Biomass and Bioenergy, 46*, 89–101.

Li, B., Gao, M.-H., Chu, X.-M., Teng, L., Lv, C.-Y., Yang, P., et al. (2015). The synergistic antitumor effects of all-trans retinoic acid and C-phycocyanin on the lung cancer A549 cells in vitro and in vivo. *European Journal of Pharmacology, 749*, 107–114.

Li, P., Li, S. P., Lao, S. C., Fu, C. M., Kan, K. K. W., & Wang, Y. T. (2006). Optimization of pressurized liquid extraction for Z-ligustilide, Z-butylidenephthalide and ferulic acid in *Angelica sinensis*. *Journal of Pharmaceutical and Biomedical Analysis, 40*(5), 1073–1079.

Liau, B.-C., Hong, S.-E., Chang, L.-P., Shen, C.-T., Li, Y.-C., Wu, Y.-P., et al. (2011). Separation of sight-protecting zeaxanthin from *Nannochloropsis oculata* by using supercritical fluids extraction coupled with elution chromatography. *Separation and Purification Technology, 78*(1), 1–8.

Liau, B. C., Shen, C. T., Liang, F. P., Hong, S. E., Hsu, S. L., Jong, T. T., et al. (2010). Supercritical fluids extraction and anti-solvent purification of carotenoids from microalgae and associated bioactivity. *Journal of Supercritical Fluids, 55*(1), 169–175.

Lim, G.-B., Lee, S.-Y., Lee, E.-K., Haam, S.-J., & Kim, W.-S. (2002). Separation of astaxanthin from red yeast *Phaffia rhodozyma* by supercritical carbon dioxide extraction. *Biochemical Engineering Journal, 11*(2–3), 181–187.

Luengo, E., Condón-Abanto, S., Álvarez, I., & Raso, J. (2014). Effect of pulsed electric field treatments on permeabilization and extraction of pigments from *Chlorella vulgaris*. *Journal of Membrane Biology, 247*(12), 1269–1277.

Luengo, E., Martínez, J. M., Bordetas, A., Álvarez, I., & Raso, J. (2015). Influence of the treatment medium temperature on lutein extraction assisted by pulsed electric fields from *Chlorella vulgaris*. *Innovative Food Science and Emerging Technologies, 29*, 15–22.

Luengo, E., Martínez, J. M., Coustets, M., Álvarez, I., Teissié, J., Rols, M. P., et al. (2015). A comparative study on the effects of millisecond- and microsecond-pulsed electric field treatments on the permeabilization and extraction of pigments from *Chlorella vulgaris*. *The Journal of Membrane Biology, 248*(5), 883–891.

Machmudah, S., Shotipruk, A., Goto, M., Sasaki, M., & Hirose, T. (2006). Extraction of astaxanthin from *Haematococcus pluvialis* using supercritical CO$_2$ and ethanol as entrainer. *Industrial and Engineering Chemistry Research, 45*(10), 3652–3657.

Macías-Sánchez, M. D., Fernandez-Sevilla, J. M., Fernández, F. G. A., García, M. C. C., & Grima, E. M. (2010). Supercritical fluid extraction of carotenoids from *Scenedesmus almeriensis*. *Food Chemistry, 123*(3), 928–935.

Macı́as-Sánchez, M. D., Mantell, C., Rodrı́guez, M., Martı́nez de la Ossa, E., Lubián, L. M., & Montero, O. (2005). Supercritical fluid extraction of carotenoids and chlorophyll a from *Nannochloropsis gaditana*. *Journal of Food Engineering, 66*(2), 245–251.

Macías-Sánchez, M. D., Mantell, C., Rodríguez, M., Martínez de la Ossa, E., Lubián, L. M., & Montero, O. (2007). Supercritical fluid extraction of carotenoids and chlorophyll a from *Synechococcus* sp. *Journal of Supercritical Fluids, 39*(3), 323–329.

Macías-Sánchez, M. D., Mantell, C., Rodríguez, M., Martínez de la Ossa, E., Lubián, L. M., & Montero, O. (2009). Comparison of supercritical fluid and ultrasound-assisted extraction of carotenoids and chlorophyll a from *Dunaliella salina*. *Talanta, 77*(3), 948–952.

Macías-Sánchez, M. D., Mantell Serrano, C., Rodríguez Rodríguez, M., Martínez de la Ossa, E., Lubián, L. M., & Montero, O. (2008). Extraction of carotenoids and chlorophyll from microalgae with supercritical carbon dioxide and ethanol as cosolvent. *Journal of Separation Science, 31*(8), 1352–1362.

Macías-Sánchez, M. D., Serrano, C. M., Rodríguez, M. R., & Martínez de la Ossa, E. (2009). Kinetics of the supercritical fluid extraction of carotenoids from microalgae with CO_2 and ethanol as cosolvent. *Chemical Engineering Journal, 150*(1), 104–113.

Martínez, J. M., Delso, C., Álvarez, I., & Raso, J. (2019). Pulsed electric field permeabilization and extraction of phycoerythrin from *Porphyridium cruentum*. *Algal Research, 37*(June 2018), 51–56. https://doi.org/10.1016/j.algal.2018.11.005.

Martínez, J. M., Delso, C., Álvarez, I., & Raso, J. (2020). Pulsed electric field-assisted extraction of valuable compounds from microorganisms. *Comprehensive Reviews in Food Science and Food Safety*, 530–552.

Martínez, J. M., Gojkovic, Z., Ferro, L., Maza, M., Álvarez, I., Raso, J., et al. (2019). Use of pulsed electric field permeabilization to extract astaxanthin from the Nordic microalga *Haematococcus pluvialis*. *Bioresource Technology, 289* (June), 121694. https://doi.org/10.1016/j.biortech.2019.121694.

Martínez, J. M., Luengo, E., Saldaña, G., Álvarez, I., & Raso, J. (2017). C-phycocyanin extraction assisted by pulsed electric field from *Artrospira platensis*. *Food Research International, 99*, 1042–1047. Available from: (2017). https://doi.org/10.1016/j.foodres.2016.09.029.

Memije-Lazaro, I. N., Blas-Valdivia, V., Franco-Colín, M., & Cano-Europa, E. (2018). Arthrospira maxima (*Spirulina*) and C-phycocyanin prevent the progression of chronic kidney disease and its cardiovascular complications. *Journal of Functional Foods, 43*(2017), 37–43. https://doi.org/10.1016/j.jff.2018.01.013.

Mendes, R. L., Nobre, B. P., Cardoso, M. T., Pereira, A. P., & Palavra, A. F. (2003). Supercritical carbon dioxide extraction of compounds with pharmaceutical importance from microalgae. *Inorganica Chimica Acta, 356*, 328–334.

Meullemiestre, A., Breil, C., Abert-Vian, M., & Chemat, F. (2015). *Innovative techniques and alternative solvents for extraction of microbial oils*: (pp. 19–42). Springer International Publishing.

Mojaat, M., Foucault, A., Pruvost, J., & Legrand, J. (2008). Optimal selection of organic solvents for biocompatible extraction of beta-carotene from *Dunaliella salina*. *Journal of Biotechnology, 133*(4), 433–441.

Molina Grima, E., Belarbi, E. H., Acién Fernández, F. G., Robles Medina, A., & Chisti, Y. (2003). Recovery of microalgal biomass and metabolites: Process options and economics. *Biotechnology Advances, 20*(7–8), 491–515.

Molino, A., Larocca, V., Di Sanzo, G., Martino, M., Casella, P., Marino, T., et al. (2019). Extraction of bioactive compounds using supercritical carbon dioxide. *Molecules, 24*(4).

Molino, A., Mehariya, S., Di Sanzo, G., Larocca, V., Martino, M., Leone, G. P., et al. (2020). Recent developments in supercritical fluid extraction of bioactive compounds from microalgae: Role of key parameters, technological achievements and challenges. *Journal of CO_2 Utilization, 36*(September 2019), 196–209. https://doi.org/10.1016/j.jcou.2019.11.014.

Molino, A., Mehariya, S., Iovine, A., Larocca, V., Di Sanzo, G., Martino, M., et al. (2018). Extraction of astaxanthin and lutein from microalga haematococcus pluvialis in the red phase using CO_2 supercritical fluid extraction technology with ethanol as co-solvent. *Marine Drugs, 16*(11).

Monteiro, M., Santos, R. A., Iglesias, P., Couto, A., Serra, C. R., Gouvinhas, I., et al. (2020). Effect of extraction method and solvent system on the phenolic content and antioxidant activity of selected macro- and microalgae extracts. *Journal of Applied Phycology 32*, 349–362.

Montero, O., Macías-Sánchez, M. D., Lama, C. M., Lubián, L. M., Mantell, C., Rodríguez, M., et al. (2005). Supercritical CO_2 extraction of β-carotene from a marine strain of the *Cyanobacterium Synechococcus* Species. *Journal of Agricultural and Food Chemistry, 53*(25), 9701–9707.

Mulchandani, K., Kar, J. R., & Singhal, R. S. (2015). Extraction of lipids from Chlorella saccharophila using high-pressure homogenization followed by three phase partitioning. *Applied Biochemistry and Biotechnology, 176*(6), 1613–1626.

Mustafa, A., & Turner, C. (2011). Pressurized liquid extraction as a green approach in food and herbal plants extraction: A review. *Analytica Chimica Acta, 703*(1), 8–18.

Nair, G. R., Divya, V. R., Prasannan, L., Habeeba, V., Prince, M. V., & Raghavan, G. S. V. (2014). Ohmic heating as a pre-treatment in solvent extraction of rice bran. *Journal of Food Science and Technology, 51*(10).

Nezammahalleh, H., Ghanati, F., Adams, T. A., Nosrati, M., & Shojaosadati, S. A. (2016). Effect of moderate static electric field on the growth and metabolism of *Chlorella vulgaris*. *Bioresource Technology*, *218*, 700–711.

Nobre, B., Marcelo, F., Passos, R., Beirão, L., Palavra, A., Gouveia, L., et al. (2006). Supercritical carbon dioxide extraction of astaxanthin and other carotenoids from the microalga *Haematococcus pluvialis*. *European Food Research and Technology*, *223*(6), 787–790.

Nobre, B. P., Villalobos, F., Barragán, B. E., Oliveira, A. C., Batista, A. P., Marques, P. A. S. S., et al. (2013). A biorefinery from *Nannochloropsis* sp. microalga—Extraction of oils and pigments. Production of biohydrogen from the leftover biomass. *Bioresource Technology*, *135*, 128–136.

Olaizola, M. (2000). Commercial production of astaxanthin from *Haematococcus pluvialis* using 25,000-liter outdoor photobioreactors. *Journal of Applied Phycology*, *12*, 499–506.

Olmstead, I. L. D., Kentish, S. E., Scales, P. J., & Martin, G. J. O. (2013). Low solvent, low temperature method for extracting biodiesel lipids from concentrated microalgal biomass. *Bioresource Technology*, *148*, 615–619.

Ong ES, Cheong JSH, Goh D. Pressurized hot water extraction of bioactive or marker compounds in botanicals and medicinal plant materials. Journal of Chromatography. A 2006;1112(1–2):92–102.

Osorio-Tobón, J. F., Meireles, M. A. A., Osorio-Tobón, J. F., & Meireles, M. A. A. (2013). Recent applications of pressurized fluid extraction: Curcuminoids extraction with pressurized liquids. *Food Public Health*, *3*(6), 289–303.

Ota, M., Watanabe, H., Kato, Y., Watanabe, M., Sato, Y., Smith, R. L., et al. (2009). Carotenoid production from *Chlorococcum littorale* in photoautotrophic cultures with downstream supercritical fluid processing. *Journal of Separation Science*, *32*(13), 2327–2335.

Parniakov, O., Apicella, E., Koubaa, M., Barba, F. J., Grimi, N., Lebovka, N., et al. (2015). Ultrasound-assisted green solvent extraction of high-added value compounds from microalgae *Nannochloropsis* spp. *Bioresource Technology*, *198*, 262–267. https://doi.org/10.1016/j.biortech.2015.09.020.

Parniakov, O., Barba, F. J., Grimi, N., Marchal, L., Jubeau, S., Lebovka, N., et al. (2015a). Pulsed electric field and pH assisted selective extraction of intracellular components from microalgae nannochloropsis. *Algal Research*, *8*, 128–134. https://doi.org/10.1016/j.algal.2015.01.014.

Parniakov, O., Barba, F. J., Grimi, N., Marchal, L., Jubeau, S., Lebovka, N., et al. (2015b). Pulsed electric field assisted extraction of nutritionally valuable compounds from microalgae *Nannochloropsis* spp. using the binary mixture of organic solvents and water. *Innovative Food Science and Emerging Technologies*, *27*, 79–85. https://doi.org/10.1016/j.ifset.2014.11.002.

Pasquet, V., Chérouvrier, J.-R., Farhat, F., Thiéry, V., Piot, J.-M., Bérard, J.-B., et al. (2011). Study on the microalgal pigments extraction process: Performance of microwave assisted extraction. *Process Biochemistry*, *46*(1), 59–67.

Pataro, G., Barca, G. M. J., Pereira, R. N., & Vicente, A. A. (2014). Author's personal copy: Quantification of metal release from stainless steel electrodes during conventional and pulsed ohmic heating. *Innovative Food Science and Emerging Technologies*, *21*, 66–73.

Pataro, G., Carullo, D., & Ferrari, G. (2019). PEF-assisted supercritical CO_2 extraction of pigments from microalgae nannochloropsis oceanica in a continuous flow system. *Chemical Engineering Transactions*, *74*(July 2018), 97–102.

Pataro, G., Goettel, M., Straessner, R., Gusbeth, C., Ferrari, G., & Frey, W. (2017). Effect of PEF treatment on extraction of valuable compounds from microalgae *C. vulgaris*. *Chemical Engineering Transactions*, *57*, 67–72.

Plaza, M., Santoyo, S., Jaime, L., Avalo, B., Cifuentes, A., Reglero, G., et al. (2012). Comprehensive characterization of the functional activities of pressurized liquid and ultrasound-assisted extracts from *Chlorella vulgaris*. *LWT-Food Science and Technology*, *46*(1), 245–253.

Plaza, M., & Turner, C. (2015). Pressurized hot water extraction of bioactives. *TrAC Trends in Analytical Chemistry*, *71*, 39–54.

Poojary, M. M., Barba, F. J., Aliakbarian, B., Donsì, F., Pataro, G., Dias, D. A., et al. (2016). Innovative alternative technologies to extract carotenoids from microalgae and seaweeds. *Marine Drugs*, *14*(11), 1–34.

Poojary, M. M., & Passamonti, P. (2015a). Optimization of extraction of high purity all-trans-lycopene from tomato pulp waste. *Food Chemistry*, *188*, 84–91.

Poojary, M. M., & Passamonti, P. (2015b). Extraction of lycopene from tomato processing waste: Kinetics and modelling. *Food Chemistry*, *173*, 943–950.

Postma, P. R., Pataro, G., Capitoli, M., Barbosa, M. J., Wijffels, R. H., Eppink, M. H. M., et al. (2016). Selective extraction of intracellular components from the microalga *Chlorella vulgaris* by combined pulsed electric field-temperature treatment. *Bioresource Technology*, *203*, 80–88.

Qv, X. Y., Zhou, Q. F., & Jiang, J. G. (2014). Ultrasound-enhanced and microwave-assisted extraction of lipid from *Dunaliella tertiolecta* and fatty acid profile analysis. *Journal of Separation Science*, *37*(20), 2991–2999.

Raja, R., Hemaiswarya, S., Ganesan, V., & Carvalho, I. S. (2016). Recent developments in therapeutic applications of *Cyanobacteria*. *Critical Reviews in Microbiology*, *42*(3), 394–405.

Raso, J., Frey, W., Ferrari, G., Pataro, G., Knorr, D., Teissie, J., et al. (2016). Recommendations guidelines on the key information to be reported in studies of application of PEF technology in food and biotechnological processes. *Innovative Food Science and Emerging Technologies*, *37*, 312–321.

Rocha, C. M. R., Genisheva, Z., Ferreira-Santos, P., Rodrigues, R., Vicente, A. A., Teixeira, J. A., et al. (2018). Electric field-based technologies for valorization of bioresources. *Bioresources Technology*, *254*(November 2017), 325–339. https://doi.org/10.1016/j.biortech.2018.01.068.

Rodríguez-Meizoso, I., Jaime, L., Santoyo, S., Cifuentes, A., García-Blairsy Reina, G., Señoráns, F. J., et al. (2008). Pressurized fluid extraction of bioactive compounds from *Phormidium* species. *Journal of Agricultural and Food Chemistry*, *56*(10), 3517–3523.

Roh, M. K., Uddin, M. S., & Chun, B. S. (2008). Extraction of fucoxanthin and polyphenol from *Undaria pinnatifida* using supercritical carbon dioxide with co-solvent. *Biotechnology and Bioprocess Engineering*, *13*(6), 724–729.

Román, R. B., Alvárez-Pez, J. M., Fernández, F. G. A., & Grima, E. M. (2002). Recovery of pure b-phycoerythrin from the microalga porphyridium cruentum. *Journal of Biotechnology*, *93*(1), 73–85.

Ruen-ngam, D., Shotipruk, A., & Pavasant, P. (2011). Comparison of extraction methods for recovery of astaxanthin from *Haematococcus pluvialis*. *Separation Science and Technology*, *46*(1), 64–70.

Ruen-ngam, D., Shotipruk, A., Pavasant, P., Machmudah, S., & Goto, M. (2012). Selective extraction of lutein from alcohol treated *Chlorella vulgaris* by supercritical CO_2. *Chemical Engineering and Technology*, *35*(2), 255–260.

Ruiz, J., Olivieri, G., De Vree, J., Bosma, R., Willems, P., Reith, J. H., et al. (2016). Towards industrial products from microalgae. *Energy & Environmental Science*, *9*(10), 3036–3043.

Safafar, H., Van Wagenen, J., Møller, P., & Jacobsen, C. (2015). Carotenoids, phenolic compounds and tocopherols contribute to the antioxidative properties of some microalgae species grown on industrial wastewater. *Marine Drugs*, *13*(12), 7339–7356.

Samarasinghe, N., Fernando, S., Lacey, R., & Faulkner, W. B. (2012). Algal cell rupture using high pressure homogenization as a prelude to oil extraction. *Renewable Energy*, *48*.

Santos, D. T., Veggi, P. C., & Meireles, M. A. A. (2012). Optimization and economic evaluation of pressurized liquid extraction of phenolic compounds from jabuticaba skins. *Journal of Food Engineering*, *108*(3), 444–452.

Saravana, P. S., Getachew, A. T., Cho, Y.-J., Choi, J. H., Park, Y. B., Woo, H. C., et al. (2017). Influence of co-solvents on fucoxanthin and phlorotannin recovery from brown seaweed using supercritical CO_2. *Journal of Supercritical Fluids*, *120*, 295–303.

Sarkar, C. R., Das, L., Bhagawati, B., & Goswami, B. C. (2012). A comparative study of carotenoid extraction from algae in different solvent systems. *Asian Journal of Plant Science and Research*, *2*(4), 546–549.

Schoenbach, K. H., Beebe, S. J., & Buescher, E. S. (2001). Intracellular effect of ultrashort electrical pulses. *Bioelectromagnetics*, *22*(6), 440–448.

Scholz, M. J., Weiss, T. L., Jinkerson, R. E., Jing, J., Roth, R., Goodenough, U., et al. (2014). Ultrastructure and composition of the *Nannochloropsis gaditana* cell wall. *Eukaryotic Cell*, *13*(11), 1450–1464.

Shang, Y. F., Kim, S. M., Lee, W. J., & Um, B. H. (2011). Pressurized liquid method for fucoxanthin extraction from *Eisenia bicyclis* (Kjellman) Setchell. *Journal of Bioscience and Bioengineering*, *111*(2), 237–241.

Shene, C., Monsalve, M. T., Vergara, D., Lienqueo, M. E., & Rubilar, M. (2016). High pressure homogenization of *Nannochloropsis oculata* for the extraction of intracellular components: Effect of process conditions and culture age. *European Journal of Lipid Science and Technology*, *118* (4), 631–639.

Shynkaryk, M. V., Ji, T., Alvarez, V. B., & Sastry, S. K. (2010). Ohmic heating of peaches in the wide range of frequencies (50 Hz to 1 MHz). *Journal of Food Science*, *75*(7).

Spiden, E. M., Yap, B. H. J., Hill, D. R. A., Kentish, S. E., Scales, P. J., & Martin, G. J. O. (2013). Quantitative evaluation of the ease of rupture of industrially promising microalgae by high pressure homogenization. *Bioresource Technology*, *140*, 165–171.

Stirk, W. A., Bálint, P., Vambe, M., Lovász, C., Molnár, Z., van Staden, J., et al. (2020). Effect of cell disruption methods on the extraction of bioactive metabolites from microalgal biomass. *Journal of Biotechnology*, *307*, 35–43.

Tang-Bin, Z., Jia, Q., Li, H. W., Wang, C. X., & Wu, H. F. (2013). Response surface methodology for ultrasound-assisted extraction of astaxanthin from *Haematococcus pluvialis*. *Marine Drugs*, *11*(5), 1644–1655.

Taucher, J., Baer, S., Schwerna, P., Hofmann, D., Hümmer, M., Buchholz, R., et al. (2016). Cell disruption and pressurized liquid extraction of carotenoids from microalgae. *Thermodynamics and Catalysis*, *7*(1), 1–7.

Thana, P., Machmudah, S., Goto, M., Sasaki, M., Pavasant, P., & Shotipruk, A. (2008). Response surface methodology to supercritical carbon dioxide extraction of astaxanthin from *Haematococcus pluvialis*. *Bioresource Technology*, *99* (8), 3110–3115.

Töpfl, S. (2006). Pulsed electric fields (PEF) for permeabilization of cell membranes in food- and bioprocessing. In *Applications, process and equipment design and cost analysis*. Technological University of Berlin.

Valderrama, J. O., Perrut, M., & Majewski, W. (2003). Extraction of astaxantine and phycocyanine from microalgae with supercritical carbon dioxide. *Journal of Chemical & Engineering Data*, *48*(4), 827–830.

Vanthoor-Koopmans, M., Wijffels, R. H., Barbosa, M. J., & Eppink, M. H. M. (2013). Biorefinery of microalgae for food and fuel. *Bioresource Technology*, *135*, 142–149.

Viskari, P. J., & Colyer, C. L. (2003). Rapid extraction of phycobiliproteins from cultured cyanobacteria samples. *Analytical Biochemistry, 319*(2), 263–271.

Vorobiev, E., & Lebovka, N. (2010). Enhanced extraction from solid foods and biosuspensions by pulsed electrical energy. *Food Engineering Reviews, 2*(2), 95–108.

Wang, D., Li, Y., Hu, X., Su, W., & Zhong, M. (2015). Combined enzymatic and mechanical cell disruption and lipid extraction of green alga *Neochloris oleoabundans. International Journal of Molecular Sciences, 16*(4), 7707–7722.

Wang, L., & Weller, C. L. (2006). Recent advances in extraction of nutraceuticals from plants. *Trends in Food Science and Technology, 17*(6), 300–312.

Wang, L., Yang, B., Yan, B., & Yao, X. (2012). Supercritical fluid extraction of astaxanthin from *Haematococcus pluvialis* and its antioxidant potential in sunflower oil. *Innovative Food Science and Emerging Technologies, 13*, 120–127.

Wiltshire, K. H., Boersma, M., Möller, A., & Buhtz, H. (2000). Extraction of pigments and fatty acids from the green alga *Scenedesmus obliquus* (Chlorophyceae). *Aquatic Ecology, 34*(2), 119–126.

Xiao, X., Si, X., Yuan, Z., Xu, X., & Li, G. (2012). Isolation of fucoxanthin from edible brown algae by microwave-assisted extraction coupled with high-speed countercurrent chromatography. *Journal of Separation Science, 35*(17), 2313–2317.

Xie, Y., Ho, S. H., Chen, C. Y. C. N. N., Chen, C. Y. C. N. N., Jing, K., Ng, I. S., et al. (2016). Disruption of thermotolerant *Desmodesmus* sp. F51 in high pressure homogenization as a prelude to carotenoids extraction. *Biochemical Engineering Journal, 109*, 243–251.

Yamamoto, K., King, P. M., Wu, X., Mason, T. J., & Joyce, E. M. (2015). Effect of ultrasonic frequency and power on the disruption of algal cells. *Ultrasonics Sonochemistry, 24*, 165–171.

Yap, B. H. J., Dumsday, G. J., Scales, P. J., & Martin, G. J. O. (2015). Energy evaluation of algal cell disruption by high pressure homogenisation. *Bioresource Technology, 184*, 280–285.

Yen, H. W., Chiang, W. C., & Sun, C. H. (2012). Supercritical fluid extraction of lutein from *Scenedesmus* cultured in an autotrophical photobioreactor. *Journal of the Taiwan Institute of Chemical Engineers, 43*(1), 53–57.

Zhang, R., Grimi, N., Marchal, L., Lebovka, N., & Vorobiev, E. (2019). Effect of ultrasonication, high pressure homogenization and their combination on efficiency of extraction of bio-molecules from microalgae *Parachlorella kessleri. Algal Research, 40*.

Zhang, R., Parniakov, O., Grimi, N., Lebovka, N., Marchal, L., & Vorobiev, E. (2019). Emerging techniques for cell disruption and extraction of valuable bio-molecules of microalgae *Nannochloropsis* sp. *Bioprocess and Biosystems Engineering, 42*(2), 173–186. Available from: (2019). https://doi.org/10.1007/s00449-018-2038-5.

Zhang, J., Sun, Z., Sun, P., Chen, T., & Chen, F. (2014). Microalgal carotenoids: Beneficial effects and potential in human health. *Food & Function, 5*(3), 413.

Zhao, L., Chen, G., Zhao, G., & Hu, X. (2009). Optimization of microwave-assisted extraction of astaxanthin from *Haematococcus pluvialis* by response surface methodology and antioxidant activities of the extracts. *Separation Science and Technology, 44*(1), 243–262.

Food applications

Marco Garcia-Vaquero

School of Agriculture and Food Science, University College Dublin, Dublin 4, Ireland

1 Introduction

The United Nations predicts a continuous growth of the World's population which is estimated to reach 8.6 billion people by 2030, and continue to increase up to 11.2 billion by 2100 (https://www.un.org/development/desa/en/news/population/world-population-prospects-2017.html, Retrieved 3 January 2020). Meeting the future demands of food of this population will require to double the current food production by 2050 (https://www.un.org/press/en/2009/gaef3242.doc.htm, Retrieved 3 January 2020), with increase in 70% of more animal production and the need of 235% more animal feed (Herman & Schmidt, 2016). If this production trend materializes, it will ultimately lead to a scarcity of natural resources, as the current reservoirs of water and arable land cannot support this growth (Chaudhary, Gustafson, & Mathys, 2018). Moreover, the current food production system accounts for 20%-30% of the total environmental impacts (Tukker & Jansen, 2006) and anthropogenic greenhouse gas emissions, estimated in 9800-16,900 megatonnes of carbon dioxide (CO_2) equivalents in 2008 (Vermeulen, Campbell, & Ingram, 2012). It is also worth mentioning that agricultural production and indirect emissions represented $\sim$80%-86% of the total

food system emissions (Vermeulen et al., 2012). Thus, meeting the future demands of food of this growing population while reducing the environmental impacts of the current food production systems constitute a major challenge to be addressed by the food industry. Within the current scenario with global threats, such as climate change, global warming, and depletion of resources, the exploitation of underutilized natural resources such as microalgae and the development of efficient and green exploitation models have recently attracted the interest of the scientific community.

Microalgal biotechnology has developed rapidly over the last two decades aiming to improve the efficiency of microalgal cultivation as an alternative natural source for the production of biofuels (Garrido-Cardenas, Manzano-Agugliaro, Acien-Fernandez, & Molina-Grima, 2018). Despite the unexpected low efficiency of microalgae for their initial intended purpose, the technological advances in this field made it possible to cultivate this biomass at the industrial level, expanding the potential applications of algae. The use of microalgae for food and animal feed purposes offers certain advantages over terrestrial crops. Microalgae are extremely efficient photosynthetic organisms, able to transform atmospheric CO_2 into a wide variety

Microalgae. https://doi.org/10.1016/B978-0-12-821218-9.00008-6

of high-value products, including carbohydrates, lipids, proteins, and pigments such as carotenoids, with promising applications in the food industry. The possibility to use microalgae as bio-sequesters of CO_2 from power plants offers a promising scenario for the reduction of anthropogenic greenhouse gas emissions when integrating microalgal production with biorefineries (Chaudhary et al., 2018; Cheah, Show, Chang, Ling, & Juan, 2015). Microalgae can grow in nonagricultural lands such as seashore lands, desert, or semiarid regions as the cultivation requires saltwater (saline or brackish waters) or wastewater from agricultural, domestic, municipal, or industrial origins (Chaudhary et al., 2018). The cultivation strategies and water supply of choice to produce microalgae will also influence the final price of the biomass. The culture of microalgae at small-scale for specialized applications can increase the cost of production of the biomass reaching 5 € per kg of microalgae (Garrido-Cardenas et al., 2018) while culturing microalgae using wastewater combined with CO_2 capture from industrial flue gases can reduce the price of microalgae production up to less than 1 € per kg of biomass (Acién, Fernández, Magán, & Molina, 2012). Moreover, microalgal cultures have a high growth rate with peaks of 3-5 days and short recovery times after harvesting cycles (8-10 days), allowing the harvesting of this biomass multiple times compared to other feedstock that can only be harvested once or twice a year (Matos, 2017).

Currently, the most widely cultivated microalgae belong to *Chlorella, Dunaliella, Nannochloris, Nitzschia, Crypthecodinium, Schizochytrium, Tetraselmins*, and *Skeletonema* species and the cyanobacteria *Arthrospira* (formerly *Spirulina*) (Ejike et al., 2017). Companies commercializing microalgae or microalgal high-value products for multiple purposes are currently located in Australia, Israel, Germany, Spain, the Netherlands, and the United States (Cuellar-Bermudez et al., 2015). The application of more sustainable practices such as the biorefinery concept (see Fig. 1) is

currently being explored in both macroalgae (Garcia-Vaquero, O'Doherty, Tiwari, Sweeney, & Rajauria, 2019) and microalgae (Hayes et al., 2019).

The algal biorefinery concept aims to use processes and technological advances for the conversion of the algal biomass into energy and a wide variety of compounds such as biofuels and other high-value products that can be potentially employed by the pharmaceutical, cosmetic, food, and animal feed industries (Garcia-Vaquero, O'Doherty, et al., 2019; Hayes et al., 2019).

This chapter focuses on the composition, benefits, and potential applications of microalgae or microalgal products—mainly carbohydrates, lipids, proteins, and pigments such as carotenoids—as a source of novel ingredients for the food industry. The potential of these compounds to be used as functional foods/nutraceuticals and the novel downstream processing strategies (i.e., cultivation and extraction techniques) used to increase the industrial production of these compounds are also summarized, together with the main legislation, current market, and prospects of the microalgal industry for food applications.

2 Composition of microalgae

As mentioned earlier, the most widely cultivated microalgae worldwide are *Chlorella, Dunaliella, Nannochloris, Nitzschia, Crypthecodinium, Schizochytrium, Tetraselmins, Skeletonema*, and *Arthrospira* (Ejike et al., 2017). The composition of microalgae is generally described as high in proteins (40%-70%), medium in carbohydrates (12%-30%), with considerable concentration of lipids (4%-20%), carotenoids (8%-14%), and variable amounts of vitamins and minerals (Ejike et al., 2017). However, this composition can be affected by other factors, including the microalgal species, growth phase, and cultivation conditions (i.e., temperature, light irradiation, and nutrients). The proximate composition of

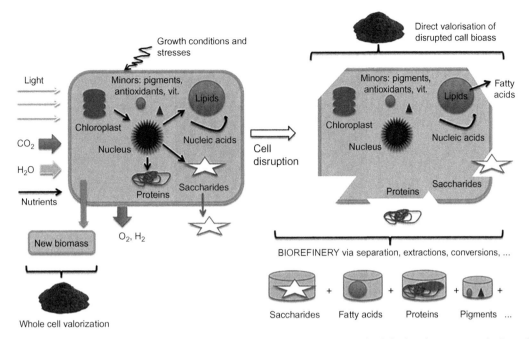

FIG. 1 Exploitation of microalgae following a biorefinery approach to recover multiple high-value compounds. *Reproduced from Hayes, M., Bastiaens, L., Gouveia, L., Gkelis, S., Skomedal, H., Skjanes, K., & Dodd, J. (2019). Microalgal bioactive compounds including protein, peptides, and pigments: Applications, opportunities, and challenges during biorefinery processes. Novel proteins for food, pharmaceuticals and agriculture: Sources, applications and advances, pp. 239–255, with permission from John Wiley & Sons.*

several commercially relevant microalgae at various stages during their life cycle is summarized in Table 1.

There are remarkable differences in the composition microalgae depending on the stage of development of the biomass (see Table 1). The protein contents of *Arthrospiraplatensis*, one of the most widely consumed microalgae as an alternative source of protein, varied between 61.3 and 40.3g protein/100g dried microalgae when collected during the exponential phase and early late stationary phases of growth, respectively (Markou et al., 2012; Tokuşoglu & üUnal, 2003).

Moreover, these variations in the composition of microalgae are also reflected in the morphology of the microalgal cells. Thereby, during the life cycle of *Haematococcus pluvialis* there are two clearly distinguishable phases depending on the number of carotenoids produced by the cells.

H. pluvialis cells in "green or vegetative" and "red or encysted" phases as described by Shah, Liang, Cheng, and Daroch (2016) are provided in Fig. 2.

These morphological changes are mainly caused by the accumulation of carotenoids in the cells. *H. pluvialis* in the green phase contains mainly lutein (75%-80%) and β-carotene (10%-20%), while during the red phase the accumulation of astaxanthin increases, representing between 80% and 99% of the total amount of carotenoids of the cells (Shah et al., 2016). Furthermore, the manipulation of several cultivation conditions including nitrogen, light intensity, carbon source, salinity, and other environmental stresses have shown to have a major effect on the production of several high-value compounds and the morphology of the cells (Chen et al., 2017; Lafarga, Clemente, & Garcia-Vaquero, 2019). For example, Abe, Hattori, and Hirano (2007)

TABLE 1 Proximate composition (protein, lipid, carbohydrate and ash contents) of selected microalgae sampled at different development stages.

Microalgae sp.	Growth phase	Protein content (%)	Lipid content (%)	Carbohydrate content (%)	Ash (%)	References
Arthrospira platensis	Exponential	61.3	8.0	15.8	10.4	Tokuşoglu and üUnal (2003)
	Early or late	40.3	16.2	7.4	NA	Markou, Chatzipavlidis, and Georgakakis (2012)
Chlorella vulgaris	Exponential	47.8	13.3	8.1	6.3	Tokuşoglu and üUnal (2003)
	Early or late	29.0	18.0	51.0	NA	Illman, Scragg, and Shales (2000)
Dunaliella salina	Exponential	26.8	24.28	14.9	8.6	Ben-Amotz, Tornabene, and Thomas (1985)
	Early or late	11.5	8.8	51.2	7.7	Ben-Amotz et al. (1985)
Nannochloropsis oculata	Exponential	45.5	17.3	18.4	NA	Lourenco, Barbarino, Mancini-Filho, Schinke, and Aidar (2002)
	Early or late	21.2	11.1	13.8	NA	Brown, Garland, Jeffrey, Jameson, and Leroi (1993)
Nitzschia paleacea	Exponential	33.2	20.8	21.1	22.9	Knuckey, Brown, Barrett, and Hallegraeff (2002)
	Early or late	17.7	18.8	7.8	38.6	Knuckey et al. (2002)
Skeletonema costatum	Exponential	32.7	11.8	29.8	NA	Lourenco et al. (2002)
	Early or late	NA	21.1	NA	NA	Rodolfi et al. (2009)
Tetraselmis suecica	Exponential	34.2	5.9	8.3	22.5	Whyte (1987)
	Early or late	12.8	11.1	48.7	6.9	Albentosa, Pérez-Camacho, Labarta, and Fernández-Reiriz (1996)

All the composition data is expressed on dry weight basis (g/100g dried microalgae). The values reported from the references cited on this table were selected from the database reported in the supplementary materials by Finkel, Z. V., Follows, M. J., Liefer, J. D., Brown, C. M., Benner, I., & Irwin, A. J. (2016). Phylogenetic diversity in the macromolecular composition of microalgae. PLoS One, 11 (5), e0155977.. The abbreviation "NA" refers to data not available.

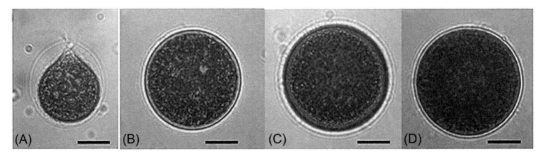

FIG. 2 Microscopic images of *H. pluvialis* showing multiple life cycle stages: "vegetative or green stage" (A) and (B) or the "red or encysted phase" (C) and (D). Scale bars: 10 μm. *Image taken from Shah, M. M. R., Liang, Y., Cheng, J. J., & Daroch, M. (2016). Astaxanthin-producing green microalga* Haematococcus pluvialis: *From single cell to high value commercial products. Frontiers in Plant Science, 7 (531). https://doi.org/10.3389/fpls.2016.00531 (web archive link) originally published by Frontiers.*

discovered the ability of microalgae *Coelastrella striolata* var. *multistriata* to increase the production of β-carotene, canthaxanthin, and astaxanthin in response to nitrogen depletion in the media, changing the appearance of the cells in a similar way than previously described for *H. pluvialis*. Thus, optimizing the cultivation conditions of microalgae and understanding the influence of multiple parameters on the metabolism and the production of high-value compounds are of key importance when targeting the production and commercialization of microalgae or microalgal high-value products.

3 Extraction of microalgal high-value compounds for food applications

Algae are a diverse group of organisms able to adapt and survive the extreme conditions of the aquatic environments (i.e., changes in salinity, temperature, UV irradiation, and CO_2) by producing unique secondary metabolites (Garcia-Vaquero, Rajauria, O'Doherty, &

Sweeney, 2017; Hayes et al., 2019). Compounds synthesized by algae include carbohydrates, lipids, proteins, and pigments such as carotenoids with multiple industrial applications (García-Vaquero, 2018; Garcia-Vaquero & Hayes, 2016; Hayes et al., 2019; Ventura et al., 2018). High-value compounds produced by microalgae together with their potential industrial applications are represented in Fig. 3.

Recently, microalgal compounds have been attributed potent biological activities, including antioxidant, antimicrobial, antiinflammatory, anticoagulant, antiproliferative, and antihypertensive among others, making microalgae an excellent source of bioactive compounds to be exploited as nutraceuticals, as seen in Fig. 3 (Hayes et al., 2019; Ventura et al., 2018). Nutraceuticals or functional foods are high-value compounds with health benefits beyond those of basic nutrition when incorporated in the diet (Garcia-Vaquero & Hayes, 2016). The market of nutraceuticals is expanding due to an increased awareness of the consumers on the health benefits of certain food products and

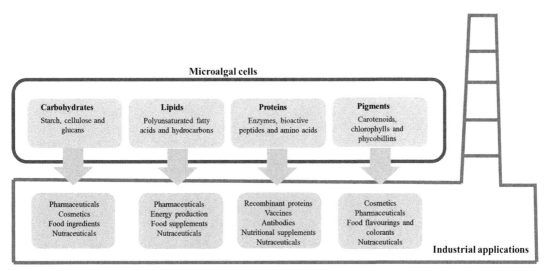

FIG. 3   Main compounds produced by microalgae and potential areas of application. The potential uses of these compounds were specified by Matos (2017) and Koller, Muhr, and Braunegg (2014).

ingredients. Recent predictions indicate that the global market of nutraceuticals will be valued in ~94 billion USD by 2023 (https://www.marketsandmarkets.com/Market-Reports/functional-food-ingredients-market-9242020.html, Retrieved 4 January 2020).

The production of nutraceuticals from microalgae comprises multiple steps. As previously mentioned, the cultivation of microalgae can significantly influence the cost of the biomass and also the yields of high-value compounds obtained from the cultures. Thereby, modifications in salinity, light irradiation, temperature, and certain additives in the culture media are able to improve the yields of certain high-value compounds and/or the growth of the microalgal biomass.

Following cultivation, the downstream processing strategies to obtain high-value compounds from microalgae include the application of cell disruption techniques. These disruption methods can be classified, based on the forces applied to the biomass and the mechanisms of action, into mechanical (i.e., bead milling, high-pressure, pulsed electric fields, microwaves, and sonication) and nonmechanical methods such as osmotic shock (Lafarga et al., 2019). The main advantages and disadvantages of these methodologies based on the recovery of high-value compounds, energy consumption, and possibilities to scale-up to industry among other factors are summarized in Table 2.

Following the disruption of the cells, the process to obtain high-value compounds includes the extraction, purification, and storage of the compounds of interest that will influence greatly the cost of the microalgal products (Khanra et al., 2018; Lafarga et al., 2019; Ventura et al., 2018). A scheme summarizing the main downstream processing strategies followed to high-value compounds from microalgae is represented in Fig. 4.

The health benefits and potential uses as functional foods or nutraceuticals of microalgal

TABLE 2 Summary elaborated by Günerken et al. (2015) comparing the main cell disruption methods used when obtaining high-value products from microalgae.

Disruption method	Mildness	Selective product recovery	Optimum dry cell weight concentration	Energy consumption	Practical scalability	Repeatability
Bead milling	Yes/no	No	Concentrated	High/medium	Yes	High
High pressure homogenization	Yes/no	No	Diluted/concentrated	High/medium	Yes	High
High speed homogenizer	No	No	Diluted	High/medium	Yes	High/medium
Ultrasound	Yes/no	No	Diluted	Medium/low	Yes/no	Medium
Microwave	Yes/no	No	Diluted	High/medium	Yes/no	Medium
Enzymatic lysis	Yes	Yes	Diluted	Low	Yes	High
Chemical treatment	Yes/no	Yes	Diluted/concentrated	Medium/low	Yes	High
Pulsed electric field	Yes/no	No	Very diluted/diluted	High/medium/low	Yes/no	Medium

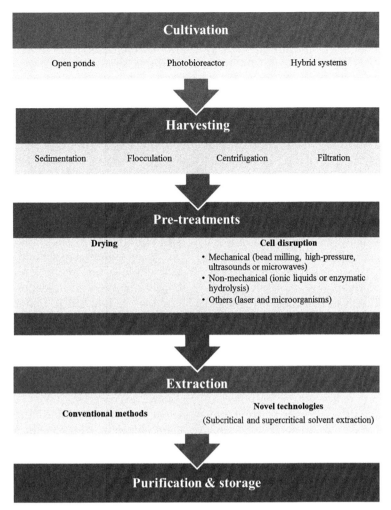

FIG. 4 Downstream processing steps to obtain high-value products from microalgae. *The content of this image was modified from Lafarga, T., Clemente, I., & Garcia-Vaquero, M. (2019). Carotenoids from microalgae. Carotenoids: Properties, processing and applications, p. 149.*

carbohydrates, lipids, proteins, and pigments are summarized below together with the main protocols and innovative extraction approaches to obtain these compounds from the biomass.

3.1 Microalgal carbohydrates

Microalgal cells have low amounts of lignin and hemicelluloses and thus, this biomass has been considered as a suitable source of carbohydrates for the production of bioethanol as an alternative to terrestrial crops such as corn and soybean (Behera et al., 2014; Chaudhary, Pradhan, Soni, Singh, & Tiwari, 2014). Recently, polysaccharides and oligosaccharides produced by microalgae have gained attention due to their wide range of biological activities and possibilities to be used as functional foods. The health benefits reported for several carbohydrates extracted from microalgae include antiviral, antibacterial, antioxidant, anti-inflammatory, immunomodulatory, antitumor, antilipidemic, antiglycemic, anticoagulant, antithrombotic, biolubricant, and antiadhesive properties (de Jesus Raposo, de Morais, & de Morais, 2015). Moreover, several carbohydrates from microalgae have shown prebiotic effects, as these compounds are nondigestible in the gastrointestinal tract, being able to influence the growth and/or activity of the bacterial population in the colon (Caporgno & Mathys, 2018). Other applications of microalgal carbohydrates in the food industry include the use of these molecules as thickeners and gelling agents to improve the quality and texture of food formulations (de Jesus Raposo et al., 2015).

The use of microalgae as a source of carbohydrates has certain advantages over other traditional sources of polysaccharides. Microalgal cell walls are rich in polysaccharides (i.e., sulphated galactan hetero-polysaccharides) and these cells are also able to produce intracellular carbohydrates (i.e., starch—amylose and amylopectin—and other sulphated compounds) and excrete extra-cellular polymers to the culture media at different stages during their biological cycle (i.e., sulphated and non-sulphated extra-cellular polysaccharides) (Clemente-Carazo, Sanchez, Condon-Abanto, & Marco, 2019). The chemical composition of microalgal carbohydrates (i.e., the degree of sulphation, monosaccharide composition and linkages, type and number of chains) will influence the physicochemical and biological properties of these compounds and thus, the final application of these molecules. Moreover, the chemical structure of algal carbohydrates can also be influenced by factors related to the biology of algae, such as the species, strain and cultivation conditions, and other processing strategies followed to obtain these high-value compounds. The composition and cell wall structure of microalgae play an important role for future technological applications of the biomass as these features will play a pivot role on

the efficacy of cell-disruption methods to recover intra-cellular compounds such as lipids and carotenoids from microalgae (Lee, Cho, Chang, & Oh, 2017).

The cell walls of *Chlamydomonas reinhardtii* are complex structures of five layers rich in glycoproteins that do not contain cellulose, one of the most common carbohydrates of microalgal cell walls (Adair, Steinmetz, Mattson, Goodenough, & Heuser, 1987; Arnold et al., 2015). Other microalgae have less complex cell wall structures with mainly two or three layers of carbohydrates. *H. pluvialis* have cell walls with three layers of variable chemical composition: an outer layer composed mainly by algaenan, a secondary layer composed of mannose and cellulose, and the internal layer consists of mainly mannose and cellulose (Praveenkumar, Lee, Lee, & Oh, 2015). Other microalgae such as *Chlorella* and *Nannochloropsis* have cells walls containing two layers, an external layer containing aliphatic polymers and an internal layer containing mainly mannose and chitin-like polysaccharides in the case of *Chlorella* sp. (Allard & Templier, 2000; Gerken, Donohoe, & Knoshaug, 2013), and cellulose in the case of *Nannochloropsis* (Scholz et al., 2014).

The extraction and purification techniques used to obtain carbohydrates from the algal biomass may influence to a significant degree the chemical properties of the extracted molecules (Clemente-Carazo et al., 2019; de Jesus Raposo et al., 2015; Garcia-Vaquero, Rajauria, et al., 2017). Owing to the wide variety of carbohydrates produced by microalgae, different extraction procedures are followed to obtain extra-cellular and cellular polysaccharides (see Table 3).

The extraction of extra-cellular carbohydrates requires the separation of the algal biomass followed by the precipitation of the extra-cellular polymers using cationic surfactants and/or organic solvents (de Jesus Raposo et al., 2015; Geresh et al., 2009). Other innovative approaches such as membrane filtration are

TABLE 3 Summary of the main methodologies used to obtain extra-cellular and cellular carbohydrates from microalgae.

Carbohydrate type	Microalgae sp.	Extraction methodology	References
Extra-cellular carbohydrates	*Porphyridium* sp.	Centrifugation of the culture and precipitation of the carbohydrates of the media using multiple solvents (NaOH, HCl, NaCl, ethanol, cetyltrimethyl ammonium bromide, and acetone)	Geresh et al. (2009)
	Amphora sp.;*Ankistrodesmus angustus; Phaeodactylum tricornutum*	Cell separation using centrifugation, followed by the fractionation of the compounds by filtration, crossed-flow ultrafiltration, and diafiltration	Chen et al. (2011)
	Porphyridium cruentum	Diafiltration using a tangential flow through a 300 kDa membrane	Patel et al. (2013)
	Porphyridium cruentum	Concentration followed by diafiltration using several membranes	Marcati et al. (2014)
Cellular carbohydrates	*Trebouxia* sp.	Microalgae was extracted with ethanol, chloroform: methanol and acetone, followed by boiling water and ethanol precipitation	Casano, Braga, Alvarez, Del Campo, and Barreno (2015)
	Pavlova viridis	Microalgae were sonicated at 65°C and the supernatant containing the carbohydrates was concentrated, treated with trichloroacetic acid, dialysed and precipitated with ethanol several times	Sun, Chu, Sun, and Chen (2016)
	Isochrysis galbana	Pretreated pellet was extracted with CaCl$_2$ (boiling temperature, three times). Pooled supernatants were ethanol precipitated and pellet washed with chloroform (remove proteins and lipophilic materials) and dialysed	Sadovskaya et al. (2014)

The content of this table was adapted from Clemente-Carazo, M., Sanchez, V., Condon-Abanto, S., & Marco, G.-V. (2019). Algal polysaccharides: Innovative extraction technologies, health benefits and industrial applications. Handbook of Algal Technologies and Phytochemicals (pp. 3–12): CRC Press.

currently being explored to extract and fractionate polymers from microalgae (Chen et al., 2011; Marcati et al., 2014; Patel et al., 2013).

In the case of cellular carbohydrates, the extraction process frequently include pretreatments of the biomass to eliminate impurities (Casano et al., 2015; Cheng, Labavitch, & Vander Gheynst, 2015; Cheng, Zheng, Labavitch, & Vander Gheynst, 2011; Sadovskaya et al., 2014) followed by cell-disruption methods including conventional processes—that is, maceration of the biomass using solvents and various temperature combinations—and innovative processes such as ultrasound or enzymatic treatments. Further details on the extraction and purification of carbohydrates produced by microalgae were recently reviewed by Clemente-Carazo et al. (2019).

Following the extraction of these carbohydrates, certain chemical features of the compounds have to be determined to evaluate their functionality and possible uses. The molecular weight can influence the biological properties of the carbohydrates as large molecules have difficulties when passing through membranes. Other parameters such as the number of monosaccharides, type of linkages and chains (linear or ramified carbohydrates) and the contents of sulphate, uronic acids, and other chemical groups (amino acids, proteins, or nucleic acids) have great influence on the biological activity and applications of these compounds (de Jesus Raposo et al., 2015). The strategies and analytical techniques currently available to evaluate the composition and conformation of carbohydrates were recently reviewed by Garcia-Vaquero (2019).

3.2 Microalgal lipids

Fat is an essential macronutrient as a source of energy in the diet, accounting for ~30% of the total energy of a well-balanced diet and as a source of antioxidants such as tocopherols and carotenoids (Bialek, Bialek, Jelinska, & Tokarz, 2017). Microalgae can accumulate high amounts of lipids ($\approx$30%-50%) depending on the cultivation conditions of the biomass (Chew et al., 2017). The manipulation of the cultures by inducing nitrogen starvation in the cells or by applying high temperatures, pH shifts, high salt concentrations, or the combination of several stresses increased the production of lipids (Kwak et al., 2016).

The health benefits of any source of dietary fat are linked to the fatty acid composition of the lipids. The consumption of polyunsaturated fatty acids (PUFAs), or hydrocarbons with two or more double bonds, has important health benefits such as protection against cardiovascular diseases, rheumatoid arthritis, and cancer (Maurer, Hatta-Sakoda, Pascual-Chagman, & Rodriguez-Saona, 2012). The European Nutritional Society determined an optimum dietary ratio of PUFAs (omega-6 (n-6) and omega-3 (n-3)) of 5:1 (n-6:n-3), being most European diets low in n-3 or high in n-6 with respect to those recommendations (Dawczynski, Schubert, & Jahreis, 2007; Holdt & Kraan, 2011). Fish oil has been used as a source of n-3 PUFAs as this product can contain levels up to 35.3 g n-3 per 100 g oil (Rubio-Rodríguez et al., 2010). However, novel sources of n-3 PUFA are currently being explored due to concerns regarding the sustainability and environmental impacts of fish oil production and safety concerns related to the presence of dioxins and other contaminants in fish oil (Maurer et al., 2012; Vannuccini, Kavallari, Bellù, Müller, & Wisser, 2019).

Microalgae are an excellent source of PUFAs as these compounds can represent up to 25%-60% of the total lipids of the biomass depending on the species and cultivation conditions (da Silva Vaz, Moreira, de Morais, & Costa, 2016). As seen in Table 4, in general microalgae are a promising source of linoleic acid (C18:2, n-6), arachidonic acid (C20:4, n-6), and other nutritionally relevant n-3 such as

TABLE 4 Fatty acid composition of selected microalgae.

Fatty acid	Arthrospira platensis	Chlorella vulgaris	Isocrysis galbana	Nannochloropsis oculata	Phaeodactylum tricornutum	Prorocentrum minimum	Seminavis gracilenta
C14:00	–	0.5	9.5	3.2	6.9	5.9	4.3
C16:00	44.2	27.6	5.2	24.1	16.9	7.2	10
C18:00	–	1	8.7	2.1	0.3	5.7	14.4
C24:00:00	1.1	nd	–	nd	nd	–	5.6
C15:01	nd	4	nd	nd	3.5	nd	nd
C16:01	6.4	2.1	10.7	26.6	19.2	9.2	3.9
C18:01	0.9	1.8	0.2	5.8	1.6	4.6	13.2
C16:4 n-3	nd	–	nd	nd	0.2	nd	nd
C18:2 n-6	18.7	14.7	0.4	5.5	3.1	0.9	4.8
C18:3 n-3	nd	40.3	9.5	nd	1.1	1.3	1.3
C18:3 n-6	23.4	4	6.4	nd	0.4	0.2	1.7
C18:4 n-3	nd	nd	nd	nd	nd	nd	nd
C20:4 n-6	nd	0.4	3.4	9.7	1	13.8	15.8
C20:5 n-3	nd	–	10.3	21.3	34.1	14	15.7
C22:6 n-3	nd	–	33.6	nd	1	28.1	0.2
Other PUFA	–	–	0.8	–	1.4	3.9	5.8
Total PUFA	42.2	59.4	64.5	36.4	42.5	62.3	45.3
References	Ramadan and Selim as'ker (2008)	Matos et al. (2016)	Suh et al. (2015)	Wei and Huang (2017)	Matos et al. (2016)	Suh et al. (2015)	Chen (2012)

Data are expressed as percent of total fatty acids, the dashes refer to nondetected and "nd" stands for nondetermined. This table was modified from Xue, Z., Wan, F., Yu, W., Liu, J., Zhang, Z., & Kou, X. (2018). Edible oil production from microalgae: A review. European Journal of Lipid Science and Technology, 120 (6), 1700428. and reproduced with permission from John Wiley &Sons.

eicosapentaenoic acid or EPA (C20:5, n-3) and docosahexaenoic acid or DHA (C22:6, n-3).

The consumption of n-3, mainly EPA and DHA, has shown promising health benefits by decreasing the risk of coronary heart disease (Adarme-Vega, Thomas-Hall, & Schenk, 2014) and acting as protective molecules in the brain and retina (Saini & Keum, 2018). Furthermore,

the recent recommendation to include DHA in infant formula has renewed the interest in the industrial production of this oil from microalgae (Ventura et al., 2018; Ward & Singh, 2005).

The cultivation and downstream processing of microalgae for the production of PUFAs has recently attracted the attention of the scientific community aiming to improve the yields of compounds produced by microalgae and the efficiency of the extraction and purification processes to obtain PUFAs while decreasing the time and production costs. The extraction of lipophilic compounds such as PUFAs and carotenoids requires the use of mechanical and nonmechanical processes to break-down the microalgal cell walls as seen in Fig. 4. Traditionally the recovery of lipophilic compounds was achieved by using a Soxhlet apparatus or by macerating the biomass in organic solvents (Chew et al., 2017). Innovative technologies based on the application of ultrasounds, microwaves, and pressure (i.e., pressurized and supercritical fluid extraction) have been successfully explored aiming to develop green and sustainable extraction processes of oil from microalgae. The main extraction and purification methods used to recover lipids from microalgae have been reviewed in detail by Lee et al. (2017) and Garcia-Vaquero and Tiwari (2020).

3.3 Microalgal proteins and peptides

Protein is an essential nutrient in a healthy diet (Garcia-Vaquero, Lopez-Alonso, & Hayes, 2017; Henchion, Hayes, Mullen, Fenelon, & Tiwari, 2017). The demands of dietary protein are growing due to the aforementioned increase of the World's population together with socio-economic changes and preferences toward an increased protein consumption influenced by a rising income, increased urbanization and aging of the population of developing countries (Henchion et al., 2017). It is projected that the demand of animal-derived protein will double by 2050 raising sustainability concerns as the production of animal protein is inefficient (i.e., the production of 1 kg of animal protein requires ~6 kg of plant protein) (Aiking, 2014) and contributes significantly to the emission of greenhouse gases related to global warming (Tilman & Clark, 2014).

In terms of protein quantity, the use of terrestrial plants and aquatic organisms is currently being considered as a more sustainable alternative to animal protein sources (Garcia-Vaquero, Lopez-Alonso, & Hayes, 2017; Tiwari, Gowen, & McKenna, 2011). Microalgae are considered promising sources of protein with variable contents depending on the microalgal species, biological stage, and cultivation conditions, as seen in Table 1. The cultivation of microalgae aiming to optimize the production of protein by these cells achieved yields per unit area of biomass 20-50 times higher than soybeans (Chronakis, 2000). Thereby, Safi et al. (2014) described protein contents ranging from 50% to 70% in *Arthrospira platensis* and 38%-58% of the protein in *Chlorella vulgaris* on a dry weight basis.

The nutritional quality of multiple sources of dietary protein is commonly evaluated on the basis of their digestibility and amino acid composition (Boisen & Eggum, 1991). Essential amino acids cannot be synthesized by humans and thus, these compounds must be supplied by the diet. Essential amino acids for humans are histidine, isoleucine, leucine, lysine, methionine, phenylalanine, threonine, tryptophan, and valine. Proteins from terrestrial plants often lack certain essential amino acids. For example, lysine and tryptophan are deficient in cereals (Ufaz & Galili, 2008); while cysteine, methionine, and tryptophan are low inpulses (Venkidasamy et al., 2019). The composition and amount of essential and nonessential amino acids of various microalgae are summarized in Table 5.

The nutritional quality of protein from microalgae and other sources is commonly evaluated by their essential amino acid index. This index is calculated by comparing the levels of essential amino acids between a novel protein and an animal protein of reference (Oser, 1951). An index >0.90 indicates high-quality protein, between 0.70 and 0.89 moderate quality and <0.70 in the case of low-quality protein (Brown & Jeffrey, 1992). *Phaeodactylum tricornutum* and *Tetraselmis chuii* had essential amino acid indices

TABLE 5 Amino acid composition of selected microalgae as described by Becker (2007).

Amino acids	Microalgae sp.					
	Chlorella vulgaris	*Dunaliella bardawil*	*Scenedesmus obliquus*	*Arthrospira maxima*	*Arthrospira platensis*	*Aphanizomenon* sp.
Ile	3.8	4.2	3.6	6	6.7	2.9
Leu	8.8	11	7.3	8	9.8	5.2
Val	5.5	5.8	6	6.5	7.1	3.2
Lys	8.4	7	5.6	4.6	4.8	3.5
Phe	5	5.8	4.8	4.9	5.3	2.5
Tyr	3.4	3.7	3.2	3.9	5.3	–
Met	2.2	2.3	1.5	1.4	2.5	0.7
Cys	1.4	1.2	0.6	0.4	0.9	0.2
Try	2.1	0.7	0.3	1.4	0.3	0.7
Thr	4.8	5.4	5.1	4.6	6.2	3.3
Ala	7.9	7.3	9	6.8	9.5	4.7
Arg	6.4	7.3	7.1	6.5	7.3	3.8
Asp	9	10.4	8.4	8.6	11.8	4.7
Glu	11.6	12.7	10.7	12.6	10.3	7.8
Gly	5.8	5.5	7.1	4.8	5.7	2.9
His	2	1.8	2.1	1.8	2.2	0.9
Pro	4.8	3.3	3.9	3.9	4.2	2.9
Ser	4.1	4.6	3.8	4.2	5.1	2.9

All the results are expressed as g amino acid per 100 g protein.
Reproduced with permission from Elsevier.

of 0.9 and *Botryococcus braunii* an index of 1, while *Nannochloropsis granulata* and *Porphyridium aerugineum* had a better quality protein than egg albumin with essential amino acid indices of 1.2 (Tibbetts, Milley, & Lall, 2015).

Apart from the nutritional role of proteins on a healthy and well-balanced diet, protein ingredients have also a major influence on the taste and texture of foods due to their functional properties (Garcia-Vaquero, Lopez-Alonso, & Hayes, 2017). Proteins can act as emulsifying agents, texture modifiers, and assist with fat and water absorption and the whipping properties of foods, influencing the final consumer's acceptability of prepared foods and food formulations (Garcia-Vaquero, Lopez-Alonso, & Hayes, 2017). Ursu et al. (2014) obtained proteins from *Chlorella vulgaris* at various pHs and the extracted compounds had emulsifying capacities and stabilities similar to commercial ingredients such as sodium caseinate. Waghmare, Salve, LeBlanc, and Arya (2016) also reported high foaming capacity and stability of proteins extracted from *Chlorella pyrenoidosa*. As seen in

Table 8, there are currently products in the market using microalgal proteins as egg replacers. Furthermore, it is worth noting that the functional properties of proteins will vary depending on the physicochemical characteristics of the compounds, including the amino acid composition, molecular weight, and structural confirmation that will be greatly influenced by the extraction procedures used to obtain these compounds (Garcia-Vaquero, Lopez-Alonso, & Hayes, 2017). In general microalgal protein, extracts can be used to develop novel food formulations with high organoleptic and nutritional quality due to their amino acid composition and functional properties. However, the extraction of protein from microalgae is a complex task and further improvements are needed in order to reduce the cost of extraction of these promising ingredients for food purposes (Ruiz et al., 2016).

Recent trends in the field of nutraceuticals and functional foods include the transformation of protein concentrates into bioactive peptides. Bioactive peptides or cryptides are sequences (2–30 amino acids) inert within their parent protein that can be released by hydrolyzing the proteins following enzymatic reactions, fermentation, and other food processing treatments (Garcia-Vaquero & Hayes, 2016). Once released, these molecules display a wide range of health-promoting benefits mainly antihypertensive (i.e., angiotensin-converting enzyme I (ACE-I) inhibitors), anti-inflammatory, antidiabetic, anticancer, and antimicrobial (Fan, Bai, Zhu, Yang, & Zhang, 2014; Garcia-Vaquero & Hayes, 2016; Garcia-Vaquero, Mora, & Hayes, 2019; Hayes & García-Vaquero, 2016). A scheme summarizing the process to generate and characterize biologically active peptides from algae is represented in Fig. 5.

The generation of protein hydrolysates from microalgae is an active field of research aiming to identify potent biologically active peptides. A summary of several peptide sequences generated from microalgae with ACE-I inhibitory activity is summarized in Table 6.

Despite the promising applications and health benefits of bioactive peptides, the high cost associated with the production of peptides is a huge challenge for the use of these compounds as food ingredients (Ejike et al., 2017). However, the generation of bioactive peptides from low-cost raw materials and the commercialization of full protein hydrolysates containing multiple peptides are promising strategies for the industrial production and future commercialization of bioactive peptides.

3.4 Microalgal pigments and carotenoids

Chlorophylls, carotenoids, anthocyanins, and betanins are responsible for the green, yellow/orange/red, red/blue, and red color of microalgae and other terrestrial plants (Rodriguez-Amaya, 2016). The market of natural colorants for food applications is currently limited due to the low stability of these compounds and their limited color range compared to chemically synthesized molecules (Rodriguez-Amaya, 2016). However, the promising health benefits of these molecules have attracted the interest of industry in natural pigments such as carotenoids in recent years (Lafarga et al., 2019). Carotenoids display potent antioxidant and pro-vitamin A activities together with a major role in other biological functions such as regulation of the immune system (Chew & Park, 2004). Moreover, the production costs and risks of toxicity of natural compounds are lower compared to chemically synthesized carotenoids (Gong & Bassi, 2016).

Carotenoids (carotenes and xanthophylls) are lipophilic compounds with a C40 backbone structure (Gong & Bassi, 2016). Depending on the function of these molecules in the cells, carotenoids can be classified as primary carotenoids (i.e., lutein) if the compounds are linked to the photosynthetic processes in the cells; or

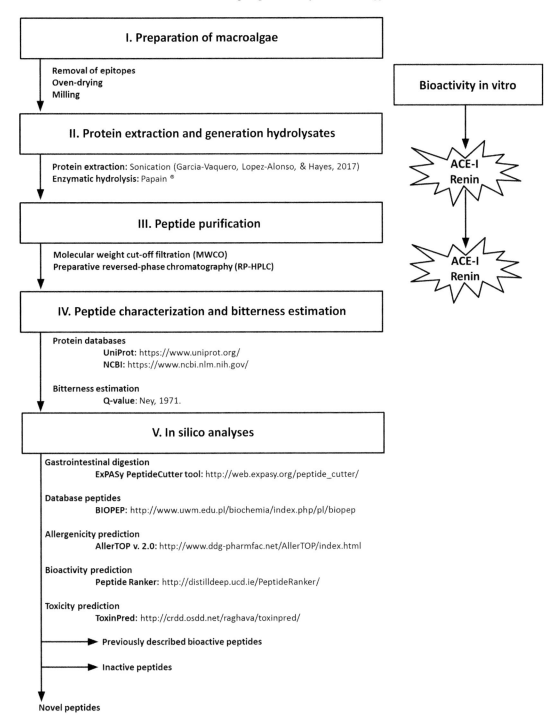

FIG. 5 Scheme showing the procedures followed to identify bioactive peptides from algae as described by Garcia-Vaquero, Mora, and Hayes (2019) published by MDPI.

TABLE 6 Summary of peptide sequences obtained from microalgae with angiotensin-converting enzyme I (ACE-I) inhibition activities.

Microalgal sources	Peptide sequences	ACE-I inhibition (IC$_{50}$)	References
	Ile-Val-Val-Glu	315.3	Suetsuna and Chen (2001)
	Ala-Phe-Leu	63.8	
Chlorella vulgaris	Phe-Ala-Leu	26.3	
	Ala-Glu-Leu	57.1	
	Val-Val-Pro-Pro-Ala	79.5	
	Ile-Ala-Glu	34.7	
	Phe-Ala-Leu	11.4	
Arthrospira platensis	Ala-Glu-Leu	11.4	
	Ile-Ala-Pro-Gly	11.4	
	Val-Ala-Phe	35.8	
Nannochloropsis oculata	Leu-Val-Thr-Val-Met	18.0	Qian et al. (2013)
Nannochloropsisoculata	Gly-Met-Asn-Asn-Leu-Thr-Pro	123	Samarakoon et al. (2013)
	Leu-Glu-Gln	173	
Chlorella ellipsoidea	Val-Glu-Gly-Tyr	128.4	Ko et al. (2012)
Chlorella vulgaris	Val-Glu-Cys-Tyr-Gly-Pro-Asn-Arg-Pro-Gln-Phe	29.6	Sheih, Fang, and Wu (2009)

The biological activities are expressed by the IC$_{50}$ value of each peptide or the concentration in μM of each compound required to inhibit by 50% the activity of ACE-I.

secondary carotenoids (astaxanthin and canthaxanthin) involved in other cell protection roles (Gong & Bassi, 2016). The production and commercialization of astaxanthin and β-carotene represent ~50% of the global carotenoid production (Gong & Bassi, 2016), a market that is estimated to be valued in over 1.5 billion USD by 2021 (Sathasivam & Ki, 2018). Currently, the most industrially relevant microalgal species for the production of carotenoids are *Dunaliella salina* and *Haematococcus pluvialis* that can accumulate high amounts of β-carotene and astaxanthin, respectively (Del Campo, García-González, & Guerrero, 2007; Wayama et al., 2013). Other microalgal pigments of industrial relevance include the photosynthetic pigments phycobilins or phycobiliproteins (allophycocyanin, phycocianin, phycoerythrin, and phycoerythrocyanin) present in the microalgal cells in complexes namely phycobilisomes (Cuellar-Bermudez et al., 2015). These compounds are responsible for the blue/green, blue, purple, and orange color of certain microalgae (Cuellar-Bermudez et al., 2015).

As in the case of microalgal lipids, the production of carotenoids from microalgae requires the application of cell disruption techniques and other downstream processing strategies (see Fig. 4). As many microalgae have thick and rigid cell walls rich in polysaccharides and other associated molecules, the release of lipophilic compounds including PUFAs and carotenoids require the application of cell disruption techniques aiming to improve the recovery of intracellular compounds without degrading the final products (Gong & Bassi, 2016). The main advantages and efficiency of different technologies currently used for the extraction of carotenoids and lipids from microalgae are summarized in Table 7.

Detailed information on the downstream processing stages of microalgae for the production of carotenoids, including cultivation and harvesting of the biomass, pretreatments, extraction techniques, purification, and storage of these products has been reviewed in detail by Lafarga et al. (2019).

4 The current market of microalgae and microalgal products

The value of the microalgal biomass varies depending on the targeted market or intended use of the algae or algal derived products, as represented in Fig. 6. The lowest revenue per unit of biomass comes from the use of microalgae as biofuel (0.3 € per kg of microalgae), while the use of the biomass as food triplicates this value, being the most promising market the production of bulk chemicals with a revenue greater than 2 € per kg (Ruiz et al., 2016).

Within this scenario, the production of multiple compounds and their correct placement within a particular market that allows achieving the highest economic benefits could increase the revenue of the overall microalgal biomass up to 30.4 € per kg (Ruiz et al., 2016). As seen in Fig. 6, natural pigments represent the most attractive market in terms of the overall revenue, while the production of proteins does not play a key role improving the value of the biomass, despite the fact of being one of the most costly ingredients to produce (Ruiz et al., 2016).

Currently, the main uses of microalgae for food purposes include the use of full microalgae. Approximately 75% of the annual microalgal biomass production is currently being processed as powders, tablets, and capsules sold as healthy food products and supplements (Chacón-Lee & González-Mariño, 2010). Moreover, regulations affecting microalgae as food highly influence the placement of products containing microalgae in the market. Thereby, the most widely consumed microalgae include *Arthrospira*, *Chlorella*, *Dunaliella*, *Haematococcus*, and *Schizochytrium* species as these species achieved the generally recognized as safe status in the United States (Hayes et al., 2017). The production of whole dried *Spirulina* sp. and *Chlorella* sp., supplied mainly by 85 producers, represents an annual turnover of ∼80 million USD (Vigani et al., 2015).

The incorporation of full microalgae has been tested as a source of coloring agents, PUFAs, or antioxidants by incorporating the biomass in food formulations including bakery products (i.e., bread, cookies, pasta, and energy bars) and beverages (Hayes et al., 2017). Baked goods have certain advantages for the delivery of healthy ingredients to the population due to the widespread consumption and integration in the diet of these products (Lafarga & Hayes, 2017). Recent research has evaluated the incorporation of multiple microalgal species at different percentages in bakery products on the sensory attributes and chemical composition of these novel food formulations. The addition of microalgal biomass to bread has significant impacts on these products as microalgae may weaken the formation of a gluten viscoelastic matrix when these proteins are mixed with water. Graça, Fradinho, Sousa, and Raymundo (2018) researched the incorporation of *Chlorella vulgaris* at percentages ranging from 1 to 5% on the rheology of wheat flour dough and texture of the final baked bread. These researchers appreciated that the addition of up to 3% of *C. vulgaris* had a positive impact on the rheology

TABLE 7 Summary of the main methods, efficiency and advantages of different technological approaches to extract carotenoids from microalgae elaborated by Gong and Bassi (2016).\

Step	Methodology	Efficiency	Advantages and disadvantages	References
	Grinding	++	Time-consuming	Hu, Wang, Wang, Liu, and Guo (2013)
	Cryogenic grinding	+++	Expensive	(Grima, Belarbi, Fernández, Medina, and Chisti (2003); Zheng et al. (2011))
	Bead milling	+++	Most efficient in some studies; not as efficient in several studies; the treatment is strain specific; generates heat	Chan et al. (2013); Halim, Danquah, and Webley (2012); Lee, Yoo, Jun, Ahn, and Oh (2010); Prabakaran and Ravindran (2011); Taucher et al. (2016)
	High pressure homogenizer	+++	Comparable with bead milling and ultrasound assisted extraction	Grima et al. (2003); Halim, Harun, Danquah, and Webley (2012); Kim et al. (2016)
	Autoclave	–	Damage to carotenoids occurs	Chan et al. (2013)
	Microwave	+++	Comparable efficiency with bead milling; low energy consumption; simple method; generates heat	Lee, Lewis, and Ashman (2012); Li, Qu, Yang, Chang, and Xu (2016); McMillan, Watson, Ali, and Jaafar (2013)
Cell disruption	Ultrasonication	++	Most efficient in some studies;	Cravotto et al. (2008); Mercer and Armenta (2011)
		+	Not efficient in other studies	Halim, Harun, et al. (2012); McMillan et al. (2013); Pasquet et al. (2011)
	Enzymatic hydrolysis	+	Highly selective; mild condition; expensive, strict condition maintenance; long treatment time	Deenu, Naruenartwongsakul, and Kim (2013); Kadam, Tiwari, and O'Donnell (2013); Zheng et al. (2011)
	Pulsed electric field	++	Highly selective; retain bioactivity of carotenoids; short treatment time; small solvent requirements	Grimi et al. (2014); Lai, Parameswaran, Li, Baez, and Rittmann (2014); Sánchez-Moreno et al. (2005); Yu, Gouyo, Grimi, Bals, and Vorobiev (2016)
	Osmotic shock,	+	Not efficient;	Halim, Danquah, and Webley (2012)
	Acid and alkaline treatment	–	Cause carotenoids to degrade	Halim, Danquah, and Webley (2012)
		–		Park, Park, Lee, and Yang (2015)

	Ionic liquids and Switchable solvent		High price; toxicity; cause carotenoids to degrade	
Solvent extraction	Conventional solvent extraction	++	Cheap and easy to scale up; long extraction time; multi-step operation; use large amount of solvents	Gil-Chavez et al., 2013; Reverchon and De Marco (2006); Taucher et al. (2016)
	Super- and subcritical solvent extraction	+++	Polarity of solvent is tunable; fast; safe; easy separation of carotenoids; expensive	Du, Schuur, Kersten, and Brilman (2015); Halim, Danquah, and Webley (2012); Reverchon and De Marco (2006); Yen, Yang, Chen, and Chang (2015)

The meaning of the symbols in the table are: "−"degradation of compounds (−), slightly efficient (+), efficient (++) and highly efficient (+++).
Reproduced with permission from Elsevier.

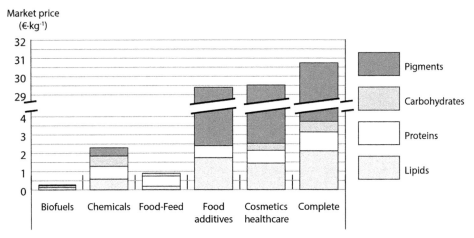

FIG. 6 Market value of microalgal carbohydrates, lipids, proteins, and pigments influencing the price of the biomass in multiple market scenarios. *Image reproduced from Ruiz, J., Olivieri, G., de Vree, J., Bosma, R., Willems, P., Reith, J. H., … Barbosa, M. J. (2016). Towards industrial products from microalgae.* Energy & Environmental Science, 9 (10), 3036–3043. *published by The Royal Society of Chemistry.*

and viscoelastic characteristics of the dough due to the strengthening of the gluten network. The addition of higher percentages of microalgal biomass had an opposite effect on the parameters evaluated, together with strong flavors and an acceleration of the process of aging of the bread (Graça et al., 2018).

Significant efforts have also been made to incorporate microalgal biomass as a source of pigments or PUFAs in pasta products and cookies. Fradique et al. (2010) studied the incorporation of different amounts of *C. vulgaris* and *Spirulina maxima* on the physicochemical and quality attributes (i.e., cooking time, losses, swelling index, and water absorption) of pasta. These researches appreciated an improvement of the quality parameters of pasta compared to control, with an increased firmness of the pasta when the microalgae wereincorporated at concentrations ranging from 0.5% to 2%. Moreover, products containing microalgae had higher acceptance scores when evaluated by a sensory panel (Fradique et al., 2010). Fradique et al. (2013) studied the incorporation of small percentages (0.5%-2%) of

Isochrysisgalbana and *Diacronemavlkianum*as a source of PUFAs. Pastas enriched with microalgae had increased contents of palmitic acid (C16:0), oleic acid (C18:1, n9), linoleic acid (C18:2 n6), EPA, and DHA with respect to control pasta. The sensorial evaluation of the products revealed a fish flavor at microalgal concentrations higher than 2% that can be useful when enriching the nutritional value of pasta for fish or seafood-based culinary applications (Fradique et al., 2013). Similarly, Gouveia et al., 2008 determined the nutritional benefits of the incorporation of *I. galbana* biomass (1%-3%) in butter biscuits. The authors reported an increased content of n3 PUFAs (EPA, docosapentaenoic acid (C22:5 n3, DPA), and DHA together with enhanced texture properties and color stability of the cookies up to 3 months of storage (Gouveia et al., 2008).

Other promising food sources for the delivery of microalgae or microalgal nutraceuticals are the elaboration of functional beverages. The inclusion of functional ingredients in beverages is a promising field of research as these products can meet multiple demands of the consumer in terms of size, shape, and appearance together with the facility of storage and distribution (Corbo, Bevilacqua, Petruzzi, Casanova, & Sinigaglia, 2014). Moreover, the design of healthy beverages is especially suited for specific sectors of the population such as the elderly that are at risk of suffering nutritional deficiencies of vitamins, minerals, and proteins. Santos, de Freitas, Moreira, Zanfonato, and Costa (2016) designed a beverage containing *Spirulina* as a source of nutrients, mainly proteins and carbohydrates, for the elderly. The shelf life of products containing *Spirulina* was 19 months, slightly lower than the same product without microalgae (24 months) and the sensorial attributes of the beverage containing microalgae were well accepted by the target population (Santos et al., 2016). *Spirulina* was also used as a source of vitamin B12 when designing smoothies targeting the dietary needs of this molecule from nonanimal sources targeting the vegetarian and vegan consumers (Castillejo et al., 2018). The recommended intakes of vitamin B12 were achieved by the consumption of 200 g of the supplemented smoothie containing *Spirulina* sp. that had also other advantages such as low vitamin C degradation and shelf life of 17 days (Castillejo et al., 2018).

Despite the promising applications of full microalgae in food formulation, the quality traits and acceptability of the food products containing microalgae are strongly influenced by the sought population or market niche. Thereby, Asian consumers appreciate the incorporation of microalgae as a favorable quality trait in food formulations, while the opposite situation occurs in Western countries (Liang, Liu, Chen, & Chen, 2004). Thus, it is necessary to evaluate the physicochemical, nutritional, and sensorial quality of new products developed containing microalgae to ensure the consumption of these foods by the target population. Several food products containing microalgae and microalgal high-value products such as lipids and proteins are summarized in Table 8.

Overall, the majority of the products currently in the market containing microalgae are designed specifically for vegetarian or vegan consumers or consumer preferences for organic food products. This is evidenced by the EU Green Leaf logo currently used by all prepackaged food products produced and sold as organic within the EU, as thisproduct must contain at least 95% of organic ingredients and strict conditions for the remaining 5% (https://ec.europa.eu/info/food-farming-fisheries/farming/organic-farming/organics-glance/organic-logo_en, Retrieved 13 January 2020). Data on the current use of microalgal derived ingredients at an industrial level is still scarce. AlgaVia, a protein-rich whole alga, is a novel ingredient marketed as a source of quality protein, insoluble fiber, lipids and pigments such as lutein and zeaxanthin. According to the manufacturers, this protein ingredient contains 63% protein, 19% carbohydrates and 11% lipids with recommended applications as protein beverages (http://algavia.com/ingredients/proteins/, Retrieved 8 January 2020).Other ingredients of industrial relevance are the n-3PUFAs, mainly DHA due to the recommendation of its inclusion in infant formula, which has an estimated world market of 10 billion USD per year (Ward & Singh, 2005). The products DHASCO and DHA Gold produced by Martek (Columbia, USA) are currently dominating the market. Other companies are also producing DHA from multiple microalgae such as *Schizochytrium* sp. (DSM-NP life's DHA plus EPA and Source-Omega Source Oil) and *Crypthecodiniumcohnii* (DSM-NP life's DHA and GCI Nutrients DHA Algae 35% Oil) and *Ulkenia*sp. (LonzaDHAid) (Ventura et al., 2018).

From all the carotenoids produced by different microalgae species, the only compounds with industrial relevance in the food industry to date are β-carotene and astaxanthin produced by *Dunaliellasalina* and *H. pluvialis*, respectively

TABLE 8 Commercialized products containing microalgal biomass or high-value products in the market.

Product	Brand	Company	Country of commercialization	Product description
Microalgal biomass				
Freeze-dried *Tetraselmis chui*	Plancton Marino Veta la Palma	Fitoplancton Marino SL, Spain	Spain	Natural product which can be used in sauces, special salts, and condiments
Dried *Tetraselmis*, *Spirulina* and *Dunaliella* species	Algaefeed	Algalimento, Spain	Spain and other EU countries	Dried microalgae cultivated in open raceway ponds with applications in aquaculture, food production, cosmetics, and as sources for pigments and antioxidants.
Organic *Spirulina* tablets	Sanatur	Sanatur, Germany	Germany	Dietary supplement featuring the EU Green Leaf, Bio, Naturland and Vegan V-Label European Vegetarian Union logos
Organic *Spirulina* microalgae tablets	Vitatrend	IQ Pharma, Germany	Germany	Dietary supplement featuring the EU Green Leaf and Naturland logos
Beverages				
Smoothie	Natura	Naturawerk, Germany	Germany	Smoothie powder with *Chlorella*, spinach, lucuma, and baobab powders. Features the EU. Green Leaf logo and the Vegan Neuform Quality
Smoothie	Be Raw!	Purella Food, Poland	Poland	Pasteurized pressed juice with added *Chlorella*
Wheat grass cocktail	Rabenhorst	HausRabenhorst, Germany	Italy and Germany	Organic functional juice drink containing young wheatgrass, green tea, agava juice and *Spirulina*
Omega 3 yogurt drink	PriégolaSimbi	GanaderíaPriégola, Spain	Spain	Enriched in DHA and omega 3 fatty acids obtained from *Schizochytrium* sp.
Sauces and spreads				
Peanut spread with Spirulina	Better & Different	Better & Different, Israel	Israel	Kosher certified low in sodium and suitable for vegans
Vegan sauce with smoked pepper	The Good Spoon	Good Spoon Foods, France	France	Free from eggs and instead made with microalgae
Original vegan supernaise sauce	The Good Spoon	Good Spoon Foods, France	France	Contains microalgae "flour" as an egg replacer. Features the V-label European Vegetarian Union
Low fat spread	St Hubert DHA Cérébral & Vision	St Hubert, France	France	Low fat spread enriched in DHA extracted from *Schizochytrium* sp.

TABLE 8 Commercialized products containing microalgal biomass or high-value products in the market—cont'd

Product	Brand	Company	Country of commercialization	Product description
Baked goods				
Sea salt algae crackers	Helga	Evasis Edibles, Germany	Austria	Green, crispy, and spicy snack containing chlorella and linseed. Features the EU Green Leaf logo
Bio matcha and Spirulina biscuits	Próvida	Próvida ProdutosNaturais, Portugal	Portugal	Protein- and fiber-rich biscuits featuring the EU Green Leaf logo
Jeju green tea Castella cake	Paris Baguette	Paris Croissant, South Korea	South Korea	Premium product made with fragrant Jeju green tea and containing *Chlorella*
Oat and rice cakes with Spirulina	Gullón Vitalday	GalletasGúllon, Spain	Spain	—
Snacks				
Super green superfood bar	OHi	OHi Foods, USA	USA	Paleo and vegan certified product free from gluten, GMO, grain, and soy and containing organic *Spirulina*
Spirulina truffles	Of the Earth Superfoods	Alara Wholefoods, UK	UK	Hand-made spirulina and lemon truffles rich in chlorophyll and phytochemicals
Smoked seaweed & sea salt organic puffs	Honest Fields	SC Honest Fields Europe, Romania	Romania	Corn-based snack containing *Chlorella* suitable for vegans and vegetarians bearing the EU Green Leaf logo
Cereal bars	All Seasons Health	All Seasons Health, UK	UK	Organic snack bars containing *Spirulina*, *Chlorella*, and blue-green algae
Baby roll	Pei Tien	Pei Tien, Taiwan	Taiwan	Algae-flavored rice roll suitable for babies from eight months onward
Green tea and Spirulina candies	CesareCarraro	Incap, Italy	Italy	Claimed to provide total well-being
Other foods				
Rice seasoning	Sengran	Woorichan, South Korea	South Korea	Sengran spinach, chlorella, spinach, cabbage, green tea, green laver and tofu green rice seasoning
Food supplement	4+ Nutrition	International Sport Nutrition, Italy	Italy	Instant powdered food supplement suitable for athletes performing intense physical exercises containing chlorella

Continued

TABLE 8   Commercialized products containing microalgal biomass or high-value products in the market—cont'd

Product	Brand	Company	Country of commercialization	Product description
Egg replacer	Follow Your Heart Vegan egg	Earth Island, USA	France and UK	Egg replacer made with whole microalgae flower and proteins
DHA-enriched cooking oil	Taisun	Taisun Enterprise, Taiwan	Taiwan	Cooking oil blend made from olive and mustard oil enriched in DHA extracted from microalgae
Baby food	Hero Baby Pedialac	Hero, Spain	Spain	Milk for premature babies from birth, containing DHA extracted from microalgae
Baby food	Tony Baby	Norbel Baby, Taiwan	Taiwan	Baby food containing DHA and AA extracted from microalgae

Table reproduced from Lafarga, T. (2019). Cultured microalgae and compounds derived thereof for food applications: Strain selection and cultivation, drying, and processing strategies. Food Reviews International, *pp. 1-26. https://doi.org/10.1080/87559129.2019.1655572 with permission from Taylor & Francis.*

(Ventura et al., 2018). These 2 carotenoids represent approximately half of the global market of carotenoids that will reach a value of 1.5 billion USD by 2021 according to recent studies (Sathasivam & Ki, 2018). Both carotenoids are valuable products; however, the market prices of β-carotene can fluctuate between 300 and 3000 USD per kg and astaxanthin is commonly priced between 2500 and 10,000 USD per kg (Taucher et al., 2016). Astaxanthin is currently sold worldwide under several trade names such as BioAstin produced by Nutrex Hawaii (https://www.nutrex-hawaii.com/pages/bioastin, Retrieved 8 January 2020) or Astapure produced by Algatech (https://www.algatech.com/algatech-product/astapure-natural-astaxanthin/, Retrieved 8 January 2020).

5 Legislation concerning microalgae as food

In the United States, the regulatory aspects of algae products and additives are under the responsibility of the Food and Drug Administration that can assign a generally recognized as safe status to a product (Hayes et al., 2017). In Canada, all-natural health products are regulated by Health Canada under the Food and Drugs Act and the Natural Health Product Regulations (Smith, Jogalekar, & Gibson, 2014). In Europe, the European Food Safety Authority established the "general food law" as the main regulatory framework for food products at all stages of production, as described in Table 9.

The regulation (EC) 258/97 described novel food and novel food ingredients as those not consumed "to a significant degree" in the EU prior to May 15, 1997. Microalgae are included as novel food/food ingredients (art. 1), Section d entitled "Foods and food ingredients consisting of or isolated from microorganisms, fungi, or algae". The novel food catalog can be accessed at https://ec.europa.eu/food/safety/novel_food/catalogue/search/public/index.cfm (Retrieved 8 January 2020). Algal oils (DHA and EPA) from *Schizochytrium* sp., β-carotene from *Dunaliella* sp., and DHA from *Crypthecodiniumcohnii* are approved as food ingredients by the EFSA following these regulations. Moreover, pesticide residues in microalgae are mentioned in Annex I of Regulation EC 396/2005 if the biomass is intended for human consumption. Microalgae as food are also regulated by the Commission Regulation

TABLE 9 European regulations applicable to algal food products.

General food law
Regulation EC 178/2002
Art. 1. Food definition
Art. 14-15. Food should not be placed in the market if they are believed unsafe
Art. 16. Presentation of the product should not mislead consumers
Art. 19-20. Withdraw a product from the market and inform authorities
Novel food regulations
Regulation EC 258/1997
Novel food/food ingredients.
Algae as Novel Food in art. 1(2) d.
Recommendation 97/618/EC. Data requirements applicants of novel food product.
Chemical food safety regulations
Regulation EC 396/2005. Maximum residue levels (MRL) pesticides food/feed of plant/animal origin.
Annex I replaced by**Regulation (EU) 212/2013.** MRL primary agricultural products (seaweed as food).
Regulation EC 1881/2006. Maximum levels for certain contaminants in foodstuffs.

Adapted from Garcia-Vaquero, M., & Hayes, M. (2016). Red and green macroalgae for fish and animal feed and human functional food development. Food Reviews International, 32 (1), 15–45. https://doi.org/10.1080/87559129.2015.1041184.

EU 420/2011 and all the food products developed using microalgae must be labeled complying with Regulation 11169/2011 since the December 13, 2014.

6 Future market and challenges of the use of microalgae as food

Microalgae represent a relatively new and untapped source of nutritionally relevant compounds, including carbohydrates, lipids, proteins-peptides, and carotenoids with promising health benefits, attracting the interest of the pharmaceutical, cosmetic, food, and animal feed industries aiming to introduce these novel products in the market. Despite the environmental and economic benefits of microalgal cultivation and the wide number of strains reported in the scientific literature, most of these species are currently not exploited at commercial scale as

the low production capacity and high production costs may limit their expansion. Thereby, recent cultivation trends aim to reduce the cultivation costs of microalgae by coupling their cultivation with wastewater treatment plants (Acién et al., 2012). More interdisciplinary studies are needed to evaluate different aspects of the production of microalgae by assessing the economic and environmental impacts of different cultivation strategies aiming to increase the productivity of the microalgal industries.

From a regulatory point of view, several conditions have to be met by microalgae or microalgal products to be placed in the European market (see Table 9). The European regulations on novel food or ingredients, food safety, nutrition, and health claims will affect the marketing of microalgal products (Hayes et al., 2019; Vigani et al., 2015). The food safety regulation requires the assessment of toxins, allergens, and other harmful compounds potentially produced by algae (Vigani et al., 2015) and thus, a correct identification of the microalgal strains is necessary to ensure the safety and future commercialization of the microalgal products (Enzing, Ploeg, Barbosa, & Sijtsma, 2014).

The production of high-value compounds from microalgae is also a promising field for researchers aiming to improve the efficiency of these cells to synthesize these molecules and the downstream processing stages to obtain the final products. The application and scale-up of promising laboratory approaches such as cultivation stresses, omics-based technologies, and molecular approaches could reduce the production costs of promising microalgal compounds. Metabolic and transcriptional engineering of microalgae is promising strategies to increase the production of carotenoids, although the negative impacts of genetically modified organisms for the human health and ecological systems are currently limiting the widespread application of these approaches (Liang, Zhu, & Jiang, 2018). Moreover, the exploration of novel exploitation approaches of algae such as the biorefinery approach and the incorporation of innovative technologies at multiple stages during the downstream processing of the biomass offer huge opportunities to improve the efficiency of the extraction of high-value compounds for future industrial use (Garcia-Vaquero, O'Doherty, et al., 2019). Future research will be needed to optimize the use of innovative technologies to improve the efficiency of the microalgal industry following a biorefinery concept.

Acknowledgments

This chapter did not receive any specific grant from funding agencies in the public, commercial, or non-for-profit sectors.

References

Abe, K., Hattori, H., & Hirano, M. (2007). Accumulation and antioxidant activity of secondary carotenoids in the aerial microalga *Coelastrella striolata* var. *multistriata*. *Food Chemistry*, *100*(2), 656–661. https://doi.org/10.1016/j.foodchem.2005.10.026.

Acién, F., Fernández, J., Magán, J., & Molina, E. (2012). Production cost of a real microalgae production plant and strategies to reduce it. *Biotechnology Advances*, *30*(6), 1344–1353.

Adair, W. S., Steinmetz, S. A., Mattson, D. M., Goodenough, U. W., & Heuser, J. E. (1987). Nucleated assembly of *Chlamydomonas* and *Volvox* cell walls. *The Journal of Cell Biology*, *105*(5), 2373–2382.

Adarme-Vega, T. C., Thomas-Hall, S. R., & Schenk, P. M. (2014). Towards sustainable sources for omega-3 fatty acids production. *Current Opinion in Biotechnology*, *26*, 14–18.

Aiking, H. (2014). Protein production: Planet, profit, plus people? *American Journal of Clinical Nutrition*, *100*(suppl_1), 483S–489S.

Albentosa, M., Pérez-Camacho, A., Labarta, U., & Fernández-Reiriz, M. J. (1996). Evaluation of live microalgal diets for the seed culture of *Ruditapes decussatus* using physiological and biochemical parameters. *Aquaculture*, *148*(1), 11–23.

Allard, B., & Templier, J. (2000). Comparison of neutral lipid profile of various trilaminar outer cell wall (TLS)-containing microalgae with emphasis on algaenan occurrence. *Phytochemistry*, *54*(4), 369–380.

Arnold, A. A., Genard, B., Zito, F., Tremblay, R., Warschawski, D. E., & Marcotte, I. (2015). Identification of lipid and saccharide constituents of whole microalgal cells by 13C solid-state NMR. *Biochimica et Biophysica Acta (BBA)-Biomembranes*, *1848*(1), 369–377.

Becker, E. (2007). Micro-algae as a source of protein. *Biotechnology Advances*, *25*(2), 207–210.

Behera, S., Singh, R., Arora, R., Sharma, N. K., Shukla, M., & Kumar, S. (2014). Scope of algae as third generation biofuels. *Frontiers in Bioengineering and Biotechnology*, *2*.

Ben-Amotz, A., Tornabene, T. G., & Thomas, W. H. (1985). Chemical profile of selected species of microalgae with emphasis on lipids 1. *Journal of Phycology*, *21*(1), 72–81.

Bialek, A., Bialek, M., Jelinska, M., & Tokarz, A. (2017). Fatty acid composition and oxidative characteristics of novel edible oils in Poland. *CyTA-Journal of Food*, *15*(1), 1–8.

Boisen, S., & Eggum, B. (1991). Critical evaluation of in vitro methods for estimating digestibility in simple-stomach animals. *Nutrition Research Reviews*, *4*(1), 141–162.

Brown, M., Garland, C., Jeffrey, S., Jameson, I., & Leroi, J. (1993). The gross and amino acid compositions of batch and semi-continuous cultures of Isochrysis sp. (clone T. ISO), Pavlova lutheri and *Nannochloropsis oculata*. *Journal of Applied Phycology*, *5*(3), 285–296.

Brown, M. R., & Jeffrey, S. (1992). Biochemical composition of microalgae from the green algal classes Chlorophyceae and Prasinophyceae. 1. Amino acids, sugars and pigments. *Journal of Experimental Marine Biology and Ecology*, *161*(1), 91–113.

Caporgno, M. P., & Mathys, A. (2018). Trends in microalgae incorporation into innovative food products with potential health benefits. *Frontiers in Nutrition*. *5*(58). https://doi.org/10.3389/fnut.2018.00058.

Casano, L. M., Braga, M. R., Alvarez, R., Del Campo, E. M., & Barreno, E. (2015). Differences in the cell walls and extracellular polymers of the two *Trebouxia* microalgae coexisting in the lichen *Ramalina farinacea* are consistent with their distinct capacity to immobilize extracellular Pb. *Plant Science*, *236*, 195–204. https://doi.org/10.1016/j.plantsci.2015.04.003.

Castillejo, N., Martínez-Hernández, G. B., Goffi, V., Gómez, P. A., Aguayo, E., Artés, F., et al. (2018). Natural vitamin B12 and fucose supplementation of green smoothies with edible algae and related quality changes during their shelf life. *Journal of the Science of Food and Agriculture*, *98*(6), 2411–2421. https://doi.org/10.1002/jsfa.8733.

Chacón-Lee, T., & González-Mariño, G. (2010). Microalgae for "healthy" foods—Possibilities and challenges. *Comprehensive Reviews in Food Science and Food Safety*, *9*(6), 655–675.

Chan, M.-C., Ho, S.-H., Lee, D.-J., Chen, C.-Y., Huang, C.-C., & Chang, J.-S. (2013). Characterization, extraction and purification of lutein produced by an indigenous microalga *Scenedesmus obliquus* CNW-N. *Biochemical Engineering Journal*, *78*, 24–31.

Chaudhary, A., Gustafson, D., & Mathys, A. (2018). Multi-indicator sustainability assessment of global food systems. *Nature Communications*, *9*(1), 848. https://doi.org/10.1038/s41467-018-03308-7.

Chaudhary, L., Pradhan, P., Soni, N., Singh, P., & Tiwari, A. (2014). Algae as a feedstock for bioethanol production: New entrance in biofuel world. *International Journal of ChemTech Research*, *6*, 1381–1389.

Cheah, W. Y., Show, P. L., Chang, J.-S., Ling, T. C., & Juan, J. C. (2015). Biosequestration of atmospheric CO_2 and flue gas-containing CO_2 by microalgae. *Bioresource Technology*, *184*, 190–201.

Chen, Y.-C. (2012). The biomass and total lipid content and composition of twelve species of marine diatoms cultured under various environments. *Food Chemistry*, *131*(1), 211–219.

Chen, C.-S., Anaya, J. M., Zhang, S., Spurgin, J., Chuang, C.-Y., Xu, C., et al. (2011). Effects of engineered nanoparticles on the assembly of exopolymeric substances from phytoplankton. *PLoS One*, *6*(7)e21865.

Chen, B., Wan, C., Mehmood, M. A., Chang, J.-S., Bai, F., & Zhao, X. (2017). Manipulating environmental stresses and stress tolerance of microalgae for enhanced production of lipids and value-added products—A review.

Bioresource Technology, *244*, 1198–1206. https://doi.org/10.1016/j.biortech.2017.05.170.

Cheng, Y. S., Labavitch, J. M., & Vander Gheynst, J. S. (2015). Elevated CO_2 concentration impacts cell wall polysaccharide composition of green microalgae of the genus *Chlorella*. *Letters in Applied Microbiology*, *60*(1), 1–7. https://doi.org/10.1111/lam.12320.

Cheng, Y.-S., Zheng, Y., Labavitch, J. M., & Vander Gheynst, J. S. (2011). The impact of cell wall carbohydrate composition on the chitosan flocculation of *Chlorella*. *Process Biochemistry*, *46*(10), 1927–1933. https://doi.org/10.1016/j.procbio.2011.06.021.

Chew, B. P., & Park, J. S. (2004). Carotenoid action on the immune response. *The Journal of Nutrition*, *134*(1), 257S–261S.

Chew, K. W., Yap, J. Y., Show, P. L., Suan, N. H., Juan, J. C., Ling, T. C., et al. (2017). Microalgae biorefinery: High value products perspectives. *Bioresource Technology*, *229*, 53–62.

Chronakis, I. S. (2000). Biosolar proteins from aquatic algae. *Developments in Food Science* (pp. 39–75). Vol. 41(pp. 39–75). Elsevier.

Clemente-Carazo, M., Sanchez, V., Condon-Abanto, S., & Marco, G.-V. (2019). Algal polysaccharides: Innovative extraction technologies, health benefits and industrial applications. In *Handbook of Algal Technologies and Phytochemicals* (pp. 3–12). CRC Press.

Corbo, M. R., Bevilacqua, A., Petruzzi, L., Casanova, F. P., & Sinigaglia, M. (2014). Functional beverages: The emerging side of functional foods. *Comprehensive Reviews in Food Science and Food Safety*, *13*(6), 1192–1206. https://doi.org/10.1111/1541-4337.12109.

Cravotto, G., Boffa, L., Mantegna, S., Perego, P., Avogadro, M., & Cintas, P. (2008). Improved extraction of vegetable oils under high-intensity ultrasound and/or microwaves. *Ultrasonics Sonochemistry*, *15*(5), 898–902.

Cuellar-Bermudez, S. P., Aguilar-Hernandez, I., Cardenas-Chavez, D. L., Ornelas-Soto, N., Romero-Ogawa, M. A., & Parra-Saldivar, R. (2015). Extraction and purification of high-value metabolites from microalgae: Essential lipids, astaxanthin and phycobiliproteins. *Microbial Biotechnology*, *8*(2), 190–209.

da Silva Vaz, B., Moreira, J. B., de Morais, M. G., & Costa, J. A. V. (2016). Microalgae as a new source of bioactive compounds in food supplements. *Current Opinion in Food Science*, *7*, 73–77.

Dawczynski, C., Schubert, R., & Jahreis, G. (2007). Amino acids, fatty acids, and dietary fibre in edible seaweed products. *Food Chemistry*, *103*(3), 891–899. https://doi.org/10.1016/j.foodchem.2006.09.041.

de Jesus Raposo, M. F., de Morais, A. M. M. B., & de Morais, R. M. S. C. (2015). Bioactivity and applications of polysaccharides from marine microalgae. In *Polysaccharides: Bioactivity and biotechnology* (pp. 1683–1727).

Deenu, A., Naruenartwongsakul, S., & Kim, S. M. (2013). Optimization and economic evaluation of ultrasound extraction of lutein from *Chlorella vulgaris*. *Biotechnology and Bioprocess Engineering, 18*(6), 1151–1162.

Del Campo, J. A., García-González, M., & Guerrero, M. G. (2007). Outdoor cultivation of microalgae for carotenoid production: Current state and perspectives. *Applied Microbiology and Biotechnology, 74*(6), 1163–1174.

Du, Y., Schuur, B., Kersten, S. R., & Brilman, D. W. (2015). Opportunities for switchable solvents for lipid extraction from wet algal biomass: An energy evaluation. *Algal Research, 11*, 271–283.

Ejike, C. E., Collins, S. A., Balasuriya, N., Swanson, A. K., Mason, B., & Udenigwe, C. C. (2017). Prospects of microalgae proteins in producing peptide-based functional foods for promoting cardiovascular health. *Trends in Food Science & Technology, 59*, 30–36.

Enzing, C., Ploeg, M., Barbosa, M., & Sijtsma, L. (2014). Microalgae-based products for the food and feed sector: An outlook for Europe. In *JRC Scientific and Policy Reports* (pp. 19–37).

Fan, X., Bai, L., Zhu, L., Yang, L., & Zhang, X. (2014). Marine algae-derived bioactive peptides for human nutrition and health. *Journal of Agricultural and Food Chemistry, 62*(38), 9211–9222. https://doi.org/10.1021/jf502420h.

Fradique, M., Batista, A. P., Nunes, M. C., Gouveia, L., Bandarra, N. M., & Raymundo, A. (2010). Incorporation of *Chlorella vulgaris* and *Spirulina maxima* biomass in pasta products. Part 1: Preparation and evaluation. *Journal of the Science of Food and Agriculture, 90*(10), 1656–1664. https://doi.org/10.1002/jsfa.3999.

Fradique, M., Batista, A. P., Nunes, M. C., Gouveia, L., Bandarra, N. M., & Raymundo, A. (2013). Isochrysis galbana and *Diacronema vlkianum* biomass incorporation in pasta products as PUFA's source. *LWT—Food Science and Technology,* *50*(1), 312–319. https://doi.org/10.1016/j.lwt.2012.05.006.

Garcia-Vaquero, M. (2019). Analytical methods and advances to evaluate dietary fiber. (chapter 6) In C. M. Galanakis (Ed.), *Dietary Fiber: Properties, Recovery, and Applications* (pp. 165–197). Academic Press.

García-Vaquero, M. (2018). Seaweed proteins and applications in animal feed. In *Novel proteins for food, pharmaceuticals and agriculture: Sources, applications and advances* (pp. 139–161).

Garcia-Vaquero, M., & Hayes, M. (2016). Red and green macroalgae for fish and animal feed and human functional food development. *Food Reviews International, 32* (1), 15–45. https://doi.org/10.1080/87559129.2015.1041184.

Garcia-Vaquero, M., Lopez-Alonso, M., & Hayes, M. (2017). Assessment of the functional properties of protein extracted from the brown seaweed *Himanthalia elongata* (Linnaeus) SF gray. *Food Research International, 99*, 971–978.

Garcia-Vaquero, M., Mora, L., & Hayes, M. (2019). In vitro and in silico approaches to generating and identifying angiotensin-converting enzyme I inhibitory peptides from green macroalga *Ulva lactuca*. *Marine Drugs, 17*(4), 204.

Garcia-Vaquero, M., O'Doherty, J. V., Tiwari, B. K., Sweeney, T., & Rajauria, G. (2019). Enhancing the extraction of polysaccharides and antioxidants from macroalgae using sequential hydrothermal-assisted extraction followed by ultrasound and thermal technologies. *Marine Drugs, 17* (8), 457.

Garcia-Vaquero, M., Rajauria, G., O'Doherty, J., & Sweeney, T. (2017). Polysaccharides from macroalgae: Recent advances, innovative technologies and challenges in extraction and purification. *Food Research International, 99*(3), 1011.

Garcia-Vaquero, M., & Tiwari, B. K. (2020). Novel sources for oil production. (chapter 3) In I. Aguilo, & T. Lafarga (Eds.), *Oil and oilseed processing, opportunities and challenges*. Wiley & Sons.

Garrido-Cardenas, J. A., Manzano-Agugliaro, F., Acien-Fernandez, F. G., & Molina-Grima, E. (2018). Microalgae research worldwide. *Algal Research, 35*, 50–60.

Geresh, S., Arad, S. M., Levy-Ontman, O., Zhang, W., Tekoah, Y., & Glaser, R. (2009). Isolation and characterization of poly- and oligosaccharides from the red microalga Porphyridium sp. *Carbohydrate Research, 344*(3), 343–349. https://doi.org/10.1016/j.carres.2008.11.012.

Gerken, H. G., Donohoe, B., & Knoshaug, E. P. (2013). Enzymatic cell wall degradation of *Chlorella vulgaris* and other microalgae for biofuels production. *Planta, 237*(1), 239–253.

Gil-Chavez, G., Villa, J., Ayala-Zavala, J., Basilio Heredia, J., Sepulveda, D., & Yahia, E. (2013). Technologies for extraction and production of bioactive compounds to be used as nutraceuticals and food ingredients: An overview. *Comprehensive Reviews in Food Science and Food Safety, 12*, 5–23.

Gong, M., & Bassi, A. (2016). Carotenoids from microalgae: A review of recent developments. *Biotechnology Advances, 34*(8), 1396–1412.

Gouveia, L., Coutinho, C., Mendonça, E., Batista, A. P., Sousa, I., Bandarra, N. M., et al. (2008). Functional biscuits with PUFA-ω3 from *Isochrysis galbana*. *Journal of the Science of Food and Agriculture, 88*(5), 891–896. https://doi.org/10.1002/jsfa.3166.

Graça, C., Fradinho, P., Sousa, I., & Raymundo, A. (2018). Impact of *Chlorella vulgaris* on the rheology of wheat flour dough and bread texture. *LWT, 89*, 466–474. https://doi.org/10.1016/j.lwt.2017.11.024.

Grima, E. M., Belarbi, E.-H., Fernández, F. A., Medina, A. R., & Chisti, Y. (2003). Recovery of microalgal biomass and metabolites: Process options and economics. *Biotechnology Advances, 20*(7–8), 491–515.

Grimi, N., Dubois, A., Marchal, L., Jubeau, S., Lebovka, N., & Vorobiev, E. (2014). Selective extraction from microalgae *Nannochloropsis* sp. using different methods of cell disruption. *Bioresource Technology*, 153, 254–259.

Günerken, E., D'Hondt, E., Eppink, M., Garcia-Gonzalez, L., Elst, K., & Wijffels, R. (2015). Cell disruption for microalgae biorefineries. *Biotechnology Advances*, 33, 243–260.

Halim, R., Danquah, M. K., & Webley, P. A. (2012). Extraction of oil from microalgae for biodiesel production: A review. *Biotechnology Advances*, 30(3), 709–732.

Halim, R., Harun, R., Danquah, M. K., & Webley, P. A. (2012). Microalgal cell disruption for biofuel development. *Applied Energy*, 91(1), 116–121.

Hayes, M., Bastiaens, L., Gouveia, L., Gkelis, S., Skomedal, H., Skjanes, K., et al. (2019). Microalgal bioactive compounds including protein, peptides, and pigments: Applications, opportunities, and challenges during biorefinery processes. In *Novel proteins for food, pharmaceuticals and agriculture: Sources, applications and advances* (pp. 239–255). John Wiley & Sons.

Hayes, M., & García-Vaquero, M. (2016). Bioactive compounds from fermented food products. In K. S. Ojha, & B. K. Tiwari (Eds.), *Novel food fermentation technologies* (pp. 293–310). Cham: Springer International Publishing.

Hayes, M., Skomedal, H., Skjånes, K., Mazur-Marzec, H., Toruńska-Sitarz, A., Catala, M., et al. (2017). Microalgal proteins for feed, food and health. (chapter 15) In C. Gonzalez-Fernandez, & R. Muñoz (Eds.), *Microalgae-based biofuels and bioproducts* (pp. 347–368). Woodhead Publishing.

Henchion, M., Hayes, M., Mullen, A. M., Fenelon, M., & Tiwari, B. (2017). Future protein supply and demand: Strategies and factors influencing a sustainable equilibrium. *Food*, 6(7), 53.

Herman, E. M., & Schmidt, M. A. (2016). The potential for engineering enhanced functional-feed soybeans for sustainable aquaculture feed. *Frontiers in Plant Science*, 7.

Holdt, S., & Kraan, S. (2011). Bioactive compounds in seaweed: Functional food applications and legislation. *Journal of Applied Phycology*, 23(3), 543–597. https://doi.org/10.1007/s10811-010-9632-5.

Hu, Y.-R., Wang, F., Wang, S.-K., Liu, C.-Z., & Guo, C. (2013). Efficient harvesting of marine microalgae Nannochloropsis maritima using magnetic nanoparticles. *Bioresource Technology*, 138, 387–390.

Illman, A., Scragg, A., & Shales, S. (2000). Increase in *Chlorella* strains calorific values when grown in low nitrogen medium. *Enzyme and Microbial Technology*, 27(8), 631–635.

Kadam, S. U., Tiwari, B. K., & O'Donnell, C. P. (2013). Application of novel extraction technologies for bioactives from marine algae. *Journal of Agricultural and Food Chemistry*, 61(20), 4667–4675.

Khanra, S., Mondal, M., Halder, G., Tiwari, O. N., Gayen, K., & Bhowmick, T. K. (2018). Downstream processing of microalgae for pigments, protein and carbohydrate in industrial application: A review. *Food and Bioproducts Processing*, 110, 60–84. https://doi.org/10.1016/j.fbp.2018.02.002.

Kim, D.-Y., Vijayan, D., Praveenkumar, R., Han, J.-I., Lee, K., Park, J.-Y., et al. (2016). Cell-wall disruption and lipid/astaxanthin extraction from microalgae: *Chlorella* and *Haematococcus*. *Bioresource Technology*, 199, 300–310.

Knuckey, R. M., Brown, M. R., Barrett, S. M., & Hallegraeff, G. M. (2002). Isolation of new nanoplanktonic diatom strains and their evaluation as diets for juvenile Pacific oysters (*Crassostrea gigas*). *Aquaculture*, 211(1–4), 253–274.

Ko, S.-C., Kang, N., Kim, E.-A., Kang, M. C., Lee, S.-H., Kang, S.-M., et al. (2012). A novel angiotensin I-converting enzyme (ACE) inhibitory peptide from a marine *Chlorella ellipsoidea* and its antihypertensive effect in spontaneously hypertensive rats. *Process Biochemistry*, 47(12), 2005–2011.

Koller, M., Muhr, A., & Braunegg, G. (2014). Microalgae as versatile cellular factories for valued products. *Algal Research*, 6, 52–63.

Kwak, H. S., Kim, J. Y. H., Woo, H. M., Jin, E., Min, B. K., & Sim, S. J. (2016). Synergistic effect of multiple stress conditions for improving microalgal lipid production. *Algal Research*, 19, 215–224.

Lafarga, T., Clemente, I., & Garcia-Vaquero, M. (2019). Carotenoids from microalgae. In *Carotenoids: Properties, processing and applications* (p. 149). Elsevier.

Lafarga, T., & Hayes, M. (2017). Bioactive protein hydrolysates in the functional food ingredient industry: Overcoming current challenges. *Food Reviews International*, 33(3), 217–246.

Lai, Y. S., Parameswaran, P., Li, A., Baez, M., & Rittmann, B. E. (2014). Effects of pulsed electric field treatment on enhancing lipid recovery from the microalga, *Scenedesmus*. *Bioresource Technology*, 173, 457–461.

Lee, S. Y., Cho, J. M., Chang, Y. K., & Oh, Y.-K. (2017). Cell disruption and lipid extraction for microalgal biorefineries: A review. *Bioresource Technology*, 244, 1317–1328. https://doi.org/10.1016/j.biortech.2017.06.038.

Lee, A. K., Lewis, D. M., & Ashman, P. J. (2012). Disruption of microalgal cells for the extraction of lipids for biofuels: Processes and specific energy requirements. *Biomass and Bioenergy*, 46, 89–101.

Lee, J.-Y., Yoo, C., Jun, S.-Y., Ahn, C.-Y., & Oh, H.-M. (2010). Comparison of several methods for effective lipid extraction from microalgae. *Bioresource Technology*, 101(1), S75–S77.

Li, H., Qu, Y., Yang, Y., Chang, S., & Xu, J. (2016). Microwave irradiation–A green and efficient way to pretreat biomass. *Bioresource Technology*, 199, 34–41.

Liang, S., Liu, X., Chen, F., & Chen, Z. (2004). *Current microalgal health food R & D activities in China*. Dordrecht.

Liang, M.-H., Zhu, J., & Jiang, J.-G. (2018). High-value bioproducts from microalgae: Strategies and progress.

Critical Reviews in Food Science and Nutrition, 01–53. https://doi.org/10.1080/10408398.2018.1455030.

Lourenco, S. O., Barbarino, E., Mancini-Filho, J., Schinke, K. P., & Aidar, E. (2002). Effects of different nitrogen sources on the growth and biochemical profile of 10 marine microalgae in batch culture: An evaluation for aquaculture. *Phycologia*, *41*(2), 158–168.

Marcati, A., Ursu, A. V., Laroche, C., Soanen, N., Marchal, L., Jubeau, S., et al. (2014). Extraction and fractionation of polysaccharides and B-phycoerythrin from the microalga *Porphyridium cruentum* by membrane technology. *Algal Research*, *5*, 258–263. https://doi.org/10.1016/j.algal.2014.03.006.

Markou, G., Chatzipavlidis, I., & Georgakakis, D. (2012). Effects of phosphorus concentration and light intensity on the biomass composition of Arthrospira (*Spirulina*) platensis. *World Journal of Microbiology and Biotechnology*, *28*(8), 2661–2670.

Matos, Â. P. (2017). The impact of microalgae in food science and technology. *Journal of the American Oil Chemists' Society*, *94*(11), 1333–1350. https://doi.org/10.1007/s11746-017-3050-7.

Matos, Â. P., Feller, R., Moecke, E. H. S., de Oliveira, J. V., Junior, A. F., Derner, R. B., et al. (2016). Chemical characterization of six microalgae with potential utility for food application. *Journal of the American Oil Chemists' Society*, *93*(7), 963–972.

Maurer, N. E., Hatta-Sakoda, B., Pascual-Chagman, G., & Rodriguez-Saona, L. E. (2012). Characterization and authentication of a novel vegetable source of omega-3 fatty acids, sacha inchi (*Plukenetia volubilis* L.) oil. *Food Chemistry*, *134*(2), 1173–1180.

McMillan, J. R., Watson, I. A., Ali, M., & Jaafar, W. (2013). Evaluation and comparison of algal cell disruption methods: Microwave, waterbath, blender, ultrasonic and laser treatment. *Applied Energy*, *103*, 128–134.

Mercer, P., & Armenta, R. E. (2011). Developments in oil extraction from microalgae. *European Journal of Lipid Science and Technology*, *113*(5), 539–547.

Oser, B. (1951). Method for integrating essential ammo acid content in the nutritional evaluation of protein. *Journal of the American Dietetic Association*, *27*, 396–402.

Park, J.-Y., Park, M. S., Lee, Y.-C., & Yang, J.-W. (2015). Advances in direct transesterification of algal oils from wet biomass. *Bioresource Technology*, *184*, 267–275.

Pasquet, V., Chérouvrier, J.-R., Farhat, F., Thiéry, V., Piot, J.-M., Bérard, J.-B., et al. (2011). Study on the microalgal pigments extraction process: Performance of microwave assisted extraction. *Process Biochemistry*, *46*(1), 59–67.

Patel, A. K., Laroche, C., Marcati, A., Ursu, A. V., Jubeau, S., Marchal, L., et al. (2013). Separation and fractionation of exopolysaccharides from *Porphyridium cruentum*. *Bioresource Technology*, *145*, 345–350. https://doi.org/10.1016/j.biortech.2012.12.038.

Prabakaran, P., & Ravindran, A. D. (2011). A comparative study on effective cell disruption methods for lipid extraction from microalgae. *Letters in Applied Microbiology*, *53*(2), 150–154.

Praveenkumar, R., Lee, K., Lee, J., & Oh, Y.-K. (2015). Breaking dormancy: An energy-efficient means of recovering astaxanthin from microalgae. *Green Chemistry*, *17*(2), 1226–1234.

Qian, Z.-J., Heo, S.-J., Oh, C. H., Kang, D.-H., Jeong, S. H., Park, W. S., et al. (2013). Angiotensin I-converting enzyme (ACE) inhibitory peptide isolated from biodiesel byproducts of marine microalgae, *Nannochloropsis oculata*. *Journal of Biobased Materials and Bioenergy*, *7*(1), 135–142.

Ramadan, M. F., & Selim asker, M. M. (2008). Functional bioactive compounds and biological activities. *Czech Journal of Food Sciences*, *26*(3), 211–222.

Reverchon, E., & De Marco, I. (2006). Supercritical fluid extraction and fractionation of natural matter. *Journal of Supercritical Fluids*, *38*(2), 146–166.

Rodolfi, L., Chini Zittelli, G., Bassi, N., Padovani, G., Biondi, N., Bonini, G., et al. (2009). Microalgae for oil: Strain selection, induction of lipid synthesis and outdoor mass cultivation in a low-cost photobioreactor. *Biotechnology and Bioengineering*, *102*(1), 100–112.

Rodriguez-Amaya, D. B. (2016). Natural food pigments and colorants. *Current Opinion in Food Science*, *7*, 20–26. https://doi.org/10.1016/j.cofs.2015.08.004.

Rubio-Rodríguez, N., Beltrán, S., Jaime, I., Sara, M., Sanz, M. T., & Carballido, J. R. (2010). Production of omega-3 polyunsaturated fatty acid concentrates: A review. *Innovative Food Science & Emerging Technologies*, *11*(1), 1–12.

Ruiz, J., Olivieri, G., de Vree, J., Bosma, R., Willems, P., Reith, J. H., et al. (2016). Towards industrial products from microalgae. *Energy & Environmental Science*, *9*(10), 3036–3043.

Sadovskaya, I., Souissi, A., Souissi, S., Grard, T., Lencel, P., Greene, C. M., et al. (2014). Chemical structure and biological activity of a highly branched (1 –> 3,1 –> 6)-beta-D-glucan from *Isochrysis galbana*. *Carbohydrate Polymers*, *111*, 139–148. https://doi.org/10.1016/j.carbpol.2014.04.077.

Safi, C., Charton, M., Ursu, A. V., Laroche, C., Zebib, B., Pontalier, P.-Y., et al. (2014). Release of hydro-soluble microalgal proteins using mechanical and chemical treatments. *Algal Research*, *3*, 55 60. https://doi.org/10.1016/j.algal.2013.11.017.

Saini, R. K., & Keum, Y.-S. (2018). Omega-3 and omega-6 polyunsaturated fatty acids: Dietary sources, metabolism, and significance—A review. *Life Sciences*, *203*, 255–267.

Samarakoon, K. W., Kwon, O.-N., Ko, J.-Y., Lee, J.-H., Kang, M.-C., Kim, D., et al. (2013). Purification and identification of novel angiotensin-I converting enzyme (ACE) inhibitory peptides from cultured marine microalgae (*Nannochloropsis oculata*) protein hydrolysate. *Journal of Applied Phycology*, *25*(5), 1595–1606.

Sánchez-Moreno, C., Plaza, L., Elez-Martínez, P., De Ancos, B., Martín-Belloso, O., & Cano, M. P. (2005). Impact of high pressure and pulsed electric fields on bioactive compounds and antioxidant activity of orange juice in comparison with traditional thermal processing. *Journal of Agricultural and Food Chemistry, 53*(11), 4403–4409.

Santos, T. D., de Freitas, B. C. B., Moreira, J. B., Zanfonato, K., & Costa, J. A. V. (2016). Development of powdered food with the addition of *Spirulina* for food supplementation of the elderly population. *Innovative Food Science & Emerging Technologies, 37*, 216–220. https://doi.org/10.1016/j.ifset.2016.07.016.

Sathasivam, R., & Ki, J. S. (2018). A review of the biological activities of microalgal carotenoids and their potential use in healthcare and cosmetic industries. *Marine Drugs 16*(1). https://doi.org/10.3390/md16010026.

Scholz, M. J., Weiss, T. L., Jinkerson, R. E., Jing, J., Roth, R., Goodenough, U., et al. (2014). Ultrastructure and composition of the *Nannochloropsis gaditana* cell wall. *Eukaryotic Cell, 13*(11), 1450–1464.

Shah, M. M. R., Liang, Y., Cheng, J. J., & Daroch, M. (2016). Astaxanthin-producing green microalga *Haematococcus pluvialis*: From single cell to high value commercial products. *Frontiers in Plant Science. 7*(531). https://doi.org/10.3389/fpls.2016.00531.

Sheih, I.-C., Fang, T. J., & Wu, T.-K. (2009). Isolation and characterisation of a novel angiotensin I-converting enzyme (ACE) inhibitory peptide from the algae protein waste. *Food Chemistry, 115*(1), 279–284.

Smith, A., Jogalekar, S., & Gibson, A. (2014). Regulation of natural health products in Canada. *Journal of Ethnopharmacology, 158*, 507–510.

Suetsuna, K., & Chen, J.-R. (2001). Identification of antihypertensive peptides from peptic digest of two microalgae, *Chlorella vulgaris* and *Spirulina platensis*. *Marine Biotechnology, 3*(4), 305–309.

Suh, S.-S., Kim, S. J., Hwang, J., Park, M., Lee, T.-K., Kil, E.-J., et al. (2015). Fatty acid methyl ester profiles and nutritive values of 20 marine microalgae in Korea. *Asian Pacific Journal of Tropical Medicine, 8*(3), 191–196.

Sun, L., Chu, J., Sun, Z., & Chen, L. (2016). Physicochemical properties, immunomodulation and antitumor activities of polysaccharide from *Pavlova viridis*. *Life Sciences, 144*, 156–161. https://doi.org/10.1016/j.lfs.2015.11.013.

Taucher, J., Baer, S., Schwerna, P., Hofmann, D., Hümmer, M., Buchholz, R., et al. (2016). Cell disruption and pressurized liquid extraction of carotenoids from microalgae. *Journal of Thermodynamics & Catalysis, 7*.

Tibbetts, S. M., Milley, J. E., & Lall, S. P. (2015). Chemical composition and nutritional properties of freshwater and marine microalgal biomass cultured in photobioreactors. *Journal of Applied Phycology, 27*(3), 1109–1119. https://doi.org/10.1007/s10811-014-0428-x.

Tilman, D., & Clark, M. (2014). Global diets link environmental sustainability and human health. *Nature, 515*(7528), 518.

Tiwari, B. K., Gowen, A., & McKenna, B. (2011). Introduction. (chapter 1) In B. K. Tiwari, A. Gowen, & B. McKenna (Eds.), *Pulse foods* (pp. 1–7). San Diego: Academic Press.

Tokuşoglu, Ö., & üUnal, M. K. (2003). Biomass nutrient profiles of three microalgae: *Spirulina platensis, Chlorella vulgaris*, and *Isochrisis galbana*. *Journal of Food Science, 68*(4), 1144–1148. https://doi.org/10.1111/j.1365-2621.2003.tb09615.x.

Tukker, A., & Jansen, B. (2006). Environmental impacts of products: A detailed review of studies. *Journal of Industrial Ecology, 10*(3), 159–182.

Ufaz, S., & Galili, G. (2008). Improving the content of essential amino acids in crop plants: Goals and opportunities. *Plant Physiology, 147*(3), 954–961.

Ursu, A.-V., Marcati, A., Sayd, T., Sante-Lhoutellier, V., Djelveh, G., & Michaud, P. (2014). Extraction, fractionation and functional properties of proteins from the microalgae *Chlorella vulgaris*. *Bioresource Technology, 157*, 134–139. https://doi.org/10.1016/j.biortech.2014.01.071.

Vannuccini, S., Kavallari, A., Bellù, L. G., Müller, M., & Wisser, D. (2019). Understanding the impacts of climate change for fisheries and aquaculture: Global and regional supply and demand trends and prospects. In *Impacts of climate change on fisheries and aquaculture* (p. 41). Food and Agriculture Organization (FAO).

Venkidasamy, B., Selvaraj, D., Nile, A. S., Ramalingam, S., Kai, G., & Nile, S. H. (2019). Indian pulses: A review on nutritional, functional and biochemical properties with future perspectives. *Trends in Food Science & Technology, 88*, 228–242. https://doi.org/10.1016/j.tifs.2019.03.012.

Ventura, S., Nobre, B., Ertekin, F., Hayes, M., Garcia-Vaquero, M., Vieira, F., et al. (2018). Extraction of added-value compounds from microalgae. In R. Muñoz, & C. Gonzalez-Fernandez (Eds.), *Microalgae-based biofuels and bioproducts*. Elsevier.

Vermeulen, S. J., Campbell, B. M., & Ingram, J. S. (2012). Climate change and food systems. *Annual Review of Environment and Resources, 37*, .

Vigani, M., Parisi, C., Rodríguez-Cerezo, E., Barbosa, M. J., Sijtsma, L., Ploeg, M., et al. (2015). Food and feed products from micro-algae: Market opportunities and challenges for the EU. *Trends in Food Science & Technology, 42*(1), 81–92. https://doi.org/10.1016/j.tifs.2014.12.004.

Waghmare, A. G., Salve, M. K., LeBlanc, J. G., & Arya, S. S. (2016). Concentration and characterization of microalgae proteins from *Chlorella pyrenoidosa*. *Bioresources and Bioprocessing, 3*(1), 16.

Ward, O. P., & Singh, A. (2005). Omega-3/6 fatty acids: Alternative sources of production. *Process Biochemistry, 40*(12), 3627–3652.

Wayama, M., Ota, S., Matsuura, H., Nango, N., Hirata, A., & Kawano, S. (2013). Three-dimensional ultrastructural study of oil and astaxanthin accumulation during encystment in the green alga *Haematococcus pluvialis*. *PLoS One*. *8*(1), e53618. https://doi.org/10.1371/journal.pone.0053618.

Wei, L., & Huang, X. (2017). Long-duration effect of multi-factor stresses on the cellular biochemistry, oil-yielding performance and morphology of Nannochloropsis oculata. *PLoS One*, *12*(3), e0174646.

Whyte, J. N. (1987). Biochemical composition and energy content of six species of phytoplankton used in mariculture of bivalves. *Aquaculture*, *60*(3–4), 231–241.

Yen, H.-W., Yang, S.-C., Chen, C.-H., & Chang, J.-S. (2015). Supercritical fluid extraction of valuable compounds from microalgal biomass. *Bioresource Technology*, *184*, 291–296.

Yu, X., Gouyo, T., Grimi, N., Bals, O., & Vorobiev, E. (2016). Pulsed electric field pretreatment of rapeseed green biomass (stems) to enhance pressing and extractives recovery. *Bioresource Technology*, *199*, 194–201.

Zheng, H., Yin, J., Gao, Z., Huang, H., Ji, X., & Dou, C. (2011). Disruption of *Chlorella vulgaris* cells for the release of biodiesel-producing lipids: A comparison of grinding, ultrasonication, bead milling, enzymatic lysis, and microwaves. *Applied Biochemistry and Biotechnology*, *164*(7), 1215–1224.

Microalgae as feed ingredients for livestock production and aquaculture

Luisa M.P. Valente[a,b], Ana R.J. Cabrita[b,c], Margarida R.G. Maia[b,c], Inês M. Valente[c,d], Sofia Engrola[e], António J.M. Fonseca[b,c], David Miguel Ribeiro[f], Madalena Lordelo[f], Cátia Falcão Martins[f], Luísa Falcão e Cunha[f], André Martinho de Almeida[f], and João Pedro Bengala Freire[f]

[a]CIIMAR, Interdisciplinary Centre of Marine and Environmental Research, University of Porto, Terminal de Cruzeiros do Porto de Leixões, Matosinhos, Portugal, [b]ICBAS, Institute of Biomedical Sciences Abel Salazar, University of Porto, Porto, Portugal, [c]REQUIMTE, LAQV, ICBAS, Abel Salazar Biomedical Sciences Institute, University of Porto, Porto, Portugal, [d]REQUIMTE, LAQV, Department of Chemistry and Biochemistry, Faculty of Sciences of the University of Porto, Porto, Portugal, [e]Centre of Marine Sciences (CCMAR), University of Algarve, Campus de Gambelas, Faro, Portugal, [f]LEAF Linking Landscape, Environment, Agriculture and Food, School of Agriculture, University of Lisbon, Lisbon, Portugal

1 Introduction

Global population growth and an increase in living standards will push up the demand for high-quality animal products in the near future (Alexandratos & Bruinsma, 2012; FAO, 2018). Consumers are becoming more aware of food beneficial effects and hazards, and care for a balanced and nutritious diet. On the other hand, concerns about the adverse effects of synthetic raw materials have increasingly favored food products containing natural ingredients. Livestock feeding requires large supplies of crops, imported protein meals (mainly soybean meal), and fodders, while aquatic species rely on fish meal and oil to satisfy the demand for cost-effective feedstuffs by the feed industry (FAO, 2018; Lum, Kim, & Lei, 2013). But crops require large areas and water supply for their cultivation, causing habitat loss, deforestation, and thereby contributing to global warming, which has severe consequences for the environment (Gibbs et al., 2015). In addition, the use of such ingredients in animal feeds competes directly with human nutrition. As in livestock production, the fast growth of aquaculture has been accompanied by aquafeed production, leading to a depletion of the fish meal and fish oil stocks and an increase of these feedstuffs' prices. This scenario puts pressure on the feed industry to search for alternative and more sustainable dietary protein and lipid sources (Tacon & Metian, 2015). In this context, the use of microalgae has been largely encouraged in the last decade in animal feeds, not only as a nutrient supply (Cardinaletti et al., 2018; Hopkins et al., 2014; Kiron et al., 2016; Lamminen et al., 2017; Lamminen, Halmemies-Beauchet-Filleau, Kokkonen, Jaakkola, & Vanhatalo, 2019; Lum et al., 2013; Qiao et al., 2014; Sørensen et al., 2017; Tibaldi et al., 2015; Valente, Custódio, Batista, Fernandes, & Kiron, 2019; Wang et al., 2017b) but also as a valuable source of bioactive compounds like fatty acids, pigments, vitamins, and minerals (Hemaiswarya, Raja, Ravi Kumar, Ganesan, & Anbazhagan, 2011; Tibbetts, 2018).

Microalgae are a diverse group of uni- or multicellular organisms composed of four major groups: diatoms (*Bacillariophyceae*), green (*Chlorophyceae*), golden (*Chrysophyceae*), and blue-green (*Cyanophyceae*) algae (Madeira et al., 2017), with the combined number of species reaching almost 100,000. These are split into autotrophic (use CO_2 and sunlight as carbon and energy sources, respectively) and heterotrophic (use organic carbon as a source of energy instead of sunlight) species; the latter grows easily in industrial bioreactors and is easily used for biomass production (Madeira et al., 2017). Microalgae are mainly composed of proteins,

carbohydrates, lipids, vitamins, minerals, and bioactive compounds, such as carotenoids. They have a variable nutrient composition, depending on several factors such as species, production conditions (e.g., nutrient availability), enzyme utilization, and biomass status (whole or defatted algae meal). The nutritional value of the main genera used in diets for livestock and aquaculture is presented in Table 1. Moreover, the capacity to use such nutrients mainly depends on the trophic level of the animal's species to be fed: herbivorous animals, particularly ruminants, use algae better than highly carnivorous monogastric species like some fish (e.g., salmonids). In this case, cost-effective processing technologies may be required to increase the nutrient bioavailability of these species.

This chapter unveils current knowledge of dietary microalgae effects on production and meat quality of livestock (ruminants, pigs, poultry, rabbits) as well as species with relevance for aquaculture (fish and shrimp).

2 Microalgae in ruminants

Among livestock animals, ruminants have the advantage of being capable to use the non-protein nitrogen and to digest the microalgae cell wall (Lum et al., 2013). In these animals, the dietary inclusion of microalgae has been evaluated mainly by their effects on milk production and composition, animal growth and meat composition, and to a lesser extent on oxidative status (Mavrommatis et al., 2018; Tsiplakou, Abdullah, Alexandros, et al., 2017; Tsiplakou et al., 2018) and animal reproduction (Senosy et al., 2017; Sinedino et al., 2017). From Tables 3–5, a selection of studies compiles major effects of dietary inclusion of microalgae on milk yield and composition, and animal growth and meat composition. The use of microalgae in animal feed dates back to 1950s, and the majority of the studies evaluated commercial products rich

TABLE 1 Proximate composition of microalgae genera considered in feeds for livestock and aquaculture (% or kJ/g dry weight basis).

Algae	Species	Form	Protein	Lipid	$\sum$PUFA (% FA)	NFE	NDF	ADF	Starch	Energy	Ash	Other	References
Arthrospira (Spirulina) (Schizochytrium)	*A. platensis* (*S. platensis*)	Whole	62.0–72.2	2.7–11.4	n.d.–44.2		0–8.7	0–1.8	3.5–6.6	n.d.–17.7		1.0% P 23.8g EAA/ 100 g DM	Lamminen et al. (2017), Lamminen, Halmemies-Beauchet-Filleau, Kokkonen, Jaakkola, et al. (2019), Lamminen, Halmemies-Beauchet-Filleau, Kokkonen, Vanhatalo, et al. (2019), Panjaitan, Quigley, Mclennan, Swain, and Poppi (2015), Costa, Quigley, Isherwood, Mclennan, and Poppi (2016), Cao, Zhang, et al. (2018), and Teimouri, Yeganeh, Mianji, Najafi, and Mahjoub (2019)
Aurantiochytrium (Schizochytrium)	*A. limacinum*	Whole	13.5–17.5	58.0–63.2	36.7–62.5		n.d.–2.0				3.8–16.5	DHA 25.7–38.5% TFA	Sinedino et al. (2017), Till et al. (2019), and Sucu, Udum, Gunes, Canbolat, and Filya (2017)
Aurantiochytrium (Schizochytrium)	*Aurantiochytrium* sp.	Whole: ALL-G-RICH, Aquagrow gold™, DHAgold™	8–17.0	40.3–70.2	36.7–64.3		4.16–36.9	0.41–20.8	n.d.–3.9	8 cal/g	2.84–9.0	DHA 17.0–42.3% TFA 44% SFA (% TL) 31% MUFA (% TL) 31% PUFA (% TL)	Moran, Morlacchini, Keegan, Warren, and Fusconi (2019), Marques et al. (2019), Ponnampalam et al. (2016), Moate et al. (2013), Flaga, Korytkowski, Górka, and Kowalski (2019), and Cortegano et al. (2019)
Chlorella	*C. pyrenoidosa*	Whole	57.4–58.0	1.0–13.6			0.4–14.6	0.0–8.8			6.4–13.1		Costa et al. (2016) and Tsiplakou, Abdullah, Alexandros, et al. (2017)

Continued

TABLE 1 Proximate composition of microalgae genera considered in feeds for livestock and aquaculture (% or kJ/g dry weight basis)—cont'd

Algae	Species	Form	Protein	Lipid	∑PUFA (% FA)	NFE	NDF	ADF	Starch	Energy	Ash	Other	References
Chlorella	*C. vulgaris*	Whole	51.5–67.7	1.1–14.0	n.d.–51.0		0.0–12.8	n.d.–4.2	4.3–5.40	20.1–20.2	5.1–15.2		Lamminen et al. (2017), Lamminen, Halmemies-Beauchet-Filleau, Kokkonen, Jaakkola, et al. (2019), Tsiplakou, Abdullah, Alexandros, et al. (2017), Tsiplakou et al. (2018), Kholif et al. (2017), and Yadav, Meena, Sahoo, Das, and Sen (2020)
Cryptecodinum	*C. cohnii*	Whole	7–15	12–15	9–11	50							2%–4% DHA or 24%–32% PUFA	Atalah et al. (2007) and Pleissner and Eriksen (2012)
Diacronema (Pavlova viridis)	*D. viridis*	Whole	39.8	19.6		30.6					23.3	10.0	EPA 25mg/mg DM EPA 10mg/mg DM	Haas et al. (2016)
Desmodesmus	*Desmodesmus sp.*	Defatted biomass (coproduct from biofuel)	27	1	—			—		17	16		Gong, Guterres, Huntley, Sørensen, and Kiron (2018)
Dunaliella	*D. salina*	Whole	7.8–53.8	10.1–18.8			n.d.–0–0	n.d.–0–0				6.51–72.0		Costa et al. (2016) and Senosy, Kassab, and Mohammed (2017)
Haematococcus	*H. pluvialis*	Defatted biomass (coproduct from biofuel)	39–40	1–3.4						4082 cal/g	13		10% crude fiber 0.8% P 0.6 Ca 19% NEAA 20% EAA	Jiang et al. (2019) and Ju, Deng, and Dominy (2012)
Isochrysis		Whole	45	27	32	—				—	10		β-Carotene 762mg/kg	Tibaldi et al. (2015)
Nannochloropsis	*N. gadiatana*	Whole	38.5	19.2	24.4	21.9			2.61		15.8		Total AA 89.0% crude protein	Lamminen, Halmemies-Beauchet-Filleau, Kokkonen, Jaakkola, et al. (2019)
Nannochloropsis	*N. oceanica*	Defatted	43.0	2.5		28.8				19.0	23.5			Sørensen et al. (2017)

	Species	Type								Reference
Nannochloropsis	*N. oceanica*	Whole	34.3	9.8	28.8		15.9	35.0	1.1% Crude fiber 20.1% EAA 12.9 % NEAA	Batista, Pereira, et al. (2020)
Nannochloropsis	*N. salina*	Defatted	44.9	10.3			17.9	26.1		Patterson and Gatlin (2013)
Nannochloropsis	*Nannochloropsis* sp.	Defatted	45.2	7.6	12.3		18.3	23.6		Valente et al. (2019)
Nanofrustulum	*Nanofrustulum* sp.	Defatted	11.9	3.14	23.1			53.1	43% EAA	Kiron, Phromkunthong, Huntley, Archibald, and De Scheemaker (2012)
Scenedesmus	*S. almeriensis*	Whole	43.2	9.6						Vizcaíno et al. (2014)
Tetraselmis	*T. suecica*	Whole	48.7	8.0	54.0	22.4		17.5	1.1 mg P/g 22% EAA; 267 mg/kg β-carotene	Tulli et al. (2012) and Cardinaletti et al. (2018)
Tisochrysis	*T. lutea*	Whole	46.3	26	43.5			11.3	0.8 mg P/g 761.7 mg/kg β-carotene	Cardinaletti et al. (2018)

AA, amino acids; *ADF*, acid detergent fiber; *DHA*, docosahexaenoic acid; *DM*, dry matter; *EPA*, eicosapentaenoic acid; *EAA*, essential amino acids; *FA*, fatty acids; *MUFA*, monounsaturated fatty acids; *n.d.*, not determined; *NDF*, neutral detergent fiber; *NEAA*, nonessential amino acids; *PUFA*, polyunsaturated fatty acids; *SFA*, saturated fatty acids; *TL*, total lipids.

in docosahexaenoic acid (DHA) derived from marine microalgae (*Aurantiochytrium* sp.), being used to a lesser extent, microalgae from the genus *Arthrospira*, *Chlorella*, *Crypthecodinium*, *Isochrysis*, *Dunaliella*, and *Nannochloropsis*. In these studies, microalgae have been used as protein sources to replace more common ones (e.g., rapeseed and soybean meals), and as energy sources either by replacing concentrate feed or to supplement the diet with lipids.

2.1 Feed intake

Inconsistent effects of dietary inclusion of microalgae on feed intake have been reported in cows, steers, sheep, lambs, goats, and kids, with most studies either describing no effect or a reduction in ingestion (Table 2). However, increased feed intake was reported in steers (Costa et al., 2016) and lactating goats (Kholif et al., 2017). This range of effects may be attributed to a variety of factors, including algae, diet, or animal-related, resulting in that no inclusion level can be pointed out for algal species, animal species, or physiological status as "safe" and above which feed intake is reduced.

Indeed, microalgae inclusion had a negative effect on total feed intake at lower inclusion levels (e.g., 595 g *Arthrospira platensis*/cow/day; Lamminen, Halmemies-Beauchet-Filleau, Kokkonen, Vanhatalo, & Jaakkola, 2019, 340 g *Aurantiochytrium* sp./cow/day; Vanbergue, Peyraud, & Hurtaud, 2018), but not at higher inclusion levels (e.g., 1248 g *A. platensis*/cow/day; Lamminen et al., 2017, 375 g *Aurantiochytrium* sp./cow/day; Moate et al., 2013). Some studies reported similar total feed intake while the concentrate intake was reduced (Lamminen et al., 2017; Lamminen, Halmemies-Beauchet-Filleau, Kokkonen, Jaakkola, et al., 2019; Papadopoulos et al., 2002). Despite presenting the diet as a total mixed ration constitutes a common strategy to reduce the sorting behavior (Schingoethe, 2017), microalgae supplementation to total mixed ration feeding did not prevent

a reduction on feed intake of cows (Lamminen, Halmemies-Beauchet-Filleau, Kokkonen, Vanhatalo, et al., 2019; Marques et al., 2019; Vanbergue, Peyraud, et al., 2018). Although less studied, microalgae had no effect milk replacer intake of goat kids and sheep lambs (Morales-De La Nuez et al., 2014) whereas on calves intake was reduced (Flaga et al., 2019).

The most pointed out explanation for the negative impact of microalgae on feed intake is their low palatability by cows (Franklin et al., 1999; Lamminen et al., 2017; Lamminen, Halmemies-Beauchet-Filleau, Kokkonen, Jaakkola, et al., 2019) and sheep (Papadopoulos et al., 2002), probably due to the fishy-like taste and odor, the finely powdered physical structure, or the prompt lipid oxidation in oil-rich algal species. Antioxidants supplementation (Ponnampalam et al., 2016) and pelleting (Lamminen et al., 2017) have been suggested, yet more research is needed on algae processing and feeding strategies.

2.2 Rumen fermentation

Effects of dietary microalgae inclusion on rumen fermentation parameters are inconsistent, with an increase, decrease, and absence of effects on pH, ammonia-N concentration, total volatile fatty acid production, and individual volatile fatty acid proportions (Table 3). However, the evaluation of the effect of microalgae on ruminal fermentation is difficult to distinguish from the effect of the composition of the whole diet provided to the animal.

Effects on ruminal pH can reflect effects on total volatile fatty acid production and ammonia-N concentrations (Lamminen et al., 2017; Marques et al., 2019; Zhu et al., 2016).

A recent in vitro study (Wild, Steingaß, & Rodehutscord, 2019) reported low ruminal fermentation of *Arthrospira*, *Chlorella*, *Nannochloropsis*, and *Phaeodactylum*. However, the studies presented in Table 3 found a decrease, no effect, or even an increase in total volatile fatty acid

TABLE 2 Effect of microalgae on ruminants feed intake.

Algae	Animal model and size	Number of animals	Animal diets	Inclusion level (%)	Trial duration	Intake	References
Arthrospira platensis	Steers	*Bos indicus* steers: 9 months old, 236 ± 2.0 kg BW	42	Ad libitum speargrass hay and 0.3 kg supplement	0, 0.08, 0.16, 0.32, and 0.48 g N/kg BW/d	70 days	Quadratic ↑ total and hay DMI ↑ Speargrass digestibility	Costa et al. (2016)
	Cows	Finnish Ayrshire cows: 112 ± 21.6 DIM, 36.2 ± 3.77 kg milk/d	4	Ad libitum grass silage plus concentrate Algae replaced 100% Soybean meal in concentrate	1238.75 g/head/d	21 days	↔ Silage, DM, OM, CP, NDF intake ↑ F:C ratio ↔ Total tract apparent digestibility (DM, OM, CP, NDF)	Lamminen, Halmemies-Beauchet-Filleau, Kokkonen, Jaakkola, et al. (2019) Spirulina vs. Chlorella
		Finnish Ayrshire cows: 190 ± 22.6 DIM, 35.8 ± 3.08 kg milk/d	8	Ad libitum grass silage plus concentrate Algae replaced 50% and 100% rapeseed protein of concentrate	0, 633.6, and 1248 g/head/d	21 days	↔ Silage, DM, OM, CP, NDF intake Linear ↑ F:C ratio ↔ Total tract apparent digestibility (DM, OM, CP, NDF)	Lamminen et al. (2017)
		Finnish Ayrshire cows: 113 ± 36.3 DIM; 33.9 ± 4.79 kg milk/d	8	TMR based on corn silage and protein feed (rapeseed meal or faba beans) Algae replaced 50% of protein feed	0% and 2.64% of TMR	21 days	↓ DM, OM, NDF and ME intake ↔ CP and EE intake ↔ Total tract apparent digestibility (DM, OM, CP, NDF)	Lamminen, Halmemies-Beauchet-Filleau, Kokkonen, Vanhatalo, et al. (2019)
Aurantiochytrium limacinum	Cows	Holstein-Friesian cows: 77 ± 17.0 DIM, 44 ± 1.9 kg milk/d	24	TMR based on corn silage	0, 50, 100, and 150 g/head/d	28 days	↔ DMI	Till et al. (2019)
Aurantiochytrium sp.	Lambs	Hu lambs: 80 ± 3 days old, 18.35 ± 1.39 kg BW	48	Standard (8.40 MJ/kg) and high (9.70 MJ/kg) energy concentrate	0% and 3% concentrate	60 days	↔ ADFI	Fan et al. (2019)

Continued

TABLE 2 Effect of microalgae on ruminants feed intake—cont'd

Algae	Animal model and size		Number of animals	Animal diets	Inclusion level (%)	Trial duration	Intake	References
Aurantiochytrium limacinum: All-G-Rich®, heterotrophically grown	Cows	Italian Friesian cows: 164 ± 65 DIM, 39 ± 8 kg milk/d	36	TMR based on corn silage and concentrate	0 and 150 g/head/d	84 days	↔ TMR intake	Moran et al. (2019)
Aurantiochytrium sp.: All-G-Rich®	Cows	Holstein cows: 130 ± 15.4 DIM, 30.8 ± 0.54 kg milk/d	24	TMR based on corn silage and concentrate (48:52 F:C) Algae replaced whole raw soybean	0, 8.00, 16.1, and 24.1 g DHA/head/d	21 days	Linear ↓ DM, OM, CP, NDF and EE intake Linear ↑ total-tract apparent digestibility (DM, OM, CP, NDF, and EE)	Marques et al. (2019)
		Holstein cows: primiparous and multiparous, average: 650 kg BW, 45 kg milk/d, 27 ± 5 DIM	739	Ad libitum TMR based on corn silage	0 and 10 g DHA/head/d	120 days	↔ DMI	Sinedino et al. (2017)
Aurantiochytrium sp.: DHA gold™	Lambs	Merino lambs: male, 5 months old, 39.1 ± 1.28 kg BW	40	Wheat grain and sunflower meal plus alfalfa hay	0 and 5 g/head/d	49 days	↔ DM, OM, CP, EE intakes	Sucu et al. (2017)
	Lambs	Canadian Arcott lambs: 22.7 ± 3.90 kg BW	44	Barley and alfalfa hay based diet	0%, 1%, 2%, and 3%	Mid December to April	↔ DMI	Meale, Chaves, He, and Mcallister (2014)
	Lambs	Poll Dorset × Border Leicester × Merino lambs: 3 months old, 34.8 ± 2.5 kg BW, 3.3 fat score	40	Oat grain based diet	0 and 12.9 g/head/d	42 days	↔ ADI	Clayton et al. (2014)
	Lambs	Manchego lambs: male, 7–8 weeks old, 15.3 ± 0.38 kg BW	24	Ad libitum grain based diet	0 and 19.3 g/head/d	Up to 26 kg BW	↓ Feed intake	Díaz et al. (2017)

Animal	n	Basal diet	Treatment	Duration	Effect on intake	Reference
Sheep	32	Alfalfa hay and concentrate	0, 23.5, 47, and 94 g/head/d (effective intake: 0, 16.9, 27.7, 51.7 g/head/d)	42 days	↔ Hay intake ↓ Concentrate intake (and consequently algae intake)	Papadopoulos, Goulas, Apostolaki, and Abril (2002)
Goats	24	Alfalfa hay and concentrate	0, 20, 40, and 60 g/head/d	60 days	↓ DMI at highest inclusion level (effective algae intake of 40 g)	Mavrommatis et al. (2018)
Cows and goats	12 and 12	Ad libitum grass hay and concentrate (45:55 F:C)	0 and 310 g/head/d (cows) or 40 g/head/d (goats)	28 days	↔ Grassland and concentrate intake ↔ DM, OM, CP, NDF intake Starch intake ↓ in cows and ↔ in goats	Fougère, Delavaud, and Bernard (2018) Algae vs. control
Calves	40	Milk replacer for 49 days and ad libitum starter mixture from day 15 onwards	0, 9, 18, and 27 g/head/d (0, 1.13, 2.26, and 3.38 g DHA/head/d)	49 days	Linear ↓ milk replacer intake, starter mixture intake	Flaga et al. (2019)
Calves		Holstein calves: females, 8.6 ± 0.8 days old, 41.1 ± 4.3 kg BW				
Cows	32	Ad libitum alfalfa hay and concentrate	0, 125, 250, and 375 g/head/d (0, 25, 50, and 75 g DHA/head/d)	30 days	Linear ↓ Total DM intake ↔ Total DM intake	Moate et al. (2013) DHA effect
Cows	32 (16 +16)	Ad libitum TMR corn silage-based	0 and 340 g/head/d	42 days	↓ Total DM, forage and concentrate intake ↓ CP intake ↔ Fat intake ↑ EPA, DHA intake	Vanbergue, Peyraud, et al. (2018) Alga vs. control only
Cows	30 (9 +21)	Ad libitum alfalfa hay, corn grain and corn silage	0, 910 g (unprotected) and 910 g xylose coated (protected)	49 days	↓ DMI vs. control	Franklin, Martin, Baer, Schingoethe, and Hippen (1999)

Continued

TABLE 2 Effect of microalgae on ruminants feed intake—cont'd

Algae	Animal model and size	Number of animals	Animal diets	Inclusion level (%)	Trial duration	Intake	References
	Holstein cows: primiparous, 46±17 DIM, rumen fistulated	4	Grass silage, corn silage, and concentrate Algae partially replaced the concentrate	0% and 2% (via rumen cannula)	21 days	↓ DMI	Boeckaert et al. (2008) Alga vs. control
	Holstein cows: 172±45 DIM, 612±32 kg of BW rumen fistulated	3	Grass silage, corn silage and concentrate	0% and 2% (via rumen cannula)	20 days	↓ DMI	Boeckaert et al. (2008) Alga vs. control
Chlorella vulgaris Goats	Damascus goats: multiparous, 1 week lactation, 44±0.8 kg BW	15	Egyptian berseem clover plus concentrate (50:50 F:C)	0, 5, and 10 g/head/d	84 days	Quadratic effect on DM, OM, CP, EE, NDF, ADF intake (highest at 10g) Quadratic effect on apparent digestibility of DM, OM, CP, EE, NDF, ADF (highest at 10g)	Kholif et al. (2017)
Chlorella vulgaris + *Arthrospira platensis* Cows	Finnish Ayrshire cows: multiparous; 212±30.7 DIM, 24.8±2.56 kg milk/d	6	Ad libitum grass silage plus concentrate Algae (1:1) replaced 50% and 100% rapeseed in concentrate	0, 511.5, and 1023 g/head/d	21 days	↔ Silage, DM, CP, NDF intake Linear ↑ CP intake ↓ F:C ↔ Total tract apparent digestibility (DM, OM, NDF, CP)	Lamminen et al. (2017)
Chlorella vulgaris and *Chlorella vulgaris* + *Nannochloropsis gaditana* Cows	Finnish Ayrshire cows: Multiparous; 112±21.6 DIM, 36.2±3.77 kg milk/d	4	Ad libitum grass silage plus concentrate Algae replaced 100% soybean meal in concentrate	1493.75 g *C. vulgaris* and 887.5 g *C. vulgaris* +877.5 g *N. gaditana*	21 days	↑ Silage intake with Chl-Nan ↔ DM, OM, CP, NDF intake, F:C ratio ↔ Total tract apparent digestibility (DM, OM, CP, NDF)	Lamminen, Halmemies-Beauchet-Filleau, Kokkonen, Jaakkola, et al. (2019)

Microalgae	Animal	Breed/detail	n	Diet	Dose	Duration	Results	Reference
Chlorella spp.		Majorera goat kids and Canarian sheep lambs	80 and 80	Ad libitum commercial milk replacer, twice a day	0% and 5.625%	60 days	↔ Milk replacer intake, FER	Morales-De La Nuez et al. (2014)
Crypthecodinium cohnii	Kids and lambs	Majorera goat kids and Canarian sheep lambs	80 and 80	Ad libitum commercial milk replacer	0% and 5.625%	60 days	↔ Milk replacer intake, FER	Morales-De La Nuez et al. (2014)
	Ewes	English mule ewes: last 9 weeks of gestation	48	Silage and commercial concentrate	0 and 12 g/d during the first 3 weeks or 9 weeks	63 days	↔ Silage intake	Pickard, Beard, Seal, and Edwards (2008)
Isochrysis galbana	Kids and lambs	Majorera goat kids and Canarian sheep lambs	80 and 80	Ad libitum commercial milk replacer	0% and 5.625%	60 days	↔ Milk replacer intake, FER	Morales-De La Nuez et al. (2014)
Chlorella pyrenoidosa, Dunalliella salina, Arthrospira platensis	Steers	Bos indicus steers: 187 ± 4.0 BW	5	Ad libitum speargrass hay plus protein supplement (9.1 N:S)	0.4 g/kg BW/d _A. platensis_, 4.7 g/kg BW/d _C. pyrenoidosa_, 4 g/kg BW/d _D. salina_	12 days	Hay intake and total DMI highest with Art, intermediate with Chl and lowest with Dun ↑ OM digestibility with Art and ↓ with Dun	Costa et al. (2016)

ADF, acid detergent fiber; _ADFI_, adjusted daily feed intake; _ADI_, average daily intake; _BW_, body weight; _CP_, crude protein; _DHA_, docosahexaenoic acid; _DIM_, days in milk; _DM_, dry matter; _DMI_, dry matter intake; _EE_, ether extract; _EPA_, eicosapentaenoic acid; _F:C_, forage-to-concentrate ratio; _FER_, feed efficiency ratio; _ME_, metabolizable energy; _NDF_, neutral detergent fiber; _OM_, organic matter; _TMR_, total mixed ration.

TABLE 3 Effect of microalgae on rumen fermentation parameters.

Algae	Animal model and size	Number of animals	Animal diets	Inclusion level (%)	Trial duration	Rumen fermentation	References
Arthrospira platensis	Steers	*Bos indicus* steers: 9 months old, 236 ± 2.0 kg BW	42	Ad libitum speargrass hay and 0.3 kg supplement	0, 0.08, 0.16, 0.32, and 0.48 g N/kg BW/d	70 days	Quadratic ↑ NH_3-N Quadratic ↑ rumen BCVFA concentration and molar proportions	Costa et al. (2016)
	Cows	Finnish Ayrshire cows: 190 ± 22.6 DIM, 35.8 ± 3.08 kg milk/d	8	Ad libitum grass silage plus concentrate Algae replaced 50% and 100% rapeseed in concentrate	0, 633.6, and 1248 g/head/d	21 days	↔ pH, NH_3-N, total VFA ↔ Acetate, propionate, butyrate, valerate, caproate Linear ↑ isobutyrate Quadratic effect on isovalerate (highest at higher level)	Lamminen et al. (2017)
		Finnish Ayrshire cows: 113 ± 36.3 DIM; 33.9 ± 4.79 kg milk/d	8	TMR based on corn silage and protein feed (rapeseed meal, RSM, or faba beans, FB) Algae replaced 50% of protein feed	0% and 2.64% of TMR	21 days	↓ pH ↔ Total VFA ↓ Acetate, acetate: propionate ratio, ↑ Propionate ↔ Butyrate ↑ Isobutyrate on RSM-diet and ↓ on FB-diet ↑ Isovalerate on RSM-diet and ↔ on FB-diet ↑ NH_4-N concentration on RSM-diet and ↓ on FB-diet	Lamminen, Halmemies-Beauchet-Filleau, Kokkonen, Vanhatalo, et al. (2019)
Aurantiochytrium sp.	Goats	Boer crossbred wether goats: 18.40 ± 0.95 kg BW	6	*Litmus chinensis* hay and concentrate	0, 6.1, and 18.3 g/head/d	21 days	↑ pH, propionate and butyrate and ↓ total VFA and acetate at highest level compared to control and lower level	Zhu, Fievez, Mao, He, and Zhu (2016)
Aurantiochytrium limacinum: All-G-Rich®, heterotrophically grown	Cows	Italian Friesian cows: 164 ± 65 DIM, 39 ± 8 kg milk yield	36	TMR based on corn silage and concentrate	0 and 150 g/head/d	84 days	↔ Rumination activity	Moran et al. (2019)

Aurantiochytrium sp.: All-G-Rich®	Cows	24	Holstein cows: 130±15.4 DIM, 30.8±0.54kg milk/d	TMR based on corn silage and concentrate (48:52 F:C) Algae replaced whole raw soybean	0, 8.00, 16.1, and 24.1g DHA/head/d	21 days	Linear ↑ pH ↓Total VFA and acetate concentration ↔ Propionate, butyrate and BCVFA concentration	Marques et al. (2019)
Aurantiochytrium sp.: DHA gold™	Lambs	40	Merino lambs: male, 5 months old, 39.1±1.28kg BW	Wheat grain and sunflower meal plus alfalfa Hay	0 and 5g/head/d	49 days	↑ total VFA concentration and propionate and valerate proportions ↓pH ↔ NH$_3$-N concentration, butyrate, isobutyrate, and isovalerate proportions, and acetate:propionate ratio	Sucu et al. (2017)
		44	Canadian Arcott lambs: 22.7±3.90kg BW	Barley and alfalfa hay based diet	0%, 1%, 2%, and 3%	Mid December to April	↔ pH, rumen weight	Meale et al. (2014)
	Cows	32	Holstein cows: 163±9.2 DIM, 20.0±3.11kg milk/d	Ad libitum alfalfa hay and concentrate	0, 125, 250, and 375g/head/d (0, 25, 50, and 75g DHA/head/d)	30 days	↓ pH ↔ Total VFA, acetate, propionate, butyrate, isobutyrate, valerate, isovalerate, caproate	Moate et al. (2013)
		4	Holstein cows: primiparous, 46±17 DIM, rumen fistulated	Grass silage, corn silage, and concentrate Algae partially replaced the concentrate	0% and 2% (via rumen cannula)	21 days	↔ pH ↔ Total VFA, acetate, propionate, isobutyrate, valerate ↓ Butyrate ↑ Isovalerate	Boeckaert et al. (2008)
		3	Holstein cows: 172±45 DIM, 612±32kg of BW, rumen fistulated	Grass silage, corn silage and concentrate	0% and 2% (via rumen cannula)	20 days	↑ pH ↓Total VFA, acetate and valerate ↔ Propionate ↑ Butyrate, isobutyrate, isovalerate	Boeckaert et al. (2008)

Continued

TABLE 3 Effect of microalgae on rumen fermentation parameters—cont'd

Algae	Animal model	Animal model and size	Number of animals	Animal diets	Inclusion level (%)	Trial duration	Rumen fermentation	References
Chlorella vulgaris	Goats	Damascus goats: multiparous, 1 week lactation, 44±0.8 kg BW	15	Egyptian berseem clover plus concentrate (50:50 F:C)	0, 5, and 10 g/head/d	84 days	↔ pH, NH_3-N, butyrate Quadratic effect on total VFA and propionate (highest at 10 g) and acetate (lowest at 10 g)	Kholif et al. (2017)
Chlorella pyrenoidosa, Dunaliella salina, Arthrospira platensis	Steers	Bos indicus steers: 187±4.0 BW	5	Ad libitum speargrass hay plus protein supplement (9.1:1 N:S)	0, 4 g/kg BW/d A. platensis, 4.7 g/kg BW/d C. pyrenoidosa, 4 g/kg BW/d D. salina	12 days	↑ NH_3-N with Art and with Dun ↓ with Dun ↔ Total VFA ↓ Acetate with Art than Chl and Dun ↑ Valerate, isovalerate, and isobutyrate with art	Costa et al. (2016)

BCVFA, branched-chain volatile fatty acids; BW, body weight; DIM, days in milk; DHA, docosahexaenoic acid; F:C, forage-to-concentrate ration; NH_3-N, ammonia-N; NH_4-N, ammonium-N; TMR, total mixed ration; VFA, volatile fatty acids.

production from microalgae supplementation. The increased digestibility in goats supplemented with *Chlorella vulgaris* (Kholif et al., 2017) might reflect an improvement in rumen fermentation kinetics, due to the growth-promoting substances present in these algae (Han, Kang, Kim, & Kim, 2002), its positive effect on some rumen bacterial species (*Butyrivibrio fibrisolvens*, *Ruminococcus albus*, and *Clostridium sticklandii*; Tsiplakou, Abdullah, Skliros, et al., 2017), and its content on β-glucan (Iwamoto, 2004), polyunsaturated fatty acids (PUFA), carotenoids, phycobiliproteins, polysaccharides, and phycotoxins (Kotrbáček, Doubek, & Doucha, 2015), all contributing to improved fermentation. Conversely, the high mineral concentration found in some of the algae species, especially Na and K, could interfere with rumen fermentation through osmolality effects (Allen, 2000).

Dietary supplementation with marine oils has been suggested to increase ruminal ammonia-N concentration (Toral et al., 2016). However, several studies reported no effect in rumen ammonia-N concentration with microalgae (Table 3). Kholif et al. (2017) observed an increased protein intake and digestibility of algae supplemented diet and no effect on rumen ammonia concentration, suggesting that the availability of peptides and amino acids along with the presence of rapidly fermentable sugars of *C. vulgaris* would have promoted a higher efficiency of N utilization by rumen microorganisms.

Studies suggest that microalgae rich in DHA affect rumen population (Boeckaert, Fievez, Van Hecke, Verstraete, & Boon, 2007; Boeckaert, Vlaeminck, Mestdagh, & Fievez, 2007), thus promoting a shift in volatile fatty acid proportions, especially acetate and propionate. Long-chain n-3 PUFA (n-3 LC-PUFA) supplementation has a toxic effect on rumen bacteria, although effects on rumen fiber digestibility are controversial (Shingfield et al., 2012; Weld & Armentano, 2017). At low levels of PUFA inclusion, rumen microbiota was suggested to have the ability to adapt to environmental changes, the inhibited fibrolytic microorganisms being replaced by less sensitive ones, with the consequent absence of effects on fiber digestibility (Toral et al., 2017). In vivo studies show inconsistent effects of DHA-rich microalgae on acetate and propionate proportions (Table 3), which can be partly explained by the volume of algae supplied. Indeed, in the study of Zhu et al. (2016), a low level of algae supplementation did not affect rumen fermentation, whereas a high level promoted a shift toward propionate at the expense of acetate, agreeing with earlier results with ewes fed diets with similar amounts of eicosapentaenoic acid (EPA) and DHA (Toral, Belenguer, Frutos, & Hervás, 2009; Toral, Shingfield, Hervás, Toivonen, & Frutos, 2010). The same microalgae species supplemented at a similar level to that having no effect on the study of Zhu et al. (2016) significantly affected rumen fermentation pattern in dairy cows (Boeckaert et al., 2008), suggesting a higher tolerance of small ruminants for dietary PUFA (Zhu et al., 2016).

Microalgae supplementation increased branched-chain volatile fatty acids in some studies, with no effects on others (Table 3). As branched-chain amino acids constitute the substrates for the rumen production of branched-chain volatile fatty acids (El-Shazly, 1952), positive effects on the later can be attributable to a higher intake of branched amino acids with microalgae supplemented diets and a higher rumen degradability of algae protein. However, it is difficult to isolate the effects of algae from the effects of the composition of amino acids on the whole diet. Indeed, in the study of Lamminen, Halmemies-Beauchet-Filleau, Kokkonen, Vanhatalo, et al. (2019), the effects of *Arthrospira platensis* inclusion on the proportions of isobutyrate and isovalerate differed with the protein source in the base diet. When *A. platensis* supplemented the rapeseed meal diet, the isobutyrate and isovalerate proportions increased in line with the increased intake of the amino acids valine and leucine, whereas when *A. platensis* was supplemented to faba beans diet, intake of leucine and valine increased, but the proportions of isobutyrate and isovalerate decreased or kept unchanged. The authors suggested that this later result might be due to the lower rumen degradable protein of *A. platensis* than that of faba beans.

2.3 Milk production and composition

The majority of the studies with cows, ewes, and goats report an absence of effect of dietary inclusion of microalgae on milk production (Table 4), even when intake was depressed (Franklin et al., 1999; Marques et al., 2019; Mavrommatis et al., 2018; Papadopoulos et al., 2002; Vanbergue, Hurtaud, et al., 2018), the latter presumably due to an increased feed efficiency (Franklin et al., 1999; Papadopoulos et al., 2002). However, some studies reported a decrease (Lamminen, Halmemies-Beauchet-Filleau, Kokkonen, Vanhatalo, et al., 2019; Reynolds et al., 2006) or an increase in milk production with the dietary inclusion of microalgae.

The comparison of results among studies is complicated given the different amounts of supplemented microalgae and the composition of the diet. Indeed, Lamminen, Halmemies-Beauchet-Filleau, Kokkonen, Vanhatalo, et al. (2019) found different results from the supplementation of *A. platensis* to diets with rapeseed meal or faba beans. When *A. platensis* was included in the rapeseed meal diet, a decrease in intake, presumably driven by palatability issues, was reflected in a decrease in milk production, whereas when *A. platensis* was included in the faba beans diet, despite the reduced intake, milk production was increased. This is most probably due to an increased feed efficiency through an improved amino acid balance or lower ruminal protein degradability or to the beneficial effects of *A. platensis* in rumen microbial population Kulpys et al. (2009).

Dietary inclusion of microalgae has differently influenced milk protein synthesis of cows, goats, and sheep (Table 4). Algae inclusion did

not affect milk protein yield and content of cows (Franklin et al., 1999; Lamminen et al., 2017; Lamminen, Halmemies-Beauchet-Filleau, Kokkonen, Jaakkola, et al., 2019; Moate et al., 2013; Moran et al., 2019; Till et al., 2019) and goats (Fougère et al., 2018; Mavrommatis et al., 2018; Póti et al., 2015; Tsiplakou, Abdullah, Alexandros, et al., 2017; Tsiplakou et al., 2018). In cows, reduced milk protein yield was observed when milk production was negatively affected (Boeckaert et al., 2008; Vanbergue, Peyraud, et al., 2018), whereas protein content was reported to reduce or to increase independently of milk production. In goats, an increase in the milk protein yield and milk production and a decrease of protein content in grazing goats compared to confined goats (Pajor et al., 2019) were reported. In sheep, the algal supplementation led to an increase in the protein content with no changes in milk yield (Papadopoulos et al., 2002). Contrasting effects of microalgae supplementation on milk protein may be related to differences in intake and/or quality of dietary protein and amino acid balance.

Contrasting results on the effects of algal supplementation on lactose have also been reported. Dietary inclusion of algae reduced milk lactose yield (Boeckaert et al., 2008) or affected differentially depending on the basal diet (Lamminen et al., 2017) and increased it in goats, being unaffected in most studies (Table 4). A decrease in lactose content was reported in cows and goats (Fougère et al., 2018; Mavrommatis et al., 2018; Pajor et al., 2019), whereas an increase was reported by Kholif et al. (2017) in goats and by Papadopoulos et al. (2002) in sheep.

Several studies also reported a decrease in milk fat secretion in cows. Milk fat reduction could reflect a higher fat content of diets with microalgae, a negative energy balance due to decreased feed intake and the accumulation of *trans* fatty acids intermediates formed during dietary PUFA rumen biohydrogenation

TABLE 4 Effect of microalgae on ruminants milk production and composition.

Algae		Animal model and size	Number of animals	Animal diets	Inclusion level (%)	Trial duration	Milk composition	Milk fatty acids	References
Arthrospira platensis	Cows	Finnish Ayrshire cows: 112±21.6 DIM, 36.2±3.77 kg milk/d	4	Ad libitum grass silage plus concentrate Algae replaced 100% soybean meal in concentrate	1238.75 g/head/d	21 days	→ Milk, ECM, fat, protein, and lactose yields ↑ Milk fat content → Milk protein, lactose, and MUN contents → ECM:DMI	↑ 6:0, 14:0, 18:3n-6 ↓ 18:2n-6, 20:0, EPA, PUFA, PUFA n-3, PUFA n-6, n-6/n-3 ratio → SFA, MUFA	Lamminen, Halmemies-Beauchet-Filleau, Kokkonen, Jaakkola, et al. (2019))
		Finnish Ayrshire cows: 190±22.6 DIM; 35.8±3.08 kg milk/d	8	Ad libitum grass silage plus concentrate Algae replaced 50% and 100% rapeseed in concentrate	0, 633.6, and 1248 g/head/d	21 days	→ Milk, ECM, fat, protein, and lactose yields → Milk fat, protein, and lactose contents Quadratic effect on MUN and ECM: DMI (highest and lowest, respectively, at intermediate level)		Lamminen et al. (2017)
		Finnish Ayrshire cows: 113±36.3 DIM; 33.9±4.79 kg milk/d	8	TMR based on corn silage and protein feed (rapeseed meal, RSM, or faba beans, FB) Algae replaced 50% CP of protein feed	0% and 2.64% of TMR	21 days	→ Milk yield and composition ↓ Milk and lactose yields ↓ on RSM-diet and ↑ on FB-diet Milk fat, protein, and MUN concentrations ↑ on RSM-diet and ↓ on FB-diet ECM:DMI ratio ↑ on FB-diet and had little effect on RSM-diet		Lamminen, Halmemies-Beauchet-Filleau, Kokkonen, Vanhatalo, et al. (2019)
		Holstein-Friesian cows: 174 DIM, 20.0 kg milk/d	16	Corn silage, alfalfa hay and grain mix	0% and 0.74% of intake	17 days	↓ Milk fat content → Milk protein, lactose, and solids without fat content	↓ 4:0, 6:0, 8:0, n-6/n-3 ratio, atherogenic index ↑ 18:0, 18t11, CLAc9t11, 18:3n-6, 20:3n-3, DHA, MUFA, PUFA n-3	Póti, Pajor, Bodnár, Penksza, and Köles (2015)

Continued

TABLE 4 Effect of microalgae on ruminants milk production and composition—cont'd

Algae	Animal model and size	Number of animals	Animal diets	Inclusion level (%)	Trial duration	Milk composition	Milk fatty acids	References
	Lithuanian black and white		Stabulated: Silage, haylage, and combined fodder Grazing: Grass, vitamin-mineral supplement and combined fodder	0 and 200 g/head/d	90 days	↑ Milk yield ↔ Milk fat, protein, and lactose contents		Kulpys, Paulauskas, Pilipavičius, and Stankevičius (2009)
Aurantiochytrium limacinum	Holstein-Friesian cows: 77±17.0 DIM, 44±1.9 kg milk/d	24	TMR based on corn silage	0, 50, 100, and 150 g/head/d	28 days	↔ Milk, protein, and lactose yields and protein content Linear ↓ ECM and fat yields, and fat and lactose contents	↔ 4:0, 14:0, 16:0, 16:1c9, 20:0, EPA Linear ↓ SFA, 6:0, 8:0, 10:0, 12:0, 18:0, 18:1c9, 22:0, n-6/n-3 ratio Linear ↑ 18:1t10, 18:1t11, 18:2n-6, 18:3n-3, CLAc9t11, CLAt10c12, 20:3n-6, 20:3n-3, DHA, MUFA, PUFA, PUFA n-3, PUFA n-6	Till et al. (2019)
Aurantiochytrium limacinum: All-G-Rich®, heterotrophically grown	Cows	36	TMR based on corn silage and concentrate	0 and 150 g/head/d	84 days	↔ Milk, fat, FCM, protein, and lactose yields ↔ Fat, protein, lactose, and MUN content	↑ EPA (at d84), DHA (>with meal than pelleted form) ↓ n-6/n-3 ratio (<with meal)	Moran et al. (2019)
Aurantiochytrium sp.: All-G-Rich®	Hungarian native goats: multiparous, 71 DIM	40	Alfalfa hay and concentrate or ad libitum pasture and concentrate	0 and 15 g/head/d	31 days	↔ Milk yield ↑ Fat content ↓ Lactose content ↓ Protein and solid nonfat content in grazing groups	↑ 10:0, 12:0, 14:0, 16:0, 18:1t11, CLAc9t11, DHA, SFA, PUFA n-3 ↓ 18:0, 18:1c9, 18:2n-6 (only in indoor groups), MUFA, n-6/n-3 ratio	Pajor, Egerszegi, Steiber, Bodnár, and Póti (2019)
	Holstein cows: 130±15.4 DIM, 30.8±0.54 kg milk/d	24	TMR based on corn silage and concentrate (48:52 F:C) Algae replaced whole raw soybean	0, 8.00, 16.1, and 24.1 g DHA/head/d	21 days	↔ Milk, lactose, and protein yields Quadratic effect on FCM, ECM and fat yields (highest at lower level) Linear ↓ Lactose, fat, protein and MUN contents Linear ↑ milk/DMI Linear ↓ FCM/	↔ MUFA, PUFA n-3, SFA/UFA, n-6/n-3 ratio Linear ↑ PUFA, CLAc9t11 Linear ↓ SFA, 18:1c9, 18:2n-6, 18:3n-3, PUFA n-6 Quadratic effect on 18:0 and 18:1t11 (lowest and highest, respectively, at higher level)	Marques et al. (2019)

						Milk	Fatty acids	
Aurantiochytrium sp.: DHA gold™	Goats	24	Goats (Alpine × local Greek): 3–4 years; 150±10 DIM; 47.6±5.9 kg BW	Alfalfa hay and concentrate	0, 20, 40, and 60 g/head/d	60 days	↔ Milk yield, FCM, and protein content ↓ Milk fat content at 40 and 60 g ↓ Milk lactose at 40 g (vs. 0 and 20 g)		Mavrommatis et al. (2018)
	Goats	12+12	Holstein cows and Alpine goats: multiparous, nonpregnant, 86±24.9 and 61±1.8 DIM, respectively	Ad libitum grass hay and concentrate (45:55 F:C)	0 and 310 g/head/d (cows) or 40 g/head/d (goats)	28 days	↔ Milk, protein, and lactose yields ↓ Fat yield and content ↓ Fat and lactose content ↔ Protein content Lactose content ↓ in cows and ↔ in goats	↑18:1t11, CLAc9t11, 20:4n-6, EPA DPA ↑ 6:0, 8:0, 14:0, 18:1t10, 20:3n-3, 22:4n-6, DHA in cows ↓ 18:0, 18:1c9, 18:3n-3 ↔ 16:0, 18:1c9, 18:3n-6, CLAt10c12, t10/t11 ratio, CLAc9t11/18:1t11 ratio	Fougère et al. (2018)
	Sheep	32	Karagouniko ewes	Alfalfa hay and concentrate	0, 23.5, 47, and 94 g/head/d (effective intake: 0, 16.9, 27.7, 51.7 g/head/d)	42 days	↔ Milk yield ↑ Protein content ↑ Fat and lactose content (at highest dose)	↑ 14:0, 16:0, EPA, 20:4n-6, 22:5n-6, DPA, DHA ↓ 18:0, 18:1c9, 18:2n-6, 18:3n-3, 20:3n-3, 22:0, n-6/n-3 ratio	Papadopoulos et al. (2002)
		48	Hampshire × Dorset ewes	Ad libitum corn silage or alfalfa pellets plus concentrate (60:30 F:C)	0 and 2.5% +2% soybean oil	42 days	↔ Milk, fat, protein, and lactose yields ↔ Fat, protein, lactose content	↔ SFA ↑ 18:1t10, 18:1t11, CLAc9t11, EPA, DHA, MUFA, PUFA ↓ 18:0 18:1c9 ↓ with corn silage but not with alfalfa 12:0, 14:0, 16:0 concentrations ↓ with alfalfa and ↑ with corn silage	Reynolds, Cannon, and Loerch (2006)
	Cows	32	Holstein cows: 163±9.2 DIM, 20.0±3.11 kg milk/d	Ad libitum alfalfa hay and concentrate	0, 125, 250, 375 g/head/d (0, 25, 50, and 75 g DHA/head/d)	30 days	↔ Milk, protein, and lactose yields, and lactose content Linear ↓ ECM, fat yield and content	↔ MUFA, 16:0, CLAc10t12, DPA ↑ 4:0, 14:0, 18:2n-6, PUFA Linear ↑ CLAc9t11, 20:4n-6, EPA, DHA, PUFA n-3, PUFA n-6 ↓ 11:0, 13:0, 15:0, 17:0, 20:0, n-6/n-3 ratio Linear ↓ 18:0, SFA	Moate et al. (2013)

Continued

TABLE 4 Effect of microalgae on ruminants milk production and composition—cont'd

Algae	Animal model and size	Number of animals	Animal diets	Inclusion level (%)	Trial duration	Milk composition	Milk fatty acids	References
	Holstein cows: primiparous and multiparous, 100±17.5 DIM	32 (16 +16)	Ad libitum TMR corn silage-based	0 and 340 g/head/d	42 days	↓ Milk, fat and protein yields ↓ Fat and protein content ↔ Lactose content	↓ SFA, 18:0, 18:1c9, 18:1c9:16:0 ratio, n-6/n-3 ratio ↑ 18:1t10, 18:1t11, CLAc9t11, 18:3n-3, DHA, BCFA, MUFA, PUFA, 14:1c9/14:0 ratio	Vanbergue, Peyraud, et al. (2018)
	Brown Swiss and Holsteins cows: primiparous and multiparous, average 145.4 DIM	30 (9 +21)	Ad libitum alfalfa hay, corn grain and corn silage	0, 910 g (unprotected) and 910 g xylose coated (protected)	49 days	↔ Milk, ECM, protein, solids yields ↓ Fat content and yield and solids content	↓ SFA, 15:0, 17:0, 18:0, 18:1c9, 18:2n-6 (when protected), 18:3n-3 (when unprotected) ↑ 14:0, 16:0, PUFA, CLA, DHA (+ when protected), 18:1t9, 18:1c11, 18:1t11	Franklin et al. (1999)
	Holstein cows: primiparous, 46±17 DIM, rumen fistulated	4	Grass silage, corn silage, and concentrate Algae partially replaced the concentrate	0% and 2% (via rumen cannula)	21 days	↓ Milk, fat, protein, and lactose yields ↔ Fat, protein, and lactose content	↓ SFA, 18:2n-6, 18:0, CLAt10c12 ↑ PUFA, 18:2t11c15, CLAc9t11, CLAc9c11, DHA, 18:1t10, 18:1t11 ↔ MUFA, OBCFA ↑ i13:0, ai13:0, ai17:0 ↓ i16:0	Boeckaert et al. (2008)
	Holstein cows: 172±45 DIM, 612±32 kg of BW	3	Grass silage, corn silage and concentrate	0% and 2% (via rumen cannula)	20 days	↓ Milk, fat, protein, and lactose yields ↓ Fat content ↔ Protein and lactose content	↓ SFA, 6:0, 8:0, 10:0, 12:0, 18:0, 18:1c9, 15:0, 17:0, i15:0 ↑ MUFA, PUFA, 18:1t10, 18:1t11, 18:2t11c15, CLAc9t11, CLAtt9c11, DHA, OBCFA, i17:0 ↔ CLAt10c12	Boeckaert et al. (2008)
Chlorella kessleri	Goats	Hungarian native goats: multiparous, 62 DIM	20	Alfalfa hay plus grain mixture	0 and 10 g/head/d	17 days	↑ Fat content ↔ Protein, lactose, and total solids without fat	↑ 4:0, 18:1t11, 18:2n-6, CLAc9t11, 18:3n-3, EPA, DHA, MUFA, PUFA, PUFA n-3 ↓ 20:4n-6, SFA, n-6/n-3 ratio ↓ Atherogenic index	Póti et al. (2015)

Algae	Animal	Animal details	n	Basal diet	Dose	Duration	Milk production	Milk FA effects	Reference
Chlorella vulgaris	Goats	Cross-bred dairy goats: 90–98 DIM, 3–4 years old, 43.0±2.3 kg BW	16	Alfalfa hay plus concentrate (53:47 F:C)	0 and 12 g/head/d	30 days	↔ Milk yield and fat, protein, and lactose content	↓ 18:1t6-9, 18:1t11, CLAc9t11 ↑ 16:0 ↔ SFA, MUFA, PUFA, SCFA, MCFA, LCFA ↓ 16:1/16:0 ratio	Tsiplakou et al. (2018)
		Damascus goats: multiparous, 1 week lactation, 44±0·8 kg BW	15	Egyptian berseem clover plus concentrate (50:50 F:C)	0, 5, and 10 g/head/d	84 days	Quadratic effect on milk, ECM, total solids, solids nonfat, fat, protein, lactose yields (highest at 10 g) Quadratic effect on solids nonfat and lactose (highest at 10 g), and fat (lowest at 10 g) contents ↑ Milk production, ECM, total solids, solids nonfat, and lactose ↔ milk/DMI, ECM/DMI	Linear ↓ 18:0 Quadratic effect on SFA, atherogenicity index (lowest at 10 g) Quadratic effect on MUFA, PUFA, CLAc9t11, 18:3n-6, 18:1c9 (highest at 10 g) ↔ n-6/n-3 ratio	Kholif et al. (2017)
Chlorella vulgaris and *Chlorella vulgaris + Nannochloropsis gaditana*	Cows	Finnish Ayrshire cows: Multiparous; 112±21.6 DIM, 36.2±3.77 kg milk/d	4	Ad libitum grass silage plus concentrate. Algae replaced 100% soybean meal in concentrate	1493.75 g C. *vulgaris* and 887.5 g C. *vulgaris* +877.5 g N. *gaditana*	21 days	↔ Milk, ECM, fat, protein, lactose yields ↔ Milk composition (fat, protein, lactose, MUN) ↔ ECM/DMI	↑ 16:0, 16:1c9, 18:2t11c15, 20:0, EPA with Chl-Nan ↓ 18:2n-6, PUFA n-6, n-6/n-3 ratio with Chl-Nan	Lamminen, Halmemies-Beauchet-Filleau, Kokkonen, Jaakkola, et al. (2019)
Chlorella pyrenoidosa	Goats	Crossbred dairy goats: 3–4 years old, 90–98 DIM, 47.4±2.3 kg BW	16	Alfalfa hay, wheat straw and concentrate	0 and 11 g/head/d	30 days	↔ Milk yield, fat, protein, and lactose content	↔ Milk FA profile, except for 16:1 ↑	Tsiplakou, Abdullah, Alexandros, et al. (2017)
Crypthecodinium cohnii	Sheep	English mule ewes: last 9 weeks of gestation	48	Silage and commercial concentrate	0 and 12 g/d during the first 3,6 or 9 weeks	63 days		↑ 14:0, 16:0, 18:0, 18:1c9 in colostrum with 9 weeks' supplementation ↑ EPA and DHA in colostrum with time	Pickard et al. (2008)

BCFA, branched-chain fatty acids; *BW*, body weight; *CLA*, conjugated linoleic acids; *DHA*, docosahexaenoic acid; *DIM*, days in milk; *DMI*, dry matter intake; *DPA*, docosapentae-noic acid; *ECM*, energy corrected milk; *EPA*, eicosapentaenoic acid; *F:C*, forage-to-concentrate ratio; *FCM*, fat-corrected milk; *LCFA*, long-chain fatty acids; *MCFA*, medium-chain fatty acids; *MUFA*, monounsaturated fatty acids; *MUN*, milk urea nitrogen; *OBCFA*, odd- and branched-chain fatty acids; *PUFA*, polyunsaturated fatty acids; *SCFA*, short-chain fatty acids; *SFA*, saturated fatty acids; *TMR*, total mixed ration; *UFA*, unsaturated fatty acids.

(Bauman & Griinari, 2003). The biohydrogenation intermediate isomers conjugated linoleic acids (CLA) trans-10,cis-12 and CLA trans-9,cis-11 were identified as potent inhibitors of milk fat synthesis (Shingfield & Griinari, 2007; Sinclair, Lock, Early, & Bauman, 2007), although other intermediates might be involved (Chilliard, Ferlay, & Doreau, 2001). Therefore dietary supplementation with microalgae rich in PUFA can inhibit the de novo synthesis of short-chain fatty acids and the uptake of fatty acids by the mammary gland (Hussein, Harvatine, Weerasinghe, Sinclair, & Bauman, 2013), although the mechanism behind milk fat depression after supplementation with marine oils is unclear. Additionally, DHA and EPA have toxic effects on rumen cellulolytic microorganisms, thus reducing the amount of acetate produced in the rumen and, consequently, de novo lipogenesis (Urrutia & Harvatine, 2017).

Microalgae supplementation had no effect on milk fat secretion or increased it (Kholif et al., 2017; Pajor et al., 2019; Papadopoulos et al., 2002; Póti et al., 2015). The increase in milk fat from microalgae supplementation has been related with higher forage intake, thus higher acetate produced in the rumen (Papadopoulos et al., 2002), and to microalgae composition, namely, its methionine content (Lamminen, Halmemies-Beauchet-Filleau, Kokkonen, Vanhatalo, et al., 2019; Póti et al., 2015).

Milk fatty acid profile is often considered unbalanced for human health (Thorning et al., 2016). Public health concerns led to a quest for its modulation, particularly toward less saturated fatty acids (SFA) and more PUFA, namely, n-3 LC-PUFA and CLA. Dairy nutrition is the most effective strategy to modulate milk fatty acid profile (Chilliard et al., 2007), particularly through the supplementation of plant and/or marine oils to the diet (Glasser, Ferlay, & Chilliard, 2008). In this context, the dietary inclusion of microalgae has been attracting interest.

Freshwater microalgae had a poor impact on cows and goats milk fatty acid profile, reflecting their low lipid content (Table 4). Total SFA, monounsaturated fatty acids (MUFA), and PUFA profile were not affected in goats milk (Tsiplakou, Abdullah, Alexandros, et al., 2017; Tsiplakou et al., 2018), whereas Lamminen, Halmemies-Beauchet-Filleau, Kokkonen, Jaakkola, et al. (2019) reported no effect on SFA and MUFA and a decrease in PUFA and n-6/n-3 ratio in cow's milk. Conversely, Póti et al. (2015) observed an increase of MUFA, PUFA, n-3 PUFA, and a decrease of SFA and n-6/n-3 ratio in goats milk fed with Chlorella kessleri. Similarly, increasing supplementation levels of C. vulgaris and A. platensis promoted a linear decrease of SFA and an increase of PUFA, with n-6/n-3 ratio decreasing in cows (Till et al., 2019) but not in goats milk (Kholif et al., 2017).

Western diets have a very high n-6/n-3 ratio, which had been associated with the increase of metabolic diseases (Simopoulos, 2004). Thus the lower n-6/n-3 ratio of milk with microalgae dietary supplementation may also improve human health. The combination of freshwater and marine microalgae may be another strategy to improve milk fat profile. Indeed, Lamminen, Halmemies-Beauchet-Filleau, Kokkonen, Jaakkola, et al. (2019) found a reduction of n-6 PUFA and n-6/n-3 ratio and an increase of EPA in the milk of cows fed a mixture of C. vulgaris and Nannochloropsis gaditana (1:1) compared to C. vulgaris alone.

Most studies aiming for the enrichment of milk n-3 content have evaluated commercial DHA-rich Aurantiochytrium sp. Dietary inclusion of marine microalgae improved the milk fat profile in cows, in sheep (Papadopoulos et al., 2002) and in goat by increasing PUFA and DHA, and decreasing SFA and n-6/n-3 ratio (Table 4). Supplementation of DHA-rich microalgae successfully promoted the health beneficial long-chain n-3 fatty acids in milk fat by increasing DHA or EPA and DHA. Interestingly, Fougère et al. (2018) observed an increase of DHA in cow's milk but not in goat's milk, suggesting a different mechanism in DHA transfer into milk between animal species.

The apparent transfer efficiency of DHA from microalgae into milk is quite low. Boeckaert et al. (2008) determined a DHA transfer efficiency of 3.1% and 5.9% when algae were supplemented, respectively, in the concentrate or through the rumen fistula, whereas Franklin et al. (1999) found a transfer efficiency of 8.4%. The low efficiency of transfer of PUFA in general, and DHA in particular, may be attributed to the extent biohydrogenation activity of the rumen microorganisms (resulting in the formation of *trans* fatty acids and SFA) and to the preferential incorporation of absorbed DHA into plasma phospholipids and cholesterol esters instead of triacylglycerols (Shingfield, Bonnet, & Scollan, 2013).

Even though the rumen biohydrogenation reduces the amount of dietary DHA reaching the small intestine and available to be transferred into the milk, it also results in the formation of health-promoting biohydrogenation intermediates, like CLA and 18:1 *trans*-11, which are also incorporated into the milk. Indeed, an increase of both of these fatty acids were reported in the milk of sheep (Papadopoulos et al., 2002), goats (Fougère et al., 2018; Pajor et al., 2019) and cows (e.g., Franklin et al., 1999; Marques et al., 2019; Moate et al., 2013; Till et al., 2019; Vanbergue, Peyraud, et al., 2018).

During particular dietary conditions (high starch and high oil), the biohydrogenation pattern is a shift from *trans*-11 toward the formation of *trans*-10, which leads to milk fat depression (Bauman & Griinari, 2003). An increase of 18:1 *trans*-10 was observed in the milk fat of cows but not in those of sheep and goats. Similar biohydrogenation shifts toward *trans*-10 formation have been reported with marine oil supplementation (Loor et al., 2005; Shingfield et al., 2012).

Although less studied, marine DHA-rich *C. cohnii* was also described to increase EPA and DHA in colostrum of ewes, linearly increasing with time (Pickard et al., 2008).

2.4 Meat production and composition

The effects of microalgae supplementation on body weight (BW) gain is controversial (Table 5). The inclusion of *A. platensis* up to 6.1 g/kg BW linearly increased daily weight gain of steers; this higher growth is explained by a higher protein supply and ultimately by an increase of forage intake (Panjaitan et al., 2015). Similarly, the inclusion of *A. platensis* at 10 g/kg BW increased the daily weight gain of lambs (El-Sabagh et al., 2014).

When lambs were fed a diet with a combination of *Aurantiochytrium* sp. (3.9% DM) and linseed (5% DM), the daily weight gain decreased, this reduction is attributed to a decreased feed intake presumably due to reduced palatability of the algae and to a decreased microbial growth and fiber digestion promoted by dietary PUFA (Urrutia et al., 2016). Díaz et al. (2017) also reported a reduction on growth rate attributed to a reduction in feed intake due to the reduced palatability of the diet with *Aurantiochytrium* sp., whereas a positive effect on daily weight gain without effects on intake was observed by Sucu et al. (2017), and no effect was reported by others (Fan et al., 2019; Meale et al., 2014). The reason for the inconsistent effects of *Aurantiochytrium* sp. on BW gain is unknown, but Meale et al. (2014) speculated that despite being the same species of microalgae, depending on the stage of growth at which the algae was harvested, the levels of antioxidative, antibacterial, or cytotoxic effects could be different, thus having different effects on intake and ruminal degradation.

Microalgae have been also studied as a strategy to decrease morbidity and mortality in young animals by benefiting from the potential antiinflammatory effects of n-3 fatty acids, mainly DHA. In this context, Flaga et al. (2019) evaluated the effect of increasing doses of *Aurantiochytrium* sp. in the milk replacer fed to calves and observed that increasing the dose of DHA in the milk replacer decreased not only

TABLE 5 Effect of microalgae on body weight gain and muscle composition of ruminants.

Algae		Animal model and size	Number of animals	Animal diets	Inclusion level (%)	Trial duration	Growth	Meat composition	Meat fatty acids	References
Arthrospira platensis	Steers	Brahman × Shorthorn steers: 18months old, 250.1±10.86kg BW	9	Mitchell grass (*Astrebla* spp.) hay based diet	0 and 0.5, 1.4, 2.5, and 6.1g/kg BW/d	30days	Linear ↑ ADG			Panjaitan et al. (2015)
	Lambs		10	Berseem hay, wheat straw and concentrate	0 and 10g/kg BW/d	35days	↑ Final BW, daily weight gain and feed conversion			El-Sabagh, Eldaim, Mahboub, and Abdel-Daim (2014)
Aurantiochytrium sp.	Lambs	Hu lambs: 80±3days old, 18.35±1.39kg BW	48	Standard (8.40MJ/kg) and high (9.70MJ/kg) energy concentrate	0% and 3% concentrate	60days		↓ Body fat	↓16:0, 18:0, 18:2n-6, 18:3n-6, n-6/n-3 ratio, SFA, MUFA, PUFA n-6 / ↑18:1t11, EPA, DHA, CLAc9t11, PUFA, PUFA n-3 / ↔18:1t10, DPA, 20:4n-6	Fan et al. (2019)
Aurantiochytrium sp.: DHA gold™	Lambs	Canadian Arcott lambs: 22.7±3.90kg BW	44	Barley and alfalfa hay based diet	0%, 1%, 2%, and 3%	Mid December to April	↔ Lambs final BW, ↔ ADG	↔ Hot carcass weight, dressing percentage, and muscle scores of the leg, back and shoulder	↔ Content of 18:1c9, 18:1t10, 18:1t11, 18:2t11c15, CLAc9t11, CLAt10c12, 18:2n-6, 18:3n-3, SFA, MUFA, PUFA / Linear ↑ EPA, DPA, DHA, PUFA n-3 content / Quadratic effect on n-6/n-3 ratio (lowest at 2 e 3%) / Linear ↓ 18:0	Meale et al. (2014)
		Merino lambs: male, 5months old, 39.1±1.28kg BW	40	Wheat grain and sunflower meal plus alfalfa hay	0 and 5g/head/d	49days	↑ Final BW, daily weight gain and growth rate			Sucu et al. (2017)
		Navarra lambs: male, weaned at 16.3±0.3kg BW and 55.1±1.5days old	33	Ad libitum concentrate feed (barley and soybean)	0% and 3.89% +5% linseed of concentrate	Slaughtered at 26.7±0.3kg BW	↓ ADG / ↑ Slaughter age	↔ Hot and cold carcass weight	↓18:0, 16:1c9, 18:1c9, n-6/n-3 ratio / ↑18:3n-3, EPA, DPA, DHA, PUFA/SFA ratio / ↔CLAc9,t11	Urrutia et al. (2016)
		Poll Dorset × Border Leicester × Merino lambs: 34.8±2.5kg BW, 3months old	40	Oat-grain-based diet	0 and 12.9g/head/d	42days	↔ Average growth rate, feed efficiency	↔ Hot carcass weight, pH	↑ EPA, DHA, 22:5n-6, PUFA n-3 / ↓ n-6/n-3 ratio / ↔ SFA, MUFA, PUFA n-6	Hopkins et al. (2014)
		Hu lambs: 80±3days old, 18.35±1.39kg BW	48	Standard (8.40MJ/kg) and high (9.70MJ/kg) energy concentrate	0% and 3% concentrate	60days	↔ Final BW, ADG			Fan et al. (2019)
			40 lambs	Oat grain based diet		42days	↔ ADG			

Microalgae	Animal	n	Dose	Diet	Duration	Performance	Carcass	Fatty acids	Reference
	Poll Dorset × Border Leicester × Merino lambs: 3 months old, 34.8±2.5 kg BW, 3.3 fat score		0 and 12.9 g/head/d						Clayton et al. (2014)
	Manchego lambs: male, 7–8 weeks old, 15.3±0.38 kg BW	24	0% and 2%	Ad libitum grain-based diet	Up to 26 kg BW	↔ Final BW ↓ ADG ↑ Fattening period	↔ Fat content	↓ 18:0, 18:1, 18:2n-6 ↑ 20:4n-6, EPA, DPA, DHA ↑ 16:0, 20:4n-6, EPA, DPA, DHA, PUFA n-3, PUFA and ↓ 18:0 in neutral lipids ↑ 14:0, 16:0, EPA, DPA, DHA, SFA, PUFA, PUFA n-3 and ↓ 18:1, 18:2n-6, 18:3n-3, CLA, MUFA, PUFA n-6 in polar lipids ↑ DHA in intramuscular lipids	Diaz et al. (2017)
	Calves Holstein calves: females, 8.6±0.8 days old, 41.1±4.3 kg BW	40	0, 9, 18, and 27 g/head/d (0, 1.13, 2.26, and 3.38 g DHA/head/d)	Milk replacer for 49 days and ad libitum starter mixture from d15 onwards	49 days	↔ Final BW and height Linear ↓ BW gain, ADG ↔ ADG:DMI			Flaga et al. (2019)
	Heifers Charolais × Limousin × Friesian heifers: 24±0.8 months old, 509±40.0 kg BW	30	0%, 1.5%, and 3.0%	Ad libitum cereal-based diet and wheat straw	95 days		↔ Carcass weight, conformation, and fat class	↑ 16:0, EPA, DHA, PUFA n-3 ↓ 18:1c9, 18:1t11, DPA, 20:3n-6, 22:4n-6, n-6/n-3 ratio ↔ 18:2n-6, 18:3n-3, 20:4n-6, SFA, MUFA, PUFA, PUFA n-6, PUFA/SFA ratio	Rodriguez-Herrera, Khatri, Marsh, Posri, and Sinclair (2018)
Chlorella spp.	Kids and lambs Majorera goat kids and Canarian sheep lambs	80 and 80	0% and 5.625%	Ad libitum commercial milk replacer, twice a day	60 days	↔ BW, growth rate			Morales-De La Nuez et al. (2014)
Crypthecodinium cohnii	Ewes English mule ewes: last 9 weeks of gestation	48	0 and 12 g/d during the first 3, 6 or 9 weeks	Silage and commercial concentrate	63 days	↔ Lamb birth weight, weight gain in first 24h, DLWG in first 5 weeks ↓ Time to stand at 6 and 9 weeks vs. 0			Pickard et al. (2008)
	Kids and lambs Majorera goat kids and Canarian sheep lambs	80 and 80	0% and 5.625%	Ad libitum commercial milk replacer	60 days	↔ BW, growth rate			Morales-De La Nuez et al. (2014)
Isochrysis galbana	Kids and lambs Majorera goat kids and Canarian sheep lambs	80 and 80	0% and 5.625%	Ad libitum commercial milk replacer	60 days	↔ BW, growth rate			Morales-De La Nuez et al. (2014)

ADG, average daily gain; BW, body weight; CLA, conjugated linoleic acid; DHA, docosahexaenoic acid; DLWG, daily live weight gain; DMI, dry matter intake; DPA, docosapentaenoic acid; EPA, eicosapentaenoic acid; MUFA, monounsaturated fatty acids; PUFA, polyunsaturated fatty acids.

its intake but also of the starter feed, thus decreasing BW gain. Authors suggested that this reduction of milk replacer reduction might result from an indirect effect of DHA on the intake regulation rather than to reduced palatability of DHA, as animals seemed to get accustomed to the specific taste of DHA with time. Similarly, the addition of *C. cohnii*, *Chlorella* spp., and *Isochrysis galbana* to milk replacer did not improve the growth of goat kids and lambs (Morales-De La Nuez et al., 2014). Absence of effects on BW gain of lambs was also observed when ewes were supplemented with *C. cohnii* during the last 9 weeks of gestation, even with a successful transfer of DHA and EPA via colostrum (Pickard et al., 2008).

Ruminant meat is a valuable source of high biological value proteins, vitamins, particularly those of vitamin B group, minerals, and other micronutrients, but also of fat and SFA. Red meat has often been pointed out as the main source of fat and SFA and *trans* fatty acids in Western diets and its intake associated with the increased incidence of metabolic diseases, including cardiovascular disease, cancer, insulin resistance, and obesity (Battaglia Richi et al., 2015). The FAO/WHO nutritional recommendations include a decrease in red meat consumption, disregarding the high nutritional value of meat. The nutritional modulation of ruminant meat toward a healthier fatty acid profile is an effective strategy to reduce the human consumption of SFA and increase that of PUFA without altering consumers' habits. In this context, marine microalgae, rich in n-3 PUFA, emerge as an interesting feed to promote healthier meat.

Few studies evaluated the effects of diets supplemented with the DHA-rich microalgae *Aurantiochytrium* sp. on body fat (Table 5), results varying from no alteration observed to an effective fat reduction. More consistently DHA-rich microalgae supplementation promoted an increase in n-3 PUFA, particularly DHA and EPA, a decrease of n-6/n-3 ratio, and of SFA.

Hopkins et al. (2014) and Rodriguez-Herrera et al. (2018) failed to modulate both SFA and PUFA of meat of animals fed with DHA-rich microalgae but successfully increased its PUFA n-3 profile. These results reflect a higher intake of PUFA (mainly DHA and EPA) from microalgae, yet the potential to manipulate the fatty acid composition of muscle lipids is also dependent on the extent of lipolysis and rumen biohydrogenation of dietary PUFA. Supplementation of DHA-rich microalgae *Aurantiochytrium* sp. strongly reduced the biohydrogenation endproduct (18:0) and promoted the content of biohydrogenation intermediates (e.g., CLA, 18:2 *trans*-11,*cis*-15, and 18:1 *trans*-11), suggesting that the rumen biohydrogenation was impaired, particularly its final step (reduction of 18:1 *trans*-11 to 18:0), more effectively than with vegetable oils (e.g., linseed) supplementation (Fan et al., 2019; Meale et al., 2014; Urrutia et al., 2016). Thus feeding microalgae is an effective way of increasing the nutritional value of ruminant meat, in terms of health claimable long-chain n-3 fatty acids, and as 20-carbon and 22-carbon n-3 fatty acids are mainly incorporated into phospholipids instead of triacylglycerols, it is possible to increase these PUFA in muscle without a further increase in body fat (Díaz et al., 2017; Fan et al., 2019).

On the negative side, *Aurantiochytrium* sp. supplementation can render the meat more susceptible to oxidation (Díaz et al., 2017; Hopkins et al., 2014) due to the increase of long-chain PUFA more prone to lipid oxidation. Oxidation reactions reduce both the nutritional value and the sensory quality, as it changes meat color (Urrutia et al., 2016) and reduces consumer ratings for odor, flavor, and overall acceptability (Rodriguez-Herrera et al., 2018; Urrutia et al., 2016). Furthermore, compounds formed during oxidation are implicated in the development of atherosclerosis, cancer, and inflammation (Domínguez et al., 2019). The dietary inclusion of antioxidants, namely, vitamin E supplementation, would ensure that diets with high PUFA levels will not lead to excessive lipid oxidation and off-flavor development in ruminant's meat.

3 Microalgae in swine

Swine are some of the main competitors with the human food chain, mostly due to the fact that their diets are usually composed of two major ingredients: cereals (e.g., corn) and soybean meal. Hence, microalgae could have an interesting role in solving this and other issues in this species' nutrition. The main effects of microalgae inclusion used in swine diets are summarized in Table 6.

3.1 Piglets

For piglets, weaning is a stressful event that is caused by social, environmental, and nutritional alterations. Hence, assuring a high dietary protein quality and maintaining and promoting gut health is of paramount importance to mitigate PWS (postweaning stress). For this purpose, antibiotics are often used to prevent PWS diarrhea (Heo et al., 2013). Many studies attempted to use microalgae as a source of n-3 PUFA to either enhance the fatty acid profiles of edible tissues or take advantage of the immunomodulatory/prebiotic properties of these microalgae to deal with PWS. Furbeyre et al. (2017) used *A. platensis* and *C. vulgaris* (both at a 1% dietary inclusion) as alternatives to antibiotic supplementation to promote piglet gut health. These authors found that, at this inclusion level, both microalgae had no effect over average daily feed intake and average daily gain, but increased villus height and villus:crypt ratio in the small intestine, compared to those fed a control diet. Moreover, *Chlorella* reduced diarrhea incidence and lowered its severity. The administration of microalgae via drinking water did not improve gut histological traits (Furbeyre et al., 2018). In a latter study, *Spirulina* supplementation reduced PWS diarrhea, which

was explained by the stabilizing effect on gut microflora after weaning. Contrary to what has been previously mentioned for *A. platensis* and *C. vulgaris*, Kibria and Kim (2019) reported that *Schizochytrium* JB5 used as an additive (1%) in postweaning diets improved feed conversion (during 8–21 days postweaning) from 0.73 to 0.67. In addition, the latter microalgae increased DM and nitrogen apparent total tract digestibility: these were 83.42%, 87.59%, and 85.21% for DM and 64.81%, 76.31%, and 73.98% for CP, respectively, for control, 0.5% and 1% dietary inclusion. The same relation was found for the ileal digestibility of these components; however, 1% inclusion increased DM and CP digestibility when compared to 0.5%. The use of *Spirulina platensis* as a prebiotic in the diets of piglets weaned with 3.7 kg LW has been reported to increase the average daily gain between the 14th and 28th days after weaning. Particularly, 0.2% inclusion had higher average daily gain (324 g/d) compared to 0.5% inclusion (291 g/d) (Grinstead, Tokach, Dritz, Goodband, & Nelssen, 2000). Hence, small inclusion levels of microalgae seem to be effective at improving animal performance through increased gut health, possibly by increasing the presence of beneficial bacteria such as *Lactobacillus*. However, further research is necessary to confirm the mechanism of action of microalgae on intestinal microflora.

Biofuel production from microalgae sources provides an opportunity for novel animal feeding strategies, by providing the defatted biomass, highly concentrated in protein and/or carbohydrates. Urriola et al. (2018) have reported that using 1%, 5%, 10%, and 20% microalgae defatted biomass (unknown species) as an energy source (partially replacing corn) did not negatively influence the growth of weaned piglets. However, it reduced the average daily feed intake when included at 20% (718 g) compared to 1% (766 g). When using 10% *Desmodesmus* sp. defatted biomass as a source of protein in weaned piglet diets, the addition of proteolytic enzymes was required to improve the final body weight of piglets from 26.7 kg to 28.1 kg (Ekmay et al., 2014). In addition, the inclusion of this biomass lowered plasma uric acid and urea nitrogen, which indicates an improvement of N efficiency and lower N excretion. Unfortunately, this feedstuff has a significant limitation, since most lipid-soluble pigments and antioxidant compounds, such as carotenoids, are removed along with oil during its extraction from intact microalgae biomass (Minhas, Hodgson, Barrow, & Adholeya, 2016). This shortcoming explains why Urriola et al. (2018) reported no differences between control and algae-fed groups regarding antioxidant metabolites in the liver of piglets. This makes defatted microalgae unsuitable to improve the oxidative status of these animals, already under heavy oxidative stress due to PWS (Guevarra et al., 2018). A possible solution would be to enrich the cultivation medium with selenium, thus enriching the microalgae with nonlipid-soluble antioxidative elements prior to oil extraction (Doucha, Lívanský, Kotrbáček, & Zachleder, 2009). However, this approach to piglet feeding is so far unexplored in the currently available literature.

Theoretically, whole microalgae biomass could provide improved antioxidant status while simultaneously supplying dietary protein and/or carbohydrates in weaned piglet diets. However, literature regarding the use of intact microalgae biomass in postweaning diets is scarce. Kalbe et al. (2019) reported the incorporation of 7% *Schizochytrium* in postweaning diets. Microalgae incorporation yielded no significant differences relative to the control diet regarding animal performance. Grinstead et al. (2000) studied the incorporation of *S. platensis* as a replacement for soybean meal in early-weaned piglets and found inconsistent effects on animal performance through the course of three different trials using 0.2%, 0.5%, and 2% inclusion levels. Contrary to what would have been expected, pelleting the feed failed to improve feed efficiency in comparison to a standard meal. Hence, further research aiming to replace soybean meal by higher levels of microalgae biomass is necessary. Also, studies aiming to evaluate the effect of the technological treatment of microalgae (pelleting, extrusion, etc.) on nutrient digestibility and animal performance would be useful.

Finally, there is a lack of information concerning intact microalgae as protein and carbohydrate sources in postweaning diets. Indeed, this could occur due to the high cost of microalgae compared to conventional feedstuffs, which makes them more appealing to use in small inclusion percentages, as prebiotics or additives. Nonetheless, with the development of increasingly efficient and high-yield production techniques, cheaper microalgae could compete with the corn or soybean meal used in postweaning diets. Hence, there are immense possibilities and opportunities for researchers to consider. Particularly, studying the digestibility of these feedstuffs at higher percentages of inclusion and the corresponding effects on metabolism using state-of-the-art -omics technology will generate both exciting results and challenges to overcome.

3.2 Growing and finishing pigs

Due to a more developed gastrointestinal tract, growing and finishing pigs could be more suitable for feeding with microalgae as protein/carbohydrate sources than weaned piglets. Indeed, since at least 1967, microalgae have been considered as protein sources for these late-stage pigs. Hintz and Heitman (1967) reported that in growing and finishing pigs, the inclusion of 10% and 5% sewage-grown microalgae (*Chlorella* and *Scendesmus*), respectively, to replace fish meal, with added B vitamins, did not negatively influence animal performance but increased backfat thickness compared to control diet. Additionally, these authors reported that the digestibility of dry matter, carbohydrates, ether extract, and energy of a diet composed of 80% barley and 20% algae were lower compared to a diet 100% barley.

Recently, Neumann, Velten, and Liebert (2018) reported that *S. platensis* have an adequate quality to be used as a protein source in grower diets (60 kg LW), when included at 13%. Soybean meal replacement of up to 100% with this microalga in the fattening diets of pigs had no detrimental effect on carcass quality or the physicochemical properties of the *Longissimus lumborum*. Although increased undesirable odors and astringent taste were perceived by a sensory panel analysis, lipid oxidation did not differ from controls (Altmann, Neumann, Rothstein, Liebert, & Mörlein, 2019). The pigs fed with this microalga also displayed increased linolenic acids (C18:3n-3 and C18:3n-6) in the subcutaneous fat.

Several authors have reported the use of microalgae as sources of LC-PUFA to enhance the nutritional properties of pork (De Tonnac et al., 2018; Marriott, Garrett, Sims, & Abril, 2002; Meadus et al., 2010; Moran, Morlacchini, et al., 2018; Sardi et al., 2006; Vossen et al., 2017). In 2002 Marriott et al. (2002) reported that finishing pigs (65–105 kg LW) were able to accumulate DHA from *Schizochytrium* sp., often with over twofold increased concentrations in various edible cuts, including bacon, with no effect on animal performance. In 2006 Sardi et al. (2006) reported that in the latest stage of fattening heavy Italian pigs (118–160 kg LW), dietary inclusion of *Schizochytrium* sp. (0.25% and 0.5%) was able to increase DHA in the muscle, regardless its inclusion level. According to these authors, 0.5% microalgae inclusion in the 4 weeks before slaughter increased 50 mg DHA/100 g of fresh *Longissimus lumborum* compared to the control diet. Microalgae inclusion had no effect over animal performance. Accordingly, Meadus et al. (2010) reported a linear increase of DHA concentration in bacon made from pigs fed 0.06%, 0.6%, and 1.6% *Schizochytrium* sp., recording 43, 122, and 339 mg/100 g of bacon, respectively. However, when pigs were fed over 0.6% microalgae, consumer acceptability was reduced due to the development of off-flavors during and after cooking the bacon. Moreover, lipid oxidation increased significantly in bacon from the same pigs. Likewise, Vossen et al. (2017) fed finishing pigs from 75 to 110 kg with 0.3%, 0.6% or 1.2% *Schizochytrium* sp. and found that dry-cured hams had increased DHA concentrations versus controls, ranging from 25 to 56 mg/100 g. However, they also reported increased lipid oxidation as a result of DHA enrichment, which was not found for uncured loins from the same pigs. Thus if the objective is to produce DHA-rich processed meat products, it is important to consider the process that will be used to transform meat from pigs fed with microalgae as DHA sources because of their oxidative nature. Indeed, in pigs fed with *Schizochytrium* sp., it was observed an upregulation of oxidation-related genes expressed in subcutaneous fat and liver, which aggravates this problem (Meadus et al., 2011). Finally, it is also relevant to consider the cut from the carcass that will be used for processing, because DHA deposition depends on tissue location (De Tonnac et al., 2018).

Tonnac and Mourot (2017) reported increased production of malondialdehyde (a product of lipid oxidation) in the subcutaneous fat of pigs fed with up to 3.7% *Schizochytrium* sp. and related it with increased fish odor reported by a sensory panel. Dietary microalgae also significantly increased C22:6n-3 in subcutaneous backfat, which corroborates the results previously mentioned. Furthermore, edible tissue oxidation will negatively influence the consumer acceptability of the final product (development of undesirable odors) and shelf-life (increased lipid oxidation). Increased dietary DHA reduced the activity of lipogenic enzymes in the liver of finisher pigs (64.6–114.9 kg) and inhibited the expression of adipogenic SREBP1 gene in the muscle and liver of these pigs, while DHA content increased (Tonnac et al., 2016). Hence, tissue DHA tissue content in the edible tissues of pigs fed with DHA-rich microalgae seems to depend on its intake. Thus further experiments are necessary to establish a threshold of microalgae inclusion that increases DHA in meat without compromising meat quality.

The use of microalgae as prebiotics during growing and finishing stages is another potential use for this novel feedstuff. Yan et al. (2012) reported increased *Lactobacillus* and lower *Escherichia coli* populations in the feces of growing pigs fed with either 0.1% or 0.2% fermented *C. vulgaris* compared to a control diet. In addition, the former algae diet had an increased average daily gain of 655 g/d compared to the control diet (624 g/d) and increased dry matter digestibility. Conversely, Saeid, Chojnacka, Korczy, and Korniewicz (2013) reported a decrease in the dry matter digestibility of diets with *Spirulina maxima* as a source of Cu. There were no effects on the animal performance of pigs.

Regarding growing and finishing pigs, there are multiple studies that report DHA enrichment of pork through the dietary inclusion of *Schizochytrium* sp. However, there are few studies aiming to evaluate the value of microalgae as feed ingredients. In the coming years, research aiming to characterize the value of microalgae as feed ingredients will be of paramount importance if the industry is to use these feedstuffs and reduce the carbon footprint of pig production. Namely, studies aiming to evaluate nutrient digestibility and those using -omics to evaluate the effects of this dietary inclusion on tissue metabolism would be of interest. Moreover, nutritionists should be cautious when using these feedstuffs due to the recalcitrant cell wall of some species as well as the mineral and amino acid profile of the microalgae that might compromise its nutritional value.

3.3 Sows and boars

Sows and boars are particular classes of animals within pig production because they play a critical role: reproduction. For boars,

microalgae have the potential of being a fine alternative to fish oil/fish meal as a source of both protein and particularly DHA. This n-3 LC-PUFA is a major component of the spermatozoid membrane and DHA-rich ingredients are routinely used to feed boars in order to enhance sperm quality. Andriola et al. (2018) supplemented boars with 150 g *Schizochytrium* sp./kg of feed to supply a total amount of 120 g DHA/kg of feed. The authors found that boars fed with increased DHA had spermatozoids with increased mobility due to increased membrane fluidity. However, feeding microalgae comes with an increased necessity to be aware of adequate n-6:n-3 PUFA ratios. If unbalanced, they can impair mitochondrial function and spermatogenesis (Andriola et al., 2018). Nonetheless, other authors have reported that feeding an algal source of DHA to boars (33.55% DHA in dietary fatty acids) increased the volume of the ejaculate and sperm count. It also increased DHA in the fatty acid profile of sperm. Increasing sperm DHA is beneficial because the tail region has the highest concentration of DHA (Murphy et al., 2017) explaining the increased sperm mobility and membrane fluidity.

In order to improve boar reproduction parameters, Otte et al. (2019) used *Schizochytrium* sp. as a source of n-3 PUFA during the postweaning phase of gilts. Supplementing F1 gilts with 3.5 g of DHA *per* day produced no tangible effects on ovarian development or the expression of LH receptors. Conversely, supplemented gilts had lower leptin and leptin receptor levels in ovarian tissue. In another study aiming to supply DHA during gestation, *Crypthercodinium cohnii* oil was shown to be as efficient as egg yolk extract in supplying DHA to piglet fetuses during the last 40 days of gestation (Gázquez, Ruíz-Palacios, & Larqué, 2017).

Briefly, the use of microalgae seems to be effective at improving the sperm quality of boars. However, further studies are required to avoid mitochondrial dysfunction in the testes.

Dietary DHA also seems to promote oxidative stress during spermatogenesis. Studies aiming to establish a threshold of microalgae inclusion in boar diets would be recommended in the near future. Also, the supplementation of antioxidant compounds such as vitamin E in these diets could be an interesting solution to lower this side effect of DHA supplementation.

Regarding sows and gilts, there is scarce information regarding how microalgae inclusion in feeds affects the reproduction cycle. However, available literature points toward increased usefulness during gestation and lactation in order to take advantage of the immunomodulatory properties of DHA to improve the immune status of piglets prior to and after birth. Further research is thus necessary to evaluate the impact of providing DHA to piglets on their oxidative status. Fig. 1 summarizes the main results found in the reviewed studies of this section (Table 6).

4 Microalgae in poultry

Poultry-derived products, such as meat and eggs, are healthy, cheap, and do not have religious restrictions; therefore they have become staples in the diet for many consumers worldwide. However, recent concerns over the environmental aspect and climate change have paved the way for increased awareness for sustainability in the poultry industry. The main conventional feedstuffs for poultry are corn and soy, which require a large amount of land to grow. In addition, a growing global population combined with an increasing overall demand for animal-derived products may lead to the unsustainable allocation of corn and soy in the globe (Madeira et al., 2017). The protein level in microalgae may be as high as 75%, while soybean meal is only around 45% crude protein. Therefore due to its high concentration in protein, microalgae may be an alternative to plant

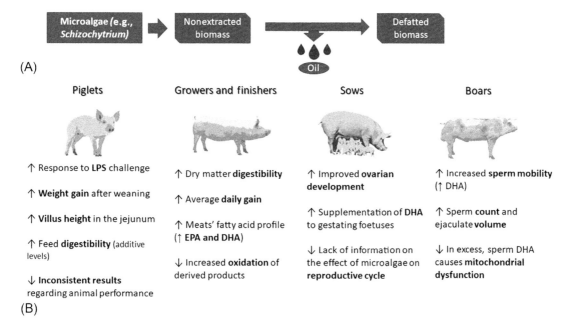

FIG. 1   (A) Microalgae forms of utilization in animal nutrition and (B) main results found for pigs fed with (*EPA*, eicosapentaenoic acid; *DHA*, docosahexaenoic acid; *LPS*, lipopolysacharide).

TABLE 6　Summary of studies evaluating different microalgae in pig diets as either ingredients (>3%) or additives (<3%).

Microalgae	Ingredient vs. additive	Inclusion level	Objective	Animal class/ live weight	Main results	References
Arthrospira platensis and *Chlorella vulgaris*	Additive	1% (both)	Mitigate postweaning stress	Weaned piglets/ 9.1–20kg	*Chlorella* increased feed:gain ratio in comparison to *Spirulina*. Both microalgae increased villus height in the jejunum	Furbeyre, Van Milgen, Mener, Gloaguen, and Labussière (2017)
	Additive	385mg/ kg LW (in drinking water)	Mitigate postweaning stress	Suckling piglets/ 4.9kg Weaned piglets/ 9.04kg	Spirulina increased weight gain in the suckling period and at weaning. Spirulina lowered diarrhea incidence. Chlorella increased diarrhea occurrence	Furbeyre, Milgen, Mener, Gloaguen, and Labussière (2018)
Aurantiochytrium limacinum	Additive	3.12%	Replace fish oil as a source of n-3 PUFA with microalgae, to boost immune stats during postweaning stress	Weaned piglets/ 6.82–22.30kg	After lipopolysaccharide challenge, algae-fed piglets had lower rectal temperature, less circulating cortisol but increased cytokine concentration. Microalgae had no effect over animal performance	Lee, You, Oh, Li, Code, et al. (2019a)
	Additive	3.12%	Enhance weaned piglet immune status through maternal supplementation	Multiparous sows/NA	Supplementation did not affect cortisol response in sows following a LPS challenge. No effect on offspring	Lee, You, Oh, Li, Fisher-Heffernan, et al. (2019b)
Aurantiochytrium limacinum	Additive	1%	Evaluate the use of the microalga as a source of DHA in finisher diets	Finisher 117.1–140.3kg	No effect over animal performance. Algae supplementation increased EPA, DHA and n-3/n-6 ratio in *Longissimus lomborum*	Moran, Morlacchini, Keegan, and Fusconi (2018)
Defatted microalgae biomass (unknown species)	Additive	0.5%–1%	Test the effect of defatted biomass supplementation on growth performance, ileal digestibility and total tract apparent digestibility	Weaned piglets 6.27kg	Defatted biomass downregulated the urea cycle in the liver. No influence of microalgae biomass dietary inclusion over animal performance	Urriola et al. (2018)
Desmodesmus sp. (defatted)	Ingredient	10%	Determine the impact of using defatted biomass in piglet diets and the effect of protease addition	Weaned piglets 9.8kg	Dietary microalgae with protease supplementation increased final live weights from 26.7–28.1kg. Biomass incorporation increased ingested N utilization	Ekmay, Gatrell, Lum, Kim, and Lei (2014)

Species	Type	Dosage	Objective	Animal/weight	Results	Reference
Chlorella vulgaris (fermented)	Additive	0.1% 0.2%	Evaluate the effect of fermented microalgae as prebiotic in the performance of growing pigs	Growers 26–53 kg	0.1% incorporation increased dry matter digestibility and ADG; Microalgae incorporation benefited the gut's *lactobacillus* population	Yan, Lim, and Kim (2012)
Chlorella and *Scendesmus*	Ingredient	10% (grower) 5% (finisher)	Test the viability of feeding sewage-grown microalgae to fattening pigs	Growers 63 kg Finisher 90 kg	Thicker backfat in diets with algae and vitamin B supplementation; Vitamin B supplementation improved ADG (866 g/d) vs. no supplementation (757 g/d) in algae-fed groups	Hintz and Heitman (1967)
Schizochytrium JB5	Additive	0.5%–1%	Test the effect of microalgae supplementation on growth performance, nutrient digestibility and blood immune profile	Weaned piglets 6.25–10.52 kg	Supplementation of 1% microalgae improved gain:feed ratio vs. 5%; Microalgae improved blood lymphocyte count; Microalgae supplementation increased dry matter and N apparent total tract digestibility and ileal digestibility	Kibria and Kim (2019)
Schizochytrium sp.	Ingredient	7% (piglets) 5% (growers)	Increase n-3 PUFA content in pig meat	Weaned piglets/9.46 kg Grower pigs/103.37 kg	Increased DHA in *Longissimus thoracis* and *semitendinosus* muscles by 7- and 14-fold, respectively, in comparison to control; Reduced feed intake; No effect over animal performance	Kalbe, Priepke, Nürnberg, and Dannenberger (2019)
	Additive	0.25% 0.5%	Evaluate the effect of microalgae as a source of DHA in the diet of Italian heavy pigs	Finisher 118 kg	No effects of microalgae dietary inclusion on animal performance or carcass quality; Feeding 0.5% microalgae over 4 weeks prior to slaughter resulted in the accumulation of 0.07 g DHA/100 g of meat vs. 0.02 g DHA/100 g in control	Sardi, Martelli, Lambertini, Parisini, and Mordenti (2006)
	Additive	0.06% 0.6% 1.6%	Evaluate DHA from *Schyzochitrium* sp. as a source to improve bacon fatty acid profile	Finisher 80 kg	Positive correlation between increased dietary microalgae and average daily gain; Over 0.6% microalgae inclusion increased lipid oxidation of bacon; Linear increase of DHA in the bacon	Meadus et al. (2010)

Continued

TABLE 6 Summary of studies evaluating different microalgae in pig diets as either ingredients (>3%) or additives (<3%)—cont'd

Microalgae	Ingredient vs. additive	Inclusion level	Objective	Animal class/live weight	Main results	References
	Additive	0.3% 0.6% 1.2%	Study the effect of adding microalgae in the diet of pigs over DHA content, meat and ham quality	Finisher 75–110 kg	Lipid oxidation increased in algae-fed groups for dry-cured hams Algae-fed pigs had increased DHA in dry-cured hams (25–56 mg/100 g) No effect on carcass characteristics	Vossen, Raes, Van Mullem, and De Smet (2017)
	Additive	0.9% 1.9% 3.7%	Study the influence of tissue location on n-3 PUFA concentration in finishing pigs	Finisher 50.7–115.2 kg	Neck subcutaneous and ham adipose tissue were the most efficient depots of n-3 and n-6 PUFA *Longissimus dorsi* and *semimembranosus* muscles had less saturated profiles compared with diaphragm	De Tomnac, Guillevic, and Mourot (2018)

ADG, average daily gain; *DHA*, docosahexaenoic acid; *EPA*, eicosapentaenoic acid; *LPS*, lipopolysaccharide; *LW*, live weight; *NA*, not available; *PUFA*, poly unsaturated fatty acids.

proteins such as soy, while serving as a means for carbon dioxide capture to produce organic compounds. Furthermore, microalgae have other advantages and may be added to poultry diets as a rich source of n-3 LC-PUFA, and as a mineral, vitamin, and antioxidant source. Furthermore, microalgae in feeds serve as a pigmentation agent for meat and egg yolks in the poultry industry. Experiments have also shown that microalgae may be used as a means of increasing beneficial microflora in the digestive tract of poultry and enhance the immune function. However, microalgae availability and price are still two of the greatest limitations preventing their wider usage.

Levels of microalgae incorporation in poultry feed depend on the objective for the intended usage of the microalga. Experiments have been done with high inclusion levels of up to 21% (Evans, Smith, & Moritz, 2015) aiming at substituting conventional protein sources. Others have tested supplementing diets with levels of microalgae as low as 0.001% (Qureshi, Garlich, & Kidd, 1996). Low levels of incorporation are used with specific goals in poultry, such as improving nutritional characteristics of meat and eggs or improve the color of meat and egg yolk. Table 7 summarizes the main results found for studies involving poultry.

4.1 Meat production

The use of different species of microalgae biomass as a raw material in poultry nutrition has been extensively studied. Microalgae *Spirulina* has the potential of replacing a large portion of the common plant-derived protein sources and thus serving as a means for decreasing soy incorporation in poultry diets. Nevertheless, when *Spirulina* is added as a feed ingredient, broiler performance results are still inconclusive. Evans et al. (2015) added 16% and 21% of dried commercial *Spirulina*, to starter diets, and concluded that up to 16% of *Spirulina* may be incorporated into diets and be utilized by broiler

chicks without causing detriment to performance or digestibility. However, when levels of *Spirulina* were as high as 21%, broilers demonstrated a decrease in performance and amino acid digestibility. The authors speculate that protein gelation during pelleting may have increased viscosity of the digestive contents, preventing endogenous enzymes access to substrates in the gastrointestinal tract, and therefore decreasing performance (Evans et al., 2015). Pestana et al. (2020) proposed a similar hypothesis suggesting that *Spirulina* proteins impair the nutritive value of microalgae for broilers through an increase in digesta viscosity and a lower digestibility, likely associated with the gelation of microalgae indigestible proteins. Enzymatic pretreatment has been tested in order to disrupt the cell wall of different microalgae, leading to better exposure of proteins and pigments to the endogenous stock of digestive enzymes (Al-Zuhair et al., 2017). Since lysozyme is an antimicrobial enzyme capable of hydrolyzing the peptidoglycan of bacterial cell walls (Cowieson & Kluenter, 2019), Coelho et al. (2019) have reported that lysozyme in combination with α-amylase is able to partially degrade *S. platensis* cell wall in vitro. However, in vivo, it has been shown that the incorporation of 15% *Spirulina*, in combination with lysozyme, decreased birds' performance, possibly due to disruption of the *Spirulina* cell wall by the lysozyme in the birds intestines, aggravating the gelation of the microalga indigestible proteins (Pestana et al., 2020).

Studies with lower incorporation levels of microalgae have also been performed. When 1% of *S. platensis* was supplemented to the diet, there was an improvement in body weight (BW) gain and feed conversion ratio (FCR) (Park et al., 2018; Shanmugapriya et al., 2015). The microalgae *Chlorella* has also been extensively studied in broiler feeding trials. Kang et al. (2013) found that dietary supplementation of 1% of fresh liquid *Chlorella* improved growth performance and the development of the digestive tract.

TABLE 7 Main effects of microalgae utilization in poultry diets.

Microalgae	Incorporation level in the diet	Animal and initial age	Major results	References
Aurantiochytrium limacinum	0.5%, 2.5%, and 5%	Broilers, 1-day old	Performance not affected. Increased omega-3 content in broiler meat	Moran, Currie, Keegan, and Knox (2018)
Chlorella vulgaris	0.1% and 0.2%	Pekin ducks, 1-day old	Increased BW gain and feed intake. Water-holding capacity and yellowness increased in breast meat	Oh et al. (2015)
	2.5%, 5%, and 7.5%	Broilers, 1-day old	Improvement in growth performance and *Lactobacillus* concentration in the intestines	Kang, Park, and Kim (2017)
	1%	Broilers, 1-day old	Improvement of BW gain, immune characteristics and production of *Lactobacillus* in the intestinal microflora	Kang et al. (2013)
Crypthecodinium cohnii	0.5%	Muscovy ducks, 50 days old males and 43 days old females	No differences in performance. DHA increase in breast meat	Schiavone et al. (2007)
Nannnochloropsis gaditana	5% and 10%	Laying hens, 25 weeks old	Yolk color shifted from yellow to orange-red. DHA enriched eggs	Bruneel et al. (2013)
Porphyridium sp.	5% and 10%	Laying hens, 30 weeks old	Reduced cholesterol levels in yolk and darker yolk colors. Layer performance not affected. Microalga incorporation decreased feed intake	Ginzberg et al. (2000)
Schizochytrium limacinum	1%, 2%, and 3%	Laying hens, 45 weeks old	Darker yolk color. Egg quality and layer performance not affected	Ao et al. (2015)
	0.5%, 1%, and 1.5%	Laying hens, 40 weeks old	Enrichment with n-3 fatty acids in eggs. No affect in performance	Kralik, Kralik, Grčević, Hanžek, and Margeta (2020)
Schizochytrium sp.	3.7% and 7.4%	Broilers, 21 days old	Increase in BW and BW gain with 7.4% incorporation of microalgae. Improvement of broiler meat with n-3 LC-PUFA	Ribeiro et al. (2013)
Spirulina platensis	6, 11, 16%, and 21%	Broilers, 1-day old	Higher digestible methionine value. Microalgae may be added up to 16% without detriment to performance	Evans et al. (2015)
	15%	Broilers, 21 days old	Increased color in muscle. Decline in broiler performance	Pestana et al. (2020)
	1.5%, 2.0%, and 2.5%	Laying hens, 63 weeks old	Increase in egg yolk color. No negative effects on performance	Zahroojian, Moravej, and Shivazad (2013)

TABLE 7 Main effects of microalgae utilization in poultry diets—cont'd

Microalgae	Incorporation level in the diet	Animal and initial age	Major results	References
	Up to 15%	Laying quails, 35 weeks old	Yolk color more intense. No differences in performance	Boiago et al. (2019)
	0.25%, 0.5%, 0.75%, and 1%	Broilers, 1-day old	Improvement in performance. Increased cecal population of *Lactobacillus*	Park, Lee, and Kim (2018)
	4% and 8%	Broilers, 21 days old	Microalgae incorporation did not affect performance. Meat pigmentation increased with microalgae.	Toyomizu, Sato, Taroda, Kato, and Akiba (2001)

BW, body weight; *DHA*, docosahexaenoic acid; *LC-PUFA*, long chain-poly unsaturated fatty acid.

Subsequently, Kang et al. (2017) tested a *Chlorella* by-product at the 2.5%, 5%, and 7.5% levels and also found an improvement in growth performance. When lower levels (0.07%, 0.14%, and 0.21%) of *Chlorella* were added to broiler diets, FCR was significantly improved with a possible enhancement in cell immune response (Rezvani, Zaghari, & Moravej, 2012).

Antibiotic growth promoters were banned in the European Union in 2006, but they continue to be used in other parts of the world to enhance growth, prevent disease, and increase the efficient use of feeds in the broiler industry. Furthermore, excessive use of antibiotics has been credited as one of the causes of the development of bacterial strains resistant to antibiotics in humans and animals. When broiler chickens were fed 1% of fresh liquid *Chlorella*, the *Lactobacillus* population in the ceca was significantly higher and the plasma IgA and IgM concentration were also significantly higher than the groups fed no *Chlorella*, and furthermore, body weight gain was improved (Kang et al., 2013). Similarly, Park et al. (2018) found an increased cecal population of *Lactobacillus* with *Spirulina* supplementation in comparison

to control groups. In light of these results, it has been suggested that microalgae may be a good alternative to the use of antibiotic growth promoters.

Color pigmentation of raw meat and skin in poultry is an important factor in the choice by the consumer (Fletcher, 1999). Venkataraman, Somasekaran, and Becker (1994) found that 14% and 17% of sun-dried *S. platensis* in the broiler diet did not affect the performance of the birds. However, pigmentation of skin, breast, and thigh muscles were deeper in those birds fed Spirulina in comparison to the control (Venkataraman et al., 1994). Similarly, Pestana et al. (2020),demonstrated that yellowness in the broiler breast muscle was affected by the supplementation with *Spirulina*. In addition, in a comprehensive study by Toyomizu et al. (2001) it was found that when 4% and 8% of *S. platensis* powder were added to the diets, it influenced both the yellowness and redness of broiler meat. Hence, it is suggested that microalgae can be safely used for the color manipulation of broiler meat.

In terms of breast meat quality, the addition of 0.5%, 0.75%, or 1% of *S. platensis* in broiler diets decreased drip loss after 7 days of storage in comparison to control diets (Park et al., 2018). Similarly, the addition of 0.2% and 1% *E. coli* fermented liquor with *Chlorella* showed increased performance and significantly less drip loss during storage in breast meat compared to control, and no differences in terms of meat color (Choi et al., 2017).

It has been well established that n-3 LC-PUFA has the potential in the prevention and treatment of several diseases and has an important role in neural development (Gogus & Smith, 2010). Additionally, consumer interest for healthier foodstuffs has been increasing the demand for functional foods with a high content of n-3 LC-PUFA. Therefore improving poultry meat quality with low doses of microalgae can be a strategy of incorporating microalgae in poultry diets. To this end, Bonos et al. (2016) added 0.5% dried *Spirulina* powder and found an increase in the concentrations of certain vital LC-PUFA such as EPA, DPA, and DHA in the thigh meat, while BW gain and FCR did not differ among groups. As previously mentioned in the swine section, *Schizochytrium* is also a type of marine microalgae with the natural capacity to produce DHA. It is used as a feed supplement under the form of DHA-Gold extract. Yan and Kim (2013) have found that 0.1% and 0.2% of *Schizochytrium* supplementation can improve the fatty acid composition of breast meat with performance parameters remaining unaltered. Conversely, Ribeiro et al. (2013) have demonstrated that the incorporation of 7.4% of DHA-Gold in broiler diets leads to an improvement in final BW and BW gain. In addition, supplementation with 3.7% of DHA-Gold may be a good strategy to improve broiler meat with n-3 LC-PUFA without affecting its quality (Ribeiro et al., 2013). Furthermore, another DHA-rich microalgae are *Aurantiochytrium limacinum* that, when supplemented at 0.5%, 2.5%, and 5%, does not negatively impact broiler health or productivity, while significantly improving the nutritional quality of the meat in terms of omega-3 content (Moran, Currie, et al., 2018).

Under high ambient temperatures, Mirzaie, Zirak-Khattab, Hosseini, and Donyaei-Darian (2018) found that supplementation with 0.5%, 1%, and 2% of the broiler chicken's diet with *S. platensis* had beneficial effect in elevating antioxidant status, decreasing stress hormones and enhancing humoral immunity response with no adverse effect on performance, therefore suggesting that supplementation with microalgae in heat stress conditions may be beneficial.

Other species of poultry have been subjected to microalgae supplementation in various studies. In Japanese quails, 0.5% of *S. platensis* in the diet improved FCR, BW gain, and breast percentage (Hajati & Zaghari, 2019). When Pekin duck feeds were supplemented with 0.1% and 0.2% of fermented *C. vulgaris*, final BW, BW gain, and feed intake improved (Oh et al., 2015). Furthermore, yellowness of duck meats, pH, and water-holding capacity of the breast meat were increased as *Chlorella* increased in the diet (Oh et al., 2015). In Muscovy ducks, dietary treatment with 0.5% of the microalgae *C. cohnii* did not induce differences in growth performances and slaughter traits (Schiavone et al., 2007). Similarly, chemical composition, color, pH, oxidative stability, and sensory characteristics of breast muscle were not influenced by the diet. However, there was a significant increase of DHA content in breast meat of ducks fed on *C. cohnii*-enriched diet (Schiavone et al., 2007).

In conclusion, according to the studies presented here, it is suggested that microalgae have the potential of largely being used in several species of poultry for the purpose of producing meat. While the addition of microalgae in large quantities (>15% of the diet) may have a detrimental effect on performance, lower doses seem to bring advantages in terms of meat quality without negatively affecting performance. Further studies are worth conducting in order to clarify why microalgae may have deleterious effects on performance when incorporated in larger doses.

4.2 Egg production

Table eggs are a rich source of high-value proteins and vitamins and can easily be manipulated with the purpose of valuable nutrient enrichment. Since food products rich in n-3 LC-PUFA are more beneficial for the metabolic and health status of the consumer, reducing the risk of chronic lifestyle diseases, microalgae supplementation in layers has been pursued with the goal of altering the fatty acid profile of the egg.

Research by Bruneel et al. (2013) established that feeding 5% and 10% of *N. gaditana* to layers significantly increased DHA in the yolk in comparison to eggs from layers from the control groups. Likewise, when levels as high as 23% of *Nannochloropsis oceanica* were added to laying hen diets, it was found a dose-dependent enrichment of DHA in the eggs, without impacting on the hen's health or performance (Manor, Derksen, Magnuson, Raza, & Lei, 2019). Similar results were demonstrated by Wu et al. (2019) with lower doses of *Nannochloropsis* sp. ranging between 1% and 8%. When hens were given levels between 0.5% and 3% of *Schizochytrium limacinum*, results revealed an enrichment with n-3 PUFA in eggs (Ao et al., 2015; Kralik et al., 2020) without negatively affecting performance. Furthermore, in a study by Ginzberg et al. (2000), lyophilized *Porphyridium* sp. biomass was added at 5% and 10% levels in the layer diet with a consequent increase in linoleic acid and arachidonic acid level in the egg yolk. Although the performance of the laying hen remained unaltered, feed intake of the hen was decreased and cholesterol levels reduced in the egg yolk (Ginzberg et al., 2000).

In fact, one of the consumer's main concerns when eating an egg is the cholesterolemic impact. Microalgae have the potential of reducing not only the saturated fatty acid but also egg yolk cholesterol content. In fact, the high contents of PUFA may be responsible for the reduction in cholesterol levels in the egg yolk (Park, Upadhaya, & Kim, 2015). DHA from microalgal sources has been found to reduce blood triglyceride levels and the triglyceride: HDL-cholesterol ratio, as well as shifts toward moderate elevations in HDL cholesterol levels (Holub, 2009). Several authors have reported a reduction in egg yolk cholesterol concentration with different levels of Spirulina ranging from 0.2% to 2% in the layer diet (Dogan et al., 2016; Mariey, Samak, & Ibrahem, 2012; Selim, Hussein, & Abou-Elkhair, 2018). On the other hand, there are also studies that indicate an absence of the effect of *Spirulina* in doses between 1.5% and 2.5% on egg yolk cholesterol (Omri et al., 2019; Zahroojian et al., 2013).

Egg yolk color is an aesthetic characteristic that is highly regarded by the consumer. Microalgae have a high concentration of beta-carotene that may intensify yolk color (Anderson, Tang, & Ross, 1991). Thus several researchers have looked into how microalgae addition in the layer feed may alter yolk coloration. To the best of our knowledge, so far, all experiments that incorporated microalgae such as *Porphyridium* sp., *S. platensis*, *S. limacinum*, *N. gaditana*, in the layer diet, revealed an alteration of the yolk color from yellow to orange-red regardless of the level of microalgae incorporation (Ao et al., 2015; Bruneel et al., 2013; Ginzberg et al., 2000; Omri et al., 2019; Zahroojian et al., 2013). Clearly, microalgae have the potential of replacing synthetic pigments in the hens' diet without detrimental effects on layer performance.

Microalgae are good biosorbents and can be enriched by binding microelements to its cell surface (Zielińska & Chojnacka, 2009). Eggs fortified in iron, zinc, and manganese were obtained from laying hens supplemented with copper and iron-enriched *S. maxima* (Saeid, Chojnacka, Opaliński, & Korczyński, 2016). In addition, eggshell strength was improved

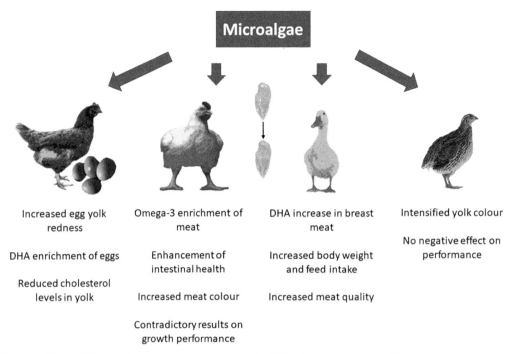

FIG. 2   The effects of dietary microalgae on poultry products (*DHA*, docosahexaenoic acid).

significantly with a lower number of cracked eggs and performance remained unaltered when enriched *Spirulina* was added to the layer diet in comparison to controls (Saeid et al., 2016).

The use of *S. platensis* has been tested in the layer Japanese quail diet with interesting results. Levels of *Spirulina* of 0.1%, 0.3%, and 0.5% in the diet decreased shell thickness, albumen height, and Haugh units in the eggs of laying quails. However, egg yolk color increased linearly and there was a significant decrease in cholesterol in the yolk (Hajati & Zaghari, 2019). Higher levels of up to 15% *S. platensis* in Japanese quail layer diets revealed a decrease in saturated fatty acids in the yolk and an intensity of the color of the yolk without altering egg production (Boiago et al., 2019).

Overall, there is clear evidence that microalgae can be safely used in the layer diet as a natural pigment agent of the egg yolk. Moreover, microalgae have the potential of improving the fatty acid profile in the egg and reducing egg cholesterol without detrimental effects on performance. There is a safe indication that microalgae may be utilized in the egg industry with clear advantages for the consumer. Fig. 2 summarizes the main results reported by reviewed studies in poultry.

5 Microalgae in rabbit

The rabbit is not only a very relevant experimental model but is also extensively used in animal production, specifically for the meat production industry in countries on the Mediterranean basin, such as Portugal, Spain, France, and Italy. Although feeding experiments evaluating alternative feedstuffs are widely available in rabbit, studies with microalgae are very scarce in the literature and mainly focus on *A. platensis*, *Chlorella*, and *Schizochytrium*, which has attracted attention due to its production of n-3 fatty acid (FA), particularly DHA (Madeira et al., 2017). Such studies are focused essentially on the use of microalgae to improve the health status and productive performances of rabbits. Additionally, studies concerning rabbit meat quality as affected by dietary microalgae are still limited, only explored in *Spirulina* by some authors. The compilation of the reviewed studies on the use of microalgae in rabbits is presented in Table 8.

Until now, the highest inclusion level of *Spirulina* in rabbit diets was studied by Peiretti and Meineri (2008, 2011). The treatments in these studies consisted of four groups of 10 rabbits aged 9 weeks, fed diets with increasing the level of *Spirulina* (0%, 5%, 10%, and 15%). They concluded that *Spirulina* diets had no detrimental effect on growth performance during the fattening period (9–13 weeks of age). This happened because of the increased feed consumption, which allowed to balance the lower digestibility of the *Spirulina*-containing diets. These authors found increased lipid content in perirenal fat and meat of rabbits with increasing dietary *Spirulina* content. For these rabbit tissues, the FA profile had significant differences (PUFA and n-6/n-3 increased; total n-3 FA decreased). The atherogenic and thrombogenic indexes were lower in the meat of the rabbits fed with *Spirulina* diets.

Gerencsér et al. (2014) studied the effect of *Spirulina* dietary supplementation using weaned rabbits, aged 35 days (42 rabbits/group). These rabbits were fed diets containing 5% of *Spirulina*, for 6 and 3 weeks before slaughter. Although without observing any effect on the productive performance or health status of rabbits, these authors concluded that the supplemented diets containing *Spirulina* reduced the nutritive value of the diets. Indeed, the authors highlight a lower total tract apparent digestibility for crude protein and mineral (Ca, P, and K) components, compared to conventional feedstuffs. Mousa et al. (2018) studied low levels of

incorporation in the rabbit's diets (0.1% and 0.2%) and concluded that growth performance and carcass traits were improved.

In the same animal experiment, Vántus et al. (2012) observed no effects on the production of volatile FA in the cecum. Dalle Zotte et al. (2014) proved that the inclusion of *Spirulina* improved the FA profile of rabbit meat, particularly of γ-linolenic acid content, which supports the results of Dal Bosco et al. (2014). The latter authors also observed no protective effect of dietary microalgae against lipid oxidation on rabbit meat. *Spirulina* constituents with antioxidant activity such as minerals and carotenoids would suggest enhanced oxidative stability of meat from rabbits fed with Spirulina. However, the authors detected no effect on the oxidative stability of rabbit meat. To explain this, authors suggest that either the *Spirulina* level in the diet was too low or that *Spirulina* antioxidants had not been absorbed from the gut (Dal Bosco et al., 2014; Dalle Zotte et al., 2014). Consistently, Mahmoud et al. (2017) report that increasing *Spirulina* content (3%, 6%, and 9%) in the diets of weaned rabbits aged between 6 and 14 weeks old generates increased PUFA content of meat, particularly regarding γ-linolenic and linoleic acids content.

Some authors used hypercholesterolemic rabbits to study the effects of 1% and 5% dietary inclusion of *Spirulina* and suggested that this microalga may be beneficial in preventing atherosclerosis, reducing risk factors for cardiovascular diseases, and protecting the cells from lipid peroxidation and oxidative DNA damage (Cheong et al., 2010; Kim et al., 2010). Moreover, these rabbits received a dose of alkaloid extract of *Spirulina* (33 or 66 mg/kg) to study its effect on the serum lipid profiles. Authors found decreasing levels of total cholesterol, triglycerides, and low-density lipoprotein and an increasing level of high-density lipoprotein (Kata et al., 2018).

The antioxidant, antiinflammatory, immunomodulatory activities of *Spirulina* were studied in rabbits by Abdel-Daim et al. (2019) and

TABLE 8 Summary of studies concerning the dietary utilization of microalgae in rabbit diets.

Microalgae	Inclusion level	Animal class/live weight/age	Objective	Main results	References
Arthrospira platensis	5%, 10%, and 15%	Growing rabbits with 2034 ± 174 g and 9 weeks old/1 week of adaptation followed by 24 days of experiment	Evaluate the effect of *Spirulina* (as a protein replacement for soybean meal) on growth performance, apparent digestibility, carcass characteristics and meat quality	Increased average daily feed intake (diet with 10%); digestibility coefficients decreased comparatively with control diet; lipid content of meat lower in control group; reducing in atherogenic and thrombogenic indexes in groups fed with *Spirulina* diets	Peiretti and Meineri (2008, 2011)
	5%	Weaned rabbits/35 days old to 11 weeks old	Evaluate the effect of *Spirulina* dietary supplementation on productive performance, health status, total tract apparent digestibility of nutrients, microbial diversity in cecum and cecal fermentation, meat quality, and protection against oxidative stress	*Spirulina* supplementation in growing rabbits reduced the nutritive value of the diets and significantly increased γ-linolenic acid content	Dal Bosco et al. (2014), Dalle Zotte et al. (2014), Gerencsér et al. (2014), and Vántus et al. (2012)
	3%, 6%, and 9%	Weaned rabbits/6–14 weeks old	Evaluate the effect of *Spirulina* dietary supplementation on performance, blood parameters, and meat quality of growing rabbits	No effect on growth performance, kidney and liver functions and immune response. The increasing *Spirulina* inclusion increased the protein and polyunsaturated fatty acids of meat	Mahmoud, Naguib, Higazy, Sultan, and Marrez (2017)
	0.1% and 0.2%	Weaned rabbits with 5 weeks old/8 weeks of experiment	Evaluate the effect of *Spirulina* on growth performance and carcass traits of growing rabbits	Growth performance and carcass traits were significantly enhanced in rabbits fed *Spirulina* compared to the control group	Mousa, Abdel-Monem, and Bazid (2018)
	1% and 5%	Male growing rabbits between 2.0% and 2.5 kg/6 weeks of adaptation followed by 8 of experiment	Evaluate the effect of *Spirulina* on the prevention of atherosclerosis, tissue lipid peroxidation and oxidative DNA damage in hypercholesterolemic rabbits	Dietary *Spirulina* reduced blood cholesterol levels and protected the cells from lipid peroxidation and oxidative DNA damage	Cheong et al. (2010), Kim et al. (2010)

Species	Dose	Animal/duration	Objective	Findings	Reference
	33 and 66mg/kg (alkaloid extract of Spirulina)	Male rabbits with an initial weight of 1500–1600 g/4 weeks of experiment	Evaluate the effect of the alkaloid extract of Spirulina on the serum lipid profile of hypercholesterolemic rabbits	Alkaloid extract of Spirulina decreased the levels of cholesterol, triglycerides and low-density lipoprotein and increased the level of high-density lipoprotein in the serum of hypercholesterolemic rabbits	Kata, Athbi, Manwar, Al-Ashoor, Abdel-Daim, and Aleya (2018)
	300 and 600 mg/rabbit/d (water supplemented)	Nulliparous female rabbit at 16–18 weeks old/4 week of experiment	Evaluate the effect of Spirulina on reproductive performance, hematological and biochemical parameters, antioxidant activity, and liver and kidney histogenesis of Egyptian APRI Rabbit Line females	Spirulina improved reproductive performance, blood constituents, antioxidative status without adverse effects on liver and kidney functions	El-Ratel (2017)
	700 mg/rabbit/d (water supplemented)	Bucks with 6–7 month/5 weeks of experiment	Evaluate the effect of oral administration of Spirulina on reproductive performance of buck rabbits	Spirulina improved semen production of buck rabbits and fertility of inseminated doe rabbits	Fouda and Ismail (2017)
	500 mg/kg LW	8 weeks old/1 week of adaptation/7 days of experiment	Evaluate the efficiency of Spirulina in protecting against amikacin (AMK)-induced nephrotoxicity in rabbits	Spirulina minimized the nephrotoxic effects of AMK through its antioxidant and anti-inflammatory activities	Abdel-Daim et al. (2019)
	0.05%, 0.1%, 0.15%	Weaned rabbit with 35 days old/8 weeks of experiment	Evaluate the effect of dietary Spirulina against subchronic lead toxicity in rabbits	In Pb intoxicated rabbits Spirulina improved lipid profile, oxidative parameters, growth performance, immune status and reduced the Pb residual concentration in muscle tissues	Aladaileh et al. (2020)
Chlorella vulgaris	0.5%, 1.0%, and 1.5%	Weaned rabbits/5–13 weeks old	Evaluate the effect of dietary Chlorella on growth performance, carcass traits, hematobiochemical variables and immunity responses of growing rabbits	No effect on growth performances; Dietary supplementation with Chlorella up to 1% had the potential to enhance immune responses and antioxidant status, as well as reduce blood lipid accumulation	Abdelnour et al. (2019)

Continued

TABLE 8 Summary of studies concerning the dietary utilization of microalgae in rabbit diets —cont'd

Microalgae	Inclusion level	Animal class/live weight/age	Objective	Main results	References
	200, 300, 400, and 500 mg/kg LW/d	Female growing/ 14 days of adaptation	Evaluate the effect of dietary supplementation of Chlorella on growth and lactating performances and serum oxidative stress in rabbits	Supplementation of Chlorella improved growth and lactating performances of rabbits through reduction of oxidative stress	Sikiru, Arangasamy, Alemede, et al. (2019), Sikiru, Arangasamy, Ijaiya, et al. (2019), and Sikiru, Arangasamy, Alemede, Guvvala, et al. (2019)
Arthrospira platensis and Chlorella vulgaris	0.075% and 0.15%	Weaned rabbits with 5 weeks old/7 weeks of experiment	Evaluate the effect of dietary supplementation of Spirulina and Chlorella on growth performance, digestibility, and serum biochemical parameters of rabbit	Both levels of Spirulina in diets improved growth performance and reduced liver enzyme activities, cholesterol, and total lipids concentration in blood comparatively to control and chlorella diets	Hassanein, Arafa, Abo Warda, and Abd-Elall (2014)
Schizochytrium sp.	180, 600, and 1800 mg/kg LW/d (oral gavage)	Rabbits with 6 months old/13 days of experiment	Evaluated the toxicity of Schizochytrium sp. in rabbits	No signals of toxicities were observed at any dose levels studied	Hammond et al. (2001)
	0.4%	Primiparous does and their 1184 kits/ 35 d of gestation + 28 d of lactation + 28 d of fattening	Evaluated the use of Schizochytrium spp. in the diets of does during gestation and lactation on reproductive and zootechnical parameters and the nutritional properties of the meat obtained from their kits	In primiparous does, the presence of microalga in the diet negatively influenced zootechnical performances. In lactating and growing-fattening rabbits the dietary supplementation influenced positively the quality of muscle lipids (loin and thigh)	Mordenti et al. (2010)

Aladaileh et al. (2020). The former authors suggested that *Spirulina* (500 mg/kg LW) minimized the nephrotoxic effect of amikacin. The latter authors noted that 0.05%, 0.1%, and 0.15% inclusion of *Spirulina* in the diets promoted antioxidant, antiinflammatory, and immunestimulatory properties against lead-induced toxicity in rabbits.

Regarding the effect of dietary microalgae on reproductive performance, there are two studies that study the oral administration of a solution containing *Spirulina*. El-Ratel (2017) reported the potential of this microalga as a natural antioxidant for protecting the doe against free radical-induced cellular damage under oxidative stress, resulting in improved reproductive performance. Fouda and Ismail (2017), with the same administration method and a similar level of *Spirulina*, found that semen production improved in rabbit bucks.

Studies using *C. vulgaris* as a supplement in diets reported that 0.5%, 1%, and 1.5% inclusion improved growth performance and protected growing rabbit against oxidative stress damage (Abdelnour et al., 2019). The remaining authors described that 200, 300, 400, and 500 mg/kg LW incorporation levels improved the lactating performance of the does and the growth performance of their litters (Sikiru, Arangasamy, Alemede, Egena, & Bhatta, 2019; Sikiru, Arangasamy, Alemede, Guvvala, et al., 2019; Sikiru, Arangasamy, Ijaiya, Ippala, & Bhatta, 2019).

Hassanein et al. (2014) studied the effect of dietary supplementation of *Spirulina* and *C. vulgaris* at two different levels (0.075% and 0.15%) in growing rabbits. Authors found that *Spirulina* supplemented groups had improved growth performances and reduced liver enzyme activities, as well as cholesterol and total lipids concentration in blood of *Chlorella*-fed rabbits compared to control groups.

Lastly, *Schizochytrium* sp. was studied in rabbits with the purpose of developing toxicity studies by Hammond et al. (2001). These authors observed that there were no toxicity effects at any inclusion level studied (180, 600, and 1800 mg/kg LW). Mordenti et al. (2010) evaluated the 0.4% inclusion of this microalga in the diets of rabbits does during gestation and lactation.

Most studies concerning the use of microalgae in rabbit diets justify their use as functional feeds to feed these animals (Fig. 3). Moreover, several studies have reported an enhanced antioxidative status as well as an enhanced inflammatory response as a result of dietary microalgae. Thus these feeds provide a means to improve the meat fatty acid profile while improving animal health, which is particularly important in this sensitive species.

6 Microalgae in diets for relevant species for aquaculture

Aquaculture is the fastest-growing animal food production sector (annual growth rate of 5.8% between 2001 and 2016) and is becoming the main source of aquatic animal food (fish, crustaceans, and mollusks). The main groups of species farmed worldwide are carps and cyprinids, tilapia, shrimp and prawns, salmon, and trout. In Europe, salmonids (Atlantic salmon and rainbow trout), bivalves (oysters and mussels), and Mediterranean marine species (European seabass and gilthead seabream) have the largest production volume and economic value (EUMOFA, 2018). Fish and seafood are a highly nutritious food, rich in protein, minerals, and vitamins, being the main natural source of long-chain omega-3 fatty acids (n-3 LC-PUFA) with recognized health benefits. In 2015 global food fish consumption peaked at about 20.2 kg per capita, and it is expected to increase even further (FAO, 2018). The growth of the world population with concomitant economic development has increased the demand for fish products and driven aquaculture impressive growth during the last decades.

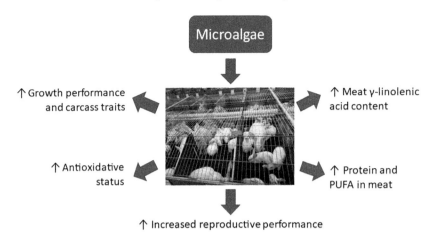

FIG. 3   Main effects found for dietary microalgae on rabbit diets (*PUFA*, poly-unsaturated fatty acids).

Compared to other livestock production sectors, aquaculture represents one of the most efficient methods to convert feed into edible protein (Fry, Mailloux, Love, Milli, & Cao, 2018). However, its fast growth is heavily dependent on the global feed market. Fish meal (FM) and fish oil (FO) have traditionally been the major protein and oil sources for fish and shrimp diets, but due to overexploitation of wild stocks, their reduction in aquafeeds has been a priority goal for the further expansion and sustainability of the aquaculture sector (FAO, 2018). The feed industry has searched for suitable alternative ingredients to partially or totally replace unsustainable fish-based ingredients without compromising animal growth and welfare (Naylor et al., 2009). These mainly include conventional vegetable sources (e.g., soybean, corn, wheat, canola) and rendered animal by-products (e.g., feather meal, blood meal, poultry meal, seafood byproducts), even though single-cell products (bacteria, yeast, protists, and microalgae) are emerging at a fast pace. Plant substitutes are widely available in the market, but besides directly competing with other feed and food production sectors, they have antinutritional factors, lower digestibility, are deficient in certain essential amino acids, and totally lack

HUFAs, ultimately affecting the growth potential, nutritional quality, and fish health (Gatlin et al., 2007). Rendered by-products have high nutritional value and lower environmental impact than FM and FO (Campos, Valente, Matos, Marques, & Freire, 2020), but safety and regulatory constraints, as well as consumer acceptance, limit their inclusion levels in aquafeeds (Tacon, Hasan, & Metian, 2011). The nutritional value of single-cell products is highly dependent on their growth conditions (autotrophy, heterotrophy, and mixotrophy), which can be manipulated to obtain high-valued biomass (Jones, Karpol, Friedman, Maru, & Tracy, 2020). However, their high production costs, limited production capacity, low digestibility, and regulatory aspects still hamper their inclusion as a macronutrient source, being mostly used as a dietary supplement ($<3\%$ inclusion) in commercial feeds for farmed fish and shrimp (Naylor et al., 2009; Shah et al., 2018). Major technological advances in production systems, including large-scale photobioreactors and fermenters, are underway and may soon result in increased cost-effectiveness of production potentiating such as "next-generation" ingredients into the market. On the other hand, increasing consumer demand for products containing food ingredients from natural sources encourages the use of environ-friendly alternatives to FM and FO able to produce a final product that retains adequate levels of HUFAs with multiple benefits to human health (Simopoulos, 1999).

Microalgae are primary producers of HUFAs (i.e., eicopentaenoic acid [EPA] and decosahexaenoic acid [DHA]) and carotenoids, namely, astaxanthin, being part of the natural diets of various fish species. Several strains have been identified as the most promising alternative sources to FO due to their fatty acid content, being also rich sources of protein with a suitable amino acid profile able to partially replace FM (Shah et al., 2018). Moreover, the large-scale cultivation of microalgae for bioenergy purposes generates high amounts of defatted coproducts, which are valuable protein sources for animal feed (Lum et al., 2013). The most relevant and recent developments concerning the use of microalgae in diets for larvae and juveniles, of species with importance for the aquaculture sector, have been summarized in Table 9.

6.1 Microdiets for larvae

Nowadays, microalgae are utilized regularly in hatcheries for several purposes, as a larval feed (Helm, Bourne, & Lovatelli, 2004) or broodstock conditioning (Anjos et al., 2017) for bivalves and penaeid shrimp; to improve live preys (rotifers and Artemia) quality since their nutritional composition is considered suboptimal for fish and shrimp larvae growth (Conceição, Yúfera, Makridis, Morais, & Dinis, 2010; Hamre et al., 2013); or to improve marine fish larvae production protocols through the "green water" technique (Nicolaisen & Bolla, 2016; Rocha, Ribeiro, Costa, & Dinis, 2008). The addition of microalgae to larval rearing tanks improved positive phototaxis in cod *Gadus morhua* larvae at early developmental stage (Nicolaisen & Bolla, 2016), and feeding ability in Senegalese sole *Solea senegalensis* and gilthead seabream *Sparus aurata* larvae (Rocha et al., 2008).

The inclusion of microalgae as a whole in hatchery feeds is still in its infancy, mainly because most marines reared species are altricial and unable to digest a microdiet due to biological constraints like an immature digestive system (Canada, Engrola, Conceição, & Valente, 2019). Contrarily to freshwater fish species, which are precocial and have the capability of fatty acid elongation and desaturation from shorter chain fatty acids (LA and LNA), marine fish larvae require HUFAs (DHA, EPA, and ARA) in their diet for proper development (Sargent, Mcevoy, & Bell, 1997). Many microalgae species are able to synthesize these HUFAs, so their inclusion as algal biomass or microalgal lipids in microdiet formulation mainly to replace FO is quite appealing. In addition, there is a lack of studies focused on FM replacement in microdiets, probably due to the high protein content in larval diets (>55%) and lower digestive capacity as already mentioned. The partial (15%–29%) replacement of FO with *C. cohnii* in microdiets for gilthead seabream larvae improved survival (Atalah et al., 2007), while total (100%) replacement of FO had a detrimental effect on survival and body weight (Ganuza et al., 2008). In shrimp larvae, an inclusion level of 4% *Nannochloropsis* (Adissin et al., 2020) or 2% *Schizochytrium* (Wang et al., 2017b) promoted growth (Table 9). In the latter study, the positive effect of microalgae was absent in higher inclusion levels. Nevertheless, these few studies in gilthead seabream (Atalah et al., 2007;Eryalçin et al., 2015; Ganuza et al., 2008), goldfish *Carassius auratus* (Coutinho et al., 2006), and shrimp *Marsupenaeus japonicos* and *Litopenaeus vannamei* (Adissin et al., 2020; Wang et al., 2017a, 2017b) provided insights regarding a positive modulation in larvae quality, mostly addressing survival and growth rates, after microalgae inclusion in microdiets. In gilthead seabream larvae, the dietary inclusion of *C. cohnii* (Eryalçin et al., 2015; Ganuza et al., 2008), *Nannochloropsis* (Eryalçin et al., 2015), or *Schizochytrium* (Ganuza et al., 2008) for total replacement of FO

TABLE 9 ajor dietary effects of microalgae biomass (MA) on growth, nutrient utilization, and body composition of fish and shrimp.

Algae	Animal model and size	Form: Whole algae vs extract	Inclusion level (%)	Protein/oil replacement (%)	Trial duration	Biological effect			References
						Growth	Nutrient utilization	Whole body composition; flesh quality	
Arthrospira (Spirulina)	Gibel carp (C. auratus); 15–31 g	Whole A. platensis	0–3–7–14	0%–25%–50%–100% FM replacement	46 days	↑Final weight, FI, SGR in MA3 and MA7 ↔FE, Survival	↔PRE (% intake)	–	Cao, Zhang, et al. (2018)
	Gibel carp (C. auratus); 5–12 g	Whole A. platensis	0–10–15–20	0%–50%–75%–100% FM replacement	60 days	↔Final weight, SGR, FI, K ↑FE in MA10	↔PER ↑Digestibility of DM, Protein, energy and P in MA diets ↔Trypsin, amylase & lipase activity in MA groups	↔Protein and lipid content in whole body and muscle ↑Whole body moisture in MA15 and MA20	Cao, Zou, et al. (2018)
	Nile tilapia (O. niloticus); 30 g	Whole A. platensis	0–15	–	9 weeks	–	–	↑Whole body fat and ash content in MA15	El-Habashi et al. (2019)
	Pacific white shrimp (L. vannamei); 3–6 g	Whole A. platensis	0–8–16–23–30	0%–25%–50%–75%–100% FM replacement	8 weeks	↔Weight gain, SGR, FCR and survival ↑Survival after an hypoxia challenge in MA groups	↔Amylase and lipase ↑Trypsin and chymotrypsin activity in MA16 ↓Chymotrypsin in MA30	↔Whole body protein and lipid content ↑PUFA and n-3 PUFA in whole body of MA groups ↑DHA in whole body of MA 8 and MA16	Pakravan, Akbarzadeh, Sajjadi, Hajimoradloo, and Noori (2017)
	Rainbow trout (Oncorhynchus mykiss); 101–235 g	Whole A. platensis	0%, 2.5%, 5%, 7.5% and 10%	0%, 6%, 12%, 18%, 24% FM replacement	10 weeks	↔Final weight, weight gain, SGR, FCR	–	↓Whole body protein ↓Whole body lipids in MA5 and MA 10 ↓Fillet Luminosity (L*) ↑Fillet a* (redness) and b* (yellowness) in MA2, MA3 and MA4 ↓Fillet Hue in MA groups ↑Carotenoid concentration in fillets and skin ↔Skin luminosity ↑Skin a* (redness) in MA	Teimouri, Amirkolaie, and Yeganeh (2013) and Teimouri et al. (2019)
	Red seabream (Pagrus major); 18–32 g	Whole Spirulina meal	3	0% (used to replace cellulose)	41 days	↔Final weight, FCR, HSI	↓Serum total lipids; ↔Serum protein and glucose ↑Carnitine Palmitoyltransferase; ↔Fatty acid synthase; arginase & Glucose 6-phosphate	↔Muscle moisture, protein & lipid ↔Intraperitoneal fat Liver lipids	Nakagawa, Mustafa, Takii, Umino, and Kumai (2000)

Genus	Species	Form	Inclusion (%)	FM replacement	Duration	Growth	Feed utilization	Body composition	Reference
	Siberian sturgeon (Acipenser baerii); 92–605 g	Whole Spirulina	0-40-50-60	0%–63%–76%–89% FM replacement	12 weeks	↓FCR in MA groups; ↑TGC in MA 40 and MA50; ↑Biomass gain in MA groups; ↔HSI, liver lipids	↑PER in MA groups; ↔Dry matter and protein ADC	↔Fillet composition (protein, lipids, energy, n-3 PUFA, SFA); ↑PUFA in MA groups; ↓MUFA	Palmegiano et al. (2005)
Chlorella	Common carp (Cyprinus carpio): 8–18 g	Whole Chlorella sp.	0-15	–	60 days	↔Final weight, SGR, HIS & VSI in all groups	↔PER	–	Yadav et al. (2020)
	Crucian carp (Carassius auratus); 3–15 g	Whole Chlorella meal	0-53-71	0%–75%–100% FM replacement	8 weeks	↑Final weight, SGR, FI in MA groups; ↓FCR in MA groups; ↓HSI and Myog in MA71; ↓K in MA 53; ↑Liver lipids in MA71; ↑Muscle Myod mRNA in MA71; ↑Muscle mRNA of Mrf4 and myf5 in MA53	↑PER in MA groups; ↑Amylase activity in MA71	↔Fillet dry matter, protein, lipid and ash	Shi et al. (2017)
	Nile tilapia, (Oreochromis niloticus); 30 g	Whole C. vulgaris	0-15	–	9 weeks	–	–	↑Whole body protein and lipids in MA15; ↓Whole body moisture in MA15	El-Habashi et al. (2019)
	Red drum (Sciaenops ocellatus); 5 g	Defatted C. vulgaris	0-10-20-40-45	0%–2.5%–5%–10%–12.5% FM protein	7 weeks	↓Weight gain in MA40 and MA45; ↓survival in MA45; ↓FE in MA20, MA40 and MA45; ↔HIS	↓PER in MA20, MA40 and MA45	↔Whole body composition	Patterson and Gatlin (2013)
Cryptecodinum	Gilthead seabream (Sparus aurata) postlarvae; 73 mg	Whole, broken cells C. cohnii	2-4	15%–29% FO replacement	57 days	↑Survival in all MA groups; ↔Final weight, SGR, Biomass in all MA groups	–	↔PUFA, DHA, EPA, ARA in all MA groups	Atalah et al. (2007)
	Gilthead seabream (S. aurata) larvae; 0.06 mg	Whole C. cohnii	8	100% FO replacement	17 days	↔TL, FBW, Survival	–	↔Whole body composition	Eryalçin, Ganuza, Atalah, and Cruz (2015)
	Gilthead seabream (S. aurata) larvae; 0.5 mg	Whole, broken cells C. cohnii	2	100% FO replacement	15 days	↔Survival, FBW, TL in all MA groups	–	↔DHA, ARA, DHA: EPA, DHA;ARA; ↓total PUFA, EPA	Ganuza et al. (2008)
	Gilthead seabream (S. aurata) larvae; 0.75 mg	Whole, broken cells C. cohnii	16	100% FO replacement	21 days	↓Survival in all MA groups; ↔Final weight, total length	–	↔DHA, DHA:EPA, DHA;ARA; ↓EPA, ARA, total PUFA	Ganuza et al. (2008)

Continued

TABLE 9 ajor dietary effects of microalgae biomass (MA) on growth, nutrient utilization, and body composition of fish and shrimp—cont'd

Algae	Animal model and size	Form: Whole algae vs extract	Inclusion level (%)	Protein/oil replacement (%)	Trial duration	Biological effect — Growth	Biological effect — Nutrient utilization	Biological effect — Whole body composition; flesh quality	References
Desmodesmus	Atlantic salmon (Salmo salar); 165–360 g	Defatted biomass Desmodesmus sp.	0–10–20	0%–9%–18% FM replacement	70 days	↔ SGR, HSI and VSI in all groups; ↓ Final weight in MA10; ↑ FCR in MA20	↔ PER; ↔ Protein and Lipid ADC; ↓ Energy ADC in MA10	↔ Whole body composition; ↓ Fillet lipids in MA10; ↓ Fillet ash in MA20	Kiron et al. (2016)
Haematococcus	Longfin yellowtail (Seriola rivoliana); 2.5–83 g	Defatted H. pluvialis	0–15–10–6–2	0%–25%–40%–60%–80% FM replacement	9 weeks	↔ Final weight, SGR, FCR, HSI, Survival rate	↓ Retained N in MA6 and MA2 where FM was replaced by 60% and 80%, respectively	↔ Whole body protein; ↓ SFA, PUFA, n-3 PUFA and DHA in fillet of MA fish; ↑ MUFA in fillet in MA	Kissinger, García-Ortega, and Trushenski (2016)
	Pacific white shrimp (Litopenaeus vannamei); 1–11 g	Defatted H. pluvialis	0–3–6–9–12	0%–13%–26%–38%–58% FM replacement	8 weeks	↑Final weigh in MA3; ↑ SGR in MA3 and MA9	↑ PER in MA3, MA6 and MA12	↔ Whole body protein and lipid composition; ↔ Whole body micro and macro-minerals composition; ↑ Astaxanthin content in MA groups	Ju et al. (2012)
	Rainbow trout (Oncorhynchus mykiss); 96–250 g	Pressure-disrupted H. pluvialis	50 ppm astaxanthin from distinct sources (Carophyll Pink® vs MA)	—	10 weeks	↔ SGR, FCR	↔ ADC and net apparent retention of dietary carotenoids and dietary astaxanthin	↔ Total carotenoid content in muscle; ↓ Astaxanthin concentration in muscle of fish fed MA	Young, Pritchard, White, and Davies (2017)
	Yellow perch (Perca flavescens); 13–38 g	Defatted H. pluvialis	0–5–10–15	0%–25%–50% FM replacement by MA + Soy protein (1:1)	8 weeks	↓ Final weight, HSI, VSI in MA10 and MA15; ↑ FCR in MA15; ↔ VFI	↓ PER in MA10 and MA15; ↔ PRE (% intake)	↔ Whole body protein and lipid content in MA groups; ↓ Ash in MA15	Jiang et al. (2019)
Isochrysis	European seabass (Dicentrarchus labrax); 140–290 g	Whole Isochrysis sp.	0–7–14	0%–10%–20% FM protein; 0%–18%–36% FO	121 days	↔ Final weight, SGR, FCR; ↑ FI in MA14; ↔ HIS, VSI	↔ PER; ↓ Lipid & energy ADC in fish fed MA14	↔ Fillet lightness (L*), yield and hardness; ↑ SFA in fillet MA7 and MA14; ↓ n-3 PUFA in fillet MA14; ↑ Redness (a*) and b* in Fillet MA; ↓ a* (increased greenness) in skin A14; ↑ Hue in skin A14	Tibaldi et al. (2015)

Fish species; size	Microalgae form/species	Inclusion level (%)	Replacement	Duration	Growth performance	Feed utilization / physiology	Body composition	Reference
(C. auratus) larvae; 3.4 mg	I. galbana		replacement		groups ↓ SL, Biomass in all MA groups			(2006)
Nannochloropsis								
Atlantic salmon (S. salar L.); 200–400 g	Defatted N. oceanica	0–10–20	0%–10%–20% FM replacement	84 days	↔ Final weight, SGR, FI, FCR and hematocrit in MA20; ↓ K in MA20; ↔ HSI	↑ PER; ↓ Lipid and energy retention in MA20; ↓ Protein, Lipid & Energy ADC; ↑ Ash ADC in MA 10 & 20	↔ Whole body and fillet composition	Sørensen et al. (2017)
Atlantic salmon (S. salar L.); 230–420 g	Extruded N. oceanica	0–10	0%–7.5% FM replacement	68 days	↔ Final weight, SGR, FCR, ↔ K; ↔ HSI, VSI	↔ PER; ↑ Dry matter & ash ADC in MA; ↔ Protein ADC; ↓ Lipid ADC in MA; ↑ Intestinal cell proliferation index in MA; ↔ Villi height & width	↔ Whole body composition; ↔ Fillet PUFA and EPA+DHA	Gong et al. (2020)
European seabass (D. labrax); 30–83 g	Whole N. oceanica.	0–8	0%–20.6 % FM replacement	106 days	↔ Final weight, SGR, FCR, VFI; ↓ HSI in MA8; ↔ Intestinal morphology	↔ Nutrient and energy ADCs; ↔ Nutrient and energy retention efficiency & gain; ↓ Sucrose-isomaltase & alkaline phosphatase activities in MA8 pyloric caeca; ↓ Maltase & sucrose-isomaltase activities in MA8 anterior intestine; ↓ Plasma triglycerides	↔ Whole body composition and fillet protein and fat content; ↔ Skin and fillet colour; ↑ Fillet cohesiveness; ↓ Fillet springiness	Batista, Pereira, et al. (2020)
European seabass (D. labrax); 20–50 g	Defatted Nannochloropsis sp.	0–5–10–15	0%–3.1%–6.2%–9.4% FM replacement	93 days	↔ Final weight, DGI, FI; ↑ FCR MA15; ↔ HSI and VSI	↓ Energy ADC in fish fed MA; ↔ Nutrient retention & gain	↔ Whole body composition	Valente et al. (2019)
European seabass (D. labrax); 13–36 g	Whole Nannochloropsis sp.	0–4–8	0%–50%–100% FO replacement	56 days	↔ Final weight, SGR, K; ↓ FCR in MA8; ↔ HSI and liver histology	↔ PER	↔ Whole body lipid and protein composition; ↑ WB energy in MA8; ↔ MUFA, PUFA and DHA in fillets; ↑ EPA in MA fillets	Haas et al. (2016)
Gilthead seabream (S. aurata); 100 g	Whole N. gaditana	0–5–10	–	4 weeks	↔ SGR	–	–	Cerezuela, Guardiola, Meseguer, and Esteban (2012)

Continued

TABLE 9 ajor dietary effects of microalgae biomass (MA) on growth, nutrient utilization, and body composition of fish and shrimp—cont'd

Algae	Animal model and size	Form: Whole algae vs extract	Inclusion level (%)	Protein/oil replacement (%)	Trial duration	Biological effect			References
						Growth	Nutrient utilization	Whole body composition; flesh quality	
	Gilthead seabream (S. aurata) larvae; 0.06 mg	Whole N. gaditana	11	100% FO replacement	17 days	↔ Survival ↓ TL, FBW	–	↓ Protein, Lipid	Eryalçin et al. (2015)
	Kuruma shrimp (Marsupenaeus japonicus) Zoea I	Whole Nannochloropsis sp.	0–1–4–7	0%–4%–14%–24% FO replacement	14 days	↔ Survival, TL ↑ FBW in MA4 and MA 7 ↓ DS, PL in MA 4 and MA 7	–	–	Adissin et al. (2020)
	Kuruma shrimp (Marsupenaeus japonicus) Zoea I	Nannochloropsis sp. lipids	0–1–3–6	0%–17%–58%–97% FO replacement	14 days	↔ Survival, TL, DS ↓ FBW in MA groups ↓ PL in MA1 and MA6			Adissin et al. (2020)
	Kuruma shrimp (M. japonicus) postlarvae; 0.07 g	Whole Nannochloropsis sp.	0–1–4–7	0%–3%–14%–24% FO replacement	37 days	↔ Survival, FE, FI ↓ FBW in MA1 and MA7 ↓ SGR in MA1		↔ Whole body protein, lipid, SFA ↓ Whole body MUFA in neutral lipid fraction in MA4 and MA7 ↑ Whole body PUFA and n-3 PUFA in neutral lipid fraction of MA groups	Adissin et al. (2020)
	Kuruma shrimp (M. japonicus) postlarvae; 0.07 g	Nannochloropsis sp. lipids	0–2–8–14	0%–3%–14%–24% FO replacement	37 days	↔ Survival, FE, FI ↓ Final weight in MA8 and MA14 ↓ SGR in MA8	–	↔ Whole body protein, lipid, and SFA ↓ MUFA in neutral lipid fraction in MA groups ↑ Whole body PUFA and n-3 PUFA in neutral lipid fraction of MA groups	Adissin et al. (2020)
	Red drum (S. ocellatus); 2g	Defatted N. salina	0–5–7.5–10–15	0%–2.5%–3.8%–5%–7.5% FM and protein replacement	8 weeks	↓ Weight gain in MA5, MA10, MA15 ↑ Weight gain in MA 7.5 ↓ Survival and FE in MA15 ↔ HSI	↓ PER in MA10 and MA15	↔ Whole body moisture and protein composition ↓ Whole body Lipids in MA15	Patterson and Gatlin (2013)

Microalgae	Species	Form	Inclusion (%)	Replacement	Duration	Growth	PER/ADC	Composition	Reference
Nanofrustulum	Atlantic salmon (*S. salar* L.); 62–270 g	Defatted *Nanofrustulum* sp.	0–9–17	0%–5%–10% FM protein replacement	12 weeks	↔ Final weight, SGR, FI, FCR	↔ PER	↔ Whole body and fillet composition	Kiron et al. (2012)
	Common carp, (*C. carpio*); 10–74 g	Defatted *Nanofrustulum* sp.	0–20–32	0%–25%–40% FM protein replacement	12 weeks	↔ Final weight, SGR, FI, FCR	↔ PER	↔ Whole body and fillet composition	Kiron et al. (2012)
	Whiteleg shrimp (*L. vannamei*); 2–11 g	Defatted *Nanofrustulum* sp.	0–22–36	0%–25%–40% FM protein replacement	9 weeks	↔ Final weight, SGR, FI, FCR	↔ PER	↔ Whole body composition	Kiron et al. (2012)
Navicula	Red drum (*Sciaenops ocellatus*); 13 g	Whole *Navicula* sp.	0–10–20	0%–2.5%–5% of FM and protein	7 weeks	↔ Weight gain, FCR, HSI	↔PER, protein retention and energy ADC; ↓ Protein ADC in MA10	↔ Whole body composition	Patterson and Gatlin (2013)
	Red drum (*S. ocellatus*); 13 g	Defatted *Navicula* sp.	0–15–30	0%–2.5%–5% of FM and protein	7 weeks	↔ Weight gain, FCR, HSI	↔PER, ↓ Protein and energy retention in fish fed MA30; ↓ Protein ADC in MA fed fish; ↓ energy ADC in MA30	↔ Whole body composition	Patterson and Gatlin (2013)
Pavlova (*Diacronema viridis*)	European seabass (*D. labrax*); 13–36 g	Whole *P. viridis*	0–10–20	0%–50%–100% FO replacement	56 days	↑ Final weight and SGR in MA groups; ↓ FCR in MA groups; ↔ HSI and liver histology	↔ PER	↔ Whole body composition; ↓ SFA in fillets of MA10; ↔ Fillet MUFA and DHA; ↑Fillet PUFA and EPA in MA groups	Haas et al. (2016)
Phaeodactylum	Atlantic salmon (*S. salar* L.); 325–560 g	Whole *P. tricornutum*	0–3–6–12	3%–7%–14% FM replacement	82 days	↔ Final weight, SGR, FI, FCR	↓ Starch ADC in MA6; ↔ Lipid, N, MUFA, PUFA and energy ADC; ↔ N, lipid and energy retention	↔ Whole body Lipid, Fatty acids, protein and energy	Sorensen, Berge, Reitan, and Ruyter (2016)
	Gilthead seabream (*S. aurata*); 100 g	Whole *P. tricornutum*	0–5–10	—	4 weeks	↔ SGR	—	—	Cerezuela et al. (2012)
	Gilthead seabream (*S. aurata*) postlarvae; 73 mg	Whole, Broken cells *P. tricornutum*	13	76% FO replacement	57 days	↔ Survival, Final weight, SGR. Biomass	—	↔ Whole body composition	Atalah et al. (2007)

Continued

TABLE 9 ajor dietary effects of microalgae biomass (MA) on growth, nutrient utilization, and body composition of fish and shrimp—cont'd

Algae	Animal model and size	Form: Whole algae vs extract	Inclusion level (%)	Protein/oil replacement (%)	Trial duration	Biological effect — Growth	Nutrient utilization	Whole body composition; flesh quality	References
Scenedesmus	Gilthead seabream (S. aurata); 8–25 g	Whole S. almeriensis	0–12–20–25–40	16%–23%–31%–47% FM 6%–13%–16%–25% FO replacement	45 days	↑ Final weight, SGR in MA20 ↔ Final weight, SGR in MA12-25-40 ↔ FCR, FI, K ↓ HSI in MA40	↔ PER ↑ Trypsin activity in MA12 ↑ Absorptive capacity (Leucine aminopeptidase & alkaline phosphatase) in MA20	–	Vizcaíno et al. (2014)
Schizochytrium	Atlantic salmon (S. salar); 213–850 g	Whole Scizochytrium sp. + yeast extract	0–1–6–15	0%–7%–40%– 100% FO replacement	12 weeks	↔ Final weight, K, FI, FCR, TGC, HSI	↔ PER ↓ ADC of lipid in MA6 and MA15 ↓ Energy ADC and ADC of SFA and PUFA in MA15	↔ Gaping score, Firmness and Liquid loss in fillet ↔ Total lipids in fillet, PUFA and DHA+EPA	Kousoulaki et al. (2015)
	Gilthead seabream (S. aurata) larvae; 0.5 mg	Whole, Broken cells Schizochytrium sp.	2.5	100% FO replacement	15 days	↔ Survival, Final weight, TL	–	↔ DHA, ARA, total PUFA DHA:EPA, DHA:ARA ↓EPA	Ganuza et al. (2008)
	Gilthead seabream (S. aurata) larvae; 0.75 mg	Whole, Broken cells Schizochytrium sp.	21	100% FO replacement	21 days	↓ Survival, Final weight, TL	–	↔ DHA, DHA:EPA, DHA:ARA ↓ EPA, total PUFA ↑ARA	Ganuza et al. (2008)
	Pacific white shrimp (L. vannamei) larvae; 4.2 mg	Schizochytrium sp.meal	2–4–6	29%–61%–92% FO replacement	24 days	↑ Final weight and SGR in MA2 ↔ TL, S	–	↔ C16:0, C18:0, total saturated fatty acids, C18:2n-6, DHA, total n-3 fatty acids, total n-6 fatty acids, total PUFA and n-3/n-6	Wang et al. (2017b)
	Pacific white shrimp (L. vannamei); 4.2 mg	Schizochytrium sp.meal	0.6–1.2–1.8–2.3–3.5	6%–11%–17%–23%–34% Lipids (FO + Soy oil) replacement	24 days	↔ Survival, Final weight, TL, SGR	–	↔ Lipid, Moisture, EPA ↑ Protein in 2%, ARA, total PUFA in 2% and 4%, DHA in 4%	Wang et al. (2017a)
	Tambaqui (Colossoma macropomum); 490–880 g	Whole Schizochytrium sp.	0–5		90 days	↑ Weight gain ↔ K	–	↔ SFA, MUFA and PUFA in muscle ↑ Muscle n-3 and n3/n6 ratios. and DHA in MA groups	Cortegano et al. (2019)

Microalgae	Fish species	Form	Microalgae sp.	DS	Replacement	Duration	Growth	Digestibility	Body composition	Reference
Tetraselmis	Common carp (*C. carpio* L.); 10–80 g	Whole	*Tetraselmis* sp.	0-11-17	0%–25%–40% FM replacement	12 weeks	↔ Final weight, SGR, FI, FCR	↔ PER	↔ Whole body and fillet composition	Kiron et al. (2012)
	European seabass (*D. labrax*); 70–120 g	Whole	*Tetraselmis* sp.	0-8-16	0%–10%–20% FM replacement	63 days	↔ Final weight, SGR, FI, FCR, VSI; ↓ HSI in MA16	↓ Protein, lipid and organic matter ADC values in fish fed MA16	↔ Carcass and fillet yield; ↔ Fillet moisture, protein, total lipids and PUFA	Tulli et al. (2012)
	Gilthead seabream (*S. aurata*); 100 g	Whole *T. chuii*		0-5-10	–	4 weeks	↔ SGR	–	–	Cerezuela et al. (2012)
	Whiteleg shrimp (*L. vannamei*); 2–11 g	Whole	*Tetraselmis* sp.	0-11-17	0%–25%–40% FM replacement	9 weeks	↔ Final weight, SGR, FI, FCR	↔ PER	↓ Whole body protein level in MA17	Kiron et al. (2012)
Microalgae mixture										
Nannochloropsis + *Isochrysis* sp.	Atlantic Cod (*Gadus morhua*); 40–78 g	Whole		0-14-28	0%–15%–30% FM replacement	84 days	↔ Survival, FCR, VSI, VFI; ↑ FBW and growth in MA14 and MA28; ↓ K and HSI in MA30	–	↔ Muscle and liver composition; ↔ n-3 and n-6 fatty acids in the muscle	Walker and Berlinsky (2011)
Tisochrysis lutea and *Tetraselmis suecica* (2:1)	European seabass (*D. labrax*); 200–400 g	Whole		0-6-12-18	0%–14%–28%–43% FM protein replacement + 0%–12%–24%–36% fish lipid replacement	105 days	↔ Final weight, SGR, K, FCR, HSI, MFI; ↑ FI in MA 18	↔ PER; ↓ DM ADC in MA fed fish; ↓ Protein and Energy ADC in fish fed MA 12 and MA43	↔ Carcass yield, fillet yield and fillet lipid content; ↔ SFA, PUFA and DHA (% FA) in fillets; ↑ protein content in fillet of MA18; ↓ DHA fillet content (g/100 g) in fish fed MA18; ↑ a* (↑ redness) in dorsal skin of fish fed MA; ↓ Hue in dorsal skin of fish fed MA	Cardinaletti et al. (2018)
Schizochytrium + *Nannochloropsis*	Oliver flounder, (*Paralichthys olivaceus*); 16–40 g	Whole		–	0%–50%–100% FO replacement	56 days	↔ SGR, VFI, VSI, HSI, K	–	↔ Whole body composition to CTRL; ↔ Muscle and liver lipid content; ↔ n-3 PUFA content	Qiao et al. (2014)

↔ Without differing from control; ↑ significantly higher than the control; ↓ significantly lower than the control; *ADC*, apparent digestible coefficients; *ARA*, arachidonic acid; *DGI*, daily growth index; *DHA*, docosahexaenoic acid; *DM*, dry matter; *DS*, developmental stage; *FM*, fish meal; *FO*, fish oil; *FCR*, feed conversion ratio; *EPA*, eicosapentaenoic acid; *FE*, feed efficiency; *FI*, feed intake; *HSI*, Hepatossomatic index; *K*, Condition factor; *MUFA*, monounsaturated fatty acids; *PL*, larvae metamorphosed to the first postlarval stage; *PRE*, protein retention efficiency; *PUFA*, polyunsaturated fatty acids; *SFA*, saturated fatty acids; *SGR*, specific growth rate; *SL*, standard length; *TGC*, thermal growth coefficient; *TL*, total body length; *VSI*, viscerossomatic index.

did not promote a higher resistance to stress (Table 9). In the case of *Schizochytrium*, a higher inclusion level (21%) had a detrimental effect on larvae survival (Ganuza et al., 2008).

6.2 Feeds for juvenile

Several microalgae species have been evaluated in fish and shrimp juveniles, either as protein sources replacing dietary FM, as lipid sources replacing dietary FO, or simply as sources of bioactive compounds as supplements for aquafeeds (Table 9). The most recent studies in fish and shrimp tested a wide range of algae representing various taxonomic groups: *Arthrospira* (Cianobacteria), *Chlorella* (Chlorophyceae), *Cryptecodinum*, *Diacronema* (*Pavlova*, Pavlovophyceae), *Desmodesmus* (Scenedesmaceae), *Haematococcus* (Haematococcaceae), *Isochrysis* (Isochrysidaceae), *Nannochloropsis* (Eustigmatophyceae), *Nanofrustulum* (Bacillariophyceae), *Navícula* (Diatomacea), *Phaeodactylum* (Phaeodactylaceae), *Scenedesmus* Chlorophyceae, *Schizochytrium* (Labyrinthulomycetes), and *Tetraselmis* (Chlorodendrales). The nutritional value of each microalgae species is mainly dependent on its protein and lipid content, in particular, the polyunsaturated fatty acids (Table 1), which will determine the inclusion level and the ability to replace either FM or FO in aquafeeds. The dietary inclusion level of microalgae biomass seems to be highly dependent on the trophic level of fish species; most studies tested inclusion levels varying between 10% and 20%, which correspond to an FM replacement between 24% and 64%, and FO replacement between 46% and 33%, in carnivorous species (e.g., salmonids and marine species) and noncarnivorous (e.g., cyprinids), respectively. The evaluation of dietary inclusion of microalgae as either additive or supplements relied on low inclusion levels (<3%) and mainly focused on their functional properties (Table 10).

Before any novel ingredient can be considered for dietary inclusion in commercial feeds, besides knowing its nutrient composition, it is crucial to evaluate its nutrient digestibility and bioavailability to formulate balanced diets for each farmed species. Not many studies evaluated the digestibility of microalgae in fish species and results are highly variable, depending on algal and fish species, but also on technological methodologies applied to algal biomass. Atlantic salmon and Arctic charr *Salvelinus alpinus* appear to digest macronutrients from *Spirulina* sp. equally well, but with apparent digestible coefficients (ADC) for protein (82%–85%) and energy (83%) below the range of values observed for some plant ingredients (Burr, Barrows, Gaylord, & Wolters, 2011). In Atlantic salmon, it was shown that *Nannochloropsis* sp. was more digestible than *Desmodesmus* sp. (DM digestibility: 48–63% vs. 32%–47%; protein ADC: 73% vs. 54%–67%; energy ADC: 61% vs. 51%, respectively), and extrusion processing could be used to improve digestibility of certain nutrients (Gong et al., 2018). However, both algae resulted in decreased nutrient digestibility of the diets when compared to the reference diet. Likewise, ADC values of the same defatted *Nannochloropsis* sp. biomass fed to European seabass *Dicentrarchus labrax* resulted in protein and energy ADC values of 85% and 68%, respectively, which are still lower than values reported for fish meal or other plant protein sources fed to this species (Valente et al., 2019). It was recently demonstrated that European seabass is able to better digest *N. oceanica* and *Chlorella vulgaris* than *Tetraselmis* sp., and enzymatic and physical-mechanical technological processes increase protein and energy ADC values of those species (Batista, Pintado, et al., 2020). Nile tilapia *Oreochromis niloticus* seems to have a good capacity to digest microalgae. According to Sarker, Gamble, Kelson, and Kapuscinski (2016), digestibility of crude protein of *Arthrospira* sp. was 86%, but values for *Chlorella* sp. and *Schizochytrium* sp. were lower (80%–82%, respectively). Even though, results are within the range of values reported for FM and other traditional plant protein sources used in Nile tilapia. The low ADC of nutrients and energy in *Chlorella* sp. was

TABLE 10 Major dietary effects of microalgae biomass (MA) on antioxidant and immunological status of fish and shrimp.

Algae	Animal model and size	Form: Whole algae vs. extract	Inclusion level (%)	Protein/Oil replacement (%)	Trial duration	Biological effect		References
						Antioxidant status	Immunological status	
Arthrospira (formerly known as Spirulina)	Gibel carp (*C. auratus*); 15–31 g	Whole *A. platensis*	0–3–7–14	0%–25%–50%–100% FM replacement	46 days	↑ Plasma SOD activity in MA14 post bacterial challenge ↑ Plasma MDA in MA3 and MA7 ↑ *TLR2, TIRAP,* and *MyD88* expression in spleen of MA groups	↑ Lysozyme in MA groups ↑ Phagocyte activity in MA groups post challenge ↑ TNF-α1 in MA groups post challenge ↑ Survival rate post challenge in MA3	Cao, Zhang, et al. (2018)
	Gibel carp (*C. auratus*); 5–12 g	Whole *A. platensis*	0–10–15–20	0%–50%–75%–100% FM replacement	60 days	↔ Plasma and hepatopancreas SOD activity ↑ Plasma MDA in MA groups ↑ Plasma AKP activity in MA groups ↑ TAC in hepatopancreas of MA groups	↑ Lysozyme in MA 15 and MA20	Cao, Zou, et al. (2018)
	Nile tilapia (*O. niloticus*); 30–40 g	Whole *A. platensis*	0–15	–	9 weeks	↓ Serum MDA in MA15 ↑ Serum CAT and SOD activities in MA15	↑ Lysozyme and bactericidal activity in MA15 ↑ Survival rate after infection with *A. hydrophila* in MA15	El-Habashi et al. (2019)
	Rainbow trout (*Oncorhynchus mykiss*); 101–235 g	Whole *A. platensis*	0%, 2.5%, 5%, 7.5%, and 10%	0%, 6%, 12%, 18%, and 24% FM replacement	10 weeks	↓ Serum and liver MDA in MA5–7.5–10 ↑ TAC in serum and liver of MA5–7.5–10 ↑ mRNA of SOD and CAT in liver of MA7.5 and MA 10	–	Teimouri et al. (2013), Teimouri et al. (2019)
		3			41 days	–	↔ Hematocrit	

Continued

TABLE 10 Major dietary effects of microalgae biomass (MA) on antioxidant and immunological status of fish and shrimp—cont'd

Algae	Animal model and size	Form: Whole algae vs. extract	Inclusion level (%)	Protein/Oil replacement (%)	Trial duration	Biological effect		References
						Antioxidant status	Immunological status	
	Red seabream (*Pagrus major*) 18–32 g	Whole Spirulina meal		0% (used to replace cellulose)				Nakagawa et al. (2000)
Chlorella	Nile tilapia, (*Oreochromis niloticus*); 30 g	Whole C. vulgaris	0–15	–	9 weeks	↓ Serum MDA in MA15 ↑ Serum CAT and SOD activities in MA15	↑ Lysozyme and bactericidal activity in MA15 ↑ Survival rate after infection with *A. hydrophila* in MA15	El-Habashi et al. (2019)
Cryptecodinium	Gilthead seabream (*S. aurata*) larvae; 0.06 mg	Whole *C. cohnii*	8	100% FO replacement	17 days	↔ Stress test survival (air exposure)	–	Eryalçin et al. (2015)
	Gilthead seabream (*S. aurata*) larvae; 0.5 mg	Whole, broken cells *C. cohnii*	2	100% FO replacement	15 days	↔ Stress test survival (air exposure)		Ganuza et al. (2008)
	Gilthead seabream (*S. aurata*) larvae; 0.75 mg	Whole, broken cells *C. cohnii*	16	100% FO replacement	21 days	↓ Stress test survival (air exposure)	–	Ganuza et al. (2008)
Desmodesmus	Atlantic salmon (*Salmo salar*); 165–360 g	Defatted biomass *Desmodesmus* sp.	0–10–20	0%–9%–18% FM replacement	70 days	↔ Serum TAC, superoxide and CAT activities	↔ mRNA of intestinal sod1, nrf2, igt, il17d, tgfb, cathl1, cathl2	Kiron et al. (2016)
Haematococcus	Yellow perch (*Perca flavescens*); 13–38 g	Defatted *H. pluvialis*	0–5–10–15	0%–25%–50%–75% FM replacement by MA+Soy protein (1:1)	8 weeks	↔ Serum biochemical indexes (albumin, AKP, ALT, amylase, Ca, globulin, glucose, Na, P, total protein)	–	Jiang et al. (2019)

Microalgae	Fish species	Form & microalgae	Inclusion levels	Replacement	Duration	Physiological effect	Immune effect	Reference
Nannochloropsis	Atlantic salmon (S. salar L.); 200–400 g	Defatted *N. oceania*	0–10–20	0%–10%–20% FM replacement	84 days	↔ Total antioxidant capacity ↔ CA activity ↑ SOD activity in MA10	—	Sørensen et al. (2017)
	European seabass (D. labrax); 30–83 g	Whole *N. oceanica.*	0–8	0%–20.6 % FM replacement	106 days	↔ Total antioxidant capacity, total glutathione, glutathione peroxidase, glutathione s-transferase, glutathione reductase, CAT, lipid peroxidation	↔ Lysozyme, Peroxidase and ACH50	Batista, Pereira, et al. (2020)
	European seabass (D. labrax); 20–50 g	Defatted *Nannochloropsis* sp.	0–5–10–15	0%–3.1%–6.2%–9.4% FM replacement	93 days	—	ACH50: MA10>MA15 ↔ Lysozyme and peroxidase	Valente et al. (2019)
	Gilthead seabream (S. aurata); 100 g	Whole *N. gaditana*	0–5–10	—	4 weeks	—	↔ IgM, respiratory burst at 4 weeks ↑ Hemolytic complement activity in MA groups at 4 weeks ↑ Phagocytic capacity in MA5 ↑ β-Defensine gene in MA10	Cerezuela et al. (2012)
	Gilthead seabream (S. aurata) larvae; 0.06 mg	Whole *N. gaditana*	11	100% FO replacement	17 days	↔ Stress test (air exposure)	—	Eryalçin et al. (2015)
	Kuruma shrimp (*Marsupenaeus japonicus*) Zoea I	Whole *Nannochloropsis* sp.	0–1–4–7	0%–4%–14%–24% FO replacement	14 days	↓ LT50 in MA 1 and MA4	—	Adissin et al. (2020)
	Kuruma shrimp (*M. japonicus*) Zoea I	*Nannochloropsis* sp. lipids	0–1–3–6	0%–17%–58%–97% FO replacement	14 days	↓ LT50 in MA 1	—	Adissin et al. (2020)
Phaeodactylum	Gilthead seabream (S. aurata); 100 g	Whole *P. tricornutum*	0–5–10	—	4 weeks	—	↔ IgM, respiratory burst at 4 weeks ↑ Hemolytic complement activity in MA groups at 2 and 4 weeks	Cerezuela et al. (2012)

Continued

TABLE 10 Major dietary effects of microalgae biomass (MA) on antioxidant and immunological status of fish and shrimp—cont'd

Algae	Animal model and size	Form: Whole algae vs. extract	Inclusion level (%)	Protein/Oil replacement (%)	Trial duration	Antioxidant status	Immunological status	References
						Biological effect		
Schizochytrium	Atlantic salmon (*S. salar*); 213–850 g	Whole *Scizochytrium* sp. + yeast extract	0–1–6–15	0%–7%–40%–100% FO replacement	12 weeks	↑ Inducible nitric oxide synthase (iNOS) activity in MA fed fish	↑ Phagocytic capacity in MA 10	Kousoulaki et al. (2015)
							Microarrays analysis revealed no sign of inflammation or stress	
	Gilthead seabream (*S. aurata*) larvae; 0.5 mg	Whole, broken cells *Schizochytrium* sp.	2.5	100% FO replacement	15 days	↔ Stress test survival (air exposure)	–	Ganuza et al. (2008)
	Gilthead seabream (*S. aurata*) larvae; 0.75 mg	Whole, broken cells *Schizochytrium* sp.	21	100% FO replacement	21 days	↓ Stress test survival (air exposure)	–	Ganuza et al. (2008)
Tetraselmis	Gilthead seabream (*S. aurata*); 100 g	Whole *T. chuii*	0–5–10	–	4 weeks	–	↔ IgM, respiratory burst at 4 weeks	Cerezuela et al. (2012)
							↑ Hemolytic complement activity in MA10 at 4 weeks	
							↑ Phagocytic capacity in MA groups	
							↑ MHCII α and CSF-1R genes in MA5 after 2 weeks	
Mixture								
Schizochytrium + *Nannochloropsis*	Olive flounder, (*Paralichthys olivaceus*); 16–40 g	Whole	–	0%–50%–100% FO replacement	56 days	–	↔ Triglyceride and cholesterol in plasma	Qiao et al. (2014)

↔ Without differing from control; ↓ significantly lower than the control; ↑ significantly higher than the control; *ACH50*, alternative complement; *AKP*, alkaline phosphatase; *ALT*, alanine amino transferase; *CAT*, catalase; *FM*, fish meal; *FO*, fish oil; *LT50*, time required for 50% mortality to occur; *MDA*, malondialdehyde; *TAC*, total antioxidant capacity; *SOD*, superoxide dismutase.

attributed to its high fiber content that might have inhibited proteolytic enzymatic activity. The reduction of ADC values in algae, compared to other plant ingredients, is probably related to the complex structure of the microalgae cell wall and to the presence of indigestible compounds. *Schizochytrium* sp. had the highest lipid and n-3 PUFA content, resulting in significantly higher ADCs for total PUFA (98% vs. 79%–91%) and n-3 PUFA (97% vs. 39%) compared to *Arthrospira* and *Chlorella* (Sarker et al., 2016). These results suggest that *Arthrospira* is a high-quality candidate to replace FO in tilapia feed.

According to Table 9, microalgae biomass is able to replace variable levels of FM. In a broad sense, we can affirm that lower trophic species with herbivorous and/or omnivorous feeding regimes can tolerate higher FM replacement levels than carnivorous fish species (with protein requirements often above 40%). According to Tibbetts (2018), salmonids can only tolerate low inclusion levels (below 20% of the diet) without compromising growth, nutrient utilization, or muscle quality. The inclusion of 10% defatted *Desmodesmus* sp. biomass (coproduct from the biofuel) increased the feed conversion ratio and reduced energy digestibility, final body weight, and muscle lipid deposition in Atlantic salmon (Kiron et al., 2016), while the same inclusion level of defatted *N. oceania* significantly reduced nutrient digestibility without impairing fish growth or fillet composition (Sørensen et al., 2017). In marine carnivorous fish species like European seabass, the dietary inclusion of microalgae biomass up to 18% had no major effects on fish growth but generally resulted in reduced nutrient and energy digestibility. The inclusion of defatted *Nannochloropsis* sp. biomass, up to 15%, resulted in lower energy digestibility without affecting European seabass growth performance or whole-body composition (Valente et al., 2019). Such results should be interpreted with caution, as most studies were short-termed (63–121 days) and the observed reduced nutrient bioavailability might compromise fish growth in the long term. In some species, like Atlantic cod *Gadus morhua*, palatability problems associated with microalgae inclusion (mix of *Nannochloropsis* sp. and *Isochrysis* sp.) decreased feed intake during the first weeks leading to lower growth performance, even at modest inclusion levels (Walker & Berlinsky, 2011). Contrarily to carnivorous fish species, *Chlorella* sp. could totally replace FM in diets for crucian carp, resulting in increased feed intake, growth, and feed utilization (Shi et al., 2017). In Gibel carp *C. auratus gibelio* total FM substitution by *A. platensis* was successfully achieved without affecting growth performance while improving nutrient and energy digestibility (Cao, Zou, et al., 2018). This very same microalga could also totally replace FM in diets for white shrimp, *L. vannamei*, a species with low protein requirement, without impacting growth and increasing survival rate after a hypoxia challenge (Pakravan et al., 2017).

The use of microalgae to replace FO has been mainly addressed during larval stages, but a few studies have also considered juveniles. *Aurantiochytrium* (*Schizochytrium*) is a heterotrophic microalga already produced by large-scale fermentation technology that can reach quite a high lipid content (>60%) and up to 42% DHA (Table 1). Atlantic salmon smolt-fed diets with graded levels of this algae to replace up to 100% FO had good growth performance with similar fillet quality among treatments in terms of n-3 HUFAs (Kousoulaki et al., 2015). In the European seabass the inclusion of 14% *Isochrysis* sp. to replace 36% FO resulted in a fillet with lower n-3 PUFA content, without negative effects on fish growth (Tibaldi et al., 2015). A 50% FO replacement by *Nannochloropsis* sp. and a total replacement by *P. viridis* meal were possible without negative effects on the growth performance and nutrient utilization of juvenile seabass (Haas et al., 2016). Moreover, fish fed the *Pavlova*-rich diet had significantly higher levels of PUFA and EPA in the fillet, highlighting the great potential of this microalga to replace FO.

Reduced growth and digestibility can be associated with gut morphological alterations, with a direct impact on intestinal absorptive capacity. In gilthead seabream postlarvae, lower survival rates in fish fed *P. tricornutum* were related to an epithelial degeneration observed in the anterior intestine (Atalah et al., 2007). But increased microvilli length and enterocyte area in the anterior intestinal region of juveniles fed 20% *Scenedesmus almeriensis* was reported by Vizcaíno et al. (2014). In another study, the dietary inclusion of 5% *Tetraselmis chuii* significantly increased the expression of CSF-1R and MHCIIα in the gut (Cerezuela et al., 2012). The CSF-1R gene encodes the receptor for colony-stimulating factor 1, a cytokine that controls the production, differentiation, and function of macrophages, suggesting that *T. chuii* can help develop an innate immune response against pathogens. In Atlantic salmon midgut, significant effects on slime, goblet cells' production, and inducible nitric oxide synthase (iNOS) activity were observed with increasing levels of dietary *Schizochytrium* sp. supplementation using histological fluorescence staining and immunofluorescence analysis (Kousoulaki et al., 2015). In Nile tilapia, the inclusion of either 15% *S. platensis* or *Chlorella* sp. mitigated the necrotic and degenerative changes induced by *A. hydrophila* infection without affecting intestinal morphology (El-Habashi et al., 2019). Likewise, in most studies evaluating increasing levels of microalgal biomass in aquafeeds, the morphology of the intestine was well preserved and showed no signs of inflammation (Table 9).

Microalgae have various bioactive compounds that have been shown to affect fish pigmentation, enhance defense activity and health status, and confer tissue protection and antioxidant effects. Several studies highlighted the ability of microalgae to modulate the immunological status of aquatic species (Table 10). Microarray analysis did not reveal any signs of toxicity, stress, inflammation, or any other negative effects from *Schizochytrium* sp. supplementation in diets for Atlantic salmon (Kousoulaki et al., 2015). *Chlorella* sp. increased the lysozyme and bactericidal activity in Nile tilapia when

added at 15% in diets, also resulting in an increased survival rate after infection with *A. hydrophila* (El-Habashi et al., 2019). The administration of *N. gaditana* and *T. chuii* in gilthead seabream resulted in a significant increase in hemolytic complement activity, phagocytic capacity, and the expression of different immune-associated genes (β-defensin, MHCIIα, and CSF-1R), while the *P. tricornutum*-supplemented diet led to an immunostimulation, with very little effect on gene expression (Cerezuela et al., 2012). Low dietary inclusion (<4%) of *A. platensis* significantly enhanced the immune response and disease resistance of Gibel carp, partly through TLR2 (Toll-like receptor 2) signaling pathway (Cao, Zhang, et al., 2018). The dietary inclusion of 3% *Dunaliella* sp. improved the survival of L. *vannamei* infected with *Vibrio parahaemolyticus* which causes significant economic losses in the shrimp industry (Medina Félix et al., 2017). In summary, these studies suggest that microalgae can be beneficial to improve the immune system of fish and shrimp, which was particularly evident after subjecting animals to a bacterial or environmental challenge.

The high content of LC-PUFAs and pigments in microalgae is one of the advantages supporting the inclusion of microalgae in animal diets because these molecules have a known positive effect on animal health and well-being. However, since inclusions levels are not yet optimized for each species, the results on the animal antioxidant status are quite heterogeneous (Table 10). A positive response in superoxide dismutase (SOD) and catalase (CAT) activity was observed with an inclusion level around 15% with *Arthrospira* sp. and with *Chlorella* sp. in Nile tilapia (El-Habashi et al., 2019). On the other hand, the inclusion of 10%–20% *Desmodesmus* sp. in diets for Atlantic salmon was unable to alter serum antioxidant capacity (Kiron et al., 2016), while 10% inclusion of *Nannochloropsis* sp. in this species had a positive impact in SOD (Sørensen et al., 2017). Overall, the data suggest that microalgae have a positive effect on the antioxidant status of fish, but species used and inclusion levels need further refinement to improve the value of this novel resource.

Exogenous feeding in aquaculture unlocks the possibility to tailor fish composition toward enhanced health-promoting properties, without compromising its sensory attributes and consumer's acceptance. Many studies showed that algal-meal rich diets did not alter whole body composition (Kiron et al., 2012; Kiron et al., 2016; Pakravan et al., 2017; Sørensen et al., 2017; Valente et al., 2019; Walker & Berlinsky, 2011). However, decreased whole-body lipid content was reported in rainbow trout *Oncorhynchus mykiss* fed *Tetraselmis or S. platensis* (Teimouri et al., 2013; Teimouri et al., 2019), and in red drum, *Sciaenops ocellatus* fed 15% defatted *N. salina* (Patterson & Gatlin, 2013). Most studies evaluating the inclusion of microalgae did not have a significant impact on fillets' nutritional value, compared to those fed FO/FM rich diets. Increased PUFA levels were reported in Pacific white shrimp fed *Arthrospira* sp. to replace FM (Pakravan et al., 2017) and in European seabass fed *Diacronema* sp. to replace FO (Haas et al., 2016). Microalgal pigments may be of great value to enhance pigmentation in fish and shrimp species. Of particular interest is astaxanthin, a carotenoid commonly used by the salmonid feed industry. Although most diets include synthetic pigments due to their significantly lower cost, there is a societal-driven opportunity to use natural sources. The dietary inclusion of microalgae in diets for Atlantic salmon has been extensively revised by Tibbetts (2018) that evidenced that, although most studies relied on *Haematococcus* sp. as the natural source of astaxanthin, *Arthrospira* sp. and *Chlorella* sp. also promoted muscle carotenoid deposition. Moreover, the inclusion of a by-product of astaxanthin production from *H. pluvialis* in diets for Pacific white shrimp still resulted in increased astaxanthin content (Ju et al., 2012). In rainbow trout fed with *Arthrospira* biomass, Teimouri et al. (2013) reported lower

fillet luminosity and increased redness associated with higher carotenoid concentration, both in fillet and skin. In white fish species like European seabass, diets with 14% *Isochrysis* resulted in increased fillet redness (a^*) coupled with decreased hue values (Tibaldi et al., 2015), but these differences could not be perceived in cooked fillets evaluated by a sensory panel. A greenish pigmentation of the skin was also reported in those fish. Likewise, feeding seabass a microalgae blend (*Tisochrysis lutea* and *Tetraselmis suecica*) resulted in a greenish pigmentation of the skin, with a slight tendency toward redness and diminished lightness and hue (Cardinaletti et al., 2018). In conclusion, some algae seem to be able to affect muscle and skin color of several species, although in some cases that can be positively evaluated by consumers (e.g., salmonids and shrimp); in other cases, the impact of such alterations is less clear and remains to be clarified.

7 Conclusion and perspectives

This chapter unveils the increasing interest in the use of microalga in diets for livestock and species with economic value for aquaculture. Selected strains for most studies were used as alternative protein, lipid, and carbohydrate sources for animal feeds, replacing traditional ingredients (crops, soybean meal and fish meal, and oil) and contributing to the environmental sustainability of animal production systems. The vast number of microalgae species and versatile characteristics opens many exciting possibilities in the coming years. It is important, however, to keep in mind that microalgae are uncharted territory in animal nutrition as researchers are still scratching the surface on the use of microalgae in animal diets. Several problems must be addressed before their large-scale use by the feed industry, as information is fragmented and results often inconsistent. The primary focus should be given at establishing the nutritive value of microalgae species, as analytics are highly variable are the bioavailability of nutrients largely unknown. More research is needed on algae processing and feeding strategies to mitigate problems related to low palatability in some animals, and to tackle the low digestibility problems of some species related to the existence of the recalcitrant cell wall (e.g., in *Spirulina*). The use of chemicals (e.g., enzymes) and mechanical processing technologies, and additives able to enable the use of algae nutrients merits further consideration. Moreover, the ash contents of microalgae may cause electrolyte imbalances highlighting the importance of studying the physiology behind animal performance and feed digestibility. Using omics (transcriptomics, proteomics, metabolomics) to zoom in on the effect of these novel feed ingredients on animal metabolism would generate useful information. It already seems that microalgae play an important role in animal production, by affecting growth, meat, and milk quality, while enhancing defense activity and health status in several species, but a large number of well-designed studies to further evaluate the potential of this "next-generation" ingredients are still required.

Finally, the large discrepancy in microalgae production cost and global supply compared to commodity feedstuffs stills hampers the large-scale use of this raw material in animal feeds. Moreover, current knowledge and technology projections suggest that third-generation biofuels from microalgae could progressively substitute a significant proportion of fossil fuels. This will consequently increase the generation of defatted coproducts which are valuable nutrients sources for feeds and can contribute to the economic and environmental sustainability of both the biofuel and the animal production sectors.

Acknowledgments

This work was supported by the project "MARINAL-GAE4AQUA—Improving bioutilization of marine algae as sustainable feed ingredients to increase efficiency and quality of aquaculture production," financed by FCT (Fundação para a Ciência e a Tecnologia), Portugal, through the research project ERA-NET COFASP/004/2015. Financial support from FCT in the form of infrastructural funding to CIIMAR (UIDB/04423/2020, UIDP/04423/2020), LEAF (UID/AGR/04129), CCMAR (UIDB/04326/2020), and REQUIMTE (UID/QUI/50006/2019) is also acknowledged. Margarida R.G. Maia and Inês M. Valente acknowledge FCT for funding through program DL 57/2016—Norma transitória. David Ribeiro acknowledges a PhD grant (SFRH/BD/143992/2019) from FCT. S. Engrola was supported by FCT investigator grant IF/00482/2014/CP1217/CT0005 funded by the European Social Fund, the Operational Programme Human Potential and the Foundation for Science and Technology of Portugal (FCT).

References

Abdel-Daim, M. M., Ahmed, A., Ijaz, H., Abushouk, A. I., Ahmed, H., Negida, A., et al. (2019). Influence of *Spirulina platensis* and ascorbic acid on amikacin-induced nephrotoxicity in rabbits. *Environmental Science and Pollution Research, 26,* 8080–8086.

Abdelnour, S. A., Sheiha, A. M., Taha, A. E., Swelum, A. A., Alarifi, S., Alkahtani, S., et al. (2019). Impacts of enriching growing rabbit diets with *Chlorella vulgaris* microalgae on growth, blood variables, carcass traits, immunological and antioxidant indices. *Animals, 9,* 788.

Adissin, T. O. O., Manabu, I., Shunsuke, K., Saichiro, Y., Moss, A. S., & Dossou, S. (2020). Effects of dietary *Nannochloropsis* sp. powder and lipids on the growth performance and fatty acid composition of larval and postlarval kuruma shrimp, *Marsupenaeus japonicus. Aquaculture Nutrition, 26,* 186–200.

Aladaileh, S. H., Khafaga, A. F., Abd El-Hack, M. E., Al-Gabri, N. A., Abukhalil, M. H., Alfwuaires, M. A., et al. (2020). *Spirulina platensis* ameliorates the sub chronic toxicities of lead in rabbits via anti-oxidative, anti-inflammatory, and immune stimulatory properties. *Science of the Total Environment, 701,* 134879.

Alexandratos, N., & Bruinsma, J. (2012). *World agriculture towards 2030/2050: The 2012 revision.* ESA Working paper No. 12-03Rome: FAO.

Allen, M. S. (2000). Effects of diet on short-term regulation of feed intake by lactating dairy cattle. *Journal of Dairy Science, 83,* 1598–1624.

Altmann, B. A., Neumann, C., Rothstein, S., Liebert, F., & Mörlein, D. (2019). Do dietary soy alternatives lead to pork quality improvements or drawbacks? A look into micro-alga and insect protein in swine diets. *Meat Science, 153,* 26–34.

Al-Zuhair, S., Ashraf, S., Hisaindee, S., Darmaki, N. A., Battah, S., Svistunenko, D., et al. (2017). Enzymatic pre-treatment of microalgae cells for enhanced extraction of proteins. *Engineering in Life Sciences, 17,* 175–185.

Anderson, D. W., Tang, C.-S., & Ross, E. (1991). The xanthophylls of *Spirulina* and their effect on egg yolk pigmentation. *Poultry Science, 70,* 115–119.

Andriola, Y. T., Moreira, F., Anastácio, E., Camelo, F. A., Silva, A. C., Varela, A. S., et al. (2018). Boar sperm quality after supplementation of diets with omega-3 polyunsaturated fatty acids extracted from microalgae. *Andrologia, 50,* e12825.

Anjos, C., Baptista, T., Joaquim, S., Mendes, S., Matias, A. M., Moura, P., et al. (2017). Broodstock conditioning of the Portuguese oyster (*Crassostrea angulata,* Lamarck, 1819): Influence of different diets. *Aquaculture Research, 48,* 3859–3878.

Ao, T., Macalintal, L. M., Paul, M. A., Pescatore, A. J., Cantor, A. H., Ford, M. J., et al. (2015). Effects of supplementing microalgae in laying hen diets on productive performance, fatty-acid profile, and oxidative stability of eggs. *Journal of Applied Poultry Research, 24,* 394–400.

Atalah, E., Cruz, C. M. H., Izquierdo, M. S., Rosenlund, G., Caballero, M. J., Valencia, A., et al. (2007). Two microalgae *Crypthecodinium cohnii* and *Phaeodactylum tricornutum* as alternative source of essential fatty acids in starter feeds for seabream (*Sparus aurata*). *Aquaculture, 270,* 178–185.

Batista, S., Pereira, R., Oliveira, B., Baião, L. F., Jessen, F., Tulli, F., Messina, M., Silva, J. L., Abreu, H., & Valente, L. M. P. (2020). Exploring the potential of seaweed Gracilaria gracilis and microalga *Nannochloropsis oceanica,* single or blended, as natural dietary ingredients for European seabass *Dicentrarchus labrax. Journal of Applied Phycology, 32,* 2041–2059.

Batista, S., Pintado, M., Marques, A., Abreu, H., Silva, J. L., Jessen, F., Tulli, F., & Valente, L. M. P. (2020). Use of technological processing of seaweed and microalgae as strategy to improve their apparent digestibility coefficients in European seabass (Dicentrarchus labrax) juveniles. *Journal of Applied Phycology.* https://doi.org/10.1007/s10811-10020-02185-10812.

Battaglia Richi, E., Baumer, B., Conrad, B., Darioli, R., Schmid, A., & Keller, U. (2015). Health risks associated with meat consumption: A review of epidemiological studies. *International Journal for Vitamin and Nutrition Research, 85,* 70–78.

Bauman, D. E., & Griinari, J. M. (2003). Nutritional regulation of milk fat synthesis. *Annual Review of Nutrition, 23,* 203–227.

Boeckaert, C., Fievez, V., Van Hecke, D., Verstraete, W., & Boon, N. (2007). Changes in rumen biohydrogenation intermediates and ciliate protozoa diversity after algae supplementation to dairy cattle. *European Journal of Lipid Science and Technology, 109,* 767–777.

Boeckaert, C., Vlaeminck, B., Dijkstra, J., Issa-Zacharia, A., Van Nespen, T., Van Straalen, W., et al. (2008). Effect of dietary starch or micro algae supplementation on rumen fermentation and milk fatty acid composition of dairy cows. *Journal of Dairy Science, 91,* 4714–4727.

Boeckaert, C., Vlaeminck, B., Mestdagh, J., & Fievez, V. (2007). *In vitro* examination of Dha-edible micro algae. 1. Effect on rumen lipolysis and biohydrogenation of linoleic and linolenic acids. *Animal Feed Science and Technology, 136,* 63–79.

Boiago, M. M., Dilkin, J. D., Kolm, M. A., Barreta, M., Souza, C. F., Baldissera, M. D., et al. (2019). *Spirulina platensis* in Japanese quail feeding alters fatty acid profiles and improves egg quality: Benefits to consumers. *Journal of Food Biochemistry, 43,* e12860.

Bonos, E., Kasapidou, E., Kargopoulos, A., Karampampas, A., Christaki, E., Florou-Paneri, P., et al. (2016). Spirulina as a functional ingredient in broiler chicken diets. *South African Journal of Animal Science, 46,* 94.

Bruneel, C., Lemahieu, C., Fraeye, I., Ryckebosch, E., Muylaert, K., Buyse, J., et al. (2013). Impact of microalgal feed supplementation on omega-3 fatty acid enrichment of hen eggs. *Journal of Functional Foods, 5,* 897–904.

Burr, G. S., Barrows, F. T., Gaylord, G., & Wolters, W. R. (2011). Apparent digestibility of macro-nutrients and phosphorus in plant-derived ingredients for Atlantic salmon, Salmo salar and Arctic charr, Salvelinus alpinus. *Aquaculture Nutrition, 17,* 570–577.

Campos, I., Valente, L. M. P., Matos, E., Marques, P., & Freire, F. (2020). Life-cycle assessment of animal feed ingredients: Poultry fat, poultry by-product meal and hydrolyzed feather meal. *Journal of Cleaner Production, 252,* 119845.

Canada, P., Engrola, S., Conceição, L. E. C., & Valente, L. M. P. (2019). Improving growth potential in Senegalese sole (*Solea senegalensis*) through dietary protein. *Aquaculture, 498,* 90–99.

Cao, S., Zhang, P., Zou, T., Fei, S., Han, D., Jin, J., et al. (2018). Replacement of fishmeal by spirulina Arthrospira platensis affects growth, immune related-gene expression in gibel carp (*Carassius auratus gibelio* var. Cas III), and its challenge against *Aeromonas hydrophila* infection. *Fish & Shellfish Immunology, 79,* 265–273.

Cao, S. P., Zou, T., Zhang, P. Y., Han, D., Jin, J. Y., Liu, H. K., et al. (2018). Effects of dietary fishmeal replacement with *Spirulina platensis* on the growth, feed utilization, digestion and physiological parameters in juvenile gibel carp (*Carassis auratus gibelio* var. Cas III). *Aquaculture Research, 49,* 1320–1328.

Cardinaletti, G., Messina, M., Bruno, M., Tulli, F., Poli, B. M., Giorgi, G., et al. (2018). Effects of graded levels of a blend of *Tisochrysis lutea* and *Tetraselmis suecica* dried biomass on growth and muscle tissue composition of European sea bass (*Dicentrarchus labrax*) fed diets low in fish meal and oil. *Aquaculture, 485,* 173–182.

Cerezuela, R., Guardiola, F. A., Meseguer, J., & Esteban, M. Á. (2012). Enrichment of gilthead seabream (*Sparus aurata* L.) diet with microalgae: Effects on the immune system. *Fish Physiology and Biochemistry, 38,* 1729–1739.

Cheong, S. H., Kim, M. Y., Sok, D.-E., Hwang, S.-Y., Kim, J. H., Kim, H. R., et al. (2010). Spirulina prevents atherosclerosis by reducing hypercholesterolemia in rabbits fed a high-cholesterol diet. *Journal of Nutritional Science and Vitaminology*, *56*, 34–40.

Chilliard, Y., Ferlay, A., & Doreau, M. (2001). Effect of different types of forages, animal fat or marine oils in cow's diet on milk fat secretion and composition, especially conjugated linoleic acid (CLA) and polyunsaturated fatty acids. *Livestock Production Science*, *70*, 31–48.

Chilliard, Y., Glasser, F., Ferlay, A., Bernard, L., Rouel, J., & Doreau, M. (2007). Diet, rumen biohydrogenation and nutritional quality of cow and goat milk fat. *European Journal of Lipid Science and Technology*, *109*, 828–855.

Choi, H., Jung, S. K., Kim, J. S., Kim, K. W., Oh, K. B., Lee, P. Y., et al. (2017). Effects of dietary recombinant chlorella supplementation on growth performance, meat quality, blood characteristics, excreta microflora, and nutrient digestibility in broilers. *Poultry Science*, *96*, 710–716.

Clayton, E. H., Lamb, T. A., Refshauge, G., Kerr, M. J., Bailes, K. L., Ponnampalam, E. N., et al. (2014). Differential response to an algae supplement high in DHA mediated by maternal periconceptional diet: Intergenerational effects of n-6 fatty acids. *Lipids*, *49*, 767–775.

Coelho, D., Lopes, P. A., Cardoso, V., Ponte, P., Brás, J., Madeira, M. S., et al. (2019). A two-enzyme constituted mixture to improve the degradation of *Arthrospira platensis* microalga cell wall for monogastric diets. *Journal of Animal Physiology and Animal Nutrition*, 310–321.

Conceição, L. E. C., Yúfera, M., Makridis, P., Morais, S., & Dinis, M. T. (2010). Live feeds for early stages of fish rearing. *Aquaculture Research*, *36*, 1–16.

Cortegano, C. A. A., De Alcântara, A. M., Da Silva, A. F., Epifânio, C. M. F., Bentes, S. P. C., Dos Santos, V. J., et al. (2019). Finishing plant diet supplemented with microalgae meal increases the docosahexaenoic acid content in *Colossoma macropomum* flesh. *Aquaculture Research*, *50*, 1291–1299.

Costa, D. F. A., Quigley, S. P., Isherwood, P., Mclennan, S. R., & Poppi, D. P. (2016). Supplementation of cattle fed tropical grasses with microalgae increases microbial protein production and average daily gain. *Journal of Animal Science*, *94*, 2047–2058.

Coutinho, P., Rema, P., Otero, A., Pereira, O., & Fabregas, J. (2006). Use of biomass of the marine microalga Isochrysis galbana in the nutrition of goldfish (*Carassius auratus*) larvae as source of protein and vitamins. *Aquaculture Research*, *37*, 793–798.

Cowieson, A. J., & Kluenter, A. M. (2019). Contribution of exogenous enzymes to potentiate the removal of antibiotic growth promoters in poultry production. *Animal Feed Science and Technology*, *250*, 81–92.

Dal Bosco, A., Gerencsér, Z., Szendro, Z., Mugnai, C., Cullere, M., Kovàcs, M., et al. (2014). Effect of dietary supplementation of Spirulina (*Arthrospira platensis*) and

Thyme (*Thymus vulgaris*) on rabbit meat appearance, oxidative stability and fatty acid profile during retail display. *Meat Science*, *96*, 114–119.

Dalle Zotte, A., Cullere, M., Sartori, A., Szendro, Z., Kovàcs, M., Giaccone, V., et al. (2014). Dietary Spirulina (*Arthrospira platensis*) and Thyme (*Thymus vulgaris*) supplementation to growing rabbits: Effects on raw and cooked meat quality, nutrient true retention and oxidative stability. *Meat Science*, *98*, 94–103.

De Tonnac, A., Guillevic, M., & Mourot, J. (2018). Fatty acid composition of several muscles and adipose tissues of pigs fed n-3 PUFA rich diets. *Meat Science*, *140*, 1–8.

Díaz, M. T., Pérez, C., Sánchez, C. I., Lauzurica, S., Cañeque, V., González, C., et al. (2017). Feeding microalgae increases omega 3 fatty acids of fat deposits and muscles in light lambs. *Journal of Food Composition and Analysis*, *56*, 115–123.

Dogan, S. C., Baylan, M., Erdogan, Z., Akpinar, G. C., Kucukgul, A., & Duzguner, V. (2016). Performance, egg quality and serum parameters of Japanese quails fed diet supplemented with *Spirulina platensis*. *Fresenius Environmental Bulletin*, *25*, 5857–5862.

Domínguez, R., Pateiro, M., Gagaoua, M., Barba, F. J., Zhang, W., & Lorenzo, J. M. (2019). A comprehensive review on lipid oxidation in meat and meat products. *Antioxidants (Basel, Switzerland)*, *8*, 429.

Doucha, J., Lívanský, K., Kotrbáček, V., & Zachleder, V. (2009). Production of Chlorella biomass enriched by selenium and its use in animal nutrition: A review. *Applied Microbiology and Biotechnology*, *83*, 1001–1008. https://doi.org/10.1007/s00253-009-2058-9.

Ekmay, R., Gatrell, S., Lum, K., Kim, J., & Lei, X. G. (2014). Nutritional and metabolic impacts of a defatted green marine microalgal (*Desmodesmus* sp.) biomass in diets for weanling pigs and broiler chickens. *Journal of Agricultural and Food Chemistry*, *62*, 9783–9791.

El-Habashi, N., Fadl, S. E., Farag, H. F., Gad, D. M., Elsadany, A. Y., & El Gohary, M. S. (2019). Effect of using *Spirulina* and *Chlorella* as feed additives for elevating immunity status of Nile tilapia experimentally infected with *Aeromonas hydrophila*. *Aquaculture Research*, *50*, 2769–2781.

El-Ratel, I. T. (2017). Reproductive performance, oxidative status and blood metabolites of doe rabbits administrated with *Spirulina* alga. *Egyptian Poultry Science Journal*, *37*, 1153–1172.

El-Sabagh, M. R., Eldaim, M. A. A., Mahboub, D. H., & Abdel-Daim, M. (2014). Effects of *Spirulina platensis* algae on growth performance, antioxidative status and blood metabolites in fattening lambs. *Journal of Agricultural Science*, *6*, 92–98.

El-Shazly, K. (1952). Degradation of protein in the rumen of the sheep. II. The action of rumen micro-organisms on amino acids. *The Biochemical Journal*, *51*, 647–653.

Eryalçin, K. M., Ganuza, E., Atalah, E., & Cruz, M. C. H. (2015). *Nannochloropsis gaditana* and *Crypthecodinium cohnii*, two microalgae as alternative sources of essential fatty acids in early weaning for gilthead seabream. *Hidrobiologica*, 25, 193–202.

EUMOFA (2018). *The EU fish market, European Market Observatory for Fisheries and Aquaculture Products*. Publications Office of the European Union.

Evans, A. M., Smith, D. L., & Moritz, J. S. (2015). Effects of algae incorporation into broiler starter diet formulations on nutrient digestibility and 3 to 21 d bird performance. *Journal of Applied Poultry Research, 24*, 206–214.

Fan, Y., Ren, C., Meng, F., Deng, K., Zhang, G., & Wang, F. (2019). Effects of algae supplementation in high-energy dietary on fatty acid composition and the expression of genes involved in lipid metabolism in Hu sheep managed under intensive finishing system. *Meat Science, 157*.

FAO (2018). *The state of the world fisheries and aquaculture—Meeting the sustainable development goals*. Rome, Italy: FAO, Fisheries and Aquaculture Department.

Flaga, J., Korytkowski, Ł., Górka, P., & Kowalski, Z. M. (2019). The effect of docosahexaenoic acid-rich algae supplementation in milk replacer on performance and selected immune system functions in calves. *Journal of Dairy Science, 102*, 8862–8873.

Fletcher, D. L. (1999). Broiler breast meat color variation, pH, and texture. *Poultry Science, 78*, 1323–1327.

Fouda, S. F., & Ismail, R. F. S. A. (2017). Effect of *Spirulina platensis* on reproductive performance of rabbit bucks. *Egyptian Journal of Nutrition and Feeds, 20*, 55–66.

Fougère, H., Delavaud, C., & Bernard, L. (2018). Diets supplemented with starch and corn oil, marine algae, or hydrogenated palm oil differentially modulate milk fat secretion and composition in cows and goats: A comparative study. *Journal of Dairy Science, 101*, 8429–8445.

Franklin, S. T., Martin, K. R., Baer, R. J., Schingoethe, D. J., & Hippen, A. R. (1999). Dietary marine algae (*Schizochytrium* sp.) increases concentrations of conjugated linoleic, docosahexaenoic and transvaccenic acids in milk of dairy cows. *Journal of Nutrition, 129*, 2048–2052.

Fry, J. P., Mailloux, N. A., Love, D. C., Milli, M. C., & Cao, L. (2018). Feed conversion efficiency in aquaculture: Do we measure it correctly? *Environmental Research Letters, 13*, 024017.

Furbeyre, H., Milgen, J. V., Mener, T., Gloaguen, M., & Labussière, E. (2018). Effects of oral supplementation with *Spirulina* and *Chlorella* on growth and digestive health in piglets around weaning. *Animal*, 1–10.

Furbeyre, H., Van Milgen, J., Mener, T., Gloaguen, M., & Labussière, E. (2017). Effects of dietary supplementation with freshwater microalgae on growth performance, nutrient digestibility and gut health in weaned piglets. *Animal, 11*, 183–192.

Ganuza, E., Benítez-Santana, T., Atalah, E., Vega-Orellana, O., Ganga, R., & Izquierdo, M. S. (2008). *Crypthecodinium cohnii* and *Schizochytrium* sp. as potential substitutes to fisheries-derived oils from seabream (*Sparus aurata*) microdiets. *Aquaculture, 277*, 109–116.

Gatlin, D. M., Barrows, F. T., Brown, P., Dabrowski, K., Gaylord, T. G., Hardy, R. W., et al. (2007). Expanding the utilization of sustainable plant products in aquafeeds: A review. *Aquaculture Research, 38*, 551–579.

Gázquez, A., Ruíz-Palacios, M., & Larqué, E. (2017). Dha supplementation during pregnancy as phospholipids or Tag produces different placental uptake but similar fetal brain accretion in neonatal piglets. *British Journal of Nutrition, 118*, 981–988.

Gerencsér, Z., Szendro, Z., Matics, Z., Radnai, I., Kovács, M., Nagy, I., et al. (2014). Effect of dietary supplementation of spirulina (*Arthrospira platensis*) and thyme (*Thymus vulgaris*) on apparent digestibility and productive performance of growing rabbits. *World Rabbit Science, 22*, 1.

Gibbs, H. K., Rausch, L., Munger, J., Schelly, I., Morton, D. C., Noojipady, P., et al. (2015). Brazil's soy moratorium. *Science, 347*, 377–378.

Ginzberg, A., Cohen, M., Sod-Moriah, U. A., Shany, S., Rosenshtrauch, A., & Arad, S. (2000). Chickens fed with biomass of the red microalga *Porphyridium* sp. have reduced blood cholesterol level and modified fatty acid composition in egg yolk. *Journal of Applied Phycology, 12*, 325–330.

Glasser, F., Ferlay, A., & Chilliard, Y. (2008). Oilseed lipid supplements and fatty acid composition of cow milk: A meta-analysis. *Journal of Dairy Science, 91*, 4687–4703.

Gogus, U., & Smith, C. (2010). N-3 omega fatty acids: A review of current knowledge. *International Journal of Food Science & Technology, 45*, 417–436.

Gong, Y., Guterres, H. A. D. S., Huntley, M., Sørensen, M., & Kiron, V. (2018). Digestibility of the defatted microalgae *Nannochloropsis* sp. and *Desmodesmus* sp. when fed to Atlantic salmon, Salmo salar. *Aquaculture Nutrition, 24*, 56–64.

Gong, Y., Sørensen, S. L., Dahle, D., Nadanasabesan, N., Dias, J., Valente, L. M. P., Sørensen, M., & Kiron, V. (2020). Approaches to improve utilization of *Nannochloropsis oceanica* in plant-based feeds for Atlantic salmon. *Aquaculture*, 735122.

Grinstead, G. S., Tokach, M. D., Dritz, S. S., Goodband, R. D., & Nelssen, J. L. (2000). Effects of *Spirulina platensis* on growth performance of weanling pigs. *Animal Feed Science and Technology, 83*, 237–247.

Guevarra, R. B., Hong, S. H., Cho, J. H., Kim, B.-R., Shin, J., Lee, J. H., et al. (2018). The dynamics of the piglet gut microbiome during the weaning transition in association with health and nutrition. *Journal of Animal Science and Biotechnology, 9*, 54.

Haas, S., Bauer, J. L., Adakli, A., Meyer, S., Lippemeier, S., Schwarz, K., et al. (2016). Marine microalgae *Pavlova viridis* and *Nannochloropsis* sp. as n-3 Pufa source in diets for juvenile European sea bass (*Dicentrarchus labrax* L.). *Journal of Applied Phycology, 28*, 1011–1021.

Hajati, H., & Zaghari, M. (2019). Effects of *Spirulina platensis* on growth performance, carcass characteristics, egg traits and immunity response of Japanese quails. *Iranian Journal of Applied Animal Science, 9,* 347–357.

Hammond, B. G., Mayhew, D. A., Holson, J. F., Nemec, M. D., Mast, R. W., & Sander, W. J. (2001). Safety assessment of DHA-rich microalgae from *Schizochytrium* sp.: II. Developmental toxicity evaluation in rats and rabbits. *Regulatory Toxicology and Pharmacology, 33,* 205–217.

Hamre, K., Yúfera, M., Rønnestad, I., Boglione, C., Conceição, L. E. C., & Izquierdo, M. (2013). Fish larval nutrition and feed formulation: Knowledge gaps and bottlenecks for advances in larval rearing. *Reviews in Aquaculture, 5,* S26–S58.

Han, J. G., Kang, G. G., Kim, J. K., & Kim, S. H. (2002). The present status and future of chlorella. *Food Science and Industry, 6,* 64–69.

Hassanein, H., Arafa, M. M., Abo Warda, M. A., & Abd-Elall, A. (2014). Effect of using *Spirulina platensis* and *Chlorella vulgaris* as feed additives on growing rabbit performance. *Egyptian Journal of Rabbit Science, 24,* 413–431.

Helm, M. M., Bourne, N., & Lovatelli, A. (2004). *Hatchery culture of bivalves, a practical manual:* (p. 471). FAO Fisheries Technical Paper.

Hemaiswarya, S., Raja, R., Ravi Kumar, R., Ganesan, V., & Anbazhagan, C. (2011). Microalgae: A sustainable feed source for aquaculture. *World Journal of Microbiology and Biotechnology, 27,* 1737–1746.

Heo, J. M., Opapeju, F. O., Pluske, J. R., Kim, J. C., Hampson, D. J., & Nyachoti, C. M. (2013). Gastrointestinal health and function in weaned pigs: A review of feeding strategies to control post-weaning diarrhoea without using in-feed antimicrobial compounds: Feeding strategies without using in-feed antibiotics. *Journal of Animal Physiology and Animal Nutrition, 97,* 207–237.

Hintz, H. F., & Heitman, H. (1967). Sewage-grown algae as a protein supplement for swine. *Animal Production, 9,* 135–140.

Holub, B. J. (2009). Docosahexaenoic acid (DHA) and cardiovascular disease risk factors. *Prostaglandins, Leukotrienes and Essential Fatty Acids, 81,* 199–204.

Hopkins, D. L., Clayton, E. H., Lamb, T. A., Van De Ven, R. J., Refshauge, G., Kerr, M. J., et al. (2014). The impact of supplementing lambs with algae on growth, meat traits and oxidative status. *Meat Science, 98,* 135–141.

Hussein, M., Harvatine, K. H., Weerasinghe, W. M. P. B., Sinclair, L. A., & Bauman, D. E. (2013). Conjugated linoleic acid-induced milk fat depression in lactating ewes is accompanied by reduced expression of mammary genes involved in lipid synthesis. *Journal of Dairy Science, 96,* 3825–3834.

Iwamoto, H. (2004). Industrial production of microalgal cell mass and secondary products—Major industrial species. Chlorella. In A. Richmond (Ed.), *Handbook of microalgal culture: Biotechnology and applied phycology.* Blackwell Science: Oxford, UK.

Jiang, M., Zhao, H. H., Zai, S. W., Shepherd, B., Wen, H., & Deng, D. F. (2019). A defatted microalgae meal (*Haematococcus pluvialis*) as a partial protein source to replace fishmeal for feeding juvenile yellow perch Perca flavescens. *Journal of Applied Phycology, 31,* 1197–1205.

Jones, S. W., Karpol, A., Friedman, S., Maru, B. T., & Tracy, B. P. (2020). Recent advances in single cell protein use as a feed ingredient in aquaculture. *Current Opinion in Biotechnology, 61,* 189–197.

Ju, Z. Y., Deng, D.-F., & Dominy, W. (2012). A defatted microalgae (*Haematococcus pluvialis*) meal as a protein ingredient to partially replace fishmeal in diets of Pacific white shrimp (*Litopenaeus vannamei*, Boone, 1931). *Aquaculture, 354-355,* 50–55.

Kalbe, C., Priepke, A., Nürnberg, G., & Dannenberger, D. (2019). Effects of long-term microalgae supplementation on muscle microstructure, meat quality and fatty acid composition in growing pigs. *Journal of Animal Physiology and Animal Nutrition, 103,* 574–582.

Kang, H. K., Park, S. B., & Kim, C. H. (2017). Effects of dietary supplementation with a chlorella by-product on the growth performance, immune response, intestinal microflora and intestinal mucosal morphology in broiler chickens. *Journal of Animal Physiology and Animal Nutrition, 101,* 208–214.

Kang, H. K., Salim, H. M., Akter, N., Kim, D. W., Kim, J. H., Bang, H. T., et al. (2013). Effect of various forms of dietary chlorella supplementation on growth performance, immune characteristics, and intestinal microflora population of broiler chickens. *Journal of Applied Poultry Research, 22,* 100–108.

Kata, F. S., Athbi, A. M., Manwar, E. Q., Al-Ashoor, A., Abdel-Daim, M. M., & Aleya, L. (2018). Therapeutic effect of the alkaloid extract of the cyanobacterium *Spirulina platensis* on the lipid profile of hypercholesterolemic male rabbits. *Environmental Science and Pollution Research, 25,* 19635–19642.

Kholif, A. E., Morsy, T. A., Matloup, O. H., Anele, U. Y., Mohamed, A. G., & El-Sayed, A. B. (2017). Dietary *Chlorella vulgaris* microalgae improves feed utilization, milk production and concentrations of conjugated linoleic acids in the milk of Damascus goats. *Journal of Agricultural Science, 155,* 508–518.

Kibria, S., & Kim, I. H. (2019). Impacts of dietary microalgae (*Schizochytrium* JB5) on growth performance, blood profiles, apparent total tract digestibility, and ileal nutrient digestibility in weaning pigs. *Journal of the Science of Food and Agriculture, 99,* 6084–6088.

Kim, M. Y., Cheong, S. H., Lee, J. H., Kim, M. J., Sok, D. E., & Kim, M. R. (2010). Spirulina improves antioxidant status by reducing oxidative stress in rabbits fed a high-cholesterol diet. *Journal of Medicinal Food, 13,* 420–426.

Kiron, V., Phromkunthong, W., Huntley, M., Archibald, I., & De Scheemaker, G. (2012). Marine microalgae from biorefinery as a potential feed protein source for Atlantic salmon, common carp and whiteleg shrimp. *Aquaculture Nutrition, 18,* 521–531.

Kiron, V., Sørensen, M., Huntley, M., Vasanth, G. K., Gong, Y., Dahle, D., et al. (2016). Defatted biomass of the microalga, *Desmodesmus* sp., can replace fishmeal in the feeds for Atlantic salmon. *Frontiers in Marine Science, 3.*

Kissinger, K. R., García-Ortega, A., & Trushenski, J. T. (2016). Partial fish meal replacement by soy protein concentrate, squid and algal meals in low fish-oil diets containing *Schizochytrium limacinum* for longfin yellowtail *Seriola rivoliana. Aquaculture, 452,* 37–44.

Kotrbáček, V., Doubek, J., & Doucha, J. (2015). The chlorococcalean alga Chlorella in animal nutrition: A review. *Journal of Applied Phycology, 27,* 2173–2180.

Kousoulaki, K., Østbye, T.-K. K., Krasnov, A., Torgersen, J. S., Mørkøre, T., & Sweetman, J. (2015). Metabolism, health and fillet nutritional quality in Atlantic salmon (*Salmo salar*) fed diets containing n-3-rich microalgae. *Journal of Nutritional Science, 4,* e24.

Kralik, Z., Kralik, G., Grčević, M., Hanžek, D., & Margeta, P. (2020). Microalgae *Schizochytrium limacinum* as an alternative to fish oil in enriching table eggs with n-3 polyunsaturated fatty acids. *Journal of the Science of Food and Agriculture, 100,* 587–594.

Kulpys, J., Paulauskas, E., Pilipaviþius, V., & Stankeviþius, R. (2009). Influence of cyanobacteria Arthrospira (*Spirulina*) platensis biomass additives towards the body condition of lactation cows and biochemical milk indexes. *Agronomy Research, 7,* 823–835.

Lamminen, M., Halmemies-Beauchet-Filleau, A., Kokkonen, T., Jaakkola, S., & Vanhatalo, A. (2019). Different microalgae species as a substitutive protein feed for soya bean meal in grass silage based dairy cow diets. *Animal Feed Science and Technology, 247,* 112–126.

Lamminen, M., Halmemies-Beauchet-Filleau, A., Kokkonen, T., Simpura, I., Jaakkola, S., & Vanhatalo, A. (2017). Comparison of microalgae and rapeseed meal as supplementary protein in the grass silage based nutrition of dairy cows. *Animal Feed Science and Technology, 234,* 295–311.

Lamminen, M., Halmemies-Beauchet-Filleau, A., Kokkonen, T., Vanhatalo, A., & Jaakkola, S. (2019). The effect of partial substitution of rapeseed meal and faba beans by *Spirulina platensis* microalgae on milk production, nitrogen utilization, and amino acid metabolism of lactating dairy cows. *Journal of Dairy Science, 102,* 7102–7117.

Lee, A., You, L., Oh, S.-Y., Li, Z., Code, A., Zhu, C., et al. (2019a). Health benefits of supplementing nursery pig diets with microalgae or fish oil. *Animals, 9,* 80.

Lee, A. V., You, L., Oh, S. Y., Li, Z., Fisher-Heffernan, R. E., Regnault, T. R. H., et al. (2019b). Microalgae supplementation to late gestation sows and its effects on the health status of weaned piglets fed diets containing high- or low-quality protein sources. *Veterinary Immunology and Immunopathology, 218,* 109937.

Loor, J. J., Doreau, M., Chardigny, J. M., Ollier, A., Sebedio, J. L., & Chilliard, Y. (2005). Effects of ruminal or duodenal supply of fish oil on milk fat secretion and profiles of *trans*-fatty acids and conjugated linoleic acid isomers in dairy cows fed maize silage. *Animal Feed Science and Technology, 119,* 227–246.

Lum, K. K., Kim, J., & Lei, X. G. (2013). Dual potential of microalgae as a sustainable biofuel feedstock and animal feed. *Journal of Animal Science and Biotechnology, 4,* 53.

Madeira, M. S., Cardoso, C., Lopes, P. A., Coelho, D., Afonso, C., Bandarra, N. M., & Prates, J. A. M. (2017). Microalgae as feed ingredients for livestock production and meat quality: A review. *Livestock Science, 205,* 111–121.

Mahmoud, A. E., Naguib, M. M., Higazy, A. M., Sultan, Y. Y., & Marrez, D. A. (2017). Effect of substitution Soybean by blue green alga *Spirulina platensis* on performance and meat quality of growing rabbits. *American Journal of Food Technology, 12,* 51–59.

Manor, M. L., Derksen, T. J., Magnuson, A. D., Raza, F., & Lei, X. G. (2019). Inclusion of dietary defatted microalgae dose-dependently enriches ω-3 fatty acids in egg yolk and tissues of laying hens. *Journal of Nutrition, 149,* 942–950.

Mariey, Y. A., Samak, H. R., & Ibrahem, M. A. (2012). Effect of using *Spirulina platensis* algae as a feed additive for poultry diets: 1-Productive and reproductive performances of local laying hens. *Egyptian Poultry Science, 32,* 201–215.

Marques, J. A., Del Valle, T. A., Ghizzi, L. G., Zilio, E. M. C., Gheller, L. S., Nunes, A. T., et al. (2019). Increasing dietary levels of docosahexaenoic acid-rich microalgae: Ruminal fermentation, animal performance, and milk fatty acid profile of mid-lactating dairy cows. *Journal of Dairy Science, 102,* 5054–5065.

Marriott, N. G., Garrett, J. E., Sims, M. D., & Abril, J. R. (2002). Performance characteristics and fatty acid composition of pigs fed a diet with docosahexaenoic acid. *Journal of Muscle Foods, 13,* 265–277.

Mavrommatis, A., Chronopoulou, E. G., Sotirakoglou, K., Labrou, N. E., Zervas, G., & Tsiplakou, E. (2018). The impact of the dietary supplementation level with *Schizochytrium* sp. on the oxidative capacity of both goats' organism and milk. *Livestock Science, 218,* 37–43.

Meadus, W. J., Duff, P., Rolland, D., Aalhus, J. L., Uttaro, B., & Dugan, M. E. R. (2011). Feeding docosahexaenoic acid to pigs reduces blood triglycerides and induces gene expression for fat oxidation. *Canadian Journal of Animal Science, 91,* 601–612.

Meadus, W. J., Duff, P., Uttaro, B., Aalhus, J. L., Rolland, D. C., Gibson, L. L., et al. (2010). Production of docosahexaenoic acid (DHA) enriched bacon. *Journal of Agricultural and Food Chemistry, 58,* 465–472.

Meale, S. J., Chaves, A. V., He, M. L., & Mcallister, T. A. (2014). Dose-response of supplementing marine algae (*Schizochytrium* spp.) on production performance, fatty acid profiles, and wool parameters of growing lambs. *Journal of Animal Science, 92*, 2202–2213.

Medina Félix, D., López Elías, J. A., Campa Córdova, Á. I., Martínez Córdova, L. R., Luna González, A., Cortes Jacinto, E., et al. (2017). Survival of *Litopenaeus vannamei* shrimp fed on diets supplemented with *Dunaliella* sp. is improved after challenges by *Vibrio parahaemolyticus*. *Journal of Invertebrate Pathology, 148*, 118–123.

Minhas, A. K., Hodgson, P., Barrow, C. J., & Adholeya, A. (2016). A review on the assessment of stress conditions for simultaneous production of microalgal lipids and carotenoids. *Frontiers in Microbiology, 7*.

Mirzaie, S., Zirak-Khattab, F., Hosseini, S. A., & Donyaei-Darian, H. (2018). Effects of dietary *Spirulina* on anti-oxidant status, lipid profile, immune response and performance characteristics of broiler chickens reared under high ambient temperature. *Asian-Australasian Journal of Animal Sciences, 31*, 556–563.

Moate, P. J., Williams, S. R. O., Hannah, M. C., Eckard, R. J., Auldist, M. J., Ribaux, B. E., et al. (2013). Effects of feeding algal meal high in docosahexaenoic acid on feed intake, milk production, and methane emissions in dairy cows. *Journal of Dairy Science, 96*, 3177–3188.

Morales-De La Nuez, A., Moreno-Indias, I., Sanchez-Macias, D., Hernandez-Castellano, L. E., Suarez-Trujillo, A., Assuncao, P., et al. (2014). Effects of *Crypthecodinium cohnii*, *Chlorela* spp. and *Isochrysis galbana* addition to milk replacer on goat kids and lambs growth. *Journal of Applied Animal Research, 42*, 213–216.

Moran, C., Currie, D., Keegan, J., & Knox, A. (2018). Tolerance of broilers to dietary supplementation with high levels of the DHA-rich microalga, *Aurantiochytrium limacinum*: Effects on health and productivity. *Animals, 8*, 180.

Moran, C. A., Morlacchini, M., Keegan, J. D., & Fusconi, G. (2018). Dietary supplementation of finishing pigs with the docosahexaenoic acid-rich microalgae, *Aurantiochytrium limacinum*: Effects on performance, carcass characteristics and tissue fatty acid profile. *Asian-Australasian Journal of Animal Sciences, 31*, 712–720.

Moran, C. A., Morlacchini, M., Keegan, J. D., Warren, H., & Fusconi, G. (2019). Dietary supplementation of dairy cows with a docosahexaenoic acid-rich thraustochytrid, *Aurantiochytrium limacinum*: Effects on milk quality, fatty acid composition and cheese making properties. *Journal of Animal and Feed Sciences, 28*, 3–14.

Mordenti, A. L., Sardi, L., Bonaldo, A., Pizzamiglio, V., Brogna, N., Cipollini, I., et al. (2010). Influence of marine algae (*Schizochytrium* spp.) dietary supplementation on doe performance and progeny meat quality. *Livestock Science, 128*, 179–184.

Mousa, Y. I. M., Abdel-Monem, U. M., & Bazid, A. I. (2018). Effect of *Spirulina* and prebiotic (Inmunair 17.5®) on New-Zealand white rabbits performance. *Animal Poultry and Fish Production Research, 45*, 385–393.

Murphy, E. M., Stanton, C., Brien, C. O., Murphy, C., Holden, S., Murphy, R. P., et al. (2017). The effect of dietary supplementation of algae rich in docosahexaenoic acid on boar fertility. *Theriogenology, 90*, 78–87.

Nakagawa, H., Mustafa, M. G., Takii, K., Umino, T., & Kumai, H. (2000). Effect of dietary catechin and *Spirulina* on vitamin C metabolism in red sea bream. *Fisheries Science, 66*, 321–326.

Naylor, R. L., Hardy, R. W., Bureau, D. P., Chiu, A., Elliott, M., Farrell, A. P., et al. (2009). Feeding aquaculture in an era of finite resources. *Proceedings of the National Academy of Sciences of the United States of America, 106*, 15103–15110.

Neumann, C., Velten, S., & Liebert, F. (2018). N balance studies emphasize the superior protein quality of pig diets at high inclusion level of algae meal (*Spirulina platensis*) or insect meal (*Hermetia illucens*) when adequate amino acid supplementation is ensured. *Animals, 8*, 172.

Nicolaisen, O., & Bolla, S. (2016). Behavioural responses to visual environment in early stage Atlantic cod *Gadus morhua* L. larvae. *Aquaculture Research, 47*, 189–198.

Oh, S. T., Zheng, L., Kwon, H. J., Choo, Y. K., Lee, K. W., Kang, C. W., et al. (2015). Effects of dietary fermented *Chlorella vulgaris* (Cbt®) on growth performance, relative organ weights, cecal microflora, tibia bone characteristics, and meat qualities in Pekin ducks. *Asian-Australasian Journal of Animal Sciences, 28*, 95–101.

Omri, B., Amraoui, M., Tarek, A., Lucarini, M., Durazzo, A., Cicero, N., et al. (2019). *Arthrospira platensis* (Spirulina) supplementation on laying hens' performance: Eggs physical, chemical, and sensorial qualities. *Food, 8*.

Otte, M. V., Moreira, F., Bianchi, I., Oliveira, J., Mendes, R. E., Haas, C. S., et al. (2019). Effects of supplying omega-3 polyunsaturated fatty acids to gilts after weaning on metabolism and ovarian gene expression. *Journal of Animal Science, 97*, 374–384.

Pajor, F., Egerszegi, I., Steiber, O., Bodnár, Á., & Póti, P. (2019). Effect of marine algae supplementation on the fatty acid profile of milk of dairy goats kept indoor and on pasture. *Journal of Animal and Feed Sciences, 28*, 169–176.

Pakravan, S., Akbarzadeh, A., Sajjadi, M. M., Hajimoradloo, A., & Noori, F. (2017). Partial and total replacement of fish meal by marine microalga *Spirulina platensis* in the diet of Pacific white shrimp *Litopenaeus vannamei*: Growth, digestive enzyme activities, fatty acid composition and responses to ammonia and hypoxia stress. *Aquaculture Research, 48*, 5576–5586.

Palmegiano, G. B., Agradi, E., Forneris, G., Gai, F., Gasco, L., Rigamonti, E., et al. (2005). *Spirulina* as a nutrient source in diets for growing sturgeon (*Acipenser baeri*). *Aquaculture Research, 36,* 188–195.

Panjaitan, T., Quigley, S. P., Mclennan, S. R., Swain, A. J., & Poppi, D. P. (2015). Spirulina (*Spirulina platensis*) algae supplementation increases microbial protein production and feed intake and decreases retention time of digesta in the rumen of cattle. *Animal Production Science, 55,* 535–543.

Papadopoulos, G., Goulas, C., Apostolaki, E., & Abril, R. (2002). Effects of dietary supplements of algae, containing polyunsaturated fatty acids, on milk yield and the composition of milk products in dairy ewes. *Journal of Dairy Research, 69,* 357–365.

Park, J. H., Lee, S. I., & Kim, I. H. (2018). Effect of dietary *Spirulina* (Arthrospira) platensis on the growth performance, antioxidant enzyme activity, nutrient digestibility, cecal microflora, excreta noxious gas emission, and breast meat quality of broiler chickens. *Poultry Science, 97,* 2451–2459.

Park, J. H., Upadhaya, S. D., & Kim, I. H. (2015). Effect of dietary marine microalgae (*Schizochytrium*) powder on egg production, blood lipid profiles, egg quality, and fatty acid composition of egg yolk in layers. *Asian-Australasian Journal of Animal Sciences, 28,* 391–397.

Patterson, D., & Gatlin, D. M. (2013). Evaluation of whole and lipid-extracted algae meals in the diets of juvenile red drum (*Sciaenops ocellatus*). *Aquaculture, 416-417,* 92–98.

Peiretti, P. G., & Meineri, G. (2008). Effects of diets with increasing levels of *Spirulina platensis* on the performance and apparent digestibility in growing rabbits. *Livestock Science, 118,* 173–177.

Peiretti, P. G., & Meineri, G. (2011). Effects of diets with increasing levels of *Spirulina platensis* on the carcass characteristics, meat quality and fatty acid composition of growing rabbits. *Livestock Science, 140,* 218–224.

Pestana, J. M., Puerta, B., Santos, H., Lopes, P. A., Madeira, M. S., Alfaia, C. M., et al. (2020). Impact of dietary incorporation of Spirulina (*Arthrospira platensis*) and exogenous enzymes on broiler performance, carcass traits and meat quality. *Poultry Science, 99,* 2519–2532.

Pickard, R. M., Beard, A. P., Seal, C. J., & Edwards, S. A. (2008). Neonatal lamb vigour is improved by feeding docosahexaenoic acid in the form of algal biomass during late gestation. *Animal, 2,* 1186–1192.

Pleissner, D., & Eriksen, N. T. (2012). Effects of phosphorous, nitrogen, and carbon limitation on biomass composition in batch and continuous flow cultures of the heterotrophic dinoflagellate *Crypthecodinium cohnii*. *Biotechnology and Bioengineering, 109,* 2005–2016.

Ponnampalam, E. N., Burnett, V. F., Norng, S., Hopkins, D. L., Plozza, T., & Jacobs, J. L. (2016). Muscle antioxidant (vitamin E) and major fatty acid groups, lipid oxidation and retail colour of meat from lambs fed a roughage based diet with flaxseed or algae. *Meat Science, 111,* 154–160.

Póti, P., Pajor, F., Bodnár, Á., Penksza, K., & Köles, P. (2015). Effect of micro-alga supplementation on goat and cow milk fatty acid composition. *Chilean Journal of Agricultural Research, 75,* 259–263.

Qiao, H., Wang, H., Song, Z., Ma, J., Li, B., Liu, X., et al. (2014). Effects of dietary fish oil replacement by microalgae raw materials on growth performance, body composition and fatty acid profile of juvenile olive flounder, *Paralichthys olivaceus*. *Aquaculture Nutrition, 20,* 646–653.

Qureshi, M. A., Garlich, J. D., & Kidd, M. T. (1996). Dietary *Spirulina platensis* enhances humoral and cell-mediated immune functions in chickens. *Immunopharmacology and Immunotoxicology, 18,* 465–476.

Reynolds, C. K., Cannon, V. L., & Loerch, S. C. (2006). Effects of forage source and supplementation with soybean and marine algal oil on milk fatty acid composition of ewes. *Animal Feed Science and Technology, 131,* 333–357.

Rezvani, M., Zaghari, M., & Moravej, H. (2012). A survey on *Chlorella vulgaris* effect's on performance and cellular immunity in broilers. *International Journal of Agricultural Science and Research, 3,* 9–15.

Ribeiro, T., Lordelo, M. M., Alves, S. P., Bessa, R. J. B., Costa, P., Lemos, J. P. C., et al. (2013). Direct supplementation of diet is the most efficient way of enriching broiler meat with n-3 long-chain polyunsaturated fatty acids. *British Poultry Science, 54,* 753–765.

Rocha, R. J., Ribeiro, L., Costa, R., & Dinis, M. T. (2008). Does the presence of microalgae influence fish larvae prey capture? *Aquaculture Research, 39,* 362–369.

Rodriguez-Herrera, M., Khatri, Y., Marsh, S. P., Posri, W., & Sinclair, L. A. (2018). Feeding microalgae at a high level to finishing heifers increases the long-chain n-3 fatty acid composition of beef with only small effects on the sensory quality. *International Journal of Food Science and Technology, 53,* 1405–1413.

Saeid, A., Chojnacka, K., Korczy, M., & Korniewicz, D. (2013). Effect on supplementation of *Spirulina maxima* enriched with Cu on production performance, metabolical and physiological parameters in fattening pigs. *Journal of Applied Phycology, 25,* 1607–1617.

Saeid, A., Chojnacka, K., Opaliński, S., & Korczyński, M. (2016). Biomass of *Spirulina maxima* enriched by biosorption process as a new feed supplement for laying hens. *Algal Research, 19,* 342–347.

Sardi, L., Martelli, G., Lambertini, L., Parisini, P., & Mordenti, A. (2006). Effects of a dietary supplement of DHA-rich marine algae on Italian heavy pig production parameters. *Livestock Science, 103,* 95–103.

Sargent, J. R., Mcevoy, J., & Bell, J. G. (1997). Requirements, presentation and sources of polyunsaturated fatty acids in marine fish larval feeds. *Aquaculture, 155,* 117–127.

Sarker, P. K., Gamble, M. M., Kelson, S., & Kapuscinski, A. R. (2016). Nile tilapia (*Oreochromis niloticus*) show high digestibility of lipid and fatty acids from marine *Schizochytrium* sp. and of protein and essential amino acids from freshwater *Spirulina* sp. feed ingredients. *Aquaculture Nutrition, 22,* 109–119.

Schiavone, A., Chiarini, R., Marzoni, M., Castillo, A., Tassone, S., & Romboli, I. (2007). Breast meat traits of Muscovy ducks fed on a microalga (*Crypthecodinium cohnii*) meal supplemented diet. *British Poultry Science, 48,* 573–579.

Schingoethe, D. J. (2017). A 100-year review: Total mixed ration feeding of dairy cows. *Journal of Dairy Science, 100,* 10143–10150.

Selim, S., Hussein, E., & Abou-Elkhair, R. (2018). Effect of *Spirulina platensis* as a feed additive on laying performance, egg quality and hepatoprotective activity of laying hens. *European Poultry Science, 82.*

Senosy, W., Kassab, A. Y., & Mohammed, A. A. (2017). Effects of feeding green microalgae on ovarian activity, reproductive hormones and metabolic parameters of Boer goats in arid subtropics. *Theriogenology, 96,* 16–22.

Shah, M. R., Lutzu, G. A., Alam, A., Sarker, P., Kabir Chowdhury, M. A., Parsaeimehr, A., et al. (2018). Microalgae in aquafeeds for a sustainable aquaculture industry. *Journal of Applied Phycology, 30,* 197–213.

Shanmugapriya, B., Babu, S. S., Hariharan, T., Sivaneswaran, S., Anusha, M. B., & College, C. N. (2015). Research article dietary administration of *Spirulina platensis* as probiotics on growth performance and histopathology in broiler chicks. *International Journal of Recent Scientific Research, 6,* 2650–2653.

Shi, X., Luo, Z., Chen, F., Wei, C.-C., Wu, K., Zhu, X.-M., et al. (2017). Effect of fish meal replacement by Chlorella meal with dietary cellulase addition on growth performance, digestive enzymatic activities, histology and myogenic genes' expression for crucian carp *Carassius auratus. Aquaculture Research, 48,* 3244–3256.

Shingfield, K. J., Bonnet, M., & Scollan, N. D. (2013). Recent developments in altering the fatty acid composition of ruminant-derived foods. *Animal, 7,* 132–162.

Shingfield, K. J., & Griinari, J. M. (2007). Role of biohydrogenation intermediates in milk fat depression. *European Journal of Lipid Science and Technology, 109,* 799–816.

Shingfield, K. J., Kairenius, P., Arola, A., Paillard, D., Muetzel, S., Ahvenjarvi, S., et al. (2012). Dietary fish oil supplements modify ruminal biohydrogenation, alter the flow of fatty acids at the omasum, and induce changes in the ruminal *Butyrivibrio* population in lactating cows. *Journal of Nutrition, 142,* 1437–1448.

Sikiru, A. B., Arangasamy, A., Alemede, I. C., Egena, S. S. A., & Bhatta, R. (2019). Dietary supplementation effects of *Chlorella vulgaris* on performances, oxidative stress status and antioxidant enzymes activities of prepubertal New

Zealand White rabbits. *Bulletin of the National Research Centre, 43,* 162.

Sikiru, A. B., Arangasamy, A., Alemede, I. C., Guvvala, P. R., Egena, S. S. A., Ippala, J. R., et al. (2019). *Chlorella vulgaris* supplementation effects on performances, oxidative stress and antioxidant genes expression in liver and ovaries of New Zealand White rabbits. *Heliyon, 5,* e02470.

Sikiru, A., Arangasamy, A., Ijaiya, A., Ippala, R., & Bhatta, R. (2019). Effects of *Chlorella vulgaris* supplementation on performances of lactating nulliparous New Zealand white rabbits does and their kits. *International Journal of Livestock Research, 9,* 37–45.

Simopoulos, A. P. (1999). Essential fatty acids in health and chronic disease. *The American Journal of Clinical Nutrition, 70,* 560S–569S.

Simopoulos, A. P. (2004). Omega-6/Omega-3 essential fatty acid ratio and chronic diseases. *Food Reviews International, 20,* 77–90.

Sinclair, L. A., Lock, A. L., Early, R., & Bauman, D. E. (2007). Effects of Trans-10, Cis-12 conjugated linoleic acid on ovine milk fat synthesis and cheese properties1. *Journal of Dairy Science, 90,* 3326–3335.

Sinedino, L. D. P., Honda, P. M., Souza, L. R. L., Lock, A. L., Boland, M. P., Staples, C. R., et al. (2017). Effects of supplementation with docosahexaenoic acid on reproduction of dairy cows. *Reproduction, 153,* 707–723.

Sørensen, M., Berge, G. M., Reitan, K. I., & Ruyter, B. (2016). Microalga *Phaeodactylum tricornutum* in feed for Atlantic salmon (*Salmo salar*)—Effect on nutrient digestibility, growth and utilization of feed. *Aquaculture, 460,* 116–123.

Sørensen, M., Gong, Y., Bjarnason, F., Vasanth, G. K., Dahle, D., Huntley, M., et al. (2017). *Nannochloropsis oceania*-derived defatted meal as an alternative to fishmeal in Atlantic salmon feeds. *PLoS One, 12,* e0179907.

Sucu, E., Udum, D., Gunes, N., Canbolat, O., & Filya, I. (2017). Influence of supplementing diet with microalgae (*Schizochytrium limacinum*) on growth and metabolism in lambs during the summer. *Turkish Journal of Veterinary & Animal Sciences, 41,* 167–174.

Tacon, A. G. J., Hasan, M. R., & Metian, M. (2011). *Demand and supply of feed ingredients for farmed fish and crustaceans—Trends and prospects.* FAO Fisheries and Aquaculture Technical Paper(Vol. 564), 87 pp.

Tacon, A. G. J., & Metian, M. (2015). Feed matters: Satisfying the feed demand of aquaculture. *Reviews in Fisheries Science & Aquaculture, 23,* 1–10.

Teimouri, M., Amirkolaie, A. K., & Yeganeh, S. (2013). The effects of *Spirulina platensis* meal as a feed supplement on growth performance and pigmentation of rainbow trout (*Oncorhynchus mykiss*). *Aquaculture, 396–399,* 14–19.

Teimouri, M., Yeganeh, S., Mianji, G. R., Najafi, M., & Mahjoub, S. (2019). The effect of *Spirulina platensis* meal on antioxidant gene expression, total antioxidant capacity, and lipid peroxidation of rainbow trout (*Oncorhynchus mykiss*). *Fish Physiology and Biochemistry*, 45, 977–986.

Thorning, T. K., Raben, A., Tholstrup, T., Soedamah-Muthu, S. S., Givens, I., & Astrup, A. (2016). Milk and dairy products: Good or bad for human health? An assessment of the totality of scientific evidence. *Food & Nutrition Research*, 60, 32527.

Tibaldi, E., Chini Zittelli, G., Parisi, G., Bruno, M., Giorgi, G., Tulli, F., et al. (2015). Growth performance and quality traits of European sea bass (*D. labrax*) fed diets including increasing levels of freeze-dried *Isochrysis* sp. (T-ISO) biomass as a source of protein and n-3 long chain PUFA in partial substitution of fish derivatives. *Aquaculture*, 440, 60–68.

Tibbetts, S. M. (2018). The potential for 'Next-Generation', microalgae-based feed ingredients for salmonid aquaculture in context of the blue revolution. In E. Jacob-Lopes, L. Q. Zepka, & M. I. Queiroz (Eds.), *Microalgal biotechnology*. IntechOpen.

Till, B. E., Huntington, J. A., Posri, W., Early, R., Taylor-Pickard, J., & Sinclair, L. A. (2019). Influence of rate of inclusion of microalgae on the sensory characteristics and fatty acid composition of cheese and performance of dairy cows. *Journal of Dairy Science*, 102, 10934–10946.

Tonnac, A. D., Labussière, E., Vincent, A., Mourot, J., Pegase, U. M. R., & Ouest, A. (2016). Effect of α-linolenic acid and DHA intake on lipogenesis and gene expression involved in fatty acid metabolism in growing-finishing pigs. *The British Journal of Nutrition*, 116(1), 7–18.

Tonnac, A. D., & Mourot, J. (2017). Effect of dietary sources of n-3 fatty acids on pig performance and technological, nutritional and sensory qualities of pork. *Animal*, 1–9. https://doi.org/10.1017/S1751731117002877.

Toral, P. G., Belenguer, A., Frutos, P., & Hervás, G. (2009). Effect of the supplementation of a high-concentrate diet with sunflower and fish oils on ruminal fermentation in sheep. *Small Ruminant Research*, 81, 119–125.

Toral, P. G., Bernard, L., Belenguer, A., Rouel, J., Hervás, G., Chilliard, Y., et al. (2016). Comparison of ruminal lipid metabolism in dairy cows and goats fed diets supplemented with starch, plant oil, or fish oil. *Journal of Dairy Science*, 99, 301–316.

Toral, P. G., Hervás, G., Carreño, D., Leskinen, H., Belenguer, A., Shingfield, K. J., et al. (2017). In vitro response to EPA, DPA, and DHA: Comparison of effects on ruminal fermentation and biohydrogenation of 18-carbon fatty acids in cows and ewes. *Journal of Dairy Science*, 100, 6187–6198.

Toral, P. G., Shingfield, K. J., Hervás, G., Toivonen, V., & Frutos, P. (2010). Effect of fish oil and sunflower oil on rumen fermentation characteristics and fatty acid composition of digesta in ewes fed a high concentrate diet. *Journal of Dairy Science*, 93, 4804–4817.

Toyomizu, M., Sato, K., Taroda, H., Kato, T., & Akiba, Y. (2001). Effects of dietary *Spirulina* on meat colour in muscle of broiler chickens. *British Poultry Science*, 42, 197–202.

Tsiplakou, E., Abdullah, M. A. M., Alexandros, M., Chatzikonstantinou, M., Skliros, D., Sotirakoglou, K., et al. (2017). The effect of dietary *Chlorella pyrenoidosa* inclusion on goats milk chemical composition, fatty acids profile and enzymes activities related to oxidation. *Livestock Science*, 197, 106–111.

Tsiplakou, E., Abdullah, M. A. M., Mavrommatis, A., Chatzikonstantinou, M., Skliros, D., Sotirakoglou, K., et al. (2018). The effect of dietary *Chlorella vulgaris* inclusion on goat's milk chemical composition, fatty acids profile and enzymes activities related to oxidation. *Journal of Animal Physiology and Animal Nutrition*, 102, 142–151.

Tsiplakou, E., Abdullah, M. A. M., Skliros, D., Chatzikonstantinou, M., Flemetakis, E., Labrou, N., et al. (2017). The effect of dietary *Chlorella vulgaris* supplementation on micro-organism community, enzyme activities and fatty acid profile in the rumen liquid of goats. *Journal of Animal Physiology and Animal Nutrition*, 101, 275–283.

Tulli, F., Chini Zittelli, G., Giorgi, G., Poli, B. M., Tibaldi, E., & Tredici, M. R. (2012). Effect of the inclusion of dried *Tetraselmis suecica* on growth, feed utilization, and fillet composition of European Sea Bass Juveniles fed organic diets. *Journal of Aquatic Food Product Technology*, 21, 188–197.

Urriola, P. E., Mielke, J. A., Mao, Q., Hung, Y.-T., Kurtz, J. F., Johnston, L. J., et al. (2018). Evaluation of a partially de-oiled microalgae product in nursery pig diets. *Translational Animal Science*, 2, 169–183.

Urrutia, N. L., & Harvatine, K. J. (2017). Acetate dose-dependently stimulates milk fat synthesis in lactating dairy cows. *Journal of Nutrition*, 147, 763–769.

Urrutia, O., Mendizabal, J. A., Insausti, K., Soret, B., Purroy, A., & Arana, A. (2016). Effects of addition of linseed and marine algae to the diet on adipose tissue development, fatty acid profile, lipogenic gene expression, and meat quality in lambs. *PLoS ONE*, 11.

Valente, L. M. P., Custódio, M., Batista, S., Fernandes, H., & Kiron, V. (2019). Defatted microalgae (*Nannochloropsis* sp.) from biorefinery as a potential feed protein source to replace fishmeal in European sea bass diets. *Fish Physiology and Biochemistry*, 45, 1067–1081.

Vanbergue, E., Hurtaud, C., Peyraud, J. L., Beuvier, E., Duboz, G., & Buchin, S. (2018). Effects of n-3 fatty acid sources on butter and hard cooked cheese; technological properties and sensory quality. *International Dairy Journal*, 82, 35–44.

Vanbergue, E., Peyraud, J. L., & Hurtaud, C. (2018). Effects of new n-3 fatty acid sources on milk fatty acid profile and milk fat properties in dairy cows. *Journal of Dairy Research*, 85, 265–272.

Vántus, V., Bónai, A., Zsolnai, A., Dal Bosco, A., Szendro, Z., Tornyos, G., et al. (2012). Single and combined effect of dietary Thyme (*Thymus vulgaris*) and Spirulina (*Arthrospira platensis*) on bacterial community in the caecum and caecal fermentation of rabbits. *Acta Agriculturae Slovenica, 100,* 77–81.

Venkataraman, L. V., Somasekaran, T., & Becker, E. W. (1994). Replacement value of blue-green alga (*Spirulina platensis*) for fishmeal and a vitamin-mineral premix for broiler chicks. *British Poultry Science, 35,* 373–381.

Vizcaíno, A. J., López, G., Sáez, M. I., Jiménez, J. A., Barros, A., Hidalgo, L., et al. (2014). Effects of the microalga *Scenedesmus almeriensis* as fishmeal alternative in diets for gilthead sea bream, *Sparus aurata,* juveniles. *Aquaculture, 431,* 34–43.

Vossen, E., Raes, K., Van Mullem, D., & De Smet, S. (2017). Production of docosahexaenoic acid (DHA) enriched loin and dry cured ham from pigs fed algae: Nutritional and sensory quality. *European Journal of Lipid Science and Technology, 119,* 1600144.

Walker, A. B., & Berlinsky, D. L. (2011). Effects of partial replacement of fish meal protein by microalgae on growth, feed intake, and body composition of Atlantic Cod. *North American Journal of Aquaculture, 73,* 76–83.

Wang, Y., Li, M., Filer, K., Xue, Y., Ai, Q., & Mai, K. (2017a). Replacement of fish oil with a Dha-rich *Schizochytrium* meal on growth performance, activities of digestive enzyme and fatty acid profile of Pacific white shrimp (*Litopenaeus vannamei*) larvae. *Aquaculture Nutrition, 23,* 1113–1120.

Wang, Y. Y., Li, M. Z., Filer, K., Xue, Y., Ai, Q. H., & Mai, K. S. (2017b). Evaluation of Schizochytrium meal in microdiets of Pacific white shrimp (*Litopenaeus vannamei*) larvae. *Aquaculture Research, 48,* 2328–2336.

Weld, K. A., & Armentano, L. E. (2017). The effects of adding fat to diets of lactating dairy cows on total-tract neutral detergent fiber digestibility: A meta-analysis. *Journal of Dairy Science, 100,* 1766–1779.

Wild, K. J., Steingaß, H., & Rodehutscord, M. (2019). Variability of in vitro ruminal fermentation and nutritional value of cell-disrupted and nondisrupted microalgae for ruminants. *GCB Bioenergy, 11,* 345–359.

Wu, Y. B., Li, L., Wen, Z. G., Yan, H. J., Yang, P. L., Tang, J., et al. (2019). Dual functions of eicosapentaenoic acid-rich microalgae: Enrichment of yolk with n-3 polyunsaturated fatty acids and partial replacement for soybean meal in diet of laying hens. *Poultry Science, 98,* 350–357.

Yadav, G., Meena, D. K., Sahoo, A. K., Das, B. K., & Sen, R. (2020). Effective valorization of microalgal biomass for the production of nutritional fish-feed supplements. *Journal of Cleaner Production, 243,* 118697.

Yan, L., & Kim, I. H. (2013). Effects of dietary ω-3 fatty acid-enriched microalgae supplementation on growth performance, blood profiles, meat quality, and fatty acid composition of meat in broilers. *Journal of Applied Animal Research, 41,* 392–397.

Yan, L., Lim, S. U., & Kim, I. H. (2012). Effect of fermented Chlorella supplementation on growth performance, nutrient digestibility, blood characteristics, fecal microbial and fecal noxious gas content in growing pigs. *Asian-Australasian Journal of Animal Sciences, 25,* 1742–1747.

Young, A. J., Pritchard, J., White, D., & Davies, S. (2017). Processing of astaxanthin-rich *Haematococcus* cells for dietary inclusion and optimal pigmentation in Rainbow trout, *Oncorhynchus mykiss* L. *Aquaculture Nutrition, 23,* 1304–1311.

Zahroojian, N., Moravej, H., & Shivazad, M. (2013). Effects of dietary marine algae (*Spirulina platensis*) on egg quality and production performance of laying hens. *Journal of Agricultural Science and Technology, 15,* 1353–1360.

Zhu, H., Fievez, V., Mao, S., He, W., & Zhu, W. (2016). Dose and time response of ruminally infused algae on rumen fermentation characteristics, biohydrogenation and *Butyrivibrio* group bacteria in goats. *Journal of Animal Science and Biotechnology, 7.*

Zielińska, A., & Chojnacka, K. (2009). The comparison of biosorption of nutritionally significant minerals in single- and multi-mineral systems by the edible microalga *Spirulina* sp. *Journal of the Science of Food and Agriculture, 89,* 2292–2301.

Cosmetics applications

Andressa Costa de Oliveira[a], Ana Lucía Morocho-Jácome[a], Cibele Ribeiro de Castro Lima[b], Gabriela Argollo Marques[a], Maíra de Oliveira Bispo[a], Amanda Beatriz de Barros[a], João Guilherme Costa[c], Tânia Santos de Almeida[c], Catarina Rosado[c], João Carlos Monteiro de Carvalho[d], Maria Valéria Robles Velasco[a], and André Rolim Baby[a]

[a]Department of Pharmacy, Faculty of Pharmaceutical Sciences, University of São Paulo, São Paulo, Brazil, [b]Physics Institute, University of São Paulo, São Paulo, Brazil, [c]CBIOS—Universidade Lusófona's Research Center for Biosciences and Health Technologies, Lisbon, Portugal, [d]Department of Biochemical and Pharmaceutical Technology, Faculty of Pharmaceutical Sciences, University of São Paulo, São Paulo, Brazil

1 Introduction

Skin is considered the major organ in the human body and its principal function is protecting body against the external environment (e.g., microorganisms and other harmful invasive agents). The more superficial skin layer is stratified to act as a protective barrier to retain water and prevent dehydration and dryness. When the *stratum corneum* is compromised, it results in transepidermal water loss (TEWL) and gradual skin dehydration (Berardesca, Loden, Serup, Masson, & Rodrigues, 2018).

The appeal for products from natural origins has been growing dramatically in the last decade, leading the cosmetic industry to look for more natural and sustainable ways to develop cosmetic products. Thus algae extracts are promising and valuable in the cosmetic market (Khanra et al., 2018). Moreover, microalgae have compounds that can be used in the manufacture and development of cosmetics with particular beneficial properties for the prevention or treatment of skin aging (Ariede et al., 2017).

Only about 10% of microalgae species are identified and described, but a few species are produced at an industrial level (i.e., *Dunaliella salina, Haematococcus, Arthrospira, Chlorella, Aphanizomenon*) (Couteau & Coiffard, 2018). Moreover, only a very few macroalgae species have been cultivated for industrial purposes

(De Jesus Raposo, De Morais, & De Morais, 2015; Hudek, Davis, Ibbini, & Erickson, 2014; Wei, Quarterman, & Jin, 2013; Yun, Choi, & Kim, 2015) due to their real potential as cosmetic ingredients, like polysaccharides, for example. Still, such organisms (microorganisms) have basic nutritional needs and their cultivation does not compete with agriculture when using closed cultivation systems. Interestingly, numerous species are able to live in deserts, producing metabolites that provide them protection against excessive exposure to sunlight and desiccation (Derikvand, Liewellyn, & Purton, 2017; Quintana, Van der Kooy, Van de Rhee, Voshol, & Verpoorte, 2011). These compounds can be cogitated as actives for cosmetics and personal care products, considering their potential as sunscreen and moisturizer, respectively.

Biotechnology advances have resolved many challenges in algae cultivation to increase productivity, as well as quality of raw material (Carvalho, Matsudo, Bezerra, Ferreira-Camargo, & Sato, 2014). However, the extraction methods require more attention to improve green methodologies with better results. Even when various technological obstacles are currently discussed and studied (Chen et al., 2017; Koller, Muhr, & Braunegg, 2014; Kroumov et al., 2017; Rizwan, Mujtaba, Memon, Lee, & Rashid, 2018), other economic problems have emerged in microalgae biorefineries that have induced an increment in the final cost of the microalgae-derived compounds (e.g., astaxanthin from *Haematococcus pluvialis*) when compared with the synthetic one (Panis & Carreon, 2016).

It is important to mention that algae compounds have the ability to improve the sensory characteristics of cosmetic products and even offer additional advantages to achieve an increasing market tendency for industrialists and consumers considering the demand for healthier products or with natural ingredients. This chapter summarizes the recent scientific advances of new compounds from algae species to be considered as cosmetic ingredients.

2 The necessity of products environmentally sustainable in cosmetics

Algae applications in cosmetic products have several favorable properties, such as promoting blood circulation, provide moisture to the skin, activate metabolism and cell renewal, antiinflammatory effect, increase skin resistance, and even drain skin tissues (Khanra et al., 2018; Morone, Alfeus, Vasconcelos, & Martins, 2019; Pereira, 2018).

Algae have benefits in relation to seasonality, allowing the continuous supply, regardless of harvest, as they are considered an almost inexhaustible natural source. They also offer reproducibility of composition when cultivated under controlled conditions and chain traceability (Carvalho et al., 2014; Fu et al., 2016). It was also demonstrated that their controlled conditions during cultivation have increased biomass productivity while reducing the environmental impact of CO_2 emissions (Ferreira, Rodrigues, Converti, Sato, & Carvalho, 2012), water consumption (Mejia-Da-Silva, Matsudo, Morocho-Jacome, Carlos, & De Carvalho, 2018; Morocho-Jácome, Mascioli, Sato, & de Carvalho, 2015a), and even production costs (Morocho-Jácome, Mascioli, Sato, & De Carvalho, 2015b). Furthermore, many algae and microalgae also have been successfully cultivated in saline waters (Chen et al., 2017; Ishika et al., 2017) offering an advantage regarding conventional agriculture that is limited by the use of high amounts of fresh water and arable land (Hu, Nagarajan, Zhang, Chang, & Lee, 2018). In addition to this, many algae and microalgae produce valuable oils, like omega 3, with low-quality deviation, although with specification and proven origin (Lowrey, Armenta, & Brooks, 2016; Priyadarshani & Rath, 2012).

The main disadvantage in the production process is the challenge to achieve the best optimal performance and efficiency to extract compounds without environmental damages and, yet, with financial gains. Furthermore, it is necessary to consider a key procedure to promote disruption in the microalgae cell wall (Günerken et al., 2015) to release the targeted compounds to be used as active ingredients in cosmetics. We also highlight the necessity to inform the consumer, the technology, and innovation that is used to obtain these raw materials with cleaner methodologies than in the past.

On the other hand, compounds extracted from plants are the most common ingredients used in natural cosmetics due to the increasing number of concerns about the consumption of environment-friendly products. Nowadays, compounds from natural sources are becoming more common in cosmetic formulations. As a new source, algae gained attention from cosmetic industries due to their particular composition with interesting properties to maintain skin health (Ariede et al., 2017). However, the cosmetic attributes of these compounds have to be tested not only in vitro but also in vivo to guarantee the desirable cosmetic safety and efficacy.

The use of plant extracts in cosmetics has been reported due to their antioxidant effects on the skin (Anunciato & da Rocha Filho, 2012; Wang, Jónsdóttir, & Ólafsdóttir, 2009) and even their photoprotective activity (Cefali, Ataide, Moriel, Foglio, & Mazzola, 2016; Radice et al., 2016). However, extraction technologies need to be evaluated in order to diminish the use of pollutants. In this context, microalgae biorefineries have been created with the main purpose to offer more green extraction technologies to avoid environmental harm (Fresewinkel et al., 2014; Guihéneuf & Stengel, 2015; Kroumov et al., 2017; Ruiz et al., 2016; Wang, Li, Hu, Su, & Zhong, 2015).

Moreover, due to the toxic effects of various synthetic colorants, many industries are involved in the development of new algal derivatives to test a new generation of products obtained from different natural sources, such as microalgae derivatives, including natural dyes, since they are nontoxic and noncarcinogenic (Cezare-Gomes et al., 2019; Paliwal et al., 2016).

3 Skin structure

Skin is considered an organ that suffers the most evident transformations with aging, thus leading to changes in the integumentary system as well as cellular immunity, with concomitant effects in both physical and emotional states. Skin provides a dynamic barrier between the human body and the environment, performing important functions as the preservation of water and electrolytes; protection; regulation of water and temperature; sensorial perception of touch, heat and cold, as well as it synthesizes vitamin D (Gilaberte, Prieto-Torres, Pastushenko, & Juarranz, 2016).

Skin characteristics, such as topology, pH, temperature, humidity, and microbiology, differ according to different body locations. For instance, the thickness in the specific area such as face may vary from 0.1 mm on the greater eyelid to 1 mm on the upper nasal lip. Skin structure may differ according to race; even gender dissimilarities in the cutaneous tissue may be partly due to hormonal differences between both that regulate the body and facial hair distribution, sweating, sebum production, and pH (Kolarsick, Kolarsick, & Goodwin, 2011). Skin is structurally denser in men than women, and the loss of estrogen at menopause induces thin skin. This skin profile can be reversed by estrogen therapy. Clinical features of aging skin include melanocytic hyperplasia, xerosis, telangiectasia, and decreased elasticity (Wong, Geyer, Weninger, Guimberteau, & Wong, 2016).

A healthy skin surface can be characterized by acid pH (4.0–6.0) and it includes three

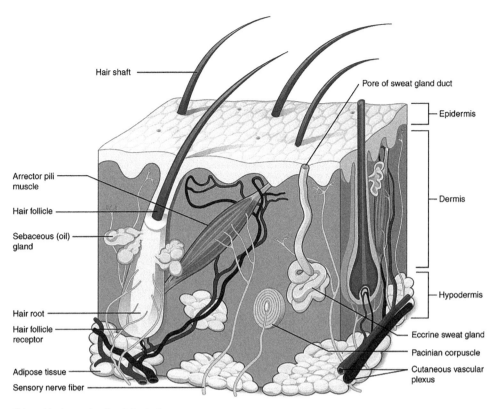

FIG. 1   Skin with the main physiological structures.

distinct layers, the epidermis, dermis, and hypodermis (Gilaberte et al., 2016). Fig. 1 shows the main structures in normal skin (OpenStax College, 2013).

The impermeable barrier is named epidermis, which is the outermost layer. It is formed by a stratified squamous epithelium, mainly composed of melanocytes and keratinocytes. Its layers include the basal layer (germ layer), which is the deepest portion of the epidermis, a *spinous stratum*, a *granular stratum*, *lucid stratum*, and *stratum corneum*, which is the most superficial portion of the epidermis with an important defensive function against environmental changes. The epidermis is a tissue that regenerates perpetually with cells continuously subjected to terminal differentiation followed by death. The total renewal time is about 2 months.

The basement membrane separates the epidermis from the dermis and contains no blood vessels (Gilaberte et al., 2016; Kolarsick et al., 2011).

The dermis is the largest part of the skin and confers flexibility, elasticity, strength, and resistance. It protects the body from mechanical injury and includes sensory stimulus receptors. It also contains resistant connective tissue, sweat glands, hair follicles, sensory neurons, and blood vessels. Its main component is collagen, an important structural protein for the whole body (Gilaberte et al., 2016; Kolarsick et al., 2011; Wong et al., 2016). The infantile dermis is composed of small collagen bundles while the adult dermis contains thicker collagen bundles. Many fibroblasts are present in the infantile dermis, but fewer persist in adults (Wong et al., 2016).

The water proportion in the dermis is maintained due to the hydrophilic properties of hyaluronic acid. Some smaller amounts of this acid can also be found in the intracellular matrix of the corneal layer (Kolarsick et al., 2011).

The hypodermis is the deepest tissue, consisting of fat, connective tissue, and elastin. It contains adipose lobes, along with some skin appendages, such as hair follicles, fibroblasts, macrophages that play a role in adipocyte homeostasis in obesity, sensory neurons, and blood vessels (Wong et al., 2016).

Knowledge about the skin layers with their components becomes more important in the prevention of skin aging, through various nutritional, environmental, mechanical factors and even with the proper use of cosmetics.

4 Property of algae in skincare products

Algae are commonly classified into microalgae and macroalgae. The great natural compositions of such living organisms reveal the interesting and increasing potential of applications in several industries, such as food, feed, cosmetics, and even pharmaceutics. Compounds extracted from algae biomass belong to the following categories: pigments (i.e., chlorophylls, carotenoids), carbohydrates (agar, alginate, carrageenan, fucoidan, polysaccharides, biopolymers), nutritional supplements (polyunsaturated fatty acids), proteins, and lipids (biodiesel) (Khanra et al., 2018).

4.1 Microalgae

Microalgae have been used as food since the last century due to their high protein content. Nowadays, *Chlorella* sp. and *Scenedesmus* sp. are produced and commercialized as a human supplement (Khanra et al., 2018).

Chlorella sp. is widely used in natural food and considered healthy and a dietary supplement (Ursu et al., 2014). It has health benefits, such as treatment of gastric ulcers, wounds, and constipation (De Jesus Raposo, De Morais, & De Morais, 2013a; Sathasivam, Radhakrishnan, Hashem, & Abd_Allah, 2017). It can also be used to prevent diseases, such as atherosclerosis, hypercholesterolemia activities, and tumors (Kulkarni & Nikolov, 2018; Sui, Gizaw, & Bemiller, 2012).

D. salina, *Chlorella vulgaris*, and *Nannochloropsis oculata* are used in the manufacture of skin, hair, and sun protection products. *C. vulgaris* extract is responsible for the production of collagen, which regenerates skin tissues, reducing the formation of wrinkles (Ariede et al., 2017). Cosmetic industries in France and Switzerland have invested in the creation of products containing carotenoids derived from *C. vulgaris* and *D. salina*, which have protective activity against UV radiation and the potential oxidative damage that can cause premature skin aging. Moreover, carotenoids are natural fat-soluble pigments extracted from algae and plants generally essential for the photosynthesis process, in which antioxidant activity plays an important role to protect human cells and tissues against oxidative stress (Stahl & Sies, 2012). Another antioxidant compound from green algae is a sulfated polysaccharide (sPS), whose main mechanism of action is the elimination of excess free radicals that can cause skin aging. In addition, it could chelate and reduce ferrous ions (Cunha & Grenha, 2016; De Jesus Raposo et al., 2015).

Another species, *H. pluvialis* from freshwater, is the largest natural resource of astaxanthin, a pigment used for the manufacture of cosmetics, nutraceuticals, and both human and animal foods. It has a high antioxidant capacity, provides defense against UV radiation and increased immunity being a source of provitamin A, and even has antiinflammatory activity (Rao et al., 2013; Wang, Zarka, Trebst, & Boussiba, 2003).

4.2 Macroalgae

Macroalgae, known as seaweeds, are large multicellular, visible to the naked eye, while microalgae are microscopic cells of single-celled or single-celled species found in various environments. Macroalgae is a natural source of proteins, polysaccharides, minerals, lipids, and secondary metabolites such as halogenated compounds, phenolic compounds, terpenoids, and nitrogen derivatives, among others, which could be used as cosmetic ingredients (Guillerme, Couteau, & Coiffard, 2017).

4.2.1 Chlorophyta (green algae)

Sulfated polysaccharides from *Ulva pertusa*, *Enteromorpha linza*, and *Bryopsis plumosa* have demonstrated antioxidant activities. Moreover, the antioxidant potential was dependent on the type and concentration of such compounds (Zhang et al., 2010). Substances from *Rhizoclonium hieroglyphicum* have demonstrated the moisturizing effect in both pig skin model and skin human. It increased the moisture effect when compared to hyaluronic acid (Pimentel, Alves, Rodrigues, & Oliveira, 2018).

4.2.2 Phaeophyta (brown seaweed)

Some extracts from brown algae can be used in cosmetic formulations. For instance, alginate is a structural phytocolloid that contains two monomeric units (α-L-guluronic acid and β-D-mannuronic acid) and may be used as a thickening agent in the food industries, as a thickening agent and moisture retainer in the cosmetic area, and in the pharmaceutical field as alginic acid. Some brown algae species such as *Macrocystis pyrifera*, *Laminaria hyperborea*, *Ascophyllum nodosum*, *Laminaria japonica*, and *Lessonia flavicans* are used for the production of alginate by industries (Khanra et al., 2018).

Another polysaccharide derived from brown algae is fucoidan. Fucoidans are cell wall constituents. Their yield depends on the seasonal variation, the species, and the marine depth of growth. Fucoidans have anticoagulant, antitumor, antiviral, and antioxidant activities, which make them interesting ingredients in the pharmaceutical and cosmetic industries (Cunha & Grenha, 2016; De Jesus Raposo, De Morais, & De Morais, 2016).

Natural polysaccharides from *Sargassum fusiforme* showed viscosity, good film-forming, moisturizing effect, as well as free radical scavenging activity (Zhang, Zhang, Tang, & Mao, 2020).

4.2.3 Rhodophyta (red seaweed)

Some red microalgae species, such as *Gracilaria gracilis*, *Gracilaria dura*, and *Pterocladia* sp. among others, have in their cell walls a mixture of polysaccharide molecules called agar that is composed of two sugar molecules: agaropectin and agarose (Yun et al., 2015). It has a gelatinous consistency and it is used in the cosmetic industry as a thickening agent. However, its most important use is as a gelatinous culture medium for the growth of microorganisms for scientific studies and medical examinations. Two known algae resources to produce agar at industries are *Gracilaria* sp. and *Gelidium* sp. (Khanra et al., 2018).

Furthermore, *Porphyridium* sp. is cultured for the commercial production of exopolysaccharides and pigments as β-phycoerythrin, which are high-value compounds in the cosmetic industries (Marcati et al., 2014; Soanen, Da Silva, Gardarin, Michaud, & Laroche, 2016; Tannin-Spitz, Bergman, Van-Moppes, Grossman, & Arad, 2005).

Another group of polysaccharides named carrageenans is derived from red algae, such as *Furcellaran* sp., *Chondrus* sp., *Hypnea* sp., *Eucheuma* sp., and *Gigartina* sp. Carrageenans contain sulfated galactans capable of forming a gel and increase viscosity (De Jesus Raposo et al., 2016; Sathasivam et al., 2017). Carrageenan is also used in the pharmaceutical and food industry in aqueous gel applications, such as fruit, jelly beans, and marmalades, among other applications (Khanra et al., 2018).

Some studies have shown that red algae *Porphyra* sp. contains high concentrations of polysaccharides (Holdt & Kraan, 2011; Zhang et al., 2010), proteins (Barbarino & Lourenço, 2005; Tibbetts, Milley, & Lall, 2015), and mycosporine-like amino acids (MAAs) that can absorb sunlight and can act as a natural sunscreen (Lalegerie et al., 2019).

4.3 Cyanobacteria

Cyanobacteria are bacteria with endosymbiotic plastids that perform oxygenic photosynthesis. They contain a variety of natural pigments, such as chlorophylls, carotenoids, and phycobiliproteins with different colors that are used as dyes in food and cosmetics (Saini, Pabbi, & Shukla, 2018).

Arthrospira platensis (*Spirulina*), a cyanobacteria species, are widely studied to produce compounds with demonstrated applications for food and health (Bernaerts et al., 2018; Sotiroudis & Sotiroudis, 2013), such as phycocyanin and sulfated polysaccharides for cosmetics purposes (De Jesus Raposo et al., 2013a; Gunes, Tamburaci, Dalay, & Deliloglu, 2017; Priyadarshani & Rath, 2012; Trabelsi, M'sakni, Ouada, Bacha, & Roudesli, 2009). It also produces a protein-rich extract that can be used in the manufacture of antiaging cosmetics (Ariede et al., 2017). Moreover, phycocyanin produced for another cyanobacterium *Geitlerinema* sp. H8DM has been studied due to its antioxidant potential. It can be used as a colorant and therapeutic agent against oxidative stress due to its stability, availability, as well as antioxidant activity (Patel, Rastogi, Trivedi, & Madamwar, 2018).

5 Natural dyes

Among the natural dyes of different algae are chlorophylls, greenish fat-soluble natural pigment; phycocyanins, a blue pigment from *Arthrospira*; astaxanthin, a yellow-to-red pigment from *Haematococcus*; and β-carotene, a yellow pigment from *Dunaliella* (Cezare-Gomes et al., 2019; Koller et al., 2014; Rodrigues, de Castro, de Santiago-Aguiar, & Rocha, 2018; Saini et al., 2018).

Phycobiliproteins are soluble, deep-colored fluorescent proteins in water, which can be produced by Rhodophyta, e.g., *Galdieria sulphuraria*, and cyanobacteria, e.g., *A. platensis*. Recent interest in phycobiliproteins is increasing due to its antioxidant activity. They can reduce oxidative stress due to their chemical structures and chelating properties (Pallela, Na-Young, & Kim, 2010). Chlorophyll was used as a deodorant in cosmetic products while carotenoids were incorporated as antioxidant compounds (Khanra et al., 2018; Saini et al., 2018).

6 Moisturizer agents

Moisturizers are the main component in daily care to maintain the healthy appearance of the skin. When the barrier function is compromised, transepidermal water loss (TEWL) could be increased and can induce some dermatological disorders, such as skin dryness, or xerosis, acne vulgaris, atopic dermatitis, irritating retinoid-induced dermatitis, rosacea, and psoriasis (Leite e Silva et al., 2009).

Topical moisturizers can restore the integrity of the *stratum corneum*, replace the skin lipids, reduce TEWL as a barrier function to water loss, and help prevent the appearance of dry skin, loss of elasticity, reduce fine lines, smooth, and maintain skin plasticity (Purnamawati, Indrastuti, Danarti, & Saefudin, 2017). There are four main classes of moisturizers, such as occlusive, humectant, emollient, and rejuvenating.

Occlusive agents are the most effective and commonly used. They are indicated for the prevention of contact dermatitis, atopic dermatitis, and xerosis (Draelos, 2018). The effect is most significant when applied to moist skin because they create hydrophobic barriers contributing to the intercorneocyte matrix. They slow the evaporation and loss of TEWL by forming a hydrophobic film on the skin surface and in the interstitial space between keratinocytes (Purnamawati et al., 2017). Occlusive agents include mineral oil, petroleum jelly, silicones, stearic acid, and shea butter (Draelos, 2018). Humectants are hygroscopic components that retain water and moisture the skin. Formulations with humectant properties can be combined with occlusive agents that prevent water loss. Humectants are indicated for xerosis and ichthyosis. Moisturizing ingredients include propylene glycol, glycerin, and urea (Draelos, 2018).

Emollients are indicated for dry and rough skin due to filling the spaces between the corneocytes and lubricating the skin, providing a feeling of softness and plasticity, contributing successfully to consumer satisfaction (Purnamawati et al., 2017). Some common emollients include octyl octanoate, isopropyl myristate, diisopropyl dilinoleate, and jojoba oil (Draelos, 2018).

The rejuvenating ones are intended to replace essential proteins to the skin. Since these proteins are large, a limitation for rejuvenators is the difficulty in penetrating the *stratum corneum*, thus requiring small molecular weight proteins. Although rejuvenators cannot replace proteins in the dermis, they can play a similar role than emollients improving skin appearance by creating a film that aesthetically softens the skin and stretches and even fills fine lines. Rejuvenators often contain skin proteins keratin, collagen, and elastin (Purnamawati et al., 2017).

There are several pharmaceutical/cosmetic forms to obtain moisturizers. Lotions are defined as "emulsified liquids intended to be applied topically on the skin and share many characteristics with creams but are more fluid and pourable" (United States PC, 2016). As an advantage, they can be more easily applied to large surfaces of the skin than viscous preparations and they are suggested for face, body, and normal skin. The serum is a highly concentrated product based on water or oil and it has a property of rapid absorption, i.e., an evanescent characteristic. Furthermore, they could be nongreasy. Serums act locally upon different body parts: face, neck, decollate, and eyelids and can be used irrespective of skin age (Draelos, 2018).

Creams are "semisolid emulsion dosage forms that often contain more than 20% water and volatiles and/or less than 50% hydrocarbons, waxes or polyols as the vehicle for the drug substance. They are generally intended to be applied topically to the skin or mucous membrane." They also develop higher viscosity than lotions. They are suitable for face, body, hands, feet, and even dry skin. Ointments are semisolid vehicles intended to be applied topically over the skin or mucous membranes. They have sticky and greasy characteristics (United States PC, 2016). The gel is a "semisolid system consisting either of suspensions of small inorganic particles or of organic molecules interpenetrated by a liquid" (United States PC, 2016). Gels are used to not let the oily feeling over the skin, according to their aqueous composition (Draelos, 2018).

Regardless of the pharmaceutical/cosmetic forms, an ideal moisturizer must be aesthetically elegant and acceptable, and suitable for all types of skin (hypoallergenic, nonsensitizing, fragrance-free, and noncomedogenic). Moreover, they should offer an affordable, long-lasting, rapidly absorbed, automatically hydrated skin (Kraft & Lynde, 2005; Lynde, 2001). In a recent study, topical application of ceramide creams could increase skin hydration and improve the skin barrier function with a suitable use even in dry skin. It was also demonstrated that such formulations were decreased TEWL values with no evidence of skin sensitization in both adults and children (Spada, Barnes, & Greive, 2018).

Some microalgae species have compounds with moisturizing effect. For instance,

Thalassiosira sp. is rich in amino acids, mainly serine, while the *Monodus subterraneus* is rich in polyunsaturated fatty acids (Daniel Jouvance, 2019). Sold in over 20 countries, *Chlorella* biomass is used for the production of water-soluble extracts, made up of a variety of compounds, such as peptides, essential amino acids, proteins, sugars, vitamins, and nucleic acids (Borowitzka, 2013; Ursu et al., 2014). It was demonstrated that it improves the expression of genes responsible for skin health, nourishing the region and stimulating the fibroblasts. It is also indicated to improve elasticity and firmness and to decrease stretch marks by increasing the presence of elastin and elfin (Nutrifarm, 2019). *Chlorella* extracts are commercialized to use in cosmetic formulations at 1%. It inhibits collagenase and elastase activities with an increment in collagen synthesis and density of the extracellular matrix to induce an increment in the elasticity and firming (Ariede et al., 2017).

The effects of the facial application of a gel formulation containing 1% *Fucus vesiculosus* extract on the thickness and mechanical properties of human skin were investigated. The results indicated that *F. vesiculosus* extract was sufficient to cause changes in skin thickness and elastic properties due to its effects on skin tightening, antisagging, and wrinkle smoothing (Fujimura, Tsukahara, Moriwaki, & Kitahara, 2002). In fact, there is a cream claiming the diminution of aging effects on the skin (Cicatricure, 2019).

N. oculata extracts have been used commercially as they have been shown to have skin-tightening properties. *N. oculata* extract contains a large amount of linolenic acid (omega 6), a polyunsaturated fatty acid that has potential as a moisturizing agent, which may help to reduce inflammation of skin cells, promote tissue regeneration, and improve the skin appearance. Moreover, *Nannochloropsis* sp. is considered a great source of omega-3 fatty acids, mainly eicosapentaenoic acid (EPA) as well as polysaccharides (Bernaerts et al., 2018).

Nannochloropsis gaditana could also be considered as a new cosmetic ingredient. The extracts demonstrated in vitro skin protection activity against induced oxidative stress. As an interesting finding, collagen production has been observed to aid both skin hydration and skin aging (Letsiou et al., 2017).

Some polysaccharides from *Saccharina japonica* (brown algae) have been identified with a higher humectant property than hyaluronic acid (HA), particularly fucoidan that can act against skin dehydration due to its higher moisture retention capacity (Wang et al., 2013).

7 Antiaging agents

Skin aging can be triggered by both intrinsic and extrinsic factors (Wang, Chen, Huynh, & Chang, 2015). Intrinsic aging is natural and chronological, less aggressive than extrinsic aging, and occurs by flattening the dermo-epidermal junction, reducing the contact surface between the epidermis and dermis (Cavinato et al., 2017), and consequently affecting the aging of fibroblasts, which effect is collagen loss. It is also important to mention ethnicity; black skin is more resistant to aging than Caucasian skin and has a higher intracellular lipid content, which increases the resistance to aging. In addition, some anatomical variations and hormonal changes in the overall skin tissues that occur mainly in women could influence skin aging, such as menopause (Zouboulis, Chen, Thornton, Qin, & Rosenfield, 2007).

On the other hand, extrinsic aging is due to several factors, including the influence of the external environment and lifestyle, the effects of smoking and nicotine, and mainly the effects of the ultraviolet (UV) radiation to the skin, which can induce the skin deterioration as well as dermal extracellular matrix damage (Chen & Wang, 2016).

The UV radiation is subdivided into three regions: UVA (320–400 nm), UVB (290–320 nm), and UVC (100–290 nm). UVA is able to penetrate the dermal region, inducing wrinkle formation and causing related skin symptoms, such as sagging and roughness, and may cause mild erythema. UVB is cancerous, penetrates only the epidermis, and is responsible for burns and erythema. UVC radiation is blocked by the ozone layer (Oliveira et al., 2015; Wang, Chen, et al., 2015).

The daily use of sunscreens can mitigate the effects of UV radiation, avoiding sunburn and photoaging, and reducing the chances of skin cancer derived from UV exposition. The incorporation of antioxidant compounds in sunscreens can benefit the skin by reducing the production of formed reactive oxygen species (ROS). It was found that the presence of antioxidants in the sunscreen formulation improved the skin's protective barrier function (Chen, Hu, & Wang, 2012).

At normal conditions, the skin produces antioxidant agents that block the ROS, thus preventing the cellular imbalance (Wang, Chen, et al., 2015). However, the overexposure to UV radiation could increase the ROS production and cause an imbalance between both oxidizing and antioxidant substances with consequent oxidative stress. Such processes cause damage to cells, mainly affecting protein and lipid membranes by the excess of free radicals that induce cell death (apoptosis), with concomitant formation of wrinkles and skin dryness (Anunciato & da Rocha Filho, 2012).

Furthermore, the excess of free radicals generated in the epidermis can accelerate the photoaging process and even induce skin inflammation, melanoma, and skin cancer (Pallela et al., 2010). Thus applying topical antioxidants into the skin favors free radical inhibition by retarding intrinsic and extrinsic factors in the aging process due to their ability to prevent the formation of ROS (Chen et al., 2012).

Microalgae can be a natural source of vitamin E in its most potent form, α-tocopherol. It is a potent natural antioxidant used as an antiaging and antifree radical compound in the cosmetic industry. *Tetraselmis suecica* and *Dunaliella tertiolecta*, both microalgae species, are used as food for fish and shellfish larvae with high α-tocopherol content and other vitamins. Therefore they also could be considered for cosmetic use, mainly the antiaging ones (Mourelle, Gómez, & Legido, 2017).

C. vulgaris extracts are used as collagen-repairing agents, regenerating the skin and making it look younger, while it acts as an aid in antiaging treatment (Koller et al., 2014). Sulfated polysaccharides from marine microalgae *Porphyridium* and *Rhodella reticulata* have shown the ability to prevent the accumulation of free radicals and chemically reactive species due to their high antioxidant power (Mourelle et al., 2017). Moreover, the extracts of *Arthrospira* sp. have demonstrated to slow down the signs of premature skin aging as well as the formation of stretch marks (Ariede et al., 2017).

The extract from microalgae *Phaeodactylum tricornutum* was launched by a Swiss company as an antioxidant and antiaging agent. It is a lipid extract capable of detoxifying cells from oxidative proteins by the proteasome, which is a specific enzymatic system, preventing premature and chronological aging of the skin, avoiding the accumulation of free radicals and harmful proteins (Mourelle et al., 2017).

The use of antioxidants is important for extrinsic aging, mainly caused by the effects of UV radiation. Some polysaccharides such as fucoidan, alginate, and laminarin from brown algae (*Turbinaria invite*, *F. vesiculosus*, and *Undaria pinnatifida*) also have important antioxidant properties to combat cutaneous aging (Jea et al., 2009). Thus the application of *F. vesiculosus* in cosmetics can improve eye bags and dark circles, as well as stimulate collagen production (Chavan, Sun, Litchauer, & Denis, 2014).

Moreover, *Porphyra umbilicalis*, other red algae, contains a large amount of mycosporine-like amino acids (MMAs) which can absorb UVA and UVB radiation and act as a sunscreen to prevent intrinsic aging of the skin (Wang, Chen, et al., 2015).

The extracts from seaweed *Coccoid* and *Filamentous*, even with undefined compounds, may help to prevent photoaging due to their ability to increase the protective barrier function of the skin as they stimulate terminal differentiation of keratinocytes as well as the expression of important genes in the skin barrier, thus preventing skin aging (Ariede et al., 2017).

Extracts from *Chlamydocapsa* sp. (snow algae) were used in the production of cosmetics to combat oxidative reactions of the skin, e.g., photoaging. It can decrease the skin function loss, the TEWL, and preventing the formation of wrinkles and fine lines after the skin UV exposure in a colder environment (Ariede et al., 2017).

On the other hand, phytohormones production has also been widely studied in algae species since they are accumulated and released by some microalgae, and they are linked to plant growth and development (Singh et al., 2017). *Nannochloropsis oceanica* has demonstrated molecular evidence that suggested the functionality of endogenous abscisic acid and cytokine with physiological effects similar to larger plants. This fact suggests that they can play an antiaging role under the skin, counteracting the signs of aging (Michelet et al., 2012).

The moisturizing ability of a cream containing *S. fusiforme* polysaccharides was better than that one made with glycerin, with significant enhancement of the free radical scavenging function in mice skin. Moreover, the cream containing 0.3% *S. fusiforme* polysaccharides exhibited high protection against skin stain, aging, and wrinkles (Zhang et al., 2020).

An *A. platensis* (*Spirulina*) bioactive peptide-enriched fraction (obtained by enzymatic digestion) presented moisturizing property by increasing gene expression of factors involved in water maintenance in epidermal cells (aquaporin-3, for instance). This peptide fraction was also able to eliminate ROS induction (Apone, Barbulova, & Colucci, 2019).

An extract obtained from *Botryococcus braunii* resulted in an improved biosynthesis of collagens in fibroblasts and also increased the expression of the cornified envelop proteins (involucrin and filaggrin) of the epidermis that could restore the skin barrier function, besides a potent antioxidant activity (Lorencini, Brohem, Dieamant, Zanchin, & Maibach, 2014).

8 Anticellulite agents

Gynoid lipodystrophy, popularly known as cellulitis, is an expanding acute gangrenous infection in subcutaneous tissue which is very common in postpubertal women (Terranova, Berardesca, & Maibachà, 2006). They usually appear mainly on the buttocks and thighs and give an "orange peel" appearance in such regions. About 85% of adult women develop cellulite, which can be considered common rather than a pathology, although it is still uncomfortable for many women on aesthetic issues (Draelos, 2005).

Some macroalgae species, such as *F. vesiculosus*, *Laminaria digitata*, among others, could be used in cosmetics to reduce cellulite. As cellulite is a condition nonpathologic, it requires regular use of cosmetics products to treat its sings and even to improve skin appearance (Pimentel et al., 2018).

The anticellulite effect of *F. vesiculosus* L. consists of increasing connective tissue density and stimulation of vascular flow, resulting in an improved appearance of the skin. Extracts of these algae generally stimulate peripheral blood circulation and inhibit collagen matrix fibrosclerosis around the localized fat where cellulite is present. Even the use of these extracts in topical products seemed safe (Pimentel et al., 2018); there are still few in vivo scientific studies to prove these effects against cellulite (Turati et al., 2014).

The aqueous extracts of *F. vesiculosus* and *Furcellaria lumbricalis* were incorporated in cosmetics to stimulate in vitro both lipolysis in mature adipocytes and the production of procollagen I by aged primary fibroblasts. It was also demonstrated significant improvement in cellulite grading in volunteers (35 women) after 8 and 12 weeks. Moreover, ultrasound imaging demonstrated a high reduction in fat thickness when compared with placebo after 12 weeks (Al-Bader et al., 2012).

9 Sunscreen/UV filter compounds

It is considered a challenge to the development of broad-spectrum sunscreens since they are composed of several ingredients (actives and excipients) that will be in contact with the skin and sunlight and must be safe and properly protect the cutaneous tissue against the UV radiation harmful effects. Unsafe overexposure to UVB and UVA radiations can cause sunburn, oxidative stress, photoaging, immunosuppression, and photocarcinogenesis (Daneluti et al., 2019; Peres et al., 2017; Wróblewska, Baby, Grombone, & Moreno, 2019). Natural compounds have been the focus of numerous researches due to ecological issues, sustainability, and minimum ambient impact, presumably safe utilization and multiple benefits, since several natural substances address distinct biological actions, for instance, antioxidant activity (Peres, Sarruf, Oliveira, Velasco, & Baby, 2018). Thus a sunscreen containing UV filters and natural compounds, since adequately developed, can present multifunctions that will protect the skin with more efficacy (Rosado et al., 2019; Tomazelli et al., 2018).

Sunscreens are products vehiculated in several pharmaceutical/cosmetic forms, including hydroalcoholic solutions, oils, gels (oily, aqueous, and hydroalcoholic), emulsions, sprays, powders, and sticks (Balogh, Velasco, Pedriali, Kaneko, & Baby, 2011). Additionally, to protect the skin against UVB and UVA rays, the proper and regular use of them avoids erythema formation, reduces actinic keratosis and squamous cell carcinoma, and, also, prevents extrinsic akin aging (Balogh et al., 2011; Lautenschlager, Wulf, & Pittelkow, 2007).

Many studies have described natural compounds from aquatic organisms with their photoprotective claims (Ariede et al., 2017; Bedoux, Hardouin, Burlot, & Bourgougnon, 2014; Chen et al., 2012; Morone et al., 2019; Pallela et al., 2010). The use of algae has been recently suggested for cosmetics as sunscreen products. Compounds from algae biomass were well described due to the high value-added and there are many reviews describing the health benefits of carotenoids (Aburai, Ohkubo, Miyashita, & Abe, 2013; Cezare-Gomes et al., 2019; De Jesus Raposo, De Morais, & De Morais, 2013b; Grundman, Richter, & Ini, 2018; Stahl & Sies, 2012). However, research relating to their cosmetic properties is still scarce (Ariede et al., 2017; Wang, Chen, et al., 2015).

A *Chlorella*-derived peptide fraction was investigated as a potential active against UVB damages. This peptide fraction (430–1350 Da) reduced UVB-induced metalloproteinase-1 gene expression in irradiated fibroblasts and also presented an effect against UVC radiation (Apone et al., 2019).

Nowadays, there is limited information about MAAs and their photoprotective potential due to the huge variety in nature as well as the different environmental conditions to grow algae (Lalegerie et al., 2019; Oren & Gunde-Cimerman, 2007). Furthermore, their production is limited to an industrial scale with economic benefits since their biosynthesis pathway is not completely understood (Lawrence, Long, & Young, 2019). Therefore an appealing claim linked to the use of more eco-friendly and safe compounds developed the use of sunscreens with natural UV compounds with lower amounts of chemical filters on cosmetic formulations.

9.1 Carotenoids

Microalgae are linked by producing different carotenoids in nature (Cezare-Gomes et al., 2019; Saini et al., 2018; Sathasivam & Ki, 2018; Singh & Das, 2011). Carotenoids exhibit different mechanisms to protect from UV radiation. They have the capacity to absorb both UV radiation and visible light due to their conjugated double bonds (Singh et al., 2017) to avoid skin damage. Carotenoids mediated photoprotection by energy dissipation in both thylakoids and UV filter layer at the chloroplast periphery or cell body. Furthermore, carotenoids may act as provitamin A in the skin protection against UV radiation as physical quenching (Stahl & Sies, 2012).

Carotenoids can scavenge the ROS (singlet molecular oxygen and peroxyl radicals) producing carotenoid radicals (Chisté, Freitas, Mercadante, & Fernandes, 2014). For instance, β-carotene could protect cells from free radicals and UV radiation. It was reported that the largest natural resource of such pigment was *D. salina*, which has produced more than 10.0% β-carotene on the dry biomass (Lamers, Janssen, De Vos, Bino, & Wijffels, 2008).

Fucoxanthin could collect and transfer energy as light to chlorophyll-protein complexes. It has greater antioxidant activity than β-carotene and it acts in photoprotection (Ayalon, 2017; De Jesus Raposo et al., 2013b; Sathasivam et al., 2017). In addition, it was reported as skin protective from photoaging (Peng, Yuan, Wu, & Wang, 2011).

As a practical application, rosemary oil and vitamin C were used to stabilize fucoxanthin from microalgae biomass of *P. tricornutum* (Grundman et al., 2018) to facilitate cosmetic application.

H. pluvialis is the highest source of astaxanthin (1.5%–3.0% dry biomass). A marketed product named AstaTROL containing astaxanthin from *H. pluvialis* is produced and marketed by the Japanese company Fuji Chemical Industry for applications in personal care products and cosmetic formulations (Fuji Chemical Industries Co., Ltd, 2019). Moreover, some studies about the benefits of these compounds in human health (Tominaga, Hongo, Karato, & Yamashita, 2012; Wang, Chen, et al., 2015), as well as the daily supplementation of astaxanthin to avoid the risk of skin cancer (Rao et al., 2013), have demonstrated their importance in the market (Tominaga, Fujishita, & Nobuko, 2018).

Zeaxanthin, another natural pigment, could be act as a colorant in cosmetic products (Bhosale & Bernstein, 2005) while canthaxanthin can induce the tan color in the skin (Gong & Bassi, 2016).

9.2 Mycosporine-like amino acids

Some species of microalgae, protozoa, seaweeds, corals, and even fishes, which suffer high sunlight exposition, could produce MAAs (Řezanka, Temina, Tolstikov, & Dembitsky, 2004; Shick & Dunlap, 2002; Sinha, Singh, & Häder, 2007). They are water-soluble, colorless (Chrapusta et al., 2017; Lawrence et al., 2019), and can absorb the UV radiation between absorption wavelength peaks in the UVB and UVA spectrum (309–362 nm) (Shick & Dunlap, 2002; Singh, Kumari, Rastogi, Singh, & Sinha, 2008). Among the most studied MMAs, we remark mycosporine-glycine, Porphyra-334, and shinorine due to their photostability. They could dissipate approximately 98% of the energy absorbed as heat, without generating ROS (Conde, Churio, & Previtali, 2004). *Porphyra umbilicalis* extract was used to produce Helioguard 365 that contains Porphyra-334 and shinorine and has claimed antiaging and skin protectant (Chrapusta et al., 2017).

9.3 Scytonemin

Scytonemin was found in some terrestrial cyanobacteria species (Rastogi, Sonani, Madamwar, & Incharoensakdi, 2016; Singh et al., 2017). The unique use of scytonemin has diminished the risk of damage provoked by the most lethal UVC radiation (Rastogi & Incharoensakdi, 2014; Rastogi et al., 2016).

10 Skin-whitening agents

The demand for skin-whitening products in the cosmetic market has increased to achieve a clearer appearance and treat pigmentary disorders, mainly melasma or postinflammatory hyperpigmentation, among others. Hyperpigmentation is usually caused by changes in melanogenesis with an increment in melanin synthesis. As a result, hyperpigmentation causes dark in a given skin region. Active ingredients from natural sources are very popular in these cases, due to their ability to act at different levels of melanogenesis (Guillerme et al., 2017; Wang, Chen, et al., 2015).

Skin hyperpigmentation can be caused by intrinsic factors. Genetic hyperpigmentation appears as freckles and spots on mature skin. Metabolic hyperpigmentation is known as Wilson's disease that includes problems with copper assimilation. Endocrine hyperpigmentation causes by melisma, chloasma, and pregnancy spots. Nutritional deficiency hyperpigmentation causes by a lack of vitamins. Photosensitivity that induces hyperpigmentation is in response to dermatitis derived from perfumes. Hyperpigmentation causes due to the use of drugs such as amiodarone, phenytoin, or overdosage of hydroquinone, among others (Couteau & Coiffard, 2018).

Melasma is a pigment disorder that affects all skin types, most often types IV–VI according to Fitzpatrick in contact with high ultraviolet radiation, pregnants, African descent, Asian, and Hispanic women. This change in skin pigmentation is related to genetic predisposition, hormonal factors, and sun exposure because they cause an increase in tyrosinase activity (Bagherani, Gianfaldoni, & Smoller, 2015). Since melasma is a chronic and recurrent condition,

controlled clinical studies have indicated alternatives as photoprotection and the use of whitening products to mitigate the darkness in the skin (Bagherani et al., 2015; Couteau & Coiffard, 2018).

Regarding extrinsic factors, excessive exposure to the sun might have induced the concomitant increment in melanin synthesis. In addition, since tyrosinase is responsible for melanin synthesis, depigmenting agents are used as tyrosinase inhibitors to provide skin uniformity and limit the speed in the pigmentation process (Wang, Chen, et al., 2015). Thus the search for natural compounds with tyrosinase inhibitory activity has increased (Zeitoun et al., 2019) and only a few compounds with significant effect in response to clinical trials have been found (Couteau & Coiffard, 2018; Espinosa-Leal & Garcia-Lara, 2019).

Algae have an interesting role in the research for these whitening agents. Among microalgae species, extracts from *N. oculata* containing zeaxanthin, a natural pigment with potential tyrosinase inhibitory activity, was reported (Ariede et al., 2017). Moreover, as a whitening agent, *Chlorella* extract has also been proposed with activity to reduce skin pigmentation in amounts greater than 10% (Couteau & Coiffard, 2018).

Some species from the brown alga *Ecklonia* have been studied to discover new compounds to treat illnesses related to melanin. Dieckol, a phlorotannin derived from *Ecklonia cava*, has demonstrated inhibitory effects on melanin synthesis. Furthermore, the phlorotannins dieckol, eckol, phloroglucinol, dioxinodehydroeckol, and phlorofucofuroeckol-A from *Ecklonia stolonifera* have shown more antityrosinase activity than arbutin and kojic acid (Manandhar, Paudel, Seong, Jung, & Choi, 2019).

Peptide extract from brown algae *U. pinnatifida* was recently reported as a whitening agent (Pimentel et al., 2018) because it inhibits the tyrosinase activity and limits melanin production. It also permits a fast reduction of skin color (ex-vivo tests). Moreover, Hikiji extract made from *Hizikia fusiforme* is also commercialized as a whitening and lightening agent for cosmetic formulations (Pimentel et al., 2018).

Fucoxanthin of ethanolic extract from *L. japonica*, commonly known as Kombu, inhibited tyrosinase activity, melanogenesis in melanoma, and UVB-induced skin pigmentation. It was demonstrated that a cosmetic with 1.0% fucoxanthin and even oral supplementation of such pigment have antipigmentary activity and could be used in cosmetics with this claim (Shimoda, Tanaka, Shan, & Maoka, 2010).

11 Haircare products: The benefits of algae

Haircare procedures involve the use of both shaft and scalp products. Hair is a biopolymer consisting primarily of amino acids bounded forming the keratin protein. The hair shaft has two main distinct regions: the cuticle and the cortex (Feughelman, 1959; Robbins, 2012). Fig. 2 shows the scanning electron microscopy (SEM) images of an oriental virgin hair sample,

highlighting the cuticle (outer portion 3.45 µm of thick) and the cortex (inner part). It is also possible to observe the cell membrane complex (CMC) layers between the cuticle layers. The cuticle is responsible for the protection of the cortex and brightness of the hair surface. Fig. 3 shows the surface of a sample of oriental hair. The cuticle layers are juxtaposed and undamaged, conditions that provide brightness and combability. The cortex is composed of keratin helix intermediate filaments (IFs) incorporated into an amorphous matrix, in which melanin granules are distributed throughout. It is in this portion that the treatments involving chemical transformation (bleaching, dyeing, straightening, etc.) act. Cuticle and cortical cells are bound by continuous multiple layers of interconnected proteins and lipids named CMC. In the cuticle CMC, there are free fatty acid among 18-methyl eicosanoic acid (18-MEA) molecules (Coderch, Méndez, Martí, Pons, & Parra, 2007; Robbins, 2012) and free fatty acids. This multilayer enhances the barrier function, which prevents external materials from penetrating the hair fibers. Free fatty acids are an important

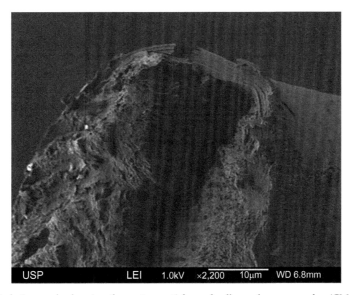

FIG. 2 Oriental virgin hair sample showing the cortex, cuticle, and cell membrane complex (CMC).

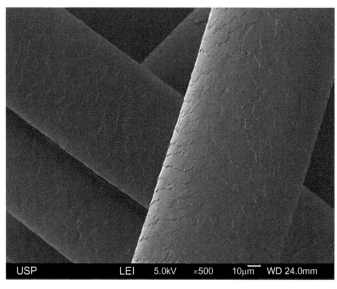

FIG. 3   Sample of oriental virgin hair.

component of the hair surface, as they play a significant role in the isoelectric point of the fibers and in the interaction of the fiber with cosmetic products (Capablanca & Watt, 1986). Changes in these structures (cuticle, CMC, and cortex) may modify the penetration of cosmetic products into the hair matrix as well as their mechanical properties (Robbins, 2012).

In the last years, products that contain algae and microalgae have been used to repair damage in both skin and hair (Pereira, 2018). Algae became an interesting source due to their wide distribution and valuable chemical composition with a high potential as active ingredients, among them, proteins, essential amino acids, polyunsaturated fatty acids; carbohydrates; vitamins and minerals (magnesium and calcium); alginates, agar, and carrageenans (polysaccharides); phytosterols and polyphenols (Messyasz et al., 2018; Pereira, 2018; Pimentel et al., 2018), for instance, lipids from Black Sea algae (a brown algae *Cystoseira barbata* from Phaeophyta and a red algae *Gelidium crinale* from Rhodophyta) containβ-carotene, α-tocopherol, polyunsaturated fatty acids, and antioxidants (Panayotova, Merzdhanova, Dobreva, Zlatanov, & Makedonski, 2017). Microalgae, though not as popular as macroalgae, are valuable sources of biomolecules, proteins, carbohydrates, lipids, vitamins, minerals, pigments, and other small molecules (Hudek et al., 2014; Skjånes, Rebours, & Lindblad, 2012). Many of these components, derived from animals or plants, have an affinity for the hair or skin and they could be used as innovative ingredients in green cosmetics.

Regarding the care of the hair shaft, specifically, some actives could act both in the hair matrix (cortex) and cuticle, improving some properties like conditioning, resistance, elasticity, and shine of fiber. Proteins, silicones, etc. could show affinity with hair proteins and/or filling pores in damaged fiber (Robbins, 2012). In this sense, some algae and their extracts are available in cosmetic raw materials commercially available, like *Pelvetia canaliculata* (Xylishine), claimed as moisturizing and repairing agent (SEPPIC, 2019a), and association of *P. canaliculata* and *L. digitata* extracts (Bioenergizer BG) is claimed as a hair moisturizer that enhances the shine of all hair types, however is no longer available in the market (SEPPIC, 2019b). Compositions containing dried *Porphyridium* biomass (red microalgae) were used topically on skin, scalp, nail, and hair to reduce hydrogen peroxide concentration and provide beneficial effects like slowing, delaying, and reducing the progression of hair graying. Other red microalgae suggested with similar properties, including the unicellular algae *Bangiophyceae*, *Florideophyceae*, *Goniotrichales*, *Dixoniellagrisea*, or members of Rhodophyta (Seiberg, 2017). In various embodiments of the method that uses microalgal biomass composition to soften and impart pliability to skin and hair, the composition usually comprises cells containing at least 15% oil by dry weight. Additionally, these authors reported that biomass cosmetic compositions can be derived from a culture of various species, among them, some of the genus *Chlorella*, like *Chlorella antarctica*, *Chlorella anitrata*, *Chlorella aureoviridis*, *Chlorella capsulata*, *Chlorella candida*, *Chlorella desiccata*, among others. In some recent inventions, algae extracts were used to moisturize, remove residue, dirt, oil, grease from skin, and hair, among others (Ariede et al., 2017). Photoprotective compositions of topical applicable (against the damaging effects of UVA and UVB radiation) containing pigments from some microalgae species were also claimed care of the human skin and/or hair (Lalegerie et al., 2019; Radice et al., 2016).

Moreover, it was described that extracts from algae biomass that contain unsaturated fatty acids, mainly omega-3, omega-6, and omega-9, are a valuable source in cosmetic formulations, as they are natural and environment-friendly (Messyasz et al., 2018). Algae extracts (alginates and carrageenans) may use as thickeners for surfactant preparation of shampoos for scalp and hair hygiene (Pimentel et al., 2018). The protein present in *Aphanizomenon flos-aquae* (blue-green algae) has been used to help to strengthen the hair and prevention of breakage and also split ends (Aubrey) (Aubrey, 2019).

On the other hand, the scalp also requires care because it also suffers aggressions caused by cosmetic chemical treatments and even inflammatory pathologies. The scalp has a big amount of large hair follicles that produce long hair fibers with sebaceous glands attached to the follicle. The hair follicle is modified during cyclical distinct phases of growth that involves three stages: anagen (growth), catagen (regression), and telogen (rest). These steps are orderly by molecular regulators that act first on stem cells and then on the newly formed cells in the bulb with concomitant differentiation of the growing fiber. During anagen, the dermal papilla near the center of the bulb is involved in some growth functions (Robbins, 2012).

Currently, an extensive variety of causes, mainly aging, have induced hair loss disturbances in many millions of people. The most diagnosed hair loss known as androgenetic alopecia is derived from genetic predisposition with an exacerbated response of the hair follicles cells to androgens intensified by scalp inflammation and extrinsic factors. Many studies have focused on compounds to stimulate hair growth and even to prevent or delay their loss (Katzer, Leite Junior, Beck, & da Silva, 2019). In this sense, the interest in algae and microalgae applied to hair care has recently grown, and some patents and patent applications containing formulations with these actives have been proposed (Ariede et al., 2017; Kang et al., 2012, 2017). Among these researches, there are some that indicate some types of algae could enhance human hair growth. In a recent study, it was related to the significant increase in the hair-fiber lengths of mice hair using *Undariopsis peterseniana*, edible brown algae. The extracts from *U. peterseniana* have decreased 5-α-reductase activity and significantly increased the proliferation of dermal papilla cells, which is a central regulator of the hair cycle (Kang et al., 2017). It was compared the promoting effect of hair growth by the mixture of the brown algae *U. pinnatifida* and *L. japonica* (U-L mixture) and 3% minoxidil and a U-L mixture. The results evidenced that the U-L mixture could induce hair growth in mice, with a similar effect than minoxidil, thus suggesting a novel application in the treatments for hair loss (Park & Park, 2016). Other seaweed extracts were studied to prevent hair loss such as *Grateloupia elliptica*. This red seaweed native to Korea was reported as promising in the hair loss in vitro (Kang et al., 2012). *U. pinnatifida* and *L. japonica* (brown algae) association demonstrated have substantial antiinflammatory activities, once one of the causes of hair loss is scalp inflammation. Extracts of microalgae of *Thalassiosira* sp., *Monodus* sp., *Chlorococcum* sp., and *Chaeloceros* sp. showed excellent properties in modulating the metabolism of human skin and hair follicles (Ariede et al., 2017).

12 Formulation adjuvants

Algae have been used as thickening agents, moisturizers binding to water, and antioxidants in cosmetic formulations. Each species could have more than one function to improve physical and organoleptic characteristics in the formulation, for instance, viscosity, stabilization, and sensory (Wang, Chen, et al., 2015).

Microalgae are promising sources of excipients due to their high protein content (50%–70%), amino acids, mineral salts, fatty acids, and triacylglycerides (TAG) (Borowitzka, 2013). The emulsifier and surfactant functions are due to proteins, lipids, polysaccharides, and nucleic acids present in the microalgae external cellular walls. However, the composition of the walls of most of microalgae species is highly undefined (Baudelet, Ricochon, Linder, & Muniglia, 2017) and must to be more studied for industrial purposes.

12.1 Thickening agents

The thickening purpose in cosmetic formulations requires the rigorous viscosity control to offer an attractive sensorial aspect to the consumers. The main thickening agents used in cosmetics are extracted from red algae (Rhodophyceae) and brown algae (*Phaenophyceae*). Although alginates are obtained from both algae, algenic acids are specifically isolated from brown algae, which provide rapid water absorption, and they are good hydrocolloids that increase viscosity and aid in the stabilization of emulsions (Khanra et al., 2018). Optimizing the thickening effect of alginates, such as alginic acid, it is essential to control the cosmetic viscosity with quelants, as ethylenediaminetetraacetic acid (EDTA) with consequent control of the metal ions (Wang, Chen, et al., 2015).

Agar is a polysaccharide extracted from red macroalgae such as *Gelidium*, *Pterocladiella*, and *Pterocladia*. However, Rhodophyceae *G. gracilis* and *Gracilaria chilensis* are the most relevant producers. Agar is used as an emulsifier, thickener, stabilizer, mucoadhesive as well as promoting the modified release of active compounds according to the conditions in the preparation of the emulsion, e.g., concentration of agar and homogenization speed in cosmetic formulations (Wang, Chen, et al., 2015).

Carrageenans are other polysaccharides isolated from red algae with anionic substituents (Khanna, Kaur, & Goyal, 2019). Thermal stimulation is required to form double helix bonds in carrageenans because the double-helix structure in the presence of cations and anionic groups of sulfates creates cross-links that increase viscosity. However, since the cation concentration increases, liquid viscosity also increases and becomes similar to a gel. Another polysaccharide that increases viscosity by the entanglement of bonds is λ-carrageenan (Wang, Chen, et al., 2015).

12.2 Surfactants

The most used surfactants are fossil fuel derivatives; however, they have been replaced by biodegradable surfactants that reduce CO_2 production and contribute to green chemistry

(Khan, Shin, & Kim, 2018). Glycolipids are non-ionic biosurfactants, with good superficial tension, and are cleaning and biocompatible products, mainly with enzymes (Olmstead et al., 2013). They are extracted from algae oils with low environmental impact and high content of stable and renewable carbon isotopes (Anyanwu, Rodriguez, Durrant, & Olabi, 2018).

12.3 Preservatives

Marine algae have bactericidal and bacteriostatic characteristics, keeping the skin microbiota in the balance as a preservative in the formulations (Fu et al., 2017). Extracts of *Isochrysis galbana*, *D. salina*, *C. vulgaris*, *N. oculata*, and *Pavlova lutheri* have shown inhibitory effects in gram-negative bacteria, mainly *Pseudomonas aeruginosa* and *Klebsiella pneumonia* (Couteau & Coiffard, 2018).

Biosynthesized nanoparticles of marine macroalgae named *Turbinaria conoides* have inhibitory activity in the formation of biofilms with *Escherichia coli* and *Salmonella* spp. They also exhibit excellent compatibility in emulsifier formulations for topical use (Khanna et al., 2019). Alginic acids from the cell wall of macroalgae *Phaeophyta* quickly absorb water preventing the proliferation of microorganisms in formulations; thus their use is mostly recommended for cosmetics (Wijesinghe & Jeon, 2011). Moreover, *Rhodomela confervoides* (red macroalgae) and *Padina pavonica* (brown macroalgae) have inhibited fungi, mainly of *Candida albicans* (Saidani, Bedjou, Benabdesselam, & Touati, 2012).

A sesquiterpenoid named laurinterol and isolated from Rhodophyta *Laurencia intermedia* and *Aplysia kurodai* has exhibited antibacterial activity against *Staphylococcus aureus* in the skin microbiota. Laurinterol can also be applied against staphylococci enterotoxins that cause pimples, boils, impetigo, abscesses, infection, and scalding skin syndrome, among other conditions (Joshi, 2018).

Algae *Himanthalia elongata* and microalgae *Synechocystis* sp. have functional components with antimicrobial action in microorganisms such as *E. coli*, *S. aureus*, *C. albicans*, and *Aspergillus brasiliensis*, being efficient preservatives (Wang, Chen, et al., 2015).

Finally, in Table 1, we describe some extracts from algal species that are currently used in the cosmetic industries.

13 Conclusions and perspectives

Compounds from algae species are described in the literature due to their particular applications in cosmetic fields. The proper extraction methods for the desired compound are currently studied for some algal species with the aid of green technologies to avoid environmental damage with successfully yields for industries. Besides several benefits of algae for human health have been addressed, the safety of them must be established. Algal cells can develop strategies to defend themselves against the environment, including the production of toxins, known as phycotoxins (Caruana & Zouher, 2018; Rhodes & Wood, 2014).

The addition of algal extracts in cosmetics has become more necessary mainly due to their safety and improvements in formulations that claim the enhancement of skin appearance and multipurpose benefits. In addition to all cosmetic attributes discussed here, we suggest the incorporation of algae-derived compounds with future possible applications for dental care. The reduction in the effects of skin aging through the continuous use of cosmetic products demands the incorporation of more natural products compatible with the natural physiology of the skin. Many algae-derived compounds are available in the cosmetic market. However, for all the marketed products there are no in vivo studies to confirm each claimed activity.

It is necessary to focus future studies on the concentration of each ingredient in the

TABLE 1   Some commercial products containing algae extracts.

Algae specie	INCI name	Cosmetic form	Claims for use in cosmetics	Reference
Thalassiosira pseudonana	*Thalassiosira pseudonana* extract	Cream gel Velvet finish	Smooth, firms, wrinkle Bust firming	Daniel Jouvance (2019)
Asterionella sp.	*Asterionella* extract	Cream	Stimulates the fat-burning trigger signal. Potentialize the anticellulite effect of caffeine	Daniel Jouvance (2019)
Chlamydomonas reinhardtii	*Chlamydomonas reinhardtii* extract	Shampoo Conditioner	Purification	Daniel Jouvance (2019)
Fucus vesiculosus	*Fucus vesiculosus* extract	Cream	Moisturizing, protect the skin	Cicatricure (2019)
Fragilaria pinnata	*Fragilaria pinnata* extract	Cream	Moisturizes, antiwrinkling	Daniel Jouvance (2019)
Tetraselmis suecica	*Tetraselmis suecica* extract	Fluid	Detoxifies, regenerates	Daniel Jouvance (2019)
Padina pavonica	*Padina pavonica* Thallus extract	Cream	Antiaging	Lancome (2019)

formulations, the physicochemical properties of each active and mainly the availability of the active ingredient at the action site according to the desired effect. We suggest that the algal extracts used for cosmetics must be standardized to allow each active compound to reach the same effect anywhere worldwide. In addition to this, new scientific research is highly necessary to verify in vivo (in volunteers) the efficacy and the ideal proportions of these unique compounds in new innovative cosmetics.

References

Aburai, N., Ohkubo, S., Miyashita, H., & Abe, K. (2013). Composition of carotenoids and identification of aerial microalgae isolated from the surface of rocks in mountainous districts of Japan. *Algal Research*, *2*, 237–243.

Al-Bader, T., Byrne, A., Gillbro, J., Mitarotonda, A., Metois, A., Vial, F., et al. (2012). Effect of cosmetic ingredients as anticellulite agents: Synergistic action of actives with in vitro and in vivo efficacy. *Journal of Cosmetic Dermatology*, *11*, 17–26.

Anunciato, T. P., & da Rocha Filho, P. A. (2012). Carotenoids and polyphenols in nutricosmetics, nutraceuticals, and cosmeceuticals. *Journal of Cosmetic Dermatology*, *11*, 51–54.

Anyanwu, R. C., Rodriguez, C., Durrant, A., & Olabi, A. G. (2018). Micro-macroalgae properties and applications. In S. Hashmi (Ed.), *Reference module in materials science and materials engineering* (pp. 1–28). New York: Elsevier.

Apone, F., Barbulova, A., & Colucci, M. G. (2019). Plant and microalgae derived peptides are advantageously employed as bioactive compounds in cosmetics. *Frontiers in Plant Science*, *10*, 756.

Ariede, M. B., Candido, T. M., Jacome, A. L. M., Velasco, M. V. R., de Carvalho, J. C. M., & Baby, A. R. (2017). Cosmetic attributes of algae—A review. *Algal Research*, *25*, 483–487.

Aubrey (2019). *Intensive hair repair mask. (2019).* https://aubreyorganics.com/blue-green-algae-hair-mask.html/ (Accessed 02 December 2019).

Ayalon, O. (2017). Improved process for producing fucoxanthin and/or polysaccharides from microalgae. WO2017037692A1.

Bagherani, N., Gianfaldoni, S., & Smoller, B. (2015). An overview on melasma. *Journal of Pigmentary Disorders*, *2*, 1000216.

Balogh, T. S., Velasco, M. V. R., Pedriali, C. A., Kaneko, T. M., & Baby, A. R. (2011). Proteção à radiação ultravioleta: recursos disponíveis na atualidade em fotoproteção. *Anais Brasileiros de Dermatologia, 86*, 732–742.

Barbarino, E., & Lourenço, S. O. (2005). An evaluation of methods for extraction and quantification of protein from marine macro- and microalgae. *Journal of Applied Phycology, 17*, 447–460.

Baudelet, P. H., Ricochon, G., Linder, M., & Muniglia, L. (2017). A new insight into cell walls of Chlorophyta. *Algal Research, 25*, 333–371.

Bedoux, G., Hardouin, K., Burlot, A. S., & Bourgougnon, N. (2014). Bioactive components from seaweeds: Cosmetic applications and future development. N. Bourgougnon (Ed.), *Advances in botanic research* (pp. 346–367). Vol. 71(pp. 346–367). San Diego, CA: Academic Press Inc.

Berardesca, E., Loden, M., Serup, J., Masson, P., & Rodrigues, L. M. (2018). The revised EEMCO guidance for the *in vivo* measurement of water in the skin. *Skin Research and Technology, 24*, 351–358.

Bernaerts, T. M. M., Gheysen, L., Kyomugasho, C., Jamsazzadeh Kermani, Z., Vandionant, S., Foubert, I., et al. (2018). Comparison of microalgal biomasses as functional food ingredients: Focus on the composition of cell wall related polysaccharides. *Algal Research, 32*, 150–161.

Bhosale, P., & Bernstein, P. S. (2005). Microbial xanthophylls. *Applied Microbiology and Biotechnology, 68*, 445–455.

Borowitzka, M. A. (2013). High-value products from microalgae—Their development and commercialisation. *Journal of Applied Phycology, 25*, 743–756.

Capablanca, J. S., & Watt, I. C. (1986). Factors affecting the zeta potential at wool fiber surfaces. *Textile Research Journal, 56*, 49–55.

Caruana, A. M. N., & Zouher, A. (2018). Microalgae and toxins. In I. A. Levine, & J. Fleurence (Eds.), *Microalgae in health and disease prevention* (pp. 263–305). London: Elsevier/Academic Press.

Carvalho, J. C. M., Matsudo, M. C., Bezerra, R. P., Ferreira-Camargo, L. S., & Sato, S. (2014). Microalgae bioreactors. R. Bajpai, A. Prokop, & M. Zappi (Eds.), *Algal biorefineries: Cultivation of cells and products* (pp. 83–126). Vol. 1 (pp. 83–126). Dordrecht: Springer.

Cavinato, M., Waltenberger, B., Baraldo, G., Grade, C. V. C., Stuppner, H., & Jansen-Dürr, P. (2017). Plant extracts and natural compounds used against UVB-induced photoaging. *Biogerontology, 18*, 499–516.

Cefali, L. C., Ataide, J. A., Moriel, P., Foglio, M. A., & Mazzola, P. G. (2016). Plant-based active photoprotectants for sunscreens. *International Journal of Cosmetic Science, 38*, 346–353.

Cezare-Gomes, E. A., Mejia-da-Silva, L. d. C., Pérez-Mora, L. S., Matsudo, M. C., Ferreira-Camargo, L. S., Singh, A. K., et al. (2019). Potential of microalgae carotenoids for industrial application. *Applied Biochemistry and Biotechnology, 188*, 602–634.

Chavan, M., Sun, Y., Litchauer, J., & Denis, A. (2014). Fucus extract: Cosmetic treatment for under-eye dark circles. *Journal of Cosmetic Science, 14*, 934.

Chen, L., Hu, J. Y., & Wang, S. Q. (2012). The role of antioxidants in photoprotection: A critical review. *Journal of the American Academy of Dermatology, 67*, 1013–1024.

Chen, B., Wan, C., Mehmood, M. A., Chang, J. S., Bai, F., & Zhao, X. (2017). Manipulating environmental stresses and stress tolerance of microalgae for enhanced production of lipids and value-added products—A review. *Bioresource Technology, 244*, 1198–1206.

Chen, L. L., & Wang, S. Q. (2016). Nanotechnology in photoprotection. M. R. Hamblin, P. Avci, & T. W. Prow (Eds.), *Nanoscience in dermatology* (pp. 229–236). Vol. 1(pp. 229–236). London: Academic Press.

Chisté, R. C., Freitas, M., Mercadante, A. Z., & Fernandes, E. (2014). Carotenoids are effective inhibitors of *in vitro* hemolysis of human erythrocytes, as determined by a practical and optimized cellular antioxidant assay. *Journal of Food Science, 79*, H1841–H1847.

Chrapusta, E., Kaminski, A., Duchnik, K., Bober, B., Adamski, M., & Bialczyk, J. (2017). Mycosporine-like amino acids: Potential health and beauty ingredients. *Marine Drugs, 15*, 326.

Cicatricure (2019). *Cicatricure creme antissinais*. (2019). https://www.cicatricure.com.br/produtos/antissinais/cicatricure-creme-antissinais/ (Accessed 4 December 2019).

Coderch, L., Méndez, S., Martí, M., Pons, R., & Parra, J. L. (2007). Influence of water in the lamellar rearrangement of internal wool lipids. *Colloids and surfaces B: Biointerfaces, 60*, 89–94.

Conde, F. R., Churio, M. S., & Previtali, C. M. (2004). The deactivation pathways of the excited-states of the mycosporine-like amino acids shinorine and porphyra-334 in aqueous solution. *Photochemical & Photobiological Sciences, 3*, 960.

Couteau, C., & Coiffard, L. (2018). Microalgal application in cosmetics. I. A. Levine, & J. Fleurence (Eds.), *Microalgae in health and disease prevention* (pp. 317–323). Vol. 1(pp. 317–323). London: Academic Press.

Cunha, L., & Grenha, A. (2016). Sulfated seaweed polysaccharides as multifunctional materials in drug delivery applications. *Marine Drugs, 14*, 1–42.

Daneluti, A. L. M., Mariano, N. F., Ruscinc, N., Lopes, I., Velasco, M. V. R., Matos, J. R., et al. (2019). Using ordered mesoporous silica SBA-15 to limit cutaneous penetration and transdermal permeation of organic UV filters. *International Journal of Pharmaceutics, 570*, 118633.

Daniel Jouvance (2019). *Daniel Jouvance*. (2019). https://www.danieljouvance.com/fr-fr/ (Accessed 4 December 2019).

De Jesus Raposo, M. F., De Morais, R. M. S. C., & De Morais, A. M. M. B. (2013a). Bioactivity and applications of sulphated polysaccharides from marine microalgae. *Marine Drugs*, *11*, 233–252.

De Jesus Raposo, M. F., De Morais, R. M. S. C., & De Morais, A. M. M. B. (2013b). Health applications of bioactive compounds from marine microalgae. *Life Sciences*, *93*, 479–486.

De Jesus Raposo, M. F., De Morais, A. M. B., & De Morais, R. M. S. C. (2015). Marine polysaccharides from algae with potential biomedical applications. *Marine Drugs*, *13*, 2967–3028.

De Jesus Raposo, M. F., De Morais, A. M. M. B., & De Morais, R. M. S. C. (2016). Emergent sources of prebiotics: Seaweeds and microalgae. *Marine Drugs*, *14*, 2.

Derikvand, P., Liewellyn, C. A., & Purton, S. (2017). Cyanobacterial metabolites as a source of sunscreens and moisturizers: A comparison with current synthetic compounds. *European Journal of Phycology*, *52*, 43–56.

Draelos, Z. D. (2005). The disease of cellulite. *Journal of Cosmetic Dermatology*, *4*, 221–222.

Draelos, Z. D. (2018). The science behind skincare: Moisturizers. *Journal of Cosmetic Dermatology*, *17*, 138–144.

Espinosa-Leal, C. A., & Garcia-Lara, S. (2019). Current methods for the discovery of new active ingredients from natural products for cosmeceutical applications. *Planta Medica*, *85*, 535–551.

Ferreira, L. S., Rodrigues, M. S., Converti, A., Sato, S., & Carvalho, J. C. M. (2012). *Arthrospira (Spirulina) platensis* cultivation in tubular photobioreactor: Use of no-cost CO_2 from ethanol fermentation. *Applied Energy*, *92*, 379–385.

Feughelman, M. (1959). A two-phase structure for keratin fibers. *Textile Research Journal*, *29*, 223–228.

Fresewinkel, M., Rosello, R., Wilhelm, C., Kruse, O., Hankamer, B., & Posten, C. (2014). Integration in microalgal bioprocess development: Design of efficient, sustainable, and economic processes. *Engineering in Life Sciences*, *14*, 560–573.

Fu, W., Chaiboonchoe, A., Khraiwesh, B., Nelson, D. R., Al-Khairy, D., Mystikou, A., et al. (2016). Algal cell factories: Approaches, applications, and potentials. *Marine Drugs*, *14*, 255.

Fu, W., Nelson, D., Yi, Z., Xu, M., Khraiwesh, B., Jijakli, K., et al. (2017). Bioactive compounds from microalgae: Current development and prospects. Atta-ur-Rahman (Ed.), *Studies in natural products chemistry* (pp. 199–225). Vol. 54(pp. 199–225). London: Elsevier.

Fuji Chemical Industries Co., Ltd (2019). *The astaxanthin of choice AstaReal®*. *(2019)*. http://www.fujichemical.co.jp/english/life_science/about_astareal/index.html/; (Accessed 20 May 2019).

Fujimura, T., Tsukahara, K., Moriwaki, S., & Kitahara, T. (2002). Treatment of human skin with an extract of *Fucus vesiculosus* chanoes its thickness and mechanical properties. *Journal of Cosmetic Science*, *53*, 1–9.

Gilaberte, Y., Prieto-Torres, L., Pastushenko, I., & Juarranz, Á. (2016). Anatomy and function of the skin. M. R. Hamblin, P. Avci, & T. W. Prow (Eds.), *Nanoscience in dermatology* (pp. 1–14). Vol. 1(pp. 1–14). London: Academic Press.

Gong, M., & Bassi, A. (2016). Carotenoids from microalgae: A review of recent developments. *Biotechnology Advances*, *34*, 1396–1412.

Grundman, O., Richter, H., Ini, S. (2018). Compositions comprising carotenoids and use thereof. US20180078521A1.

Guihéneuf, F., & Stengel, D. B. (2015). Towards the biorefinery concept: Interaction of light, temperature and nitrogen for optimizing the co-production of high-value compounds in *Porphyridium purpureum*. *Algal Research*, *10*, 152–163.

Guillerme, J.-B., Couteau, C., & Coiffard, L. (2017). Applications for marine resources in cosmetics. *Cosmetics*, *4*, 35.

Günerken, E., D'Hondt, E., Eppink, M. H. M., Garcia-Gonzalez, L., Elst, K., & Wijffels, R. H. (2015). Cell disruption for microalgae biorefineries. *Biotechnology Advances*, *33*, 243–260.

Gunes, S., Tamburaci, S., Dalay, M. C., & Deliloglu, G. I. (2017). *In vitro* evaluation of *Spirulina platensis* extract incorporated skin cream with its wound healing and antioxidant activities. *Pharmaceutical Biology*, *55*, 1824–1832.

Holdt, S. L., & Kraan, S. (2011). Bioactive compounds in seaweed: Functional food applications and legislation. *Journal of Applied Phycology*, *23*, 543–597.

Hu, J., Nagarajan, D., Zhang, Q., Chang, J. S., & Lee, D. J. (2018). Heterotrophic cultivation of microalgae for pigment production: A review. *Biotechnology Advances*, *36*, 54–67.

Hudek, K., Davis, L. C., Ibbini, J., & Erickson, L. (2014). Commercial products from algae. R. Bajpai, A. Prokop, & M. Zappi (Eds.), *Algal biorefineries: Cultivation of cells and products* (pp. 275–295). Vol. 1(pp. 275–295). Dordrecht: Springer.

Ishika, T., Moheimani, N. R., Bahri, P. A., Laird, D. W., Blair, S., & Parlevliet, D. (2017). Halo-adapted microalgae for fucoxanthin production: Effect of incremental increase in salinity. *Algal Research*, *28*, 66–73.

Jea, J. Y., Park, P. J., Kim, E. K., Park, J. S., Yoon, H. D., & Kim, K. R. (2009). Antioxidant activity of enzymatic extracts from the brown seaweed *Undaria pinnatifida* by electron spin resonance spectroscopy. *LWT—Food Science and Technology*, *42*, 874–878.

Joshi, R. K. (2018). Role of natural products against microorganisms. *American Journal of Clinical Microbiology and Antimicrobials*, *1*, 1005.

Kang, J., Kim, S. C., Han, S. C., Hong, H. J., Jeon, Y. J., Kim, B., et al. (2012). Hair-loss preventing effect of *Grateloupia elliptica*. *Biomolecules & Therapeutics*, *20*, 118–120.

Kang, J., Kim, M. K., Lee, J. H., Jeon, Y. J., Hwang, E. K., Koh, Y. S., et al. (2017). *Undariopsis peterseniana* promotes hair growth by the activation of Wnt/β-catenin and ERK pathways. *Marine Drugs, 15,* 5.

Katzer, T., Leite Junior, A., Beck, R., & da Silva, C. (2019). Physiopathology and current treatments of androgenetic alopecia: Going beyond androgens and anti-androgens. *Dermatologic Therapy, 32,* e13059.

Khan, M., Shin, J., & Kim, J. (2018). The promising future of microalgae: Current status, challenges, and optimization of a sustainable and renewable industry for biofuels, feed, and other products. *Microbial Cell Factories, 17,* 36.

Khanna, P., Kaur, A., & Goyal, D. (2019). Algae-based metallic nanoparticles: Synthesis, characterization and applications. *Journal of Microbiological Methods, 163,* 105656.

Khanra, S., Mondal, M., Halder, G., Tiwari, O. N., Gayen, K., & Bhowmick, T. K. (2018). Downstream processing of microalgae for pigments, protein and carbohydrate in industrial application: A review. *Food and Bioproducts Processing, 110,* 60–84.

Kolarsick, P. A. J., Kolarsick, M. A., & Goodwin, C. (2011). Anatomy and physiology of the skin. *Journal of the Dermatology Nurses' Association, 3,* 203–213.

Koller, M., Muhr, A., & Braunegg, G. (2014). Microalgae as versatile cellular factories for valued products. *Algal Research, 6,* 52–63.

Kraft, J., & Lynde, C. (2005). Moisturizers: What they are and a practical approach to product selection. *Skin Therapy Letter, 10,* 1–8.

Kroumov, A. D., Scheufele, F. B., Trigueros, D. E. G., Modenes, A. N., Zaharieva, M., & Najdenski, H. (2017). Modeling and technoeconomic analysis of algae for bioenergy and coproducts. R. Rastogi, D. Madamwar, & A. Pandey (Eds.), *Algal green chemistry: Recent progress biotechnology* (pp. 201–241). Vol. 1(pp. 201–241). Amsterdam: Elsevier.

Kulkarni, S., & Nikolov, Z. (2018). Process for selective extraction of pigments and functional proteins from *Chlorella vulgaris. Algal Research, 35,* 185–193.

Lalegerie, F., Lajili, S., Bedoux, G., Taupin, L., Stiger-Pouvreau, V., & Connan, S. (2019). Photo-protective compounds in red macroalgae from Brittany: Considerable diversity in mycosporine-like amino acids (MAAs). *Marine Environmental Research, 147,* 37–48.

Lamers, P. P., Janssen, M., De Vos, R. C. H., Bino, R. J., & Wijffels, R. H. (2008). Exploring and exploiting carotenoid accumulation in *Dunaliella salina* for cell-factory applications. *Trends in Biotechnology, 26,* 631–638.

Lancome (2019). *Absolue jour precious cells. (2019).* https://www.lancome.com.br/absolue-jour-precious-cells/p; (Accessed 4 December 2019).

Lautenschlager, S., Wulf, H. C., & Pittelkow, M. R. (2007). Photoprotection. *Lancet, 370,* 528–537.

Lawrence, K. P., Long, P. F., & Young, A. R. (2019). Mycosporine-like amino acids for skin photoprotection. *Current Medicinal Chemistry, 25,* 5512–5527.

Leite e Silva, V. R., Schulman, M. A., Ferelli, C., Gimenis, J. M., Ruas, G. W., Baby, A. R., et al. (2009). Hydrating effects of moisturizer active compounds incorporated into hydrogels: *In vivo* assessment and comparison between devices. *Journal of Cosmetic Dermatology, 8,* 32–39.

Letsiou, S., Kalliampakou, K., Gardikis, K., Mantecon, L., Infante, C., Chatzikonstantinou, M., et al. (2017). Skin protective effects of *Nannochloropsis gaditana* extract on H_2O_2-stressed human dermal fibroblasts. *Frontiers in Marine Science, 4,* 15.

Lorencini, M., Brohem, C. A., Dieamant, G. C., Zanchin, N. I. T., & Maibach, H. I. (2014). Active ingredients against human epidermal aging. *Ageing Research Reviews, 15,* 100–115.

Lowrey, J., Armenta, R. E., & Brooks, M. S. (2016). Nutrient and media recycling in heterotrophic microalgae cultures. *Applied Microbiology and Biotechnology, 100,* 1061–1075.

Lynde, C. (2001). Moisturizers: What they are and how they work. *Skin Therapy Letter, 6,* 3–5.

Manandhar, B., Paudel, P., Seong, S. H., Jung, H. A., & Choi, J. S. (2019). Characterizing eckol as a therapeutic aid: A systematic review. *Marine Drugs, 17,* 361.

Marcati, A., Ursu, A. V., Laroche, C., Soanen, N., Marchal, L., Jubeau, S., et al. (2014). Extraction and fractionation of polysaccharides and B-phycoerythrin from the microalga *Porphyridium cruentum* by membrane technology. *Algal Research, 5,* 258–263.

Mejia-Da-Silva, L. D. C., Matsudo, M. C., Morocho-Jacome, A. L., Carlos, J., & De Carvalho, M. (2018). Application of physicochemical treatment allows reutilization of *Arthrospira platensis exhausted medium* an investigation of reusing medium in *Arthrospira platensis* cultivation. *Applied Biochemistry and Biotechnology, 186,* 1–14.

Messyasz, B., Michalak, I., Łęska, B., Schroeder, G., Górka, B., Korzeniowska, K., et al. (2018). Valuable natural products from marine and freshwater macroalgae obtained from supercritical fluid extracts. *Journal of Applied Phycology, 30,* 591–603.

Michelet, J. F., Olive, C., Rieux, E., Fagot, D., Simonetti, L., Galey, J. B., et al. (2012). The anti-ageing potential of a new jasmonic acid derivative (LR2412): *In vitro* evaluation using reconstructed epidermis episkin™. *Experimental Dermatology, 21,* 398–400.

Morocho-Jácome, A. L., Mascioli, G. F., Sato, S., & de Carvalho, J. C. M. (2015a). Ferric chloride flocculation plus carbon adsorption allows to reuse spent culture medium of *Arthrospira platensis. Engineering in Life Sciences, 15,* 208–219.

Morocho-Jácome, A. L., Mascioli, G. F., Sato, S., & De Carvalho, J. C. M. (2015b). Continuous cultivation of *Arthrospira platensis* using exhausted medium treated with granular activated carbon. *Journal of Hydrology, 522,* 467–474.

Morone, J., Alfeus, A., Vasconcelos, V., & Martins, R. (2019). Revealing the potential of cyanobacteria in cosmetics and cosmeceuticals—A new bioactive approach. *Algal Research*, *41*, 1015–1041.

Mourelle, M., Gómez, C., & Legido, J. (2017). The potential use of marine microalgae and cyanobacteria in cosmetics and thalassotherapy. *Cosmetics*, *4*, 1–14.

Nutrifarm (2019). *Dermochlorella D. Nutrifarm*. (2019). https://www.nutrifarm.com.br/Arquivos/Insumo/0e289d72-7bd6-43c9-9308-0e2ed03fc559.pdf/ (Accessed 4 December 2019).

Oliveira, C. A., Peres, D. D., Rugno, C. M., Kojima, M., Pinto, C. A. S. O., Consiglieri, V. O., et al. (2015). Functional photostability and cutaneous compatibility of bioactive UVA sun care products. *Journal of Photochemistry and Photobiology B: Biology*, *148*, 154–159.

Olmstead, I., Hill, D., Dias, D., Jayasinghe, N., Callahan, D., Kentish, S., et al. (2013). A quantitative analysis of microalgal lipids for optimization of biodiesel and omega-3 production. *Biotechnology and Bioengineering*, *110*, 2096–2104.

OpenStax College (2013). *Skin structure. (2013).* https://commons.wikimedia.org/wiki/File:501_Structure_of_the_skin.jpg/; (Accessed 4 December 2019).

Oren, A., & Gunde-Cimerman, N. (2007). Mycosporines and mycosporine-like amino acids: UV protectants or multipurpose secondary metabolites? *FEMS Microbiology Letters*, *1*, 1–10.

Paliwal, C., Ghosh, T., George, B., Pancha, I., Maurya, R., Chokshi, K., et al. (2016). Microalgal carotenoids: Potential nutraceutical compounds with chemotaxonomic importance. *Algal Research*, *15*, 24–31.

Pallela, R., Na-Young, Y., & Kim, S. K. (2010). Antiphotoaging and photoprotective compounds derived from marine organisms. *Marine Drugs*, *8*, 1189–1202.

Panayotova, V., Merzdhanova, A., Dobreva, D. A., Zlatanov, M., & Makedonski, L. (2017). Lipids of Black Sea algae: Unveiling their potential for pharmaceutical and cosmetic applications. *Journal of IMAB*, *23*, 1747–1751.

Panis, G., & Carreon, J. R. (2016). Commercial astaxanthin production derived by green alga *Haematococcus pluvialis*: A microalgae process model and a techno-economic assessment all through production line. *Algal Research*, *18*, 175–190.

Park, K. S., & Park, D. H. (2016). Comparison of *Saccharina japonica–Undaria pinnatifida* mixture and minoxidil on hair growth promoting effect in mice. *Archives of Plastic Surgery*, *43*, 498–505.

Patel, H. M., Rastogi, R. P., Trivedi, U., & Madamwar, D. (2018). Structural characterization and antioxidant potential of phycocyanin from the cyanobacterium *Geitlerinema* sp. H8DM. *Algal Research*, *32*, 372–383.

Peng, J., Yuan, J. P., Wu, C. F., & Wang, J. H. (2011). Fucoxanthin, a marine carotenoid present in brown seaweeds and diatoms: Metabolism and bioactivities relevant to human health. *Marine Drugs*, *9*, 1806–1828.

Pereira, L. (2018). Seaweeds as source of bioactive substances and skincare therapy-cosmeceuticals, algotheraphy, and thalassotherapy. *Cosmetics*, *5*, 1–4.

Peres, D. D., Ariede, M. B., Candido, T. M., Almeida, T. S., Lourenco, F. R., Consiglieri, V. O., et al. (2017). Quality by design (QbD), Process Analytical Technology (PAT), and design of experiment applied to the development of multifunctional sunscreens. *Drug Development and Industrial Pharmacy*, *2*, 246–256.

Peres, D. D., Sarruf, F. D., Oliveira, C. A., Velasco, M. V. R., & Baby, A. R. (2018). Ferulic acid photoprotective properties in association with UV filters: Multifunctional sunscreen with improved SPF and UVA-PF. *Journal of Photochemistry and Photobiology B: Biology*, *185*, 46–49.

Pimentel, F. B., Alves, R. C., Rodrigues, F., & Oliveira, M. B. P. P. (2018). Macroalgae-derived ingredients for cosmetic industry—An update. *Cosmetics*, *5*, 2–20.

Priyadarshani, I., & Rath, B. (2012). Commercial and industrial applications of micro algae—A review. *Journal of Algal Biomass Utilization*, *3*, 89–100.

Purnamawati, S., Indrastuti, N., Danarti, R., & Saefudin, T. (2017). The role of moisturizers in addressing various kinds of dermatitis: A review. *Clinical Medicine & Research*, *15*, 75–87.

Quintana, N., Van der Kooy, F., Van de Rhee, M. D., Voshol, G. P., & Verpoorte, R. (2011). Renewable energy from cyanobacteria: Energy production optimization by metabolic pathway engineering. *Applied Microbiology and Biotechnology*, *91*, 471–490.

Radice, M., Manfredini, S., Ziosi, P., Dissette, V., Buso, P., Fallacara, A., et al. (2016). Herbal extracts, lichens and biomolecules as natural photo-protection alternatives to synthetic UV filters. A systematic review. *Fitoterapia*, *114*, 144–162.

Rao, A. R., Sindhuja, H. N., Dharmesh, S. M., Sankar, K. U., Sarada, R., & Ravishankar, G. A. (2013). Effective inhibition of skin cancer, tyrosinase, and antioxidative properties by astaxanthin and astaxanthin esters from the green alga *Haematococcus pluvialis*. *Journal of Agricultural and Food Chemistry*, *61*, 3842–3851.

Rastogi, R. P., & Incharoensakdi, A. (2014). Characterization of UV-screening compounds, mycosporine-like amino acids, and scytonemin in the cyanobacterium *Lyngbya* sp. CU2555. *FEMS Microbiology Ecology*, *87*, 244–256.

Rastogi, R. P., Sonani, R. R., Madamwar, D., & Incharoensakdi, A. (2016). Characterization and antioxidant functions of mycosporine-like amino acids in the cyanobacterium *Nostoc* sp. R76DM. *Algal Research*, *16*, 110–118.

Řezanka, T., Temina, M., Tolstikov, A. G., & Dembitsky, V. M. (2004). Natural microbial UV radiation filters—Mycosporine-like amino acids. *Folia Microbiologia (Praha)*, *49*, 339–352.

Rhodes, L., & Wood, S. (2014). Micro-algal and cyanobacterial producers of biotoxins. G. P. Rossini (Ed.), *Toxins and biologically active compounds from microalgae* (pp. 21–50). Vol. 2(pp. 21–50). Florida: CRC Press.

Rizwan, M., Mujtaba, G., Memon, S. A., Lee, K., & Rashid, N. (2018). Exploring the potential of microalgae for new biotechnology applications and beyond: A review. *Renewable and Sustainable Energy Reviews*, *92*, 394–404.

Robbins, C. R. (2012). Chemical composition of different hair types. C. R. Robbins (Ed.), *Chemical and physical behavior of human hair* (pp. 105–176). Vol. 5(pp. 105–176). Berlin: Springer.

Rodrigues, R. D. P., de Castro, F. C., de Santiago-Aguiar, R. S., & Rocha, M. V. P. (2018). Ultrasound-assisted extraction of phycobiliproteins from *Spirulina* (Arthrospira) platensis using protic ionic liquids as solvent. *Algal Research*, *31*, 454–462.

Rosado, C., Tokunaga, V. K., Sauce, R. S., Oliveira, C. A., Sarruf, F. D., Parise-Filho, R., et al. (2019). Another reason for using caffeine in dermocosmetics: Sunscreen adjuvant. *Frontiers in Physiology*, *10*, 1.

Ruiz, J., Olivieri, G., De Vree, J., Bosma, R., Willems, P., Reith, J. H., et al. (2016). Towards industrial products from microalgae. *Energy & Environmental Science*, *9*, 3036–3043.

Saidani, K., Bedjou, F., Benabdesselam, F., & Touati, N. (2012). Antifungal activity of methanolic extracts of four Algerian marine algae species. *African Journal of Biotechnology*, *11*, 9496–9500.

Saini, D. K., Pabbi, S., & Shukla, P. (2018). Cyanobacterial pigments: Perspectives and biotechnological approaches. *Food and Chemical Toxicology*, *120*, 616–624.

Sathasivam, R., & Ki, J. S. (2018). A review of the biological activities of microalgal carotenoids and their potential use in healthcare and cosmetic industries. *Marine Drugs*, *16*, 1–31.

Sathasivam, R., Radhakrishnan, R., Hashem, A., & Abd_Allah, E. F. (2017). Microalgae metabolites: A rich source for food and medicine. *Saudi Journal of Biological Sciences*, *26*, 709–722.

Seiberg, M. (2017). Compositions containing natural extracts and use thereof for skin and hair. WO2017/087177A2.

SEPPIC (2019a). *Xylishine.* SEPPIC Launches. (2019a). https://www.seppic.com/xylishine/ (Accessed 02 December 2019).

SEPPIC (2019b). *Bioenergizer BG.* SEPPIC Launches. (2019b). https://cosmetics.specialchem.com/product/i-biotech-marine-seppic-group-bioenergizer-bg/ (Accessed 02 December 2019).

Shick, J. M., & Dunlap, W. C. (2002). Mycosporine-like amino acids and related gadusols: Biosynthesis, accumulation, and UV-protective functions in aquatic organisms. *Annual Review of Physiology*, *64*, 223–262.

Shimoda, H., Tanaka, J., Shan, S. J., & Maoka, T. (2010). Antipigmentary activity of fucoxanthin and its influence on skin mRNA expression of melanogenic molecules. *The Journal of Pharmacy and Pharmacology*, *62*, 1137–1145.

Singh, S., & Das, S. (2011). Screening, production, optimization and characterization of cyanobacterial polysaccharide. *World Journal of Microbiology and Biotechnology*, *27*, 1971–1980.

Singh, S. P., Kumari, S., Rastogi, R. P., Singh, K. L., & Sinha, R. P. (2008). Mycosporine-like amino acids (MAAs): Chemical structure, biosynthesis and significance as UV-absorbing/screening compounds. *Indian Journal of Experimental Biology*, *46*, 7–17.

Singh, R., Parihar, P., Singh, M., Bajguz, A., Kumar, J., Singh, S., et al. (2017). Uncovering potential applications of cyanobacteria and algal metabolites in biology, agriculture and medicine: Current status and future prospects. *Frontiers in Microbiology*, *8*, 1–37.

Sinha, R. P., Singh, S. P., & Häder, D.-P. (2007). Database on mycosporines and mycosporine-like amino acids (MAAs) in fungi, cyanobacteria, macroalgae, phytoplankton and animals. *Journal of Photochemistry and Photobiology B: Biology*, *89*, 29–35.

Skjånes, K., Rebours, C., & Lindblad, P. (2012). Potential for green microalgae to produce hydrogen, pharmaceuticals and other high value products in a combined process. *Critical Reviews in Biotechnology*, *33*, 1–44.

Soanen, N., Da Silva, E., Gardarin, C., Michaud, P., & Laroche, C. (2016). Improvement of exopolysaccharide production by *Porphyridium marinum*. *Bioresource Technology*, *213*, 231–238.

Sotiroudis, T. G., & Sotiroudis, G. T. (2013). Health aspects of *Spirulina* (Arthrospira) microalga food supplement. *Journal of the Serbian Chemical Society*, *78*, 395–405.

Spada, F., Barnes, T. M., & Greive, K. (2018). Skin hydration is significantly increased by a cream formulated to mimic the skin's own natural moisturizing systems. *Clinical, Cosmetic and Investigational Dermatology*, *11*, 491–497.

Stahl, W., & Sies, H. (2012). β-Carotene and other carotenoids in protection from sunlight. *The American Journal of Clinical Nutrition*, *96*, 1179–1184.

Sui, Z., Gizaw, Y., & Bemiller, J. N. (2012). Extraction of polysaccharides from a species of *Chlorella*. *Carbohydrate Polymers*, *90*, 1–7.

Tannin-Spitz, T., Bergman, M., Van-Moppes, D., Grossman, S., & Arad, S. (2005). Antioxidant activity of the polysaccharide of the red microalga *Porphyridium* sp. *Journal of Applied Phycology*, *17*, 215–222.

Terranova, F., Berardesca, E., & Maibachà, H. (2006). Cellulite: Nature and aetiopathogenesis. *International Journal of Cosmetic Science*, *28*, 157–167.

Tibbetts, S. M., Milley, J. E., & Lall, S. P. (2015). Chemical composition and nutritional properties of freshwater and marine microalgal biomass cultured in photobioreactors. *Journal of Applied Phycology*, *27*, 1109–1119.

Tomazelli, L. C., Ramos, M. M. A., Sauce, R. S., Candido, T. M., Sarruf, F. D., Pinto, C. A. S. O., et al. (2018). SPF enhancement provided by rutin in a multifunctional sunscreen. *International Journal of Pharmaceutics*, *552*, 401–406.

Tominaga, K., Fujishita, M., Nobuko, H. (2018). Use of astaxanthin for reducing progression of damage caused to human skin. WO062427.

Tominaga, K., Hongo, N., Karato, M., & Yamashita, E. (2012). Cosmetic benefits of astaxanthin on humans subjects. *Acta Biochimica Polonica*, *59*, 2012.

Trabelsi, L., M'sakni, N. H., Ouada, H. B., Bacha, H., & Roudesli, S. (2009). Partial characterization of extracellular polysaccharides produced by cyanobacterium *Arthrospira platensis*. *Biotechnology and Bioprocess Engineering*, *14*, 27–31.

Turati, F., Pelucchi, C., Marzatico, F., Ferraroni, M., Decarli, A., Gallus, S., et al. (2014). Efficacy of cosmetic products in cellulite reduction: Systematic review and meta-analysis. *Journal of the European Academy of Dermatology and Venereology*, *28*, 1–15.

United States PC (2016). *Nomenclature guidelines*. U.S. Pharmacopeial Conv.. (2016). https://www.usp.org/sites/default/files/usp/document/about/expert-volunteers/expert-committees/nomenclature-guideline.pdf; (Accessed 04 December 2019).

Ursu, A. V., Marcati, A., Sayd, T., Sante-Lhoutellier, V., Djelveh, G., & Michaud, P. (2014). Extraction, fractionation and functional properties of proteins from the microalgae *Chlorella vulgaris*. *Bioresource Technology*, *157*, 134–139.

Wang, H. M. D., Chen, C. C., Huynh, P., & Chang, J. S. (2015). Exploring the potential of using algae in cosmetics. *Bioresource Technology*, *184*, 355–362.

Wang, J., Jin, W., Hou, Y., Niu, X., Zhang, H., & Zhang, Q. (2013). Chemical composition and moisture-absorption/retention ability of polysaccharides extracted from five algae. *International Journal of Biological Macromolecules*, *57*, 26–29.

Wang, T., Jónsdóttir, R., & Ólafsdóttir, G. (2009). Total phenolic compounds, radical scavenging and metal chelation of extracts from Icelandic seaweeds. *Food Chemistry*, *116*, 240–248.

Wang, D., Li, Y., Hu, X., Su, W., & Zhong, M. (2015). Combined enzymatic and mechanical cell disruption and lipid extraction of green alga *Neochloris oleoabundans*. *International Journal of Molecular Sciences*, *16*, 7707–7722.

Wang, B., Zarka, A., Trebst, A., & Boussiba, S. (2003). Astaxanthin accumulation in *Haematococcus pluvialis* (Chlorophyceae) as an active photoprotective process under high irradiance. *Journal of Phycology*, *39*, 1116–1124.

Wei, N., Quarterman, J., & Jin, Y. (2013). Marine macroalgae: An untapped resource for producing fuels and chemicals. *Trends in Biotechnology*, *31*, 70–77.

Wijesinghe, W., & Jeon, Y. (2011). Biological activities and potential cosmeceutical applications of bioactive components from brown seaweeds: A review. *Phytochemistry Reviews*, *10*, 431–443.

Wong, R., Geyer, S., Weninger, W., Guimberteau, J. C., & Wong, J. K. (2016). The dynamic anatomy and patterning of skin. *Experimental Dermatology*, *25*, 92–98.

Wróblewska, K. B., Baby, A. R., Grombone, G. M. T., & Moreno, P. R. H. (2019). *In vitro* antioxidant and photoprotective activity of five native Brazilian bamboo species. *Industrial Crops and Products*, *130*, 208–215.

Yun, E. J., Choi, I. G., & Kim, K. H. (2015). Red macroalgae as a sustainable resource for bio-based products. *Trends in Biotechnology*, *33*, 247–249.

Zeitoun, H., Michael-Jubeli, R., El Khoury, R., Baillet-Guffroy, A., Tfayli, A., Salameh, D., et al. (2019). Skin lightening effect of natural extracts coming from Senegal botanical biodiversity. *International Journal of Dermatology*, *1*, 6.

Zhang, Z., Wang, F., Wang, X., Liu, X., Hou, Y., & Zhang, Q. (2010). Extraction of the polysaccharides from five algae and their potential antioxidant activity *in vitro*. *Carbohydrate Polymers*, *82*, 118–121.

Zhang, R., Zhang, X., Tang, Y., & Mao, J. (2020). Composition, isolation, purification and biological activities of *Sargassum fusiforme* polysaccharides: A review. *Carbohydrate Polymers*, *228*, 115381.

Zouboulis, C. C., Chen, W. C., Thornton, M. J., Qin, K., & Rosenfield, R. (2007). Sexual hormones in human skin. *Hormone and Metabolic Research*, *39*, 85–95.

Microalgal applications toward agricultural sustainability: Recent trends and future prospects

Kshipra Gautam, Meghna Rajvanshi,
Neera Chugh, Rakhi Bajpai Dixit, G. Raja Krishna Kumar,
Chitranshu Kumar, Uma Shankar Sagaram, and Santanu Dasgupta

Reliance Technology Group, Reliance Industries Limited, Reliance Corporate Park, Ghansoli,
Maharashtra, India

1 Introduction

With the onset of the green revolution, agricultural productivity has increased tremendously due to the introduction of better yielding varieties and the use of various agricultural inputs. In general, agricultural inputs are chemical and biological materials used in crop production.

Fertilizers and pesticides attract major attention with respect to inputs in increasing agricultural production. Fertilizer application provides nutrients required for crop growth while pesticide application can significantly reduce plant diseases or insect pests or weeds thus indirectly contributing to an increase in agricultural production. Gradually, the dependence on chemical inputs, mainly the use of chemical fertilizers and pesticides has increased significantly, in all modern agriculture practices. However, in a disturbing trend, the utilization rate of agriculture chemicals is only ~35% and the unutilized fertilizers and pesticides are most likely to contaminate soil and water bodies (Zhang, Yan, Guo, Zhang, & Ruiz-Menjivar, 2018). As a result, an alarming level of residues of agricultural chemicals, which are likely to be the result of runoff or unused chemical inputs, were reported to be present in the soil, water, air, and agricultural products in several parts of the world. For example, the buildup of metal contaminants, such as arsenic, cadmium, fluorine, lead, and mercury in agricultural soils was reported to be associated with the vast use of inorganic fertilizers (Udeigwe et al., 2015). Similarly, pesticides were detected in almost all stream water samples at multiple agricultural sites in the USA (Gilliom, 2007) and the residential environments of agricultural communities in Japan (Kawahara, Horikoshi, Yamaguchi, Kumagai, & Yanagisawa, 2005).

In the last century, the use of agricultural chemicals has aided in doubling the production; however, the current need to increase food production keep pressure on the intensive use of fertilizers and pesticides (Carvalho, 2017). There is still a mounting pressure on agriculture to meet the demands of the growing population. As per the United Nations, the world's population will increase by 2.2 billion, reaching around 9.7 billion by 2050 (https://www.un.org/en/development/desa/news/population/2015-report.html). To meet the growing demand for food, excessive and imbalanced use of pesticides and fertilizers continued, and this trend has caused adverse effects on the environment. Although harmful organic pesticides have been replaced by biodegradable chemicals to a large extent, contamination by historical residues and ongoing accumulation still impact the quality of food, water, and environment (Carvalho, 2017). It is essential to develop and adopt sustainable and environmentally friendly agriculture practices, which not only enhance yield but also crop quality and environmental sustainability. With respect to agricultural inputs, pollution impact assessment and pollution prevention/reduction strategies are the most researched areas in the past 3 decades (Zhang et al., 2018), and significant efforts are being continually taken to use harmless sustainable agriculture inputs such as natural fertilizers and biopesticides.

Microalgae can be a great value to agriculture. Many studies indicate the use of microalgae in sustainable and organic agricultural practices (Priyadarshani & Rath, 2012; Sharma, Khokhar, Jat, & Khandelwal, 2012) and still, extensive research is being carried out.

Microalgae are a diverse group of microorganisms that are ubiquitous and found in almost every habitat on earth be it soil, oceans, hot springs or in dessert lands. Microalgae are unicellular or multicellular eukaryotic organisms, however, cyanobacteria that are commonly called blue-green algae (BGA) are also interchangeably referred to as microalgae in this chapter. Microalgae can perform photosynthesis by capturing CO_2 from the atmosphere and energy from sunlight. They have a high growth rate and hence produce higher biomass per unit area as compared to other microbes (Gautam, Pareek, & Sharma, 2013, 2015; Hu et al., 2008).

Microalgae are known to possess several functional properties that can make agriculture more sustainable. For example, microalgae have plant growth promoting, insecticidal, and pesticidal activities. Biostimulants produced by microalgae result in improved plant growth and hence enhanced crop performance. Further, microalgae act as biofertilizers and enhance nutrient availability by fixing nitrogen and improving the soil fertility/soil structure. Several microalgae symbiotically interact with higher plants, bacteria, fungi, mycorrhiza, etc., resulting in enhanced growth of the interacting species. Microalgae also find an application in crop protection and combating environmental stress by eliciting defense mechanisms in the plant and suppressing diseases by controlling the growth of pathogens. These beneficial qualities if further exploited in a judicial manner, microalgae can act as a sustainable alternative for wide applications in agriculture (Richmond, 2003). In this review, a detailed overview of a wide range of applications of microalgae, especially as alternatives to synthetic chemicals, in improving the agricultural sustainability has been presented.

2 Biofertilizers

The health of the soil is one of the primary concerns to farmers and the global community, whose livelihoods depend on agriculture. Most chemical fertilizers employ the use of harsh chemicals prolonged use of which leads to soil degradation and contamination. The chemicals are also leached out from the soil and reach the water bodies, which contaminates the water and leads to an imbalance in aquatic flora and fauna (Savci, 2012). Human activities such as deforestation, grazing, mining that has worsened the soil quality resulting in a reduction of soil fertility and deterioration of the soil's structural condition making it unsuitable for agriculture use. Hence, the use of new modes of organic and sustainable farming techniques using biofertilizers was devised to reduce environmental degradation and sustainably enhance the productive capacity of the soil.

Looking further on the disadvantages of using conventional fertilizers, especially for N and P are the use of nonrenewable sources such as petroleum and rock phosphate minerals as raw material or feedstock for the production of such fertilizers. The conventional fertilizers also come with a heavy cost. Therefore, to bridge the gap between demand and supply of conventional fertilizers and to protect the environment from negative effects of chemical fertilizers, the agriculture sector is focusing on research methods to find effective biofertilizers and soil conditioners, which promote crop yield. Round (1973) reported the use of marine algae as biofertilizers on farms near the wetland areas.

Biofertilizersis a term used for a group of microbes such as fungi, bacteria, BGA (cyanobacteria) and green algae, which are used directly in soil as whole-cell biomass and indirectly as cell extracts to improve soil fertility and growth/yield of crops (Prasanna et al., 2017; Renuka, Guldhe, Prasanna, Singh, & Bux, 2018). *Rhizobium* spp., bacteria, which fix nitrogen symbiotically, also falls in the category of biofertilizers. Biofertilizers can be either liquid or carrier-based, and their application is targeted either to the soil, the seeds or as a foliar spray. The various types of biofertilizers can aid in enhancing soil fertility and nutrient recycling by solubilization and mobilization of phosphate potash or sulfur apart from other chemicals such as zinc and silica.

Microalgal biofertilizers containing cyanobacteria are increasingly becoming popular owing to their ability to grow at high rates in diverse temperature conditions. Cyanobacteria or BGA have a history of use as biofertilizers and are well-known for improving the productivity of rice plants (Prasanna, Jaiswal, Singh, & Singh, 2008; Prasanna, Nain, et al., 2008). Algae can grow as solitary form or in association with fungi as lichens, which helps in improving the soil quality, carbon sequestration, fertilizer use efficiency in plants (Mandal, Vlek, & Mandal, 1999). It has been reported that biofertilizers in the form of microalgal biomass can improve the seed quality, seed germination, crop yield, grain yield, number of branches, flowers, and fresh weight of the plant (Karthikeyan, Prasanna, Nain, & Kaushik, 2007; Prasanna, Kanchan, Kaur, et al., 2016; Prasanna, Kanchan, Ramakrishnan, et al., 2016; Bhalamurugan, Valerie, & Mark, 2018). In addition, microalgal fertilizers are advantageous to the farmers as they are comparatively cheaper than chemical fertilizers and can give better results (Sahu, Priyadarshani, & Rath, 2012). Further, cyanobacterial fertilizers may be applied with some carriers such as agricultural straw or multani mitti (Fuller's earth) in case of rice cultivation and has shown to give better growth as compared to direct inoculation of microalgae in soil, referred to as "algalization" (Dhar, Prasanna, & Singh, 2007).

2.1 Enhancing soil fertility

Soil fertility is understood as the intrinsic capacity of the soil to supply nutrients in suitable amounts for crop growth and crop yield. Several green algae and cyanobacteria that are natural nitrogen fixers are known to play a crucial role in plant growth and soil fertility by producing natural substances that positively impact the growth (Saadatnia & Riahi, 2009). Algae-based bio-fertilizers or soil conditioners are used in two ways in agriculture. These either function in a symbiotic association between algae and the microorganism when added as inoculums in soil or in a combination of algae extracts with inorganic nutrients.

Microalgal biofertilizer maintains the soil nutrients and improves the soil structure by improving the microbial activity of the soil. Microalgae facilitate microbial interactions and support the growth of beneficial microbes that further improve the soil quality (Leloup, Nicolau, Pallier, Yéprémian, & Feuillade-Cathalifaud, 2013; Yan-Gui, Xin-Rong, Ying-Wu, Zhi-Shan, & Yan, 2013). Microalgae as biofertilizers are applied directly to the soil or as foliar spray. It has been observed in many experiments that nutrient absorption by plants is more rapid with foliar applications (Ronga et al., 2019). Table 1 summarizes the application, formulation and growth effects of microalgal/cyanobacteria biofertilizer on growth and yield various crops.

2.2 Nitrogen uptake by microalgae

Soil macronutrients constitute nitrogen, potassium, and phosphorous of which nitrogen is the most crucial nutrient for plant growth. Cyanobacteria are the most well-studied for its use as microalgal biofertilizer. Cyanobacteria have specialized cells called heterocysts, which harbor the nitrogenase enzyme and act as a site for nitrogen fixation (Babu, Prasanna, Bidyarani, & Singh, 2015; Karthikeyan et al., 2007). The nitrogen from the atmosphere is fixed by cyanobacteria growing in association with agricultural crop plants. Nitrogen molecule in the presence of hydrogen gets reduced to ammonia by the action of the nitrogenase enzyme. Ammonia uptake in cyanobacteria is either direct as ammonium ions or by diffusion in an energy-independent process (Goyal et al., 1997; Mishra & Pabbi, 2004). Another advantage of biofertilizer is that the nitrogen is not leached out from the soil as much as in the case of chemical fertilizers. This is because the nitrogen once fixed by the heterocysts does not remain in the soil in the free form and reduces the chance of leaching. However, the mineralization of the fixed nitrogen may result in leaching the nitrogen out and reducing the N content of the soil (Mager & Thomas, 2011). The diazotrophic cyanobacteria do not offer any competition to the crops for soil nitrogen; instead, the nitrogen content of the soil is improved (Nilsson, Bhattacharya, Rai, & Bergman, 2002; Renuka et al., 2016). Such soil inoculations of cyanobacteria in paddy resulted in atleast a 50% reduction in the use of chemical fertilizers while maintaining the grain quality and yield (Pereira et al., 2009). Cyanobacterial inoculations are now being used in a variety of crops such as maize, wheat, cotton, tomato (Osman et al., 2010; Prasanna et al., 2015). Fig. 1 shows the number of studies available in the published literature on the use of microalgal biofertilizer in plants.

2.3 Maintenance of soil structure and quality by microalgae

It is important to maintain the soil structure and optimal levels of soil organic matter to maintain the soil fertility that will determine the crop yield. The physical properties of soil are determined by its porosity, water holding capacity, bulk density, water infiltration, and hydraulic conductivity (Herencia, Garcia-Galavis, & Maqueda, 2011; Li et al., 2018). A good soil structure will have stable soil aggregates that will increase the porosity of the soil, which is important for plant and microbial health. Soils with good aggregate stability will have a higher water infiltration rate. Physical forces like wind, water, or anthropogenic activities and excessive farming have deteriorated the soil quality and structure. Soil structure quality can be measured by its soil aggregate stability (Six, Elliott, & Paustian, 2000). Excessive flooding or wind erosion causes the breakdown of soil aggregates into smaller particles, which gets re-deposited between aggregates close to the surface forming soil crust. This results in sealing of the surface and limits infiltration

TABLE 1 Application, formulation and mode of action of microalgal fertilizer on various crops and vegetable plants.

Species	Crop	Mode of action/ formulation	Application	Improvements and % reduction in chemical fertilizer	References
Microalgae					
Chlorella sp.	Not mentioned	Suspensions of microalgae culture and sterile filtrates from wastewater treatment	Spread to agricultural soil	Enhancement in microbial fauna and their activity	Marks et al. (2017)
Chlorella vulgaris	Tomato and Cucumber	Seed germination *Chlorella* culture suspension	—	Faster seed germination, enhanced root and shoot length	Bumandalai and Tserennadmid (2019)
Chlorella vulgaris and *Spirulina platensis*	Maize (*Zea mays* L.)	Microalgal biomass along with cow dung	Applied to soil	Growth performance enhancement at early stage of growth, improved grain yield, mineral content, macro and micronutrients in soil. Enhanced carbohydrates, Lipids, fiber and sugar content	Dineshkumar et al. (2019)
Chlorella	Wheat and Barley	Seed germination in media containing microalgae	Cell suspension applied to seed	Enhanced seed germination, Increase in root and shoot length	Odgerel and Tserendulam (2016)
Chlorella vulgaris	Not mentioned	Cells digestate of anaerobic reactors after growth in wastewater	—	No effect of the microalgae as biofertilizer was observed	Doğan-Subaşı and Demirer (2016)
Chlorella vulgaris	Tomato	Dry or liquid microalgae biomass	Spread to agricultural soil; foliar spray	Improved growth, yield and quality of fruits	Özdemir, Sukatar, and Bahar (2016)
Chlorella pyrenoidosa	Rice	Cellular biomass after immobilizing for dairy effluent treatment	Spread on rice seeds	Enhanced growth due to improved nitrogen fixation	Yadavalli and Heggers (2013)

Continued

TABLE 1 Application, formulation and mode of action of microalgal fertilizer on various crops and vegetable plants.—cont'd

Species	Crop	Mode of action/formulation	Application	Improvements and % reduction in chemical fertilizer	References
Acutodesmus dimorphus	Tomato	Cellular extracts, in distilled water and dry Biomass	Applied to seeds, spread to agricultural soil; foliar spray	Enhanced seed germination, growth and flowering	Garcia-Gonzalez and Sommerfeld (2016)
Microcystis aeruginosa MKR 0105 *Anabaena* PCC 7120 *Chlorella* sp.	Willow (*Salix viminalis* L.)	Intact cells of BGA mixed with microalgae with limited use of YaraMila Complex synthetic fertilizer	Triple foliar	Enhanced physiological performance and plant growth	Grzesik, Romanowska-Duda, and Kalaji (2017)
Chlorella vulgaris Scenedesmus obliquus "Consortium C"	Not mentioned	Possible use of cellular bio mass growth in wastewater	Extract application on the seeds	Not mentioned	Gouveia et al. (2016)
Polyculture of *Chlorella vulgaris* and *Scenedesmus dimorphus*	Rice	Microalgal biomass from airlift photobioreactor used as biofertilizer	Water dissolved Biomass applied to soil	Improved plant height	Jochum, Moncayo, and Jo (2018)
Phaeodactylum tricornutum, Pavlova lutheri	Not mentioned	Possible use of cellular biomass growth in mineral medium and agro-industrial ultrafiltrate	Not mentioned	Not mentioned	Veronesia, Ida, and D'Imporzano (2015)
Chlorella vulgaris	Lettuce	*C. vulgaris* biomass used as biofertilizer	Not mentioned	Seed germination and soil fertility was enhanced	Faheed and Fattah (2008)
Microalgal sludge from biogas production	Maize	Microalgal sludge with cellulose degrading enzymes	Improved growth of maize plant	Applied to soil	Kompi, Mekbib, and George (2018)
Cyanobacteria (BGA)					
Spirulina platensis	Egg plant	Intact cells (Spirufert® biofertilizer)	Foliar treatment	Improved eggplant yield and vegetative growth	Dias, Rocha, Araújo, De Lima, and Guedes (2016)

Species/Organism	Crop	Form/Method	Application	Effect	Reference
Spirulina platensis (partial)	leafy vegetable	intact cells after using them for nitrogen fixation in aquaculture and wastewater treatment	spraying biomass on leaves of plant	enhancement in plant growth and seedling dry weight	Wang, Yang, Cretu, and Luo (2016)
Consortium ZOB1	Rice	Intact cells	Not mentioned	Improved growth and seed germination	Zayadan et al. (2014)
Anabaena sp., *Aulosira* sp., *Cylindrospermum* sp., *Nostoc* sp., *Tolypothrix* sp.	Rice	Intact cells for N-fixation	Wet land cultivation	Improved growth and seed germination	Ashok, Ravi, and Saravanan (2017)
Nitrogen fixing cyanobacteria	Rice	Intact cells	Spread on soil	Reduced heavy metal load	Padhy, Nayak, Dash-Mohini, Rath, and Sahu (2016)
Anabaena and *Nostoc* Consortia	Rice	Consortium biofertilizer directly applied to soil-termed as Algalization		enhanced crop yields and soil fertility	Kaushik (1998), Goyal, Singh, Nagpal, and Marwaha (1997)
N-fixing cyanobacteria	Rice	Direct application in soil		Same grain yield and quality with 50% reduction in chemical fertilizer	Pereira et al. (2009)
Cylindrospermum sphaerica HH-202, *Anabaena doliolum* HH-209, and *Nostoc calcicola* HH-201	Pearl millet and wheat	—	Direct application to semi-arid clay-loam soil	Enhanced plant growth and yield; Improved carbon and nitrogen mineralization	Nisha, Kaushik, and Kaushik (2007)
Biofilms of *Anabaena* with *Trichoderma*/*Mesorhizobium*/*Azotobacter*/*Anabaena*-*Nostoc* consortium	Maize Cotton Chickpea (Desi and Kabuli chana) Chrysanthemum	Vermiculite and paddy straw (1:1) used as carrier; Vermiculite and compost carrier (1:1); Vermiculite and Paddy straw compost (1:1); Compost and vermiculite (1:1) with *Anabaena*-*Trichoderma* biofilm	Applied to soil/artificial soil at 60% water holding capacity	Plant growth characteristics and yield was improved; Improved uptake of macro and micronutrients	Prasanna et al. (2015), Prasanna, Kanchan, Kaur, et al. (2016), Prasanna, Kanchan, Ramakrishnan, et al. (2016), Prasanna et al. (2017)

Continued

TABLE 1 Application, formulation and mode of action of microalgal fertilizer on various crops and vegetable plants.—cont'd

Species	Crop	Mode of action/formulation	Application	Improvements and % reduction in chemical fertilizer	References
Calothrix elenkinii	Rice	Not mentioned	Cyanobacterial culture inoculated in agar media/soil	Enhancement in plant growth and plant hormones; Defense and hydrolytic enzymes activity increased; Changes in soil microbiota	Priya et al. (2015), Ranjan et al. (2016)
Calothrix ghosei, Hapalosiphon intricatus, Nostoc sp.	Wheat	Not mentioned	Applied to the soil	Increase in plant growth Enhanced grain yield	Karthikeyan et al. (2007)
Cyanobacterial consortia and their consortia with eubacteria	Wheat	Consortia mixed with Charcoal:soil (3:1)	Applied to soil	Plant growth and yield was enhanced Uptake of nutrients was improved The enzymes in soil showed enhanced activity	Nain et al. (2010), Rana, Joshi, Prasanna, Shivay, and Nain (2012); Rana et al. (2015)
Inoculation with *Anabaena* biofilm	Wheat	Not mentioned	Inoculated in soil	40% and 57% increase in soil N with 50% and 100% doses of chemical fertilizer, respectively	Swarnalakshmi, Dhar, Senthilkumar, and Singh (2013), Swarnalakshmi et al., 2013
Nostoc strain 2S9B and *Anabaena* strains LC2, C5	*Zea mays, Beta vulgaris, Phaseolus vulgaris, Triticum Vulgare*	Not mentioned	Biofilm of cyanobacteria was grown on the sand and then the crop was planted	Increase in plant length and dry weight	Svircev et al. (1997)
N-fixing cyanobacteria	Rice	Not mentioned	Not mentioned	Increased straw and grain yield with 25% N reduction by chemical fertilizer	Jha and Prasad (2006)
Vegetative cells an Akinetes of *Nostoc* sp. VICCRI	Rice	Not mentioned	Spread on the field	Improved grain yield	Innok, Chunleuchanon, Boonkerd, and Teaumroong (2009)
Nostoc entophytum and *Oscillatoria augustissima*	Pea Plant	Not mentioned	Applied to soil	Increased nutritional value of pea seed with 50% reduction in chemical fertilizer	Osman, El-Sheekh, El-Naggar, and Gheda (2010)

Organism	Plant	Formulation/Carrier	Application	Effect	Reference
Acutodesmus dimorphus	Tomato var. Roma	Not mentioned	50 g and 100 g of dry algal biomass used with soil	Enhanced germination rate and plant growth, increase in soil organic matter	Gomiero, Pimentel, and Paoletti (2011)
Chlorella-bacteria	Maize (*Zea mays* L.)	Not mentioned	The liquid cultures were applied to soil	Increase in organic carbon in soil Enhancement in aggregate stability at micro and macro scale	Yilmaz and Sönmez (2017)
Chlorella vulgaris	Maize and Wheat	Not mentioned	Applied to soil in the pots	Increase in seed germination rate and plant growth	Uysal, Uysal, and Ekinci (2015)
Nannochloropsis	Tomato	Organic slow release biofertilizer	Not mentioned	Increase carotenoid and sugar levels in tomato	Coppens et al. (2016)
Consortium of Cyanobacteria and Green Algae	Wheat	Consortium was added with carriers, vermiculite and compost (1:1)	Applied to soil	25% N savings, Improvement in soil microbial activity Increase in soil organic carbon, plant growth and yield	Renuka et al. (2016, 2017)
Mixture of *Nostoc* sp., *Anabaena doliolum*, *Calothrix* sp., *Westiellopsis* sp. and *Phormidium papyraceum*	Rice	Mixed with fly ash and garden soil	Soil application	Enhanced growth with cyanobacterial supplements, decrease in stress inducible components in soil such as nonprotein thiols (NP-SH) and cysteine	Tripathi et al. (2008)
Frankia Hsli10	Not mentioned	Intact cells application on saline soil	Not mentioned	Possible application as biofertilizer for nitrogen fixation	Srivastava and Mishra (2014)

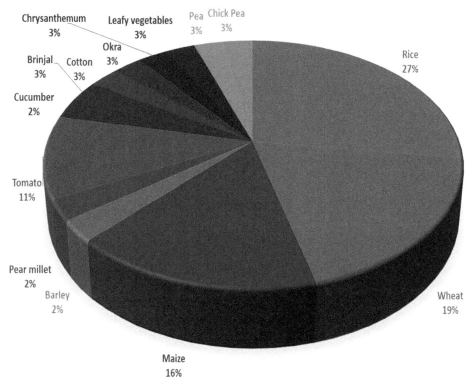

FIG. 1 Number of studies in published literature on application of microalgal fertilizer across various crops and vegetable plants. Data from more than 50 publications was used to derive at the percentage of studies conducted on the crops till date using microalgal/cyanobacterial biofertilizers.

and increases runoff. Several strategies are implemented in managing soil structure like tillage, the addition of organic materials, crop rotations, harvesting methods that can impact both aggregate size distribution and stability. However, extensive tillage and physical disturbances can further damage the soil and would require a lot of organic amendments and fertilizers to improve the soil structure.

Microalgal fertilizers increase the porosity, water-holding capacity thus reducing the bulk density of soil, which improves water infiltration and avoids runoff. The hydraulic conductivity of the soil is improved by microalgal fertilizers as it renders water and nutrients available to plant roots such that they are easily absorbed (Almendro-Candel, Lucas, Navarro-Pedreño, & Zorpas, 2018). Moreover, microalgae improve the soil organic matter by secretion of exopolysaccharides (EPS), which is one of the principal sources of soil carbon, which in turn enhances the population of agriculturally beneficial microbes and their activity in the soil. The microbial population comprises of photosynthetic organisms such as algae, diatoms, cyanobacteria, bacteria (phototrophic and heterotrophic), and fungi (Bharti, Velmourougane, & Prasanna, 2017). The high molecular weight natural polymers EPS are secreted during bio floc and biofilm formation in plant rhizosphere (Weiss et al., 2012;

Xiao & Zheng, 2016). Microalgal EPS comprises organic and inorganic components such as extracellular monosaccharides, polysaccharides, proteins, nucleic acids, carbonate, and silica (Flemming & Wingender, 2010; Bondoc, Heuschele, Gillard, Vyverman, & Pohnert, 2016). It is also a source of some noncarbohydrates such as acetate phosphate, succinate, and pyruvate. Some of the numerous reports of enhancement of microbial activity in soil due to the improved microflora and fauna, as given in Table 1. The structural and mechanical integrity of the soil is also preserved owing to the ability of EPS that allows the formation of a hydrated matrix rich in microbial colonies, which holds the different microbial colonies together. This renders a better flow of nutrients in the microsphere due to the exoenzyme activity of microalgae and acts as a defense mechanism against grazers and predators. The phototrophic cyanobacterial and green algal layers provide carbon and nitrogen to the heterotrophic partners in form of dead cellular mass and photosynthetic by-products, which sequentially recycle the nutrients in the biofilm (Drever & Stillings, 1997; Roeselers, Van Loosdrecht, & Muyzer, 2008). Apart from nitrogen and carbon, microalgae also provide vitamins and proteins in the form of amino acids, which are synthesized and are also secreted outside in the soil (Goyal et al., 1997). Some recent studies have shown that cyanobacterial consortium results in strengthened structure of microbial communities in the root rhizosphere which helps in nutrient availability by biomineralization and solubilization of macro and micronutrients (Manjunath et al., 2016; Priya et al., 2015; Ranjan et al., 2016) and also provides oxygen in the submerged rhizosphere.

2.4 Stabilization of soil aggregates

A stable soil aggregate reflects a healthy soil structure. Soil aggregation plays an essential function in protecting soil from wind and water erosion. In addition, it also helps in increasing porosity, soil carbon sequestration, and retention of moisture and nutrients (Le Bissonnais, 1996). Various biotic and abiotic factors help in building up soil aggregates. Mineral particles such as silt, clays, metal oxides, etc., including cation complexes flocculate result in the formation of small microaggregates (<0.05 mm). These microparticles get bound and cemented to form larger microaggregates (0.05–0.25 mm) by the biochemical action of exopolymeric substances and soil organic matter secreted by plants and microorganisms (Puget, Angers, & Chenu, 1998; Six, Bossuyt, Degryze, & Denef, 2004). Many green algae and cyanobacteria are known to form temporary photosynthetic micro biotic soil crusts and therefore prevent the topsoil from erosion (Knapen, Poesen, Galindo-Morales, Baets, & Pals, 2007). In various other reports, the use of algae and cyanobacteria to promote soil aggregation by direct inoculation of these strains in the soil is described (Bailey, Mazurak, & Rosowski, 1973). Inoculation of cyanobacteria strains in the soil leads to the formation of organo-mineral aggregates comprising of filamentous structures and EPS, which enhanced the stability of aggregates in comparison to un-inoculated control (Issa et al., 2007). Yilmaz and Sönmez (2017) reported that the inoculation of *Chlorella* spp. alone or in combination with a carrier, vermiculite, increased the stability of soil micro aggregates (0.25–0.050 mm) in comparison to the control.

2.5 Nutrient recycling in soil

The microalgal biofertilizers are also involved in the insolubilization and mineralization of complex micro and macronutrients. This is a significant linkage between the biogeochemical cycles and the nutrient pool. The process of breaking down of complex nutrients into simpler mineral forms for various cellular processes by microbes is referred to as biomineralization (Coppens et al., 2016; Goldman, Lammers, Berman, & Sanders-Loehr, 1983; Yilmaz &

Sönmez, 2017). Nisha et al. (2007) reported an in-depth study on the use of microalgae as biofertilizer in semiarid land. The study showed that not only the nitrogen and carbon content in the soil is enhanced, microalgal biofertilizer also led to the betterment of cation exchange potential and water holding potential. The soil structure was also improved as the bulk density reduced and the hydraulic conductivity was enhanced.

Biomineralization by microalgae or cyanobacteria occurs mainly by two means, by the production of organic acids and side rophores (McKnight & Morel, 1980).

2.5.1 Biomineralization by organic acids

It is known that many bacterial species produce humic acid that causes weathering of rocks. Similar substances are also reported to be produced by microalgae (Renuka et al., 2018; Sharma, Sayyed, Trivedi, & Gobi, 2013). A green microalga, *Microcystis aeruginosa* is reported to produce EPS, which has humic acid-like properties and results in the breakdown of phenanthrene (Bai, Xu, Wang, Deng, & Jiang, 2016). Some biological mats consisting of cyanobacterial species mainly *Lyngbya* sp. have also been reported to mineralize and precipitate magnesium carbonates (dypingite) in an alkaline wetland soil (Power, Wilson, Thom, Dipple, & Southam, 2007). Phosphorous in the soil is ~0.05%-0.1% (w/w), abundantly in organic and organic states. However, its availability is restricted as it occurs mostly in insoluble forms (Illmer & Schinner, 1995; Sharma et al., 2013). It is important that phosphorous remains in soil in the unbound form for it to absorbed and utilized by plants. There are three ways in which phosphorous solubilization occurs by various microbes—(i) by release of mineral dissolving compounds (e.g., protons, organic acid anions, hydroxyl ions, carbon dioxide, siderophores), (ii)secretion of extracellular enzymes (biochemical method), and (iii) liberation of P during substrate breakdown (biological method) (McGill &

Cole, 1981; Sharma et al., 2013). Cyanobacterial strains such as *Westiellopsis prolifica* and *Anabaena variabilis* are reported to solubilize tricalcium phosphate and Mussorie rocks (Yandigeri, Yadav, Meena, & Pabbi, 2010). It was also concluded from this study that P solubilization by BGA is strain and species-specific. In a study by Mukherjee, Chowdhury, and Ray (2015), *Chlorella* sp., *Lyngbya* sp., *Cyanobacterium* sp., and *Anabaena* sp. were used to sequester phosphorous from contaminated water in the form of polyphosphate. The polyphosphate was slowly mineralized and acted as a slow-release P fertilizer and improved plant growth. The slow release of phosphorous from polyphosphate was comparable to the release of chemical fertilizers-NPK and superphosphate. The release of organic acids and phosphatases by the microbes in the rhizospheric region resulted in the release of plant absorbable phosphorous.

2.5.2 Biomineralization by siderophores

Siderophores are chemical compounds formed by microalgae that help in the growth of microbes and plants by chelating iron in iron-deficient conditions. This allows the plants or microbes to utilize the iron under iron stress conditions. *Chlorella* spp. and *Scenedesmus incrassatulus* have been studied to produce iron siderophores for iron chelation (Benderliev, Ivanova, & Pilarski, 2003). There are reports of chelation of copper and iron by cyanobacterial strains *Anabaena flos-aquae* and *Anabaena cylindrica* by forming siderophores (Goldman et al., 1983; McKnight & Morel, 1980). Similar reports are also available for the production of Fe, Zn, M, and Cu chelators by cyanobacterial consortium with bacteria and green algae. There are many types of siderophores produced by cyanobacteria such as hydroxamate-type, citrate type, and catechol-type (Jaiswal, Das, Koli, & Pabbi, 2018). A citrate derivative siderophore, namely Schizokinen, produced from the *Anabaena* PCC 7120 is one of the most well-studied systems (Gademann & Portmann, 2008). Brown and

Trick (1992) reported that *Oscillatoria tenuis* is the only cyanobacterium to date that produces catechol-type siderophores. In a recent study by Jaiswal et al. (2018), *Nostoc muscorum* (CCC442) showing the maximum siderophore production along with *Anabaena variabilis* (CCC441) and closely followed by *Aulosira fertilissima* (CCC444), *Tolypothrix tenuis* (CCC443), and *Westiellopsis prolifica* (CCC128) (Umamaheswari, Madhavi, & Venkateswarlu, 1997). It was concluded from this study that variations in siderophore production in these species are because of cyanobacterial species and strain diversity. *Calothrix* sp., used as biofertilizer in sweet corn under iron-deficient conditions enhanced algal iron by 11-fold and nitrogenase activity by 9-fold (Mahasneh & Tiwari, 1992). This was due to the production of siderophores by alga as compared to the non-siderophore producing strains which resulted in the death of the plant. In a recent study, a diatom, *Navicula pelliculosa*, was found to induce Cupriachelin, a photoreactive lipopeptide siderophore in a freshwater bacterium *Cupriavidus necator* H16 (Kurth, Wasmuth, Wichard, Pohnert, & Nett, 2019). Productivity enhancement using new-age fertilizers that are organic and sustainable has given a new dimension to agricultural technology. It has been demonstrated by a plethora of studies that microalgal fertilizers are more effective in enhancing plant growth as compared to chemical fertilizers. In addition, biofertilizers ensure more efficient use of waste and degraded land. Biofertilizer application using various carriers such as soil, clay, etc., and renewable carriers such as agricultural straw and fuller's earth are more efficient in improving plant growth and yields as it increases the retention time of algal biomass in soil. Further, biorefinery approaches employing the use of microalgae in wastewater treatment followed by biofertilizer application renders the process renewable and more sustainable. Biofertilizer market is on a surge as many countries have imposed a ban on the usage of chemicals fertilizer for agricultural crops. The biofertilizer market is at USD 2.0 billion which is expected to increase to USD 3.8 billion by 2025. The escalation in demand for organic or environmentally sustainable products is driving the biofertilizers market globally (https://www.prnewswire.com/news-releases/the-biofertilizers-market-is-estimated-to-be-valued-at-usd-2-0-billion-in-2019-and-is-projected-to-grow-at-a-cagr-of-11-2-recording-a-value-of-usd-3-8-billion-by-2025–300917109.html).

3 Plant biostimulants

Plant biostimulants (PBs) (commonly called "Biostimulants") are another group of compounds that are gaining importance as an eco-friendly solution to address the critical challenges of agricultural sustainability. Researchers have defined PBs in many ways; a detailed chronological order of evolution of definitions is presented by Yakhin, Lubyanov, Yakhin, and Brown (2017). Based on European Biostimulant Industry Council, "plant biostimulants contain substance(s) and/or microorganisms, whose function when applied to plants or the rhizosphere is to stimulate natural processes to enhance/benefit nutrient uptake, nutrient efficiency, tolerance to abiotic stress, and crop quality" (EBIC, 2019). Thus, PBs contribute toward sustainable agriculture by triggering physiological and molecular processes, which lead to enhanced crop yield and quality, help in making plants more resilient to abiotic stresses, improve nutrient assimilation and their utilization, thereby enabling better returns on the investment. PBs improve quality attributes of produce by improving nutrient content and color. They also promote the growth of beneficial soil microorganisms leading to improved soil health and in turn better crop yield (EBIC, 2019). European Commission has categorized PBs as organic non-microbial plant biostimulants (humic acids, protein hydrolysates (PHs), and seaweed extracts) and microbial plant biostimulants.

3.1 Types of plant biostimulants

3.1.1 Microbial PBs

This category primarily includes plant growth-promoting rhizobacteria (PGPR) belonging to genera *Azospirillum, Azotobacter,* and *Rhizobium* and endophytic fungi such as mycorrhizal fungi and *Trichoderma sp* (Rouphael & Colla, 2018). PGPRs and mycorrhizal fungi improve plant growth by enhanced uptake and translocation of macro (N, P) and micronutrients, improved root architecture, which includes an increased surface area of roots, number of lateral roots, and ultimately higher root biomass. These also help in the production of enzymes like phosphatase that converts phosphate source into ion form, production of growth-promoting substances like amino acids, phenolics, sugars, proteins, organic acids, and regulation of plant growth-promoting hormones (Korir, Mungai, Thuita, Hamba, & Masso, 2017; Rouphael & Colla, 2018).

3.1.2 Humic substances

These are the substances formed through the biological and chemical conversion of dead organic matter. Biostimulation action of humic substances is attributed to improved soil structure, neutralization of soil pH, increased cation exchange capacity of soil resulting in better nutrient availability. Humic substances also form soluble complexes with micronutrients like iron, which prevents their leaching and makes them more bioavailable. Humic substances also save plants from negative effects of heavy metal uptake as heavy metals bind to the carboxylic and phenolic hydroxy group of the humic substances, resulting in reduced uptake by the plant (Canellas et al., 2015; De Pascale, Rouphael, & Colla, 2017; Du Jardin, 2015).

3.1.3 Protein hydrolysates

They primarily comprise a mixture of polypeptides, oligopeptides, and free amino acids

such as glutamate, proline, glutamine, and glycine betaine (Colla et al., 2015). They are derived from both animal and plant residues through a chemical, enzymatic or thermal hydrolysis processes. Common plant sources for protein hydrolysate preparation include alfalfa residue, algal protein, *Nicotiana* cell wall glycoproteins, and wheat germ PHs, whereas, animal sources include epithelium, animal collagen, and elastine (Cavani, Ter Halle, Richard, & Ciavatta, 2006). There are several commercial PBs available, which consist of protein hydrolysates. Siapton (animal-based product), ILSATOP, Megafol, and Macro-Sorb Foliar (animal-based product) are some of the examples of commercially available protein hydrolysates. Protein hydrolysates are mainly applied as a foliar spray (Umemiya & Furuya, 2001; Stiegler, Richardson, Karcher, Roberts, & Norman, 2013). They work by improving nitrogen uptake and assimilation through modulation of enzyme activity, signaling pathway in roots and by increasing root system in terms of root hair diameter, a number of lateral roots and length as more lateral roots result in greater water and nutrient uptake. Through the chelating activity of proline, the availability of micro and macronutrients improve significantly. Proline and glycine betaine also act as free radical scavengers and osmoprotectants and thus help in alleviating abiotic stress. Stress-relieving effects of PHs have been demonstrated on many crop plants (Van Oosten, Pepe, De Pascale, Silletti, & Maggio, 2017). For example, salt tolerance in *Zeamays* (Ertani, Schiavon, Muscolo, & Nardi, 2013), *Lactuca sativa* (Lucini et al., 2015), *Diospyros kaki and D. lotus* (Visconti et al., 2015), heavy metal tolerance in *Triticum aestivum* (Zhu, Zhou, & Qian, 2006), cold tolerance in *L. sativa* (Botta, 2012), and heat tolerance in *Lolium perenne* (Botta, 2012) to name a few.

3.1.4 Algal extracts

Macroalgae (seaweeds) and their extracts are used as fertilizer and soil conditioner for ages but its new-found application is as PBs

(Sharma, Fleming, Selby, Rao, & Martin, 2014). Although close to 9000 macroalgal species are known, only a few are commercially exploited (Khan et al., 2009). These belong to Phaeophyta, Rhodophyta, and Chlorophyta group. Common macroalgae used for seaweed extract preparation are *Ascophylum nodosum*, *Ecklonia maxima*, *Durvillea potatorum*, *Durvillea antarctica*, *Fucus serratus*, *Himanthalia elongate*, *Laminaria digitata*, *L. hyperborea*, *Macrocystis pyrifera*, and *Sargassum* sp. (Sharma et al., 2014). Out of this *Ascophylum nodosum* (brown algae) is used in the majority of commercial formulations (Van Oosten et al., 2017). Seaweed extracts are rich source of phytohormones, osmoprotectants, polysaccharides (*Phaeophyta*: alginate, fucoidan, laminarin, *Rhodophyta*: agar, carrageenan, porphyrin, *Chlorophyta*: xylan, cellulose, ulvan), vitamins, and phenolic compounds, which offers beneficial response in target plant by triggering various physiological and biochemical pathways (Khan et al., 2009; Michalak & Chojnacka, 2014). However, obtaining consistent composition of seaweed extract is always a challenge because of variable environmental conditions, time of harvesting, nutrient availability to seaweed and the age of seaweed tissue (Chiaiese, Corrado, Colla, Kyriacou, & Rouphael, 2018). Apart from this, seaweed harvesting is regulated in order to avoid the destruction of natural habitat and sustained production of macroalgae. Once harvested, the area has to be left untouched for the next 4-6 years based on the country's regulations (Sharma et al., 2014). Moreover, the cultivation of macroalgae in controlled conditions is not yet in practice. Keeping these challenges in mind, biostimulant extraction from microalgae gained momentum, as their products can be standardized and controlled. Microalgal species that have been commercially utilized for various purposes are *Chlorella, Dunaliella, Haematococcus, Isochrysis, Nannochloropsis, Porphyridium, and Spirulina* (Chiaiese et al., 2018). Microalgal biomass is rich in terms of macro and microelements, which are essential for plant growth. Their ability to produce phytohormones (auxins, gibberelins, and cytokinins)

(Tarakhovskaya, Maslov, & Shishova, 2007) and other bioactive molecules antioxidants like polyphenols, tocopherols, and pigments make them lucrative source for exploitation as PBs (Michalak & Chojnacka, 2014). Table 2 summarizes the biostimulatory action of microalgal extracts derived from various microalgal strains.

3.2 Cell lysis and extraction methods

The process of biostimulant extraction varies depending on the source. However, the first step in the preparation of macro or microalgal extracts is pretreatment, which involve, washing, drying, and grinding of algal biomass. Washing helps in removing biomass adhered impurities. This biomass is then subjected to either freeze-drying or low temperature ($\approx 35°C$) drying to minimize the degradation of bioactive molecules. Dried biomass is then blended and sieved to reduce particle size, increase contact surface area with solvent and make the powder homogeneous (Michalak & Chojnacka, 2014). To improve accessibility to bioactive molecules, efficient cell disruption is a critical step. There are multiple methods available in the literature, which include mechanical/physical, chemical, and enzymatic cell disruption. The selection of extraction method largely depends on the complexity of biomass, target molecules and scale of operation. Mechanical methods include high-pressure homogenization, ultrasonication, bead milling, microwave-assisted extraction, pulsed electric field, and autoclaving. Heating however, can degrade heat-labile components and degrade the quality of extract. Chemical means of cell rupturing include primary treatment with acid (HCl, H_2SO_4, nitric acid, acetic acid), alkali (NaOH, KOH, $CaCO_3$, Na_2CO_3, K_2CO_3) or osmotic shock (NaCl). Multiple studies are reported on the extraction of biomolecules from microalgae using these chemicals either alone or in combinations that have been used (Günerken et al., 2015; Michalak & Chojnacka, 2014; Sharma et al., 2014). In addition to chemicals, enzymatic treatment for cell rupturing has been used, which is

TABLE 2 Plant response to bio-stimulatory action of microalgae.

Microalgae	Crop	Extraction and application method	Effect on plant	References
Anabaena sp. *Leptolyngbya* sp. *Neochloris* sp. *Klebsormidium flaccidum*	*Zea mays* L.	Another culture medium supplemented with dried algal biomass	Auxin and cytokinin like activities Improved anther induction and regeneration frequency	Jäger, Bartók, Ördög, and Barnabás (2010)
Spirulina maxima Chlorella ellipsoida	Wheat	Aqueous extraction and application through irrigation	Total carotenoid, tocopherol, phenolic, and protein (PC) contents in whole grains increased	Abd El-Baky, El-Baz, and El Baroty (2010)
Nannochloris sp. 424-1	Tomato	Cell rupture through homogenization followed by hydrolysis and aqueous extraction of phytohormones, osmo-protectants in permeate, enzymatic hydrolysis of defatted biomass for protein hydrolysate preparation. Algal biostimulant consist of phytohormones, betaines, soluble carbohydrates, and micro-algal protein hydrolysates	Increased plant height, root length, leaf number and leaf area Partial alleviation of water stress	Oancea, Velea, Fatu, Mincea, and Ilie (2013)
Calothrix elenkinii (RPC1)	Rice cv. *Pusa Sugandh 5*	Cyanobacterial cells applied to rice seedlings	Increase in shoot and root length Increase in fresh weight Increase of number of rod-shaped bacteria in microbiome associated to plant Increase in nitrogenase activity, IAA production and activity of many defense enzymes measured in roots and leafs	Priya et al. (2015)
Acutodesmus dimorphus (LARB-0414)	*Solanum lycopersicum* var. Roma	Aqueous extraction through microfluidizer Application on seeds Foliar spray	Seed priming Faster germination Greater lateral root development Greater plant biomass and crop yield Foliar spray Greater shoot length, number of branches and flower buds	Garcia-Gonzalez and Sommerfeld (2016)

Microalgae	Plant	Application	Effects	References
Spirulina platensis	*Solanum lycopersicum* *Capsicum annum*	Total polysaccharide aqueous extract used as plant biostimulant Application by spraying	Improved plant size, root weight, size and number of nodes, leaf area	Elarroussia et al. (2016)
Spirulina plantensis	Winter Wheat (variety Akteur)	Super critical CO_2 extraction	Number of grains in ear and shank length were higher in treated plant	Manjunath et al. (2016)
Azotobacter sp. *Anabaena* sp.– *Azotobacter* sp. *Anabaena* sp.– *Providencia* sp. *Calothrix* sp.	*Abelmoschus esculentum* (L.)	Seed coating with *Cyanobacterial* inoculants	Higher root weight and increased crop yield Increased alkaline phosphatase and dehydrogenase activity in soil Increased availability of micro (Fe, Zn, Cu) and macronutrients (N, P, C)	Manjunath et al. (2016)
Chlorella vulgaris *Scenedesmus quadricauda*	*Beta vulgaris* L.	Extracts added to Hoagland solution used for seedling growth	Increased number of root tips, length and diameter Improved nutrient uptake Up-regulation of genes related to nutrient uptake	Barone et al. (2018)
Chlorella infusionum	Tomato	Cultivation in hydroponic culture system along with tomato plant	Root weight higher Fruit yield more	Zhang, Wang, and Zhou (2017)
Dunaliella salina	*Solanum lycopersicum*	Sulfated exopolysaccharides in water, application through spraying	Alleviated the salt stress and mitigated the decrease in length and dry weight of the plant's shoot and root systems	Elarroussia et al. (2016)

gentle, specific and has a less adverse effect on the compounds being extracted. Proteolytic enzymes are used for the preparation of protein hydrolysates, which are one of the categories of biostimulants. Apart from cell rupturing, enzymes like glycosidases, glucanases, peptidases, cellulase, and lipases have been used to extract pigments, fermentable sugars, lipids, etc. (Günerken et al., 2015; Ronga et al., 2019). Oancea et al. (2013) have used a combination of mechanical and enzymatic methods to prepare the algal extract, which was shown to improve the yield of tomato plants. The cascade process developed by them includes microalgae cell rupture by homogenization, hydrolysis with a mixture of enzymes, recovery of major components of biostimulants through tangential flow filtration and combining this fraction with protein hydrolysate prepared through enzymatic treatment (Oancea et al., 2013).

3.3 Mode of application

Biostimulants can be applied to soil or directly to plants through foliar application. Usually, humic substances, compounds containing high nitrogen content and algal dry biomass are applied through the soil, whereas, extracts from algae are applied through foliar spray (Drobek, Frąc, & Cybulska, 2019). Foliar spray is more effective compared to soil application, as it enables rapid nutrient utilization and rectification in nutrient deficiencies. However, determining adequate supply through the foliar application is critical and is governed by many factors. For example, temperature, light intensity, surfactants, frequency of application, time of application (morning is preferred, where stomata are open, and rate of assimilation is fast), and nutrient concentration. These parameters must be optimized to have an efficient outcome of foliar spray (Fernández & Eichert, 2009; Garcia-Gonzalez & Sommerfeld, 2016). Biostimulants can also be applied through the irrigation system (Drobek et al., 2019).

3.4 Composition and mode of action

The active ingredients present in seaweeds and microalgal extracts, which promote plant growth are phytohormones (cytokinin, auxins, gibberellins, brassinosteriods, ethylene, and abscisic acids) (Xu & Geelen, 2018), amino acids (Colla et al., 2015), polyamine, vitamins precursors, fatty acids, mineral nutrients, and polysaccharides (De Pascale et al., 2017; Sharma et al., 2014).

Microalgal extracts provide plant biostimulatory action by increasing root growth, and nutrient uptake and by enhancing the activity of PGPR and mycorrhizae (Rouphael & Colla, 2018). Also, polysaccharides present in the microalgae extract help in water retention and soil aeration by forming gel (Du Jardin, 2015). Moreover, it is reported that laminarin stimulates defense response in plants by inducing various genes involved in synthesizing proteins related to antimicrobial properties (Khan et al., 2009). Likewise, the presence of amino acids like tryptophan and arginine enhances growth as they are precursors of phytohormones (auxin and salicylic acid), aromatic secondary compounds and polyamines. Proline and glutamate help in reducing salinity stress and oxidative stress in the plant by regulating reactive oxygen species (Oancea et al., 2013).

3.5 Application of microalgal PBs for crop yield improvement

Microalgal extracts from various algal and cyanobacterial species are evaluated to improve crop yield. Table 2 has a compilation of some key studies, describing extract preparation, its application and significant outcome on the treated plants.

Finally, algal biostimulants can be eco-friendly and cost-effective ways to meet growing agriculture needs and improve the nutritional quality of plants and thus reduce malnutrition. Considering the advantages that PBs offer, their market potential is growing significantly. The global biostimulant market was valued at USD 2.6 billion in 2019 and is expected to reach $5.23 billion by 2026 growing at a CAGR of 11.9% from 2018 to

2026 (Marketsandmarkets.com, 2019; Research and Markets, 2019). Europe holds a major share in the biostimulant market, followed by North America, Asia Pacific, and South America (Marketsandmarkets.com, 2019).

Although the area of microalgal PBs is growing rapidly, a concerted effort is needed to gain a better understanding of the interactive effect of environment, algal species, and target plant species. Optimization of parameters like time of PBs application, mode, frequency of application, and plant growth stage is most effective for application and needs to be understood well. Microalgal strain selection and improvement to enhance bioactive compounds, identification of synergistic algal and cyanobacterial strains with complementary traits, a complete characterization of extract, formulation, and stability studies are other key areas that need concentrated focus for successful application in agriculture.

4 Biopesticides

Conventionally used pesticides are synthetic chemicals used to manage insects, fungi, bacteria, viruses or weeds affecting crop yield. Pesticides not only inhibit the growth of organisms infecting crop plants, but they also cause deleterious effects on animals and humans by consuming pesticides treated crops. Long usage of pesticides in the agriculture field and their discharge in water bodies deteriorate the environment and decrease crop productivity. Most of the agricultural pesticides that are in the market today can be toxic to the plant and environment, especially due to misuse. There is a need to look for alternative sources that can sustainably protect crop plants against diseases. Biopesticides are considered better alternatives to pesticides, as they are biodegradable and harmless to animals and the environment. United States Environment Protection Agency (US-EPA) has registered 299 active ingredients of biopesticides and 1401 biopesticide products that are available in the market. US-EPA guidelines (2016) have classified biopesticides into three main classes, i.e., biochemicals, microbial, and plant-incorporated protectants (PIPs).

Biochemical pesticides control pests by a natural non-toxic mechanism. Biochemical pesticides include plant extracts or biomolecules that restrict insect mating or attract insect pests to traps. Microbial pesticides consist of microorganisms (algae, bacteria, fungus, virus, or protozoa) as an active ingredient that can control different types of plant pathogenic pests. The commonly used microbial biopesticides are active against plant pathogenic organisms such as *Trichoderma, Phytopthora, Bacillus thuringiensis,* and*Bacillus sphaericus* (Gupta & Dikshit, 2010). PIPs are another class of biopesticides that includes the incorporation of foreign genetic material into plant genome that makes the crop resistant against pest infection. For example, bacterium *Bacillus thuringiensis* (Bt) produces insecticidal toxins that are active against lepidopteran larvae, which is the main pest in commercial cotton crop. Lepidopteran larvae were killed naturally by eating genetically modified cotton and reduce the usage of large amounts of insecticides to kill lepidopteran pests. Thus, PIPs are produced by the genetically modified plants itself that protect them from the attack of insects (Lu, Wu, Jiang, Guo, & Desneux, 2012).

Synthetic pesticides and fertilizers cause long-lasting risk for the environment and human health. Hence, microalgae may be considered as a useful tool to reduce the input of chemicals for pest management. The European Regulation (EC 1107/2009) and repealing Council Directives (79/117/EEC and 91/414/EEC) recommends that priority should be given to non-chemical and natural alternatives of plant protection products (Righini & Roberti, 2019).

4.1 Microalgae as a sustainable source of biopesticides

Microalgal biopesticides are sustainable, non-phytotoxic and biodegradable in the environment (Righini & Roberti, 2019). Microbial pesticides act on crop plants through several mechanisms.

Biopesticides protect the crop against pest by inhibiting feeding, reproduction, growth, or developmental process of the pest (Mnif & Ghribi, 2015). Bacteria-derived pesticides are presently the major players in the biopesticide market (Weinzierl, Henn, Koehler, & Tucker, 2007).

Microalgae based biopesticides are unique as they could be grown with minimal nutritional requirements as compared to other microorganisms. Microalgae could be grown in mass culture and are suitable candidates for exploitation as biocontrol agents of plant pathogenic bacteria, fungi and nematodes (Righini & Roberti, 2019). Microalgae, especially cyanobacteria are the main biological agents for the control of pests in the crop plants. Microalgae produce biologically active molecules such as tannins, phenols, tocopherols, terpenes, steroids, allelochemicals, sugars, peptides and saponins with biocidal, antifungal, antibacterial, and toxic activities against pests. Microalgae biopesticides not only showed an inhibitory effect on larval stages but also inhibit the pupation processes and formation of the adult stage. Microalgae can disrupt the cellular membrane, deactivate enzymes, and inhibit protein synthesis in the targeted pest (Kulik, 1995).

4.2 Microalgae against plant pathogenic bacteria, fungi, and nematodes

Microalgae produce bioactive chemicals that have antibacterial, antifungal, and antinematodal properties. The first report of microalgal biopesticidal activity was reported from green alga *Chlorella*, due to secretion of chlorellin inhibiting bacterial growth (Pratt, Oneto, & Pratt, 1945). Microalgae inhibited the growth of pathogenic fungi (*Chaetomium globosum*, *Cunninghamella blakesleeana*, and *Aspergillus oryzae*) and bacteria (*Rhizoctonia solani* and *Sclerotinia sclerotiorum*) (Kulik, 1995). Abbassy, Marei, and Rabia (2014) reported that ethanolic extracts of microalgae *Spirulina platensis* and *Scendesmus* sp. showed growth inhibitory effect against soil-borne bacteria *Agrobacterium tumefaciens* and *Erwinia carotovora* with minimum inhibitory concentration (MIC) of 2500-7500 mg/L. Similarly, another cyanobacterial species *Fischerella muscicola* releases antifungal compound Fischerellin A that completely inhibits the infection of *Uromyces appendiculatus* and *Erysiphe graminis* on the growth of *Phasealus vulgaris* and *Hordeum vulgare* at 250 and 1000 ppm concentration, respectively. Fischerellin A also inhibits 80% infection of *Phytophthora infestans* on tomatoes and *Pericularia oryzae* on rice at 1000 ppm concentration (Hagmann & Jüttner, 1996). Similarly, crude ethanolic extract of *Anabaenalaxa* contains antifungal compound Laxaphycins that inhibits the growth of plant pathogenic fungi *Aspergillus oryzae* and *Penicillum notatum* (Frankmölle, Knübel, Moore, & Patterson, 1992).

Kulik (1995) reported that there is greater success with the various formulation of microalgae applied directly to the seeds as protectants rather than spraying on the leaves against damping-off disease caused by pathogenic bacteria and fungi (*Fusarium* sp., *Pythium* sp. and *Rhizoctonia solani*). Yanni and Osman (1990) found that soil application of microalgae *Anabaena cylindrica*, *Anabaena oryzae*, *Nostoc muscorum*, and *Tolypothrix tenuis* reduced the incidence of blast fungus infestation by *Pyricularia oryzae* in rice crop (Yanni & Osman, 1990). Damping off the disease of soybean could be controlled with the biochemicals released by *Nostoc muscorum* and it could be a promising biofungicides (De Caire, De Cano, De Mule, & De Halperin, 1990). Plant pathogenic fungi, that is, *Alternaria alternate*, *Botrytis fabae*, *Fusarium oxysporum*, *Fusarium solani*, *Macrophomina phaseolina*, *Rhizooctonia solani*, *Sclerotium rolfsii* were isolated from infected leaves and stem of faba bean that cause leaf spot, damping off and wilt disease. It was found that a combination of *Nostoc muscorum* and *Oscillatoria angusta* culture filtrates completely control the infection of several pathogenic fungi in faba bean (Abo-Shady, Al-ghaffar, Rahhal, & Abd-El Monem, 2007).

Fungicidal activity in extracellular filtrates of *Anabaena* and *Calothrix* strains revealed the presence of several hydrolytic enzymes. It was found that microalgal extract improves plant immunity by eliciting plant defense enzymes activity (Prasanna, Jaiswal, et al., 2008; Prasanna, Nain, et al., 2008). Biocontrol potential of *Calothrix* spp. culture filtrates were found to be active in controlling damping-off disease in vegetables belonging to solanaceae family (Manjunath et al., 2010). Novel microbe amended composts consisting of *Anabaena oscillarioides* and *Bacillus subtillis* could be utilized in eliciting disease resistance in tomato plants against pathogenic fungi (*Fusarium oxysporum, Pythium debaryanum* and *Pythium aphanidermatum*) and bacterium (*Rhizoctonia solani*) (Dukare et al., 2011).

Microalgae were also found to produce several nematocidal compounds and toxins that control the nematode infection in crop plants (Khan & Park, 1999). Nematicidal activity of three microalgae *Nostoc calcicola, Anabaena oryzae,* and *Spirulina* sp. tested against nematode *Meloidogyne incognita* infecting cowpea plants. *Nostoc calcicola* showed maximum nematicidal activity and highest plant productivity compared to *Anabaena oryzae,* and *Spirulina* sp. (Youssef & Ali, 1998). The addition of microalga *Microcoleus vaginatus* to soil prevented the growth of nematode *Meloidogyne incognita* population in the soil, hence protects the root gall formation in tomato crop, and hence improved crop yields. Nematicidal properties were tested by dipping tomato seedling in culture filtrates of *Microcoleus vaginatus* that leads to 65.9% and 97.5% reduction in root galling and nematode population, respectively (Khan et al., 2005). The cyanobacterium *Oscillatoria chlorina* also showed to have nematicidal properties and with 1% formulation of *Oscillatoria chlorina* powder in soil lead to 68.9% and 97.6% reduction in gall formation and nematode number respectively, as compared to untreated soil (Khan, Kim, Kim, & Kim, 2007). *Aulosira fertilissima* culture filtrate inhibited the infection of root-knot nematode *Meloidogyne triticoryzae* (Chandel, 2009). Inoculation of microalgae into the soil prior to implanting plant seedlings showed increased plant growth and showed higher nematicidal activity (Khan et al., 2005, 2007). Glycine rich toxin of microalgae *Scytonema* MKU toxins showed nematicidal (*Helicoverpa armigera*) and insecticidal (*Stylepta derogate*) activity against at 0.01% and 0.1 % concentration, respectively (Sathiyamoorthy & Shanmugasundaram, 1996). Microalgal pesticides offer an eco-friendly approach toward the environment and improve the crop yield by protecting against plant pathogenic fungi, nematodes and bacteria (Table 3).

4.3 Smart agriculture using algal nanoparticles for pest control

Smart agriculture using algal nanoparticles could provide green and effective management of plant pathogenic pests. The synthesis of eco-friendly and efficient nanoparticles is an important aspect of nanotechnology. Application of microalgal nanotechnology in agriculture includes the usage of nanopesticides for controlling of pests and crop disease. Application of nanosilicon product of *Oscillatoria agardhii* on wheat crop reduces the infection of powdery mildew, leaf rust and spot disease in wheat crop and also increases crop yield (Haggag, Abd El-Aty, & Mohamed, 2014). Synthesis of nanoparticle using silicon is reported to be absorbed into plants quickly and it increases disease and stress resistance in the plant via promoting physiological activities (Currie & Perry, 2007). Algae provide an effective platform for the synthesis of varied nanoparticles due to the presence of bioactive compounds like pigments and antioxidants in their cell extracts that act as biocompatible reductants. Antimicrobial properties of microalgae *Chlorella* spp. and macroalgae *Sargassum* spp. are utilized for the synthesis of nanoparticles using gold and silver

TABLE 3   Microalgal pesticides against plant pathogenic fungi, nematodes and bacteria.

Microalgae	Infected Crop	Target Pathogen	References
Anabaena oscillarioides, Bacillus subtillis, Anabaena spp. and *Calothrix* spp.	Tomato	*Fusarium oxysporum, Fusarium moniliforme, Pythium debaryanum, Pythium Aphanidermatum,* and *Rhizoctonia solani*	Manjunath et al. (2010), Chaudhary et al. (2012)
Nostoc commune FA-103	Tomato	*Fusarium oxysporum f.* sp. *lycopersici*	Kim and Kim (2008)
Oscillatoria chlorine and *Microcoleus vaginatus*	Tomato	*Meloidogyne arenaria* and *Meloidogyne incognita*	Khan et al. (2005), Khan et al. (2007)
Fischerella muscicola	Tomato	*Phytophthora infestans* and *Pericularia oryzae*	Hagmann and Jüttner (1996)
Nostoc commune FK-103, Oscillatoria tenuis FK-109, Oscillatoria, Anabaena, Nostoc, Nodularia, and *Calothrix* spp.	Rice	*Phytophthora capsici* and *Alternaria alternate*	Kim (2006)
Aulosira fertilissima	Rice, Wheat and Vegetables	*Meloidogyne triticoryzae* and *Meloidogyne incognita*	Chandel (2009)
Scytonema MKU 106	Cotton	*Helicoverpa armigera, Heliothis larvae, Styleptaderogate*	Sathiyamoorthy and Shanmugasundaram (1996)
Anabaena spp.	Cotton	*Rhizoctonia solani*	Prasanna, Kanchan, Kaur, et al. (2016), Prasanna, Kanchan, Ramakrishnan, et al. (2016)
Oscillatoria redekei HUB 051	–	*Bacillus subtilis, Micrococcus flavus,* and *Staphylococcus aureus*	Mundt, Kreitlow, and Jansen (2003)
Nostoc muscorum	Rice, Potato	*Rhizoctonia solani*	De Caire et al. (1990)
Spirulina platensis and *Scendesmus* sp.	–	*Agrobacterium tumefaciens* and *Erwinia carotovora*	Moustafa et al. (2014)
Anabaenalaxa	Rice	*Aspergillus oryzae* and *Penicillum notatum*	Frankmölle et al. (1992)

that has the potential to substitute for conventional antibiotics (Khanna, Kaur, & Goyal, 2019). There are also reports on the usage of zinc nanoparticles using the macroalga *Ulva lactuca* extract as a potential eco-friendly option against antimicrobials and insecticides (Ishwarya et al., 2018).

Microalgae have been identified as a potential biocontrol agent against pests and these also provide additional nutritional benefits to plants that lead to increased crop yields and disease resistance (Prasanna, Kanchan, Kaur, et al., 2016; Prasanna, Kanchan, Ramakrishnan, et al., 2016). However, one of the main challenges in the use of microalgal biopesticides is a lack of clear understanding regarding the mechanisms of production and their mode of action on specific dose-dependent pesticidal activity. Hence, thorough trials with microalgal extract or their active ingredients to understand the mechanism of action will pave the way to meet biopesticide market needs.

5 Symbiotic interaction of microalgae with higher plants

Symbiosis is an intimate relationship and it may occur between two (Bipartite) or three partners (Tripartite). Bipartite interaction is more commonly present in agricultural crops. Diverse types of symbiotic associations are reported between cyanobacteria, bacteria, algae, fungi, gymnosperms, pteridophytes, and vascular plants.

Microalgal interaction with higher plants have several advantages such as providing growth-promoting substances and fixing atmospheric nitrogen for plant growth, enhancing seedlings germination rate, eliciting plant defense mechanism, protecting the plants against disease/pest control, reclamation of wasteland, improving soil fertility, micronutrient availability, and translocation to plants (Gautam, Tripathi, Pareek, & Sharma, 2019; Renuka et al., 2018). Counterpart microalgae get nutrition in the form of reduced carbon and moisture-rich habitat, which helps to protect from desiccation, freezing and fluctuating environmental conditions. Understanding the symbiotic relationship may help to improve crop productivity and offer other agronomical benefits.

5.1 Cyanobacteria symbiotic relationship with higher plants

Cyanobacteria have a unique ability to fix nitrogen from the atmosphere with specialized cells called heterocysts and reduce the nitrogen under microoxic environment and supply nutrients to the host plants, which in turn reciprocates

a niche for cyanobiont's growth and survival. In addition, soil fertility is also enhanced during this process, which favors the neighboring plant growth, which is not directly involved in symbiosis. In this relationship, cyanobionts like *Nostoc* and *Anabaena* are considered as obligate phototrophs and colonizes the roots of vascular plants to fix the nitrogen.

Gunnera-Nostoc symbiosis is a classic and exclusive example of angiosperm-cyanobacterial endosymbiosis. *Gunnera* is an herbaceous flowering plant makes a complex relationship with *Nostoc*. *Nostoc* invades through specialized bright red glands located on the stems of Gunnera, *which* are rich in carbohydrate and guide the cyanobacteria to symbiotic cavities by the hormonal response and eventually cyanobacteria colonize inside the host cells. In this process, *Nostoc's* heterocysts fix the nitrogen from the environment and pass it to *Gunnera* for its growth and in turn, cyanobiont gets reduced carbon as a nutrient from the host and habitat for living. Further, changes in gene expression and phenotypes also observed in both the organisms (Adams, Bergman, Nierzwicki-Bauer, Rai, & Schüßler, 2006; Renuka et al., 2018). This insight may be helpful to develop a new strategy of the artificial symbiosis of angiosperm-cyanobacteria for crop improvement.

Similarly, various symbiotic interactions have been reported to enhance plant growth. Gupta and Lata (1964) isolated three species of *Phormordium foveolarum*, *P. corium*, and *P. autumnale* from the rice field and concluded that plant growth-promoting hormones from *P. foveolarum* accelerates the germination and growth of rice seedling and conversely *P. corium* and *P. autumnale* showed a negative effect on rice. In another study on endosymbiosis of cycads-cyanobacteria, the major cyanobionts found were *Nostoc, Tolypothrix* sp., and *Leptolyngbya* sp. in 22 colonized coralloid roots of *Macrozamia, Lepidozamia, Bowenia*, and *Cycas*. Further, the rhizosphere isolates were distinct from root cyanobionts due to environmental factors (Cuddy, Neilan, & Gehringer, 2012).

5.2 Artificial symbiosis "Nature identical symbiosis" for crop improvement

Based on the insight of the above natural symbiosis, researchers have mimicked the symbiotic relationship by co-culturing selected cyanobacterial/algal strains with higher plants artificially for crop improvement. *Nostoc, Anabaena,* and *Calothrix* are major growth-promoting cyanobacteria and are found to form an intimate relationship with roots of rice, tomato and wheat plants, and enhances seed germination rate, growth and biomass. Two types of associations have been discussed in this section. Bipartite association involves the interaction of two species, that is, plant and algae/cyanobacteria. The second is tripartite which has three interaction partners including plant, microalgae/cynobacteria and bacteria. Fig. 2 shows a general interaction between plants and microalgal biofertilizer and their bipartite and tripartite associations.

In one of the earliest studies by Gantar and Elhai (1999; Gantar, 2000), wheat seedlings were co-cultivated with *Nostoc* sp. strain 2S9B in the hydroponics method and showed improved penetration and colonization of cyanobacteria in the wheat roots. Cyanobacteria were able to penetrate the intercellular spaces as well as inside the epidermal and cortex cells. Further, they emphasized that their findings paved a way for creating a model system in non-legume plant-microbe studies and crop improvement.

Later, Karthikeyan et al. (2009) have shown chlorophyll and growth enhancement in wheat seedlings when co-cultured with consortia of *C. ghosei, H. intricatus, Nostoc* sp., *Hapalosiphon* sp., *Calothrix* sp., *N. muscorum, W. prolifica, C. membranacea.* Similar interaction of bacterial and cyanobacterial interaction with wheat was observed resulting in growth enhancements. In a study by Nain et al. (2010) several bacterial and cyanobacterial isolates were screened from the wheat rhizosphere concluding their potential application as an alternative to chemical fertilizers and pesticide and promising integrated nutrient management for the wheat crop with nitrogen savings of 40–80 kg N/ha. These

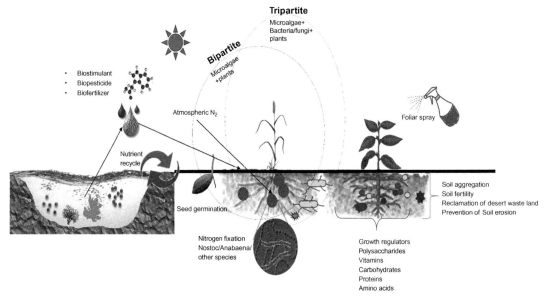

FIG. 2   Association of microalgal/cyanobacterial biofertilizer with higher plants to impart better growth and yield in the plants by acting as biofertilizers, biostimulants, biopesticides, and bioremediating agents. Interaction of microalgae with plants (bipartite) and interaction of microalgae with bacteria and plants (tripartite).

findings demonstrated the feasibility of environment-friendly organic farming and the low use of expensive chemical fertilizers. An association of a consortium of filamentous cyanobacteria *Phormidium, Anabaena, Westiellopsis, Fischerella, Spirogyra* with wheat plant showed an enhancement of 51% in organic carbon content in grains over control (Renuka et al., 2017).

Association of *Chroococcidiopsis* sp. Ck4 and *Anabaena* sp. Ck1. with wheat release, cytokinin (22.7 pmol/mg Chl-a) and indole-3-acetic acid (38 pmol/mg Chl-a) which resulted in an increase in grain weight (43%), seed germination, shoot length, tillering, number of lateral roots, and spike length under natural conditions (Hussain & Hasnain, 2011). Swarnalakshmi, Dhar, et al. (2013) and Swarnalakshmi, Prasanna, et al. (2013) have developed a different combination of biofilms using co-inoculation of *Anabaena torulosa* with agriculturally useful bacteria (*Azotobacter, Mesorhizobium, Serratia, and Pseudomonas*) mixed with wheat seeds and measured the nitrogen-fixing potential of acetylene reduction activity (ARA). The findings suggest that *Anabaena-Pseudomonas* biofilm showed the highest P uptake and nitrogen fixation. This tripartite synergism demonstrated that the interrelationship of cyanobacteria which fixes the nitrogen fixation and bacteria solubilize phosphate, which eventually benefited the plants for its growth.

Calothrix elenkinii (RPC1) inoculation with soil microbiome on rice plants showed an increment of biomass upto 70%, improved activity of nitrogenase (~48%), hydrolytic and defense enzymes and enhancement of soil chlorophyll concentration from 20% to 50% in sterilized soil. In addition, *Calothrix* exhibited selective interaction with gram-negative bacteria in rhizosphere (Ranjan et al., 2016). Consortia of *Anabaena variabilis, Chlorella vulgaris,* and *Azotobacter* sp exhibited the highest germination rate in rice seeds and an increase in the length of sprouts (Zayadan et al., 2014). Rice plants inoculated with consortia of *Anabaena torulosa, Nostoc carneum, Nostoc piscinale,* and *Anabaena doliolum* have been shown

to enhance the yield in intermittent irrigation (Prasanna et al., 2013). Singh, Prabha, Yandigeri, and Arora (2011) have shown that accumulation of phenolic acids, flavonoids, growth regulators, protein, and chlorophyll in rice with consortia of different cyanobacterial strains.

Triveni et al. (2015) have studied the *Trichoderma* (fungus) and *Anabaena* (cyanobacteria) based biofilms in infected cotton crops (plants). They found this biofilm has growth promoting characteristics and biocontrol potential against *Macrophomina phaseolina*, which infects the cotton crops. In this tripartite relationship, all three partners get mutual benefits by providing growth-promoting substances, nutrition, habitat and protection against competitive organisms.

As microalgal interaction has positively affected the growth of various plants, it is imperative to utilize these interactions with cereal crops to enhance their productivity. In this regard, some reports are available which throw light on this aspect. Many interactions of cyanobacterial association with cereal plants such as wheat (Renuka et al., 2017), rice (Prasanna et al., 2017), tomato (Coppens et al., 2016), Sorghum (Ramos-Ibarra, Rubio-Ramírez, Mondragón-Cortez, Torres-Velázquez, & Choix, 2019; Trejo et al., 2012) are known for their action as biofertilizers. However, the microalgae also grow in association with higher plants and interact with it to show the improved effects on growth and yield. Hence, extensive research in artificial symbiosis and genetic modification of microalgae to support agriculture will give a cost-effective, environment-friendly green technology to improve the agronomical traits like crop quality, productivity, biocontrol against pest, disease, maintain the soil fertility, and ecosystem.

6 Microalgae in bioremediation and reclamation of degraded land

One of the most important elements in agriculture is soil. It is the most essential element that provides a medium for plant growth, maintain atmospheric gases, store carbon and act as a filtration system for surface water. With the increasing population, there has been an increased demand to convert forests and grasslands to farm fields. A major portion of the earth's topsoil has been degraded in the last 150 years causing major implications on crop productivity (Soil erosion and degradation report, WWF). Unsustainable use of land, farming practices and abandonment of land had subjected the soil to series of degradation processes like loss of soil structure, compaction, nutrient degradation, salinity, sodification, contamination with heavy metals, and pesticides, etc. (Virto et al., 2015). As reported by UN's food and agriculture organization if soil degradation continues at the same pace the world's top soil would run out within 60 years (Arsenault, Reuters- sustainability "Only 60 Years of Farming Left If Soil Degradation Continues", Dec 5 2014).

As land resources are limited, requisite measures are needed to reclaim degraded and infertile wastelands. Microalgae play a significant role in minimizing soil erosion by helping regulate fluids into the soil, enhance soil fertility and helps in soil reclamation (Sharma et al., 2012). Cyanobacteria are also known to contribute toward soil fertility, produce biologically active substances, and concentrate metal ions from their surrounding environment (Ibraheem, 2007; Samhan, 2008; Shaaban, Haroun, & Ibraheem, 2004). Altogether, green algae and BGA can potentially act as a soil conditioner and help in the reclamation of degraded lands.

6.1 Use of Algae as soil conditioners

While maintaining healthy soil is challenging, there are many innovative options available to maintain soil's fragile skin and preserve the natural biodiversity of the soil. Soil conditioners can be organic or inorganic or a combination of synthetic or natural ingredients. A wide variety of ingredients described as soil conditioners are animal manure, compost, sawdust, limestone, straw, peat, etc. The most innovative and recent

soil conditioner are microalgal supplements. The use of microalgae as a soil conditioner not only improves health but also the overall quality of the soil. While fertilizers only provide nutrients to the soil, microalgae are additives that improve the physical, chemical, and biological properties of the soil. Microalgae enhance soil structure by increasing water retention capacity, aeration, and thus the absorption of nutrients.

6.2 Microalgae as bioremediating agents

There has been a tremendous decrease in the cultivable lands due to increased environmental problems like drought, soil acidity, soil salinity, metal contamination, extreme weather conditions, and anthropogenic activities. A total of 147 million hectares (Mha) of Indian land suffers from soil degradation out of which major portion is from water erosion (94 Mha), acidification (16 Mha), flooding (14 Mha), wind erosion (9 Mha), salinity (6 Mha), and a combination of factor (7 Mha) (Bhattacharyya et al., 2015). In the past two decades, immense attention has been given on overcoming problems related to environmental pollution caused by dangerous materials like heavy metals released from industrial waste, oil contaminated sites from oil spillages, chemicals like organochlorine or organophosphate mainly coming from pesticide residues. There are numerous physiochemical methods to combat the removal of these contaminants and reclaim these degraded lands. However, the methods used are costly and pose several disadvantages. Hence, there is a need to identify sustainable, cost-effective, and efficient methods to remediate and reclaim degraded soils.

Microalgae are looked up as a potential bioremediating agent. These photosynthetic organisms can act as potential sinks for the elimination of harmful metals and other toxic contaminants present in their surroundings. Both cyanobacteria and green algae are found almost everywhere and can survive and grow under harshest environmental conditions (Megharaj, Singleton, McClure, & Naidu, 2000; Qiao, Takano, & Liu, 2015; Subramaniyam et al., 2016). Be it oil/metal-contaminated sites or salt-affected areas microalgae's ability to survive under such conditions make them the most opted and suitable candidate for bioremediation of degraded lands (Abed & Köster, 2005; Monteiro, Marques, Castro, & Malcata, 2009; Pandey, Shukla, Giri, & Kashyap, 2005; Renuka et al., 2018; Trejo et al., 2012).

6.2.1 Reclamation of heavy metals contaminated sites

Increased urbanization, industrialization, and use of fertilizers have loaded the environment with toxic heavy metals. As these heavy metals are non-biodegradable they continuously get accumulated in the environment. The industrial effluents rich in these toxic metal (PB, Hg, Cu, Zn, Cd, Cr, and Ni, etc.), when released in the water bodies, pose a potential threat to the aquatic ecosystem (Ali & Khan, 2018). When the concentration of these heavy metals exceeds the permissible limits, they accumulate in the entire food chain causing life-threatening diseases like cancer. The existing techniques for the removal of heavy metals are membrane filtration, chemical precipitation, ultrafiltration, ion-exchange, electrodialysis, etc. However, the above-mentioned methods have their own disadvantages like inefficient metal removal, intensive energy requirement, and production of toxic sludge that would require further treatment for disposal (Ahalya, Ramachandra, & Kanamadi, 2003).

Microalgae are reported to have high metal binding capacity due to the surface chemistries of their cell wall that have polysaccharides, proteins, lipids containing functional groups like sulfates, aminoacids, carboxyl, hydroxyl, etc., that act as metal-binding sites (Schiewer & Volesky, 2000; Yu, Matheickal, Yin, & Kaewsarn, 1999). Both cyanobacteria and microalgae are known to occur in metal-contaminated sites and play an

effective role in the sequestration of heavy metal (Hamed, Selim, Klöck, & AbdElgawad, 2017; Renuka et al., 2018; Subramaniyam et al., 2016).

Biosorption studies on freely suspended biomass of *Microcystis aeruginosa* suggest the potential of this organism in the removal of high quantities of Pb and Cd from contaminated aqueous solutions (Rzymski, Niedzielski, Karczewski, & Poniedziałek, 2014). *Chlorella* sp. when exposed to arsenic at a concentration of 0 to 50 µg As cm-3 showed no growth inhibition, concluding that this species can retain around 50% of arsenite from a solution (Beceiro-Gonzalez et al., 2000; Ohki, Kuroiwa, & Maeda, 1999). A study reported the use of both living and non-living *S. abundans* in Biosorption of copper and cadmium (Terry & Stone, 2002). Another in-vitro study demonstrated the use of freshwater *C. reinhardtii* and Ca-alginate beads for the removal of Cd(II), Pb(II), and Hg(II) ions from contaminated solutions. Adsorption of metal ions on these immobilized microalgae was as fast as around 60 min (Bayramoğlu, Tuzun, Celik, Yilmaz, & Arica, 2006). In view of the wide knowledge based on in vitro studies, one can conclude that both microalgae and cyanobacteria can be used as bio-sorbent material for the removal of toxic metals. In addition, proving to be the most reliable and economical approach to reclaim the metal-contaminated habitats.

6.2.2 Reclamation of desert land by the production of polysaccharides

Desertification has become a rapidly growing environmental issue; factors like anthropogenic activities and climatic variations are rapidly degrading semiarid, arid, and humid land areas (MEA, Millennium Ecosystem Assessment). Ecosystems and human well-being: Desertification synthesis, Washington, DC: World Resources Institute, 2005). Desertification of land mainly involves loss of biological soil crust which in turn leads to a decrease in soil stability, loss of productivity of the soil, loss of biodiversity, lack of nutrients in the soil, dust formation, etc. Biological soil crust is the main component of the soil and helps in aggregating unconsolidated soil particles both by releasing chemicals like polysaccharides and physical mesh or net (Abdel-Raouf, Al-Homaidan, & Ibraheem, 2012).

Various strategies have been implemented for the stabilization of desertified soil. Use of chemicals like organic polymers—PAM (polyacrylamide), PASP (poly aspartic acid), Gypsum, and PVA (polyvinyl alcohol), has shown to increase sand aggregate stability thereby preventing soil erosion (Yang, Wang, Fang, & Tan, 2007). Although this method has its own ease of applicability and might be less expensive but still cannot be considered as a safe methodology for ecological restoration (Corti, Cinelli, D'Antone, Kenawy, & Solaro, 2002). Another common and widely adapted method is the monotonous cultivation of the same species. However, this mode of planting native crops may result in widespread damage caused by various diseases and pests (Jia, Li, Liu, Gao, & Zhang, 2012). Another very conventional way of biological soil crust restoration is mulching. Covering the soil surface through a protective layer of mulch that is organic in nature. However, this method of restoration also does not last long as it requires periodic renewal (Chalker-Scott, 2007).

It is well-known that algal crusts of desert regions retard soil erosion (Chungxiang Hu, Liu, Song, & Zhang, 2002; Metting, 1981). The growth of cyanobacteria mostly causes algal crust formation in an arid region. Cyanobacterial crust mainly secretes extracellular polymeric substances and provide organic matter that helps in improving soil aggregate stability (Belnap & Gardner, 1993; Issa et al., 2007; Issa, Le Bissonnais, Défarge, & Trichet, 2001; Johansen, 1993; Rossi, Potrafka, Pichel, & De Philippis, 2012; Wu et al., 2013). Many studies reported inoculating cyanobacterial cells to enhance biological soil crust to improve soil

stability, productivity, and nutrition (Acea, Diz, & Prieto-Fernandez, 2001; Acea, Prieto-Fernández, & Diz-Cid, 2003; Maestre et al., 2006; Nisha et al., 2007). Park, Li, Zhao, Jia, and Hur (2017) reported restoration of desertified areas by combining cyanobacteria and soil fixing chemicals. Another study revealed that the addition of cyanobacterial polysaccharides (*Phormidium tenue*) at the time of growth of shrub *Caragana korshinskii* in desert soil enhanced seed germination, growth, and nutritional properties of the shrub (Xu et al., 2013).

6.2.3 Reclamation of oil-contaminated site

Petroleum hydrocarbons accumulation in the environment due to crude oil spills from tankers, drilling rigs and wells, spills of gasoline, diesel, and other by-products caused during transportation, etc., is a serious environmental concern. Oil spillage causes serious damages to the soil leading to loss of fertility of the soil. Spillage of petroleum hydrocarbon has an adverse effect on the germination capacity of the plant and retards plant growth. Crude oil spillage results in the quenching of essential nutrients like nitrogen and oxygen from the soil. The hydrocarbons are known to shield the seed by forming a film thereby preventing the entry of oxygen and water (Adam & Duncan, 2002). Another important observation in plants exposed to oil and it's by-products include the deterioration of chlorophyll (Malallah, Afzal, Kurian, Gulshan, & Dhami, 1998) causing serious damages to land and aquatic habitats where these spillages happen.

One of the most eco-friendly and economic approaches in the reclamation of these oil-contaminated sites is bioremediation. It is well known that many microorganisms possess the capability to degrade oil components or certain classes of oil hydrocarbons (Dalby, Kormas, Christaki, & Karayanni, 2008; Harayama, Kasai, & Hara, 2004). Excretion of organic substances by microalgae act as a substrate for many heterotrophic microorganisms like fungi and bacteria. These bacteria have active oxidase enzymes that have the capability to metabolize oil hydrocarbons (Iliev, Petkov, & Lukavský, 2015). Algae are known to survive under extreme conditions and this fact was supported by a study done by Iliev, Gacheva, Nikova, and Nedelcheva (2011), who reported the presence of algae of different taxa from oil-polluted soil. These strains were identified as *Nostoc* sp., *Phormidium* sp, green algae *Klebsormidium* sp., and *Chlamydomonas* sp. (Iliev et al., 2011). Another study by Abed (2010) reported the association of petroleum degrading bacteria with cyanobacteria in the degradation of n-alkanes. The cyanobacterial EPS helps in the formation of biofilms, which in turn provide a stable matrix for these heterotrophic bacteria and provide them with O_2 and fixed nitrogen. The microbes can penetrate, flourish, and degrade the surrounding oil contaminants and thereby making the surroundings feasible for cyanobacterial growth. This unique and beneficial relationship between bacteria and cyanobacteria plays a key role in the remediation of oil-contaminated sites (Abed, 2010; Abed & Köster, 2005). Studies conducted in vitro on one of the *Chlorella* sp. showed the potential to biodegrade crude oil under mixotrophic conditions. Crude oil acts as a carbon source during mixotrophic conditions and thereby stimulates algal biomass productivity and photosynthetic pigments (Romero, Cordero, & Garizado, 2018). Therefore, inoculating microalgal and cyanobacterial strains can be an effective strategy to improve the microbial consortia of the oil-contaminated site and thereby reclaim and recover the lost fertility of the land contaminated with oil and other by-products.

6.2.4 Reclamation of alkaline and saline soil

Salt or alkalinity stress is a major contributing factor that restricts the productivity of any crop. Salt affected soils are widely distributed throughout the world including dessert, river valleys, coastal areas, plains and excessively irrigated lands (Singh & Dhar, 2007). Salt affected soils are mainly of two types saline and alkaline soils. As the name indicates saline soils are formed by the influence of sodium salts like NaCl or Na_2SO_4 whereas alkaline soils are formed due to the presence of Na_2CO_3 and $NaHCO_3$ (Szabolcs, 1993; Singh & Dhar, 2010).

The reclamation of saline/alkaline soil requires complete replacement of exchangeable Na ions (Na^+) by Ca ions (Ca^+) thereby completely removing the exchanged sodium. Conventional approaches for reclamation of these alkaline and saline soil involves the use of gypsum ($CaSO_4$, $2H_2O$) or pyrite ($FeSO_4$) followed by excessive irrigation and flooding to remove excess salts (Singh & Singh, 2015). A more sustainable and equally efficient approach to reclaim these salt-affected lands is by the use of blue-green algae or cyanobacteria. There are numerous reports that are evidence in the role of cyanobacteria to reclaim saline or alkaline soil because these organisms have the capability to grow successfully under high salt and alkaline soil conditions (Jaiswal, Kashyap, Prasanna, & Singh, 2010; Murtaza et al., 2011; Pandey et al., 2005; Singh, 1961). Cyanobacteria particularly of the genera *Anabaena*, *Synechocystis* and *Aphanothece* are known to predominate in the saline soils (Pandhal, Snijders, Wright, & Biggs, 2008; Reed, 1988). Reports suggest that a large number of cyanobacterial strains have the capability to quench toxic sodium (Na^+) from the soil and aids in the improvement of the soil properties (Apte & Thomas, 1983; Kratz & Myers, 1955). A study conducted by Singh and Singh (2015) reported a decrease in pH, electrical conductivity, and Na^+ of saline-alkaline soil upon inoculation of cyanobacterium *Nostoc calcicola* with gypsum (Singh & Singh, 2015) indicating that this combination can be an effective approach toward the reclamation of such degraded lands. On the similar lines, another study showed improved soil aggregation upon inoculation of cyanobacteria in user soil by reducing the pH and electrical conductivity, which further increased the hydraulic conductivity of the soils (Kaushik & Subhashini, 1985).

Driving toward sustainable agricultural practices, utilization of green algae and cyanobacteria in the form of soil conditioners and bioremediation agents can be a step toward revolutionizing the current agriculture system as a sustainable and efficient means for land reclamation.

7 Conclusion and future prospects

The agriculture that sustainably fed humans for last 100 years or so was mainly driven by the use of chemical fertilizers and pesticides. Mankind is faced with the daunting task of not only increasing agricultural production to match the population growth but also doing it in a sustainable manner. Algae, mainly seaweeds and cyanobacteria have been used since ages in agriculture to boost productivity and crop health. This chapter has described the enormous potential of microalgae in agriculture-related applications such as bioremediation, biofertilizers, biopesticides and bio-stimulants to enhance soil/crop health and productivity. Clearly, several studies have demonstrated the potential of these green cell factories to sustain agriculture and boost productivity beyond doubt. For microalgae to be commercially viable as agriculture inputs, a reliable year-round supply with consistent quality and economic production process is a prerequisite. This has been the biggest bottleneck in the microalgae field. Considerable progress was made in microalgae cultivation, harvesting and downstream processing during biofuel research but this has been largely demonstrated at the lab scale. At the pilot scale, so far, the microalgae species have been mainly cultivated in open ponds where they are highly susceptible to biotic and abiotic stress and hence prone to crashes and loss of crop. The cost-effective scalability of alternative cultivation

systems like PBRs remains to be demonstrated. Further, there are challenges with a lack of scalable, efficient and cost-effective technologies for downstream processing of microalgal biomass like dewatering and lysis of cells to extract compounds of interest. However, these lacunae or gaps at the pilot scale will be soon filled by the findings of several very recent studies. These studies have shown that with the implementation of integrated pest management approaches and tailor-made nutrient formulations (Subhash et al, 2019; McBride et al., 2014; Karuppasamy et al., 2018; Rajvanshi et al., 2019), microalgae can indeed be successfully grown in open ponds at a commercial scale on a consistent basis. Complimenting these studies are also recent new technical developments in the harvesting of biomass and breaking of microalgal cells using mechanical or enzymatic technologies. Finally, the ace in the shot appears to be the biorefinery approach where the focus is on generating multiple products from the same microalgal biomass thus reducing the cost of production significantly and thereby enhancing the economic viability of the whole pipeline. Microalgal biomass or extracts as niche products to sustain agriculture and boost crop health might be on the anvil and this would just be the beginning of a long success story.

References

Abbassy, M. A., Marei, G. I. K., & Rabia, S. M. (2014). Antimicrobial activity of some plant and algal extracts. *International Journal of Plant & Soil Science, 3*, 1366–1373.

Abd El-Baky, H. H., El-Baz, F. K., & El Baroty, G. S. (2010). Enhancing antioxidant availability in wheat grains from plants grown under seawater stress in response to microalgae extract treatments. *Journal of the Science of Food and Agriculture, 90*(2), 299–303.

Abdel-Raouf, N., Al-Homaidan, A. A., & Ibraheem, I. B. M. (2012). Agricultural importance of algae. *African Journal of Biotechnology, 11*(54), 11648–11658.

Abed, R. M. (2010). Interaction between cyanobacteria and aerobic heterotrophic bacteria in the degradation of hydrocarbons. *International Biodeterioration & Biodegradation, 64*(1), 58–64.

Abed, R. M., & Köster, J. (2005). The direct role of aerobic heterotrophic bacteria associated with cyanobacteria in the degradation of oil compounds. *International Biodeterioration & Biodegradation, 55*(1), 29–37.

Abo-Shady, A. M., Al-ghaffar, B. A., Rahhal, M. M. H., & Abd-El Monem, H. A. (2007). Biological control of faba bean pathogenic fungi by three cyanobacterial filtrates. *Pakistan Journal of Biological Sciences, 10*, 3029–3038.

Acea, M. J., Diz, N., & Prieto-Fernandez, A. (2001). Microbial populations in heated soils inoculated with cyanobacteria. *Biology and Fertility of Soils, 33*(2), 118–125.

Acea, M. J., Prieto-Fernández, A., & Diz-Cid, N. (2003). Cyanobacterial inoculation of heated soils: Effect on microorganisms of C and N cycles and on chemical composition in soil surface. *Soil Biology and Biochemistry, 35*(4), 513–524.

Adam, G., & Duncan, H. (2002). Influence of diesel fuel on seed germination. *Environmental Pollution, 120*(2), 363–370.

Adams, D. G., Bergman, B., Nierzwicki-Bauer, S. A., Rai, A. N., & Schüßler, A. (2006). Cyanobacterial-plant symbioses. *The prokaryotes* (pp. 331–363)*Symbiotic associations, biotechnology, applied microbiology: Vol. 1*, (pp. 331–363).

Ahalya, N., Ramachandra, T. V., & Kanamadi, R. D. (2003). Biosorption of heavy metals. *Research Journal of Chemistry and Environment, 7*(4), 71 79.

Ali, H., & Khan, E. (2018). Bioaccumulation of non-essential hazardous heavy metals and metalloids in freshwater fish. Risk to human health. *Environmental Chemistry Letters, 16*(3), 903–917.

Almendro-Candel, M. B., Lucas, I. G., Navarro-Pedreño, J., & Zorpas, A. A. (2018). Physical properties of soils affected by the use of agricultural waste. In *Agricultural waste and residues* (p. 9).

Apte, S. K., & Thomas, J. (1983). Sodium transport in filamentous nitrogen fixing cyanobacteria. *Journal of Biosciences, 5*(3), 225–233.

Ashok, A. K., Ravi, V., & Saravanan, R. (2017). Influence of cyanobacterial auxin on sprouting of taro (Colocasia esculenta var. antiquorum) and corm yield. *Indian Journal of Agricultural Sciences, 87*(11), 1437–1444.

Babu, S., Prasanna, R., Bidyarani, N., & Singh, R. (2015). Analysing the colonisation of inoculated cyanobacteria in wheat plants using biochemical and molecular tools. *Journal of Applied Phycology, 27*(1), 327–338.

Bai, L., Xu, H., Wang, C., Deng, J., & Jiang, H. (2016). Extracellular polymeric substances facilitate the biosorption of phenanthrene on cyanobacteria Microcystis aeruginosa. *Chemosphere, 162*, 172–180.

Bailey, D., Mazurak, A. P., & Rosowski, J. R. (1973). Aggregation of soil particles by algae 1. *Journal of Phycology, 9*(1), 99–101.

Barone, V., Baglieri, A., Stevanato, P., Broccanello, C., Bertoldo, G., Bertaggia, M., et al. (2018). Root morphological and molecular responses induced by microalgae extracts in sugar beet (*Beta vulgaris* L.). *Journal of Applied Phycology, 30*(2), 1061–1071.

Bayramoğlu, G., Tuzun, I., Celik, G., Yilmaz, M., & Arica, M. Y. (2006). Biosorption of mercury (II), cadmium (II) and lead (II) ions from aqueous system by microalgae Chlamydomonas reinhardtii immobilized in alginate beads. *International Journal of Mineral Processing, 81*(1), 35–43.

Beceiro-Gonzalez, E., Taboada-de la Calzada, A., Alonso-Rodrıguez, E., López-Mahıa, P., Muniategui-Lorenzo, S., & Prada-Rodrıguez, D. (2000). Interaction between metallic species and biological substrates: Approximation to possible interaction mechanisms between the alga Chlorella vulgaris and arsenic (III). *TrAC Trends in Analytical Chemistry, 19*(8), 475–480.

Belnap, J., & Gardner, J. S. (1993). Soil microstructure in soils of the Colorado Plateau: The role of the cyanobacterium Microcoleus vaginatus. *Great Basin Naturalist, 53*(1), 6.

Benderliev, K. M., Ivanova, N. I., & Pilarski, P. S. (2003). Singlet oxygen and other reactive oxygen species are involved in regulation of release of iron-binding chelators from Scenedesmus cells. *Biologia Plantarum, 47*(4), 523–526.

Bhalamurugan, G. L., Valerie, O., & Mark, L. (2018). Valuable bioproducts obtained from microalgal biomass and their commercial applications: A review. *Environmental Engineering Research, 23*(3), 229–241.

Bharti, A., Velmourougane, K., & Prasanna, R. (2017). Phototrophic biofilms: Diversity, ecology and applications. *Journal of Applied Phycology, 29*(6), 2729–2744.

Bhattacharyya, R., Ghosh, B. N., Mishra, P. K., Mandal, B., Rao, C. S., Sarkar, D., et al. (2015). Soil degradation in India: Challenges and potential solutions. *Sustainability, 7*(4), 3528–3570.

Bondoc, K. G. V., Heuschele, J., Gillard, J., Vyverman, W., & Pohnert, G. (2016). Selective silicate-directed motility in diatoms. *Nature Communications, 7*, 10540.

Botta, A. (2012). Enhancing plant tolerance to temperature stress with amino acids: An approach to their mode of action. In: *I World Congress on the Use of Biostimulants in Agriculture 1009*, pp. 29–35.

Brown, C. M., & Trick, C. G. (1992). Response of the cyanobacterium, Oscillatoria tenuis, to low iron environments: The effect on growth rate and evidence for siderophore production. *Archives of Microbiology, 157*(4), 349–354.

Bumandalai, O., & Tserennadmid, R. (2019). Effect of *Chlorella vulgaris* as a biofertilizer on germination of tomato and cucumber seeds. *International Journal of Aquatic Biology, 7*(2), 95–99.

Canellas, L. P., Olivares, F. L., Aguiar, N. O., Jones, D. L., Nebbioso, A., Mazzei, P., et al. (2015). Humic and fulvic acids as biostimulants in horticulture. *Scientia Horticulturae, 196*, 15–27.

Carvalho, F. P. (2017). Pesticides, environment, and food safety. *Food and Energy Security, 6*(2), 48–60.

Cavani, L., Ter Halle, A., Richard, C., & Ciavatta, C. (2006). Photosensitizing properties of protein hydrolysate-based fertilizers. *Journal of Agricultural and Food Chemistry, 54* (24), 9160–9167.

Chalker-Scott, L. (2007). Impact of mulches on landscape plants and the environment—A review. *Journal of Environmental Horticulture, 25*(4), 239–249.

Chandel, S. T. (2009). Nematicidal activity of the cyanobacterium, *Aulosira fertilissima* on the hatch of Meloidogynetriticoryzae and Meloidogyneincognita. *Archives of Phytopathology and Plant Protection, 42*(1), 32–38.

Chaudhary, V., Prasanna, R., Nain, L., Dubey, S. C., Gupta, V., Singh, R., et al. (2012). Bioefficacy of novel cyanobacteria-amended formulations in suppressing damping off disease in tomato seedlings. *World Journal of Microbiology and Biotechnology, 28*, 3301–3310.

Chiaiese, P., Corrado, G., Colla, G., Kyriacou, M. C., & Rouphael, Y. (2018). Renewable sources of plant biostimulation: Microalgae as a sustainable means to improve crop performance. *Frontiers in Plant Science, 9*, 1782.

Colla, G., Nardi, S., Cardarelli, M., Ertani, A., Lucini, L., Canaguier, R., et al. (2015). Protein hydrolysates as biostimulants in horticulture. *Scientia Horticulturae, 196*, 28–38.

Coppens, J., Grunert, O., Van Den Hende, S., Vanhoutte, I., Boon, N., Haesaert, G., et al. (2016). The use of microalgae as a high-value organic slow-release fertilizer results in tomatoes with increased carotenoid and sugar levels. *Journal of Applied Phycology, 28*(4), 2367–2377.

Corti, A., Cinelli, P., D'Antone, S., Kenawy, E. R., & Solaro, R. (2002). Biodegradation of poly (vinyl alcohol) in soil environment: Influence of natural organic fillers and structural parameters. *Macromolecular Chemistry and Physics, 203*(10-11), 1526–1531.

Cuddy, W. S., Neilan, B. A., & Gehringer, M. M. (2012). Comparative analysis of cyanobacteria in the rhizosphere and as endosymbionts of cycads in drought-affected soils. *FEMS Microbiology Ecology, 80*(1), 204–215.

Currie, H. A., & Perry, C. C. (2007). Silica in plants: Biological, biochemical and chemical studies. *Annals of Botany, 100*, 1383–1389.

Dalby, A. P., Kormas, K. A., Christaki, U., & Karayanni, H. (2008). Cosmopolitan heterotrophic microeukaryotes are active bacterial grazers in experimental oil-polluted systems. *Environmental Microbiology, 10*(1), 47–56.

De Caire, G. Z., De Cano, M. S., De Mule, M. C. Z., & De Halperin, D. R. (1990). Antimycotic products from the Cyanobacterium *Nostoc muscorum* against *Rhizoctonia solani*. *Phyton, Buenos Aires, 51*, 1–4.

De Pascale, S., Rouphael, Y., & Colla, G. (2017). Plant biostimulants: Innovative tool for enhancing plant nutrition in organic farming. *European Journal of Horticultural Science, 82*(6), 277–285.

Dhar, D. W., Prasanna, R., & Singh, B. V. (2007). Comparative performance of three carrier based blue green algal biofertilizers for sustainable rice cultivation. *Journal of Sustainable Agriculture, 30*(2), 41–50.

Dias, G. A., Rocha, R. H. C., Araújo, J. L., De Lima, J. F., & Guedes, W. A. (2016). *Growth, yield, and postharvest quality in eggplant produced under different foliar fertilizer (*Spirulina platensis*) treatments*. Semina: Ciências Agrárias.

Dineshkumar, R., Subramanian, J., Gopalsamy, J., Jayasingam, P., Arumugam, A., Kannadasan, S., et al. (2019). The impact of using microalgae as biofertilizer in maize (Zea mays L.). *Waste and Biomass Valorization, 10*(5), 1101–1110.

Doğan-Subaşı, E., & Demirer, G. N. (2016). Anaerobic digestion of microalgal (*Chlorella vulgaris*) biomass as a source of biogas and biofertilizer. *Environmental Progress & Sustainable Energy, 35*(4), 936–941.

Drever, J. I., & Stillings, L. L. (1997). The role of organic acids in mineral weathering. *Colloids and Surfaces A: Physicochemical and Engineering Aspects, 120*(1-3), 167–181.

Drobek, M., Frąc, M., & Cybulska, J. (2019). Plant biostimulants: Importance of the quality and yield of horticultural crops and the improvement of plant tolerance to abiotic stress—A review. *Agronomy, 9*(6), 335.

Du Jardin, P. (2015). Plant biostimulants: Definition, concept, main categories and regulation. *Scientia Horticulturae, 196*, 3–14.

Dukare, A. S., Prasanna, R., Chandra Dubey, S., Nain, L., Chaudhary, V., Singh, R., et al. (2011). Evaluating novel microbe amended composts as biocontrol agents in tomato. *Crop Protection, 30*, 436–442.

Elarroussia, H., Elmernissia, N., Benhimaa, R., El Kadmiria, I. M., Bendaou, N., Smouni, A., et al. (2016). Microalgae polysaccharides a promising plant growth biostimulant. *Journal of Algal Biomass Utilization, 7*(4), 55–63.

Ertani, A., Schiavon, M., Muscolo, A., & Nardi, S. (2013). Alfalfa plant-derived biostimulant stimulate short-term growth of salt stressed Zea mays L. plants. *Plant and Soil, 364*(1-2), 145–158.

Faheed, F. A., & Fattah, Z. A. (2008). Effect of Chlorella vulgaris as bio-fertilizer on growth parameters and metabolic aspects of lettuce plant. *Journal of Agriculture and Social Sciences (Pakistan) 4*, 165–169.

Fernández, V., & Eichert, T. (2009). Uptake of hydrophilic solutes through plant leaves: Current state of knowledge and perspectives of foliar fertilization. *Critical Reviews in Plant Sciences, 28*(1-2), 36–68.

Flemming, H. C., & Wingender, J. (2010). The biofilm matrix. *Nature Reviews Microbiology, 8*(9), 623.

Frankmölle, W. P., Knübel, G., Moore, R. E., & Patterson, G. M. (1992). Antifungal cyclic peptides from the terrestrial blue-green alga Anabaena laxa. II. Structures of laxaphycins A, B, D and E. *The Journal of Antibiotics, 45*(9), 1458–1466.

Gademann, K., & Portmann, C. (2008). Secondary metabolites from cyanobacteria: Complex structures and powerful bioactivities. *Current Organic Chemistry, 12*(4), 326–341.

Gantar, M. (2000). Mechanical damage of roots provides enhanced colonization of the wheat endorhizosphere by the dinitrogen-fixing cyanobacterium Nostoc sp. strain 2S9B. *Biology and Fertility of Soils*, *32*(3), 250–255.

Gantar, M., & Elhai, J. (1999). Colonization of wheat paranodules by the N 2-fixing cyanobacterium Nostoc sp. strain 2S9B. *The New Phytologist*, *141*(3), 373–379.

Garcia-Gonzalez, J., & Sommerfeld, M. (2016). Biofertilizer and biostimulant properties of the microalga Acutodesmus dimorphus. *Journal of Applied Phycology*, *28*(2), 1051–1061.

Gautam, K., Pareek, A., & Sharma, D. K. (2013). Biochemical composition of green alga Chlorella minutissima in mixotrophic cultures under the effect of different carbon sources. *Journal of Bioscience and Bioengineering*, *116*(5), 624–627.

Gautam, K., Pareek, A., & Sharma, D. K. (2015). Exploiting microalgae and macroalgae for production of biofuels and biosequestration of carbon dioxide—A review. *International Journal of Green Energy*, *12*(11), 1122–1143.

Gautam, K., Tripathi, J. K., Pareek, A., & Sharma, D. K. (2019). Growth and secretome analysis of possible synergistic interaction between green algae and cyanobacteria. *Journal of Bioscience and Bioengineering*, *127*(2), 213–221.

Gilliom, R. J. (2007). Pesticides in US streams and groundwater. *Environmental Science and Technology*, *41*(10), 3408–3414.

Goldman, S. J., Lammers, P. J., Berman, M. S., & Sanders-Loehr, J. (1983). Siderophore-mediated iron uptake in different strains of Anabaena sp. *Journal of Bacteriology*, *156*(3), 1144–1150.

Gomiero, T., Pimentel, D., & Paoletti, M. G. (2011). Environmental impact of different agricultural management practices: Conventional vs. organic agriculture. *Critical Reviews in Plant Sciences*, *30*(1-2), 95–124.

Gouveia, L., Graça, S., Sousa, C., Ambrosano, L., Ribeiro, B., Botrel, E. P., et al. (2016). Microalgae biomass production using wastewater: Treatment and costs: Scale-up considerations. *Algal Research*, *16*, 167–176.

Goyal, S. K., Singh, B. V., Nagpal, V., & Marwaha, T. S. (1997). An improved method for production of algal biofertilizer. *Indian Journal of Agricultural Sciences*, *67*, 314–315.

Grzesik, M., Romanowska-Duda, Z., & Kalaji, H. M. (2017). Effectiveness of cyanobacteria and green algae in enhancing the photosynthetic performance and growth of willow (Salix viminalis L.) plants under limited synthetic fertilizers application. *Photosynthetica*, *55*(3), 510–521.

Günerken, E., D'Hondt, E., Eppink, M. H. M., Garcia-Gonzalez, L., Elst, K., & Wijffels, R. H. (2015). Cell disruption for microalgae biorefineries. *Biotechnology Advances*, *33*(2), 243–260.

Gupta, S., & Dikshit, A. K. (2010). Biopesticides: An eco-friendly approach for pest control. *Journal of Biopesticides*, *3*, 186–188.

Gupta, A. B., & Lata, K. (1964). Effect of algal growth hormones on the germination of paddy seeds. *Hydrobiologia*, *24*(1-2), 430–434.

Haggag, W., Abd El-Aty, A. M., & Mohamed, A. A. (2014). The potential effect of two cyanobacterial species; Anabaena sphaerica and Oscillatoria agardhii against grain storage fungi. *European Scientific Journal*, *10*, 1857–7881.

Hagmann, L., & Jüttner, F. (1996). Fischerellin A, a novel photosystem-II-inhibiting allelochemical of the cyanobacterium Fischerella muscicola with antifungal and herbicidal activity. *Tetrahedron Letters*, *37*, 6539–6542.

Hamed, S. M., Selim, S., Klöck, G., & AbdElgawad, H. (2017). Sensitivity of two green microalgae to copper stress: Growth, oxidative and antioxidants analyses. *Ecotoxicology and Environmental Safety*, *144*, 19–25.

Harayama, S., Kasai, Y., & Hara, A. (2004). Microbial communities in oil-contaminated seawater. *Current Opinion in Biotechnology*, *15*(3), 205–214.

Herencia, J. F., Garcia-Galavis, P. A., & Maqueda, C. (2011). Long-term effect of organic and mineral fertilization on soil physical properties under greenhouse and outdoor management practices. *Pedosphere*, *21*(4), 443–453.

Hu, C., Liu, Y., Song, L., & Zhang, D. (2002). Effect of desert soil algae on the stabilization of fine sands. *Journal of Applied Phycology*, *14*(4), 281–292.

Hu, Q., Sommerfeld, M., Jarvis, E., Ghirardi, M., Posewitz, M., Seibert, M., et al. (2008). Microalgal triacylglycerols as feedstocks for biofuel production: Perspectives and advances. *The Plant Journal*, *54*(4), 621–639.

Hussain, A., & Hasnain, S. (2011). Phytostimulation and biofertilization in wheat by cyanobacteria. *Journal of Industrial Microbiology & Biotechnology*, *38*(1), 85–92.

Ibraheem, I. B. (2007). Cyanobacteria as alternative biological conditioners for bioremediation of barren soil. *Egyptian Journal of Phycology*, *8*(100).

Iliev, I., Gacheva, G., Nikova, I., & Nedelcheva, D. (2011). Algal diversity of oil polluted Vertisol. In: *Proceedings of Youth Scientific Conference "Kliment's Days"*, pp. 22–23.

Iliev, I., Petkov, G., & Lukavský, J. (2015). An approach to bioremediation of mineral oil polluted soil. *Genetics and Plant Physiology*, *5*(2), 162–169.

Illmer, P., & Schinner, F. (1995). Solubilization of inorganic calcium phosphates—Solubilization mechanisms. *Soil Biology and Biochemistry*, *27*(3), 257–263.

Innok, S., Chunleuchanon, S., Boonkerd, N., & Teaumroong, N. (2009). Cyanobacterial akinete induction and its application as biofertilizer for rice cultivation. *Journal of Applied Phycology*, *21*(6), 737.

Ishwarya, R., Vaseeharan, B., Kalyani, S., Banumathi, B., Govindarajan, M., Alharbi, N. S., et al. (2018). Facile green synthesis of zinc oxide nanoparticles using Ulva lactuca seaweed extract and evaluation of their photocatalytic, antibiofilm and insecticidal activity. *Journal of Photochemistry and Photobiology B: Biology*, *178*, 249–258.

Issa, O. M., Défarge, C., Le Bissonnais, Y., Marin, B., Duval, O., Bruand, A., et al. (2007). Effects of the inoculation of cyanobacteria on the microstructure and the structural stability of a tropical soil. *Plant and Soil*, *290*(1-2), 209–219.

Issa, O. M., Le Bissonnais, Y., Défarge, C., & Trichet, J. (2001). Role of a cyanobacterial cover on structural stability of sandy soils in the Sahelian part of western Niger. *Geoderma*, *101*(3-4), 15–30.

Jäger, K., Bartók, T., Ördög, V., & Barnabás, B. (2010). Improvement of maize (Zea mays L.) anther culture responses by algae-derived natural substances. *South African Journal of Botany*, *76*(3), 511–516.

Jaiswal, A., Das, K., Koli, D. K., & Pabbi, S. (2018). Characterization of cyanobacteria for IAA and siderophore production and their effect on rice seed germination. *International Journal of Current Microbiology and Applied Sciences*, *5*, 212–222.

Jaiswal, P., Kashyap, A. K., Prasanna, R., & Singh, P. K. (2010). Evaluating the potential of N. calcicola and its bicarbonate resistant mutant as bioameleorating agents for 'usar'soil. *Indian Journal of Microbiology*, *50*(1), 12–18.

Jha, M. N., & Prasad, A. N. (2006). Efficacy of new inexpensive cyanobacterial biofertilizer including its shelf-life. *World Journal of Microbiology and Biotechnology*, *22*(1), 73–79.

Jia, R. L., Li, X. R., Liu, L. C., Gao, Y. H., & Zhang, X. T. (2012). Differential wind tolerance of soil crust mosses explains their micro-distribution in nature. *Soil Biology and Biochemistry*, *45*, 31–39.

Jochum, M., Moncayo, L. P., & Jo, Y. K. (2018). Microalgal cultivation for biofertilization in rice plants using a vertical semi-closed airlift photobioreactor. *PLoS One*, *13*(9), e0203456.

Johansen, J. R. (1993). Cryptogamic crusts of semiarid and arid lands of North America. *Journal of Phycology*, *29*(2), 140–147.

Karthikeyan, N., Prasanna, R., Nain, L., & Kaushik, B. D. (2007). Evaluating the potential of plant growth promoting cyanobacteria as inoculants for wheat. *European Journal of Soil Biology*, *43*(1), 23–30.

Karthikeyan, N., Prasanna, R., Sood, A., Jaiswal, P., Nayak, S., & Kaushik, B. D. (2009). Physiological characterization and electron microscopic investigation of cyanobacteria associated with wheat rhizosphere. *Folia Microbiologica*, *54*(1), 43–51.

Karuppasamy, S., Musale, A. S., Soni, B., Bhadra, B., Gujarathi, N., Sundaram, M., et al. (2018). Integrated grazer management mediated by chemicals for sustainable cultivation of algae in open ponds. *Algal Research*, *35*, 439–448.

Kaushik, B. D. (1998). Use of cyanobacterial biofertilizers in rice cultivation: A technology improvement. In *Cyanobacterial Biotechnology* (pp. 211–222).

Kaushik, B. D., & Subhashini, D. (1985). Amelioration of salt-affected soils with blue-green algae. II. Improvement in

soil properties. *Proceedings of the Indian National Science Academy*, *51*, 380–389.

Kawahara, J., Horikoshi, R., Yamaguchi, T., Kumagai, K., & Yanagisawa, Y. (2005). Air pollution and young children's inhalation exposure to organophosphorus pesticide in an agricultural community in Japan. *Environment International*, *31*(8), 1123–1132.

Khan, Z., Kim, Y. H., Kim, S. G., & Kim, H. W. (2007). Observations on the suppression of root knot nematode (*Meloidogyne arenaria*) on tomato by incorporation of cyanobacterial powder (*Oscillatoria chlorina*) into potting field soil. *Bioresource Technology*, *98*, 69–73.

Khan, Z., & Park, S. D. (1999). Effects of inoculum level and time of *Microcoleus vaginatus* on control of *Meloidogyne incognita* on tomato. *Journal of Asia-Pacific Entomology*, *2*, 93–96.

Khan, Z., Park, S. D., Shin, S. Y., Bae, S. G., Yeon, I. K., & Seo, Y. J. (2005). Management of *Meloidogyne incognita* on tomato by root-dip treatment in culture filtrate of the blue green alga, *Microcoleus vaginatus. Bioresource Technology*, *96*, 1338–1341.

Khan, W., Rayirath, U. P., Subramanian, S., Jithesh, M. N., Rayorath, P., Hodges, D. M., et al. (2009). Seaweed extracts as biostimulants of plant growth and development. *Journal of Plant Growth Regulation*, *28*(4), 386–399.

Khanna, P., Kaur, A., & Goyal, D. (2019). Algae-based metallic nanoparticles: Synthesis, characterization and applications. *Journal of Microbiological Methods*, *163*, 105656.

Kim, J. D. (2006). Screening of cyanobacteria (blue-green algae) from rice paddy soil for antifungal activity against plant pathogenic fungi. *Mycobiology*, *34*, 138–142.

Kim, J., & Kim, J.-D. (2008). Inhibitory effect of algal extracts on mycelial growth of the tomato-wilt pathogen, Fusariumoxysporum f. sp. lycopersici. *Mycobiology*, *36*, 242–248.

Knapen, A., Poesen, J., Galindo-Morales, P., Baets, S. D., & Pals, A. (2007). Effects of microbiotic crusts under cropland in temperate environments on soil erodibility during concentrated flow. *Earth Surface Processes and Landforms: The Journal of the British Geomorphological Research Group*, *32*(12), 1884–1901.

Kompi, M., Mekbib, S. B., & George, M. J. (2018). Evaluation of microalgae sludge as biofertilizer for growth of Maize under greenhouse trials. In: *Proceedings of National University of Lesotho International Science, Technology and Innovation Conference and Expo. January 23–25, 2018, Maseru, Lesotho*, pp. 10–12.

Korir, H., Mungai, N. W., Thuita, M., Hamba, Y., & Masso, C. (2017). Co-inoculation effect of rhizobia and plant growth promoting rhizobacteria on common bean growth in a low phosphorus soil. *Frontiers in Plant Science*, *8*, 141.

Kratz, W. A., & Myers, J. (1955). Photosynthesis and respiration of three blue-green algae. *Plant Physiology*, *30*(3), 275.

Kulik, M. M. (1995). The potential for using cyanobacteria (blue-green algae) and algae in the biological control of plant pathogenic bacteria and fungi. *European Journal of Plant Pathology, 101*, 585–599.

Kurth, C., Wasmuth, I., Wichard, T., Pohnert, G., & Nett, M. (2019). Algae induce siderophore biosynthesis in the freshwater bacterium Cupriavidus necator H16. *Biometals, 32*(1), 77–88.

Le Bissonnais, Y. L. (1996). Aggregate stability and assessment of soil crustability and erodibility: I. Theory and methodology. *European Journal of Soil Science, 47*(4), 425–437.

Leloup, M., Nicolau, R., Pallier, V., Yéprémian, C., & Feuillade-Cathalifaud, G. (2013). Organic matter produced by algae and cyanobacteria: Quantitative and qualitative characterization. *Journal of Environmental Sciences, 25*(6), 1089–1097.

Li, Z., Schneider, R. L., Morreale, S. J., Xie, Y., Li, C., & Li, J. (2018). Woody organic amendments for retaining soil water, improving soil properties and enhancing plant growth in desertified soils of Ningxia, China. *Geoderma, 310*, 143–152.

Lu, Y., Wu, K., Jiang, Y., Guo, Y., & Desneux, N. (2012). Widespread adoption of Bt cotton and insecticide decrease promotes biocontrol services. *Nature, 487*, 362–365.

Lucini, L., Rouphael, Y., Cardarelli, M., Canaguier, R., Kumar, P., & Colla, G. (2015). The effect of a plant-derived biostimulant on metabolic profiling and crop performance of lettuce grown under saline conditions. *Scientia Horticulturae, 182*, 124–133.

Maestre, F. T., Martín, N., Díez, B., Lopez-Poma, R., Santos, F., Luque, I., et al. (2006). Watering, fertilization, and slurry inoculation promote recovery of biological crust function in degraded soils. *Microbial Ecology, 52*(3), 365–377.

Mager, D. M., & Thomas, A. D. (2011). Extracellular polysaccharides from cyanobacterial soil crusts: A review of their role in dryland soil processes. *Journal of Arid Environments, 75*(2), 91–97.

Mahasneh, I. A., & Tiwari, D. N. (1992). The use of biofertilizer of Calothrix sp. M103, enhanced by addition of iron and siderophore production. *Journal of Applied Bacteriology, 73*(4), 286–289.

Malallah, G., Afzal, M., Kurian, M., Gulshan, S., & Dhami, M. S. I. (1998). Impact of oil pollution on some desert plants. *Environment International, 24*(8), 919–924.

Mandal, B., Vlek, P. L. G., & Mandal, L. N. (1999). Beneficial effects of blue-green algae and *Azolla*, excluding supplying nitrogen, on wetland rice fields: A review. *Biology and Fertility of Soils, 28*(4), 329–342.

Manjunath, M., Kanchan, A., Ranjan, K., Venkatachalam, S., Prasanna, R., Ramakrishnan, B., et al. (2016). Beneficial cyanobacteria and eubacteria synergistically enhance bioavailability of soil nutrients and yield of okra. *Heliyon, 2*(2), e00066.

Manjunath, M., Prasanna, R., Nain, L., Dureja, P., Singh, R., Kumar, A., et al. (2010). Biocontrol potential of cyanobacterial metabolites against damping off disease caused by *Pythium aphanidermatum* in solanaceous vegetables. *Archives of Phytopathology and Plant Protection, 43*, 666–677.

Marketsandmarkets.com (2019). *Biostimulants market by active ingredient (humic substances, amino acids, seaweed extracts, microbial amendments), crop type (fruits & vegetables, cereals, turf & ornamentals), application method, form, and region—Global Forecast to 2025.*

Marks, E. A., Miñón, J., Pascual, A., Montero, O., Navas, L. M., & Rad, C. (2017). Application of a microalgal slurry to soil stimulates heterotrophic activity and promotes bacterial growth. *Science of the Total Environment, 605*, 610–617.

McBride, R. C., Lopez, S., Meenach, C., Burnett, M., Lee, P. A., Nohilly, F., et al. (2014). Contamination management in low cost open algae ponds for biofuels production. *Industrial Biotechnology, 10*(3), 221–227.

McGill, W. B., & Cole, C. V. (1981). Comparative aspects of cycling of organic C, N, S and P through soil organic matter. *Geoderma, 26*(4), 267–286.

McKnight, D. M., & Morel, F. M. (1980). Copper complexation by siderophores from filamentous blue-green algae 1. *Limnology and Oceanography, 25*(1), 62–71.

Megharaj, M., Singleton, I., McClure, N. C., & Naidu, R. (2000). Influence of petroleum hydrocarbon contamination on microalgae and microbial activities in a long-term contaminated soil. *Archives of Environmental Contamination and Toxicology, 38*(4), 439–445.

Metting, B. (1981). The systematics and ecology of soil algae. *The Botanical Review, 47*(2), 195–312.

Michalak, I., & Chojnacka, K. (2014). Algal extracts: Technology and advances. *Engineering in Life Sciences, 14*(6), 581–591.

Mishra, U., & Pabbi, S. (2004). Cyanobacteria: A potential biofertilizer for rice. *Resonance, 9*(6), 6–10.

Mnif, I., & Ghribi, D. (2015). Potential of bacterial derived biopesticides in pest management. *Crop Protection, 77*, 52–64.

Monteiro, C. M., Marques, A. P., Castro, P. M., & Malcata, F. X. (2009). Characterization of Desmodesmus pleiomorphus isolated from a heavy metal-contaminated site: Biosorption of zinc. *Biodegradation, 20*(5), 629–641.

Mukherjee, C., Chowdhury, R., & Ray, K. (2015). Phosphorus recycling from an unexplored source by polyphosphate accumulating microalgae and cyanobacteria—A step to phosphorus security in agriculture. *Frontiers in Microbiology, 6*, 1421.

Mundt, S., Kreitlow, S., & Jansen, R. (2003). Fatty acids with antibacterial activity from the cyanobacterium *Oscillatoria redekei* HUB 051. *Journal of Applied Phycology, 15*, 263–267.

Murtaza, B., Murtaza, G., Zia-ur-Rehman, M., Ghafoor, A., Abubakar, S., & Sabir, M. (2011). Reclamation of salt-affected soils using amendments and growing wheat crop. *Soil & Environment*, *30*(2).

Nain, L., Rana, A., Joshi, M., Jadhav, S. D., Kumar, D., Shivay, Y. S., et al. (2010). Evaluation of synergistic effects of bacterial and cyanobacterial strains as biofertilizers for wheat. *Plant and Soil*, *331*(1-2), 217–230.

Nilsson, M., Bhattacharya, J., Rai, A. N., & Bergman, B. (2002). Colonization of roots of rice (Oryza sativa) by symbiotic Nostoc strains. *New Phytologist*, *156*(3), 517–525.

Nisha, R., Kaushik, A., & Kaushik, C. P. (2007). Effect of indigenous cyanobacterial application on structural stability and productivity of an organically poor semi-arid soil. *Geoderma*, *138*(1-2), 49–56.

Oancea, F., Velea, S., Fătu, V., Mincea, C., & Ilie, L. (2013). Micro-algae based plant biostimulant and its effect on water stressed tomato plants. *Romanian Journal of Plant Protection*, *6*, 104–117.

Odgerel, B., & Tserendulam, D. (2016). Effect of chlorella as a biofertilizer on germination of wheat and barley grains. *Proceedings of the Mongolian Academy of Sciences*, 26–31.

Ohki, A., Kuroiwa, T., & Maeda, S. (1999). Arsenic compounds in the freshwater green microalga *Chlorella vulgaris* after exposure to arsenite. *Applied Organometallic Chemistry*, *13*(2), 127–133.

Osman, M. E. H., El-Sheekh, M. M., El-Naggar, A. H., & Gheda, S. F. (2010). Effect of two species of cyanobacteria as biofertilizers on some metabolic activities, growth, and yield of pea plant. *Biology and Fertility of Soils*, *46*(8), 861–875.

Özdemir, S., Sukatar, A., & Bahar, G. (2016). Production of Chlorella vulgaris and its effects on plant growth, yield and fruit quality of organic tomato grown in greenhouse as biofertilizer. *The Journal of Agricultural Science 22*, 596–605.

Padhy, R. N., Nayak, N., Dash-Mohini, R. R., Rath, S., & Sahu, R. K. (2016). Growth, metabolism and yield of rice cultivated in soils amended with fly ash and cyanobacteria and metal loads in plant parts. *Rice Science*, *23*(1), 22–32.

Pandey, K. D., Shukla, P. N., Giri, D. D., & Kashyap, A. K. (2005). Cyanobacteria in alkaline soil and the effect of cyanobacteria inoculation with pyrite amendments on their reclamation. *Biology and Fertility of Soils*, *41*(6), 451–457.

Pandhal, J., Snijders, A. P., Wright, P. C., & Biggs, C. A. (2008). A cross-species quantitative proteomic study of salt adaptation in a halotolerant environmental isolate using 15N metabolic labelling. *Proteomics*, *8*(11), 2266–2284.

Park, C. H., Li, X. R., Zhao, Y., Jia, R. L., & Hur, J. S. (2017). Rapid development of cyanobacterial crust in the field for combating desertification. *PLoS One*, *12*(6), e0179903.

Pereira, I., Ortega, R., Barrientos, L., Moya, M., Reyes, G., & Kramm, V. (2009). Development of a biofertilizer based on filamentous nitrogen-fixing cyanobacteria for rice crops in Chile. *Journal of Applied Phycology*, *21*(1), 135–144.

Power, I. M., Wilson, S. A., Thom, J. M., Dipple, G. M., & Southam, G. (2007). Biologically induced mineralization of dypingite by cyanobacteria from an alkaline wetland near Atlin, British Columbia, Canada. *Geochemical Transactions*, *8*(1), 13.

Prasanna, R., Babu, S., Bidyarani, N., Kumar, A., Triveni, S., Monga, D., et al. (2015). Prospecting cyanobacteria-fortified composts as plant growth promoting and biocontrol agents in cotton. *Experimental Agriculture*, *51*(1), 42–65.

Prasanna, R., Jaiswal, P., Singh, Y., & Singh, P. (2008). Influence of biofertilizers and organic amendments on nitrogenase activity and phototrophic biomass of soil under wheat. *Acta Agronomica Hungarica*, *56*(2), 149–159.

Prasanna, R., Kanchan, A., Kaur, S., Ramakrishnan, B., Ranjan, K., Singh, M. C., et al. (2016). Chrysanthemum growth gains from beneficial microbial interactions and fertility improvements in soil under protected cultivation. *Horticultural Plant Journal*, *2*(4), 229–239.

Prasanna, R., Kanchan, A., Ramakrishnan, B., Ranjan, K., Venkatachalam, S., Hossain, F., et al. (2016). Cyanobacteria-based bioinoculants influence growth and yields by modulating the microbial communities favourably in the rhizospheres of maize hybrids. *European Journal of Soil Biology*, *75*, 15–23.

Prasanna, R., Nain, L., Tripathi, R., Gupta, V., Chaudhary, V., Middha, S., et al. (2008). Evaluation of fungicidal activity of extracellular filtrates of cyanobacteria—Possible role of hydrolytic enzymes. *Journal of Basic Microbiology*, *48*, 186–194.

Prasanna, R., Ramakrishnan, B., Simranjit, K., Ranjan, K., Kanchan, A., Hossain, F., et al. (2017). Cyanobacterial and rhizobial inoculation modulates the plant physiological attributes and nodule microbial communities of chickpea. *Archives of Microbiology*, *199*(9), 1311–1323.

Prasanna, R., Sharma, E., Sharma, P., Kumar, A., Kumar, R., Gupta, V., et al. (2013). Soil fertility and establishment potential of inoculated cyanobacteria in rice crop grown under non-flooded conditions. *Paddy and Water Environment*, *11*(1-4), 175–183.

Pratt, R., Oneto, J. F., & Pratt, J. (1945). Studies on *Chlorella Vulgaris*. X. influence of the age of the culture on the accumulation of chlorellin. *American Journal of Botany*, *32*, 405–408.

Priya, H., Prasanna, R., Ramakrishnan, B., Bidyarani, N., Babu, S., Thapa, S., et al. (2015). Influence of cyanobacterial inoculation on the culturable microbiome and growth of rice. *Microbiological Research*, *171*, 78–89.

Priyadarshani, I., & Rath, B. (2012). Commercial and industrial applications of micro algae–A review. *Journal of Algal Biomass Utilization*, *3*(4), 89–100.

Puget, P., Angers, D. A., & Chenu, C. (1998). Nature of carbohydrates associated with water-stable aggregates of two cultivated soils. *Soil Biology and Biochemistry*, 31(1), 55–63.

Qiao, K., Takano, T., & Liu, S. (2015). Discovery of two novel highly tolerant NaHCO3 Trebouxiophytes: Identification and characterization of microalgae from extreme saline–alkali soil. *Algal Research*, 9, 245–253.

Rajvanshi, M., Gautam, K., Manjre, S., Kumar, G. R. K., Kumar, C., Govindachary, S., et al. (2019). Stoichiometrically balanced nutrient management using a newly designed nutrient medium for large scale cultivation of Cyanobacterium aponinum. *Journal of Applied Phycology*, 31(5), 2779–2789.

Ramos-Ibarra, J. R., Rubio-Ramírez, T. E., Mondragón-Cortez, P., Torres-Velázquez, J. R., & Choix, F. J. (2019). Azospirillum brasilense-microalga interaction increases growth and accumulation of cell compounds in Chlorella vulgaris and Tetradesmus obliquus cultured under nitrogen stress. *Journal of Applied Phycology*, 1–13.

Rana, A., Joshi, M., Prasanna, R., Shivay, Y. S., & Nain, L. (2012). Biofortification of wheat through inoculation of plant growth promoting rhizobacteria and cyanobacteria. *European Journal of Soil Biology*, 50, 118–126.

Rana, A., Kabi, S. R., Verma, S., Adak, A., Pal, M., Shivay, Y. S., et al. (2015). Prospecting plant growth promoting bacteria and cyanobacteria as options for enrichment of macro-and micronutrients in grains in rice–wheat cropping sequence. *Cogent Food & Agriculture*, 1(1), 1037379.

Ranjan, K., Priya, H., Ramakrishnan, B., Prasanna, R., Venkatachalam, S., Thapa, S., et al. (2016). Cyanobacterial inoculation modifies the rhizosphere microbiome of rice planted to a tropical alluvial soil. *Applied Soil Ecology*, 108, 195–203.

Reed, R. H. (1988). The responses of cyanobacteria to salt stress. In *Biochemistry of the algae and cyanobacteria* (pp. 217–231).

Renuka, N., Guldhe, A., Prasanna, R., Singh, P., & Bux, F. (2018). Microalgae as multi-functional options in modern agriculture: Current trends, prospects and challenges. *Biotechnology Advances*, 36(4), 1255–1273.

Renuka, N., Prasanna, R., Sood, A., Ahluwalia, A. S., Bansal, R., Babu, S., et al. (2016). Exploring the efficacy of wastewater-grown microalgal biomass as a biofertilizer for wheat. *Environmental Science and Pollution Research*, 23(7), 6608–6620.

Renuka, N., Prasanna, R., Sood, A., Bansal, R., Bidyarani, N., Singh, R., et al. (2017). Wastewater grown microalgal biomass as inoculants for improving micronutrient availability in wheat. *Rhizosphere*, 3, 150–159.

Research and Markets (2019). *Global biostimulants market report 2019: Market is expected to reach $5.23 billion by 2026, growing at a CAGR of 11.9% during 2018 to 2026, 2019*. Available online.

Richmond, A. (2003). *Handbook of microalgal culture*: (p. 239). Oxford, UK: Blackwell Publishing Ltd.

Righini, H., & Roberti, R. (2019). Algae and cyanobacteria as biocontrol agents of fungal plant pathogens. In *Plant microbe interface* (pp. 219–238). Cham: Springer.

Roeselers, G., Van Loosdrecht, M. C. M., & Muyzer, G. (2008). Phototrophic biofilms and their potential applications. *Journal of Applied Phycology*, 20(3), 227–235.

Romero, D. V., Cordero, A. P., & Garizado, Y. O. (2018). Biodegradation activity of crude oil by Chlorella sp. under mixotrophic conditions. *Indian Journal of Science and Technology*. 11(29). https://doi.org/10.17485/ijst/2018/v11i29/127832.

Ronga, D., Biazzi, E., Parati, K., Carminati, D., Carminati, E., & Tava, A. (2019). Microalgal biostimulants and biofertilisers in crop productions. *Agronomy*, 9(4), 192.

Rossi, F., Potrafka, R. M., Pichel, F. G., & De Philippis, R. (2012). The role of the exopolysaccharides in enhancing hydraulic conductivity of biological soil crusts. *Soil Biology and Biochemistry*, 46, 33–40.

Round, F. E. (1973). *The biology of the algae* (2nd ed.). Edward Arnold Ltd.

Rouphael, Y., & Colla, G. (2018). Synergistic biostimulatory action: Designing the next generation of plant biostimulants for sustainable agriculture. *Frontiers in Plant Science*, 9, 1655.

Rzymski, P., Niedzielski, P., Karczewski, J., & Poniedziałek, B. (2014). Biosorption of toxic metals using freely suspended Microcystis aeruginosa biomass. *Open Chemistry*, 12(12), 1232–1238.

Saadatnia, H., & Riahi, H. (2009). Cyanobacteria from paddy fields in Iran as a biofertilizer in rice plants. *Plant, Soil and Environment*, 55(5), 207–212.

Sahu, D., Priyadarshani, I., & Rath, B. (2012). Cyanobacteria–as potential biofertilizer. *CIBTech Journal of Microbiology*, 1, 20–26.

Samhan, A. F. (2008). *Assessment of the ability of microalgae in removal of some industrial wastewater pollutants*: (pp. 1–16). (Doctoral dissertation, M. Sc. Thesis). Egypt: Department of Botany, Faculty of Science, Beni-Suef University.

Sathiyamoorthy, P., & Shanmugasundaram, S. (1996). Preparation of cyanobacterial peptide toxin as a biopesticide against cotton pests. *Applied Microbiology and Biotechnology*, 46, 511–513.

Savci, S. (2012). Investigation of effect of chemical fertilizers on environment. *APCBEE Procedia*, 1, 287–292.

Schiewer, S., & Volesky, B. (2000). Biosorption processes for heavy metal removal. In *Environmental microbe-metal interactions* (pp. 329–362). American Society of Microbiology.

Shaaban, A. M., Haroun, B. M., & Ibraheem, I. B. M. (2004). Assessment of impact of Microcystis aeruginosa and Chlorella vulgaris in the uptake of some heavy metals from culture media. In: *Proc. 3rd Int. Conf. Biol. Sci. Fac. Sci., Tanta Univ.*Vol. 3, (pp. 433–450). , pp. 433–450.

Sharma, H. S., Fleming, C., Selby, C., Rao, J. R., & Martin, T. (2014). Plant biostimulants: A review on the processing of macroalgae and use of extracts for crop management to reduce abiotic and biotic stresses. *Journal of Applied Phycology, 26*(1), 465–490.

Sharma, R., Khokhar, M. K., Jat, R. L., & Khandelwal, S. K. (2012). Role of algae and cyanobacteria in sustainable agriculture system. *Wudpecker Journal of Agricultural Research, 1*(9), 381–388.

Sharma, S. B., Sayyed, R. Z., Trivedi, M. H., & Gobi, T. A. (2013). Phosphate solubilizing microbes: Sustainable approach for managing phosphorus deficiency in agricultural soils. *Springerplus, 2*(1), 587.

Singh, R. N. (1961). *Role of blue-green algae in nitrogen economy of Indian agriculture.*

Singh, N. K., & Dhar, D. W. (2007). Nitrogen and phosphorous scavenging potential in microalgae. *Indian Journal of Biotechnology, 6*, 52–56.

Singh, N. K., & Dhar, D. W. (2010). Cyanobacterial reclamation of salt-affected soil. In *Genetic engineering, biofertilisation, soil quality and organic farming* (pp. 243–275). Dordrecht: Springer.

Singh, D. P., Prabha, R., Yandigeri, M. S., & Arora, D. K. (2011). Cyanobacteria-mediated phenylpropanoids and phytohormones in rice (Oryza sativa) enhance plant growth and stress tolerance. *Antonie Van Leeuwenhoek, 100*(4), 557–568.

Singh, V., & Singh, D. V. (2015). Cyanobacteria modulated changes and its impact on bioremediation of saline-alkaline soils. *Bangladesh Journal of Botany, 44*(4), 653–658.

Six, J., Bossuyt, H., Degryze, S., & Denef, K. (2004). A history of research on the link between (micro) aggregates, soil biota, and soil organic matter dynamics. *Soil and Tillage Research, 79*(1), 7–31.

Six, J., Elliott, E. T., & Paustian, K. (2000). Soil structure and soil organic matter II. A normalized stability index and the effect of mineralogy. *Soil Science Society of America Journal, 64*(3), 1042–1049.

Srivastava, A., & Mishra, A. K. (2014). *Regulation of nitrogen metabolism in salt tolerant and salt sensitive Frankia strains.*

Stiegler, J. C., Richardson, M. D., Karcher, D. E., Roberts, T. L., & Norman, R. J. (2013). Foliar absorption of various inorganic and organic nitrogen sources by creeping bentgrass. *Crop Science, 53*(3), 1148–1152.

Subhash, G. V., Rangappa, M., Raninga, S., Prasad, V., Dasgupta, S., & Kumar, G. R. K. (2019). Electromagnetic stratagem to control predator population in algal open pond cultivation. *Algal Research, 37*, 133–137.

Subramaniyam, V., Subashchandrabose, S. R., Thavamani, P., Chen, Z., Krishnamurti, G. S. R., Naidu, R., et al. (2016). Toxicity and bioaccumulation of iron in soil microalgae. *Journal of Applied Phycology, 28*(5), 2767–2776.

Svircev, Z., Tamas, I., Nenin, P., & Drobac, A. (1997). Co-cultivation of N2-fixing cyanobacteria and some agriculturally important plants in liquid and sand cultures. *Applied Soil Ecology, 6*(3), 301–308.

Swarnalakshmi, K., Dhar, D. W., Senthilkumar, M., & Singh, P. K. (2013). Comparative performance of cyanobacterial strains on soil fertility and plant growth parameters in rice. *Vegetos, 26*(2), 227–236.

Swarnalakshmi, K., Prasanna, R., Kumar, A., Pattnaik, S., Chakravarty, K., Shivay, Y. S., et al. (2013). Evaluating the influence of novel cyanobacterial biofilmed biofertilizers on soil fertility and plant nutrition in wheat. *European Journal of Soil Biology, 55*, 107–116.

Szabolcs, I. (1993). Soils and salinaization. M. Pessarakli (Ed.), *Handbook of plant and crop stress* (pp. 344–346). Vol. 32(pp. 344–346). New York: Marcel Dekker.

Tarakhovskaya, E. R., Maslov, Y. I., & Shishova, M. F. (2007). Phytohormones in algae. *Russian Journal of Plant Physiology, 54*(2), 163–170.

Terry, P. A., & Stone, W. (2002). Biosorption of cadmium and copper contaminated water by Scenedesmus abundans. *Chemosphere, 47*(3), 249–255.

Trejo, A., De-Bashan, L. E., Hartmann, A., Hernandez, J. P., Rothballer, M., Schmid, M., et al. (2012). Recycling waste debris of immobilized microalgae and plant growth-promoting bacteria from wastewater treatment as a resource to improve fertility of eroded desert soil. *Environmental and Experimental Botany, 75*, 65–73.

Tripathi, R. D., Dwivedi, S., Shukla, M. K., Mishra, S., Srivastava, S., Singh, R., et al. (2008). Role of blue green algae biofertilizer in ameliorating the nitrogen demand and fly-ash stress to the growth and yield of rice (Oryza sativa L.) plants. *Chemosphere, 70*(10), 1919–1929.

Triveni, S., Prasanna, R., Kumar, A., Bidyarani, N., Singh, R., & Saxena, A. K. (2015). Evaluating the promise of Trichoderma and Anabaena based biofilms as multifunctional agents in Macrophomina phaseolina-infected cotton crop. *Biocontrol Science and Technology, 25*(6), 656–670.

Udeigwe, T. K., Teboh, J. M., Eze, P. N., Stietiya, M. H., Kumar, V., Hendrix, J., et al. (2015). Implications of leading crop production practices on environmental quality and human health. *Journal of Environmental Management, 151*, 267–279.

Umamaheswari, A., Madhavi, D. R., & Venkateswarlu, K. (1997). Siderophore production in two species of Nostoc as influenced by the toxicity of nitrophenolics. *Bulletin of Environmental Contamination and Toxicology, 59*(2), 306–312.

Umemiya, Y., & Furuya, S. (2001). The influence of chemical. Forms on foliar-applied nitrogen absorption for peach trees. In: *International Symposium on Foliar Nutrition of Perennial Fruit Plants, Vol. 594*, pp. 97–103.

Uysal, O., Uysal, F. O., & Ekinci, K. (2015). Evaluation of microalgae as microbial fertilizer. *European Journal of Sustainable Development, 4*(2), 77–82.

Van Oosten, M. J., Pepe, O., De Pascale, S., Silletti, S., & Maggio, A. (2017). The role of biostimulants and bioeffectors as alleviators of abiotic stress in crop plants. *Chemical and Biological Technologies in Agriculture, 4*(1), 5.

Veronesia, D., Ida, A., & D'Imporzano, G. (2015). Microalgae cultivation: Nutrient recovery from digestate for producing algae biomass. *Chemical Engineer, 43*, 1201–1206.

Virto, I., Imaz, M. J., Fernández-Ugalde, O., Gartzia-Bengoetxea, N., Enrique, A., & Bescansa, P. (2015). Soil degradation and soil quality in Western Europe: Current situation and future perspectives. *Sustainability, 7*(1), 313–365.

Visconti, F., de Paz, J. M., Bonet, L., Jordà, M., Quinones, A., & Intrigliolo, D. S. (2015). Effects of a commercial calcium protein hydrolysate on the salt tolerance of Diospyros kaki L. cv."Rojo Brillante" grafted on Diospyros lotus L. *Scientia Horticulturae, 185*, 129–138.

Weinzierl, R., Henn, T., Koehler, P. G., & Tucker, C. L. (2007). *Microbial insecticides 1 agricultural entomology.* University of Illinois at Urbana-Champaign. ENY-275 IN081.

Weiss, T. L., Roth, R., Goodson, C., Vitha, S., Black, I., Azadi, P., et al. (2012). Colony organization in the green alga Botryococcus braunii (Race B) is specified by a complex extracellular matrix. *Eukaryotic Cell, 11*(12), 1424–1440.

Wu, Y., Rao, B., Wu, P., Liu, Y., Li, G., & Li, D. (2013). Development of artificially induced biological soil crusts in fields and their effects on top soil. *Plant and Soil, 370*(1-2), 115–124.

Wuang, S. C., Khin, M. C., Chua, P. Q. D., & Luo, Y. D. (2016). Use of Spirulina biomass produced from treatment of aquaculture wastewater as agricultural fertilizers. *Algal Research, 15*, 59–64.

Xiao, R., & Zheng, Y. (2016). Overview of microalgal extracellular polymeric substances (EPS) and their applications. *Biotechnology Advances, 34*(7), 1225–1244.

Xu, L., & Geelen, D. (2018). Developing biostimulants from agro-food and industrial by-products. *Frontiers in Plant Science, 9*, 1567.

Xu, Y., Rossi, F., Colica, G., Deng, S., De Philippis, R., & Chen, L. (2013). Use of cyanobacterial polysaccharides to promote shrub performances in desert soils: A potential approach for the restoration of desertified areas. *Biology and Fertility of Soils, 49*(2), 143–152.

Yadavalli, R., & Heggers, G. R. V. N. (2013). Two stage treatment of dairy effluent using immobilized Chlorella pyrenoidosa. *Journal of Environmental Health Science and Engineering, 11*(1), 36.

Yakhin, O. I., Lubyanov, A. A., Yakhin, I. A., & Brown, P. H. (2017). Biostimulants in plant science: A global perspective. *Frontiers in Plant Science, 7*, 2049 Zhang, J., Wang, X. and Zhou, Q., 2017. Co-cultivation of Chlorella spp and tomato in a hydroponic system. Biomass and Bioenergy, 97, pp.132-138.

Yandigeri, M. S., Yadav, A. K., Meena, K. K., & Pabbi, S. (2010). Effect of mineral phosphates on growth and nitrogen fixation of diazotrophic cyanobacteria Anabaena variabilis and Westiellopsis prolifica. *Antonie van Leeuwenhoek, 97*(3), 297–306.

Yang, J., Wang, F., Fang, L., & Tan, T. (2007). Synthesis, characterization and application of a novel chemical sand-fixing agent-poly (aspartic acid) and its composites. *Environmental Pollution, 149*(1), 125–130.

Yan-Gui, S., Xin-Rong, L., Ying-Wu, C., Zhi-Shan, Z., & Yan, L. (2013). Carbon fixation of cyanobacterial–algal crusts after desert fixation and its implication to soil organic carbon accumulation in desert. *Land Degradation & Development, 24*(4), 342–349.

Yanni, Y. G., & Osman, Z. H. (1990). Contribution of alkalization to rice growth, yield, N attributes and incidence of infestation with blast fungus Pyricularia oryzae under different fungicidal treatments. *World Journal of Microbiology and Biotechnology, 6*, 371–376.

Yilmaz, E., & Sönmez, M. (2017). The role of organic/bio–fertilizer amendment on aggregate stability and organic carbon content in different aggregate scales. *Soil and Tillage Research, 168*, 118–124.

Youssef, M. M. A., & Ali, M. S. (1998). Management of *Meloidogynein cognita* infecting cowpea by using some native blue green algae. *Journal of Pest Science, 71*, 15–16.

Yu, Q., Matheickal, J. T., Yin, P., & Kaewsarn, P. (1999). Heavy metal uptake capacities of common marine macro algal biomass. *Water Research, 33*(6), 1534–1537.

Zayadan, B. K., Matorin, D. N., Baimakhanova, G. B., Bolathan, K., Oraz, G. D., & Sadanov, A. K. (2014). Promising microbial consortia for producing biofertilizers for rice fields. *Microbiology, 83*(4), 391–397.

Zhang, J., Wang, X., & Zhou, Q. (2017). Co-cultivation of Chlorella spp and tomato in a hydroponic system. *Biomass and Bioenergy, 97*, 132–138.

Zhang, L., Yan, C., Guo, Q., Zhang, J., & Ruiz-Menjivar, J. (2018). The impact of agricultural chemical inputs on environment: global evidence from informetrics analysis and visualization. *International Journal of Low-Carbon Technologies, 13*(4), 338–352.

Zhu, K., Zhou, H., & Qian, H. (2006). Antioxidant and free radical-scavenging activities of wheat germ protein hydrolysates (WGPH) prepared with alcalase. *Process Biochemistry, 41*(6), 1296–1302.

Web references

EBIC (2019). *European Biostimulant Industry Council. Available online: (2019).* http://www.biostimulants.eu/. Accessed 30 November 2019.

Further reading

Ali, H., Khan, E., & Ilahi, I. (2019). Environmental chemistry and ecotoxicology of hazardous heavy metals: Environmental persistence, toxicity, and bioaccumulation. *Journal of Chemistry, 2019*, .

Arroussi, H. E., Benhima, R., Elbaouchi, A., Sijilmassi, B., Mernissi, N. E., Aafsar, A., et al. (2018). Dunaliella salina exopolysaccharides: A promising biostimulant for salt stress tolerance in tomato (Solanum lycopersicum). *Journal of Applied Phycology, 30*(5), 2929–2941.

Buchmann, L., & Mathys, A. (2019). Perspective on pulsed electric field treatment in the bio-based industry. *Frontiers in Bioengineering and Biotechnology, 7*, 265.

Cameron, R. E., & Devaney, J. R. (1970). Antarctic soil algal crusts: Scanning electron and optical microscope study. *Transactions of the American Microscopical Society*, 264–273.

Chris, A. (2014). *Sustainability "Only 60 Years of Farming Left If Soil Degradation Continues" December 5. (2014)*. https://www.scientificamerican.com/article/only-60-years-of-farming-left-if-soil-degradation-continues/.

Council, M.E.A (2005). *Ecosystems and human well-being: Desertification synthesis*: (pp. 1–22). Washington, DC: World Resources Institute.

Decho, A. W. (1990). Microbial exopolymer secretions in ocean environments: Their role (s) in food webs and marine processes. *Oceanography and Marine Biology, 28* (737153), 9–16.

Delbarre-Ladrat, C., Sinquin, C., Lebellenger, L., Zykwinska, A., & Colliec-Jouault, S. (2014). Exopolysaccharides produced by marine bacteria and their applications as glycosaminoglycan-like molecules. *Frontiers in Chemistry, 2*, 85.

Dominy, C., & Haynes, R. (2002). Influence of agricultural land management on organic matter content, microbial activity and aggregate stability in the profiles of two Oxisols. *Biology and Fertility of Soils, 36*(4), 298–305.

El-Zeky, M. M., El-Shahat, R. M., Metwaly, G. S., & Elham, M. A. (2005). Using Cyanobacteria or Azolla as alternative nitrogen sources for rice production. *Journal of Agricultural Science, Mansoura University, 30*(9), 5567–5577.

Falchini, L., Sparvoli, E., & Tomaselli, L. (1996). Effect of Nostoc (Cyanobacteria) inoculation on the structure and stability of clay soils. *Biology and Fertility of Soils, 23*(3), 346–352.

Grobbelaar, J. U. (1983). Availability to algae of N and P absorbed on suspended solids in turbid waters of the Amazon River. *Archiv für Hydrobiologie, 96*, 302–316.

Hamed, S. M. M. (2007). *Studies on nitrogen fixing Cyanobacteria.* (Doctoral dissertation M. Sc. Thesis Beni-Suef, Egypt: Botany Department, Faculty of Science, Beni-Suef University.

Healey, F. P. (1973). Characteristics of phosphorus deficiency in anabaena 1. *Journal of Phycology, 9*(4), 383–394.

Hu, C., Liu, Y., Paulsen, B. S., Petersen, D., & Klaveness, D. (2003). Extracellular carbohydrate polymers from five desert soil algae with different cohesion in the stabilization of fine sand grain. *Carbohydrate Polymers, 54*(1), 33–42.

Juttner, F., & Wu, J. T. (2000). Evidence of allelochemical activity in subtropical cyanobacterial biofilms of Taiwan. *Archiv für Hydrobiologie*, 505–517.

Kambourova, R., Petkov, G., & Bankova, V. (2006). Extracellular polar organic substances in cultures of the green alga Scenedesmus. *Algological Studies, 119*(1), 155–162.

Katznelson, R. (1989). Clogging of groundwater recharge basins by cyanobacterial mats. *FEMS Microbiology Ecology, 5*(4), 231–242.

Kheirfam, H., Sadeghi, S. H., Homaee, M., & Darki, B. Z. (2017). Quality improvement of an erosion-prone soil through microbial enrichment. *Soil and Tillage Research, 165*, 230–238.

Kidron, G. J., Yaalon, D. H., & Vonshak, A. (1999). Two causes for runoff initiation on microbiotic crusts: Hydrophobicity and pore clogging. *Soil Science, 164*(1), 18–27.

Koffi, K. T., Kumar, S., & Sur, D. H. (2018). Extraction of plant nutrients from freshwater algae and their role in sustainable agriculture. *International Journal of Current Biotechnology, 6*(4), 1–8.

Krings, M., Hass, H., Kerp, H., Taylor, T. N., Agerer, R., & Dotzler, N. (2009). Endophytic cyanobacteria in a 400-million-yr-old land plant: A scenario for the origin of a symbiosis? *Review of Palaeobotany and Palynology, 153* (1-2), 62–69.

Lund, J. W. G. (1962). Soil algae. In R. D. Lewin (Ed.), *Physiology and biochemistry of algae* (pp. 759–766). New York: Academic Press.

Marathe, K. V., & Chaudhari, P. R. (1975). An example of algae as pioneers in the lithosere and their role in rock corrosion. *The Journal of Ecology*, 65–69.

Mazor, G., Kidron, G. J., Vonshak, A., & Abeliovich, A. (1996). The role of cyanobacterial exopolysaccharides in structuring desert microbial crusts. *FEMS Microbiology Ecology, 21*(2), 121–130.

Metting, B. (1986). Population dynamics of Chlamydomonas sajao and its influence on soil aggregate stabilization in the field. *Applied and Environmental Microbiology, 51*(6), 1161–1164.

Metting, B. (1987). Dynamics of wet and dry aggregate stability from a three-year microalgal soil conditioning experiment in the field. *Soil Science, 143*(2), 139–143.

Michalak, I., Chojnacka, K., Dmytryk, A., Wilk, R., Gramza, M., & Rój, E. (2016). Evaluation of supercritical extracts of algae as biostimulants of plant growth in field trials. *Frontiers in Plant Science, 7*, 1591.

Mishra, A., & Jha, B. (2009). Isolation and characterization of extracellular polymeric substances from micro-algae Dunaliellasalina under salt stress. *Bioresource Technology, 100*(13), 3382–3386.

Narayana, D. V., & Babu, R. (1983). Estimation of soil erosion in India. *Journal of Irrigation and Drainage Engineering, 109*(4), 419–434.

Prasanna, R., Triveni, S., Bidyarani, N., Babu, S., Yadav, K., Adak, A., et al. (2014). Evaluating the efficacy of cyanobacterial formulations and biofilmed inoculants for leguminous crops. *Archives of Agronomy and Soil Science, 60*(3), 349–366.

Rao, D. L. N., & Burns, R. G. (1990). The effect of surface growth of blue-green algae and bryophytes on some microbiological, biochemical, and physical soil properties. *Biology and Fertility of Soils, 9*(3), 239–244.

Shields, L. M., & Durrell, L. W. (1964). Algae in relation to soil fertility. *The Botanical Review, 30*(1), 92–128.

Singh, S. (2014). A review on possible elicitor molecules of cyanobacteria: Their role in improving plant growth and providing tolerance against biotic or abiotic stress. *Journal of Applied Microbiology, 117*(5), 1221–1244.

Sutherland, I. W. (1996). A natural terrestrial biofilm. *Journal of Industrial Microbiology, 17*(3-4), 281–283.

Underwood, G. J. C. (1997). Microalgal colonization in a salt-marsh restoration scheme. *Estuarine, Coastal and Shelf Science, 44*(4), 471–481.

Velu, C., Cirés, S., Alvarez-Roa, C., & Heimann, K. (2015). First outdoor cultivation of the N 2-fixing cyanobacterium Tolypothrix sp. in low-cost suspension and biofilm systems in tropical Australia. *Journal of Applied Phycology, 27*(5), 1743–1753.

Zhou, Z., Cheng, Z., & Liu, Z. (1995). Study on the ecology of algae in surface crust of desert. *Acta Ecologica Sinica, 15*(4), 385–391.

Microalgae biofilms for the treatment of wastewater

Hassimi Abu Hasan, Siti Nur Hatika Abu Bakar, and Mohd Sobri Takriff

Department of Chemical & Process Engineering, Faculty of Engineering and Built Environment, Universiti Kebangsaan Malaysia, Bangi, Selangor, Malaysia

1 Introduction

Microalgae are categorized as eukaryotic or prokaryotic microorganisms that naturally grow in freshwater (pond, lake, river) or seawater. Microalgae are unicellular, colonial, and filamentous microorganisms. Structurally, microalgae contain a plasma membrane, a nucleus, chloroplasts, mitochondria, vacuoles, and starch grains. The chloroplast conducts photosynthesis through the absorption of carbon dioxide, light, nutrients, and water to produce new cells or biomass. The chloroplast is a unique characteristic that can be used in the identification of microalgae and has distinct forms such as spiral and spherical chloroplasts. Microalgae are often described by common names such as green algae (Chlorophyta), brown algae (Phaeophyta), golden-brown algae (Chrysophyta), and red algae (Rhodophyta) (Talaro, 2005).

There are many applications for microalgae such as commercial, industrial, and environmental applications. In commercial applications, microalgae have been used to produce cosmetic products (Arad & Yaron, 1992; Ariede et al., 2017; Khairuddin, Idris, & Irfan, 2019), as human nutrition (Koyande et al., 2019), and to produce the high-value molecules fatty acids, carotenoids, and phycobiliproteins (Sierra, Stoykoba, & Nikolov, 2018; Zhang, Grimi, Marchal, Lebovka, & Vorobiev, 2019; Zhang, Yu, et al., 2019). Regarding industrial and environmental applications, microalgae have mostly been used for pharmaceutical products (Manirafasha, Ndikubwimana, Zeng, Lu, & Jing, 2016; Li, Liu, et al., 2019; Li, Tong, Zhou, Huang, & Wang, 2019), animal feed (D'Este, Morales, & Angelidaki, 2017; Dineshbabu, Goswami, Kumar, Sinha, & Das, 2019; Yaakob, Ali, Zainal, Mohamad, & Takriff, 2014), biofertilizer (Renuka, Guldhe, Prasanna, Singh, & Bux, 2018; Sharma et al., 2019), biofuel (Hossain et al., 2020; Japar, Takriff, & Yasin, 2017; Shurtz, Wood, & Quinn, 2017; Wu, Lin, & Chang, 2018), bioremediation agents for the removal of pollutants in water and wastewater (Ding, Yaakob, Takriff, Salihon, & Rahaman, 2016; Hazman et al., 2018; Khalid, Yaakob,

Abdullah, & Takriff, 2019; Udaiyappan, Hasan, Takriff, & Abdullah, 2017; Udaiyappan et al., 2020) and carbon dioxide (CO_2) sequestration (Hariz, Takriff, Yasin, Ba-Abbad, & Hakimi, 2019; Verma & Srivastava, 2018; Xu et al., 2017).

In recent years, microalgae have been widely studied for use in environmental applications, especially as bioremediation agents. This process is known as phycoremediation. The treatment of water and wastewater using microalgae can be performed in four modes of growth i.e., photoautotrophic, mixotrophic, heterotrophic, and photoheterotrophic (Wayne et al., 2018). Photoautotrophic microalgae require light and CO_2. If there are other organic carbon sources in the wastewater, these organisms favor CO_2 as a carbon source. Mixotrophic microalgae consume carbon from CO_2 or wastewater and can grow under photoautotrophic or heterotrophic conditions where light intensity and organic compounds affect growth conditions (Kong et al., 2012; Wayne et al., 2018). Heterotrophic microalgae favor only organic carbon sources from wastewater in limited light conditions or the absence of light, while photoheterotrophic microalgae favor organic carbon sources from wastewater in the presence of light.

Advantages of the application of microalgae in wastewater treatment are: less production of activated sludge compared to bacteria-based wastewater treatment, a low energy requirement for growth, potential reduction of greenhouse gases through CO_2 sequestration, and algae biomass can be used to produce valuable by-products such as biofertilizers, biofuels, and bioactive compounds (Udaiyappan et al., 2017). Without knowledge and proper maintenance, the application of microalgae has the potential to be disastrous for an operator instead of providing benefits. The drawbacks and challenges of using microalgae are difficult to separate from the water and wastewater, the turbidity of wastewater will prevent the light to penetrate thus inhibiting the growth (Udaiyappan et al., 2017), and algal bloom

problems. In this book chapter, various subtopics are discussed including species of microalgae involved in wastewater treatment, affecting factors of microalgae biofilms formation and adhesion, microalgae biofilms reactors, and reactor performance including growth and removal kinetics.

2 Species of microalgae applicable for wastewater treatment

Many microalgae species, including species of freshwater and marine microalgae, have been used to remediate various types of wastewater. Among these classes of microalgae, freshwater microalgae have been used most often due ease of growth, in addition to a lower salinity requirement compared to marine microalgae. *Scenedesmus*, *Chlorella*, *Desmodesmus*, and *Chlamydomonas* are favorable genera used in the treatment of wastewater. Fig. 1 depicts examples of each of these genera, while Table 1 shows details application of microalgae in wastewater treatment.

Scenedesmus is a genus in the class Chlorophyceae. This genus consists of species of green microalgae with 4, 8, 16, or 32 cells arranged in a row (Encyclopaedia Britannica, 2015). A strain of *S. obliquus* was investigated for the treatment of leachate, food processing wastewater, aquaculture wastewater, and piggery wastewater. Strains of *Scenedesmus* species are utilized for the treatment of a wide variety of industrial forms of wastewater such as palm oil mill effluent (POME), power plant wastewater, textile, sewage, tannery, dairy, swine, and pulp and paper mill wastewater. Specific species such as *S. quadricauda*, *S. platensis*, and *S. abundans* show potential for the remediation of dairy, piggery, and food processing wastewater, respectively.

Chlorella is a genus of green, single-celled microalgae under the phylum Chlorophyta and class Trebouxiophyceae. This genus has a high capacity for photosynthesis. The species

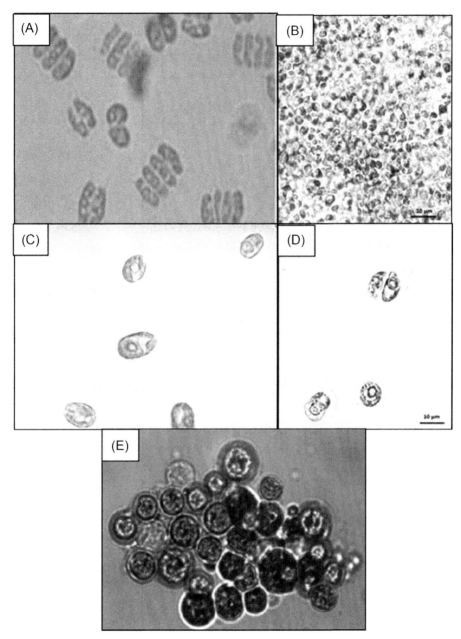

FIG. 1 (A) *Scenedesmus*, (B) *Chlorella*, (C) *Desmodesmus*, (D) *Chlamydomonas*, (E) *Neochloris* ((A) Vidyashankar et al., 2015; (B and D) Mondal et al., 2016; (C) Rios, Kleina, Luz Jrb, Maciela, & Filho, 2015; (E) Gopalakrishnan & Ramamurthy, 2014).

TABLE 1   Species of microalgae utilized as bioremediation agents.

Species	Wastewater type	Reference
Scenedesmus		
Scenedesmus obliquus	Leachate wastewater	Hernández-García et al. (2019)
	Food processing wastewater	Gupta and Pawar (2018)
	Aquaculture wastewater	Gao et al. (2016)
	Piggery wastewater	Wang et al. (2016)
Scenedesmus sp.	Palm oil mill effluent	Hariz et al. (2019)
	Power plant wastewater	Mohammadi, Mowla, Esmaeilzadeh, and Ghasemi (2018)
	Textile wastewater	Kumar, Huy, Bakonyi, Bélafi-Bakó, and Kim (2018)
	Sewage wastewater	González-Camejo et al. (2017)
	Tannery wastewater	da Fontoura, Rolim, Farenzena, and Gutterres (2017)
	Dairy wastewater	Labbe, Ramos-Suárez, Hernández-Pérez, Baeza, and Hansen (2017)
	Swine wastewater	Prandini et al. (2016)
	Pulp and paper mill	Usha, Chandra, Sarada, and Chauhan (2016)
Scenedesmus quadricauda	Dairy wastewater	Daneshvar, Zarrinmehr, Hashtjin, Farhadian, and Bhatnagar (2018)
Scenedesmus platensis	Piggery wastewater	Wang et al. (2016)
Scenedesmus abundans	Food processing wastewater	Gupta and Pawar (2018)
Chlorella		
Chlorella vulgaris	Electroplating wastewater	Ajitha et al. (2019)
	Municipal wastewater	Mujtaba and Lee (2017)
	Industrial wastewater	Al-Momani and Örmeci (2016)
	Aquaculture wastewater	Gao et al. (2016)
	Piggery wastewater	Franchino, Tigini, Varese, Sartor, and Bona (2016) and Wang et al. (2016)
	Saline wastewater	Shen et al. (2015)
	Urban wastewater	Caporgno et al. (2015)
	Swine wastewater	Wang et al. (2015)
Chlorella sp.	Palm oil mill effluent	Hariz et al. (2019)
	Power plant wastewater	Mohammadi et al. (2018)

TABLE 1 Species of microalgae utilized as bioremediation agents—cont'd

Species	Wastewater type	Reference
	Textile wastewater	Kumar et al. (2018)
	Sewage wastewater	González-Camejo et al. (2017)
	Dairy wastewater	Labbe et al. (2017)
Chlorella sorokiniana	Piggery wastewater	Leite, Hoffmann, and Daniel (2019)
Chlorella sorokiniana L3	Fermentation wastewater	Qi et al. (2018)
	Food processing wastewater	Gupta and Pawar (2018)
C. pyrenoidosa	Piggery wastewater	Wang et al. (2016)
Chlorella kessleri	Urban wastewater	Caporgno et al. (2015)
Desmodesmus		
Desmodesmus sp.	Leachate wastewater	Hernández-García et al. (2019)
Desmodesmus sp.	Domestic wastewater	Eze, Orta, García, Ramírez, and Ledesma (2018)
Desmodesmus sp.	Sewage wastewater	Mehrabadi, Farid, and Craggs (2017)
Chlamydomonas		
Chlamydomonas sp.	Power plant wastewater	Mohammadi et al. (2018)
Chlamydomonas incerta	Palm oil mill effluent	Kamyab et al. (2015)
Other		
Fischerella sp.	Power plant wastewater	Mohammadi et al. (2018)
Haematococcus pluvialis	Piggery wastewater	Wang et al. (2016)
Micractinium pusillum	Sewage wastewater	Mehrabadi et al. (2017)
Mucidosphaerium pulchellum	Sewage wastewater	Mehrabadi et al. (2017)
Nannochloropsis oculata	Urban wastewater	Caporgno et al. (2015)
Neochloris oleoabundans	Industrial wastewater	Al-Momani and Örmeci (2016)
Oocystis sp.	Power plant wastewater	Mohammadi et al. (2018)
Pediastrum boryanum	Sewage wastewater	Mehrabadi et al. (2017)
Porphyridium cruentum	Piggery wastewater	Wang et al. (2016)
Tribonema sp.	Petrochemical	Huo et al. (2019)
Coelastrella sp.	Swine wastewater	Li et al. (2018)
Coelastrum sp.	Sewage wastewater	Mehrabadi et al. (2017)
(*Tetraselmis suecica*, Ts)	Dairy wastewater	Daneshvar et al. (2018)
Aphanothece microscopica	Rice parboiling wastewater	Bastos, Bonini, Zepka, Lopes, and Queiroz (2015)

of *Chlorella* that have been used for the bioremediation are *Chlorella vulgaris, Chlorella sorokiniana, Chlorella pyrenoidosa,* and *Chlorella kessleri. Chlorella vulgaris* has been widely applied to treat various kinds of wastewater such as electroplating wastewater, piggery wastewater, aquaculture wastewater, and municipal wastewater. Another study demonstrated the use of *Chlorella* species for the treatment of power plant wastewater, POME, textile wastewater, sewage wastewater, and dairy wastewater. A report by Bich, Yaziz, and Bakti (1999) showed that *Chlorella vulgaris* could remove 93.4% chemical oxygen demand (COD) and 79.3% total Kjeldahl nitrogen in rubber latex processing wastewater. Additional findings by Pathak, Kothari, Chopra, and Singh (2015) demonstrated that the cultivation of *C. pyrenoidosa* in textile wastewater could reduce the nitrate, phosphate, and biochemical oxygen demand (BOD) by up to 82%, 87%, and 63%, respectively.

Besides, the genus *Desmodesmus* has also been studied for the treatment of leachate and domestic wastewater. Other species of microalgae that showed potential for the treatment of wastewater were *Haematococus pluvialis* (piggery wastewater), *Coelastrella* sp. (swine wastewater), *Aphanothece microscopica* (rice parboiling wastewater), *Oocystis* sp. (power plant wastewater), and *Tribonema* sp. (petrochemical wastewater).

3 Microalgae biofilms

3.1 Definition

Biofilms are communities of microorganisms that adhere to surfaces due to sticky binding by an extracellular polymeric substance (EPS) matrix which is produced by the cells. Microalgae biofilms are composed of a consortium of microalgae (including cyanobacteria) in addition to heterotrophic microorganisms (bacteria, fungi, and protozoa) and other microorganisms which are embedded in EPS matrix (Mantzorou & Ververidis, 2019), as shown in

Fig. 2. According to Choudhary, Prajapati, Kumar, Malik, and Pant (2017), microalgae biofilms are also known as autotrophic biofilms. The five processes involved in microalgae biofilm development and formation are: (1) the presence of microalgae and other microorganisms, (2) the initial adhesion onto a surface via adsorption, (3) adhesion via EPS formation, (4) biofilm development, (5) biofilm maturation, and (6) loss of biofilm integrity.

3.2 Characterization of microalgae biofilms

3.2.1 Chemical composition

Microalgae biofilms consist of various forms of chemical compositions which are influenced by a consortium of microalgae, bacteria, fungi, protozoa, and other microorganisms. In general, microalgae biofilms contain carbon, nitrogen, and phosphorus. According to de Assis, Calijuri, do Couto, and Assemany (2017), microalgae biofilms used for the treatment of domestic sewage treatment contained carbon, nitrogen, phosphorus, neutral lipids, and proteins. A species of *S. obliquus* that grew without a CO_2 supplement contained only 3.8% nitrogen and 0.83% phosphorus (Zamalloa, Boon, & Verstraete, 2013). In an additional study by de Assis et al. (2019), microalgae biofilms adhered to cotton, nylon, and polyester surfaces that contained proteins (23.8%–25.6%), carbohydrates (20.7%–21.1%), and neutral lipids (3.7%–10.6%).

3.2.2 Surface structures

Surface structures of microalgae biofilms can be observed using a traditional microscope, a scanning electron microscope (SEM), or a transmission electron microscope (TEM). For observation under SEM, a sample of microalgae biofilm needs to be fixed in glutaraldehyde and a phosphate buffer before dehydration using serial ethanol concentrations of 30%, 50%, 70%, 90%, and 100% (v/v). Fig. 3(A) shows the diversity of *Chlorella* (CL) and *Phormidium* sp. (PH) along with EPS after 6 days of growth

Microalgal biofilm formation

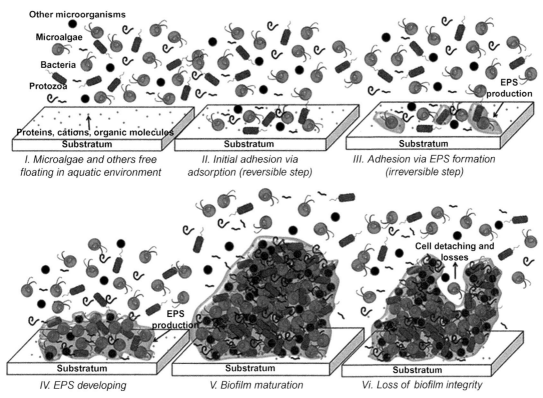

FIG. 2 Microalgae biofilms. *Reproduced from Mantzorou, A., Ververidis, F. (2019). Microalgal biofilms: A further step over current microalgal cultivation techniques.* Science of the Total Environment, 651(2), 3187–3201.

(Choudhary et al., 2017). Fig. 3(B), clearly shows the growth of PH and *Oscillatoria* sp., and *S. obliquus* on the biofilm matrix (Zamalloa et al., 2013), while the TEM micrograph clearly shows that the microalgae were surrounded by EPS matrix as indicated by the arrows in Fig. 3(C) (Choudhary et al., 2017).

3.3 Factors affecting microalgae biofilm formation and adhesion

A study by Irving and Allen (2011), demonstrated that there was a lack of information regarding the factors affecting microalgae formation and adhesion. However, as time

progressed, many studies have been conducted on microalgae biofilms to improve the quality of industrial wastewater treatment. As shown in Fig. 4, some of these factors affecting microalgae formation and adhesion are physicochemical properties of microalgae and surface materials, nutrients, light availability, pH, CO_2, temperature, other microorganisms, and biofilm carriers.

3.3.1 Surface physicochemical properties of microalgae

Surface physicochemical properties of microalgae influence the formation and adhesion of microalgae biofilms to surface materials. The surface properties of the microorganisms can

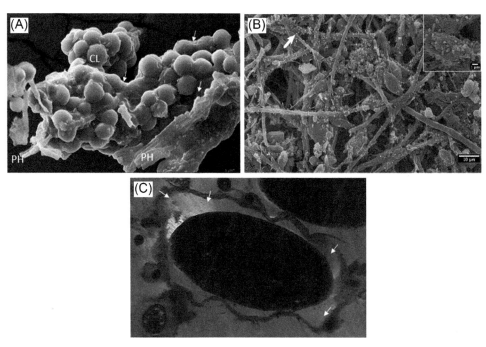

FIG. 3   Micrograph of microalgae biofilm (A) SEM image of *Chlorella* sp. and *Phormidium* sp. biofilms (Choudhary et al., 2017), (B) *Phormidium* sp., *Oscillatoria* sp. and *S. obliquus* biofilms (Zamalloa et al., 2013), and (C) TEM image of microalgae biofilms (Choudhary et al., 2017).

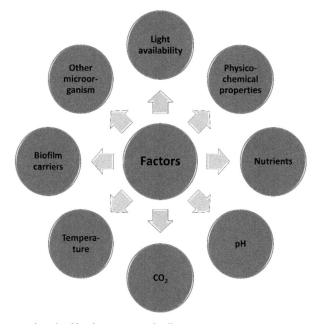

FIG. 4   Factors affecting microalgae biofilm formation and adhesion.

be measured through the absolute degree of hydrophobicity of any surface in comparison with their interaction with water as demonstrated by Van (1995). The degree of hydrophobicity of microalgae between surface materials was stronger if the $\triangle G_{sws}^{TOT} < 0$, but the surface of microalgae was hydrophilic if the $\triangle G_{sws}^{TOT} > 0$. The surface properties of *Pseudokirchneriella subcapitata* and *Chlorella vulgaris* were hydrophilic where $\triangle G_{sws}^{TOT}$ was reported at 13.8 and 52.5 mJ m^{-2} (Gonçalves, Ferreira, Loureiro, Pires, & Simões, 2015), and 24.5 ± 14.4 and 28.1 ± 13.0 mJ m^{-2} (Barros, Goncalves, & Simões, 2019), respectively. At the lowest $\triangle G_{sws}^{TOT}$, the formation and adhesion of microalgae biofilms on the surface of materials were easier.

3.3.2 Surface physicochemical properties of adhesion materials

Surface properties of materials influence growth and formation rates of microalgae biofilms. According to Schnurr and Allen (2015), an understanding of the properties of materials is critical to predict the formation and growth of biofilms. The hydrophobic properties of surface materials are ideal for biofilm adhesion. Similar to microalgae, material surfaces with $\Delta G_{sws}^{TOT} < 0$ will allow for easier formation and adhesion of

biofilms. Table 2 shows the $\triangle G_{sws}^{TOT}$ values as obtained by various researchers (Barros et al., 2019; Oliveira et al., 2001; Teixeira et al., 2005; Van, 2005)

Irving and Allen (2011) showed that the hydrophobicity interaction between microalgae and materials was weakly correlated in the adhesion process. Barros et al. (2019) reported that the adhesion of microalgae on surfaces could be predicted through the free energy of adhesion ($\triangle G_{adhesion}$). The $\triangle G_{adhesion}$ of the microalgae on the surface of materials is shown in Table 3. The adhesion of biofilms is favored if the $\triangle G_{adhesion} < 0$ and vice versa. From Table 3, as reported by Barros et al. (2019), *P. subcapitata*, and *Chlorella vulgaris* were expected to adhere to the surface of copper and glass. Besides, other properties that have been shown to affect the adhesion of microalgae biofilms are surface roughness, and texture (Cao et al., 2009; Katarzyna, Sai, & Singh, 2015; Sekar, Venugopalan, Satpathy, Nair, & Rao, 2004).

3.3.3 Media composition

Media composition is a crucial factor for microalgae growth and biofilm formation. The media should have sufficient nutrients such as carbon, nitrogen, and phosphorus. In the

TABLE 2 The ΔG_{sws}^{TOT} of surface materials.

Materials	ΔG_{sws}^{TOT} (mJ m^{-2})			
	Barros et al. (2019)	Teixeira, Lopes, Azeredo, Oliveira, and Vieira (2005)	Van (2005)	Oliveira, Azeredo, Teixeira, and Fonseca (2001)
Polystyrene	−18.5	–	−29.3	–
Poly(methyl methacrylate)	−19.2	−18.9	–	–
Polyvinyl chloride	−22.6	–	–	−22.0
Glass	−22.6	−13.8	–	–
Copper	−62.1	−79.6	–	–
Stainless steel	12.3	–	–	–

TABLE 3 $\Delta G_{adhesion}$ of microalgae on the surface of materials.

Materials	$\Delta G_{adhesion}$ (mJ m^{-2})	
	Pseudokirchneriella subcapitata	*Chlorella vulgaris*
Polystyrene	11.4	9.4
Poly(methyl methacrylate)	10.3	10.1
Polyvinyl chloride	7.9	7.9
Glass	−10.5	−8.8
Copper	−14.7	−11.3
Stainless steel	21.1	22.1

present day, many researchers use wastewater as media for microalgae growth. The abundance of wastewater produced (which contains beneficial nutrients) is an ideal alternative to commercial media such as Bold Basal Medium (BBM). The composition of the media will influence the secretion of EPS which subsequently influences the adhesion of microalgae (Bellinger, Abdullahi, & Gretz, 2005). In a heterotrophic-assisted photoautotrophic biofilm (HAPB) of *Chlorella vulgaris*, the optimum media composition of total inorganic carbon (CO_2) to total organic carbon (glucose) was 20:1 and the molar ratio of total carbon to total nitrogen was 72:1 (Ye et al., 2018).

3.3.4 CO$_2$

Phototrophic and mixotrophic microalgae require CO_2 as a carbon source for metabolism. However, photoheterotrophic and heterotrophic microalgae can obtain a supply of carbon from wastewater or water containing organic carbon. Suspended microalgae required approximately 5%–7% (v/v) CO_2 for growth (Yun, Park, & Yang, 1996), while microalgae biofilm will need a higher concentration of CO_2 to cross the film (Schnurr & Allen, 2015). Previous reports showed that a *Chlorella vulgaris* biofilm required approximately 10% (v/v) CO_2 for growth (Huang et al., 2016).

3.3.5 Light availability/intensity

Microalgae biofilms may have the challenge to adapt to light intensity. Microalgae cells on the surface of a biofilm may be inhibited due to excessive light, while cells far from the surface may be light-limited (Murphy, Macon, & Berberoglu, 2012). The cells begin to be photo-limited at a biofilm thickness of 20 µm and become dark at a thickness of 200–350 µm (Cheng, Wang, & Liu, 2014; Li et al., 2016; Zhuang et al., 2018). As summarized by Zhuang et al. (2018), the photon flux density of the microalgae biofilm culture ranges from 60 to 300 µmol m^{-2}s^{-1} for different microalgae species and cultures. A strategy is required to optimize the light intensity to get more diversity within microalgae biofilms. The light intensity can be strategized through light dilution. As was reported by Zhuang et al. (2018), two light dilutions were strategized by placing the algal biofilm in vertical, multiple planar arrays, and intermittently illuminating the biofilm with an appropriate frequency to produce time-averaged irradiation around 100–150 µmol (Wang et al., 2016). The intensity of light penetrating the microalgae biofilm also depended on the level of wastewater turbidity or color. High turbidity wastewater required high light intensity and vice versa.

3.3.6 The presence of indigenous microorganisms

The presence of indigenous microorganisms in industrial wastewater such as bacteria will facilitate the formation of microalgae biofilms. Researches have shown that there are symbiotic relationships between bacteria and microalgae (Boelee, Temmink, Janssen, Buisman, & Wijffels, 2014; Church et al., 2018; Holmes, 1986; Katam & Bhattacharyya, 2019; Schnurr & Allen, 2015; Wang et al., 2020; Yang et al., 2018). During the growth process, microalgae release O_2 which is consumed by bacteria during respiration. Then the CO_2 released by bacteria will be used as a carbon source for the microalgae during photosynthesis. The presence of bacteria which produces the EPS will condition the surface of the material for initial microalgae biofilm formation. Previous studies have reported that the presence of bacteria affected the thickness of *Chlorella vulgaris* (Irving &

Allen, 2011) and the growth of *Botryococcus braunii* (Rivas, Vargas, & Riquelme, 2010), *Nitzschia paleacea,* and *Cylindrotheca fusiformis* (Mieszkin, Callow, & Callow, 2013). Example of the symbiotic interaction for the industrial wastewater treatment are symbiotic between effective microorganism-*Chlorella* sp. (Lananan et al., 2014), activated sludge-*Scenedesmus* sp. 336 (Chen, Hu, Qi, Song, & Chen, 2019), and yeast-*Scenedesmus* sp. (Laura et al., 2019)

4 Microalgae biofilm photobioreactors

The design of a microalgae biofilm photobioreactor (PBR) depends on whether the biofilm carriers are permanently submerged, periodically submerged, or nonsubmerged as summarized in Fig. 5. Examples of permanently submerged carriers are flat-plate biofilm PBRs, turf scrubbers, closed biofilm PBRs, while

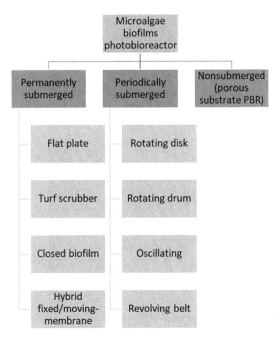

FIG. 5   Categories of microalgae biofilm photobioreactors.

examples of periodically submerged carriers are rotating disk PBRs, rotating drum PBRs, oscillating PBRs, and revolving belt PBRs.

4.1 Flat-plate microalgae biofilm photobioreactors

In the last decade, microalgae biofilm photobioreactors (MBPBRs) did not receive much intention from researchers. However, during the past 5 years, this technology has received more interest. MBPBRs are an innovation for the growth of microalgae and wastewater treatment. In contrast to conventional PBRs, the microalgae in MBPBRs grow and adhere to the material surface supported by EPS secretion by other microorganisms such as bacteria. The

advantages of the MBPBR is that the dewatering requirement can be completely omitted while achieving maximum extraction efficiency (Choudhary et al., 2017), great mass transfer of CO_2 to the microalgae (Gross, 2015), high solid retention time over hydraulic retention time (SRT/HRT) ratio for high-efficiency wastewater treatment, biofilm detachment/regrowth phases which allows for serial growth/harvesting cycles (Osorio et al., 2019), time-saving, and low costs during biomass harvesting (Guzzon, Di Pippo, & Congestri, 2019).

The MBPBR can be designed to be vertical or horizontal as shown in Fig. 6, where flow direction is a key to the design. The operational mode of the MBPBR can either be batch, continuous, or semibatch. The biofilm carriers that have been

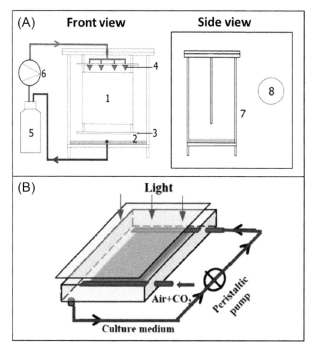

FIG. 6   Schematic of MBPBR (A) vertical MBPBR: (1) cotton carrier (for attachment of microalgae), (2) synthetic wastewater, (3) carrier holder, (4) dropper, (5) reservoir of synthetic growth medium, (6) pump, (7) PBR side view, and (8) LED lamp. (B) Horizontal MBPBR. *(A) Reproduced from Osorio, J.H.M., Pinto, G., Pollio, A., Frunzo, L., Lens, P.N.L., Esposito, G. (2019). Start-up of a nutrient removal system using Scenedesmus vacuolatus and Chlorella vulgaris biofilms.* Bioresources and Bioprocessing, *6, 27; (B) Adapted from Cheng, P., Wang, J., Liu, T. (2014). Effects of nitrogen source and nitrogen supply model on the growth and hydrocarbon accumulation of immobilized biofilm cultivation of B. braunii.* Bioresource Technology, *166, 527–533.*

used for microalgae biofilm adhesion are concrete slabs, foamed-glass beads, grid plastics, cotton cloth or fabrics, glass fiber, gauze, and baffle, and glass plates (Table 4). Some researchers have used suspended carriers in photobioreactors which are known as microalgae biofilm airlift photobioreactors (Tao et al., 2017). An additional study demonstrated that

TABLE 4 Conditions of microalgae biofilm photobioreactors.

Wastewater	Microalgae species	Biofilms carrier	Light intensity	HRT	Pollutants	References
Treated sewage	*Chlorella vulgaris*	C-1 (fiber) C-2 (plastic) C-3 (terylene)	$120.8\,\mu mol\,m^{-2}\,s^{-1}$	10 days	Dissolved inorganic nitrogen: ~62% dissolved inorganic phosphorus: ~71%	Tao et al. (2017)
Municipal wastewater	*Stenocranus acutus* *Pseudanabaena* sp. *Phormidium autumnale*	Concrete slab $(1 \times 2 \times 0.02\,m)$	$90 \pm 15\,\mu mol\,m^{-2}\,s^{-1}$	1 day	TP: ~97%	Sukačová, Trtílek, and Rataj (2015)
Industrial wastewater	*Chlorella sorokiniana* strain 211/8k	Poraver foamed-glass beads	$180\,Em^{-2}\,s^{-1}$	–	BOD: 74-100%	Muñoz, Köllner, and Guieysse (2009)
Hog manure wastewater	*Chlorella vulgaris* UTEX 2714	Grid plastic, gauze +baffle, cotton cloth, glass fiber, and a standard cotton cloth	$200 \pm 10\,\mu mol\,m^{-2}\,s^{-1}$	5 days	COD: 95.67% TP: 64.40% TN: 69.55% NH_4^+-N: 91.24%	Cen et al. (2019)
Municipal wastewater	*Scenedesmus rubescens* and *Chlorella vulgaris*	Bioreactor wall	$129.5\,\mu mol\,m^{-2}\,s^{-1}$ (7000 lux)	8 days	PO_4^{3-}-P: 98% NH_4^+-N: ~73%	Su, Mennerich, and Urban (2016)
Synthetic wastewater	*Scenedesmus vacuolatus* ACUF_053 and *Chlorella vulgaris* ACUF_809	Cotton fabric $(0.10 \times 0.10\,m)$	$90\,\mu mol\,m^{-2}\,s^{-1}$	1 day	PO_4^{3-}-P: 100%	Osorio et al. (2019)
Swine Wastewater	*Chlorella pyrenoidosa*	Glass plate and attached algal biofilm disks	$100\,\mu mol\,m^{-2}\,s^{-1}$	8 days	NH_4^+-N: 75.9% TP: 68.4% COD: 74.8% Zn^{2+}: 65.7% Cu^{2+}: 53.6% Fe^{2+}: 58.9%	Cheng, Wang, Liu, and Defu Liu (2017)

the photobioreactor wall could also be used as a surface material for biofilm adhesion (Su et al., 2016). Various types of wastewater were efficiently treated by MBPBR, for example, swine wastewater (Cheng et al., 2017), municipal wastewater (Sukačová et al., 2015), hog manure wastewater (Cen et al., 2019), and treated sewage (Tao et al., 2017).

In a study by Tao et al. (2017), fiber carriers were used for *Chlorella vulgaris* growth and adhesion in the removal of dissolved inorganic nitrogen (DIN) and dissolved inorganic phosphorous (DIP) from treated sewage. The photobioreactor was continuously airlifted at hydraulic loading of $0.1 \, \text{day}^{-1}$ where the carriers suspend and move by assisting aeration and fluid circulation. The aeration rate required to circulate the fiber carriers was approximately $1.5 \, \text{L} \, \text{min}^{-1}$. In the study, the volumetric biomass productivity at the end of the cultivation was $15.9 \, \text{mg} \, \text{L}^{-1} \, \text{day}^{-1}$ with removal performances of DIN and DIP at 62% and 71%, respectively. Sukačová et al. (2015) operated the MBPBR continually with artificial and real municipal wastewater cultivated with cyanobacteria (*Pseudanabaena* sp., *Phormidium autumnale*) and microalgae (*Stenocranus acutus*). The MBPBR was designed with recirculation flow. The test using artificial wastewater operated under continual artificial light intensity at $90 \pm 15 \, \mu\text{mol} \, \text{m}^{-2} \, \text{s}^{-1}$ while for the real wastewater test, under a regime of day-night, sunlight up to an intensity of $833 \, \mu\text{mol} \, \text{m}^{-2} \, \text{s}^{-1}$ was supplied to the microalgae and cyanobacteria with a regime of 12 h of light and 12 h of darkness. The results of the test showed that the MBPBR did not perform well for total phosphorus (TP) removal in real municipal wastewater under sunlight radiation. In contrast, TP in artificial wastewater was removed up to 97% and the effluent of $0.082 \, \text{mg} \, \text{L}^{-1}$ within only 1 day HRT.

In a study by Cheng et al. (2017), a horizontal MBPBR designed with a glass plate and attached biofilm disks containing *C. pyrenoidosa* which was illuminated at $100 \, \mu\text{mol} \, \text{m}^{-2} \, \text{s}^{-1}$ for 8 days

performed well for real swine wastewater treatment containing various constituents such as $NH_4^+ - N$, TP, metals (Zn, Cu, Fe), and other organic compounds. The innovation of conventional PBR operation through growth and adhesion of microalgae biofilms on the bioreactor wall would be one of the alternatives in designing the MBPBR. However, the problem would arise to get the high surface area for the biofilm attachment. Comparison of a conventional PBR with and without a microalgae biofilm (*Scenedesmus rubescens* and *Chlorella vulgaris*) showed that the bioreactor with microalgae biofilm on its wall was better at removing $PO_4^3 - P$ and $NH_4^+ - N$, where 98% and 73% of the removal was achieved, respectively (Su et al., 2016).

4.2 Rotating microalgae biofilm photobioreactor

A rotating microalgae biofilm photobioreactor (RABPBR), otherwise known as a rotating algal biofilm (RAB) is similar in principle to a rotating biological contactor. The RAB is categorized into four types i.e., disk, drum, oscillating, and revolving. The RAB is a rotating biological contactor (RBC) concept. Even though the RBC is widely used for bacterial biofilms in the treatment of wastewater, the technology is still limited for microalgae biofilms. Very few articles have been published on the RAB. Studies have been conducted by Woolsey (2011), Christenson and Sims (2012), Orandi, Lewis, and Moheimani (2012), Gross, Henry, Michael, and Wen (2013), Blanken et al. (2014), Sebestyén et al. (2016), Shayan, Agblevor, Bertin, and Sims (2016), Zhao, Kumar, Gross, Kunetz, and Wen (2018), Bara, Bonnefond, and Bernard (2019), and Lamare et al. (2019).

The RAB consists of a set of rotating disks mounted on a horizontal shaft as shown in Fig. 7. The disks are partially submerged in a tank containing wastewater, and the other parts are exposed to the air. The disks are rotated by a

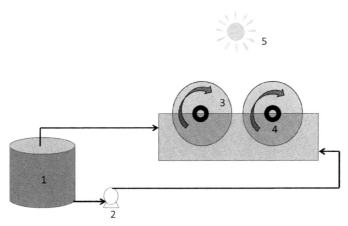

FIG. 7 Schematic of the RAB photobioreactor (RABPBR); (1) Feeding tank, (2) feeding pump, (3) disk/drum, (4) mounted shaft, (5) sunlight.

motor or a compressed air assembly with a shaft. The disks provide a surface area for the microalgae biofilm growth, and also provide contact between the wastewater and the air (Patwardhan, 2003). Various studies have been conducted to investigate the feasibility of the RAB, either for algal cultivation or wastewater treatment. Some studies focused on the type of surface material, hydraulic retention time, temperature and organic carbon concentration effect, the gradient of microalgae exposure to light, mathematical modeling, and upscaling of the RAB.

The advantages of microalgae RAB design compared to other microalgae biofilm PBRs are that it provides for a low ratio between footprint and cultivation surface, a simple disk rotation with frequent contact of microalgae biofilm with wastewater, the regulation of the intensity of light per disk by manipulating the size of the disk and distances between disks, and a large biofilm area (Blanken et al., 2014; Orandi et al., 2012; Wijffels & Barbosa, 2010). The RBC offers a low microalgae harvesting cost and produces a high microalgae yield (Bara et al., 2019). Some of the design drawbacks of the RBC system are rotation speed and spatial separation of CO_2 and light from nutrients in wastewater. A high rotation speed will enhance the mass transfer

and shear, but the mass transfer will decrease at low speed and may contribute to the drying of the microalgae biofilm (Blanken et al., 2014; Gross et al., 2013).

As illustrated in Table 5, the disks can be rotated at speeds between 4 and 11 rpm using a shaft that is connected to a motor. Disk materials such as polyvinyl chloride (PVC) (Orandi et al., 2012; Sebestyén et al., 2016), plastic (Christenson & Sims 2012; Shayan et al., 2016), aramid fiberglass, chamois cloth, and cotton duct (Gross et al., 2013), and steel mesh (Blanken et al., 2014) have been widely investigated. The disk diameter can be designed within a range of 10–40 cm depending on the scale of the RAB. As investigated by Sebestyén et al. (2016), the disk diameter was designed at 25 cm for laboratory RAB and 130 cm for the pilot scale.

The RAB has not been widely used for the treatment of wastewater. The RAB has been applied to treat synthetic acid mine drainage wastewater (Orandi et al., 2012) and municipal wastewater (Shayan et al., 2016). The photobioreactor is well-known for algal cultivation using commercial media to harvest the biomass for other end products. In one study, wastewater treatment using *Ulothrix zonata* biofilms eliminated heavy metals (Cu, Ni, Mn, Zn, Sb, Se) by

TABLE 5   Operating conditions of the RAB.

Wastewater	Microalgae species	Material of disk	Rotating speed	Light intensity	HRT	Pollutants	References
Synthetic acid mine drainage	*Ulothrix zonata*	PVC (25cm)	5rpm	$756\,\mu mol\,m^{-2}\,s^{-1}$	24	Heavy metals (Cu, Ni, Mn, Zn, Sb, Se): 50%	Orandi et al. (2012)
Municipal wastewater	–	Plactic drum (41cm)	5.4	$170\,\mu mol\,m^{-2}\,s^{-1}$	4.8days	—	Christenson and Sims (2012)
M8-amedium	*Chlorella sorokiniana*	PVC+sand paper (25cm)	11	$40\pm15\,\mu mol\,m^{-2}\,s^{-1}$	—	—	Sebestyén et al. (2016)
	–	PVC+sand paper (130cm)	4	—			
Municipal wastewater	Not mentioned	plastic cylindrical (10cm)	4.8	$200\pm20\,\mu mol\,m^{-2}\,s^{-1}$	2-day HRT 6-day HRT	TN: 93.7 TP:80.6 TN: 81.3 TP:99.5	Shayan et al. (2016)
M8-a media + Urea	*Chlorella sorokiniana*	Steel mesh (~27cm)	11rpm	$422\,\mu mol\,m^{-2}\,s^{-1}$	7days	–	Blanken et al. (2014)

up to 50% with a rotating speed of 5rpm using PVC disks (Orandi et al., 2012). Another study by Shayan et al. (2016), reported that total nitrogen and phosphorus were removed by more than 80% and 99%, respectively at rotation speeds of 4.8rpm and 6days HRT.

4.3 Microalgae biofilm membrane photobioreactors

A microalgae biofilm membrane photobioreactor (BMPBR) consists of a few key components such as biofilm carriers, a membrane module, an air compressor, and a separator. This reactor is considered to be an innovation in the field of photobioreactors. The BMPBR involves a combination of a biofilm membrane supported by light for microalgae growth. The carriers act as attachment materials to adhere to the microalgae biofilms, while the role of the membrane module is to separate solids and liquids thus achieving isolation of the microalgae from the effluent. The biofilm carriers in the BMPBR can be designed either as fixed (Fig. 8A) or moving (Fig. 8B) throughout the PBR. Examples of carriers are flexible fiber bundles (Gao et al., 2015), and kenaf fiber (Derakhshan et al., 2018). The BMPBR was studied using polyvinylidene fluoride (PVDF) hollow-fiber microfiltration membrane with a pore size of 0.1μm that was submerged in the PBR (Derakhshan et al., 2018; Gao et al., 2015).

The BMPBR with cultivated *Chlorella vulgaris* demonstrated good performance for the removal of nitrogen and phosphate in simulated municipal wastewater with a removal average of 83% and 85%, respectively, and produced

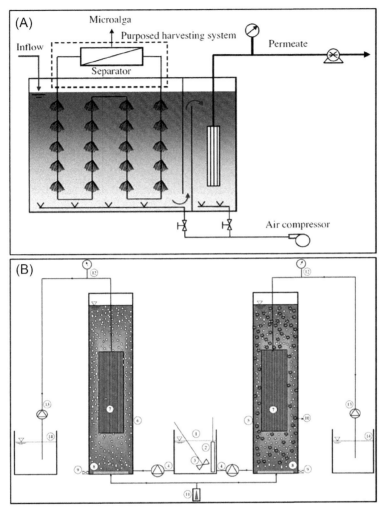

FIG. 8   BMPBR schematic diagram (A) with the fixed carrier (Gao et al., 2015), (B) with moving carrier (Derakhshan et al., 2018)—(1) reservoir of feedstock, (2) electric heater, (3) electric mixer, (4) peristaltic pump, (5) HMPBR reactor, (6) MPBR reactor, (7) submerged membrane, (8) diffuser, (9) discharge sludge port, (10) solid biofilm carriers, (11) air compressor, (12) gauge, (13) suction pump, (14) reservoir of outlet.

$72\,\mathrm{mg\,L^{-1}\,day^{-1}}$ of biomass (Gao et al., 2015). In the study by Gao et al. (2015), the BMPBR was equipped with submerged membrane and solid carriers under operational conditions of HRT for 48h, with an aeration rate of $2\,\mathrm{L\,min^{-1}}$ (4% CO_2). Another study reported that the removal of 99% of COD was achieved using a mixed culture microalgae biofilm under operational conditions of HRT of 24h, and aeration rate of $4–5\,\mathrm{L\,h^{-1}}$ and a 12:12 (light:dark) cycle (Derakhshan et al., 2018). The BMPBR was also reported to perform well for the removal of up to 95% xenobiotic atrazine within 12h HRT and 12h of light (~8500 Lux) (Derakhshan et al., 2019).

5　Kinetics of microalgae biofilm photobioreactors

The study if kinetics through mathematical modeling is important since it can be used to predict the behavior of the system towards a variety of key factors. Moreover, the study of kinetics can also provide precious response information when determining which parameter the system is most sensitive to, especially in flat-plate MBPBR, hybrid BMPBR, and RABPBR. For each system, the kinetic study focused on the growth and removal of pollutants by microalgae biofilms. Pollutant removal via microalgae uptake was also considered as it has become accepted as a source of nutrients for the growth of the microalgae biofilms. Furthermore, inhibitory factors were also included in the derivation of the kinetic model.

5.1　Kinetics in flat-plate MBPBR

Flat-plate MBPBR was normally adopted for mass production of photoautotrophic microorganisms. For mass production to be successful, the nutrient uptake by the microalgae was monitored often and was expressed in the form of a kinetic study. Lin, Leu, Lan, Lin, and Chang (2003) showed that microalgae biofilms used bicarbonate (HCO_3^-) as an inorganic carbon at a pH from 8 to 9. The development of the kinetic model in Eqs. (1), (2) (Table 6) by Lin et al. (2003) was conducted with several assumptions, as follows: (a) the microalgae biofilms was covered with a stagnant layer, (b) the concentrations of inorganic carbon within the microalgae biofilms were assumed to vary only in the normal direction to the microalgae biofilms surface, (c) molecular diffusion was involved in the transportation of inorganic carbon from bulk liquid to biofilm phase through the stagnant liquid layer, and (d) the flow pattern of liquid in a flat-plate MBPBR was not affected by the microalgae biofilm growth. By combining Fick's law for

diffusion and Monod kinetics that represent biological reactions that occur within microalgae biofilms, a differential equation as shown in Eq. (1) was developed. However, to specifically describe the growth of microalgae biofilms, Eq. (2) could be used under the assumption that the density of the attached microalgae biomass was constant.

In a study by Li, Liu, et al. (2019) and Li, Tong, et al. (2019), the influence of light was included in the growth kinetic model. This was since they claimed that light had a significant effect on the overall performance of microalgae cultivation in flat-plate MBPBR. According to Pruvost, Cornet, and Legrand (2008), the light was the limiting factor for the performance of the photobioreactors, therefore, light input had to either be increased using high photon flux density (PFD) or the illuminated surface had to be maximized for a given culture volume. Eq. (3) was developed by considering the light influence on the overall performance of flat-plate MBPBR. The overall concentration of microalgae biomass could be calculated based on Eq. (3) over cultivation time (t). However, to solve the equation, Eq. (4) with light intensity (I) had to initially be solved to obtain the value of the average of the specific growth rate of the microalgae.

5.2　Kinetics in hybrid BMPBR

Another novel configuration of the photobioreactor is the hybrid BMPBR with low HRT and minimal biomass washout. A study by Najm, Jeong, and Leiknes (2017) expressed the growth rate of *Chlorella vulgaris* in a simple growth rate kinetic model. See Eq. (5) in Table 7. A linear regression analysis was initially developed between the value of optical density at a wavelength of 683 nm (OD_{683}) and cell number per microliter (N) to obtain initial and final cell concentrations before applying Eq. (5). In addition to the growth of the microalgae, many researchers were also interested in the reaction

TABLE 6 The kinetic model involved in flat-plate MBPBR.

Kinetic model	Equation number	Equation	Parameter	References
Fick's law and Monod kinetics	1	$\dfrac{dS_f}{dt} = D_f \dfrac{d^2 S_f}{dz_f^2} - \dfrac{k S_f}{K_S + S_f} X_f$	S_f: concentration of HCO_3^- in microalgal biofilm (M_s/L^3); D_f: diffusion coefficient of HCO_3^- in the microalgal biofilm (L^2/T); k: Monod maximum utilization rate of HCO_3^- by microalgae $(1/T)$; K_s: Monod half-velocity coefficient of HCO_3^- by microalgae (M_s/L^3); X_f: density of microalgal biofilm (M_x/L^3); z_f: radical distance in microalgal biofilm (L); L_f: microalgal biofilm thickness (L); Y: yield coefficient of microalgal biofilm (M_x/M_S); b: decay coefficient of microalgal biofilm $(1/T)$; b_s: shear-loss coefficient of microalgal biofilm $(1/T)$	Lin et al. (2003)
	2	$\dfrac{dL_f}{dt} = \int_0^{L_f} \left(\dfrac{Y_k S_f}{K_s + S_f} - b - b_s \right) dz_f$		
Biomass growth model	3	$\dfrac{dC}{dt} = \mu_{avg} C$	C: biomass concentration of microalgae; μ_{avg}: average specific growth rate; $\mu(I)$: specific growth rate with light influence; μ_m: parameter in the growth model, (h^{-1}); I: light intensity, $(\mu mol\, m^{-2}\, s^{-1})$; k_1: parameter in the growth model, $(\mu mol\, m^{-2}\, s^{-1})$; k_2: parameter in the growth model, $((\mu mol\, m^{-2}\, s^{-1})$; μ_s: parameter in the growth model, (h^{-1})	Li, Liu, et al. (2019) and Li, Tong, et al. (2019)
	4	$\mu(I) = \mu_m \dfrac{I}{I + k_1 + \dfrac{I^2}{k_2}} - \mu_s$		

kinetics of biofilms in pollutant removal. Kinetic models such as modified a Stover-Kincannon model can be used to express the reaction kinetics of biofilms in pollutant removal (such as COD and atrazine in hybrid BMPBR (HMPBR)) as shown in Eq. (6) (Derakhshan et al., 2018). Eq. (6) which was used to determine substrate removal rate can also be expressed in linear form as shown in Eq. (7).

In addition to the modified Stover-Kincannon model, the Grau second-order model is often used to express nutrient utilization in microalgae biofilms. The Grau second-order model can be observed in Eq. (8). A simplified version can be observed in Eq. (9) where $\frac{S_0 - S}{S_0}$ can be expressed with E, representing the substrate removal efficiency. Modified models such as the modified Stover-Kincannon and Grau models usually consist of a small number of variables and make it easier to determine the reaction kinetics. Moreover, kinetic models have been used in many studies to predict the performance of the reactor and to optimize scaling up the construction of wastewater treatment plants (Babaei et al., 2013; Derakhshan et al., 2018).

TABLE 7　Kinetic models that are often used in hybrid BMPBR.

Kinetic model	Equation number	Equation	Parameter	References
Growth rate	5	$$\mu = \frac{\ln X_2 - \ln X_1}{t_2 - t_1}$$	μ: specific growth rates; X_2: final cell concentration at time t_2; X_1: initial cell concentration at time t_1	Najm et al. (2017)
Modified Stover-Kincannon reaction kinetics	6	$$\frac{dS}{dt} = \frac{Q(S_0 - S_e)}{V} = \frac{U_{max}\left(\frac{QS_0}{V}\right)}{K_B + \left(\frac{QS_0}{V}\right)}$$	$\frac{dS}{dt}$: substrate removal rate $(mg\,L^{-1}\,day^{-1})$; Q: flow rate $(L\,day^{-1})$; V: reactor working volume (L); S_0: influent substrate concentration $(mg\,L^{-1})$;	Derakhshan et al. (2018) and Babaei, Azadi, Jaafarzadeh, and Alavi (2013)
	7	The linear form of Stover-Kincannon reaction kinetics: $$\frac{V}{Q(S_0 - S_e)} = \frac{K_B}{U_{max}} \times \frac{V}{QS_0} + \frac{1}{U_{max}}$$	S_e: effluent substrate concentration $(mg\,L^{-1})$; U_{max}: maximum removal rate of substrate $(g\,L^{-1}\,day^{-1})$; K_B: constant saturation value $(g\,L^{-1}\,day^{-1})$	
Grau second-order substrate removal model	8	$$\frac{S_0 HRT}{S_0 - S} = a + bHRT$$	S_0: influent substrate concentrations $(mg\,L^{-1})$; S: effluent substrate concentration $(mg\,L^{-1})$	Babaei et al. (2013)
	9	Simplified Grau second-order substrate removal model: $$\frac{S_0 - S}{S_0} = E, therefore$$ $$\frac{HRT}{E} = a + bHRT$$	HRT: hydraulic retention time (h); a: Grau kinetic constant b: Grau kinetic constant E: substrate removal efficiency	

5.3 Kinetics in RABPBRs

RABPBRs can reduce the costs of harvesting and provide high productivity due to light dilution (Bara et al., 2019). However, long-term exposure of microalgae cells to high light intensity caused the cells to become photo-saturated and resulted in growth inhibition. For this reason, light inhibition factors are often expressed together in kinetic models since these factors affect microalgae growth. In a study by Bara et al. (2019), a Han model was applied by placing the limelight on the dynamics of photosystem II to describe the complex process of photosynthesis. The equation related to the previously mentioned photosystem dynamics is shown in Eq. (10) as illustrated in Table 8. The Han equation considers light intensity (I) and an inhibitory factor (C).

TABLE 8 Kinetic model for RABPBR.

Kinetic model	Equation number	Equation	Parameter	References
Han model	10	$$\frac{dC}{dt} = -\left(k_d\tau\,\frac{(\sigma I)^2}{\tau\sigma I + 1} + kr\right)C + \frac{\tau k_d(\sigma I)^2}{\tau\sigma I + 1}$$	σ: effective absorption cross-section per unit of photosynthetic units (m^2 µE^{-1}); I: light intensity (µmol m^{-2} s^{-1}); τ: turnover time of the electron transport chain (s); k_d: damage rate; k_r: repair rate; C: inhibition factor	Bara et al. (2019)
General logistic growth	11	$$\frac{dN}{dt} = \mu N\left(1 - \frac{N}{K}\right)$$	N: current population; K: carrying capacity (maximum number of organisms that can be grown in a given space) (g m^{-2}); μ: maximum growth rate (1 day^{-1})	Woolsey (2011)
Theta-logistic growth	12	$$\frac{dN}{dt} = \mu N\left(1 - \frac{N}{K}\right)^\theta$$	θ: inhibitory effect of populations; N: current population; K: carrying capacity (maximum number of organisms that can be grown in a given space) (g m^{-2}); μ: maximum growth rate (1 day^{-1})	
Biomass growth	13	$$\frac{dB}{dt} = \mu B\left(1 - \frac{B}{K}\right)^\theta - DB$$	B: biomass (g m^{-2}); D: death rate (1 day^{-1}); K: carrying capacity (maximum number of organisms that can be grown in a given space) (g m^{-2}); μ: maximum growth rate (1 day^{-1})	
Nitrogen uptake	14	$$\frac{dN}{dt} = \mu g N B \times \frac{S}{V}\left(1 - \frac{B}{K}\right)^\theta + F(N_{in} - N)$$	g: nitrogen uptake (gN gB^{-1}); S: surface area (m^3); V: volume (L); F: dilution rate (1 day^{-1}); N: nitrogen (mg L^{-1}); μ: maximum growth rate (1 day^{-1})	
Phosphorus uptake	15	$$\frac{dP}{dt} = \mu r P B \times \frac{S}{V}\left(1 - \frac{B}{K}\right)^\theta + F(P_{in} - P)$$	S: surface area (m^3); V: volume (L); F: dilution rate (1 day^{-1}); N: nitrogen (mg L^{-1}); μ: maximum growth rate (1 day^{-1}); r: phosphorus uptake (gP gB^{-1}); P: phosphorus (mg L^{-1})	

In addition to the Han model, a General Logistic Growth model (GLG), as shown in Eq. (11), was used to describe the growth of microalgae cells in the RABPBR (Woolsey, 2011).

The GLG model describes biomass population behavior, however, there are inhibition factors that occur inside the system. To overcome these inhibition factors, a modification of the GLG model was needed. Thus, the Theta-Logistic Growth (TLG) model (Eq. 12) was developed with consideration of θ, which represented the inhibitory phenomenon that occurred inside the system (Woolsey, 2011). It must be noted that the value of θ when greater than 1 ($>$1) showed a significant inhibitory effect towards the growth cycle of microalgae and θ values less than 1 ($<$1) showed fast growth of microalgae at the initial stage but a gradual decrease in growth rate to time.

From Eqs. (11), (12), a further derivation was made for biomass growth (Eq. 13) where the biomass (B) and the death rate (D) were considered. In addition to the growth rate, nutrient uptake was also considered. Both nitrogen and phosphorus uptake is crucial for microalgae cell metabolism (Liu et al., 2017). Using Eq. (12), a modification was made to generate nitrogen uptake (Eq. 14) and phosphorus uptake (Eq. 15). However, it must be noted that the biomass, nitrogen, and phosphorus changed with time whereas volume, surface area, and dilution rate remained constant based on the physical parameters of the rotating algal biofilm (Woolsey, 2011).

6 Conclusion

The development of wastewater treatment technologies is increasing from year to year. A broad range of technologies, from chemical to biological methods, with applications for primary to tertiary treatment have been explored. Microalgae biofilms for wastewater treatment have received attention during the last decade. Due to high treatment efficiency and a high yield of microalgae biomass, this method has become one of the best alternatives for low-cost wastewater treatment. Various species of microalgae, including freshwater and marine microalgae, are capable of forming biofilms and have been shown to act as bioremediation agents depending on environmental factors (light, pH, CO_2, temperature), surface materials, and nutrient

availability. Microalgae biofilm photobioreactors can be designed in three ways: permanently submerged, periodically submerged, or nonsubmerged, and each bioreactor has its advantages and disadvantages.

Acknowledgments

This research was financially supported by the Universiti Kebangsaan Malaysia with Grant No. DIP-2018-022.

References

Ajitha, V., Sreevidya, C. P., Kim, J. H., Bright, I. S., Mohandas, S. A., Lee, J. S., et al. (2019). Effect of metals of treated electroplating industrial effluents on antioxidant defense system in the microalga *Chlorella vulgaris*. *Aquatic Toxicology*, *217*, 105317.

Al-Momani, F. A., & Örmeci, B. (2016). Performance Of *Chlorella vulgaris*, *Neochloris oleoabundans*, and mixed indigenous microalgae for treatment of primary effluent, secondary effluent and centrate. *Ecological Engineering*, *95*, 280–289.

Arad, S. M., & Yaron, A. (1992). Natural pigments from red microalgae for use in foods and cosmetics. *Trends in Food Science and Technology*, *3*, 92–97.

Ariede, M. B., Candido, T. M., Jacome, A. L. M., Velasco, M. V. R., de Carvalho, J. C. M., & Baby, A. R. (2017). Cosmetic attributes of algae—A review. *Algal Research*, *25*, 483–487.

Babaei, A. A., Azadi, R., Jaafarzadeh, N., & Alavi, N. (2013). Application and kinetic evaluation of upflow anaerobic biofilm reactor for nitrogen removal from wastewater by Anammox process. *Iranian Journal of Environmental Health Science & Engineering*, *10*(1), 20.

Bara, Q., Bonnefond, H., & Bernard, O. (2019). Model development and light effect on a rotating algal biofilm. *IFAC-PapersOnLine*, *52*, 376–381.

Barros, A. C., Goncalves, A. L., & Simões, M. (2019). Microalgal/cyanobacterial biofilm formation on selected surfaces: The effects of surface physicochemical properties and culture media composition. *Journal of Applied Phycology*, *31*(1), 375–387.

Bastos, R. G., Bonini, M. A., Zepka, L. Q., Lopes, E. J., & Queiroz, M. I. (2015). Treatment of rice parboiling wastewater by cyanobacterium *Aphanothece microscopica Nägeli* with potential for biomass products. *Desalination and Water Treatment*, *56*(3), 608–614.

Bellinger, A. S. B. J., Abdullahi, M. R., & Gretz, G. J. C. (2005). Underwood biofilm polymers: relationship between carbohydrate biopolymers from estuarine mudflats and unialgal cultures of benthic diatoms. *Aquatic Microbial Ecology*, *38*(2), 169–180.

Bich, N. N., Yaziz, M. I., & Bakti, N. A. K. (1999). Combination of *Chlorella vulgaris* and *Eichhornia crassipes* for wastewater nitrogen removal. *Water Research*, 33(10), 2357–2362.

Blanken, W., Janssen, M., Cuaresma, M., Libor, Z., Bhaiji, T., & Wijffels, R. H. (2014). Biofilm growth of *Chlorella Sorokiniana* in a rotating biological contactor based photobioreactor. *Biotechnology and Bioengineering*, 111(12), 2436–2445.

Boelee, N. C., Temmink, H., Janssen, M., Buisman, C. J. N., & Wijffels, R. H. (2014). Balancing the organic load and light supply in symbiotic microalgal–bacterial biofilm reactors treating synthetic municipal wastewater. *Ecological Engineering*, 64, 213–221.

Cao, J., Yuan, W., Pei, Z. J., Davis, T., Cui, Y., & Beltran, M. (2009). A preliminary study of the effect of surface texture on algae cell attachment for a mechanical–biological energy manufacturing system. *Journal of Manufacturing Science and Engineering*, 131, 1–4.

Caporgno, M. P., Taleb, A., Olkiewicz, M., Font, J., Pruvost, J., Legrand, J., et al. (2015). Microalgae cultivation in urban wastewater: Nutrient removal and biomass production for biodiesel and methane. *Algal Research*, 10, 232–239.

Chen, X., Hu, Z., Qi, Y., Song, C., & Chen, G. (2019). The interactions of algae-activated sludge symbiotic system and its effects on wastewater treatment and lipid accumulation. *Bioresource Technology*, 292, 122017.

Cheng, P., Wang, J., & Liu, T. (2014). Effects of nitrogen source and nitrogen supply model on the growth and hydrocarbon accumulation of immobilized biofilm cultivation of *B. braunii*. *Bioresource Technology*, 166, 527–533.

Cheng, P., Wang, Y., Liu, T., & Defu Liu, D. (2017). Biofilm attached cultivation of *Chlorella pyrenoidosa* is a developed system for swine wastewater: Treatment and lipid production. *Frontiers in Plant Science*, 8, 1594.

Choudhary, P., Prajapati, S. K., Kumar, P., Malik, A., & Pant, K. K. (2017). Development and performance evaluation of an algal biofilm reactor for treatment of multiple wastewaters and characterization of biomass for diverse applications. *Bioresource Technology*, 224, 276–284.

Christenson, L. B., & Sims, R. C. (2012). Rotating algal biofilm reactor and spool harvester for wastewater treatment with biofuels by-products. *Biotechnology and Bioengineering*, 109(7), 1674–1684.

Church, J., Ryu, H., Anwar Sadmani, A. H. M., Randall, A. A., Domingo, J. S., & Lee, W. H. (2018). Multiscale investigation of a symbiotic microalgal-integrated fixed film activated sludge (MAIFAS) process for nutrient removal and photo-oxygenation. *Bioresource Technology*, 268, 128–138.

D'Este, M., Morales, M. A., & Angelidaki, I. (2017). *Laminaria digitata* as potential carbon source in heterotrophic microalgae cultivation for the production of fish feed supplement. *Algal Research*, 26, 1–7.

da Fontoura, J. T., Rolim, G. S., Farenzena, M., & Gutterres, M. (2017). Influence of light intensity and tannery wastewater concentration on biomass production and nutrient removal by microalgae *Scenedesmus sp*. *Process Safety and Environment Protection*, 111, 355–362.

Daneshvar, E., Zarrinmehr, M. J., Hashtjin, A. M., Farhadian, O., & Bhatnagar, A. (2018). Versatile applications of freshwater and marine water microalgae in dairy wastewater treatment, lipid extraction and tetracycline biosorption. *Bioresource Technology*, 268, 523–530.

de Assis, L. R., Calijuri, M. L., Assemany, P. P., Berg, E. C., Febroni, L. V., & Bartolome, T. A. (2019). Evaluation of the performance of different materials to support the attached growth of algal biomass. *Algal Research*, 39, 101440.

de Assis, L. R., Calijuri, M. L., do Couto, E. A., & Assemany, P. P. (2017). Microalgal biomass production and nutrients removal from domestic sewage in a hybrid high-rate pond with biofilm reactor. *Ecological Engineering*, 106, 191–199.

Derakhshan, Z., Ehrampoush, M. H., Mahvid, A. H., Dehghani, M., Faramarzian, M., & Eslami, H. (2019). A comparative study of hybrid membrane photobioreactor and membrane photobioreactor for simultaneous biological removal of atrazine and CNP from wastewater: A performance analysis and modelling. *Chemical Engineering Journal*, 355, 428–438.

Derakhshan, Z., Mahvid, A. H., Ehrampoush, M. H., Ghaneian, M. T., Yousefinejad, S., Faramarzian, M., et al. (2018). Evaluation of kenaf fibers as moving bed biofilm carriers in algal membrane photobioreactor. *Ecotoxicology and Environmental Safety*, 152, 1–7.

Dineshbabu, G., Goswami, G., Kumar, R., Sinha, A., & Das, D. (2019). Microalgae–nutritious, sustainable aqua- and animal feed source. *Journal of Functional Foods*, 62, 103545.

Ding, G. T., Yaakob, Z., Takriff, M. S., Salihon, J., & Rahaman, M. S. A. (2016). Biomass production and nutrients removal by a newly-isolated microalgal strain *Chlamydomonas sp* in palm oil mill effluent (POME). *International Journal of Hydrogen Energy*, 41(8), 4888–4895.

Eze, V. C., Orta, S. B. V., García, A., Ramírez, I. M., & Ledesma, M. T. O. (2018). Kinetic modelling of microalgae cultivation for wastewater treatment and carbon dioxide sequestration. *Algal Research*, 32, 131–141.

Franchino, M., Tigini, V., Varese, G. C., Sartor, R. M., & Bona, F. (2016). Microalgae treatment removes nutrients and reduces ecotoxicity of diluted piggery digestate. *Science of the Total Environment*, 569–570, 40–45.

Gao, F., Li, C., Yang, Z. H., Zeng, G. M., Feng, L. J., Liu, J. Z., et al. (2016). Continuous microalgae cultivation in aquaculture wastewater by a membrane photobioreactor for biomass production and nutrients removal. *Ecological Engineering*, 92, 55–61.

Gao, F., Yang, Z. H., Li, C., Zeng, G. M., Ma, D. H., & Zhou, L. (2015). A novel algal biofilm membrane photobioreactor for attached microalgae growth and nutrients removal from secondary effluent. *Bioresource Technology*, 179, 8–12.

Gonçalves, A. L., Ferreira, C., Loureiro, J. A., Pires, J. C. M., & Simões, M. (2015). Surface physicochemical properties of selected single and mixed cultures of microalgae and cya-nobacteria and their relationship with sedimentation kinetics. *Bioresources and Bioprocessing*, 2, 1–10.

González-Camejo, J., Serna-García, R., Viruela, A., Pachés, M., Durán, F., Robles, A., et al. (2017). Short and long-term experiments on the effect of sulphide on microalgae cultivation in tertiary sewage treatment. *Bioresource Technology*, 244(1), 15–22.

Gopalakrishnan, V., & Ramamurthy, D. (2014). Dyeing industry effluent system as lipid production medium of *Neochloris sp.* for biodiesel feedstock preparation. *BioMed Research International*, 2014, 1–7.

Gross, M. A. (2015). *Development and optimization of biofilm based algal cultivation.* (thesis). 14850Iowa State University.

Gross, M., Henry, W., Michael, C., & Wen, Z. (2013). Development of a rotating algal biofilm growth system for attached microalgae growth with in situ biomass harvest. *Bioresource Technology*, 150C, 195–201.

Gupta, S., & Pawar, S. B. (2018). An integrated approach for microalgae cultivation using raw and anaerobic digested wastewaters from food processing industry. *Bioresource Technology*, 269, 571–576.

Guzzon, A., Di Pippo, F., & Congestri, R. (2019). Wastewater biofilm photosynthesis in photobioreactors. *Microorganisms*, 7, 252.

Hariz, H. B., Takriff, M. S., Yasin, N. H. M., Ba-Abbad, M. M., & Hakimi, N. I. N. M. (2019). Potential of the microalgae-based integrated wastewater treatment and CO$_2$ fixation system to treat palm oil mill effluent (POME) by indigenous microalgae; *Scenedesmus sp.* and *Chlorella sp. Journal of Water Process Engineering*, 32, 100907.

Hazman, N. A. S., Mohd Yasin, N. H., Takriff, M. S., Hasan, H. A., Kamarudin, K. F., & Mohd Hakimi, N. I. N. (2018). integrated palm oil mill effluent treatment and CO$_2$ sequestration by microalgae. *Sains Malaysiana*, 47, 1455–1464.

Hernández-Garcia, A., Velásquez-Orta, S. B., Novelo, E., Yánez-Noguez, I., Monje-Ramirez, I., & Orta Ledesma, M. T. (2019). Wastewater-leachate treatment by microalgae: Biomass, carbohydrate and lipid production. *Ecotoxicology and Environmental Safety*, 174, 435–444.

Holmes, P. E. (1986). Bacterial enhancement of vinyl fouling by algae. *Applied and Environmental Microbiology*, 52(6), 1391–1393.

Hossain, M., Akter, Z. S., Yun, J., Zhang, G., Zhang, Y., & Qi, X. (2020). Biogas from microalgae: Technologies, challenges and opportunities. *Renewable and Sustainable Energy Reviews*, 117, 109503.

Huang, Y., Xiong, W., Liao, Q., Ao, Q. F., Zhu, X. X., & Sun, Y. (2016). Comparison of *Chlorella vulgaris* biomass productivity cultivated in biofilm and suspension from the aspect of light transmission and microalgae affinity to carbon dioxide. *Bioresource Technology*, 222, 367–373.

Huo, S., Chen, J., Zhu, F., Zou, B., Chen, X., Basheer, S., et al. (2019). Filamentous microalgae Tribonema sp.

cultivation in the anaerobic/oxic effluents of petrochemical wastewater for evaluating the efficiency of recycling and treatment. *Biochemical Engineering Journal*, 145, 27–32.

Irving, T. E., & Allen, D. G. (2011). Species andmaterial considerations in the formation and development of microalgal biofilms. *Applied Microbiology and Biotechnology*, 92, 283–294.

Japar, A. S., Takriff, M. S., & Yasin, N. H. M. (2017). Harvesting microalgal biomass and lipid extraction for potential biofuel production: A review. *Journal of Environmental Chemical Engineering*, 5, 555–563.

Kamyab, H., Din, M. F. M., Keyvanfar, A., Majid, M. Z. A., Talaiekhozani, A., Shafaghat, A., et al. (2015). Efficiency of microalgae chlamydomonas on the removal of pollutants from palm oil mill effluent (POME). *Energy Procedia*, 75, 2400–2408.

Katam, K., & Bhattacharyya, D. (2019). Simultaneous treatment of domestic wastewater and bio-lipid synthesis using immobilized and suspended cultures of microalgae and activated sludge. *Journal of Industrial and Engineering Chemistry*, 69, 295–303.

Katarzyna, L., Sai, G., & Singh, O. A. (2015). Non-enclosure methods for non-suspended microalgae cultivation: literature review and research needs. *Renewable and Sustainable Energy Reviews*, 42, 1418–1427.

Khairuddin, N. F. M., Idris, A., & Irfan, M. (2019). Towards efficient membrane filtration for microalgae harvesting: A review. *Jurnal Kejuruteraan SI*, 2(1), 103–112.

Khalid, A. A. H., Yaakob, Z., Abdullah, S. R. S., & Takriff, M. S. (2019). Analysis of the elemental composition and uptake mechanism of *Chlorella sorokiniana* for nutrient removal in agricultural wastewater under optimized response surface methodology (RSM) conditions. *Journal of Cleaner Production*, 210, 673–686.

Kong, W. B., Hua, S. F., Cao, H., Mu, Y. W., Yang, H., & Song, H. (2012). Optimization of mixotrophic medium components for biomass production and biochemical composition biosynthesis by *Chlorella vulgaris* using response surface methodology. *Journal of the Taiwan Institute of Chemical Engineers*, 43(3), 360–367.

Koyande, A. K., Chew, K. W., Rambabu, K., Tao, Y., Chu, D. T., & Show, P. L. (2019). Microalgae: A potential alternative to health supplementation for humans. *Food Science and Human Wellness*, 8(1), 16–24.

Kumar, G., Huy, M., Bakonyi, P., Bélafi-Bakó, K., & Kim, S. H. (2018). Evaluation of gradual adaptation of mixed microalgae consortia cultivation using textile wastewater via fed batch operation. *Biotechnology Reports*, 20, e00289.

Labbe, J. I., Ramos-Suárez, J. L., Hernández-Pérez, A., Baeza, A., & Hansen, F. (2017). Microalgae growth in polluted effluents from the dairy industry for biomass production and phytoremediation. *Journal of Environmental Chemical Engineering*, 5(1), 635–643.

Lamare, P. O., Aguillon, N., Sainte-Marie, J., Grenier, J., Bonnefond, H., & Bernard, O. (2019). Gradient-based optimization of a rotating algal biofilm process. *Automatica*, *105*, 80–88.

Lananan, F., Hamid, S. H. A., Sakinah, W. N., Khatoon, H., Jusoh, A., & Endut, A. (2014). Symbiotic bioremediation of aquaculture wastewater in reducing ammonia and phosphorus utilizing effective microorganism (EM-1) and microalgae (*Chlorella sp.*). *International Biodeterioration & Biodegradation*, *95*(A), 127–134.

Laura, E., Sharon, B. W., Orta, V., Frasca, E. R., Leary, P., Noguez, I. Y., et al. (2019). Non-sterile heterotrophic cultivation of native wastewater yeast and microalgae for integrated municipal wastewater treatment and bioethanol production. *Biochemical Engineering Journal*, *151*, 107319.

Leite, L. S., Hoffmann, M. T., & Daniel, L. A. (2019). Microalgae cultivation for municipal and piggery wastewater treatment in Brazil. *Journal of Water Process Engineering*, *31*, 100821.

Li, X., Liu, J., Chen, G., Zhang, J., Wang, C., & Liu, B. (2019). Extraction and purification of eicosapentaenoic acid and docosahexaenoic acid from microalgae: A critical review. *Algal Research*, *43*, 101619.

Li, T., Piltz, B., Podola, B., Dron, A., de Beer, D., & Melkonian, M. (2016). Microscale profiling of photosynthesis-related variables in a highly productive biofilm photobioreactor. *Biotechnology and Bioengineering*, *113*(5), 1046–1055.

Li, M. J., Tong, Z. X., Zhou, Z. J., Huang, D., & Wang, R. L. (2019). A numerical model coupling bubble flow, light transfer, cell motion and growth kinetic for real timescale microalgae cultivation and its applications in flat plate photobioreactor. *Algal Research*, *44*, 101727.

Li, X., Yang, W. L., He, H., Wu, S., Zhou, Q., Yang, C., et al. (2018). Responses of microalgae *Coelastrella sp.* to stress of cupric ions in treatment of anaerobically digested swine wastewater. *Bioresource Technology*, *251*, 274–279.

Lin, Y. H., Leu, J. Y., Lan, C. R., Lin, P.-H., & Chang, F. L. (2003). Kinetics of inorganic carbon utilization by microalgal biofilm in a flat plate photoreactor. *Chemosphere*, *53*, 779–787.

Liu, J., Wu, Y., Wu, C., Muylaert, K., Vyverman, W., Yu, H. Q., et al. (2017). Advanced nutrient removal from surface water by a consortium of attached microalgae and bacteria: A review. *Bioresource Technology*, *241*, 1127–1137.

Manirafasha, E., Ndikubwimana, T., Zeng, X., Lu, Y., & Jing, K. (2016). Phycobiliprotein: Potential microalgae derived pharmaceutical and biological reagent. *Biochemical Engineering Journal*, *109*, 282–296.

Mantzorou, A., & Ververidis, F. (2019). Microalgal biofilms: A further step over current microalgal cultivation techniques. *Science of the Total Environment*, *651*(2), 3187–3201.

Mehrabadi, A., Farid, M. M., & Craggs, R. (2017). Potential of five different isolated colonial algal species for wastewater treatment and biomass energy production. *Algal Research*, *21*, 1–8.

Mieszkin, S., Callow, M. E., & Callow, J. A. (2013). Interactions between microbial biofilms and marine fouling algae: A mini review. *Biofouling*, *29*, 1097–1113.

Mohammadi, M., Mowla, D., Esmaeilzadeh, F., & Ghasemi, Y. (2018). Cultivation of microalgae in a power plant wastewater for sulfate removal and biomass production: A batch study. *Journal of Environmental Chemical Engineering*, *6*(2), 2812–2820.

Mondal, M., Ghosh, A., Sharma, A. S., Tiwari, O. N., Gayen, K., Mandal, M. K., et al. (2016). Mixotrophic cultivation of *Chlorella sp.* BTA 9031 and *Chlamydomonas sp.* BTA 9032 isolated from coal field using various carbon sources for biodiesel production. *Energy Conversion and Management*, *124*, 297–304.

Mujtaba, G., & Lee, K. (2017). Treatment of real wastewater using co-culture of immobilized Chlorella vulgaris and suspended activated sludge. *Water Research*, *120*, 174–184.

Muñoz, R., Köllner, C., & Guieysse, B. (2009). Biofilm photobioreactors for the treatment of industrial wastewaters. *Journal of Hazardous Materials*, *161*(1), 29–34.

Murphy, T. E., Macon, K., & Berberoglu, H. (2012). An image processing technique to recover the biomass concentration in algae biofilm photobioreactors. In: *ASME proc. heat mass transf. biotechnol.*

Najm, Y., Jeong, S., & Leiknes, T. (2017). Nutrient utilization and oxygen production by Chlorella vulgaris in a hybrid membrane bioreactor and algal membrane photobioreactor system. *Bioresource Technology*, *237*, 64–71.

Oliveira, R., Azeredo, J., Teixeira, P., & Fonseca, A. (2001). *The role of hydrophobicity in bacterial adhesion*: (Vol. 1). BioLine. (pp. 11–22).

Orandi, S., Lewis, D. M., & Moheimani, N. R. (2012). Biofilm establishment and heavy metal removal capacity of an indigenous mining algal-microbial consortium in a photo-rotating biological contactor. *Journal of Industrial Microbiology & Biotechnology*, *39*(9), 1321–1331.

Osorio, J. H. M., Pinto, G., Pollio, A., Frunzo, L., Lens, P. N. L., & Esposito, G. (2019). Start-up of a nutrient removal system using *Scenedesmus vacuolatus* and *Chlorella vulgaris* biofilms. *Bioresources and Bioprocessing*, *6*, 27.

Pathak, V. V., Kothari, R., Chopra, A. K., & Singh, D. P. (2015). Experimental and kinetic studies for phycoremediation and dye removal by *Chlorella pyrenoidosa* from textile wastewater. *Journal of Environmental Management*, *1631*, 270–277.

Patwardhan, A. W. (2003). Rotating biological contactors: A review. *Industrial and Engineering Chemistry Research*, *42*, 2035–2051.

Prandini, J. M., Silva, M. L. B., Mezzari, M. P., Pirolli, M., Michelon, W., & Soares, H. M. (2016). Enhancement of nutrient removal from swine wastewater digestate coupled to biogas purification by microalgae *Scenedesmus sp. Bioresource Technology*, *202*, 67–75.

Pruvost, J., Cornet, J.-F., & Legrand, J. (2008). Hydrodynamics influence on light conversion in photobioreactors: An energetically consistent analysis. *Chemical Engineering Science, 63*, 3679–3694.

Qi, W., Mei, S., Yuan, Y., Li, X., Tang, T., Zhao, Q., et al. (2018). Enhancing fermentation wastewater treatment by co-culture of microalgae with volatile fatty acid- and alcohol-degrading bacteria. *Algal Research, 31*, 31–39.

Renuka, N., Guldhe, A., Prasanna, R., Singh, P., & Bux, F. (2018). Microalgae as multi-functional options in modern agriculture: Current trends, prospects and challenges. *Biotechnology Advances, 36*(4), 1255–1273.

Rios, L. F., Kleina, B. C., Luz Jrb, L. F., Maciela, M. R. W., & Filho, R. M. (2015). Influence of culture medium on *Desmodesmus sp.* growth and lipid accumulation for biodiesel production. *Chemical Engineering Transactions, 43*, 601–606.

Rivas, M. O., Vargas, P., & Riquelme, C. E. (2010). Interactions of *Botryococcus braunii* cultures with bacterial biofilms. *Microbial Ecology, 60*, 628–635.

Schnurr, P. J., & Allen, D. G. (2015). Factors affecting algae biofilm growth and lipid production: A review. *Renewable and Sustainable Energy Reviews, 52*, 418–429.

Sebestyén, P., Blanken, W., Bacsa, I., Tóth, G., Martin, A., Bhaiji, T., et al. (2016). Upscale of a laboratory rotating disk biofilm reactor and evaluation of its performance over a half-year operation period in outdoor conditions. *Algal Research, 18*, 266–272.

Sekar, R., Venugopalan, V. P., Satpathy, K. K., Nair, K. V. K., & Rao, V. N. R. (2004). Laboratory studies on adhesion of microalgae to hard substrates. *Hydrobiologia, 512*, 109–116.

Sharma, G. K., Malla, F. A., Kumar, A., Rashmi, & Gupta, N. (2019). Microalgae based biofertilizers: A biorefinery approach to phycoremediate wastewater and harvest biodiesel and manure. *Journal of Cleaner Production, 211*, 1412–1419.

Shayan, S. I., Agblevor, A. F., Bertin, L., & Sims, R. C. (2016). Hydraulic retention time effects on wastewater nutrient removal and bioproduct production via rotating algal biofilm reactor. *Bioresource Technology, 211*, 527–533.

Shen, Q. H., Gong, Y. P., Fang, W. Z., Bi, Z. C., Cheng, L. H., Xu, X. H., et al. (2015). Saline wastewater treatment by Chlorella vulgaris with simultaneous algal lipid accumulation triggered by nitrate deficiency. *Bioresource Technology, 193*, 68–75.

Shurtz, B. K., Wood, B., & Quinn, J. C. (2017). Nutrient resource requirements for large-scale microalgae biofuel production: Multi-pathway evaluation. *Sustainable Energy Technologies and Assessments, 19*, 51–58.

Sierra, L. S., Stoykoba, P., & Nikolov, Z. L. (2018). Extraction and fractionation of microalgae-based protein products. *Algal Research, 36*, 175–192.

Su, Y., Mennerich, A., & Urban, B. (2016). The long-term effects of wall attached microalgal biofilm on algae-based wastewater treatment. *Bioresource Technology, 218*, 1249–1252.

Sukačová, K., Trtílek, M., & Rataj, T. (2015). Phosphorus removal using a microalgal biofilm in a new biofilm photobioreactor for tertiary wastewater treatment. *Water Research, 71*, 55–63.

Talaro, K. P. (2005). *Foundations in microbiology* (5th ed.). Boston, MA: McGraw-Hill Inc.

Tao, Q., Gao, F., Qian, C. Y., Guo, X. Z., Zheng, Z., & Yang, Z. H. (2017). Enhanced biomass/biofuel production and nutrient removal in an algal biofilm airlift photobioreactor. *Algal Research, 21*, 9–15.

Teixeira, P., Lopes, Z., Azeredo, J., Oliveira, R., & Vieira, M. J. (2005). Physicochemical surface characterization of a bacterial population isolated from a milking machine. *Food Microbiology, 22*, 247–251.

The Editors of Encyclopaedia Britannica (2015). *Scenedesmus.* Encyclopædia Britannica, Inc. (2015). https://www.britannica.com/science/Scenedesmus. Accessed 2 February 2020.

Udaiyappan, A. F. M., Hasan, H. A., Takriff, M. S., & Abdullah, S. R. S. (2017). A review of the potentials, challenges and current status of microalgae biomass applications in industrial wastewater treatment. *Journal of Water Process Engineering, 20*, 8–21.

Udaiyappan, A. F. M., Hasan, H. A., Takriff, M. S., Abdullah, S. R. S., Maeda, T., Mustapha, N. A., et al. (2020). Microalgae-bacteria interaction in palm oil mill effluent treatment. *Journal of Water Process Engineering, 35*, 101203.

Usha, M. T., Chandra, T. S., Sarada, R., & Chauhan, V. S. (2016). Removal of nutrients and organic pollution load from pulp and paper mill effluent by microalgae in outdoor open pond. *Bioresource Technology, 214*, 856–860.

Van, O. C. (1995). Hydrophobicity of biosurfaces—Origin, quantitative determination and interaction energies. *Colloids and Surfaces, B: Biointerfaces, 5*, 91–110.

Van, O. C. (2005). *Interfacial forces in aqueous media.* New York: Marcel Decker Inc.

Verma, R., & Srivastava, A. (2018). Carbon dioxide sequestration and its enhanced utilization by photoautotroph microalgae. *Environmental Development, 27*, 95–106.

Vidyashankar, S., VenuGopal, K. S., Swarnalatha, G. V., Kavitha, M. D., Chauhan, V. S., Ravi, R., et al. (2015). Characterization of fatty acids and hydrocarbons of chlorophycean microalgae towards their use as biofuel source. *Biomass and Bioenergy, 77*, 75–91.

Wang, Y., Gua, W., Yen, H. W., Ho, S. H., Lo, Y. C., Cheng, C. L., et al. (2015). Cultivation of *Chlorella vulgaris* JSC-6 with swine wastewater for simultaneous nutrient/COD removal and carbohydrate production. *Bioresource Technology, 198*, 619–625.

Wang, M., Yang, Y., Chen, Z., Chen, Y., Wen, Y., & Chen, B. (2016). Removal of nutrients from undiluted anaerobically treated piggery wastewater by improved microalgae. *Bioresource Technology, 222*, 130–138.

Wang, M., Zhang, S. C., Tang, Q., Shi, L. D., Tao, X. M., & Tian, G. M. (2020). Organic degrading bacteria and nitrifying bacteria stimulate the nutrient removal and biomass accumulation in microalgae-based system from piggery digestate. *Science of the Total Environment, 707*, 134442.

Wayne, C. K., Reen, C. S., Show, P. L., Jiun, Y. Y., Chuan, L. T., & Shu, C. J. (2018). Effects of water culture medium, cultivation systems and growth modes for microalgae cultivation: A review. *Journal of the Taiwan Institute of Chemical Engineers*, *91*, 332–344.

Wijffels, R. H., & Barbosa, M. J. (2010). An outlook on microalgal biofuels. *Science*, *329*(5993), 796–799.

Woolsey, P. A. (2011). *Rotating algal biofilm reactors: Mathematical modelling and lipid production.* (M.Sc. thesis) Logan, UT: Biological Engineering. Utah State University.

Wu, X., Cen, Q., Addy, M., Zheng, H., Luo, S., Liu, Y., et al. (2019). A novel algal biofilm photobioreactor for efficient hog manure wastewater utilization and treatment. *Bioresource Technology*, *292*, 121925.

Wu, W., Lin, K. H., & Chang, J. S. (2018). Economic and lifecycle greenhouse gas optimization of microalgae-to-biofuels chains. *Bioresource Technology*, *267*, 550–559.

Xu, J., Wang, Z., Yuan, T., Zhou, W., Xu, J., Liang, C., et al. (2017). Metabolic acclimation mechanism in microalgae developed for CO_2 capture from industrial flue gas. *Algal Research*, *26*, 225–233.

Yaakob, Z., Ali, E., Zainal, A., Mohamad, M., & Takriff, M. S. (2014). An overview: Biomolecules from microalgae for animal feed and aquaculture. *Journal of Biological Research*, *21*, 6.

Yang, Z., Pei, H., Hou, Q., Jiang, L., Zhang, L., & Nie, C. (2018). Algal biofilm-assisted microbial fuel cell to enhance domestic wastewater treatment: Nutrient, organics removal and bioenergy production. *Chemical Engineering Journal*, *332*, 277–285.

Ye, Y., Huang, Y., Xia, A., Fu, Q., Liao, Q., Zeng, W., et al. (2018). Optimizing culture conditions for heterotrophic-assisted photoautotrophic biofilm growth of *Chlorella vulgaris* to simultaneously improve microalgae biomass and lipid productivity. *Bioresource Technology*, *270*, 80–87.

Yun, Y. S., Park, J. M., & Yang, J. M. (1996). Enhancement of CO_2 tolerance of *Chlorella vulgaris* by gradual increase of CO_2 concentration. *Biotechnology Techniques*, *10*(9), 713–716.

Zamalloa, C., Boon, N., & Verstraete, W. (2013). Decentralized two-stage sewage treatment by chemical-biological flocculation combined with microalgae biofilm for a roof installed parallel plate reactor. *Bioresource Technology*, *130*, 152–160.

Zhang, R., Grimi, N., Marchal, L., Lebovka, N., & Vorobiev, E. (2019). Effect of ultrasonication, high pressure homogenization and their combination on efficiency of extraction of bio-molecules from microalgae *Parachlorella kessleri*. *Algal Research*, *40*, 101524.

Zhang, Q., Yu, Z., Jin, S., Zhu, L., Liu, C., Zheng, H., et al. (2019). Lignocellulosic residue as bio-carrier for algal biofilm growth: Effects of carrier physicochemical proprieties and toxicity on algal biomass production and composition. *Bioresource Technology*, *293*, 122091.

Zhao, X., Kumar, K., Gross, M. A., Kunetz, T. E., & Wen, Z. (2018). Evaluation of revolving algae biofilm reactors for nutrients and metals removal from sludge thickening supernatant in a municipal wastewater treatment facility. *Water Research*, *143*, 467–478.

Zhuang, L. L., Yu, D., Zhang, J., Liu, F. F., Wu, Y. H., Zhang, T. Y., et al. (2018). The characteristics and influencing factors of the attached microalgae cultivation: A review. *Renewable and Sustainable Energy Reviews*, *94*, 1110–1119.

Techno-economic assessment of microalgae for biofuel, chemical, and bioplastic

Yessie Widya Sari[a,b], Kiki Kartikasari[c], Widyarani[d], Iriani Setyaningsih[e], and Dianika Lestari[f]

[a]Department of Physics, Faculty of Mathematics and Natural Sciences, IPB University, Bogor, Indonesia
[b]Center for Transdisciplinary and Sustainability Sciences, Institute of Research and Community Services, IPB University, Bogor, Indonesia
[c]Carbon and Environmental Research Indonesia, Bogor, Indonesia
[d]Research Unit for Clean Technology, Indonesian Institute of Sciences, Bandung, Indonesia
[e]Department of Aquatic Product Technology, Faculty of Fisheries and Marine Sciences, IPB University, Bogor, Indonesia
[f]Department of Food Engineering, Faculty of Industrial Technology, Institut Teknologi Bandung, Bandung, Indonesia

1 Introduction

The industrial revolution has played an essential role in global changes, mainly related to the energy sector. Following the first Industrial Revolution, the energy demand for the later Industrial Revolution was fuelled by fossil-based sources. The increasing use of coal and thereof, followed by petroleum and natural gas, has made a considerable increase in the amount of fossil fuel demand. Later, concerns on fossil depletion and global warming trigger the development of biofuel.

Microalgae are expected to serve as a potential feedstock for the second generation of biofuels. This was motivated by the failure of first-generation biofuels, which are directly produced from crops to fuel the global energy demand due to the food vs. fuel concern. For the ease of the microalgae cultivation, as well as the possibility to produce a higher yield than the input energy, microalgae are considered as the third generation biofuel. This class of biofuel is explicitly given for microalgae taking advantage of low-cost, high energy, and entirely renewable feedstock.

Efforts have been made to industrialize microalgae-based biofuel. However, the high cost ranging from the cultivation (Amer, Adhikari, & Pellegrino, 2011), harvesting, and conversion (Jonker & Faaij, 2013), into downstream processing contributed to the delay on this. Through the biorefinery concept, maximizing each microalga component is, therefore, expected to realize the production of biofuel as well as high-value products from microalgae soon.

This chapter will provide current information on the techno-economic assessment (TEA) of several routes on microalgae processing for biofuel, chemicals, pharmaceuticals, and bioplastics.

2 Microalgae for biofuels

As the population increased, global energy demand also increased. Renewable energy resources, especially from various third generation feedstocks, are being investigated and developed to provide feasible alternatives energy in the form of biofuel, biodiesel, biogas, or bioethanol. In particular, biodiesel is one of the high demand biofuel types due to its liquid and energy-dense form, which used in many applications such as transportation, industrial fuel, even generate electricity. One of the promising third-generation feedstocks for biodiesel production is microalgae, which are characterized by their high lipid productivity per hectare land, the possibility to be cultivated in a nonarable land, saltwater and wastewater; and potential integration with a various waste stream or other feedstocks such as Jatropha or with biogas production facility (Giwa et al., 2018; Harun et al., 2011; Xin et al., 2016). The suitability of microalgal-oil as raw materials for biodiesel was indicated by several researchers. Several physicochemical properties of the transesterified oil-based microalgae have been summarized (Kumar, Suseela, & Toppo, 2011), which were similar to the properties of fossil-based diesel oil, such as the pH range of 6–7, the density

range of 0.85–0.89 g/cm^3, and viscosity range of 3.8–4.4 mm^2/s. Moreover, microalgae have oil productivity of 136,900 L/ha, which is much higher than the most efficient palm oil productivity of 5950 L/ha (Chisti, 2007; Kumar et al., 2011).

Microalgae offer several advantages over the earlier generations of biofuels such as a broad range of conversion pathways (biochemical, thermochemical, or focused exploitation on oil content), more efficient conversion process compared to lignocellulosic feedstocks, or less conflict of food versus fuel since algae cultivation taking place on nonarable land. Microalgae can grow in various environmental conditions, including wastewater. This will both answer the demand for biofuel as well as contribute to waste management and pollution control. Potential products of biofuels from microalgae are shown in Table 1. However, production with current technologies yet remains at the research and development level. The idea of larger-scale production or start serving a niche market remains unviable unless the production costs can compromise (Davis, Aden, & Pienkos, 2011; Davis et al., 2014; Kumar, Singh, & Sharma, 2017). The potential window for cost reduction is mainly from technological options, e.g., low-cost engineering equipment and maximizing lipid content extraction (Ansari, Shekh, Gupta, & Bux, 2017; Davis et al., 2011; Shriwastav & Gupta, 2017). Further, the constraints are also coming from the supply chains and distributions, including infrastructure, logistical issues of transportation from and to processing facilities and distribution points, end-users. Fuel shortages are common in remote areas, while it is also unlikely to establish processing facilities in those areas as it will require very high investment.

There have been multiple on-going studies, researches, and process developments aimed to create commercially available microalgal biofuels. Researchers and technologists have always kept the effort on updating various

TABLE 1 Biofuel product targets from various microalgal strains.

Products	Pretreatment	Microalgae	Reference
Biobutanol	Acid + alkali	*Chlorella vulgaris*	Wang et al. (2017), Amiri & Karimi (2018), and Gao, Orr, & Rehnmann (2016)
Biomethane	*Milling, anaerobic digestion with waste sludge, organic compound e.g., N-methylmorpholine-N-oxide, sonication*	*Nannochloropsis oculata, Tertraselmis* sp., *Tertraselmis* spp.	Hernandez, Solana, Riaño, García-González, & Bertucco (2014), Ward & Lewis (2015), and Zhao et al. (2014)
Bioethanol	Acid or diluted acid	*Scenedesmus abundans PKUAC 12, Chlorella vulgaris FSP-E, Chlorella* sp. *KR-1, Scenedesmus obliquus CNW-N*	Guo et al. (2013), Ho et al. (2013), and Lee, Oh, & Lee (2015)
Biodiesel	Biocatalyst: immobilized *Aspergillus niger*	*Scenedesmus obliquus*	Guldhe, Singh, Kumari, & Rawat (2016) and Misra, Guldhe, Singh, Rawat, & Bux (2014)
Biogas (syngas + biohydrogen)	Thermal, HCl hydrolysis, catalytic pyrolysis	*Scenedesmus* sp., *Chlorella vulgaris*	Beneroso, Bermudez, Arenillas, & Menendez (2013), Hu, Ma, Li, & Wu (2014), Liu et al. (2010), Wang & Yin (2018), and Yang, Guo, Xu, Fan, & Li (2010)

assessment tools, consisting of resource assessment, TEA, and life cycle assessment (LCA) to evaluate the progressing study on the feasibility of microalgal biofuels (Giwa et al., 2018; Harun et al., 2011; Nagarajan, Chou, Cao, Wu, & Zhou, 2013; Quinn & Davis, 2014; Xin et al., 2016). These assessment tools were conducted in various microalgae, process selection, and various range of scale, from laboratory scale to pilot-plant scale and commercial facilities (Giwa et al., 2018; Moazami, Ashori, Ranjbar, & Tangestani, 2012; Quinn & Davis, 2014; Rawat, Kumar, Mutanda, & Bux, 2013; Wen et al., 2016).

In general, biodiesel production from microalgae is conducted through several stages: cultivation and harvesting of microalgae, oil extraction, and oil conversion into biodiesel by transesterification, followed by downstream processing of microalgal biodiesel to fulfill commercial specification (Amaro, Guedes, & Malcata, 2011; Chisti, 2007; Kumar et al., 2011; Moazami et al., 2012; Pragya, Pandey, & Sahoo, 2013; Rawat et al., 2013; Scott et al., 2010).

2.1 Cultivation of microalgae

There were particular concerns regarding microalgae cultivation, such as feasibility between the two options of closed or open pond bioreactor, strategies to eliminate contamination problems, and options for CO_2 sources (Scott et al., 2010). Growth conditions must be manipulated to ensure optimum oil production. Commonly, microalgae are cultivated in two significant types of reactor growth: mainly in open pond (OP) systems and photobioreactor (PBR), also in the combination of both along with some modifications, particularly, OP methods proved to be easier to be scaled-up (Rawat et al., 2013). PBRs had high productivity and better culture control over an OP system, but the latter had lower construction and maintenance costs

TABLE 2 Summary of cost of biodiesel through various methods of microalgae cultivation (Nagarajan et al., 2013).

No	Cultivation methods	Process scale	Cost of biodiesel ($/L)	Details	References
1	Open pond	333.3 ha	1.68	Monte Carlo sampling method	Delrue et al. (2012)
		500 ha	3.55	Solvent extraction of oil is eliminated	Amer et al. (2011)
		1950.58 ha	2.73	–	Davis et al. (2011)
		400 ha	0.50–0.82	Biogas was produced as coproduct for electricity generation	Benemann & Oswald (1996)
		Not stated	3.91	Monte Carlo sampling method	Richardson, Johnson, & Outlaw (2012)
			0.45	Algal turf scrubber (ATS)	Hoffman, Pate, Drennen, & Quinn (2017)
			0.59	ORP(open pond raceway)	Hoffman et al. (2017)
2	Photobioreactor (PBR)	Not stated	2.80	Monte Carlo sampling method	Delrue et al. (2012)
		1950.58 ha	5.70	–	Davis et al. (2011)
		Not stated	9.89	Monte Carlo sampling method	Richardson et al. (2012)
		500 ha	21.72	Solar lit PBR	Amer et al. (2011)
3	Integrated system		2.69	Raceway+PBR, Monte Carlo sampling method	Delrue et al. (2012)
			49.46–75.77	Carbon credits and biogas production credits are included	Harun et al. (2011)

(Rawat et al., 2013; Resurreccion, Colosi, White, & Clarens, 2012). Giwa compared several OP systems production rates from several existing works that have been compared, and it was known that OP system would be able to provide high microalgae productivity (Giwa et al., 2018), although several previous studies showed inconsistent values of productivity. Cost analysis of algal biodiesel systems of the National Renewable Energy Laboratory (NREL) (Benemann & Oswald, 1996) was assumed by (Nagarajan et al., 2013) to be based on an OP system with paddle wheels, where the total area of the system was 4×10^6 m^2, the land cost assumed to be $0.2/

m^2, and the productivity was assumed to be 109 and 219 MT/ha/year. Summary of cultivation costs are shown in Table 2.

2.1.1 Raceway or open ponds

The basic principle of microalgae cultivation, which can be conducted in a closed-loop recirculation channel of about 0.3 m deep (Chisti, 2007). The cultivation broth is fed continuously during the daylight into the pond and circulated throughout the channel by using paddlewheel. The flow of the broth is ended just behind the paddlewheel, completing one full circulation before it is harvested. The paddlewheel should be operated continuously to prevent biomass sedimentation, and the baffle is positioned to facilitate better mixing along the raceway. During the process, water evaporation can be significant. The water evaporation can facilitate system cooling, but the water loss needs to be compensated. The options of raceway materials are concrete, compacted earth, and can be lined with white plastic. Compared to the closed system of a PBR; there may be significant losses of CO_2 supplies in raceway ponds. In general, the concentration of biomass is relatively low due to the poor mixing and difficulties in maintaining an optically dark zone. Microalgae cultivation using raceway ponds is relatively lower in cost than PBR due to the lower investment cost and low operating cost. However, biomass productivity is also lower.

2.1.2 Photobioreactor

Basic configurations of PBRs consist of arrays of transparent plastic or glass tubes to capture and collect the sunlight (Chisti, 2007). Cultivation broth is pumped to flow through the collector tubes. The diameter of the tube should be small enough (generally 0.1 m or less) to ensure that the sunlight can penetrate through the tubes into the broth. Briefly, the fresh medium broth is fed into a degassing column where it mixes with the air. The column temperature is maintained by cooling water, while the exhaust gas is released to the outside air. After it mixes with air, the medium broth is then pumped by a mechanical pump or an airlift pump to flow through the collector tubes. One of the advantages of using PBR for microalgae cultivation is that it enables the cultivation of single-species culture for extended time growth and has successfully produced microalgae biomass at a larger scale.

PBR and raceways for the capacity of 100-ton biomass, which have been used in commercial operations, have been compared (Chisti, 2007). Both methods consumed an identical amount of carbon dioxide; however, the utilization efficiency is higher for the closed system of PBR since the carbon dioxide loss is eliminated. PBR can also facilitate biomass growth for more than 13-fold higher to the raceway ponds; therefore, results on the higher oil yield per hectare of land. In terms of technical feasibility, both production methods are feasible.

2.2 Biomass harvesting

The efficient biomass harvesting from the cultivation broth is an essential step toward large-scale biodiesel production from biomass (Amaro et al., 2011). Primary microalgae harvesting techniques include centrifugation, flocculation, filtration, sedimentation, flotation, and electrophoresis, but the cost was typically high due to the low mass fraction. Microalgae harvesting technique should be chosen based on the characteristic of the biomass, such as density, size, and desired by-product to be recovered. In general, harvesting microalgae consist of two-step processes: bulk harvesting and thickening. Bulk or primary harvesting was aimed to separate biomass from the cultivation broth, while thickening steps or secondary harvesting was aimed to concentrate biomass slurry by filtration or centrifugation. The cost of biomass recovery was inversely proportional to the concentration of the cultivation broth, where the higher concentration will have a lower cost of recovery (Chisti, 2007). Therefore, the cost of biomass recovery after cultivation in PBR costs much lower than the recovery cost of biomass in raceways. Besides, the higher biomass concentration means a smaller volume of the reactor for a certain amount of biomass.

2.3 Oil extraction from microalgae

After harvesting, the oil should be extracted from microalgae using the most energy-efficient technique at low cost, for example, by avoiding extraction by large volume organic solvent. Further, biorefinery processes to optimize the utilization of the remaining cell to recover high-value products should also be considered. The lipid extraction process from microalgae was not yet fully understood, which led to difficulties in up-scaling to the industrial process (Amaro et al., 2011). Besides, oil extraction for biodiesel production should be selective to triglycerides fractions, which is the raw material for the transesterification process. The extraction process must be controlled to prevent contamination by other cellular components, such as DNA or chlorophyll. One of the approaches to minimize solvent use can be made by the cell wall decomposition using enzymes (Scott et al., 2010). Oil recovery from microalgae may consist of several processes, such as microalgae dryer, bead milling of dried microalgae, and oil decanter (Giwa et al., 2018).

2.4 Oil conversion to biodiesel (transesterification)

Oil extracted from microalgae is then converted to biodiesel. There were several methods available to convert algal oil into biodiesel, such as micro-emulsification, pyrolysis (or cracking), and transesterification (Pragya et al., 2013). However, micro-emulsification and pyrolysis or catalytic cracking are cost-intensive and produce low-quality biodiesel. Therefore, microalgal oil conversion to biodiesel is mainly conducted through the transesterification process by reacting to the oil (triglycerides) with excess methanol in the presence of alkali to produce fatty acid methyl esters and glycerol as a side product. Briefly, transesterification using an alkaline catalyst, which is carried out at a temperature of approximately 60°C under atmospheric pressure (below-boiling temperature of methanol), is completed in about 90 min. This reaction consists of two liquid phase systems, the methanol phase, and the oil phase. Saponification of oil and alkali could be prevented by

removing water from the system and minimize free fatty acid in the feed oil. The resultant biodiesel is then washed repeatedly with water to remove methanol and water. This procedure has become the standard industrial technology (Scott et al., 2010). Still, there were studies to improve the efficiency of the oil conversion process. Transesterification of microalgal oil can be done in situ or in a one-step method, where extraction and transesterification of the oil are conducted simultaneously in the reactor, which can reduce the number of unit operations, reduce time consumption, thereby lower the overall biodiesel production cost. Concisely, harvested microalgae biomass is dried and milled prior to the simultaneous one-step extraction and transesterification using methanol as solvent. Mixing methanol with a nonpolar solvent such as methylene dichloride in ratio of 3:1 proved to enhance the efficiency of the extraction (Li et al., 2011). A highly efficient simultaneous oil extraction and transesterification (SOET) process have been tested, where the slurry of microalgae after primary settling is directly mixed with methanol and alkali using high power dual-frequency ultrasonicator (Nagarajan et al., 2013). Recovery of oil from microalgae biomass followed by the conversion of the extracted oil to biodiesel is most likely similar, whether the biomass is produced in raceways or PBRs. Therefore, biomass production or cultivation cost will be the most determining cost difference between the two methods of producing biodiesel from microalgae (Chisti, 2007).

2.5 TEA of biodiesel production from microalgae

TEA mainly consists of two components: (1) capital cost or total capital investment (TCI), which includes equipment cost, engineering and contingency cost, land cost, and working capital cost, and (2) operating cost, which includes the cost of raw materials, labor and overheads, utilities, waste treatment, and disposal, maintenance, tax, and insurances. Among all the input, some criteria are highly dependent on the process selection, such as the cost for equipment and land purchase (capital cost), and the raw materials, utilities, and waste treatment cost (operating cost). Various component cost distribution of various process selection and production scale has been presented in the actual result of microalgal biodiesel TEA (Giwa et al., 2018; Harun et al., 2011; Hoffman et al., 2017; Nagarajan et al., 2013; Quinn & Davis, 2014; Xin et al., 2016). Correctly, the cost breakdown for biodiesel production from microalgae can be classified into the following:

2.5.1 Capital cost for microalgal biodiesel production

Among the components of capital cost for microalgal biodiesel production, which significantly affects the process selection, are equipment cost and land cost, while the rest of the cost (engineering and contingency and working capital) can be calculated by using a multiplication factor to the total main equipment cost. Process equipment in microalgae cultivation stages, which commonly included the TEA, were pond levees, geotextiles, mixing equipment (paddlewheels), CO_2 sumps/diffuser, CO_2 distribution system, piping for water and nutrient supply, and pump systems. Microalgae harvesting stages commonly required several process equipment such as settler and flocculation pond. Oil extraction from microalgae commonly conducted using a centrifuge or extraction tank, while oil conversion to biodiesel via transesterification reaction commonly required mixing reactor and decanter centrifuge (Hoffman et al., 2017; Nagarajan et al., 2013). The capital costs of each stage of process based on the existing study on the known processes, such as raceways, PBRs, harvesting, drying, anaerobic digestion, and gasification, have also been estimated (Delrue et al., 2012).

2.5.2 Operating cost for microalgal biodiesel production

Operating cost for biodiesel production from microalgae consists of the cost of raw materials, labor and overheads, utilities, waste treatment and disposal, maintenance, tax, and insurances. Raw materials include microalgae cultures, CO_2 supplies, and nutrition to ensure the optimum growth of microalgae (Hoffman et al., 2017; Nagarajan et al., 2013). The total annual operating cost was calculated as the sum of the operating cost (utilities, labor, and other costs) and fixed cost, which is calculated from the depreciable capital cost (Delrue et al., 2012). This cost consists of 55% capital cost for the general maintenance, storage, engineering, and spare parts costs, license fees, initial expenses at 2% of capital cost, and process commissioning at 25% of operating cost. The calculation used annuities of 20 years and a discount rate of 8% and 7% of the capital cost per year. To conduct a TEA to the two methods of microalgae cultivation, several assumptions were used as follows: the developed plant is assumed to have a 3-year start-up with capital costs occurring at 8%, 60%, and 32% in years 1, 2, and 3, respectively; the plant operates at 50% capacity during the first three months of operation; plant life is 30 years; an internal rate of return of 10% with 60% financing as a 10-year, 8% interest loan, and 40% equity; and tax rate at 35% (Hoffman et al., 2017). The production cost of biodiesel is estimated from the ratio between total annual operating cost and annual production of biodiesel.

It is very beneficial to have a feasible biodiesel production at commercial scale. However, commercial-scale can also be conducted in a range of production scales. Small-scale production and large-scale production may have advantages and disadvantages, not only in the technological aspect but also in economic and environmental aspects, which can be indicated by the component cost, which later determines the feasibility of the production process. A new process evaluation methodology of microalgae biodiesel has been proposed based on four evaluation criteria, such as net energy ratio (NER), biodiesel production costs, greenhouse gases (GHG) emission rate, and water footprint (Delrue et al., 2012). Lipid productivity, cultivation step, and downstream processes were found to have a significant effect on the sensitivity analysis of biodiesel production cost. Considering microalgae for biodiesel as an immature technology, Monte-Carlo sampling method has also been applied to account for the variability and the uncertainty of process performance values (Delrue et al., 2012), such as biomass yield, oil recovery rate, etc., where the value is defined by its average of the minimum and maximum values based on a detailed and critical literature review.

3 Microalgae for chemicals

Microalgae can be used for chemical production via either or the combination of two approaches. The first one is the extraction and isolation of high-value components from microalgae. The second one is using microalgae as feedstocks for bulk chemical production.

The first approach is based on the presence of several high-value chemicals and pharmaceuticals in considerable quantities. Lipid-based chemicals include polyunsaturated fatty acids (PUFA), carotenoids, and sterols. Carbohydrate-based chemicals include alcohols, lactic acid, and bioactive polysaccharides. Protein-based chemicals include phycobiliprotein, bioactive peptides, and amino acids (Andrade, Andrade, Dias, Nascimento, & Mendes, 2018; Borowitzka, 2013). Although these high-value components have been identified in several species of microalgae, commercial production of chemicals from microalgae is still limited. Whole-cells microalgae from *Chlorella* and *Arthrospira* (*Spirulina*) genus are

marketed as neutraceuticals by several companies such as Taiwan Chlorella Manufacturing Co. (Taiwan), Klotze (Germany), and Earthrise Nutritionals (United States) (Andrade et al., 2018). DSM (the Netherlands) is the market leader of concentrated algal oil rich in PUFA (van der Voort, Spruijt, Potters, de Wolf, & Ellissen, 2017).

The second approach offers biobased (bulk) chemicals as green alternatives to many drawbacks of fossil-based chemicals. The main advantage is the renewability of biomass, including microalgae, compared to the limited reserve of fossil resources. Besides, some functionalities that are required in bulk chemicals are often already present in microalgae biomass. However, the production of biobased chemicals may require different resources, processes, and infrastructure altogether.

3.1 Polyunsaturated fatty acids

PUFA such as eicosapentaenoic acid (EPA) and docosahexaenoic acid (DHA) are essential fatty acids that are required in the human diet. Microalgae, particularly of marine origin, are excellent sources of PUFA (Ryckebosch et al., 2014; Tokuşoglu & Ünal, 2003). Several studies have explored PUFA production from microalgae (Li et al., 2019). Besides, several industrial processes already existed (van der Voort et al., 2017). A broad outline of the process is presented in Fig. 1.

Fig. 2 shows that process cost for omega-3 EPA and DHA, two of the most critical PUFA, varies from 12 to 52 USD/kg. Each process is detailed in Table 3. Estimation of the market price of PUFA varies from 21 USD/kg (Industry, 2014) to 75 USD/kg (Barsanti & Gualtieri, 2018). This shows that given optimum condition and favorable market price, PUFA production from microalgae is economically feasible. Fig. 2 also shows that in all processes except D and E, the highest process cost comes from biomass at 16–36 USD/kg-EPA + DHA or

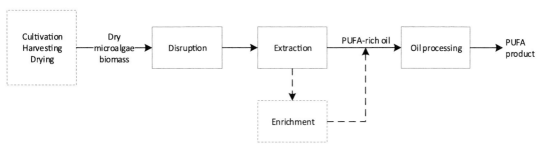

FIG. 1 PUFA production process from microalgae.

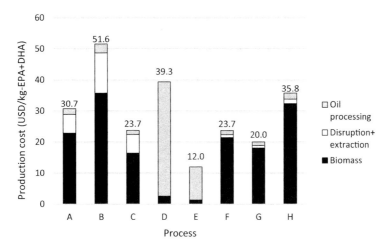

FIG. 2 PUFA production cost for several species and process conditions (explained in Table 3). Value in 2019$.

69%–90% of the total process cost. Biomass cost depends on cultivation, harvesting, and drying methods, which is also influenced by location, available technology, and labor and utility costs.

Comparison between processes B and C where different PUFA contents were assumed shows that selection of biomass with high PUFA content significantly influences the process cost. A previous study suggests that increasing PUFA content in the biomass has a more substantial influence on the overall process cost than increasing biomass production (Chauton et al., 2015).

Cell disruption is required as a pretreatment to rupture the cell and release the compounds of interest. This is followed by the lipid extraction process. Most studies optimize disruption and extraction from microalgae for biodiesel production. For biodiesel production, lipid fraction that is richer in medium- and short-chained fatty acids is more desirable since longer unsaturated fatty acids tend to yield biodiesel with low cetane number and high iodine value (Ramos, Fernandez, Casas, Rodriguez, & Perez, 2009). However, processes with high lipid yield do not always conserve the PUFA in the lipid; therefore, a balance between lipid extraction yield and PUFA recovery is required. Lipid extraction from *Nannochloropsis gaditana* was performed with hexane extraction (10% lipid yield, 42% PUFA content) and subcritical water extraction (18% lipid

TABLE 3   Process condition for PUFA production.

No.	Process	Species	Process condition	References
1	A	*Chlorella vulgaris*	Oil extraction using high-pressure homogenization and hexane extraction. Oil processing was estimated as 6% of the total cost (van der Voort et al., 2017). EPA + DHA concentration was 24% for *C. vulgaris* (Tokuşoglu & Ünal, 2003), 18% for *Nannochloropsis* sp. B (Ryckebosch et al., 2014), 40% for *Nannochloropsis* sp. C (He et al., 2019)	Kang, Heo, & Lee (2019)
	B	*Nannochloropsis* sp.		
	C	*Nannochloropsis* sp.		
2	D	*Phaeodactylum tricornutum*	Biomass production in flat panel photobioreactor, Spain (base case). Process cost is not elaborated	Chauton, Reitan, Norsker, Tveteras, & Kleivdal (2015)
3	E	*Phaeodactylum tricornutum*	Similar to D with EPA + DHA content increased from 6%-dw to 12%-dw	Chauton et al. (2015)
4	F	*Prorocentrum cassubicum*	Biomass production in flat panel photobioreactor, Spain. Process sequence: drying, ball milling, supercritical-CO_2 extraction, distillation, saponification, solvent extraction	van der Voort et al. (2017)
5	G	*Thalassiosira weissflogii*	Biomass production in flat panel photobioreactor, Spain. Process sequence: drying, ball milling, supercritical CO_2 extraction, transesterification with methanol and catalyst, filtration and distillation	van der Voort et al. (2017)
	H	*Chloridella + Raphidonema*		

yield, 21% PUFA content) suggesting that PUFA was readily extractable and additional extracted lipid mostly contained other fatty acids (Ho, Kamal, & Harun, 2018). On the other hand, while hexane extraction only yielded 6% lipid with 31% PUFA from *Nannochloropsis oculata*, both higher yield and content was achieved from combination of hexane and 2-propanol (20% lipid yield, 35% PUFA content) and 96% ethanol (36% lipid yield, 34% PUFA content) making the latter two methods attractive candidates for lipid extraction aiming for PUFA (Pieber, Schober, & Mittelbach, 2012).

After the extraction, PUFA content in the oil can be enriched via hydrolysis with lipase (Byreddy, Barrow, & Puri, 2016). Alternatively, hydrolysis can be applied on wet biomass (Jacob & Mathew, 2017). The subsequent process is purification and concentration of PUFA. Several methods that have been applied are saponification, winterization, and urea complexation (Cuellar-Bermudez et al., 2015).

3.2 Bulk commodity chemicals

Protein from microalgae can be used in either of these forms: the whole cell, concentrate, isolate, or hydrolysate (Soto-Sierra, Stoykova, & Nikolov, 2018). *Chlorella* and *Arthrospira* (*Spirulina*) have more than 50% protein content that can be used in food or nutraceutical applications (Andrade et al., 2018). Protein concentrate and isolate can be used in food applications for their functional properties such as emulsifying and gelation capabilities (Suarez Garcia et al., 2018). Hydrolysate can be added to functional food and beverages. Purified hydrolysate can yield bioactive peptides with neutraceutical or pharmaceutical properties (Suarez Garcia et al., 2018).

In addition to the abovementioned forms, specific proteins such as RuBisCo can be isolated from microalgae (Fig. 3). Microalgae proteins can also be wholly hydrolyzed into single amino acids, which subsequently can be used for producing nitrogen-containing chemicals. Fig. 3

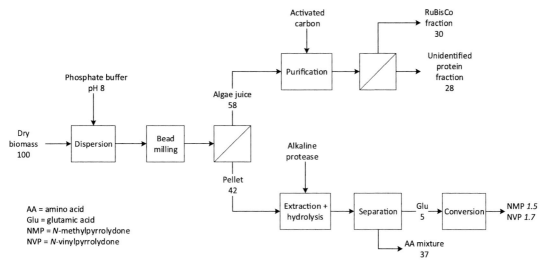

FIG. 3 RuBisCo and pyrrolidone production from microalgae. Numbers indicate protein weight equivalent.

shows a proposed process for the production of RuBisCo, *N*-methylpyrrolidone (NMP), and *N*-vinylpyrrolidone (NVP) from microalgae protein.

Microalgae biomass (*N. gaditana*) can be processed with bead milling and subsequently centrifuged to obtain algae juice and pellet fractions (Teuling, Wierenga, Schrama, & Gruppen, 2017). The juice can then be purified with activated carbon to obtain a RuBisCo fraction (Van De Velde, Alting, & Pouvreau, 2011). Meanwhile,

the pellet fraction can be hydrolyzed to obtain an amino acid mixture (Sari, Syafitri, Sanders, & Bruins, 2015; Sari, Sanders, & Bruins, 2016), which can be subsequently separated to obtain glutamic acid (Teng, Scott, & Sanders, 2012). The latter can be converted into several nitrogen-containing chemicals; in this case, NMP and NVP (Lammens, Gangarapu, Franssen, Scott, & Sanders, 2012).

Table 4 shows the processing cost and potential revenue for the production of RuBisCo,

TABLE 4 Processing cost and revenue for chemicals production from microalgae.

	Processing cost			Product			
Item	Amount processed (kg-protein)	Unit cost (USD)	Cost (USD)	Item	Product (kg)	Market price (USD	Revenue (USD
Biomass	100	2.46	246	RuBisCo	30	8.00	240
Disruption	100	0.04	4	NMP	1.5	3.38	5
Purification	58	0.17	10	NVP	1.7	3.38	6
Extraction + hydrolysis	42	0.92	38				
Separation	42	1.49	62				
Conversion	5	0.66	3				
Total cost			364	Total revenue			251

NMP, and NVP from *N. gaditana*. In total, the processing cost of 100 kg protein is USD 364, while the potential revenue is only USD 251. This suggests that the process is not economically feasible and needs to be improved.

The highest cost comes from biomass at USD 2.46/kg-protein (USD 1.28/kg for 222 kg dry biomass), which accounts for 68% of the processing cost. Dry biomass cost was assumed to be constant and only depend on the content of the compound of interest (Kang et al., 2019). In practice, the high biomass cost does not only depend on protein content, but also on the availability of the protein in either juice or pellet fractions which determine potential product and suitable required process. Furthermore, the cultivation method and location also influence biomass costs (Chauton et al., 2015).

In our calculation, the best-case scenario is referred; thus the actual cost may be higher. RuBisCo content was assumed to be 30% of the protein (Teuling et al., 2017), while the actual content might be as low as 6% (Losh, Young, & Morel, 2013). Furthermore, also it is assumed that all the RuBisCo is present in the juice fraction, which may not be the case in the actual process.

Extraction/hydrolysis, amino acid separation, and conversion, respectively account for 11%, 17%, and 1% of the processing cost. The process to economically separate amino acid from a mixture with more than three amino acids is currently still hypothetical (Widyarani, Bowden, Kolfschoten, Sanders, & Bruins, 2016). We propose using electrodialysis to separate glutamic and aspartic acids from the hydrolysate, followed by specific conversion to separate glutamic from aspartic acid (Teng, Scott, & Sanders, 2012). Electrodialysis was assumed to be the most costly process (Shen, Huang, Ruan, Wang, & van der Bruggen, 2014). The feasibility of glutamic acid conversion to NMP and NVP relies on cost-effective hydrolysis and separation of glutamic acid (Lammens et al., 2012).

In our calculation, we only considered RuBisCO, NMP, and NVP as the products. We did not consider non-RuBisCo protein fraction and hydrolysate after glutamic acid separation, which may also have economic value. RuBisCo's price was assumed to be USD 8/kg, while the actual price may vary from USD 5 to USD 11 (Ansari, Shriwastav, Gupta, Rawat, & Bux, 2017; Kamm, Hille, Schönicke, & Dautzenberg, 2010).

Using defatted biomass instead of full-fat biomass yields more protein, reduces biomass cost per unit protein, and, therefore, more is economically feasible (Sari et al., 2016). On the other hand, if the aim is to obtain intact protein, a harsh condition during lipid extraction might influence protein structure. To extract valuable products from all fractions, extraction of (1) protein, (2) lipid, and (3) carbohydrate has been suggested as the optimum sequence (Ansari, Shriwastav, et al., 2017).

3.3 Pharmaceuticals

The pharmaceutical market is the most important driver for innovation in industrial biotechnology. The pharmaceutical industry is the technology sector, with the highest added value per person employed. Over the past 30 years, medicine products based on biotechnology or biologics have been growing double in the medicine market worldwide. The global sales of biopharmaceuticals were estimated at $170 billion in 2012, accounting for 18% of the overall global market for pharmaceuticals (Scarlat, Dallemand, Monforti-Ferrario, & Nita, 2015).

Microalgae are considered as nutritious food and contain various bioactive compounds that can provide pharmaceutical benefits for human health. Generally, microalgae are considered as an essential component of primary productivity in the aquatic ecosystem. Marine microalgae have potential as the sources of novel bioactive compounds, which can provide additional physiological and pharmacological benefits (Morais, Vaz, Morais, & Costa, 2015). Microalgae are resources of natural antioxidant due to their ability to develop the defense system against photo-oxidative damage through the oxidative mechanism. Besides that, microalgae have several biological activities, such as antimicrobial (Mala, Sarojini, Saravanababu, & Umadevi, 2009), immunomodulatory effect as well as its antiangiogenic modality (Ali, Barakat, & Hassan, 2015), antidiabetic activity (Setyaningsih, Bintang, & Madina, 2015; Windari, Tarman, Safithri, & Setyaningsih, 2019), and anticancer activity (Sirait, Setyaningsih, & Tarman, 2019).

A range of pharmaceutical substances can be obtained from the extraction of biological sources. Pharmaceutical substances can be used as modern medicinal therapy. Biopharmaceuticals will be the most important bioindustry in the future. Natural products are the main target in the development of biopharmaceuticals. Microalgae are one of the aquatic products that have great potential for the development of biopharmaceutical. Active compounds in microalgae, including natural pigments, have bioactivity that are needed for human health.

Bioactive compounds derived from microalgae can be obtained from primary metabolism, including amino acids, fatty acids, vitamins, and pigments, or can be synthesized from secondary metabolism (Morais et al., 2015). Recently, pharmaceutical products derived from microalgae biomass are commercially available in the form of tablets, capsules, powder, or liquid.

The classification of Spirulina products as traditional medicine according to current government regulations can be an obstacle in spreading knowledge about Spirulina. This classification is considered too narrow and inappropriate. The development of Spirulina products is better directed as a healthy food product since it has several advantages, such as highly nutritious (in the form of protein) and beneficial for human health (Kumari et al., 2011).

The quantitative analysis is based on a TEA of harvesting and dewatering systems available at the industrial scale (Fasaei, Bitter, Slegers, & van Boxtel, 2018). However, knowledge about costs on microalgal cultivation and processing at commercial scale limited, particularly concerning PBRs. Likewise, the techno-economic aspects of biopharmaceuticals on microalgae have not thoroughly investigated.

Projections show that the production of high-value products from microalgae could be profitable nowadays, and commodities will become profitable shortly. The commercialization of different functional components requires selective biorefining of biomass that remains a challenge (Ruiz et al., 2016). For that, a strategy to adapt is needed accordingly. Culturing technique for obtaining microalgae biomass for biofuel is different from culturing for that for biopharmaceutical. The cultivation of food and biopharmaceutical has to be more aseptic, providing a challenge in the development of microalgae for biopharmaceuticals.

Large-scale cultivation in closed systems involves some constraints to consider. Overheating can be lethal to microalgae. Likewise, lighting, pH media, salinity, temperature, etc. also affect the metabolism of microalgae. These parameters are needed for further TEA study on biopharmaceuticals.

4 Microalgae for bioplastic

Demand for bioplastics is continuously increasing, boosted by the need to have a safer environment as well as a sustainable world. Microalgae have been investigated to be a natural resource for bioplastics production. Bioplastic production from microalgae may utilize their lipid, carbohydrate, or protein fractions.

Carbohydrate based bioplastics are considered as the oldest form of bioplastics. In this type of bioplastics, cellulose or starch is commonly utilized. *Chlorela* sp. may consist of 2%–11%

(dw) of starch (Cheng, Labavitch, & VanderGheynst, 2015). Most of microalgae strains have a small fraction of starch. This condition further limits the industrial application of starch-based bioplastics. Efforts have been made to increase the starch content of several microalgae strains. A high level of starch can be achieved by applying sulfur depletion growth medium, tailoring the light intensity, microelement (nitrogen, phosphorus, sulfur) limitation, and the use of a specific inhibitor of cytoplasmic protein synthesis, cycloheximide. Starch content, up to 50% of starch for *Chlamydomonas reinhardtii* strain was achieved. This indicates the potential of this strain to be used as a starch resource for bioplastics. However, the starch content was decreased into 25% (dw) as the growth medium was up the scale to pilot scale (Mathiot et al., 2019), which later limit the possible industrial commercialization for microalgae-starch-based bioplastics.

Microalgae proteins have also been investigated to be used in bioplastics production. Soybean and sunflower seed proteins are examples of the crop protein resources for bioplastics. The need to avoid food shortage in the future motivates the use of microalgae proteins instead of crop proteins. The ability of microalgae to grow in nutrient-rich wastewater provides ease in growing and added values to the environment. Chlorella and spirulina are examples of the microalgae strains having a high amount of protein content, up to 60%. Proteins are brittle; therefore, to have plastic behavior, plasticizers are required. Applying the hot press treatment, whole biomass can be directly converted into protein-based bioplastics (Figs. 4 and 5). However, the different amino acid compositions resulted in different mechanical properties of the bioplastics (Zeller, Hunt, Jones, & Sharma, 2013). In addition to direct conversion into bioplastics, microalgal protein can also be used as raw materials for the production of thermoplastic polyurethane (Kumar et al., 2014), which is usually made from petroleum feedstock.

The major obstacle in the commercial production of PHAs is the high cost of bacterial fermentation processes, which make bacterial bioplastic more expensive than petroleum-based polymers such as polypropylene and polyethylene

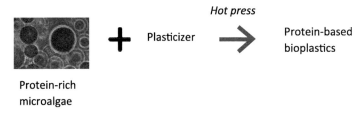

FIG. 4   Direct conversion of protein-rich microalgae into bioplastics.

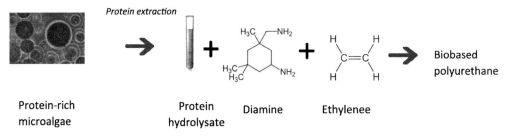

FIG. 5   Indirect conversion of protein-rich microalgae into bio-based polyurethane.

(Nagarajan et al., 2013). Although microalgae have been shown to have potency as bioplastic resources, TEA has not been conducted.

5 Microalgae: Integrated system

Although studies indicated the possibility to produce biofuels and other high-value products from microalgae, an outlook indicates that the integrated system may work well for microalgae to reduce the production cost of microalgae's main products. A recent analysis of the techno-economic of an integrated microalgae system has been conducted.

5.1 Integrated: Biodiesel from microalgae and jatropha

Microalgae are among the lists of biodiesel resources, together with jatropha. Considering the natures of this biomass, which consist of the high amount of lipid and both can be cultivated at peculiar environment conditions, scenarios for combining this biomass for biofuels and other products have been investigated (Giwa et al., 2018). In this study, microalgae and jatropha are used as feedstocks in the integrated systems for producing streams of products. Scenario 1 assumed that these feedstocks could be used for producing biodiesel, glycerol, biogas, fertilizer, and animal feed. Scenario 2 assumed that they would produce biodiesel, glycerol, and bioethanol. While Scenario 3 assumed that microalgae and jatropha would produce biodiesel, glycerol, biohydrogen, and animal feed. The analysis was made based on the model developed using Super-Pro designer indicated that scenario 1 was more profitable compared with others. Scenario 2 and

3 required a much higher investment cost compared to Scenario 1. Scenario 1 also had the lowest operating cost, among others. This was due to the production of methane in Scenario 1, which later can be utilized to supply the power for the production process. Another factor that contributes to the economic profitability of scenario 1 was that the products produced by scenario 2 and 3 have low economic value and usage to be applied in the author's regional area, which is UAE.

5.2 Integrated: Bioethanol, crude bio-oil, and biofertilizer

High carbohydrate content combined with low lignin content are two characteristics to justify the use of microalgae as a feedstock for bioethanol production. Fig. 6 illustrates the process mechanism in converting microalgal biomass into bioethanol. Cultivation and harvesting of the microalgae as a feedstock for bioethanol are like that of these processes for producing biodiesel (see the section mentioned previously).

Pretreatment is considered as the initial process for utilizing microalgae as bioethanol feedstock. The main objective of this step is to enable the microalgae cell to undergo lysis so that the intracellular compounds, particularly carbohydrates, from microalgae can be released.

Conventional pretreatment includes: (1) physical, (2) chemical, (3) enzymatic, and (4) combined methods. Physical methods aimed to disrupt the cell wall using physical interaction like a pulsed electric field (Guo et al., 2019), high-pressure homogenization (Carullo et al., 2018), ultrasonication (Skorupskaite, Makareviciene, Sendzikiene, & Gumbyte, 2019), and ball milling. Chemicals and

FIG. 6   Processing microalgae into bioethanol.

enzymatic methods are commonly applied to pretreat microalgae for bioethanol production. These methods are preferred compared to physical methods due to the conjunction in disrupting cell walls to release carbohydrates and hydrolyzing them into fermentable sugars. Furthermore, recently, enzymatic hydrolysis shows its potential over chemical hydrolysis, such as acid hydrolysis (Phwan, Ong, Chen, & Ling, 2018).

Although studies have indicated the success of bioethanol production from microalgae, this is still considered as a contemporary application that requires modern technology. As what has been learned from biodiesel production, biofuel production seems to be more economically feasible using an integrated system, and this system is also applied for TEA of bioethanol production. In this case, bioethanol would be integrated with the production of crude bio-oil liquid and biofertilizer (Hossain, Mahlia, Zaini, & Saidur, 2019). In this study, *Chlorella vulgaris* was used as the feedstock due to its high content of carbohydrates (up to 52%). Two types of cultivation methods were involved to supply the plant with total microalgae of 220 tons per year: PBRs and OP. The cell lysis was conducted using a physical method (extractor), while the hydrolysis was conducted using an enzymatic method. The calculation was conducted for the project lifetime of 20 years. With this model, the total equipment cost (TEC) for bioethanol production would be $339,600. This TEC was still considered to be reasonable, although PBR was utilized. TEC would be the significant cost contributor for fixed capital investment (FCI), which was calculated to be $1,303,724. The FCI, together with total cultivation area cost and working cost, would give the final total investment cost of $1,653,148. The major cost bearing was the operational cost (OC), which covered the operating laboratory and utility cost. The OC total was $1,796,000.

Considering the market price of bioethanol, crude bio-oil, and biofertilizer of 2.51/gal, 5.000/gal, and 3.75/kg, respectively, the model would give total bioethanol sales of $143,290 and total by-product sales of $80,000 per year. The by-products (crude bio-oil and biofertilizer) were shown to contribute significantly in providing annual total plant profit of $ 591,333, emphasizing the economic feasibility to produce bioethanol in the integrated system.

5.3 Integrated: Biodiesel and methane

The benefit of integrating biodiesel and methane production was also indicated by the techno-economic study of biodiesel and biogas production from microalgae. The current biodiesel production system involves cultivation, dewatering, extraction, and transesterification processes. Cultivation is the most energy-consuming step, among others. Self-sufficient electricity is expected to reduce the biodiesel production cost, thus improving the techno-economic feasibility of biodiesel production. Instead of biorefining, the study showed that an integrated system exhibited economic benefit (Harun et al., 2011). Higher annual electricity production was achieved when whole-cell of microalgae was used for methane production compare to the fractionated microalgae one. In this case, when using whole microalgae cells for producing methane, as much as 70.7 GW h/ year electricity was produced. Whilst, only 31.3 GW h/year was produced when methane was produced from protein and carbohydrate fractions, while lipid was only used for biodiesel production. The energy produced from microalgae converted into methane was later being used as the input energy for microalgae cultivation ponds. These microalgae biomass obtained from these ponds are further being used for biodiesel production. The energy required for cultivation varies with the type of the pond. The electrical energy consumption for the horizontal tubular reactor, external loop reactor, and raceway pond is estimated to be 153, 136.8, and 8.7 GW h/year, respectively, for producing 50 kt/year dried microalgae. Considering the methane produced from microalgae biomass and the electrical energy required for cultivation, the raceway pond is likely the preferable method for biodiesel production. With this type of cultivation, followed by the solvent extraction method for producing biodiesel and transesterification process, around 71% of electricity is still available. This excess may further be sold or used for powering the nearby houses. Overall, integrating methane and biodiesel production from microalgae can reduce biodiesel production costs and provide environmental profit in terms of carbon credit.

5.4 Integrated: Wastewater treatment and biofuel production

The use of wastewater as media has been reviewed as an option for the feasibility of microalgae extensive scale cultivation. In addition to electricity, cultivating microalgae also requires three primary nutrients: carbon, nitrogen, and phosphorous. Besides, micronutrients are also required, such as silica, calcium, magnesium, potassium, iron, manganese, sulfur, zinc, copper, and cobalt. If the primary nutrients and the micronutrients are not available in the water source, then fertilizers are added to the cultivating system to ensure the growth of microalgae. The addition of commercial fertilizers is unlikely due to the increase in production cost. Considering that wastewater somehow contains nutrients, this is then an attractive resource for microalgae growth (Christenson & Sims, 2011; DeRose, DeMill, Davis, & Quinn, 2019).

Although wastewater helps in improving the economic feasibility of extensive scare microalgae cultivation and may contribute to environmental benefit in terms of water remediation, however; the nature of wastewater may affect the character of the microalgae. As the wastewater may contain a high amount of total suspended solids, the grown microalgae may contain a high amount of ash (DeRose et al., 2019). Thus, fractionation is needed to ensure that this will not trouble the downstream processing.

Recently, TEAs have been conducted on the microalgae having low lipid while high ash content (DeRose et al., 2019). An integrated wastewater remediation and biofuel productions have been modeled. Microalgae composition used for this model was based on the data obtained from Sandia National Laboratories (US) that grow microalgae using wastewater. The average microalgae composition was protein 30%, carbohydrate 50%, and 15% other organics on the ash-free dry weight. The ash content was 75% (g ash/g harvested microalgae biomass). The production streams of biofuel involve the biochemical and thermochemical routes Fig. 7. Within this process, fractionations were conducted. Fig. 7 indicates the production rates of fusel alcohol, struvite, diesel, and naptha from microalgae cultivated in wastewater (Algae Turf Scrubber, ATS Microalgae). As a comparison, data from microalgae cultivated in clean water using open race pond is also included. As can be seen in Fig. 8, the microalgae biomass grown in wastewater are tailored for producing fusel alcohol and struvite, indicated by the higher rate of production of these two products compare to the rate of production from open race pond systems.

Economic analysis of the biofuel production from microalgae grown in wastewater indicated that ash content in this type of microalgae is the critical point for the viable biofuel production. Microalgae ash reduction requires additional capital cost, mainly related to settling the equipment for ash reduction. Unfortunately, reducing ash content has an asymptotic result, such that each extra percentage of ash removal gives diminishing returns on capital cost reduction. In terms of carbon emission, the 50% ash reduction would reduce overall process emissions to 53.81 and $-32.6\,g\,CO_{2eq}/(MJ\,fuel)$ for the biochemical and thermal-chemical pathways, respectively. However, based on the lifecycle analysis, these numbers, unfortunately, have not met the renewable fuel standard target of a 50% reduction in well-to-wheels emissions

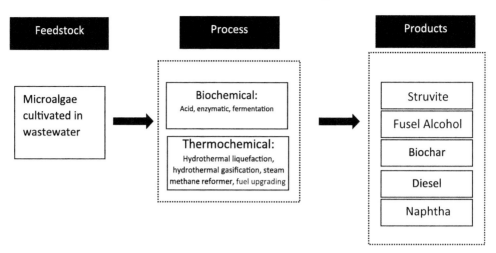

FIG. 7   A schematic integrated and fractionated microalgae system, for producing streams of biofuels. *From DeRose, K., et al. (2019). Integrated techno economic and life cycle assessment of the conversion of high productivity, low lipid algae to renewable fuels. Algal Research, 38, 101412.*

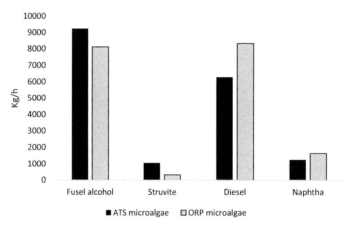

FIG. 8   Production rate of the biofuel production rate of biofuel following integrated microalgae systems. ATS (algae turf scrubber) microalgae: microalgae growth in wastewater. ORP (open race pond): microalgae growth in clean water using open race pond system. *Data from DeRose, K., et al. (2019). Integrated techno economic and life cycle assessment of the conversion of high productivity, low lipid algae to renewable fuels. Algal Research, 38, 101412.*

compared to traditional diesel. Therefore, the study on the TEA of integrated microalgae cultivated in wastewater and biofuel production recommends that an investigation of the pathways for reduction of both ash and direct carbon dioxide emission is required to meet the economic viability (DeRose et al., 2019).

5.5 Integrated: Bioethanol, biodiesel, and biogas

The high lipid and carbohydrate content, along with the incredibly high biomass productivity per hectare land, make microalgae very attractive to be used as potential oil sources to produced liquid fuels, such as biodiesel and bioethanol. However, the biofuel production process from microalgae required not only high energy and nutrient input but also conversion reaction and downstream processing to convert algal oil into market-suitable biodiesel or bioethanol (Ward, Lewis, & Green, 2014; Wu et al., 2019). Several studies stated that biogas production through anaerobic conditions might be the potential alternatives to reduce the energy footprint for the overall microalgae utilization (Passos & Ferrer, 2014; Passos, Uggetti, Carrere, & Ferrer, 2014). Anaerobic digestions have been commonly utilized in the waste treatment unit, and it is proved to be suitable to be applied for high moisture content biomass, such as microalgae, without further dewatering process (Jankowska, Sahu, & Oleskowicz-Popiel, 2017; Raheem, Prinsen, Vuppaladadiyam, Zhao, & Luque, 2018; Wu et al., 2019). In general, anaerobic digestion of microalgae consists of three stages. Firstly, macromolecules (polysaccharides, lipids, and proteins) are hydrolyzed into smaller components (simple carbohydrates, fatty acids, and amino acids) and glycerols. Secondly, the hydrolysis products are further converted into volatile fatty acids (acidogenesis), acetate compound (acetogenesis), hydrogen, and carbon dioxide. Then, the acetate, hydrogen, and carbon dioxide are converted into methane, which are called as methanogenesis steps (Gonzalez-Fernandez, Sialve, & Molinuevo-Salces, 2015). The yield and composition of biogas are affected by the composition of microalgal biomass, temperature, pH, substantial retention time, and nutrient supply rate (Raheem et al., 2018).

Several advantages of anaerobic digestion of microalgae are that it has higher energy efficiency compared to biodiesel or bioethanol production. Besides, the nutrients, such as organic nitrogen or phosphorus from methanogenesis steps, can be recycled back and used as nutrients for algae cultivation, while the methane produced from anaerobic digestion process can be burnt to produce heat and electricity. Furthermore, anaerobic digestion can utilize substrate not only from fresh microalgal biomass, but also from the microalgal biomass residue after bioethanol or biodiesel production process, which may improve profitability while reducing process waste (Jankowska et al., 2017). Anaerobic digestion of microalgae can facilitate the conversion of biomass into biogas, which mainly consist of methane and carbon dioxide, by the aid of microorganisms under anaerobic or oxygen-free conditions (Jankowska et al., 2017; Passos et al., 2014; Uggetti, Passos, Sole, Garfi, & Ferrer, 2017). Besides, not only the methane has high heating value and can be used to generate electricity, but the resulted carbon dioxide can also be used as a nutrient to grow microalgae biomass during cultivation (Harun et al., 2011). Anaerobic digestion of microalgae should be integrated with the production of biodiesel, bioethanol, or volatile fatty acids from microalgae (Jankowska et al., 2017). This integration will optimize wastewater treatment, improve profit from the side products while reducing production and utility cost of biodiesel or bioethanol production.

Despite the potential benefit, the economic viability of this process still needs to be evaluated. One of the standard tools that can be conducted to evaluate the economic feasibility of biogas production from microalgae is TEA. This tool requires an estimate on capital and operating cost profile to evaluate the potential economic viability of the technology, based on the existing experience, to enable production at commercial scale. As the TEA on anaerobic digestion of microalgae at full-scale production was still limited, The TEA has been conducted on the overall anaerobic digestion of microalgae *Cyanothece* BG0011 for biogas production, which consist of microalgae cultivation up to biogas upgrading (Wu et al., 2019). This study indicated that the cost of biomethane production from biomass productivity of 9.6 g afdw/m/day was a range between \$12.16 and 14.18/MMBTu, while electricity production cost range between 11 and 13 cents/kwh (Wu et al., 2019). The integration of biodiesel production with anaerobic digestion to produce methane proved to improve economic feasibility by enabling in situ generating electricity from the produced methane (Harun et al., 2011). This in situ generating electricity can be used to supply electricity requirements during cultivation, dewatering, oil extraction, and then the oil conversion process to biodiesel. Theoretical calculations informed that electricity generated from biomethane reduced the cost of biodiesel production by 33% while reducing carbon emissions by approximately 75% (Harun et al., 2011). The attractiveness of biogas production from microalgae via anaerobic digestion is its potential to be integrated within the microalgae biorefinery concept, where the process may provide bioenergy, nutrients (nitrogen, phosphorus, and CO_2), and also water for cultivation of microalgae (Uggetti et al., 2017).

6 Conclusion

The economic feasibility of microalgal-based products does not only depend on the investment and operational cost but also on the economic value and usage of the products in a particular region. The assessment reveals that microalgae have significant potentials for biofuels, chemicals, biopharmaceuticals, and bioplastics production. However, several routes of the current single-production approach do not show the profitability of the products. An integrated production system of microalgal-based biofuels and other value-added products is promising as the capital, and operating costs are borne to cover several target products at the same time. TEA has been a practical framework to estimate the performance of each route. There is a high variation in TEA results due to a diverse range of technical solutions for microalgae cultivation and extraction. The selection of input parameters (assumptions) are the key to the TEA.

References

Ali, E. A. I., Barakat, B. M., & Hassan, R. (2015). Antioxidant and angiostatic effect of *Spirulina platensis* suspension in complete Freund's adjuvant induced arthritis in rats. *PLoS One*, 10(4), 1–13.

Amaro, H. M., Guedes, A. C., & Malcata, F. X. (2011). Advances and perspectives in using microalgae to produce biodiesel. *Applied Energy*, 88(10), 3402–3410.

Amer, L., Adhikari, B., & Pellegrino, J. (2011). Technoeconomic analysis of five microalgae-to-biofuels processes of varying complexity. *Bioresource Technology*, 102(20), 9350–9359.

Amiri, H., & Karimi, K. (2018). Pretreatment and hydrolysis of lignocellulosic wastes for butanol production: Challenges and perspectives. *Bioresource Technology*, 270, 702–721.

Andrade, L. M., Andrade, C. J., Dias, M., Nascimento, C. A. O., & Mendes, M. A. (2018). *Chlorella* and *spirulina* microalgae as sources of functional foods, nutraceuticals, and food supplements: An overview. *MOJ Food Processing & Technology*, 6(1), 45–58.

Ansari, F. A., Shekh, A. Y., Gupta, S. K., & Bux, S. K. (2017). Microalgae for biofuels: Applications, process constraints and future needs. In S. K. Gupta, A. Malik, & F. Bux (Eds.), *Algal biofuels: Recent advances and future prospects* (pp. 58–73). Springer International Publishing.

Ansari, F. A., Shriwastav, A., Gupta, S. K., Rawat, I., & Bux, F. (2017). Exploration of microalgae biorefinery by optimizing sequential extraction of major metabolites from *Scenedesmus obliquus*. *Industrial & Engineering Chemistry Research*, 56(12), 3407–3412.

Barsanti, L., & Gualtieri, P. (2018). Is exploitation of microalgae economically and energetically sustainable? *Algal Research*, 31, 107–115.

Benemann, J., & Oswald, W. J. (1996). *Systems and economic analysis of microalgae ponds for conversion of carbon dioxide to biomass: Fourth Quarterly Technical Progress Report.* (pp. 90–156).

Beneroso, D., Bermudez, J. M., Arenillas, A., & Menendez, J. A. (2013). Microwave pyrolysis of microalgae for high syngas production. *Bioresource Technology*, 144, 240–246.

Borowitzka, M. A. (2013). High-value products from microalgae—Their development and commercialisation. *Journal of Applied Phycology*, 25(3), 743–756.

Byreddy, A. R., Barrow, C. J., & Puri, M. (2016). Bead milling for lipid recovery from thraustochytrid cells and selective hydrolysis of *Schizochytrium* DT3 oil using lipase. *Bioresource Technology*, 200, 464–469.

Carullo, D., et al. (2018). Effect of pulsed electric fields and high pressure homogenization on the aqueous extraction of intracellular compounds from the microalgae *Chlorella vulgaris*. *Algal Research*, 31, 60–69.

Chauton, M. S., Reitan, K. I., Norsker, N. H., Tveteras, R., & Kleivdal, H. T. (2015). A techno-economic analysis of industrial production of marine microalgae as a source of EPA and DHA-rich raw material for aquafeed: Research challenges and possibilities. *Aquaculture*, 436, 95–103.

Cheng, Y.-S., Labavitch, J. M., & VanderGheynst, J. S. (2015). Elevated CO_2 concentration impacts cell wall polysaccharide composition of green microalgae of the genus *Chlorella*. *Letters in Applied Microbiology*, 60(1), 1–7.

Chisti, Y. (2007). Biodiesel from microalgae. *Biotechnology Advances*, 25(3), 294–306.

Christenson, L., & Sims, R. (2011). Production and harvesting of microalgae for wastewater treatment, biofuels, and bioproducts. *Biotechnology Advances*, 29(6), 686–702.

Cuellar-Bermudez, S. P., et al. (2015). Extraction and purification of high-value metabolites from microalgae: Essential lipids, astaxanthin and phycobiliproteins. *Microbial Biotechnology*, 8(2), 190–209.

Davis, R., Aden, A., & Pienkos, P. T. (2011). Techno-economic analysis of autotrophic microalgae for fuel production. *Applied Energy*, 88(10), 3524–3531.

Davis, R. E., et al. (2014). Integrated evaluation of cost, emissions, and resource potential for algal biofuels at the national scale. *Environmental Science Technology*, 48(10), 6035–6042.

Delrue, F., et al. (2012). An economic, sustainability, and energetic model of biodiesel production from microalgae. *Bioresource Technology*, 111, 191–200.

DeRose, K., DeMill, C., Davis, R. W., & Quinn, J. C. (2019). Integrated techno economic and life cycle assessment of the conversion of high productivity, low lipid algae to renewable fuels. *Algal Research*, 38, 101412.

Fasaei, F., Bitter, J. H., Slegers, P. M., & van Boxtel, A. J. B. (2018). Techno-economic evaluation of microalgae harvesting and dewatering systems. *Algal Research*, 31, 347–362.

Gao, K., Orr, V., & Rehmann, L. (2016). Butanol fermentation from microalgae-derived carbohydrates after ionic liquid extraction. *Bioresource Technology*, 206, 77–85.

Giwa, A., Adeyemi, I., Dindi, A., Lopez, C. G. B., Lopresto, C. G., Curcio, S., et al. (2018). Techno-economic assessment of the sustainability of an integrated biorefinery from microalgae and Jatropha: A review and case study. *Renewable and Sustainable Energy Reviews*, 88, 239–257.

Gonzalez-Fernandez, C., Sialve, B., & Molinuevo-Salces, B. (2015). Anaerobic digestion of microalgal biomass: Challenges, opportunities and research needs. *Bioresource Technology*, 198, 896–906.

Guldhe, A., Singh, P., Kumari, S., & Rawat, I. (2016). Biodiesel synthesis from microalgae using immobilized *Aspergillus niger* whole cell lipase biocatalyst. *Renewable Energy*, 85, 1002–1010.

Guo, H., Daroch, M., Liu, L., Qiu, G., Geng, S., & Wang, G. (2013). Biochemical features and bioethanol production of microalgae from coastal waters of Pearl River Delta. *Bioresource Technology*, 127, 422–428.

Guo, B., Yang, B., Silve, A., Akaberi, S., Schere, D., Papachristou, I., et al. (2019). Hydrothermal liquefaction of residual microalgae biomass after pulsed electric field-assisted valuables extraction. *Algal Research*, 43, 101650.

Harun, R., Davidson, M., Doyle, M., Gopiraj, R., Danquah, M., & Forde, G. (2011). Technoeconomic analysis of an integrated microalgae photobioreactor, biodiesel and biogas production facility. *Biomass and Bioenergy*, 35(1), 741–747.

He, Y., Wang, X., Wei, H., Zhang, B. J., Chen, B., & Chen, F. (2019). Direct enzymatic ethanolysis of potential *Nannochloropsis* biomass for co-production of sustainable biodiesel and nutraceutical eicosapentaenoic acid. *Biotechnology for Biofuels*, 12(1)78.

Hernandez, D., Solana, M., Riaño, B., García-González, M. C., & Bertucco, A. (2014). Biofuels from microalgae: Lipid extraction and methane production from the residual biomass in a biorefinery approach. *Bioresource Technology*, 170, 370–378.

Ho, B. C. H., Kamal, S. M. M., & Harun, M. R. (2018). Extraction of eicosapentaenoic acid from *Nannochloropsis gaditana* using sub-critical water extraction. *Malaysian Journal of Analytical Science*, 22(4), 619–625.

Ho, S. H., Huang, S. W., Chen, C. Y., Hasunuma, T., Kondo, A., & Chang, J. S. (2013). Bioethanol production using carbohydrate-rich microalgae biomass as feedstock. *Bioresource Technology*, 135, 191–198.

Hoffman, J., Pate, R. C., Drennen, T., & Quinn, J. C. (2017). Techno-economic assessment of open microalgae production systems. *Algal Research*, 23, 51–57.

Hossain, N., Mahlia, T. M. I., Zaini, J., & Saidur, R. (2019). Techno-economics and sensitivity analysis of microalgae as commercial feedstock for bioethanol production. *Environmental Progress & Sustainable Energy*, 38(5)13157.

Hu, Z., Ma, X., Li, L., & Wu, J. (2014). The catalytic pyrolysis of microalgae to produce syngas. *Energy Conversion and Management*, 85, 545–550.

Industry, E. (2014). *Omega-3 polyunsaturated fatty acids (PUFAs)—A global market overview. (2014).* https://www.reportlinker.com/p02029889/Omega-3-Polyunsaturated-Fatty-Acids-PUFAs-A-Global-Market-Overview.html.

Jacob, J. P., & Mathew, S. (2017). Effect of lipases from *Candida cylinderacea* on enrichment of PUFA in marine microalgae. *Journal of Food Processing and Preservation*, 41(1) e12928.

Jankowska, E., Sahu, A. K., & Oleskowicz-Popiel, P. (2017). Biogas from microalgae: Review on microalgae's cultivation, harvesting and pretreatment for anaerobic digestion. *Renewable and Sustainable Energy Reviews*, 75, 692–709.

Jonker, J. G. G., & Faaij, A. P. C. (2013). Techno-economic assessment of micro-algae as feedstock for renewable bio-energy production. *Applied Energy*, 102, 461–475.

Kamm, B., Hille, C., Schönicke, P., & Dautzenberg, G. (2010). Green biorefinery demonstration plant in Havelland (Germany). *Biofuels, Bioproducts and Biorefining*, 4(3), 253–262.

Kang, S., Heo, S., & Lee, J. H. (2019). Techno-economic analysis of microalgae-based lipid Production: Considering influences of microalgal species. *Industrial & Engineering Chemistry Research*, 58(2), 944–955.

Kumar, D., Singh, B., & Sharma, Y. C. (2017). Challenges and opportunities in commercialization of algal biofuels. In S. K. Gupta, A. Malik, & F. Bux (Eds.), *Algal biofuels: Recent advances and future prospects. Springer International Publishing. (pp. 421–450).*

Kumar, P., Suseela, M. R., & Toppo, K. (2011). Physico-chemical characterization of algal oil: A potential biofuel. *Asian Journal of Experimental Biological Science*, 2(3), 493–497.

Kumar, S., Hablot, E., Moscoso, J. L. G., Obeid, W., Hatcher, P. G., DuQuette, B. M., et al. (2014). Polyurethanes preparation using proteins obtained from microalgae. *Journal of Materials Science*, 49(22), 7824–7833.

Kumari, D. J., Babitha, B., Jaffar, S. K., Prasad, M. G., Ibrahim, Md., & Khan, Md. S. A. (2011). Potential health benefits of *Spirulina platensis. Pharmanest*, 2(2–3) 2231-0541 (online).

Lammens, T. M., Gangarapu, S., Franssen, M. C. R., Scott, E. L., & Sanders, J. P. M. (2012). Techno-economic assessment of the production of bio-based chemicals from glutamic acid. *Biofuels, Bioproducts and Biorefining*, 6(2), 177–187.

Lee, O. K., Oh, Y.-K., & Lee, E. Y. (2015). Bioethanol production from carbohydrate-enriched residual biomass obtained after lipid extraction of *Chlorella* sp. KR-1. *Bioresource Technology*, 196, 22–27.

Li, Y., Lian, S., Tong, D., Song, R., Yang, W., Fan, Y., et al. (2011). One-step production of biodiesel from *Nannochloropsis* sp. on solid base Mg-Zr catalyst. *Applied Energy*, 88(10), 3313–3317.

Li, X., Liu, J., Chen, G., Zhang, J., Wang, C., & Liu, B. (2019). Extraction and purification of eicosapentaenoic acid and docosahexaenoic acid from microalgae: A critical review. *Algal Research*, 43, 101619.

Liu, J., Huang, J., Fan, K. W., Jiang, Y., Zhong, Y., Sun, Z., et al. (2010). Production potential of *Chlorella zofingienesis* as a feedstock for biodiesel. *Bioresource Technology*, 101 (22), 8658–8663.

Losh, J. L., Young, J. N., & Morel, F. M. M. (2013). Rubisco is a small fraction of total protein in marine phytoplankton. *New Phytologist*, 198(1), 52–58.

Mala, R., Sarojini, M., Saravanababu, S., & Umadevi, G. (2009). Screening for antimicrobial activity of crude extracts of *Spirulina platensis. Journal of Cell and Tissue Research*, 9(3), 1951–1955.

Mathiot, C., Ponge, P., Gallard, B., Sassi, J. F., Delrue, F., & Le Moigne, N. (2019). Microalgae starch-based bioplastics: Screening of ten strains and plasticization of unfractionated microalgae by extrusion. *Carbohydrate Polymers*, 208, 142–151.

Misra, R., Guldhe, A., Singh, P., Rawat, I., & Bux, F. (2014). Electrochemical harvesting process for microalgae by using nonsacrificial carbon electrode: A sustainable approach for biodiesel production. *Chemical Engineering Journal*, 255, 327–333.

Moazami, N., Ashori, A., Ranjbar, R., & Tangestani, M. (2012). Large-scale biodiesel production using microalgae biomass of *Nannochloropsis. Biomass and Bioenergy*, 39, 449–453.

Morais, M. G., Vaz, B. S., Morais, E. G., & Costa, J. A. V. (2015). Biologically active metabolites synthesized by microalgae. *Biomed Research International*, 2015, 835761.

Nagarajan, S., Chou, S. K., Cao, S., Wu, C., & Zhou, Z. (2013). An updated comprehensive techno-economic analysis of algae biodiesel. *Bioresource Technology*, 145, 150–156.

Passos, F., & Ferrer, I. (2014). Microalgae conversion to biogas: Thermal pretreatment contribution on net energy production. *Environmental Science and Technology*, 48 (12), 7171–7178.

Passos, F., Uggetti, E., Carrere, H., & Ferrer, I. (2014). Pretreatment of microalgae to improve biogas production: A review. *Bioresource Technology*, 172, 403–412.

Phwan, C. K., Ong, H. C., Chen, W. H., & Ling, T. C. (2018). Overview: Comparison of pretreatment technologies and fermentation processes of bioethanol from microalgae. *Energy Conversion and Management*, 173, 81–94.

Pieber, S., Schober, S., & Mittelbach, M. (2012). Pressurized fluid extraction of polyunsaturated fatty acids from the microalga *Nannochloropsis oculata. Biomass and Bioenergy*, 47, 474–482.

Pragya, N., Pandey, K. K., & Sahoo, P. K. (2013). A review on harvesting, oil extraction and biofuels production technologies from microalgae. *Renewable and Sustainable Energy Reviews*, 24, 159–171.

Quinn, J. C., & Davis, R. (2014). The potentials and challenges of algae based biofuels: A review of the techno-economic, life cycle, and resource assessment modeling. *Bioresource Technology*, 184, 444–452.

Raheem, A., Prinsen, P., Vuppaladadiyam, A. K., Zhao, M., & Luque, R. (2018). A review on sustainable microalgae based biofuel and bioenergy production: Recent developments. *Journal of Cleaner Production*, 181, 42–59.

Ramos, M. J., Fernandez, C. M., Casas, A., Rodriguez, L., & Perez, A. (2009). Influence of fatty acid composition of raw materials on biodiesel properties. *Bioresource Technology*, 100(1), 261–268.

Rawat, I., Kumar, R. R., Mutanda, T., & Bux, F. (2013). Biodiesel from microalgae: A critical evaluation from laboratory to large scale production. *Applied Energy*, 103, 444–467.

Resurreccion, E. P., Colosi, L. M., White, M. A., & Clarens, A. F. (2012). Comparison of algae cultivation methods for bioenergy production using a combined life cycle assessment and life cycle costing approach. *Bioresource Technology*, 126, 298–306.

Richardson, J. W., Johnson, M. D., & Outlaw, J. L. (2012). Economic comparison of open pond raceways to photo bio-reactors for profitable production of algae for transportation fuels in the southwest. *Algal Research*, 1(1), 93–100.

Ruiz, J., Olivieri, G., de Vree, J., Bosma, R., Willems, P., Reith, J. H., et al. (2016). Towards industrial products from microalgae. *Energy and Environmental Science*, 6(10), 3036–3043.

Ryckebosch, E., Bruneel, C., Termote-Verhalle, R., Goiris, K., Muylaert, K., & Foubert, I. (2014). Nutritional evaluation of microalgae oils rich in omega-3 long chain polyunsaturated fatty acids as an alternative for fish oil. *Food Chemistry*, 160, 393–400.

Sari, Y. W., Sanders, J. P. M., & Bruins, M. E. (2016). Techno-economical evaluation of protein extraction for microalgae biorefinery. *IOP Conference Series: Earth and Environmental Science*, 31(1), 012034.

Sari, Y. W., Syafitri, U., Sanders, J. P. M., & Bruins, M. E. (2015). How biomass composition determines protein extractability. *Industrial Crops and Products*, 70, 125–133.

Scarlat, N., Dallemand, J. F., Monforti-Ferrario, F., & Nita, V. (2015). The role of biomass and bioenergy in a future bioeconomy: Policies and facts. *Environmental Development*, 15, 3–34.

Scott, S. A., Davey, M. P., Dennis, J. S., Horst, I., Howe, C. J., Lea-Smith, D. J., et al. (2010). Biodiesel from algae: Challenges and prospects. *Current Opinion in Biotechnology*, 21 (3), 277–286.

Setyaningsih, I., Bintang, M., & Madina, N. (2015). Potentially antihyperglycemic from biomass and phycocyanin of *Spirulina fusiformis* Voronikhin by in vivo test. *Procedia Chemistry*, 14, 211–215.

Shen, J., Huang, J., Ruan, H., Wang, J., & van der Bruggen, B. (2014). Techno-economic analysis of resource recovery of glyphosate liquor by membrane technology. *Desalination*, 342, 118–125.

Shriwastav, A., & Gupta, S. K. (2017). Key issues in pilot scale production, harvesting and processing of Algal biomass for biofuels. In S. K. Gupta, A. Malik, & F. Bux (Eds.), *Algal biofuels: Recent advances and future prospects* (p. 12). Springer International Publishing.

Sirait, P. S., Setyaningsih, I., & Tarman, K. (2019). Anticancer activity of Spirulina cultivated in Walne and organic media. *Jurnal Pengolahan Hasil Perikanan Indonesia*, 22(1), 50–59.

Skorupskaite, V., Makareviciene, V., Sendzikiene, E., & Gumbyte, M. (2019). Microalgae *Chlorella* sp. cell disruption efficiency utilising ultrasonication and ultrahomogenisation methods. *Journal of Applied Phycology*, 31(4), 2349–2354.

Soto-Sierra, L., Stoykova, P., & Nikolov, Z. L. (2018). Extraction and fractionation of microalgae-based protein products. *Algal Research*, 36, 175–192.

Suarez Garcia, E., van Leeuwen, J., Safi, C., Sijtsma, L., Eppink, M. H. M., Wijffels, R. H., et al. (2018). Selective and energy efficient extraction of functional proteins from microalgae for food applications. *Bioresource Technology*, 268, 197–203.

Teng, Y., Scott, E. L., & Sanders, J. P. M. (2012). Separation of L-aspartic acid and L-glutamic acid mixtures for use in the production of bio-based chemicals. *Journal of Chemical Technology & Biotechnology*, 87(10), 1458–1465.

Teuling, E., Wierenga, P. A., Schrama, J. W., & Gruppen, H. (2017). Comparison of protein extracts from various unicellular green sources. *Journal of Agricultural and Food Chemistry*, 65(36), 7989–8002.

Tokuşoglu, O., & Ünal, M. K. (2003). Biomass nutrient profiles of three microalgae: *Spirulina platensis*, *Chlorella vulgaris*, and *Isochrisis galbana*. *Journal of Food Science*, 68(4), 1144–1148.

Uggetti, E., Passos, F., Sole, M., Garfi, M., & Ferrer, I. (2017). Recent achievements in the production of biogas from microalgae. *Waste and Biomass Valorization*, 8(1), 129–139.

Van De Velde, F., Alting, A. C., & Pouvreau, L. (2011). *Process for isolating a dechlorophylllized RuBisCo preparation from a plant material*. NIZO Food Research B.V.

van der Voort, M. P. J., Spruijt, J., Potters, J., de Wolf, P., & Ellissen, H. (2017). *Socio-economic assessment of algae-based PUFA production: The value chain from microalgae to PUFA ('PUFACHAIN')*. Göttingen: PUFAChain.

Wang, Y., Ho, S. H., Yen, H. W., Nagarajan, D., Ren, N. Q., Li, S., et al. (2017). Current advances on fermentative biobutanol production using third generation feedstock. *Biotechnology Advances*, 35(8), 1049–1059.

Wang, J., & Yin, Y. (2018). Fermentative hydrogen production using pretreated microalgal biomass as feedstock. *Microbial Cell Factories*, 17(22), 16.

Ward, A., & Lewis, D. (2015). Pre-treatment options for halophytic microalgae and associated methane production. *Bioresource Technology*, 177, 410–413.

Ward, A. J., Lewis, D. M., & Green, F. B. (2014). Anaerobic digestion of algae biomass: A review. *Algal Research*, 5 (1), 204–214.

Wen, X., Du, K., Wang, Z., Peng, X., Luo, L., Tao, H., et al. (2016). Effective cultivation of microalgae for biofuel production: A pilot-scale evaluation of a novel oleaginous microalga *Graesiella* sp. WBG-1. *Biotechnology for Biofuels*, 9(1), 1–12.

Widyarani, R., Bowden, N., Kolfschoten, R. C., Sanders, J. P. M., & Bruins, M. E. (2016). Fractional precipitation of amino acids from agro-industrial residues using ethanol. *Industrial & Engineering Chemistry Research*, 55, 7462–7472.

Windari, H. A. S., Tarman, K., Safithri, M., & Setyaningsih, I. (2019). Antioxidant activity of spirulina platensis and sea cucumber *Stichopus hermanii* in streptozotocin-induced diabetic rats. *Tropical Life Sciences Research*, 30 (2), 119–129.

Wu, N., Moreira, C. M., Zhang, Y., Doan, N., Yang, S., Phlips, E. J., et al. (2019). Techno-economic analysis of biogas production from microalgae through anaerobic digestion. In *Anaerobic Digestion*.

Xin, C., Addy, M. M., Zhao, J., Cheng, Y., Cheng, S., Mu, D., et al. (2016). Comprehensive techno-economic analysis of wastewater-based algal biofuel production: A case study. *Bioresource Technology*, 211(March), 584–593.

Yang, Z., Guo, R., Xu, X., Fan, X., & Li, X. (2010). Enhanced hydrogen production from lipid-extracted microalgal biomass residues through pretreatment. *International Journal of Hydrogen Energy*, 35(18), 9618–9623.

Zeller, M. A., Hunt, R., Jones, A., & Sharma, S. (2013). Bioplastics and their thermoplastic blends from Spirulina and Chlorella microalgae. *Journal of Applied Polymer Science*, 130(5), 3263–3275.

Zhao, B., Ma, J., Zhao, Q., Laurens, L., Jarvis, E., Chen, S., et al. (2014). Efficient anaerobic digestion of whole microalgae and lipid-extracted microalgae residues for methane energy production. *Bioresource Technology*, 161, 423–430.

Index

Note: Page numbers followed by f indicate figures and t indicate tables.

A

Acetylene reduction activity (ARA), 363
Adhesion materials, 389, 389–390t
Aeration rate, 394, 396–397
Agar, 318, 330
Aiba model, 74–75
Air-lift fermenter, 19–20, 19f
Alcohol, 130–131
Aldehydes, 130–131
Algalization, 341
AlgaVia, 227
Alginates, 330
Alginic acids, 330–331
Alkaline soil, reclamation of, 367–368
Alkalinity stress, 367
Amikacin, 279–283
Aminoclays (ACs), 93–94
Anabaena variabilis, 350
Anaerobic digestion, 161–162, 426–428
Anaerobic fermentation, 23, 158–159
Angiotensin-converting enzyme I (ACE-I), 220, 222t
Antiaging agents, 321–323
Antibiotic growth promoters, 275
Anticellulite agents, 323–324
Antioxidant compounds, from microalgae
  conventional solvent extraction methods, 176–177
  emerging extraction technologies, 177, 178–183t, 197–198
  high-pressure homogenization (HPH), cell disruption by, 196–197
  high voltage electric discharges (HVED)-assisted extraction, 177, 187–188
  microwave-assisted extraction (MAE), 192–193

moderate electric field (MEF)-assisted extraction, 177, 186–187
  pressurized liquid extraction (PLE), 188–189
  pulsed electric field (PEF)-assisted extraction, 177–186
  supercritical fluid extraction (SFE), 189–192
  ultrasound-assisted extraction (UAE), 193–195
Aphanizomenon flos-aquae, 329
Apparent digestible coefficients (ADC), 294–299
Aquaculture, dietary microalgae inclusion
  fish and shrimp, 285, 286–293t, 295–298t
  juvenile, feeds for, 294–301
  larvae, microdiets for, 285–294
  nutritional value, of microalgae genera, 240, 241–243t
Aqueous enzymatic extraction (AEE), 102
Arrhenius equation, 81, 126
Arthrospira platensis, 209
Artificial symbiosis, for crop improvement, 362–363, 362f
Astapure, 227–230
AstaTROL, 325
Astaxanthin, 92
  cosmetic applications, 317, 325
  fish diets, 285
  market prices, 227–230
Autotrophic biofilms, 386

B

Bacillus thuringiensis (Bt), 357
Bacteria, 358–359, 360t
Bakery products, microalgae in, 223–226

Basification method, 136
Bead beating (BB), 96, 98t
Beer-Lambert law, 69, 72
Beverages, microalgae in, 226–227
Bioactive peptides, 220, 221f
BioAstin, 227–230
Biocatalyst, 126–127
Biochar, 102, 161–162
Biochemical conversion process, 124
Biochemical pesticides, 357
Biodiesel production, from microalgae, 124–125, 146–153, 414, 423
  biodiesel FAME profile, 151
  capital cost for, 415
  cell disruption methods, 98t
  integrated systems
    bioethanol and biogas, 426–428
    jatropha, 423
    methane, 424–425
  operating cost for, 415–416
  production methods, 148–151, 149f
  properties, 151–153, 152t
  techno-economic assessment (TEA), 414–416
Bioelectricity, 161
Bioethanol production, from microalgae, 153–154, 154f, 423f
  advantages, 154
  characteristics, 155–157
  disadvantage, 154
  integrated systems
    biodiesel and biogas, 426–428
    crude bio-oil and biofertilizer, 423–424
  production methods
    cell disruption, 154
    fermentation, 155, 156–157t
    saccharification processes, 154–155
Biofertilizers, 340–351, 423–424, 423f
  market, 350–351

Biofertilizers (Continued)
   nitrogen uptake by microalgae, 342, 348f
   nutrient recycling in soil, 349–351
   soil fertility, 341–342, 343–347t
   soil structure, 342–349
   stabilization of soil aggregates, 349
Biofilm membrane photobioreactor (BMPBR), 396–397, 397f
Biofilms, microalgae. See Microalgae biofilms
Biofuel production
   biomass, 117–118
   engineered microalgae, 118t, 119–120
   first-generation feedstock, 117–119, 118t, 146, 146f
   from microalgae, 146, 146–147f, 410–416, 411t, 425–426, 426f
     advantages, 118t, 119–120, 125
     algae metabolism, 162–164
     algal cultivation systems, 164–165
     biochar, 161–162
     biodiesel (see Biodiesel production, from microalgae)
     bioelectricity, 161
     bioethanol (see Bioethanol production, from microalgae)
     biogas, 162
     biohydrogen, 157–160
     biomethane, 161
     bio-oil, 161
     cellulose ethanol, 125
     commercial application, 165–166
     disadvantages, 118t, 119–120, 125
     flue gas, 161
   second generation feedstock, 118t, 119, 146, 146f
Biogas, 162, 426–428
Biohydrogen production, from microalgae, 160t
   benefits, 157
   characteristics, 160
   production methods, 158–160, 158–159f
   substrates, 157–158
Biomethane, 161
Biomineralization, 349–350
Biopesticides, 357–361
   algal nanoparticles, for pest control, 359–361
   bacteria, 358–359
   fungi, 358–359
   microalgae, 357–358
   nematodes, 358–359
Biopharmaceuticals, 420–421

Biophotolysis, 159–160
Bioplastic, microalgae for, 421–423, 422f
Biorefinery, 24, 208, 209f, 413–414
Bioremediation and reclamation, of degraded land, 363–368
   bioremediating agents, 364–368
   soil conditioners, 364
Biosorption, 365
Bipartite association, 362
Bipartite interaction, 361
Bligh and Dyer's method, 101–102, 102t
Blue-green algae (BGA), 340
BMPBR. See Biofilm membrane photobioreactor (BMPBR)
Boars diet, microalgae inclusion in, 268–269, 269f, 270–272t
Bold basal medium (BBM), 389–390
Boltzmann transfer equation, 71
Botryococcus braunii, 323
Broiler chicken diets, microalgae effects on. See Poultry diets, microalgae inclusion in
Bubble column fermenter, 19–20, 19f
Bubble-induced turbulence (BIT), 77
Bulk commodity chemicals, 418–420

C
Calothrix elenkinii (RPC1) inoculation, 363
Canthaxanthin, 325
Capital cost, for biodiesel production, 415
Carbohydrates, microalgae
   advantages, 214
   cellular carbohydrates, extraction methods, 214, 215t, 216
   cell walls, 214
   chemical features, 216
   extra-cellular carbohydrates, extraction methods, 214–216, 215t
   in growing and finishing pigs diets, 266–267
   health benefits, 213–214
   market value, 226f
   prebiotic effects, 213–214
   in weaned piglet diets, 265–266
Carbon dioxide (CO_2), 390
   injection system, 67
   mass transfer, 41–42, 42f
Carbon monoxide, 130–131
Carbon nanotubes (CNTs), 93–94
Carotenoids, from microalgae
   advantages, 173–174
   biological activities, 173–174, 220

cell disruption techniques, 222–223
   cosmetic applications, 317, 319, 325
   downstream processing strategies, 222–223
   example of, 173–174
   extraction techniques, efficiency and advantages of, 222–223, 224–225t
   high-pressure homogenization (HPH), 197
   industrially relevant microalgal species, 220–222
   market prices, 227–230
   market value, 226f
   MEF-assisted extraction process, 186–187
   microwave-assisted extraction (MAE), 193
   PEF pretreatment, 184–185
   pressurized liquid extraction (PLE), 188–189
   primary and secondary carotenoids, 173, 220–222
   $scCO_2$ extraction process, 190–192
   ultrasound-assisted extraction (UAE), 194–195
Carrageenans, 318, 330
Catalysis
   catalysts, 126
     activation energy, 126
     biocatalyst, 126–127
     catalytic conversion, of microalgae oil (see Catalytic deoxygenation (DO), of microalgae oil)
     heterogeneous catalyst, 126–127
     homogeneous catalysts, 126–127
     reaction inhibitor, 125–126
     reaction mechanism, 126
     reaction pathway, 125–126
     use of, 127–128
   definition, 125–126
Catalytic deoxygenation (DO), of microalgae oil, 124, 138
   deactivation of catalyst, 137–138
   decarboxylation/decarbonylation (DCO) process, 128–130, 133–134t, 135–137
   fatty acid, deoxygenation of, 128–129, 128t
   hydrocarbon chain, breaking of, 128–129
   hydrodeoxygenation (HDO) process, 128–130, 133–134t, 135–136
   lipid content, 131–132
   methanation, 128–131

nonprecious metal-based catalysts,
132–135
precious metal-based catalysts, 132
reaction pathway, 130–131, 130*f*
water gas shift (WGS) reaction,
128–131
Cell disruption methods, of microalgae,
98*t*
bioethanol production, 154
edible bio-oil production
bead beating (BB), 96
chemical method, 97
enzymatic disruption, 97
high-pressure homogenization
(HPH), 96
microwave method, 97
pressing, 96–97
selection of, 98–100
ultrasonication, 97–98
high-pressure homogenization
(HPH), 196–197
high-value compounds, 212, 212*t*
Cell lysis, 353–356
Cell membrane complex (CMC),
327–328, 327*f*
Cellulitis. *See* Gynoid lipodystrophy
Cellulose ethanol, 125
Centrifugation, 175–176
Ceramic membranes, 106
Chemical fertilizers, 340–341
Chemical method, cell disruption, 97,
98*t*
Chemical oxygen demand (COD),
382–386
Chemicals, microalgae for, 416–421
Chemostat, 17, 18*f*
Chlamydocapsa sp., 323
Chlorofluorocarbons (CFCs), 127–128
Chlorophyll, 174, 184, 319
Chlorophyta (green algae), 318
Circular ponds, 8, 10, 12*t*, 65
Circulation time, photobioreactors,
40–41
Citric acid, 106
Coke formation, 137–138
Colloid milling (CM), 196
Computational fluid dynamics (CFD),
75
Conventional transesterification (CT)
biodiesel production
biodiesel properties, 151–153, 152*t*
disadvantages, 148
FAME yield, 149, 150*t*
Cookies, microalgae in, 226
Cortex, 327–328, 327*f*

Cosmetic products
algae applications in, 313–315, 332*t*
benefits, 314
formulation adjuvants
(*see* Formulation adjuvants,
algae applications in)
haircare products, 328–330
natural dyes, 315, 319
properties, 314
skincare products (*see* Skincare
products, algae applications in)
environment-friendly products,
necessity of, 315
C-phycocyanin, 174
Creams, 320
Crude bio-oil, 423–424, 423*f*
Cultivation techniques, microalgae,
411–413, 412*t*
cost reduction
fermenters, 23
high-value products, cultivation
for, 24, 25–26*t*
in open pond cultivation, 22
PBRs, 22–23
wastewater, cultivation in,
23–24
dark fermentation
heterotrophic cultivation
(*see* Heterotrophic microalgae
cultivation)
heterotrophic microalgae
strains, 13
history of, 2–3, 3*f*
hybrid system, 11–12
industrial cultivation techniques, 2,
13, 14*t*
laboratory cultivation, 3–5, 6*f*
open ponds (*see* Open pond (OP)
systems)
parameters, 5–6
photobioreactors (PBRs)
(*see* Photobioreactors (PBRs))
types of, 1
Cuticle, 327–328, 327*f*
Cyanobacteria, 319, 341–342, 361
Cystoseira barbata, 328

D

Dark fermentation
biohydrogen production, 159–160,
159*f*
heterotrophic cultivation
(*see* Heterotrophic microalgae
cultivation)
heterotrophic microalgae strains, 13

Decarboxylation/decarbonylation
(DCO), 104, 128–130, 133–134*t*,
135–137
Deposition/precipitation (DP) method,
136
Desertification, 365–366
Desmodesmus, 4, 196–197, 383*f*, 386
Dewatering process, 123–124
DHA. *See* Docosahexaenoic acid (DHA)
Dieckol, 326
Dimethyl sulfoxide (DMSO), 194–195
Direct transesterification (DT),
biodiesel production, 148–149,
149*f*
biodiesel properties, 151–153, 152*t*
ethanol, 149–151
FAME yield, 149, 150*t*
Dissolved inorganic nitrogen (DIN),
394
Dissolved inorganic phosphorous
(DIP), 394
Docosahexaenoic acid (DHA), 216–217,
227, 230–231, 416
growing and finishing pigs, 267–268
poultry
egg yolk, 277
meat, 276
ruminants
meat, 261–264
milk, 254–261
rumen fermentation, 253
sows and boars, 268–269, 269*f*

E

Ecklonia
E. cava, 326
E. stolonifera, 326
Edible bio-oil production, from
microalgae, 92, 103*f*, 111
bio-oil recovery, distillation, and
purification, 104–105
liquid-liquid extraction (LLE),
105–106
membrane extraction, 106
precipitation process, 106
supercritical fluid separation
(SFS), 105
cell disruption methods, 98*t*
bead beating (BB), 96
chemical method, 97
enzymatic disruption, 97
high-pressure homogenization
(HPH), 96
microwave method, 97
pressing, 96–97

Edible bio-oil production, from microalgae *(Continued)*
  selection of, 98–100
  ultrasonication, 97–98
  conversion processes, of bio-oil, 102
  decarboxylation, 104
  hydrodeoxygenation (HDO), 104
  hydrothermal liquefaction (HTL), 102–104
  slow and fast pyrolysis, 104
  environmental and socioeconomic impacts, 108–111
  integrated approach, 107–108, 107f
  lipid extraction methods, 100
    solvent extraction methods, 101–102, 102t
    solvent-free extraction, 102
    supercritical fluid extraction (SFE), 100–101
  nanotechnology application, 93–94, 95t
  suitable candidates for, 92–94, 93t
Egg production, microalgae effects on, 277–278, 278f
Eicosapentaenoic acid (EPA), 216–217, 253–261, 264, 269f, 276, 416
Electrodialysis, 420
Electroporation, 181–184, 186
Electrotechnologies, 177
  high voltage electric discharges (HVED)-assisted extraction, 187–188
  moderate electric field (MEF)-assisted extraction, 186–187
  pulsed electric field (PEF)-assisted extraction, 177–186
Emollients, moisturizers, 320
Endocrine hyperpigmentation, 326
Enzymatic disruption, 97, 98t
Enzymatic hydrolysis, 155
Enzymatic methods, 423–424
EPA. *See* Eicosapentaenoic acid (EPA)
Essential amino acids, 218–219, 219t
European Biostimulant Industry Council, 351–352
European Food Safety Authority, 230
European regulations, to algal food products, 230–232, 231t
Exopolysaccharides (EPS), 318, 348–349
Extracellular polymeric substance (EPS) matrix, 386

F
Fast pyrolysis, 104, 124
Fatty acid methyl ester (FAME), 124–125, 148
  biodiesel FAME profile, 151
  yield, in conventional and direct transesterification, 149, 150t
Feed conversion ratio (FCR), 273–275
Fermentation methods, bioethanol production, 155, 156–157t
Fermenters, heterotrophic cultivation, 18–19
  advantages, 21, 21t
  air-lift fermenter, 19–20, 19f
  batch operation mode, 17
  bubble column fermenter, 19–20, 19f
  chemostat, 17, 18f
  cost reduction, 23
  limitations, 21, 21t
  packed-bed fermenters, 19–20, 19f
  perfusion culture systems, 18, 18f
  pulsed fed-batch strategy, 17
  tank fermenters pressure cycle, 19–20, 20f
FFAs. *See* Free fatty acids (FFAs)
Fick's law, 398
Filtration, 175–176
Fischerellin A, 358
Fish and shrimp diets
  conventional vegetable sources, 284–285
  fish meal (FM), 284–285
  fish oil (FO), 284–285
  microalgae inclusion, effects of, 285, 286–293t, 295–298t
    juvenile, feeds for, 294–301
    larvae, microdiets for, 285–294
  rendered animal byproducts, 284–285
  single cell products, 284–285
Fish oil (FO), 216, 284–285
Fixed capital investment (FCI), 424
Flat-plate photobioreactors, 8, 9–10t, 66, 77, 78t, 164
Flue gas, 161
Food applications, microalgae in, 212–213
  carbohydrates, 213–216
  commercialized products, in market, 223–230, 228–230t
  future market and challenges, 231–232
  lipids, 216–218

pigments and carotenoids, 220–223
proteins and peptides, 218–220
regulatory aspects, of algae products, 230, 231t
Formulation adjuvants, algae applications in, 330
  preservatives, 331
  surfactants, 330–331
  thickening agents, 330
Fossil fuels, 117–118
Fourier transform infrared spectroscopy (FTIR) analysis, 103–104
Free fatty acids (FFAs), 129, 133–134t
Fucoidans, 318
Fucoxanthin, cosmetic applications, 325, 327
Fucus vesiculosus, 321–324
Functional foods/nutraceuticals, from microalgae, 212–213
  carbohydrates, 213–216
  lipids, 216–218
  pigments and carotenoids, 220–223
  proteins and peptides, 218–220
Fungi, 358–359, 360t
Furcellaria lumbricalis, 324

G
Gas holdup analysis, 38–39
Gasification, 124
Gelidium crinale, 328
Gels, 320
General logistic growth (GLG) model, 401–402
Genetic hyperpigmentation, 326
Glutamate, 356
Glycine betaine, 352
Glycolipids, 330–331
Grau second-order model, 399
Greenhouse gas (GHG) emissions, 117
Green hydrocarbon production, microalgae oil
  biochemical conversion process, 124
  catalytic deoxygenation (DO) process, 124, 138
    deactivation of catalyst, 137–138
    decarboxylation/decarbonylation (DCO) process, 128–130, 133–134t, 135–137
    fatty acid, deoxygenation of, 128–129, 128t
    hydrocarbon chain, breaking of, 128–129

hydrodeoxygenation (HDO)
process, 128–130, 133–134*t*,
135–136
lipid content, 131–132
methanation, 128–131
nonprecious metal-based catalysts,
132–135
precious metal-based catalysts,
132
reaction pathway, 130–131, 130*f*
water gas shift (WGS) reaction,
128–131
chemical extraction process, 123–124
dewatering process, 123–124
fatty acid composition, 120–123,
121–122*t*
lipid content, 120–123, 123*t*
pressing method, 123–124
pyrolysis, 124–125
supercritical fluid extraction,
123–124
thermochemical processes, 124
Green water technique, 285
Growth, of microalgae
energy consumption model, 82–83
factors for, 64–65
availability of nutrients, 67–68
closed photobioreactors, 66
light intensity, 68
open photobioreactors, 65–66
pH, 66–67, 67*t*
temperature, 66
gas exchange and temperature effect,
79–82, 82*f*
growth models, 73–74
Aiba model, 74–75
categories of, 73, 74*t*
Steele model, 75
irradiation models (*see* Irradiation
models)
momentum transfer models, 75–77,
78*f*
shear stress, effect of, 77–79
Guaiacol, 137–138
Gunnera-Nostoc symbiosis, 361
Gynoid lipodystrophy, 323

H

Haematococcus pluvialis, 209–211, 210*f*,
317, 325
Hair
haircare products, algae applications
in, 328–330

oriental virgin hair sample, 327–328,
327–328*f*
Han model, 400–402
HDO. *See* Hydrodeoxygenation (HDO)
Health Canada, 230
Heavy metals contaminated sites,
reclamation of, 365
Helical photobioreactors, 8
Henry's law, 80–81
Heterocysts, 342, 361
Heterogeneous catalysts, 126–127
Heterotrophic assisted
photoautotrophic biofilm
(HAPB), 389–390
Heterotrophic microalgae cultivation,
1, 13–17, 16–17*t*
aerobic products, 15
agitation, 13–15
anaerobic products, 15–17
cost, 21–22
fermenters, 18–19
advantages, 21, 21*t*
air-lift fermenter, 19–20, 19*f*
batch operation mode, 17
bubble column fermenter, 19–20,
19*f*
chemostat, 17, 18*f*
cost reduction, 23
limitations, 21, 21*t*
packed-bed fermenters, 19–20, 19*f*
perfusion culture systems, 18, 18*f*
pulsed fed-batch strategy, 17
tank fermenters pressure cycle,
19–20, 20*f*
glucose/acetate, 13–15
lipids, 15
nitrogen, source of, 13–15
organic carbon sources, 15
oxygen, 13–15
productivity figures for, 15, 16–17*t*
Hierarchical zeolite, 135–136, 138
High-pressure homogenization
(HPH), 96, 98*t*, 196–198
High voltage electric discharges
(HVED)-assisted extraction,
187–188
Hizikia fusiforme, 326–327
Homogeneous catalysts, 126–127
Humectants, moisturizers, 319–320
Humic substances, 352
Hybrid nanomaterials, 93–94
Hydrodeoxygenation (HDO), 104,
128–130, 133–134*t*, 135–136

Hydroformylation, 127
Hydrogenation, 124
Hydrophobic compounds, 192
Hydrothermal liquefaction (HTL),
102–104, 124–125
Hyperpigmentation, 326

I

Impregnation method, 136
Indigenous microorganisms, 391
Industrial cultivation, microalgae, 2, 13,
14*t*
Integrated system, microalgae, 394–396
Ion-exchange/precipitation (IP)
method, 136
Irradiation models
Beer-Lambert law, 69
interactions of microalgae, 68, 68*f*
radiative transfer equation
(RTE), 71–73
two-flux approximation
(TFA), 69–71, 70*f*
Isotropic scattering, 72–73

J

Jatropha, 423

K

Kinetic models, photobioreactors,
48–50, 49*t*
Kinetics, of microalgae biofilm
photobioreactors, 398–402
flat-plate MBPBR, 398, 399*t*
hybrid BMPBR, 398–399, 400*t*
rotating microalgae biofilm
photobioreactor (RABPBR),
400–402, 401*t*

L

Laboratory cultivation techniques, 3–5,
6*f*
Laminaria japonica, 327, 329–330
Laurinterol, 331
Legendre polynomials, 72
Light-emitting diode (LED), 46
Light intensity, 68, 390
Lipid extraction, edible bio-oil
production, 100
cell disruption process, effect of,
98–100, 99–100*t*
solvent extraction methods, 101–102,
102*t*
solvent-free extraction, 102

Lipid extraction, edible bio-oil
    production *(Continued)*
  supercritical fluid extraction (SFE),
    100–101
Lipid oxidation, 244, 264–265, 267–268,
  279
Lipids, microalgae, 120, 123*t*, 216
  cultivation and downstream
    processing, 217–218
  extraction and purification methods,
    217–218
  fatty acid composition, 216–217, 217*t*
  health benefits, 216–217
  market value, 226*f*
  storage lipids, 120–123
  structural lipids, 120–123
Liquid-liquid extraction (LLE), 105–106
Livestock animals, dietary microalgae
    inclusion
  nutritional value, of microalgae
    genera, 240, 241–243*t*
  poultry diets
    advantages, 269–273
    egg production, effects on,
      277–278, 278*f*
    inclusion levels, 273, 274–275*t*
    meat production, effects on,
      273–277, 278*f*
  rabbit diets, effects on, 278–283,
    280–282*t*, 284*f*
  ruminants, 240–244
    feed intake, effects on, 244,
      245–249*t*
    meat production and composition,
      effects on, 261–265, 262–263*t*
    milk production and composition,
      effects on, 254–261, 255–259*t*
    rumen fermentation parameters,
      effects on, 244–254, 250–252*t*
  swine diets, 265, 269*f*, 270–272*t*
    growing and finishing pigs,
      266–268
    piglets, 265–266
    sows and boars, 268–269
LLE. *See* Liquid-liquid extraction
    (LLE)
Lotions, 320

M
Macroalgae, 318–319
Macrophomina phaseolina, 363
MAE. *See* Microwave-assisted
    extraction (MAE)

Malondialdehyde, 267–268
Mariculture on Land (MCL), 3
Mass transfer coefficient, 80–81
Mathematical modeling, 40
MBPBRs. *See* Microalgae biofilm
    photobioreactors (MBPBRs)
Melasma, 326
Meloidogyne incognita, 359
Membrane extraction, 106
Metabolic hyperpigmentation, 326
Metallic nanomaterials, 93–94
Methane, 424–425
Microalgae, 382–386, 413–414
  antioxidants recovery, extraction
    technologies for (*see* Antioxidant
    compounds, from microalgae)
  applications, 340, 381–382
  biofertilizers (*see* Biofertilizers)
  biomass harvesting, 413
  bioplastic, 421–423, 422*f*
  biorefinery concept, 208, 209*f*
  blue-green algae (BGA), 340
  cell envelope structures and
    composition, 174, 175*f*, 177
  composition of, 208–211, 210*t*
  cosmetics applications (*see* Cosmetic
    products, algae applications in)
  cultivation system (*see* Cultivation
    techniques, microalgae)
  definition, 91, 240
  edible bio-oil production (*see* Edible
    bio-oil production, from
    microalgae)
  growth of (*see* Growth, of
    microalgae)
  health benefits, 317
  high-value compounds, 24, 25–26*t*
    biological activities, 211–212
    cell disruption methods, 212, 212*t*
    downstream processing
      strategies, 212, 213*f*
    for food applications (*see* Food
      applications, microalgae in)
    industrial applications, 211, 211*f*
    macro stages, for production, 63,
      64*f*
  integrated system, 423–428
  proteins, 418–419, 422
Microalgae biofilm airlift
    photobioreactors, 392–394
Microalgae biofilm photobioreactors
    (MBPBRs), 391–398, 392*f*, 399*t*,
    402

  categories of, 391*f*
  conditions of, 393*t*
  flat-plate microalgae biofilm
    photobioreactors, 392–394, 392*f*
  kinetics of, 398–402
  microalgae biofilm membrane
    photobioreactors, 396–397, 397*f*
  rotating microalgae biofilm
    photobioreactor, 394–396
Microalgae biofilms, 386–391, 387*f*
  characterization of, 386–387
    chemical composition, 386
    surface structures, 386–387
  definition, 386
  formation and adhesion, factors
    affecting, 387–391, 388*f*
    adhesion materials, surface
      physicochemical properties, 389
    carbon dioxide (CO_2), 390
    indigenous microorganisms,
      presence of, 391
    light availability/intensity, 390
    media composition, 389–390
    microalgae, surface
      physicochemical properties of,
      387–389
  micrograph of, 388*f*
Microcoleus vaginatus, 359
Microcystis aeruginosa, 350, 365
Micronutrients, 68
Microporous materials, 135
Microwave-assisted extraction (MAE),
    192–193, 197–198
Microwave method, 97, 98*t*
Milk production and composition,
    microalgae effects on, 254,
    255–259*t*
  fatty acid profile, 254–261
  lactose content, 254
  protein yield and content, 254
  rapeseed/faba beans diet, *A. platensis*
    supplementation to, 254
Mixing time (TM), photobioreactors,
    40–41
Mixotrophic microalgae, 382, 390
Moderate electric field (MEF)-assisted
    extraction, 186–187
Modified Stover-Kincannon model,
    398–399
Moisturizer agents, 319–321
Molecular distillation, 104–105
Momentum transfer models, 75–77, 78*t*
Monod equation, 74–75

Monodus subterraneus, 320–321
Monounsaturated fatty acids (MUFA), 260
Monte-Carlo sampling method, 415–416
Multiwall carbon nanotubes (MWCNT), 132
Mycosporine-like amino acids (MAAs), 319, 322–323, 325

N

Nannochloropsis
    N. gaditana, 321, 417–418
    N. oceanica, 323
    N. oculata, 321, 417–418
Nanoparticle (NP) method, 136
Nanoscale zerovalent iron (nZVI) nanoparticles, 93–94
Natural dyes, 315, 319
Navicula pelliculosa, 350–351
Navier-Stokes equations, 75, 77, 78*t*
Nematodes, 358–359, 360*t*
Net energy ratio (NER), 50–51, 50*t*
Ni/HBEA catalyst, 133–134*t*, 136–138
Nitrogen to phosphorus (N:P) ratio, 23
Nitrogen uptake, by microalgae, 342, 348*f*
Ni/ZrO$_2$ catalyst, 135, 138
Noble metal catalysts, 132
Nonprecious metal-based catalysts, 132–135
Nutritional deficiency hyperpigmentation, 326

O

Occlusive agents, moisturizers, 319–320
Oil-contaminated site, reclamation of, 366–367
Ointments, 320
Olefins, 130–131
Open pond (OP) systems, 412–413
    advantages, 2, 7
    algal turf scrubbers, 11, 12*t*
    challenges, 1–2
    circular pond, 8, 10, 12*t*
    cost reduction, 22
    disadvantages, 2
    inclined (cascade) systems, 10
    raceway ponds, 8, 10, 12*t*
    thin-layer systems, 10–11, 12*t*
Operating cost, for biodiesel production, 415–416

Organic acids, biomineralization by, 350
Oscillatoria
    O. agardhii, 359–360
    O. angusta, 358–359
    O. chlorina, 359
Oxalic acid, 106
Oxygen (O$_2$) mass transfer, 42–43

P

Packed-bed fermenters, 19–20, 19*f*
Paraffins, 130–131
Pasta products, microalgae in, 226
PBRs. *See* Photobioreactors (PBRs)
Pd/C catalyst, 132
Pelvetia canaliculata, 328–329
Perfusion-bleeding culture systems, 18, 18*f*
Petroleum hydrocarbons, 366
Phaeodactylum tricornutum, 322
Phaeophyta (brown seaweed), 318
Pharmaceuticals, 420–421
Phosphorous, 350
Photoautotrophic microalgae, 163, 382
Photobioreactors (PBRs), 1–2, 7, 35–36, 63–64, 66, 395–396, 411–413
    advances in configurations, 54–56, 55–56*t*
    advantages, 2, 7, 66
    cost reduction, 22–23
    cultivation regime, 8
    disadvantages, 2
    energy consumption, of cultivation system, 82–83
    energy requirements and net energy ratio, 50–51, 50*t*
    environmental conditions, parameters of
        light, 43–46
        pH value, 47–48
        temperature, 46–47, 47*t*
    flat-plate photobioreactors, 8, 9–10*t*, 66, 164
    gas exchange and temperature effect, 79–82, 82*f*
    helical PBRs, 8
    irradiation models (*see* Irradiation models)
    manifold/unusual, 8
    material quality and investment cost, 51–52, 51*t*
    matter and energy exchange, theories of, 64

microalgae biofilm photobioreactors (MBPBRs) (*see* Microalgae biofilm photobioreactors (MBPBRs))
momentum transfer models, 75–77, 78*t*
open photobioreactors, 65–66
performance assessment, parameters of, 48–50, 49*t*
scalability, 52–54
serpentine, 8
shear stress, effect of, 77–79
system hydrodynamics, 36
    flow regime, 39–40
    gas holdup, 38–39
    horizontal/vertical flow patterns, 37, 37*t*
    mass transfer, 41–43
    mixing, 40–41
    superficial gas and liquid velocity, 36–38, 37*f*, 38*t*
    tubular photobioreactors, 7–8, 9–10*t*, 66, 164–165
Photoheterotrophic microalgae, 382, 390
Photon flux density (PFD), 45–46
Photosynthetically active region (PAR), 44
Phototrophic cultivation, 1
Phycobiliproteins, 174, 184–186, 220–222, 319
Phycoremediation, 382
Pig diets, microalgae inclusion in. *See* Swine diets, microalgae inclusion in
Plant biostimulants (PBs), 351–357
    algal extracts, 352–353, 354–355*t*
    application, 356
    cell lysis and extraction methods, 353–356
    composition and mode of action, 356
    for crop yield improvement, 356–357
    humic substances, 352
    microbial, 352
    protein hydrolysates, 352
Plant growth promoting rhizobacteria (PGPR), 352
Plant-incorporated protectants (PIPs), 357
PLE. *See* Pressurized liquid extraction (PLE)
Polyphenols, 174, 185
Polyphosphate, 350

Polysaccharides, production of, 365–366
Polyunsaturated fatty acids (PUFAs), 174, 193, 216–218, 217t, 416–418, 417f, 418t
  fish and shrimp diets, 294–301
  poultry diets, 276
  rabbit diets, 279
  ruminants
    meat, 261, 264–265
    milk, 254–261
    rumen fermentation, 253
Polyvinyl chloride (PVC), 395
Polyvinylidene fluoride (PVDF), 396
Porphyra-334, 325
Porphyra sp., 319
Porphyra umbilicalis, 322–323
Postweaning stress (PWS), 265
Poultry diets, microalgae inclusion in
  advantages, 269–273
  egg production, effects on, 277–278, 278f
  inclusion levels, 273, 274–275t
  meat production, effects on, 273–277, 278f
Precious metal-based catalysts, 132
Precipitation process, 106
Preservatives, 331
Pressing method, 96–97, 98t, 123–124
Pressurized liquid extraction (PLE), 188–189
Proline, 352, 356
Protein, 269–273
  animal-derived protein, 218
  in growing and finishing pigs diets, 266–267
  hydrolysates, 352
  from microalgae, 218
    amino acid composition, 218, 219t
    bioactive peptides, 220, 221f
    essential amino acid index, 218–219
    for food formulations, 219–220
    market value, 226f
    peptide sequences, ACE-I inhibitory activity, 220, 222t
  terrestrial plants and aquatic organisms, use of, 218
  in weaned piglet diets, 265–266
Proteolytic enzymes, 265–266, 353–356
Pt/Al$_2$O$_3$ catalysts, 138
Pt/SiO$_2$ catalysts, 138

PUFAs. *See* Polyunsaturated fatty acids (PUFAs)
Pulsed electric field (PEF)-assisted extraction, 177–186
Pyrolysis, 104, 124–125
  biochar, 161–162
  bio-oil, 161

R
RAB. *See* Rotating algal biofilm (RAB)
Rabbit diets, microalgae inclusion in, 278–283, 280–282t, 284f
Raceway ponds, 8, 10, 12t, 65, 412–413
Radiative transfer equation (RTE), 71–73
  advantages of, 71
  Beer-Lambert's law, 72
  Boltzmann transfer equation, 71
  phase function, 71–73
  two-flux approximation, 72
Reactive oxygen species (ROS), 322
Rejuvenators, 320
Renewable hydrocarbon fuel, 117
Renewable resources, 117–118
Reproductive performance, dietary microalgae effects on
  rabbit, 283
  sows and boars, 268–269, 269f
Reynolds's number, 39
Rhizoclonium hieroglyphicum, 318
Rhodophyta (red seaweed), 318–319
Rotating algal biofilm (RAB), 394–396
  advantages of, 395
  conditions of, 396t
  design, 395
  photobioreactor, 395f
Rotating biological contactor (RBC) concept, 394–395
Rotating microalgae biofilm photobioreactor (RABPBR), 394–396, 400–402, 401t
RTE. *See* Radiative transfer equation (RTE)
RuBisCo, 418–420, 419t, 419f
Ruminants, dietary microalgae inclusion, 240–244
  feed intake, effects on, 244, 245–249t
  meat production and composition, effects on, 261–265, 262–263t
  milk production and composition, effects on, 254, 255–259t
  fatty acid profile, 254–261

  lactose content, 254
  protein yield and content, 254
  rapeseed/faba beans diet, *A. platensis* supplementation to, 254
  rumen fermentation parameters, effects on, 244–254, 250–252t

S
Saccharification processes, 154–155
Saccharina japonica, 321
Saline soil, reclamation of, 367–368
Salt stress, 367
Sandia National Laboratories (US), 425
Sargassum fusiforme, 318, 323
Saturated fatty acids (SFA), 260–261, 264
Scanning electron microscope (SEM), 386–387
Scytonemin, cosmetic applications, 325–326
Separate hydrolysis and fermentation (SHF), 155
Serums, 320
SFE. *See* Supercritical fluid extraction (SFE)
SFS. *See* Supercritical fluid separation (SFS)
Shallow sloping ponds, 65
Shear stress, 77–79
Shinorine, 325
Si/Al catalyst, 135
Siderophores, biomineralization, 350–351
Simultaneous oil extraction and transesterification (SOET) process, 414
Simultaneous saccharification and fermentation (SSF), 155
Skincare products, algae applications in
  antiaging agents, 321–323
  anticellulite agents, 323–324
  moisturizer agents, 319–321
  property of
    cyanobacteria, 319
    macroalgae, 318–319
    microalgae, 317
  skin-whitening agents, 326–327
  sunscreens (*see* Sunscreen/UV filter compounds)
Skin structure, 315–317, 316f
Skin-whitening agents, 326–327
Slow pyrolysis, 104, 124

Soil
    aggregates, stabilization of, 349
    conditioners, algae, 364
    fertility, 341–342, 343–347t
    hydraulic conductivity of, 348–349
    macronutrients, 342
    nutrient recycling in, 349–351
    structure and quality,
        maintenance of, 342–349
Solar Energy Research Institute (SERI),
    3
Solvent extraction methods, 101–102,
    102t
Sows diet, microalgae inclusion in, 269,
    269f, 270–272t
Soxhlet extraction (SE), 101, 102t
Stanford Research Institute (SRI), 2–3
Stearic acid, 130–131
Steele model, 75
Sterilization, 23
Storage lipids, 120–123
Structural lipids, 120–123
Sugarcane biorefinery, 24
Sulfated polysaccharide (sPS), 317–318,
    322
SunCHem, 161
Sunscreen/UV filter compounds, 322,
    324
    carotenoids, 325
    mycosporine-like amino acids
        (MAAs), 325
    scytonemin, 325–326
Supercritical CO_2 extraction ($scCO_2$)
    extraction process, 189–192
Supercritical fluid extraction (SFE),
    100–101, 123–124, 189–192
Supercritical fluid separation (SFS), 105

Supercritical gasification technique,
    159
Superficial gas velocity, 36–38, 37f, 38t
Superficial liquid velocity, 36–38, 37f
Superoxide dismutase (SOD), 300–301
Surface/volume (S/V) ratio, 46
Surfactants, 330–331
Swine diets, microalgae inclusion in,
    265, 269f, 270–272t
    growing and finishing pigs, 266–268
    piglets, 265–266
    sows and boars, 268–269
Symbiosis, of microalgae with higher
    plants, 361–363
Synthetic pesticides, 357

T
Tamiya's equation, 74–75
Tank fermenters, 19–20, 20f
Taylor-Couette reactor, 77, 78t
Techno-economic assessment (TEA),
    414–416, 427–428
Temperature
    microalgae growth, impact on, 66
    photobioreactors, 46–47, 47t
Thermal pretreatment, 23
Thermochemical processes, 124
Theta-logistic growth (TLG) model,
    401–402
Thickening agents, 330
Three phase model, 76
Topical moisturizers, 319
Torus reactor, 77, 78t
Total capital investment (TCI),
    414–415
Total equipment cost (TEC), 424
Total phosphorus (TP), 394

Transesterification of triacylglycerol
    (TAG), 148
Transesterification process, 107–108,
    124–125, 148–149, 414
Transmission electron microscope
    (TEM), 386–387
Tripartite association, 362
Tubular photobioreactors, 7–8, 9–10t,
    66, 164–165
Two-flux approximation
    (TFA), 69–72, 70f

U
Ulothrix zonata, 395–396
Ultrasonication, 97–98, 98t
Ultrasound-assisted extraction
    (UAE), 193–195, 197–198
Ultraviolet (UV) radiation, 321–322
Ulva lactuca, 359–360
Undaria pinnatifida, 326–327
Undariopsis peterseniana, 329–330
Unsaturated glycol di-fatty esters
    (UGDE), 130–131

W
Wastewater, microalgae cultivation in,
    23–24
Wastewater treatment, 382–386,
    384–385t, 425–426, 426f
Water gas shift (WGS), 128–131
Westiellopsis prolifica, 350
Wet-milling techniques, 196

Z
Zeaxanthin, in cosmetic
    products, 325
Zeolite, 135–136, 138